Handbook of X-Ray Crystallography

Handbook of X-Ray Crystallography

Edited by **Sharon Levine**

NY RESEARCH PRESS

New York

Published by NY Research Press,
23 West, 55th Street, Suite 816,
New York, NY 10019, USA
www.nyresearchpress.com

Handbook of X-Ray Crystallography
Edited by Sharon Levine

International Standard Book Number: 978-1-63238-284-9 (Hardback)

Printed in China.

Contents

Preface

This book has been an outcome of determined endeavour from a group of educationists in the field. The primary objective was to involve a broad spectrum of professionals from diverse cultural background involved in the field for developing new researches. The book not only targets students but also scholars pursuing higher research for further enhancement of the theoretical and practical applications of the subject.

X-ray Crystallography is a scientific technique used to demonstrate the arrangement of atoms of a crystalline solid in 3D space. This book is a collection of latest trends, using X-ray crystallography and related structural determination techniques in a diverse variety of spheres. The procedures emphasized in this book consist of various processes and spectroscopic complimentary methods. The spectrum ranges from basic organic compounds to metal complexes to proteins and includes meta-analyses of a host of weak connections.

It was an honour to edit such a profound book and also a challenging task to compile and examine all the relevant data for accuracy and originality. I wish to acknowledge the efforts of the contributors for submitting such brilliant and diverse chapters in the field and for endlessly working for the completion of the book. Last, but not the least; I thank my family for being a constant source of support in all my research endeavours.

Editor

Part 1

Small Molecules

1

Calix[8]arenes Solid-State Structures: Derivatization and Crystallization Strategies

David J. Hernández and Ivan Castillo
Universidad Nacional Autónoma de México
México

1. Introduction

The macrocycles known as calix[*n*]arenes, where *n* represents the number of phenolic units bridged by methylene groups, represent ideal building blocks in supramolecular chemistry for the development of scaffolds with a preorganized structure, a well-defined cavity size, and modifiable positions for the introduction of a variety of functional groups, as shown in Fig. 1 (Böhmer, 1995; Asfari et al., 2001). The development of novel calixarene derivatives with the capability to act as receptors, sensors, catalysts, or ion transporters designed for specific purposes has been exploited to a great extent with the smaller member of the family calix[4]arene, and to a lesser degree with calix[6]arene. In the particular case of calix[4]arenes, the ease of modification by introduction of several types of functional groups at the phenolic rim has led to the development of numerous examples of versatile compounds (Baklouti et al., 2006; Baldini et al., 2007). The variety of derivatives reported to date is related to the well established synthetic protocols, which allow the preparation of calix[4]arenes with regio- and atropisomeric control by deprotonation of the phenolic OH groups with specific alkali-metal bases. These synthetic methods have been extended to the more recently developed thiacalix[4]arenes, which feature sulfur atoms as bridging groups between the phenolic components.

The development of systems based on the larger members of the calixarene and thiacalixarene families, namely calix[8]arene and thiacalix[8]arene (from now on referred to indiscriminately as calix[8]arenes), has been slow relative to its smaller analogues. This is likely due to the number of phenolic OH and aromatic positions available for functionalization, for which the regioselective introduction of substituents remains a challenging synthetic task. As a consequence, reports on crystallographically characterized calix[8]arene derivatives are relatively sparse. While the solution structures can be determined by a variety of methods, notably NMR spectroscopy, crystallographic characterization still represents the most reliable proof of the spatial arrangement of the macrocycles, particularly when the mobility of the large calix[8]arene is concerned. The limited availability of structural data is likely related to the large number of degrees of freedom present in the larger macrocycles, which does not allow the long-range ordering required for single-crystal formation. A search of the Cambridge Structural Database affords 89 structures of methylene-bridged calix[8]arenes, compared to the numbers of the four- and six-member macrocycles (Table 1).

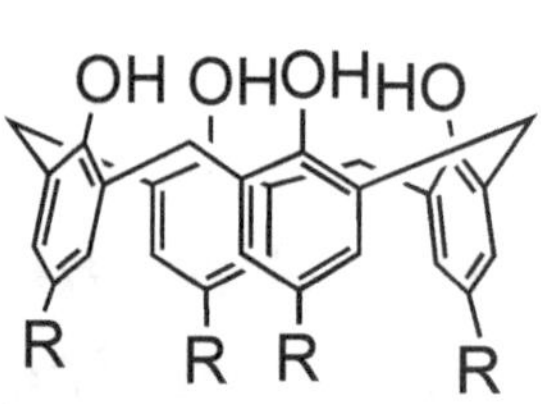

p-R-calix[4]arene

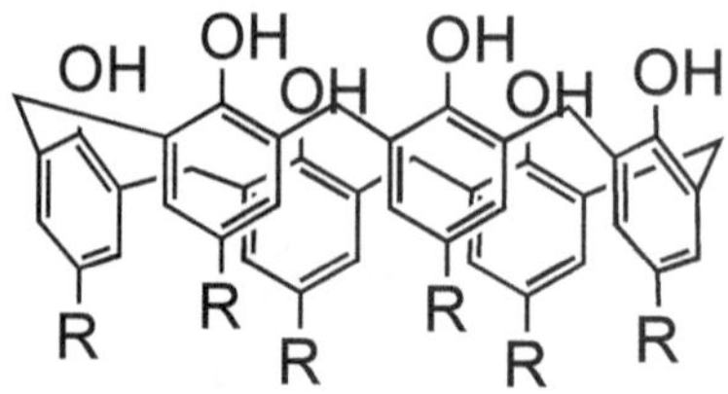

p-R-calix[6]arene

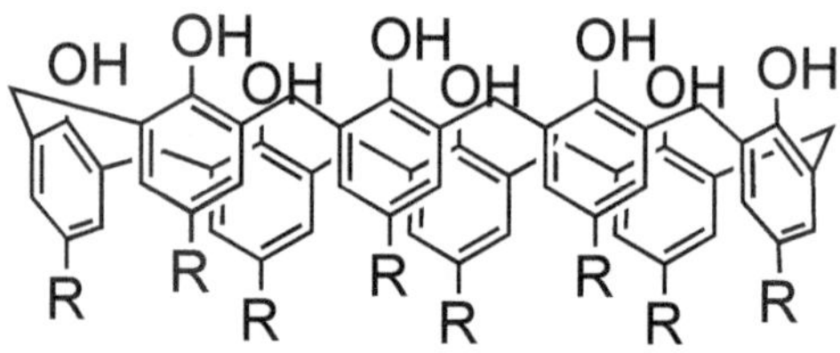

p-R-calix[8]arene

Fig. 1. Schematic representation of calix[*n*]arenes

The current chapter covers the synthetic strategies that have proven successful for the preparation of calix[8]arene derivatives amenable for structural characterization. One of them involves the formation of anionic derivatives, which are obtained by deprotonation of the phenolic OH groups, and render the calix[8]arenes as ligands towards main group and transition metals. Formally anionic phenolate derivatives have also been obtained by elimination of HCl from the reaction of calix[8]arenes with oxophilic transition-metal chlorides. These strategies have resulted in the prevalence of structural information on the 8-member macrocycles in which the calix[8]arene framework becomes rigid due to the formation of multiple oxygen-metal-oxygen bridges (Redshaw, 2003).

Compound	Number of structures
Calix[4]arene	2227
Calix[6]arene	283
Calix[8]arene	89

Table 1. Crystallographic structures reported in the CSD.

The other general strategy described in this chapter is the one involving the introduction of intramolecular covalent bridges to limit the conformational flexibility of calix[8]arenes (Geraci et al., 1995). In this respect, the use of cesium salts has allowed the regioselective introduction of covalent bridges to the 1 and 5 phenolic positions of *p-tert*-butylcalix[8]arene (Cunsolo et al., 1994). The importance of 1,5-substitution (or 1,5-3,7 substitution) resides in the high symmetry of such derivatives, relative to 1,2- or 1,4- derivatives, which appears to result in better packing interactions. In this context, we will discuss the introduction of nitrogen-containing spanning elements, which could lead to the development of new types of receptors, as well as for the binding of non-oxophilic metals within the calixarene cavity. These include our recently reported 1,5-disubstituted *p-tert*-butylcalix[8]arene by introduction of a 2,6-dimethylpyridyl group (Hernández & Castillo, 2009). A general overview on the crystallization techniques for each type of calix[8]arenes derivative accompanies the discussion.

2. Discussion

Original reports on the synthesis of the parent *p-tert*-butylcalix[8]arene date back to 1955 (Cornforth et al., 1955), where it was described as a high-melting solid with a proposed octameric structure, based on osmometry and mass spectrometry (Gutsche & Muthukrishnan, 1978; Muthukrishnan & Gutsche, 1979). Unambiguous structural assignment as an octaphenol-containing macrocycle by X-ray crystallography was initially precluded by solvent loss from the plates obtained by recrystallization from chloroform. It therefore seemed necessary to obtain calix[8]arene derivatives that did not lose solvent readily under ambient conditions, in order to afford single crystals amenable for structural characterization.

One property of calix[8]arene that was inferred from the structure of its smaller congener calix[4]arene is its large macrocyclic cavity, although crystallographic characterization was needed in order to corroborate it. Confirmation of its large cavity size in the solid state made it an attractive alternative to crown ethers for the potential binding of large cationic species. Among other possibilities, this property placed it as an ideal candidate for the selective binding of oxophilic heavy metals, including alkali, alkaline earth, lanthanide and actinide metals through the phenolic oxygen atoms. It is therefore natural that some of the first crystallographically characterized calix[8]arene derivatives consisted of metal complexes where the usually flexible structure of the macrocycle becomes relatively rigid due to the presence of multiple oxygen-metal ion-oxygen bridges. These types of derivatives were extended to *p*-block elements, including metals and non-metals such as phosphorus, germanium and bismuth.

2.1 Description of solid-state structures

2.1.1 Unfunctionalized calix[8]arenes

Although chronologically the parent *p-tert*-butylcalix[8]arene was not the first calix[8]arene to be structurally characterized due to loss of solvent molecules when crystallized from chloroform, it was obtained shortly after the first report of a calix[8]arene derivative; crystals stable enough towards solvent loss were successfully obtained from the high-boiling (115 °C) solvent pyridine (Gutsche et al., 1985). Subsequent reports include the chloroform and acetonitrile clathrates (Schatz et al., 2001; Dale et al., 2003), as well as a new determination of the pyridine-derived crystals (Huang et al., 2001); the structure of calix[8]arene with H atoms in the *para* positions also includes a molecule of the solvent pyridine (Zhang & Coppens, 2001). The aforementioned cases are described as clathrates despite the pleated loop conformation adopted by the macrocycle (Fig. 2), which is favored by the maximization of intramolecular hydrogen bonding. This configuration lacks a well-defined, deep cavity for inclusion to take place, although the incipient guest molecules may interact via hydrogen bonds, particularly in the case of pyridine.

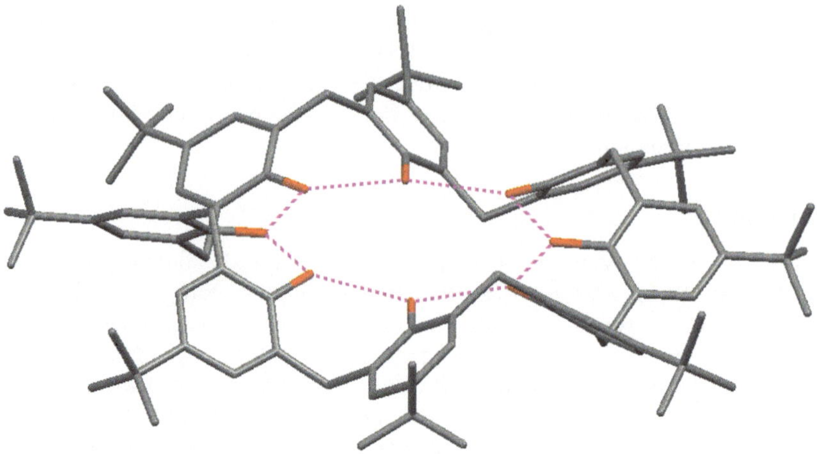

Fig. 2. Depiction of the pleated loop conformation of *p-tert*-butylcalix[8]arene with intramolecular hydrogen bonds shown in magenta (generated from Schatz et al., 2001)

Among calix[8]arenes without substituents at the phenolic positions that have been structurally characterized, the compound obtained by condensation of silyl-protected bisphenol A with formaldehyde resulted in the *p*-2-(4-hydroxyphenyl)propylcalix[8]arene (Ahn et al., 2000). Crystallization involved isopropyl ether diffusion into an acetone solution of the macrocycle, and the presence of *n*-Bu_4NBF_4 appeared to be necessary although its role is not understood. A related compound is the *p*-cumylcalix[8]arene analogue (Ettahiri et al., 2003), which was obtained by slow evaporation from dimethylsulfoxide (DMSO) solution in an alternate conformation, with two phenolic units up and two down around the macrocycle. In addition to the high boiling point of the solvent (189 °C), four molecules of DMSO form hydrogen bonds with the OH moieties, thus stabilizing the crystalline arrangement. A more recent example of an unsubstituted calix[8]arene features two deprotonated phenolic units in 1 and 3 positions of the macrocycle, and was obtained from an octasilylated precursor by fluoride attack with 2 equivalents of *n*-$Bu_4NF{\cdot}(H_2O)$. This reaction results in the formation of (*p-tert*-butylcalix[8]arene-2H)(*n*-$Bu_4N)_2$, which was crystallized from a mixture of the polar aprotic solvent dimethylformamide (DMF, b.p. 153 °C), and acetone in a 10:1 ratio (Martínez-Alanis and Castillo, 2005). The macrocycle adopts a conformation that is very similar to the pleated loop described for the parent *p-tert*-butylcalix[8]arene, with one tetrabutylammonium cation hosted within the cavity probably due to electrostatic interactions with the phenolate units.

2.1.2 Octasubstituted calix[8]arene derivatives

Initial motivation for the preparation of calix[8]arene derivatives arose from the need to substantiate its octameric structure by X-ray crystallography. Naturally, the most easily accessible derivatives are the phenolic *O*-octasubstituted compounds, which circumvent the problem of selectively introducing a limited number of functional groups at the phenolic positions. As mentioned in section 2.1.1, chronologically the first successful attempt to obtain a stable, crystalline derivative was the report of *p-tert*-butylcalix[8]arene acetylated at all the phenolic oxygen atoms (Andreetti et al., 1981). The octasubstituted derivative was

crystallized from the high-boiling acetic acid (118 °C), which likely prevented the problems associated with the loss of crystallization solvent observed for the parent *p-tert*-butylcalix[8]arene. In what would later become a recurring observation, the *p-tert*-butyl groups were disordered over at least two positions with occupancy factors close to 0.5 each.
Analogous octasubstituted compounds represent some of the first examples of structurally characterized calix[8]arene derivatives, with a growing number reported in recent years. The initial report of the completely acetylated calix[8]arene was followed by the structure of the octa-*O*-substituted macrocycle with 1,1,3,3-tetramethylbutyl substituents in the *para* phenolic positions (Ungaro et al., 1985). This compound was crystallized from the polar solvent mixture acetone/methanol in a 1:1 ratio, although no guest molecules are present in the structure; this is probably due to the self-inclusion of four of the *O*-(2-methoxy)ethyl substituents filling the macrocyclic cavity. Shortly afterwards the *para*-H methyl ether analogue (Coleman et al., 1986), in which a clathrate was obtained by ethyl ether diffusion into a deuterated chloroform solution, was also reported. The molecules of $CDCl_3$ were described as being partially hosted within the calixarene cavity, and it is important to note that data were collected with the crystals kept in a sealed capillary with the mother liquor. This likely prevented loss of the relatively volatile chloroform, which does not appear to be tightly bound to the calixarene.
O-methylated derivatives abound among structural reports, with different substituents such as *t*-Bu, Br, and NO_2 at the *para* positions of the phenol. The former was crystallized from chloroform solution as the clathrate (Bolte et al., 2002). Crystals of the *p*-bromo derivative have been obtained from CCl_4 (Baudry et al., 2003) and tetrahydrofuran (Bolte et al., 2003), in both cases by slow evaporation of solvents resulting in two molecules of each being hosted within the macrocyclic cavities. The *p*-nitro analogue crystallized from tetrahydrofuran (THF), but the solvent molecules in the structure are not included within the cavities (Podoprygorina et al., 2003); instead, two nitro groups on nitroanisole units opposite to each other in the macrocycle fill the cavity. The molecules of THF present in the structure fill the voids between stacks of the calix[8]arenes, which are stabilized by π-π interactions. In a related *p*-Br octa-*O*-butyl calix[8]arene, crystal packing appears to be promoted by Br-π interactions (Perret et al., 2007). The absence of solvent molecules in the latter structure is explained by the orientation of six butoxy groups towards the macrocyclic cavity.
The *p*-OH octapropylated calix[8]arene derivative has been crystallized from pyridine/water (Leverd et al., 2000), as well as from acetone (Leverd et al., 2000a). In both cases, the structures are stabilized by the presence of H-bonds between the solvent molecules and the *para*-hydroxy groups. The cavities are partially filled by the self-inclusion of *O*-propyl groups in both reports, with no solvent molecules hosted inside.
Two final examples of octasubstituted calix[8]arene feature the ester groups $-CH_2CO_2Et$ (Volkmer et al., 2004; Yan et al., 2009), and although it is not explicitly reported in the latter, in the former case the compound was crystallized from ethanol. These derivatives differ in the *p*-(1,1,3,3-tetramethylbutyl) and *p*-*t*-Bu substituents, with both adopting the familiar cone conformation commonly observed for calix[4]arenes, and two phenolic units on opposites ends of the macrocycle tilted towards the cavity. The ester groups attached to these phenol moieties are self-included, thus rendering the presence of ethanol within the cavity unnecessary. Nonetheless, the former structure does contain 2 hydrogen-bonded solvent molecules.

2.1.3 Calix[8]arene complexes with alkali and alkaline-earth metals

The ion-binding and transporting properties of calixarenes have been of particular interest for the development of novel derivatives analogous to the crown ethers. In this context, the oxygen-rich environment of calixarenes is ideal for the preparation of the oxophilic alkali and alkaline-earth metal complexes; in the case of calix[8]arenes, speculation on their potential to support polynuclear assemblies received confirmation from the initial solid-state characterization of a dipotassium complex (Clague et al., 1999). The macrocycle adopts a pinched conformation, with phenolic OH groups bridging the potassium ions at the pinch (Fig. 3). Surprisingly, the formally anionic oxygen atoms are located furthest from the K^+ ions; although this disposition of oxygen donors may seem counter-intuitive, it has been observed in related Cs^+ systems (Hernández & Castillo, 2009). This arrangement appears to be favored by intra-calixarene hydrogen bonds, with further stabilization by molecules of ethanol that was employed as solvent in combination with diethyl carbonate. In addition, molecules from the solvent mixture also play a role in coordinating to the cations. A related potassium complex featuring two K^+ ions sandwiched between two monoanionic calix[8]arenes (Bergougnant et al., 2005) was crystallized from the water/THF interface. In contrast to the dipotassium complex described by Clague and coworkers, in the complex reported by Bergougnant et al. the phenolate is bound directly to the K^+ ion, while the molecules of water present in the structure form H-bonded clusters.

The water/THF interfacial strategy for the crystallization of mono- and dianionic *p-tert*-butylcalix[8]arenes with alkali metal cations has been exploited by the group of Fromm (Bergougnant et al., 2007). The method consists of the dissolution of the metal carbonates in water, while the calix[8]arene is suspended in THF and then layered on top of the aqueous solutions. For the lighter alkali metals Li and Na, dianionic calix[8]arene complexes of general formula M_2(calix[8]arene-2H)$(THF)_x(H_2O)_y$ were obtained, whereas the heavier congeners K-Cs afforded monoanionic complexes of the type M(calix[8]arene-H)$(THF)_x(H_2O)_y$. In the latter, the relatively flat conformation of the macrocycles resulted in stacks that incorporate the alkali cations and water molecules aligned with the phenolic OH groups, thus generating inorganic channel-like structures.

Mixed alkali/alkaline-earth complexes have been obtained with *p-i*-Pr and *p-i*-Bu-calix[8]arenes from DMF. The crystallization method was not clearly stated, although it appears that the crystals formed on standing after 72 hours (Clague et al., 1999a). It is quite evident that the macrocycle becomes rigid upon metal complexation, since all the calixarene O atoms are involved in coordination to the four Li^+ and two Sr^{2+} cations in both structures; moreover, six of the macrocyclic oxygen donors act as Li-O-Sr bridging ligands. Bimetallic strontium complexes have also been prepared from octasubstituted calix[8]arenes (Casnati et al., 2000), with all carbonyl *O*-atoms of the eight amides present coordinating to the Sr^{2+} cations, which are additionally chelated by six of the eight phenolic oxygen atoms. Although the complexes differ in the *p*-substituents of the calix[8]arenes (*p*-OMe and *p-t*-Bu), as well as in the identity of the counter anions (picrate and chloride), the ¾ cone (or flattened partial-cone) conformations adopted by the macrocycles are very similar, likely with a similar degree of rigidity. A synergistic effect appears to be responsible for the coordination of the second strontium cation, since both reactions were initially attempted with a 1:1 molar ratio of calix[8]arene to Sr salt. In the case of the picrate, crystals were obtained from a solvent mixture that included acetic acid, which ultimately chelates the cations, fills the voids defined by the calixarene and the diethyl amide arms, and stabilizes the free picrates via H-bonds. In the latter case one chloride ligand remains coordinated to each strontium cation, while the extended structure is stabilized by H-bonded water and methanol molecules.

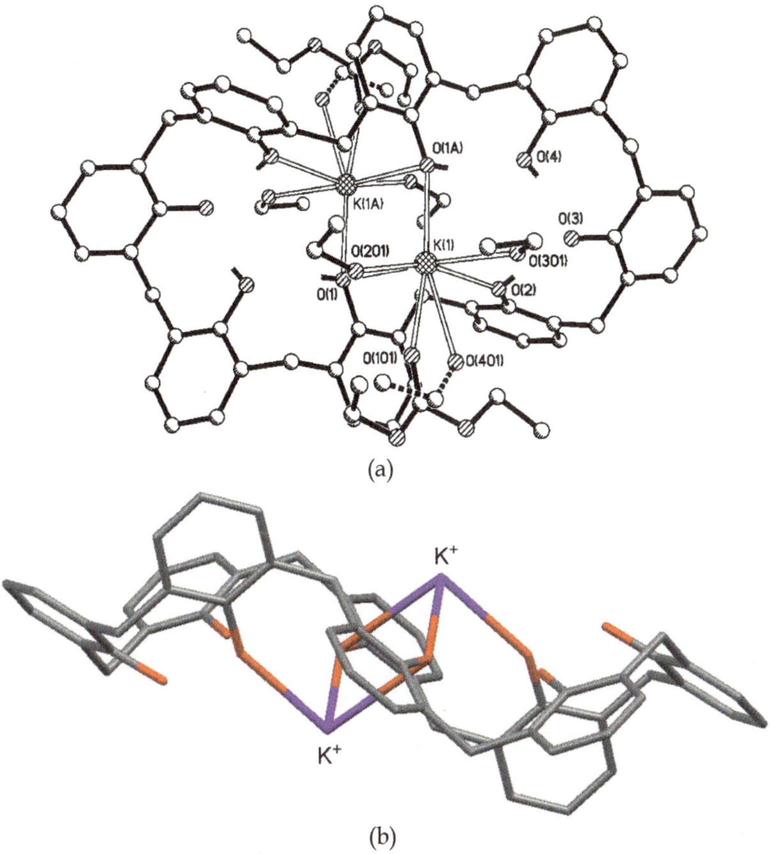

Fig. 3. a) Solid-state structure of the K^+ complex of dianionic *p-tert*-butylcalix[8]arene (Clague et al., 1999); b) Side view of the pinched conformation, *p-t*-Bu groups and solvent molecules removed for clarity

In addition to the aforementioned complexes, a monometallic structure with Ca^{2+} and doubly deprotonated *p-tert*-butylcalix[8]arene has been reported (Harrowfield et al., 1991). In this example, only two adjacent phenoxide oxygen atoms coordinate to the metal, as well as solvent molecules of DMF (crystals were obtained by cooling a hot DMF solution of the complex). This results in high mobility for the calcium cation and apparent eight-fold symmetry, as evidenced by ^1H NMR spectroscopy in solution.

2.1.4 Calix[8]arene complexes with lanthanide and actinide metals

Lanthanide derivatives are among the first structurally characterized calix[8]arene-metal complexes. As in the case of alkali and alkaline-earth metals, speculation on their potential to act as scaffolds for polymetallic assemblies received early confirmation from the solid-state characterization of a dieuropium complex (Furphy et al., 1987). The macrocycle adopts a pinched conformation similar to that observed in a dipotassium complex (Clague et al., 1999), except for the fact that in the potassium complex there are two phenolic OH groups bridging the metals, while in the europium case two phenoxide groups bridge the Eu^{3+} ions as depicted in Fig. 4a. Europium is the lanthanide with the highest representation among crystallographically characterized calix[8]arene complexes, with 4 other bimetallic examples reported. The sole exception to this general trend is the monometallic Eu^{3+} complex with a coordinated nitrate, analogous to the Ca^{2+} analogue described in the previous section (Harrowfield et al., 1991). All of the dinuclear complexes are essentially isostructural, whether they are crystallized from DMSO or DMF (Harrowfield et al., 1991a; Harrowfield et al., 1991b). Two of the europium complexes were obtained from *p*-NO_2 and *p*-H-calix[8]arenes; the former was crystallized from DMF (Bünzli et al., 1998), while the latter was obtained from DMSO solution (Fleming et al., 2003). The differences in the *para*-substituents do not affect the overall structural arrangement. In all cases the high boiling points and strongly coordinating properties of the solvents appear to be necessary to stabilize the crystal lattice, as well as to complete the coordination environment of the metal centers. Lanthanum and lutetium complexes were obtained from both DMSO and DMF, while the thulium analogue was exclusively crystallized from the former solvent. Finally, the analogous bimetallic praseodymium complex was obtained from DMF solution.

Regarding actinide metals, complexation of uranium and thorium is of particular interest due to the possibility to selectively bind the radioactive metals within the large macrocyclic cavities of calix[8]arene derivatives. Although the reports on uranium complexes far outnumber those of thorium, the latter was the first calix[8]arene-actinide complex to be structurally characterized (Harrowfield et al., 1991c). The structure of the thorium (IV) complex is unique due to the presence of two independent calix[8]arenes in the asymmetric unit with completely different conformations: one calix[8]arene ligand is in a pinched conformation, akin to that observed for the bimetallic lanthanides described above, while a second macrocycle adopts a conformation that approaches that of the free macrocyclic pleated loop; the two calixarenes assemble around Th^{4+} cations to afford a tetrameric core. Recrystallization of the complex from acetone afforded the DMSO/water solvate, with the DMSO molecules hosted within the cavities acting as terminal *O*-ligands towards the metal cations.

Polymetallic complexes have also been obtained from the reactions of uranium (IV) and *p*-H-calix[8]arene; the seemingly random reaction conditions reported (3 equivalents of UCl_4, pyridine or THF as solvents, absence or presence of NaH as base) resulted in bi-, tri- and pentauranium complexes, in one of the cases with 4 sodium ions associated. The trinuclear complex was the first to be reported, and pyridine was employed as solvent due to the poor solubility of the calix[8]arene in other solvents such as THF (Salmon et al., 2006). Pyridine likely solubilizes the macrocycle and facilitates its deprotonation upon metal complexation, resulting in an anionic complex $[U_3Cl_{11}(\text{calix[8]arene-7H})]^{6-}$ that is charge balanced by six pyridinium cations. The latter stabilize the extended structure by H-bonding to phenoxide *O*-atoms, one chloride, and lattice pyridine molecules. In the case of the bi- and pentauranium (IV) complexes, deprotonation of calix[8]arene with NaH promotes the reactions with $U(acac)_4$ and UCl_4, respectively (acac = acetylacetonate). The bimetallic complex consists of polymeric chains connected by multiple Na-O bridges through the acac ligands. The conformation of the macrocycle is described as two partial-cones, fused together in a propeller-like fashion; each partial-cone binds one U^{4+} ion through the oxygen atoms, while a Na^+ ion with a pyridine ligand is hosted within each of the cavities defined by the four phenolic units. Likewise, the conformation of the two anionic calix[8]arenes in the pentanuclear complex is described as distorted partial-cones. Each calixarene binds two uranium (IV) cations featuring additional chloride and pyridine ligands, while the fifth U^{4+} ion bridges the two calixarene ligands.

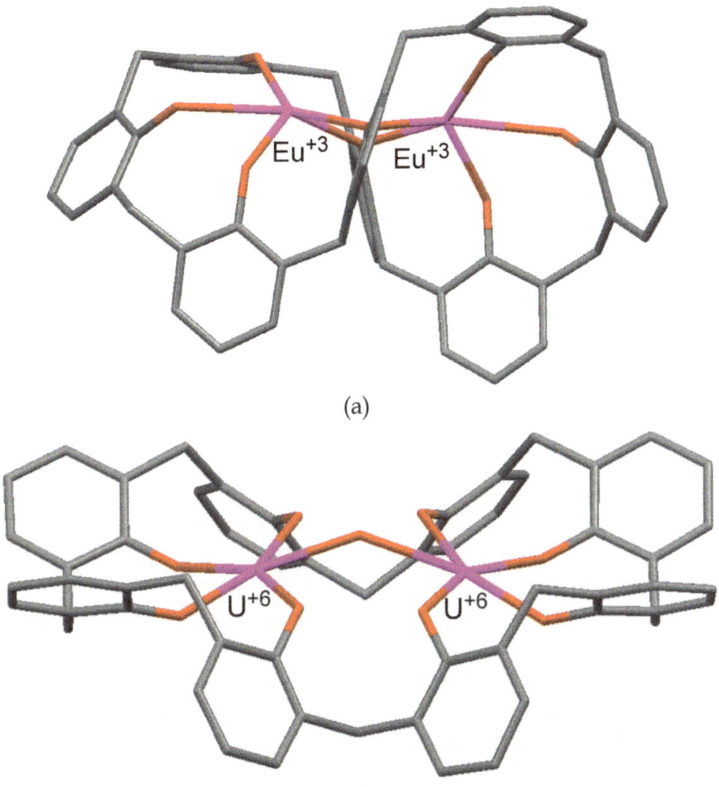

(b)

Fig. 4. a) Eu^{3+} complex of the hexaanionic *p-tert*-butylcalix[8]arene in pinched conformation (from Furphy et al., 1987); b) Pleated loop conformation of the μ-OH-bis(uranyl) complex (from Thuéry et al., 1995), *p-t*-Bu groups and exogenous ligands removed for clarity

Uranyl salts are more predictable in terms of the nuclearity of the complexes formed with calix[8]arene ligands than their uranium (IV) counterparts. A bimetallic complex was obtained by initial deprotonation of *p-tert*-butylcalix[8]arene with excess triethylamine in acetonitrile; two of the four uranyl oxo moieties interact with protonated triethylammonium cations, and the overall structure is characterized by hydrogen-bonded water and acetonitrile molecules (Thuéry et al., 1995). This diuranium (VI) species shares geometric features with one of the macrocycles observed in the structure of the thorium complex reported (Harrowfield et al., 1991c), in which the metal cations are coordinated by four phenolic oxygen atoms each, and bridged only by a hydroxyl ion. The phenolic units adopt a conformation close to that observed for the free *p-tert*-butylcalix[8]arene, which is commonly described as pleated loop, although with a distortion towards a saddle shape

(Fig. 4b). Thus, these complexes differ from the bimetallic lanthanide derivatives, in which phenolic bridges are always present. Nonetheless, the structure appears to be rigid due to the presence of the OH^- bridge between the U^{6+} ions. A related uranyl complex with a very similar structure was reported, with the main difference being the presence of two equivalents of crown ether-encapsulated KOH (Thuéry et al., 2007).

2.1.5 Calix[8]arene complexes with transition metals

As in the case of lanthanides, reactions of transition metals with calix[8]arenes give rise predominantly to bimetallic complexes. This is likely due to the extended bridging that occurs, resulting in macrocycles with reduced flexibility relative to the free calix[8]arenes, making them amenable for crystallization. For example, the first report of a transition metal complex characterized in the solid state consists of a Ti^{4+} dimer (Hofmeister et al., 1989), with a chiral propeller-like macrocyclic conformation that is very similar to that of the Eu^{3+} complexes. Molybdenum and tungsten also give rise to bimetallic complexes in most cases, irrespective of the oxidation state of the metal. The highest oxidation state attainable for molybdenum in the Mo^{6+}-imido complex gives rise to an oxophilic metal center, which coordinates to four phenolates in the complex $[(CH_3CN)Mo{=}NAr]_2$(calix[8]arene-8H), Ar = 2,6-diisopropylphenyl; the macrocycle adopts a twisted conformation with no pinch, thus no bridging phenolates are present (Gibson et al., 1995). Two molecules of acetonitrile, which was employed as solvent for recrystallization of the complex, are bound to the molybdenum ions while hosted within the two cavities defined by each half of the macrocycle. A related tungsten(VI) hydrazido complex (hydrazido = $NNPh_2^{2-}$) was structurally characterized by diffraction experiments with synchrotron radiation (Redshaw & Elsegood, 2000). The complex is characterized by a twisted macrocyclic conformation, as well as bridging phenolates in *trans*-positions relative to the hydrazido ligands; the crystals were obtained from an acetonitrile/dichloromethane mixture, both of which are present in the structure as solvate molecules. Replacement of an imido or hydrazido ligand for an oxo moiety results in a very similar macrocyclic conformation in $[(CH_3CN)W{=}O]_2$(*p-tert*-butylcalix[8]arene-8H) (Redshaw & Elsegood, 2003).

The flexibility of the *p-tert*-butylcalix[8]arene backbone was demonstrated in the report of di-, tri- and tetratungsten complexes reported from the reaction with WCl_6 (Gibson et al., 2002). The bimetallic complex is of particular interest due to the presence of a W-W triple bond, formed upon reduction of the dinuclear W^{6+} precursor to W^{3+} with sodium amalgam; moreover, the conformation adopted by the macrocycle is unique for transition metal complexes, and was described as two ¾ cones (cups) facing each other. As in the case of many other complexes, crystallization was achieved from acetonitrile solution, which stabilizes the structure by coordination to the sodium cations necessary to charge-balance the anionic ditungsten-calixarene complex.

Although chronologically developed at a later stage, the group 5 metals vanadium and niobium have also been employed in the preparation of bimetallic calix[8]arene-derived complexes. The bimetallic V^{5+}-imido (imido = N-*p*-tolyl^{2-}) complex has a structure similar to that reported for the Ti^{4+} complex described above, with bridging phenoxides in *trans*-positions relative to the imido groups, thus precluding coordination of acetonitrile solvent molecules (Gibson et al., 2001). Likewise, the analogous vanadyl derivative is characterized by bridging phenoxides *trans* to the oxo ligands (Hoppe et al., 2006); this configuration appears to preclude the presence of voids, thus favoring crystallization from acetonitrile solution upon cooling. The niobium complex reported was prepared from $NbCl_5$ as metal source, and crystallized from hot acetonitrile solution as two polymorphs, only one of which was refined as $(NbCl_2)_2$(*p-tert*-butylcalix[8]arene-6H). The macrocycle adopts a twisted conformation, and no oxygen bridges between the metal centers are observed; the two OH protons on the phenolic units are H-bonded to acetonitrile molecules in the lattice (Redshaw et al., 2007).

Not surprisingly, late transition metals are hardly represented among structurally characterized calix[8]arene complexes. This is related to the high oxophilicity of the early transition metals in high oxidation states (Ti^{4+}, V^{5+}, Mo^{6+}), compared to the reluctance of the low-valent metals in the late groups of the *d*-block to bind π-electron rich phenolate donors. Electronic repulsion of the electrons on the oxygen-centered lone pairs with the metal-centered electrons associated to a high *d*-electron count accounts for their low stability. The first complex of this type required solvothermal conditions in order to afford an anionic dicobalt complex sandwiched between two trideprotonated *p-tert*-butylcalix[8]arenes, charge balanced by two triethylammonium cations (Petit et al., 2007). In addition to the Co^{2+}, d^7 complex, the second example involves a bimetallic Cu^{2+}, d^9 species coordinated to an octasubstituted calix[8]arene derived from the octaester described in section 2.1.2 (Yan et al., 2009). Four of the eight Schiff base moieties coordinate to the cupric ions through two oxygen and one nitrogen atom each, emphasizing the importance of the nitrogen-containing substituents for the formation of metal complexes involving late transition metals; no solvent molecules (chloroform or petroleum ether) are present in the crystal structure.

2.1.6 Calix[8]arene complexes with *p*-block elements

The oxophilic nature of aluminum was exploited to obtain a trimethylaluminum complex with methylated calix[8]arene, which represents one of the first structurally characterized derivatives (Coleman et al., 1987). Both *p-t*-Bu and *p*-H-calix[8]arene methyl ethers react exothermically with 8 equivalents of $AlMe_3$ to afford the hexaaluminum complexes, which were crystallized from benzene and toluene, respectively. The conformations adopted by the macrocycles do not resemble any of the other metal complexes, probably due to the ethereal nature of the phenolic units, which precludes the formation of H-bonds and metal-oxygen-metal bridges. Steric considerations appear to be responsible for the presence of only six trimethylaluminum units. The second report on these types of derivatives with *p*-block elements involves three bridging phosphates, the central one linking the 1 and 5 phenolic positions, with the other two phosphorus atoms bound to three adjacent phenolates (Gloede et al., 2001). Although considerably flattened, the shape of the macrocycle can be described as having two ¾ cups or bowls oriented in opposite directions. Another phosphorus-containing octa-*O*-acetyl derivative features substantially disordered diethoxyphosphonate groups in the *para*-positions of the calixarene (Clark et al., 2008).

Heavy group 14 and 15 elements are represented by germanium and bismuth in terms of crystallographically characterized calix[8]arenes. One of the Ge^{2+} derivatives assembles from benzene as two rhombic Ge_2O_2 dimers with bridging oxygen donors, as well as one terminal phenolate ligand for each germanium (II) resulting in a tetragermanium-*p*-H-calix[8]arene complex that acquires a deep bowl shape (Wetherby et al., 2007). The lone pair on each Ge^{2+} allows them to act as donors towards $Fe_2(CO)_8$ fragments in a formal oxidation to Ge^{4+}; as a consequence of the oxidation process each resulting germanium (IV) is bound to only two

oxygen atoms from the calixarene. The analogous tetragermanium (II) complex with *p-tert*-butylcalix[8]arene adopts a different conformation, since the bowl shape attained in the *p*-H analogue would result in considerable steric repulsion between *p-t*-Bu substituents on the phenolic units in 1 and 5 positions (Green et al., 2009). The reaction with $Fe_2(CO)_9$ in benzene has a different outcome as well, since only two germanium (II) ions interact with $Fe(CO)_4$ fragments in the product, without any oxidation at Ge taking place.
The first bismuth derivative of *p-tert*-butylcalix[8]arene was obtained from the silylamide $Bi[N(SiMe_3)_2]_3$ by recrystallization from a toluene/THF/acetonitrile mixture (Liu et al., 2004). The solid-state structure is defined by two calix[8]arene-Bi_4 complexes bridged by μ_4-oxo ligands; the macrocycle adopts a pinched cone conformation, with the cavity filled by two molecules of toluene. When the same calix[8]arene was treated with *n*-butyllithium and subsequently with 4 equivalents of $BiCl_3$, an anionic tetrabismuth-tetralithium complex with both phenolate and chloride ligands was formed (Liu et al., 2008). Despite these differences, the macrocyclic conformation can also be described as pinched cone, with one dimethoxyethane (DME) and one THF molecules inside the cavity acting as ligands towards the lithium cations.

2.1.7 Covalently-bridged calix[8]arenes

An alternative strategy for the functionalization of calix[8]arenes that has led to crystalline derivatives is the introduction of bridging organic substituents at the phenolic rim, thus linking two (or four) oxygen atoms. The usefulness of this approach relies on the regioselectivity of the transformation, since the reactions could in principle lead to a complex mixture of isomers. Instead, the methods developed have allowed the introduction of substituents at 1,2-, 1,4- and 1,5-phenolic positions selectively, thus restricting the degrees of freedom of the macrocycle. While the reduced mobility caused by functionalization appears to be beneficial in itself for the crystallization of the organic derivatives, the preorganization seems to also result in an ideal binding pocket for large metal ions. The first type of organic-linked derivative to be structurally characterized is the doubly tetramethylene-bridged *p-tert*-butylcalix[8]arene in 1,5 and 3,7 phenolic positions (Geraci et al., 2000). The idealized symmetry of the macrocycle is D_{2d}, made up of four ¾ cone clefts and filled with one molecule of dichloromethane and two of water.
The template effect observed for cesium in the 1,5-regioselective introduction of substituents was confirmed in the singly 1,5-tetramethylene-bridged derivative of *p-tert*-butylcalix[8]arene, which crystallizes as the CsCl complex (Consoli et al., 2002). Conformationally, the macrocycle has a very similar structure to that of the 1,5:3,7-doubly bridged derivative, with the Cs^+ coordinated to all oxygen atoms, as well as two molecules of methanol (employed as solvent together with dichloromethane), one molecule of water, and the chloride counterion. Two additional molecules of water are present in the cavities of the ¾ cone clefts defined by three phenolic units. A related 1,5-tetramethylene-bridged hexa-*O*-(4-*t*-*Bu*-benzyl) derivative has also been characterized (Consoli et al., 2002a). One of the *t*-*Bu*-benzyl-substituted phenolic units is tilted towards the cavity, participating in self-inclusion in one of the ¾ cone clefts, while another cleft contains a molecule of CH_2Cl_2 from the methanol/dichloromethane solvent mixture used for crystallization.
The introduction of heteroatoms in the covalent bridges has generated interest due to the potential to bind metal ions different from the frequently observed oxophilic alkali, lanthanide, and actinide metals. A 1,4-regioisomer with a phenanthroyl group has been

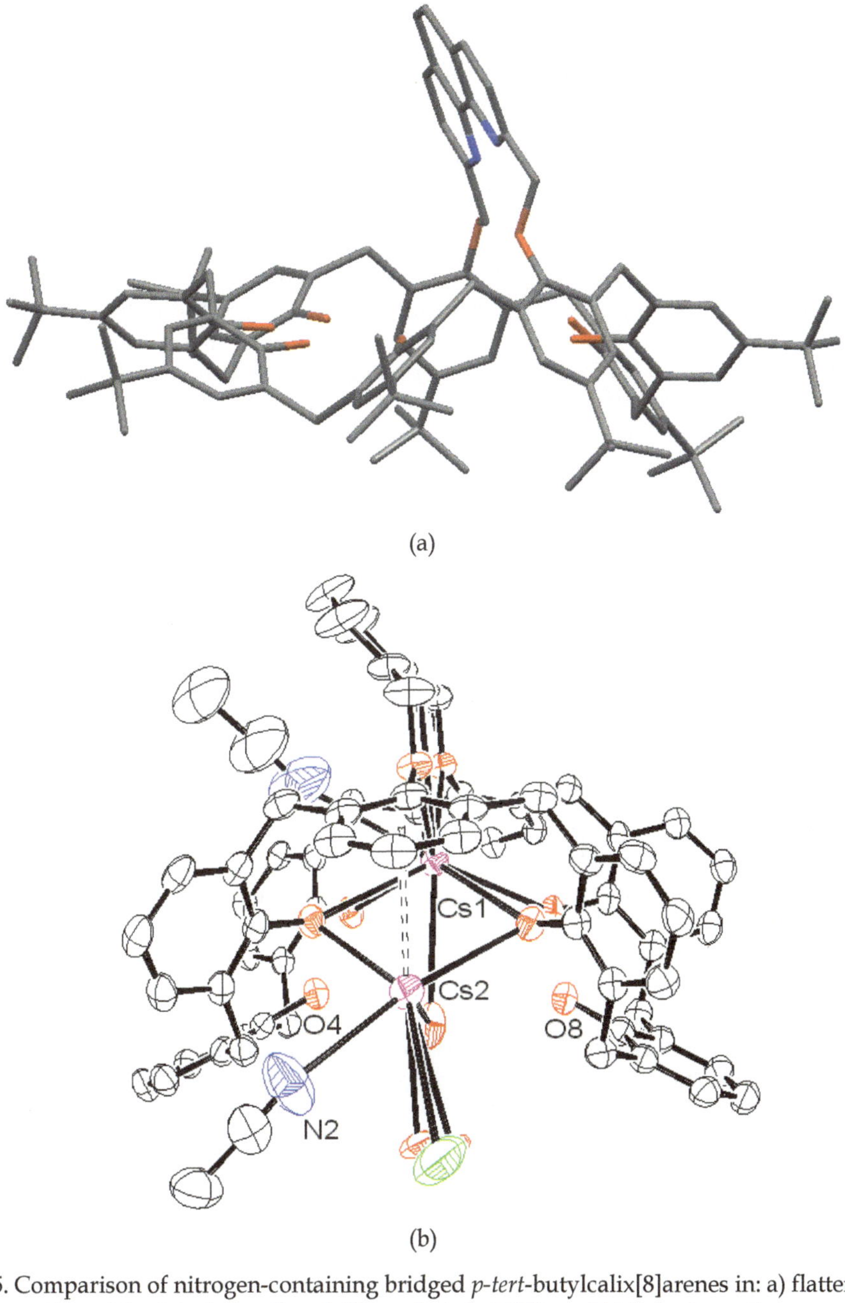

Fig. 5. Comparison of nitrogen-containing bridged *p-tert*-butylcalix[8]arenes in: a) flattened conformation of the 1,4-phenanthroyl (Konrad et al., 2005), and b) 'tub-shaped' conformation of the 1,5-pyridyl-substituted derivative (Hernández & Castillo, 2009)

crystallized from acetonitrile/dichloromethane solvent mixture, and the solid-state structure shows that the macrocycle is in an almost planar conformation stabilized by intracalixarene H-bonds (Konrad et al., 2005). The 1,4-regioselective outcome may be attributed to the use of K^+ as templating agent. In contrast, the use of cesium leads to the selective introduction of a 2,6-dimethylenepyridyl spanning element to the 1,5-phenolic positions (Hernández & Castillo, 2009). The conformational template effect of Cs^+ was also evidenced in the crystal structure, which contains two cesium cations per macrocycle, with one of them hosted deep inside the macrocyclic cavity, within bonding distance to the pyridyl N-atom. The dicesium complex was crystallized from acetonitrile/dichloromethane by slow evaporation, resulting in both molecules being present in the structure; one molecule of CH_3CN coordinates to each Cs^+ cation, while neither of the phenolate O-atoms in the 3 and 7 positions interact with the metal centers. Conformationally, the macrocycle can be described as 'tub-shaped' in an analogous fashion to the 1,5-bridged cesium complex described above (Consoli et al., 2002). Thus, the conformation of covalently-bridged calix[8]arenes is highly influenced by the regioselectivity of the substitution reaction, which is dictated by the templating cation used during the derivatization process, as evidenced in the comparison presented in Fig. 5.

2.1.8 Thia- and *para*-sulfonatocalix[8]arenes

Two novel classes of calix[8]arenes have been developed in recent years, which involve the use of sulfur in either a reduced form as thioether bridging elements between phenolic units, or in an oxidized form as *para*-sulfonate substituents. The former are commonly referred to as thiacalix[*n*]arenes, for which the octamethylated *p*-*tert*-butylthiacalix[8]arene was crystallized by slow diffusion of CH_3CN into a chloroform solution of the compound (Kon et al., 2002); the octaether adopts a 1,2,3,4-alternate conformation, with two molecules of acetonitrile hosted within the cavity. The parent thiacalix[8]arene was later characterized by X-ray crystallography, the structure being nearly identical to that of *p*-*tert*-butylcalix[8]arene in the pleated loop conformation (Kondo et al., 2005). A tetrapotassium complex of tetraanionic thiacalix[8]arene has been reported to have a zeolite-like structure, and was crystallized from methanolic solution by slow evaporation as the octasolvate (Kondo et al., 2005a).

Regarding *p*-sulfonatocalix[8]arenes, structural authentication involved the co-crystallization of the decaanion (eight sulfonate and two phenolate groups) in the presence of 4,4′-dipyridine-*N*,*N*′-dioxide and Eu^{3+} (Dalgarno et al., 2005). Despite the presence of π-stacking interactions and europium-O(sulfonate) coordination, these do not seem to perturb the macrocyclic conformation, which can be best described as the ubiquitous pleated loop. The anionic *p*-sulfonatocalix[8]arene has been crystallized with a variety of inorganic and organic cations, including combinations such as Yb^{3+} and three Ph_4P^+ cations encapsulated by two anionic *p*-sulfonatocalix[8]arenes (Makha et al., 2006). Ytterbium also favors the encapsulation of the globular $[Co(phen)_3]^{3+}$ cation (phen = 1,10-phenanthroline) by the anionic macrocycle in a chalice-like conformation (Smith et al., 2006). Likewise, in the isomorphous structures of $[M(phen)_2(H_2O)]_2[M(phen)_2]_2$-*p*-sulfonatocalix[8]arene·$2H_2O$ (M = Cu, Zn) the octaanionic calix[8]arene in a 1,2,3,4-alternate conformation encapsulates one $[M(phen)_2(H_2O)]^{2+}$ in each of the two cone subunits defined by four adjacent phenols; the $[M(phen)_2]^{2+}$ complexes join two cones from adjacent calixarenes via M-O(sulfonate) and π-π interactions (Liu et al., 2009). Among organic guests, 1,4-butanediamine and 1,2-*cis*-cyclohexanediamine have been co-crystallized from methanolic solutions by mixing with aqueous solutions of the *p*-sulfonatocalix[8]arene, followed by slow evaporation (Perret et al., 2006). The solid-state structures determined are characterized by an almost flat versus an 'inverted double partial cone' (1,2,3,4-alternate) conformation, respectively. The latter structure has a 1:2:4 stoichiometry of tetraanionic *p*-sulfonatocalix[8]arene, cyclohexanediammonium, and water, in which the ammonium cations are included deep within the macrocyclic cavity.

2.2 General overview of crystallization conditions

An initial assessment of the observations presented for the different types of calix[8]arenes points to the inherent difficulty in either obtaining adequate crystals, or being able to stabilize them in the absence of guests to fill the voids of the macrocycles. For this purpose, the use of solvents with high boiling points and the capability to form hydrogen bonds facilitates crystal formation and subsequent data collection, particularly in the case of *O*-unsubstituted calix[8]arenes. Moreover, acquisition of diffraction data at low temperature has become standard procedure for the resolution of the solid-state structures of the large macrocycles, since the commonly observed disorder in *t-Bu* groups is minimized, along with thermal motion that might be present in other parts of the molecules, for example in *p*-SO_3 groups or in solvates. These general considerations may result in the formation of stable crystals of both parent calix[8]arenes and their derivatives to be taken as a first approach towards successful crystallization and data collection.

Further analysis of the data reveals that octasubstituted derivatives lacking OH groups for hydrogen bonding can generally be crystallized from non-polar solvents. In this class of compounds, some of the phenolic *O*-substituents generally participate in self-inclusion, thus eliminating the need for high-boiling solvents; moreover, solvents with the ability to act as hydrogen bond acceptors are not essential due to the absence of phenolic OH groups available for this kind of interaction. Clathrates can indeed form, but they tend to include solvents such as chloroform, carbon tetrachloride, benzene, toluene, and tetrahydrofuran. Calix[8]arene derivatives with elements of the *p*-block that attain a closed-shell electronic configuration can also be included in this category, since they do not need solvent molecules as auxiliary ligands. This is the case with the aluminum-, phosphorus-, and germanium-containing calix[8]arenes, which were crystallized predominantly from aromatic, non-coordinating solvents. Bismuth deviates from this general rule, probably due to its larger ionic size that results in a tendency to form complexes with higher coordination numbers.

For calix[8]arene derivatives with phenolic OH groups and calix[8]arene-metal complexes with open-shell electronic configurations, solvents that can act as electron pair donors towards hydrogen atoms or electron-deficient metal centers are of key importance. Among the solvents that meet the criteria of high boiling points and electron donor capabilities, acetonitrile, dimethylformamide, dimethylsulfoxide, and in some instances pyridine, have been the most successful for the crystallization of calix[8]arenes. Their properties appear to be better suited than those of methanol or ethanol, which are only represented in a few solid-state structures, despite the comparable polarity of the alcohols and the aprotic solvents (Flick, 1998). While alcoholic solvents can act as both donors and acceptors in H-bonding interactions, the former property (with the alcohol as H-donor) does not seem to be crucial in the stabilization of most calix[8]arene crystal lattices. A disadvantage in the use of polar aprotic solvents may be the low solubility of some macrocyclic derivatives, but a co-solvent can be employed to dissolve the calix[8]arenes in appreciable quantities for crystallization.

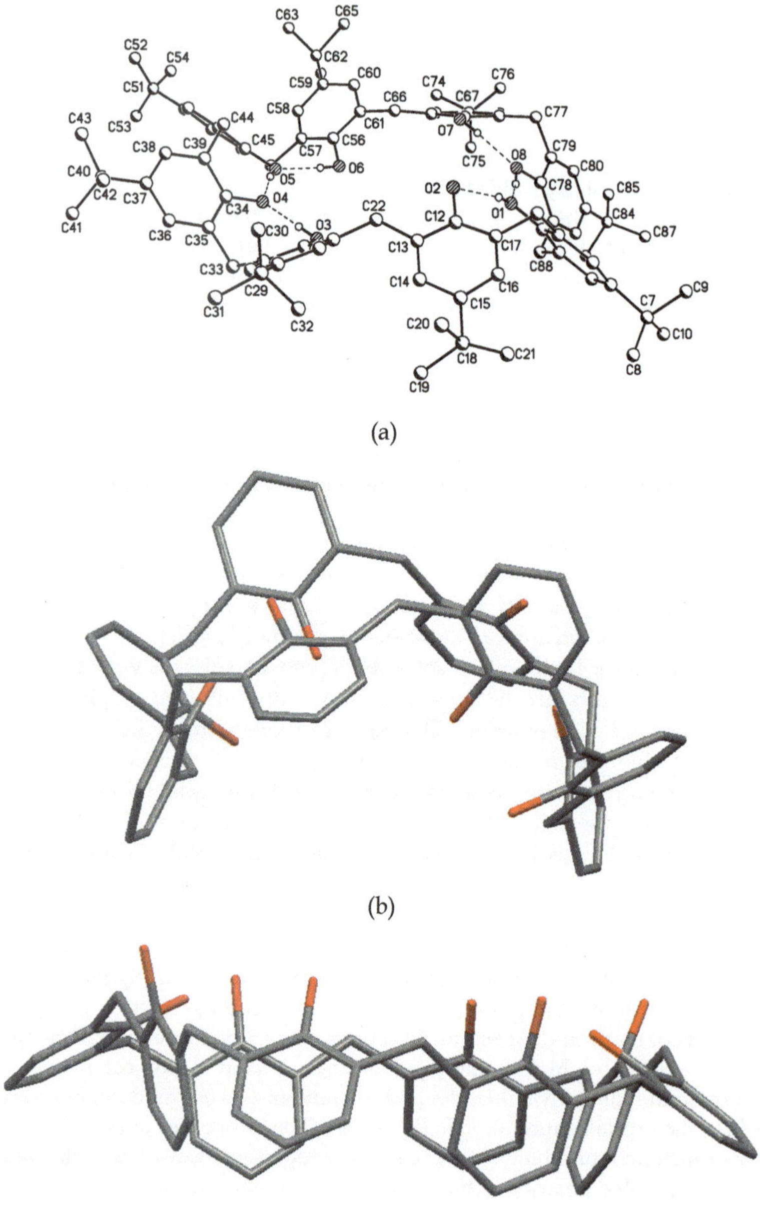

(a)

(b)

(c)

Fig. 6. Conformation adopted by the large macrocycles in the solid state: a) Pleated loop (from Martínez-Alanis and Castillo, 2005); b) 1,2-alternate (Ettahiri et al., 2003); c) cone (Volkmer et al., 2004)

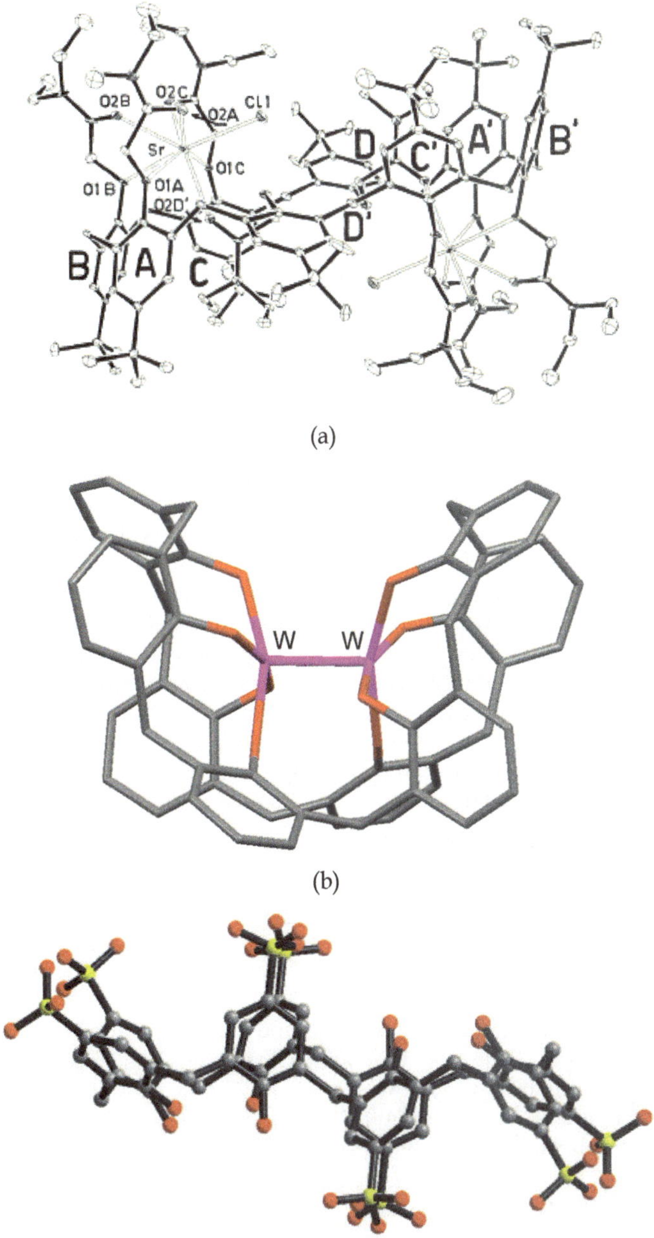

Fig. 7. Conformation adopted by the large macrocycles in the solid state: a) ¾ Cones facing away (from Casnati et al., 2000); b) ¾ cones facing each other (Gibson et al., 2002); c) 1,2,3,4-alternate (from Liu et al., 2009)

For this purpose, volatile solvents such as acetone or dichloromethane are ideal, since slow evaporation tends to result in saturated solutions of the desired compounds predominantly in the high-boiling solvent. Alternatively, heating solutions of the calix[8]arene derivatives followed by slow cooling may result in crystal formation. These guidelines apply for unsubstituted and partially substituted calix[8]arenes with phenolic OH groups, including the covalently bridged derivatives, thiacalixarenes, and complexes with electron-deficient metals. Regarding water as a solvent, most macrocycles are insoluble with the exception of *p*-sulfonatocalix[8]arene, providing the opportunity to further test the interfacial technique.
Finally, a summary of the calix[8]arene conformations determined in the solid state is presented in Figs. 6 and 7, complementing those already presented in previous sections. While it is expected that the macrocycles characterized in the future will adopt one of the conformations herein included, novel structures cannot be ruled out, including variations and intermediate structures related to those already described in the current Chapter.

3. Conclusions

Solid-state characterization of calix[8]arene derivatives involves subtle details, but general trends emerge from an analysis of the reported structures. Complete functionalization at the penolic rim with substituents capable of filling the cavity by self-inclusion may result in derivatives that can be crystallized from non-polar solvents. This method may also be applied to calix[8]arenes with light *p*-block elements. Introduction of organic bridges at the phenolic OH groups, together with complex formation of alkali, alkaline earth, and open-shell (*d*- and *f*-block) metals may result in crystalline derivatives primarily from polar aprotic solvents that can H-bond to the remaining OH functional groups, and coordinate to the electron-deficient metals. The presence of solvent molecules is stabilized by collection of diffraction data at low temperature, although crystal formation appears to be facilitated by slow evaporation of solvent, rather than by cooling. These general guidelines should serve as a first approximation for crystal growth of calix[8]arene derivatives.

4. Acknowledgment

The authors thank CONACyT (Proyecto 58408, Beca 239715) for financial support.

5. References

Ahn, K. H., Kim, S.-G., & U, J. S. (2000). Synthesis and Characterization of New Calixarenes from Bisphenol A. *Bulletin of the Korean Chemical Society*, Vol. 21, No. 8, (August 2000), pp. 813-816, ISSN 0253-2964

Andreetti, G. D. , Ungaro, R., & Pochini, A. (1981). X-Ray Crystal and Molecular Structure of the p-t-Butylphenol-Formaldehyde Cyclic Octamer Cyclo{octa[(5-t-butyl-2-acetoxy-1,3-phenylene)methylene}. *Journal of the Chemical Society, Chemical Communications*, No. 11, (June 1981), pp. 533-534, ISSN 0022-4936

Asfari, Z., Böhmer, V., Harrowfield, J., & Vicens, J. (Eds.), (2001). *Calixarenes 2001*, Kluwer, ISBN 978-079-2369-60-8, Dordrecht, The Netherlands

Baklouti, L., Harrowfield, J., Pulpoka, B., Vicens, J. (2006). 1,3-Alternate, The Smart Conformation of Calix[4]arenes. *Mini-Reviews in Organic Chemistry*, Vol. 3, No. 4, (November 2006), pp. 355-384, ISSN 1570-193X

Baldini, L., Casnati, A., Sansone, F., Ungaro, R. (2007). Calixarene-Based Multivalent Ligands. *Chemical Society Reviews*, Vol. 36, No. 2, (February 2007), pp. 254-266, ISSN 0306-0012

Baudry, R., Felix, C., Bavoux, C., Perrin, M., Vocanson, F., Dumazet-Bonnamour, I., & Lamartine, R. (2003). Biaryl Cross-Coupling Reactions: Convenient Routes to New Functionalized Calix[8]arenes. *New Journal of Chemistry*, Vol. 27, No. 10, (October 2003), pp. 1540-1543, ISSN 1144-0546

Bergougnant, R. D., Robin, A. Y., & Fromm, K. M. (2005). "Hooked-on" Calix[8]arenes: A $(H_2O)_{10}$ Cluster with an Unprecedented Structure. *Crystal Growth & Design*, Vol. 5, No. 5, (October 2005), pp. 1691-1694, ISSN 1528-7483

Bergougnant, R. D., Robin, A. Y., & Fromm, K. M. (2007). From Simple Rings to One-Dimensional Channels with Calix[8]arenes, Water Clusters, and Alkali Metal Ions. *Tetrahedron*, Vol. 63, No. 44, (October 2007), pp. 10751-10757, ISSN 0040-4020

Böhmer, V. (1995). Calixarenes, Macrocycles with (Almost) Unlimited Possibilities. *Angewandte Chemie International Edition*, Vol. 34, No. 7, (April 1995), pp. 713-745, ISSN 1433-7851

Bolte, M., Brusko, V., & Böhmer, V. (2002). Crystal Structure of *t*-Butylcalix[8]arene Octamethyl Ether. *Journal of Supramolecular Chemistry*, Vol. 2, No. 1-3, (June 2002), pp. 57-61, ISSN 1472-7862

Bolte, M., Brusko, V., & Böhmer, V. (2003). A New Pseudopolymorph of 5,11,17,23,29,35,41,47-Octabromo-49,50,51,52,53,54,55,56-octamethoxycalix[8]arene. *Acta Crystallographica*, Vol. E59, No. 11, (November 2003), pp. o1691-o1693, ISSN 1600-5368

Bünzli, J.-C. G., Ihringer, F., Dumy, P., Sager, C., & Rogers, R. D. (1998). Structural and Dynamic Properties of Calixarene Bimetallic Complexes: Solution *versus* Solid-State Structure of Dinuclear Complexes of Eu^{III} and Lu^{III} with Substituted Calix[8]arenes. *Journal of the Chemical Society, Dalton Transactions*, No. 3, (February 1998), pp. 497-503, ISSN 1472-7773

Casnati, A., Baldini, L., Pelizzi, N., Rissanen, K., Ugozzoli, F., & Ungaro, R. (2000). Strontium Complexes of Calixarene Amides in the Solid State: Structural Dependence on the Ligand Size and on the Counter Ions. *Journal of the Chemical Society, Dalton Transactions*, No. 19, (October 2000), pp. 3411-3415, ISSN 1472-7773

Clague, N. P., Clegg, W., Coles, S. J., Crane, J. D., Moreton, D. J.; Sinn, E., Teat, S. J., & Young, N. A. (1999). An Insight into Ion-Transport by Calixarenes; the Structure of the Dipotassium Complex of *p-tert*-Butylcalix[8]arene Crystallised from a Protogenic, Coordinating Solvent [Ethanol/Diethylcarbonate (10:1)]. *Chemical Communications*, No. 4, (February 1999), pp. 379-380, ISSN 1359-7345

Clague, N. P., Crane, J. D., Moreton, D. J., Sinn, E., Teat, S. J., & Young, N. A. (1999a). A Mixed Lithium-Strontium Polynuclear Complex Formed within the Hexa-Deprotonated Calix[8]arene Framework; the Synthesis and Structure $Li_4Sr_2(H_2L)(O_2CC_4H_9)_2(dmf)_8$ [H_8L = *p*-Pr^i or *p*-Bu^i-calix[8]arene]. *Journal of the Chemical Society, Dalton Transactions*, No. 20, (October 1999), pp. 3535-3536, ISSN 1472-7773

Clark, T. E., Makha, M., Sobolev, A. N., Su, D., Rohrs, H., Gross, M. L., Atwood, J. L., & Raston, C. L. (2008). Self-Organized Nano-Arrays of p-Phosphonic Acid

Functionalised Higher Order Calixarenes. *New Journal of Chemistry*, Vol. 32, No. 9, (September 2008), pp. 1478-1483, ISSN 1144-0546

Coleman, A. W., Bott, S. G., & Atwood, J. L. (1986). Preparation and Structure of (Calix[8]arene Methyl Ether)·$2CDCl_3$. *Journal of Inclusion Phenomena and Macrocyclic Chemistry*, Vol. 4, No. 3, (September 1986), pp. 247-253, ISSN 0167-7861

Coleman, A. W., Bott, S. G., & Atwood, J. L. (1987). Reaction of Trimethylaluminum with Calixarenes. I. Synthesis and Structure of [Calix[8]arene Methyl Ether]$[AlMe_3]_6$·2 Toluene and of [*p-tert*-Butylcalix[8]arene Methyl Ether]$[AlMe_3]_6$·4 Benzene. *Journal of Inclusion Phenomena and Macrocyclic Chemistry*, Vol. 5, No. 5, (October 1987), pp. 581-590, ISSN 0167-7861

Consoli, G. M. L., Cunsolo, F., Geraci, C., Gavuzzo, E., & Neri, P. (2002). Alkali Cation 'Conformational Templation' in 1,5-Bridged Calix[8]arenes: A Single Crystal X-Ray Proof. *Tetrahedron Letters*, Vol. 43, No. 7, (February 2002), pp. 1209-1211, ISSN 0040-4039

Consoli, G. M. L., Cunsolo, F., Geraci, C., Gavuzzo, E., & Neri, P. (2002a). Atropisomerism in 1,5-Bridged Calix[8]arenes. *Organic Letters*, Vol. 4, No. 16, (August 2002), pp. 2649-2652, ISSN 1523-7060

Cornforth, J. W., D'Arcy Hart, P., Nicholls, G. A., Rees, R. S. W., & Stock, J. A. (1955). Antituberculous Effects of Certain Surface-Active Polyoxyethylene Ethers. *British Journal of Pharmacology*, Vol. 10, No. 1, (May 2009), pp. 73-86, ISSN 0007-1188

Cunsolo, F., Piattelli, M., & Neri, P. (1994). Shaping Calix[8]arene Framework by Intramolecular Bridging. Synthesis of Conformationally Blocked Calix[8]arene Derivatives. *Journal of the Chemical Society, Chemical Communications*, No. 17, (September 1994), pp. 1917-1918, ISSN 0022-4936

Dale, S. H., Elsegood, M. R. J., & Redshaw, C. (2003). Polymorphism and Pseudopolymorphism in Calixarenes: Acetonitrile Clathrates of *p-Bu*^{t}-calix[n]arenes (n = 6 and 8). *CrystEngComm*, Vol. 5, No. 65, (October 2003), pp. 368-373, ISSN 1466-8033

Dalgarno, S. J., Hardie, M. J., Atwood, J. L., Warren, J. E., & Raston, C. L. (2005). A Complex 3D 'Wavy Brick Wall' Coordination Polymer Based on *p*-Sulfonatocalix[8]arene. *New Journal of Chemistry*, Vol. 29, No. 5, (May 2005), pp. 649-652, ISSN 1144-0546

Ettahiri, A., Perrin, M., & Thozet, A. (2003). Crystal Structure of the p-Cumylcalix[8]arene-Dimethylsulfoxide 1:5 Complex. *Journal of Inclusion Phenomena and Macrocyclic Chemistry*, Vol. 47, No. 1-2, (October 2003), pp. 11-13, ISSN 0923-0750

Fleming, S., Gutsche, C. D., Harrowfield, J. M., Ogden, M. I., Skelton, B. W., Stewart, D. F., & White, A. H. (2003). Calixarenes as Aryloxides: Oligonuclear Europium(III) Derivatives. *Dalton Transactions*, No. 17, (September 2003), pp. 3319-3327, ISSN 0300-9246

Flick, E. W. (1998). *Industrial Solvents Handbook*, Noyes Data Corporation, ISBN 978-081-5514-13-8, Park Ridge, United States

Furphy, B. M., Harrowfield, J. M., Kepert, D. L., Skelton, B. W., White, A. H., & Wilner, F. R. (1987). Bimetallic Lanthanide Complexes of the Calixarenes: Europium(III) and *tert*-Butylcalix[8]arene. *Inorganic Chemistry*, Vol. 26, No. 25, (December 1987), pp. 4231-4236, ISSN 0020-1669

Geraci, C., Piattelli, M., & Neri, P. (1995). Preorganization of Calix[8]arenes. Synthesis of Basket-Shaped Doubly-Crowned Calix[8]arenes. *Tetrahedron Letters,* Vol. 36, No. 30, (July 1995), pp. 5429-5432, ISSN 0040-4039

Geraci, C., Bottino, A., Piattelli, M., Gavuzzo, E., & Neri, P. (2000). Interplay Between Cone and Partial-Cone Geometry in Doubly-Bridged Calix[8]arenes Investigated by X-Ray and 2D NMR. *Journal of the Chemical Society, Perkin Transactions 2,* No. 2, (February 2000), pp. 185-187, ISSN 0300-9580

Gibson, V. C., Redshaw, C., Clegg, W., & Elsegood, M. R. J. (1995). Novel Metal Imido Calixarene Complexes. *Journal of the Chemical Society, Chemical Communications,* No. 23, (December 1995), pp. 2371-2372, ISSN 0022-4936

Gibson, V. C., Redshaw, C., & Elsegood, M. R. J. (2001). Calix[6] and [8]arene Complexes of Vanadium. *Journal of the Chemical Society, Dalton Transactions,* No. 6, (March 2001), pp. 767-769, ISSN 1472-7773

Gibson, V. C., Redshaw, C., & Elsegood, M. R. J. (2002). Novel Tungsten Calix[8]arene Complexes. *Chemical Communications,* No. 11, (June 2002), pp. 1200-1201, ISSN 1359-7345

Gloede, J., Ozegowski, S., Matt, D., & De Cian, A. (2001). Shaping Calixarene Frameworks. Synthesis and Structure of a Calix[8]arene Containing Three Bridging Phosphate Units. *Tetrahedron Letters,* Vol. 42, No. 52, (December 2001), pp. 9139-9142, ISSN 0040-4039

Green, R. A., Rheingold, A. L., & Weinert, C. S. (2009). Synthesis of the Germanium(II) Calixarene {*p*-$Bu^t{}_8$calix[8]arene}Ge_4 and its Reaction with $Fe_2(CO)_9$: Generation of the Germanium(II)/Iron(0) Complex {*p*-$Bu^t{}_8$calix[8]arene}$Ge_4[Fe(CO)_4]_2$. *Inorganica Chimica Acta,* Vol. 362, No. 9, (July 2009), pp. 3159-3164, ISSN 0020-1693

Gutsche, C. D., & Muthukrishnan, R. (1978). Analysis of the Product Mixtures Produced by the Base-Catalyzed Condensation of Formaldehyde with *para*-Substituted Phenols. *Journal of Organic Chemistry,* Vol. 43, No. 25, (December 1978), pp. 4905-4906, ISSN 0257-3717

Gutsche, C. D., Gutsche, A. E., & Karaulov, A. I. (1985). Calixarenes 11. Crystal and Molecular Structure of *p-tert*-butylcalix[8]arene. *Journal of Inclusion Phenomena,* Vol. 3, No. 4, (December 1985), pp. 447-451, ISSN 0167-7861

Harrowfield, J. M., Ogden, M. I., Richmond, W. R., & White, A. H. (1991). Lanthanide Ions as Calcium Substitutes: A Structural Comparison of Europium and Calcium Complexes of a Ditopic Calixarene. *Journal of the Chemical Society, Dalton Transactions,* No. 8, (August 1991), pp. 2153-2160, ISSN 1472-7773

Harrowfield, J. M., Ogden, M. I., & White, A. H. (1991a). Lanthanide Ion Complexes of Calixarenes. VII. Bimetallic Lanthanide Complexes of *p*-t-Butylcalix[8]arene from Dimethyl Sulfoxide Solutions. *Australian Journal of Chemistry,* Vol. 44, No. 9, (September 1991), pp. 1237-1247, ISSN 0004-9425

Harrowfield, J. M., Ogden, M. I., & White, A. H. (1991b). Lanthanide Ion Complexes of Calixarenes. VIII. Bimetallic Lanthanide Complexes of *p*-t-Butylcalix[8]arene from Dimethylformamide Solutions. *Australian Journal of Chemistry,* Vol. 44, No. 9, (September 1991), pp. 1249-1262, ISSN 0004-9425

Harrowfield, J. M., Ogden, M. I., & White, A. H. (1991c). Actinide Complexes of the Calixarenes. Part 2. Synthesis and Crystal Structure of a Novel Thorium(IV)

Complex of *p-tert*-Butylcalix[8]arene. *Journal of the Chemical Society, Dalton Transactions*, No. 10, (October 1991), pp. 2625-2632, ISSN 1472-7773

Hofmesiter, G. E., Hahn, F. E., & Pedersen, S. F. (1989). Chiral Recognition in the Synthesis of Dimetalla-4-*tert*-butylcalix[8]arene Complexes. The Incorporation of a Metal Alkoxide Ligand into a Molecular Cavity. *Journal of the American Chemical Society*, Vol. 111, No. 6, (March 1989), pp. 2318-2319, ISSN 0002-7863

Hoppe, E., Limberg, C., & Zlemer, B. (2006). Mono- and Dinuclear Oxovanadium(V)calixarene Complexes and Their Activity as Oxidation Catalysts. *Inorganic Chemistry*, Vol. 45, No. 20, (October 2006), pp. 8308-8317, ISSN 0020-1669

Hernández, D. J., & Castillo, I. (2009). Synthesis of 1,5-(2,6-Dimethylpyridyl)-Calix[8]arene: Solid-state Structure of its Dicesium Complex. *Tetrahedron Letters*, Vol. 50, No. 21, (May 2009), pp. 2548-2551, ISSN 0040-4039

Huang, R.-B., Zheng, N.-F., Xie, S. Y., & Zheng, L.-S. (2001). C-H···π Interaction and N···H-O Hydrogen Bonding in the Chair-like *p-tert*-Butylcalix[8]arene Complex Including Four Pyridine Molecules. *Journal of Inclusion Phenomena and Macrocyclic Chemistry*, Vol. 40, No. 1-2, (June 2001), pp. 121-124, ISSN 0923-0750

Kon, N., Iki, N., & Miyano, S. (2002). Synthesis of *p-tert*-Butylthiacalix[*n*]arenes ($n = 4$, 6, and 8) from a Sulfur-Bridged Acyclic Dimer of *p-tert*-Butylphenol. *Tetrahedron Letters*, Vol. 43, No. 12, (March 2002), pp. 2231-2234, ISSN 0040-4039

Kondo, Y., Endo, K., Iki, N., Miyano, S., & Hamada, F. (2005). Synthesis and Crystal Structure of *p-tert*-Butylthiacalix[8]arene: A New Member of Thiacalixarenes. *Journal of Inclusion Phenomena and Macrocyclic Chemistry*, Vol. 52, No. 1-2, (June 2005), pp. 45-49, ISSN 0923-0750

Kondo, Y., Endo, K., & Hamada, F. (2005a). Potassium-Thiacalix[8]arene Assembly: Structure and Guest Sorption Profiles. *Chemical Communications*, No. 6, (February 2005), pp. 711-712, ISSN 1359-7345

Konrad, S., Näther, C., & Lüning, U. (2005). Calix[5]- and Calix[8]arenes Bridged with Heterocycles. *European Journal of Organic Chemistry*, No. 11, (June 2005), pp. 2330-2337, ISSN 1434-193X

Leverd, P. C., Huc, V., Palacin, S., & Nierlich, M. (2000). Octa(*p*-hydroxy)octakis(propyloxy)calix[8]arene: The First Crystal Structure of a *p*-Hydroxy Calixarene. *Journal of Inclusion Phenomena and Macrocyclic Chemistry*, Vol. 36, No. 3, (March 2000), pp. 259-266, ISSN 0923-0750

Leverd, P. C., Huc, V., Palacin, S., & Nierlich, M. (2000a). Crystal Structure of Octa(*p*-hydroxy)octakis(propyloxy)calix[8]arene-Acetone (1/4). *Zeitschrift für Kristallographie-New Crystal Structures*, Vol. 215, No. 4, (December 2000), pp. NCS 1267-418, ISSN 1433-7266

Liu, L., Zakharov, L. N., Rheingold, A. L., & Hanna, T. A. (2004). Synthesis and X-Ray Crystal Structures of the First Antimony and Bismuth Calixarene Complexes. *Chemical Communications*, No. 13, (July 2004), pp. 1472-1473, ISSN 1359-7345

Liu, L., Zakharov, L. N., Golen, J. A., Rheingold, A. L., & Hanna, T. A. (2008). Synthesis and Characterization of Bismuth(III) and Antimony(III) Calixarene Complexes. *Inorganic Chemistry*, Vol. 47, No. 23, (December 2008), pp. 11143-11153, ISSN 0020-1669

Liu, Y., Liao, W., Bi, Y., Wang, M., Wu. Z., Wang, X., Su, Z., & Zhang, H. (2009). 1,2,3,4-Alternate Double Cone Conformational Extreme in the Supramolecular Assemblies of *p*-Sulfonatocalix[8]arene. *CrystEngComm*, Vol. 11, No. 9, (September 2009), pp. 1803-1806, ISSN 1466-8033

Makha, M. Sobolev, A. N., & Raston, C. L. (2006). Constructing 2D Porous Material Based on the Assembly of Large Organic Ions: *p*-Sulfonatocalix[8]arene and Tetraphenylphosphonium Ions. *Chemical Communications*, No. 5, (February 2006), pp. 511-513, ISSN 1359-7345

Martínez-Alanis, P. R., & Castillo, I. (2005). Calix[8]arene Anions: Solid State Structure of an Inclusion Compound with a Tetrabutylammonium Cation. *Tetrahedron Letters*, Vol. 46, No. 51, (December 2005), pp. 8845-8848, ISSN 0040-4039

Muthukrishnan, R., & Gutsche, C. D. (1979). Preparation of the 2,4-Dinitrophenyl and Camphorsulfonyl Derivatives of the Calix[8]arene from p-tert-Butylphenol. *Journal of Organic Chemistry*, Vol. 44, No. 22, (October 1979), pp. 3962-3964, ISSN 0257-3717

Perret, F., Bonnard, V., Danylyuk, O., Suwinska, K., & Coleman, A. W. (2006). Conformational Extremes in the Supramolecular Assemblies of para-Sulfonato-Calix[8]arene. *New Journal of Chemistry*, Vol. 30, No. 7, (July 2006), pp. 987-990, ISSN 1144-0546

Perret, F., Suwinska, K., Bertino Ghera, B., Parrot-Lopez, H., & Coleman, A. W. (2007). Synthesis, Solid State Structures and Interfacial Properties of New *para*-Phosphonato-*O*-alkoxy-calix[8]arene Derivatives. *New Journal of Chemistry*, Vol. 31, No. 6, (June 2007), pp. 893-900, ISSN 1144-0546

Petit, S., Pilet, G., Luneau, D., Chibotaru, L. F., & Ungur, L. (2007). A Dinuclear Cobalt(II) Complex of Calix[8]arene Exhibiting Strong Magnetic Anisotropy. *Dalton Transactions*, No. 40, (October 2007), pp. 4582-4588, ISSN 0300-9246

Podoprygorina, G., Zhang, J., Brusko, V., Bolte, M., Janshoff, A., & Böhmer, V. (2003). Supramolecular Structures Formed by Calix[8]arene Derivatives. *Organic Letters*, Vol. 5, No. 26, (December 2003), pp. 5071-5074, ISSN 1523-7060

Redshaw, C., & Elsegood, M. R. J. (2000). Group 6 Metallocalixarenes Bearing "Hydrazido(2-)" Groups. *Inorganic Chemistry*, Vol. 39, No. 22, (October 2000), pp. 5164-5168, ISSN 0020-1669

Redshaw, C. (2003). Coordination Chemistry of the Larger Calixarenes. *Coordination Chemistry Reviews*, Vol. 244, No. 1-2, (September 2003), pp. 45-70, ISSN 0010-8545

Redshaw, C., & Elsegood, M. R. J. (2003). Tungsten(VI) Complexes Derived from Calix[6]- and -[8]arenes: Oxo and Oxychloride Species. *European Journal of Inorganic Chemistry*, No. 11, (June 2003), pp. 2071-2074, ISSN 1434-1948

Redshaw, C., Rowan, M., Homden, D. M., Elsegood, M. R. J., Yamato, T., & Pérez-Casas, C. (2007). Niobium- and Tantalum-Based Ethylene Polymerisation Catalysts Bearing Methylene- or Dimethyleneoxa-Bridged Calixarene Ligands. *Chemistry - A European Journal*, Vol. 13, No. 36, (December 2007), pp. 10129-10139, ISSN 0947-6539

Salmon, L., Thuéry, P., & Ephritikhine, M. (2006). Uranium(IV) Complexes of Calix[*n*]arenes (*n* = 4, 6 and 8). *Chemical Communications*, No. 8, (February 2006), pp. 856-858, ISSN 1359-7345

Schatz, J., Schildbach, F., Lentz, A., Rastätter, S., & Debaerdemaeker, T. (2001). A CH/π Interaction at a Highly Solvated Macrocycle in the Solid State – the *p-tert*-Butylcalix[8]arene–Chloroform 1:8 Clathrate. *Zeitschrift für Kristallographie*, Vol. 216, No. 3, (March 2001), pp. 182-186, ISSN 0044-2968

Smith, C. B., Barbour, L. J., Makha, M., Raston, C. L., & Sobolev, A. N. (2006). Unlocking the Elusive Binding Cavity in *p*-Sulfonatocalix[8]arene. *New Journal of Chemistry*, Vol. 30, No. 7, (July 2006), pp. 991-996, ISSN 1144-0546

Thuéry, P., Keller, N., Lance, M., Vigner, J.-D., & Nierlich, M. (1995). A Bimetallic Complex Between Uranyl and *p-tert*-Butylcalix[8]arene. *Acta Crystallographica*, Vol. C51, No. 8, (August 1995), pp. 1570-1574, ISSN 0108-2701

Thuéry, P., Masci, B., Takimoto, M., & Yamato, T. (2007). Supramolecular Assemblages from Uranyl Complexes of Calixarenes and Potassium Complexes of 18-Crown-6 or Dibenzo-18-Crown-6. *Inorganic Chemistry Communications*, Vol. 10, No. 7, (July 2007), pp. 795-799, ISSN 1387-7003

Ungaro, R., Pochini, A., Andreetti, G. D., & Ugozzoli, F. (1985). Molecular Inclusion in Functionalized Macrocycles. Part 12. Crystal and Molecular Structure of a *p*-(1,1,3,3)-Tetramethylbutylcalix[8]arene Octapodand. *Journal of Inclusion Phenomena*, Vol. 3, No. 4, (December 1985), pp. 409-420, ISSN 0167-7861

Volkmer, D., Fricke, M., Agena, C., & Mattay, J. (2004). Interfacial Electrostatics Guiding the Crystallization of $CaCO_3$ Underneath Monolayers of Calixarenes and Resorcarenes. *Journal of Materials Chemistry*, No. 14, (April 2004), pp. 2249-2259, ISSN 0959-9428

Wetherby, A. E. Jr., Goeller, L. R., DiPasquale, A. G., Rheingold, A. L., & Welnert, C. S. (2007). Synthesis and Structures of an Unusual Germanium(II) Calix[4]arene Complex and the First Germanium(II) Calix[8]arene Complex and Their Reactivity with Diiron Nonacarbonyl. *Inorganic Chemistry*, Vol. 46, No. 18, (September 2007), pp. 7579-7586, ISSN 0020-1669

Yan, C.-G., Han, J., Li, L., & Liu, D.-M. (2009). Dicopper Complex of *p-tert*-Butylcalix[8]arene Bearing Acylhydrazone Pendant Domains. *Journal of Coordination Chemistry*, Vol. 62, No. 5, (March 2009), pp. 825-832, ISSN 0095-8972

Zhang, Y., & Coppens, P. (2001). Calix[8]arene/6 Pyridine, in: *Private Communications to the CSD, Deposition number 162605*, 13.05.2011, Available from http://www.ccdc.cam.ac.uk/cgi-bin/catreq.cgi

2

Polycyclic Aromatic Ketones – A Crystallographic and Theoretical Study of Acetyl Anthracenes

Sergey Pogodin, Shmuel Cohen, Tahani Mala'bi and Israel Agranat
Organic Chemistry, Institute of Chemistry, The Hebrew University of Jerusalem, Edmond J. Safra Campus, Jerusalem
Israel

1. Introduction

"Acylation differs from alkylation in being virtually irreversible" [Olah, 1973], free of rearrangements and isomerizations [Wang, 2009; Norman & Taylor, 1965]. This authoritative exposition of the state of the art of Friedel-Crafts chemistry in 1973 close to the centennial of the invention of the Friedel-Crafts reaction has been long recognized and not without reason. The difference in behavior between Friedel-Crafts acylation and Friedel-Crafts alkylation was attributed to the resonance stabilization existing between the acyl group and the aromatic nucleus [Buehler & Pearson, 1970], which may serve as a barrier against rearrangements and reversible processes. However, if the acyl group is tilted out of the plane of the aromatic nucleus, e.g., by bulky substituents, the resonance stabilization is reduced and the pattern of irreversibility of Friedel-Crafts acylation may be challenged [Buehler & Pearson, 1970; Pearson & Buehler, 1971; Gore, 1974]. Under these conditions deacylations and acyl rearrangements become feasible [Buehler & Pearson, 1970; Pearson & Buehler, 1971; Gore, 1974].

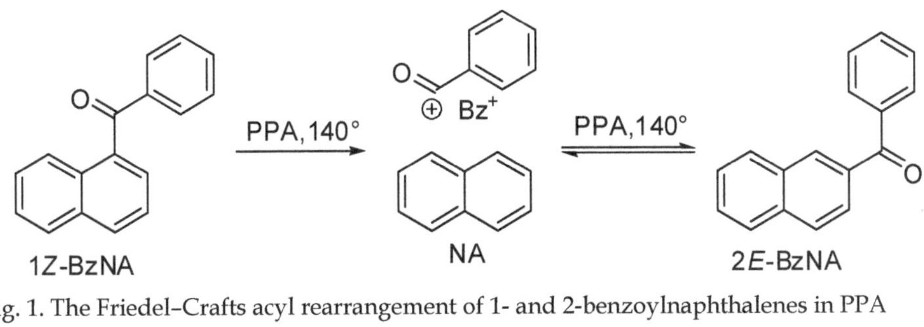

Fig. 1. The Friedel-Crafts acyl rearrangement of 1- and 2-benzoylnaphthalenes in PPA

The concept of reversibility in Friedel-Crafts acylations [Gore, 1955, 1964] was put forward in 1955 by Gore, who proposed that "the Friedel-Crafts acylation reaction of reactive hydrocarbons is a reversible process" [Gore, 1955]. Gore concluded that "Reversibility is an important factor in acylation reactions" [Gore, 1955]. The reversibility studies have been focused mainly on unusual aspects of selectivity, including deacylations, one-way rearrangements and kinetic versus thermodynamic control [Gore, 1974]. Under classical Friedel-Crafts conditions (e. g., $AlCl_3$ and a trace of water), the pattern of irreversibility (e. g., in the naphthalene series) has been highlighted [Gore, 1964, 1974; Andreou et al., 1978; Dowdy et al., 1991].

The incursion of reversibility in Friedel-Crafts acylations was revealed by Agranat, et al. in the benzoylation of naphthalene in polyphosphoric acid (PPA) at elevated temperatures (Fig. 1) [Agranat et al., 1974]. The kinetically controlled 1-benzoylnaphthalene rearranged to the thermodynamically controlled 2-benzoylnaphthalene (PPA, 140 °C) (*vide infra*). The reversibility concept was then applied to the synthesis of linearly annelated polycyclic aromatic ketones by intramolecular Friedel-Crafts rearrangements of their angularly annelated constitutional isomers [Agranat & Shih, 1974a; Heaney, 1991]. The Haworth synthesis of PAHs, which previously had allowed access to angularly annelated PAHs could thus be applied to the synthesis of linearly annelated PAHs [Agranat & Shih, 1974b]. Further experimental evidence in support of true reversibility of Friedel-Crafts acylation is limited [Frangopol et al., 1964; Balaban, 1966; Nenitzescu & Balaban, 1964; Effenberger et al., 1973; Levy et al., 2007; Mala'bi et al,. 2009; Titinchi et al., 2008; Adams et al., 1998; Okamoto & Yonezawa, 2009]. Notable cases are the report by Balaban [Frangopol et al., 1964; Balaban, 1966; Nenitzescu & Balaban, 1964] on the reversibility of Friedel-Crafts acetylation of olefins to β-chloroketones, the report by Effenberger [Effenberger et al., 1973] of the retro-Fries rearrangement of phenyl benzoates (CF_3SO_3H, 170 °C) and the reversible ArS_E aroylation of naphthalene derivatives [Okamoto & Yonezawa, 2009]. Additional examples are the acyl rearrangements of acetylphenanthrenes [Levy et al., 2007] and acetylanthracenes [Mala'bi et al., 2009] in PPA, the acetylation of fluorene [Titinchi et al., 2008], the disproportionation of 9-acetylanthracene into 1,5- and 1,8-diacetylanthracenes in an ionic liquid systems [Adams et al., 1998]. Complete reversibility of Friedel-Crafts acylation was established in the intramolecular *para* ⇌ *ortho* acyl rearrangements of fluorofluorenones in PPA (Fig. 2) [Agranat et al., 1977]. Friedel-Crafts acyl rearrangement of polycyclic aromatic ketones (PAKs) has been referred to as the Agranat-Gore rearrangement [Levy et al., 2007; Mala'bi et al., 2009]. The Friedel-Crafts acylation can be adjusted to give a kinetically controlled ketone or a thermodynamically controlled ketone [Buehler & Pearson, 1970]. Acyl rearrangements and reversibility in Friedel-Crafts acylations have been associated with thermodynamic control [Pearson & Buehler, 1971; Andreou et al., 1978; Agranat et al., 1977]. The contributions of kinetic control vs. thermodynamic control in Friedel-Crafts acyl rearrangements remain an open question, in spite of the rich chemistry of Friedel-Crafts acylations. We have recently shown that kinetic control wins out over thermodynamic control in the Friedel-Crafts acyl rearrangement of diacetylanthracenes in PPA [Mala'bi et al., 2011].

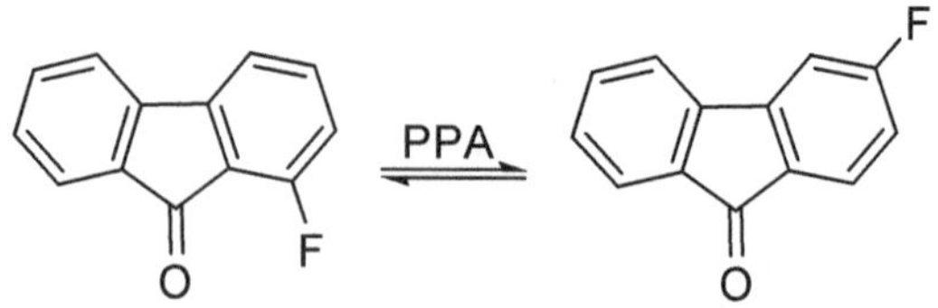

Fig. 2. The Friedel-Crafts intramolecular acyl rearrangements of fluorofluorenones in PPA

A plausible mechanism of the Friedel-Crafts acyl rearrangement of 1-benzoylnaphthalene (1-BzNA) into 2-benzoylnaphthalene (2-BzNA) in PPA, is presented in Fig. 3. The mechanism involves the two dibenzoylnaphthalenes, their *O*-protonates and their σ-complexes. In the kinetically controlled pathway 1σ-BzNAH$^+$ is more stable than 2σ-BzNAH$^+$ and by virtue of the Hammond-Leffler postulate [Muller, 1994] the transition state leading to 1σ-BzNAH$^+$ is lower in energy than the transition state leading to 2σ-BzNAH$^+$. Thus, 1-BzNA is the kinetically controlled product. By contrast, in the thermodynamically controlled pathway, 1-BzNAH$^+$ and 1-BzNA are less stable than 2-BzNAH$^+$ and 2-BzNA, respectively. Therefore, 2-BzNA is the thermodynamically controlled product. Under conditions of thermodynamic control, the relative stabilities of the constitutional isomers of a given PAK are detrimental to the products of the Friedel-Crafts acyl rearrangement of the PAK and of the Friedel-Crafts acylation of the corresponding PAH.

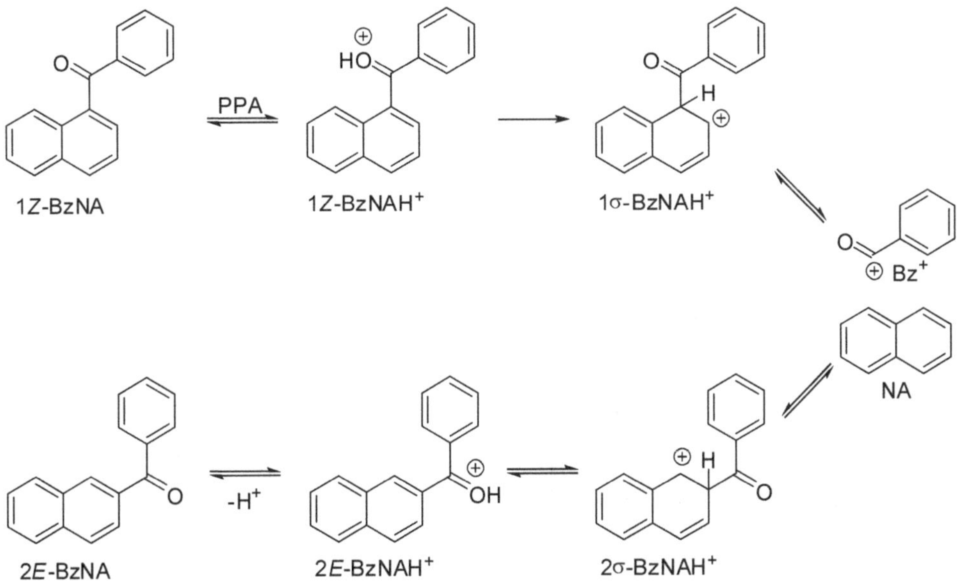

Fig. 3. The mechanism of the Friedel-Crafts acyl rearrangement of representative PAKs 1- and 2-benzoylnaphthalenes.

2. Structures of monoacetylanthracenes and diacetylanthracenes

Anthracene (AN) is essentially a planar PAH. Due to its D_{2h} symmetry, three constitutional isomers of acetylanthracenes (AcAN) are possible: 1-acetylanthracene (1-AcAN), 2-acetylanthracene (2-AcAN), and 9-acetylanthracene (9-AcAN) (see Fig. 4). These isomers differ in the position of the acetyl substituent at the anthracene ring system. The three constitutional isomers of AcAN can be categorized, depending on the degree of their overcrowding: (i) the non-overcrowded isomer 2-AcAN, in which the acetyl group is flanked by two *ortho*-hydrogens (H^1, H^3); (ii) the overcrowded isomer 1-AcAN, in which the overcrowding is due to the presence of one hydrogen atom (H^9) *peri* to the acetyl group; (iii) the severely overcrowded isomer 9-AcAN (assuming the planar conformation), in which the overcrowding is due to the presence of two *peri*-hydrogens (H^1, H^8) to the acetyl group. The overcrowding in 1-AcAN and 9-AcAN should result in significant deviations of the acetyl groups from the plane of the anthracene nucleus, which is expected to encourage reversibility and rearrangements.

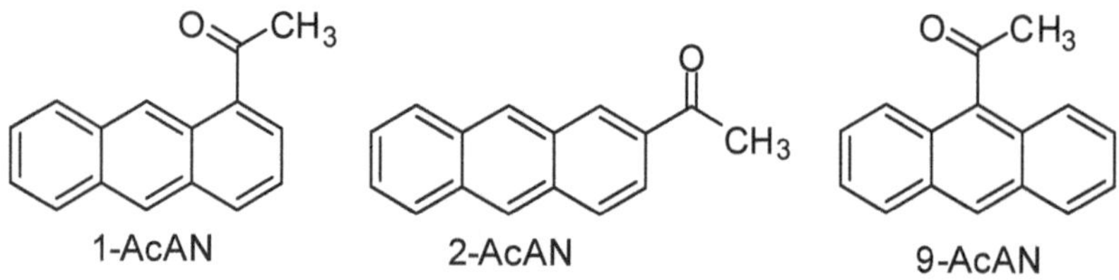

Fig. 4. Constitutional isomers of monoacetylanthracenes (*E* and *Z* stereodescriptors are omitted)

There are 15 constitutional isomers of diacetylanthracenes (Ac_2AN), shown in Fig. 5. These isomers differ in the position of the acetyl substituents at the anthracene ring system. The present study encompasses the three monoacetylanthracenes 1-AcAN, 2-AcAN, 9-AcAN and the following eleven diacetylanthracenes: 1,5-Ac_2AN, 1,6-Ac_2AN, 1,7-Ac_2AN, 1,8-Ac_2AN, 1,9-Ac_2AN, 1,10-Ac_2AN, 2,6-Ac_2AN, 2,7-Ac_2AN, 2,9-Ac_2AN, 2,10-Ac_2AN and 9,10-Ac_2AN. The remaining diacetylanthracenes, 1,2-Ac_2AN, 1,3-Ac_2AN, 1,4-Ac_2AN and 2,3-Ac_2AN were not included in the present study. These constitutional isomers are not expected to be formed in the Friedel–Crafts acylations of 1-AcAN and 2-AcAN, due to the deactivation effect of the electron-withdrawing acetyl group towards further acetylation. This effect is not necessarily operating with respect to the "remote" unsubstituted benzene ring.

In 1-AcAN and 2-AcAN, the *E*- and *Z*-diastereomers should be considered. *E* and *Z* are the stereodescriptors applied to monoacetylanthracenes and diacetylanthracenes with a fractional bond order of the bond between the carbonyl carbon and the corresponding aromatic carbon [Moss, 1996]. In diacetylanthracenes, four diastereomers should be considered: *ZZ*, *ZE*, *EZ* and *EE*. Depending on the symmetry of a given diacetylanthracene, *ZE* and *EZ* diastereomers could be equivalent. 9,10-AcAN is a special case: only one stereodescriptor, *Z* or *E*, is required. In this case, *Z* or *E* refers to whether the carbonyl bonds lie on the same or on the opposite sides of the plane containing the C^9–C^{11} and C^{10}–C^{12} bonds and perpendicular to the aromatic plane.

Acetylanthracene 2-AcAN and diacetylanthracenes 1,5-Ac_2AN, 1,6-Ac_2AN, 1,7-Ac_2AN, 1,8-Ac_2AN, 2,7-Ac_2AN and 9,10-Ac_2AN have been synthesized in the present study and their crystal structures have been determined. Ketones 2-AcAN, 1,5-Ac_2AN and 1,8-Ac_2AN have been prepared by the Friedel–Crafts acetylation of anthracene. Ketones 1,6-Ac_2AN, 1,7-Ac_2AN and 2,7-Ac_2AN have been prepared by the Friedel–Crafts acetylation of 2-AcAN. Ketone 9,10-Ac_2AN has been synthesized by methylation (MeLi) of 9,10-dicarbomethoxyanthracene (prepared from 9,10-dibromoanthracene via 9,10-anthracenedicarboxylic acid). Ketones 1,7-Ac_2AN and 2,7-Ac_2AN are reported here for the first time.

The present study encompasses the crystal and molecular structures of monoacetylanthracenes (AcANs) and diacetylanthracenes (Ac_2ANs), the results of a systematic DFT study of the structures and the conformational spaces of AcANs and Ac_2ANs, as well as the comparison between the calculated and the experimental structures of these PAKs.

2.1 Molecular and crystal structure of monoacetylanthracenes and diacetylanthracenes

Of the three monoacetylanthracenes and eleven diacetylanthracenes included in the present study, only the crystal structures of 1-AcAN [Langer & Becker, 1993], 9-AcAN [Anderson et al., 1984; Zouev et al., 2011] and 1,5-AcAN [Li & Jing, 2006] have previously been described. The molecular and crystal structures of 2-AcAN, 1,6-Ac_2AN, 1,7-Ac_2AN, 1,8-Ac_2AN, 2,7-Ac_2AN and 9,10-Ac_2AN are reported here for the first time, along with the previously reported structures.

2.1.1 Geometries of monoacetylanthracenes and diacetylanthracenes

Table 1 shows the crystallographic data for 2-AcAN, 1,5-Ac_2AN, 1,6-Ac_2AN, 1,7-Ac_2AN, 1,8-Ac_2AN, 2,7-Ac_2AN and 9,10-Ac_2AN.[1] The ORTEP diagrams of 2-AcAN and of the six diacetylanthracenes as determined by X-ray crystallography are presented in Fig. 6–10. Ketones 2-AcAN, 1,5-Ac_2AN, 1,8-Ac_2AN and 2,7-Ac_2AN crystallize in the monoclinic space groups $P2_1/n$, $P2_1/c$, $P2/n$ and $I2/a$, respectively. The unit cell dimensions of the crystal structure of 1,5-Ac_2AN are essentially identical to those reported in the literature [Li & Jing, 2006]. Ketones 1,6-Ac_2AN and 1,7-Ac_2AN crystallize in the triclinic space group *P*-1. Ketone 9,10-Ac_2AN crystallizes in the orthorhombic space group $Pna2_1$. Table 2 gives selected geometrical parameters derived from the X-ray crystal structures of the mono- and diacetylanthracenes under study. The following geometrical parameters were considered: the twist angles $\tau(C_{arom}–C_{arom}–C_{carb}–O)$ (divided into τ_1, τ_2 and τ_9 depending on the position of the acetyl group) and $\upsilon(C_{arom}–C_{arom}–C_{carb}–O)$ around the anthracenyl-carbonyl bond; the dihedral angle θ between the least-square planes of the carbonyl group and the anthracene system; the dihedral angle φ between the least-square planes of two side rings of the anthracene system; the pyramidalization angles χ at C_{arom} and C_{carb}. Table 3 gives the bond lengths in the mono- and diacetylanthracenes under study, as compared with the parent anthracene.

The data presented in Table 3 indicate the considerable variation in bond lengths in mono- and diacetylanthracenes. The bond lengths may be classified into several types: four C^1–C^2, or α-β, bonds (134.2–137.7 pm), two C^2–C^3, or β-β, bonds (138.7–144.4 pm), four C^1–C^{9a} bonds (141.8–145.5 pm), four C^{9a}–C^9 bonds (138.3–140.9 pm), and two C^{4a}–C^{9a} bonds (142.8–145.3 pm). These values are in the same range as the respective bond lengths in the X-ray crystal structure of anthracene, which are 136.1, 142.8, 143.4, 140.1 and 143.6 pm [Brock & Dunitz, 1990]. It has previously been shown that the bond lengths in anthracene are in agreement with the superposition of its four Kekulé structures and with the free valence numbers [Sinclair et al., 1950]. Table 3 also shows that the bonds adjacent to the acetyl group are elongated as compared to the respective bonds in anthracene, e.g. the C^2–C^3 bond in 2-AcAN (143.2 pm vs. 142.8 pm), the C^1–C^2 bonds in 1,5-Ac_2AN and 1,8-Ac_2AN (137.5 pm vs. 136.1 pm), the C^5–C^6 bonds in 1,6-Ac_2AN (137.0 pm vs. 136.1 pm), the C^7–C^8 bond in 1,7-Ac_2AN (137.4 pm vs. 136.1 pm) and the C^2–C^3 bond in 2,7-Ac_2AN (144.4 pm vs. 142.8 pm). This elongation effect stems from the contributions of dipolar Kekulé structures, in which the anthracene bonds adjacent to the acetyl group are necessarily single. This effect, however, is not noticeable in 9,10-Ac_2AN, because the carbonyl groups are almost perpendicular to the aromatic plane and are hardly conjugated with the anthracene system.

---

[1] CCDC 839159 – 839165 contains the supplementary crystallographic data for this paper. These data can be obtained free of charge from The Cambridge Crystallographic Data Centre via www.ccdc.cam.ac.uk/data_request/cif.

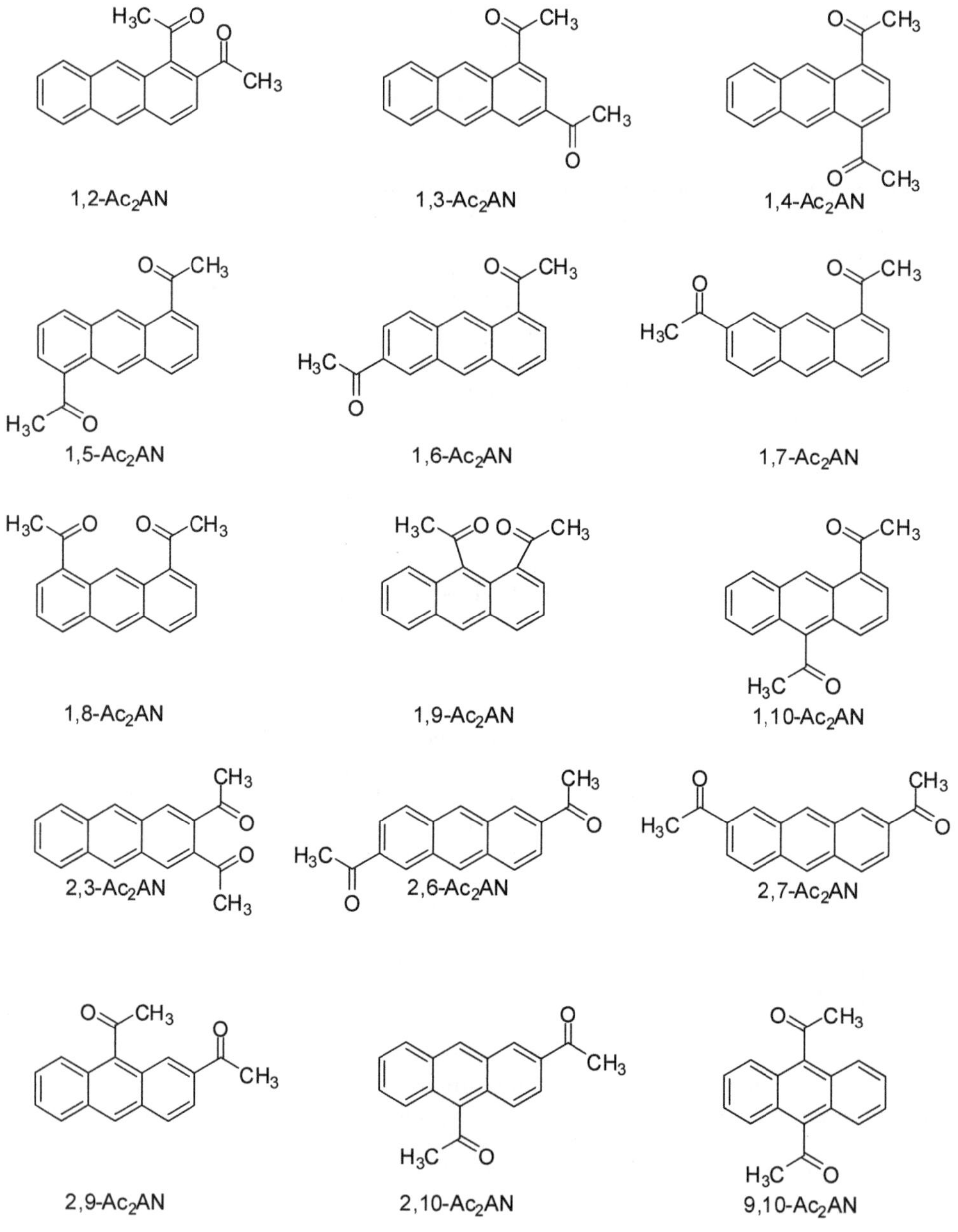

Fig. 5. Constitutional isomers of diacetylanthracenes (*E* and *Z* stereodescriptors are omitted)

	2-AcAN	1,5-Ac_2AN	1,6-Ac_2AN	1,7-Ac_2AN	1,8-Ac_2AN	2,7-Ac_2AN	9,10-Ac_2AN
Formula	$C_{16}H_{12}O$	$C_{18}H_{14}O_2$	$C_{18}H_{14}O_2$	$C_{18}H_{14}O_2$	$C_{18}H_{14}O_2$	$C_{18}H_{14}O_2$	$C_{18}H_{14}O_2$
Temp, K	173(1)	173(1)	173(1)	123(2)	173(1)	173(1)	173(1)
Crystal system	Monoclinic	Monoclinic	Triclinic	Triclinic	Monoclinic	Monoclinic	Orthorhombic
Space group	$P2_1/n$	$P2_1/c$	P-1	P-1	$P2/n$	$I2/a$	$Pna2_1$
a, Å	6.031(2)	9.8394(9)	7.5776(12)	7.9765(6)	12.6504(9)	17.003(4)	10.4235(14)
b, Å	7.394(3)	6.2073(6)	8.8581(14)	8.2802(6)	8.5008(6)	5.8390(13)	7.9835(10)
c, Å	24.847(9)	10.8876(10)	10.1653(16)	11.2321(8)	12.6756(9)	26.395(6)	16.164(2)
α, deg	90.0	90.0	92.062(3)	96.661(1)	90.0	90.0	90.0
β, deg	90.051	107.630(1)	94.348(3)	96.863(1)	108.605(1)	94.592(4)	90.0
γ, deg	90.0	90.0	110.604(3)	115.295(1)	90.0	90.0	90.0
Volume, Å^3	1108.07(7)	633.7(1)	637.9(2)	654.28(8)	1291.9(2)	2612.1(10)	1345.1(3)
Z	4	2	2	2	4	8	4
Calc density Mg/m^3	1.320	1.375	1.365	1.331	1.349	1.334	1.295
Crystal size							
max, mm	0.16	0.27	0.40	0.25	0.42	0.37	0.27
mid, mm	0.14	0.26	0.20	0.22	0.40	0.18	0.24
min, mm	0.06	0.23	0.15	0.13	0.28	0.09	0.17
Reflections collected	6443	6860	5077	7510	14485	14453	13845
Independent	2581	1494	2875	3033	3089	3124	2936
reflections	R_{int}=0.0676	R_{int}=0.0231	R_{int}=0.0181	R_{int}=0.0181	R_{int}=0.0257	R_{int}=0.0390	R_{int}=0.0263
Reflections with I>2σ(I)	1307	1429	2336	2689	2841	2275	2865
Final R indices	R_1=0.0880	R_1=0.0496	R_1=0.0583	R_1=0.0470	R_1=0.0705	R_1=0.0771	R_1=0.0510
[I>2σI]	wR_2=0.1803	wR_2=0.1253	wR_2=0.1498	wR_2=0.1312	wR_2=0.1847	wR_2=0.1747	wR_2=0.1242
R indices	R_1=0.1691	R_1=0.0518	R_1=0.0710	R_1=0.0523	R_1=0.0748	R_1=0.1051	R_1=0.0520
(all data)	wR_2=0.2185	wR_2=0.1271	wR_2=0.1588	wR_2=0.1362	wR_2=0.1900	wR_2=0.1896	wR_2=0.1248

Table 1. Crystallographic data for acetylanthracenes 2-AcAN, 1,5-Ac_2AN, 1,6-Ac_2AN, 1,7-Ac_2AN, 1,8-Ac_2AN, 2,7-Ac_2AN and 9,10-Ac_2AN.

Compound			τ^a	υ^b	θ	φ	C_{carb}–C_{an}^c	$\chi(C_{arom})$	$\chi(C_{carb})$	O···H		CH_3···H	
			deg	deg	deg	deg	pm	deg	deg	pm		pm	
1-AcAN	*Z*	C_i	27.1	-152.7	28.6	3.2	149.3	-0.1	-0.2	223.4	H^9	223.4	H^2
2-AcAN	*E*	C_1	173.1	-5.3	5.9	0.4	148.6	-1.6	0.4	249.2	H^3	244.3	H^1
9-AcAN	--	C_1	87.9	-91.8	89.2	5.8	150.4	0.3	-0.8	294.0	H^1	263.1	H^8
1,5-Ac_2AN	*ZZ*	C_i	20.0	-156.8	22.7	0.0	149.4	-3.2	3.4	226.9	H^9	243.7	H^2
1,6-Ac_2AN	*ZE*	C_1	30.0	-147.1	32.2	1.3	150.1	-2.9	3.0	228.8	H^9	230.0	H^2
			178.6	-0.7	1.9		149.3	-0.7	-0.3	249.9	H^7	230.1	H^5
1,7-Ac_2AN	*ZE*	C_1	-15.2	162.9	16.0	2.3	149.8	1.9	-1.9	221.3	H^9	229.2	H^2
			-176.6	3.7	4.5		149.0	0.3	-0.6	247.9	H^6	229.5	H^8
1,8-Ac_2AN	*ZZ*	C_2	-34.0	145.4	36.0	0.3	149.3	0.6	0.3	228.2	H^9	231.1	H^2
			-32.4	144.9	35.4	3.4	148.9	2.7	0.2	225.9	H^9	226.7	H^2
2,7-Ac_2AN	*EZ*	C_1	171.9	-3.4	9.9	2.9	149.0	-4.7	3.0	253.9	H^3	239.0	H^1
			0.9	-178.5	2.0		148.9	0.7	-1.3	246.0	H^8	226.4	H^3
9,10-Ac_2AN	*E*	C_1	-85.0	94.0	86.7	1.6	151.3	-1.0	-1.2	288.2	H^8	257.4	H^1
			87.0	-93.7	86.5		151.5	-0.6	0.1	290.5	H^4	264.4	H^5

[a] $\tau_1(C^{9a}$–C^1–C^{11}–$O^{15})$ for 1-AcAN, 1,5-Ac_2AN, 1,6-Ac_2AN, 1,7-Ac_2AN and 1,8-Ac_2AN, $\tau_2(C^1$–C^2–C^{11}–$O^{15})$ for 2-AcAN and 2,7-Ac_2AN, $\tau_2(C^5$–C^6–C^{13}–$O^{16})$ for 1,6-Ac_2AN, $\tau_2(C^8$–C^7–C^{13}–$O^{16})$ for 1,7-Ac_2AN, $\tau_9(C^{9a}$–C^9–C^{11}–$O^{15})$ for 9-AcAN and 9,10-Ac_2AN.

[b] $\upsilon_1(C^2$–C^1–C^{11}–$O^{15})$ for 1-AcAN, 1,5-Ac_2AN, 1,6-Ac_2AN, 1,7-Ac_2AN and 1,8-Ac_2AN, $\upsilon_2(C^3$–C^2–C^{11}–$O^{15})$ for 2-AcAN and 2,7-Ac_2AN, $\upsilon_2(C^7$–C^6–C^{13}–$O^{16})$ for 1,6-Ac_2AN, $\upsilon_2(C^6$–C^7–C^{13}–$O^{16})$ for 1,7-Ac_2AN, $\upsilon_9(C^{8a}$–C^9–C^{11}–$O^{15})$ for 9-AcAN and 9,10-Ac_2AN.

[c] C^1–C^{11} for 1-AcAN, 1,5-Ac_2AN, 1,6-Ac_2AN, 1,7-Ac_2AN and 1,8-Ac_2AN, C^2–C^{11} for 2-AcAN and 2,7-Ac_2AN, C^6–C^{13} for 1,6-Ac_2AN, C^7–C^{13} for 1,7-Ac_2AN, C^9–C^{11} for 9-AcAN and 9,10-Ac_2AN.

Table 2. Selected geometrical parameters of the X-ray molecular structures of acetylanthracenes 2-AcAN, 1,5-Ac_2AN, 1,6-Ac_2AN, 1,7-Ac_2AN, 1,8-Ac_2AN, 2,7-Ac_2AN and 9,10-Ac_2AN.

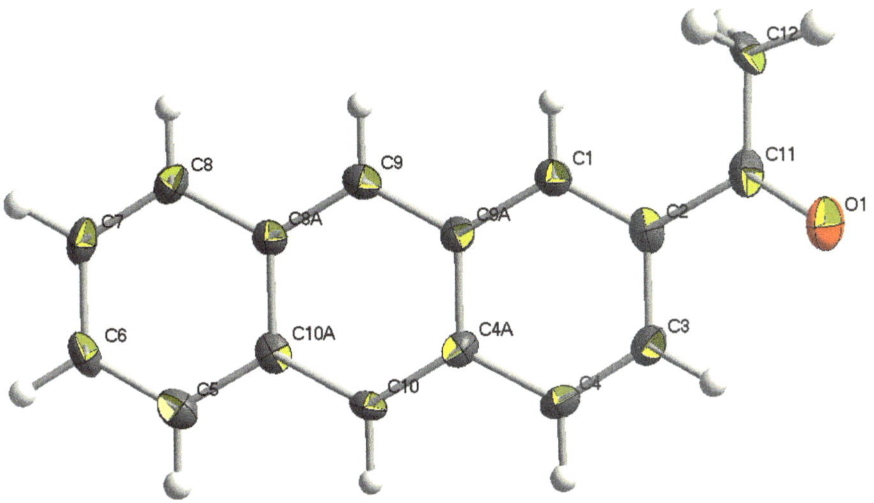

Fig. 6. ORTEP drawing of the crystal structure of 2-AcAN, scaled to enclose 50% probability

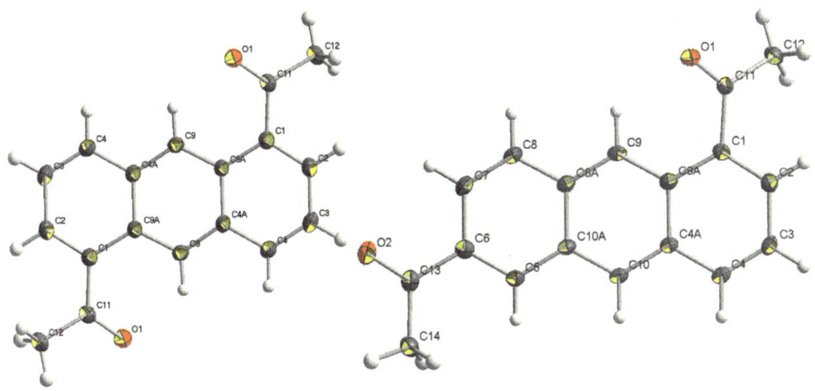

Fig. 7. ORTEP drawings of the crystal structures of 1,5-Ac_2AN (left) and 1,6-Ac_2AN (right), scaled to enclose 50% probability

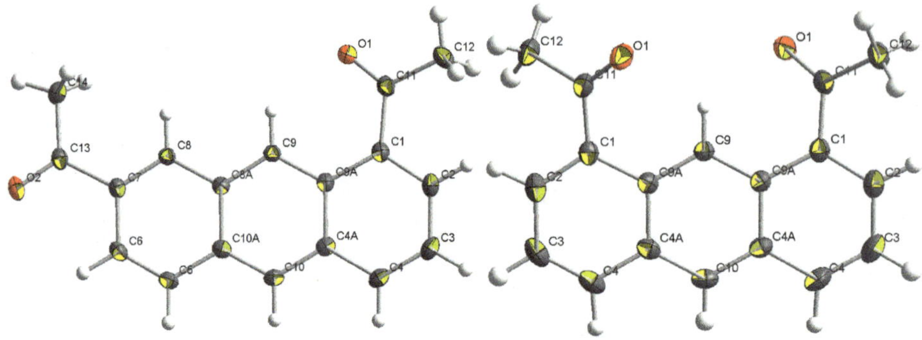

Fig. 8. ORTEP drawings of the crystal structures of 1,7-Ac$_2$AN (left) and 1,8-Ac$_2$AN (right), scaled to enclose 50% probability

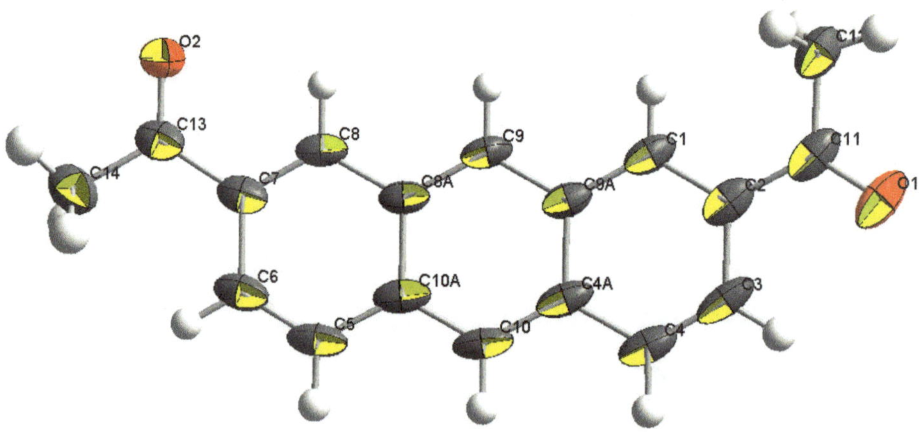

Fig. 9. ORTEP drawings of the crystal structure of 2,7-Ac$_2$AN, scaled to enclose 50% probability

	C^1–C^2	C^2–C^3	C^3–C^4	C^4–C^{4a}	C^{4a}–C^{9a}	C^{4a}–C^{10}
AN[a]	136.1	142.8	136.1	143.4	143.6	140.1
1-AcAN[b]	137.6(3)	138.7(3)	134.6(3)	143.1(3)	143.1(3)	139.2(3)
2-AcAN	136.4(4)	143.2(4)	134.6(4)	144.10(4)	143.0(4)	138.7(4)
9-AcAN[c]	134.9(3)	141.2(3)	135.0(3)	142.9(3)	143.2(3)	138.8(3)
1,5-Ac$_2$AN	137.5(2)	141.4(2)	135.6(2)	142.9(2)	143.8(2)	139.9(2)
1,6-Ac$_2$AN	136.9(2)	142.0(2)	135.5(2)	142.7(2)	144.1(2)	140.0(2)
1,7-Ac$_2$AN	137.7(2)	141.5(2)	135.6(2)	143.0(2)	144.0(2)	140.0(2)
1,8-Ac$_2$AN	137.3(2)	141.4(3)	135.5(2)	142.8(2)	143.7(2)	139.2(2)
1,8-Ac$_2$AN	137.5(2)	141.2(3)	135.2(3)	142.9(2)	143.7(2)	139.2(2)
2,7-Ac$_2$AN	137.1(3)	144.4(3)	134.2(4)	142.5(3)	144.6(3)	138.3(3)
9,10-Ac$_2$AN	136.3(3)	141.8(4)	134.6(4)	142.7(3)	143.8(3)	140.9(3)

	C^{10}–C^{10a}	C^{10a}–C^{8a}	C^{10a}–C^5	C^5–C^6	C^6–C^7	C^7–C^8	C^8–C^{8a}
AN[a]	140.1	143.6	143.4	136.1	142.8	136.1	143.4
1-AcAN[b]	139.5(3)	142.8(3)	143.0(3)	134.6(3)	140.8(3)	135.1(3)	142.2(3)
2-AcAN	139.3(4)	143.6(4)	142.4(4)	136.1(4)	141.3(4)	135.0(4)	143.0(4)
9-AcAN[c]	139.1(3)	142.9(3)	142.4(3)	134.5(3)	140.4(3)	135.6(3)	142.7(3)
1,5-Ac$_2$AN	139.7(2)	143.8(2)	145.0(2)	137.5(2)	141.4(2)	135.6(2)	142.9(2)
1,6-Ac$_2$AN	139.2(2)	143.7(2)	142.5(2)	137.0(2)	143.9(2)	135.5(2)	143.1(2)
1,7-Ac$_2$AN	139.0(2)	142.8(2)	143.3(2)	135.6(2)	142.8(2)	137.4(1)	142.7(1)
1,8-Ac$_2$AN	139.2(2)	143.7(2)	142.8(2)	135.5(2)	141.4(3)	137.3(2)	144.4(2)
1,8-Ac$_2$AN	139.2(2)	143.7(2)	142.9(2)	135.2(3)	141.2(3)	137.5(2)	144.9(2)
2,7-Ac$_2$AN	139.4(3)	144.3(3)	142.3(3)	134.9(3)	142.8(3)	136.6(3)	141.8(3)
9,10-Ac$_2$AN	139.4(3)	143.8(3)	143.3(3)	135.2(4)	141.1(4)	136.5(4)	143.0(3)

	C^{8a}–C^9	C^9–C^{9a}	C^{9a}–C^1	C_{ar}–C^{11}	C^{11}–C^{12}	C^{11}–O
AN[a]	140.1	140.1	143.4	-	-	-
1-AcAN[b]	139.0(3)	138.9(3)	144.8(3)	149.3(3)	148.8(3)	121.7(3)
2-AcAN	139.3(4)	139.7(4)	142.3(4)	148.6(4)	149.5(4)	121.5(3)
9-AcAN[c]	140.3(3)	140.2(3)	142.4(3)	150.3(3)	148.5(3)	120.8(2)
1,5-Ac$_2$AN	139.9(2)	139.7(2)	145.0(2)	149.4(2)	150.9(2)	121.8(2)
1,6-Ac$_2$AN	139.5(2)	140.0(2)	144.6(2)	150.1(2)	150.5(2)	121.7(2)
				149.3(2)	150.6(2)	121.8(2)
1,7-Ac$_2$AN	140.5(1)	140.1(1)	145.5(1)	149.8(2)	151.6(2)	121.7(1)
				149.0(2)	150.1(2)	122.1(1)
1,8-Ac$_2$AN	140.1(2)	140.1(2)	144.4(2)	149.3(2)	151.2(2)	121.5(2)
1,8-Ac$_2$AN	139.9(2)	139.9(2)	144.9(2)	148.9(2)	151.4(2)	121.4(2)
2,7-Ac$_2$AN	139.3(3)	139.6(3)	141.9(3)	149.0(4)	150.0(4)	122.1(3)
				148.9(3)	149.5(3)	121.0(3)
9,10-Ac$_2$AN	140.0(3)	140.1(3)	142.9(3)	151.4(3)	149.4(3)	121.3(3)
				151.5(3)	149.4(4)	120.0(3)

[a] Brock & Dunitz, 1990; averaged bonds lengths

[b] Langer & Becker, 1993

[c] Zouev et al., 2011

Table 3. Bond lengths (pm) in the X-ray structures of anthracene, monoacetylanthracenes and diacetylanthracenes.

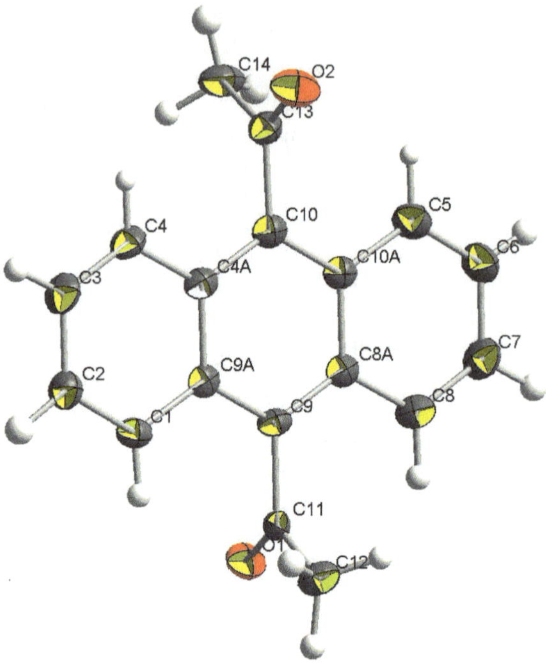

Fig. 10. ORTEP drawings of the crystal structure of 9,10-Ac_2AN, scaled to enclose 50% probability

Ketone 2-AcAN crystallizes in the *E* conformation with a small twist angle of $\tau_2(C^1\text{-}C^2\text{-}C^{11}\text{-}O^{13})=173.1°$ and a small dihedral angle $\theta=5.9°$. There are two pairs of enantiomeric molecules in the unit cell of 2-AcAN. The structure does not contain any short contact distances. Ketone 1,5-Ac_2AN adopts the *Z,Z* conformation with large twist angles, $\tau_1(C^{9a}\text{-}C^1\text{-}C^{11}\text{-}O^{15})=20.0°$, $\tau_1(C^{10a}\text{-}C^5\text{-}C^{13}\text{-}O^{16})=-20.0°$. The $O^{15}...H^9$ and $O^{16}...H^{10}$ contact distances are 226.9 pm, which is slightly shorter (7% penetration) than the sum of the respective van der Waals radii of hydrogen (115 ppm) and oxygen (129ppm) [Zefirov, 1997]. There are two identical molecules of 1,5-Ac_2AN in the unit cell, each posessing the C_i symmetry. Ketone 1,6-Ac_2AN crystallizes in the *Z,E* conformations, with a large twist angle of the *Z* carbonyl group, $\tau_1(C^{9a}\text{-}C^1\text{-}C^{11}\text{-}O^{15})=30.0°$ and a small twist of the *E* carbonyl group, $\tau_2(C^5\text{-}C^6\text{-}C^{13}\text{-}O^{16})=178.6°$. The $O^{15}...H^9$ contact distance is 228.8 pm (6% penetration), while the $O^{16}...H^7$ contact distance is 249.9 pm. There are two enantiomeric molecules in the unit cell of 1,6-Ac_2AN. Ketone 1,7-Ac_2AN also crystallizes in the *Z,E* conformations, with the twist angles of $\tau_1(C^{9a}\text{-}C^1\text{-}C^{11}\text{-}O^{15})=-15.2°$ and $\tau_2(C^8\text{-}C^7\text{-}C^{13}\text{-}O^{16})=-176.6°$. The $O^{15}...H^9$ contact distance is 221.3 pm (9% penetration). There are two enantiomeric molecules in the unit cell of 1,7-Ac_2AN. Ketone 1,8-Ac_2AN adopts the *Z,Z* conformation with two large twist angles, due to the repulsive *peri*-interactions $O^{15}...H^9$ and $O^{16}...H^9$ (225.9 and 228.2 pm) between two carbonyl oxygens and the same aromatic hydrogen. There are two enantiomeric pairs of non-equivalent molecules, A and B, in the unit cell of 1,8-Ac_2AN, each of them posessing the C_2 symmetry. The respective twist angles are $\tau_1(C^{9a}\text{-}C^1\text{-}C^{11}\text{-}O^{15})=-34.0°$ (A) and $\tau_1(C^{9a}\text{-}C^1\text{-}C^{11}\text{-}O^{15})=-32.4°$ (B). Ketone 2,7-Ac_2AN adopts the *E,Z* conformation, which is similar to those of 1,6-Ac_2AN and 1,7-Ac_2AN but with a larger twist of the *E*-acetyl group, $\tau_2(C^1-C^2-C^{11}-O^{15})$=171.9°, and a smaller twist of the *Z*-acetyl group, $\tau_2(C^6-C^7-C^{12}-O^{16})$=0.9°. There are four pairs of enantiomeric molecules in the unit cell of 2,7-Ac_2AN. Ketone 9,10-Ac_2AN crystallizes in the *E* conformation with the twist angles of $\tau_9(C^{9a}-C^9-C^{11}-O^{15})$=-85.0° and $\tau_9(C^{10a}-C^{10}-C^{13}-O^{16})$=87.0°. There are two pairs of enantiomeric molecules in the unit cell of 9,10-Ac_2AN. According to the literature structure [Langer & Becker, 1993], 1-AcAN crystallizes in the *Z*-conformation with a twist angle of $\tau_1(C^{9a}-C^1-C^{11}-O^{13})$=27.1°. The carbonyl group of ketone 9-AcAN [Zouev2011] is nearly orthogonal to the aromatic plane, $\tau_9(C^1-C^2-C^{11}-O^{13})$=87.9°.

None of the mono- and diacetylanthracenes under study adopts a planar conformation in their crystal structures. The absolute values of the twist angles in the mono- and diacetylanthracenes vary, depending on the position of substitution and on the conformation of the acetyl groups: $|\tau_1|$=15.2–34.0° for the 1*Z*-acetyl groups, $|\tau_2|$=0.9° for the 2*Z*-acetyl group, $|\tau_2|$=171.9–178.6° for the 2*E*-acetyl groups, and $|\tau_9|$=85.0–87.9°. The higher twist angles of the 1*Z*-acetyl groups are caused by the repulsive interactions between the carbonyl oxygen and the respective *peri*-hydrogen. The acetyl groups themselves are nealry planar (excluding the methyl hydrogens), and the pyramidalization angles χ at the carbonyl carbon atom are small, 0.1–3.4°. The dihedral angles θ between the plane of the carbonyl group and the aromatic plane are very close to the respective twist angles τ (Table 2). The anthracene systems in the mono- and diacetylanthracenes under study are also essentially planar: the dihedral fold angles φ between the side six-membered rings of the anthracene unit are 0.0–5.8°. The pyramidalization angles χ at the carbon atom bonded to the acetyl substituent are small, 0.1–4.7°. Thus, the twist of the acetyl group(s) is the main feature of non-planarity in the mono- and diacetylanthracenes.

The diacetylanthracenes under study can be arranged in the order of decreasing twist angles τ:

9,10-Ac_2AN>>1,8-Ac_2AN>1,6-Ac_2AN>1,5-Ac_2AN>1,7-Ac_2AN>2,7-Ac_2AN.

The magnitude ot the twist angle of the acetyl group is important. It has been shown that if an acyl group is tilted out of the plane of the aromatic ring of an aromatic ketone by neighboring bulky groups, the resonance stabilization is reduced and the pattern irreversibility of Friedel-Crafts acylation may be challenged, allowing deacylation, transacylation and acyl rearrangments [Buehler & Pearson, 1970; Gore, 1974; Mala'bi, et al., 2011]. Thus, the twist angle τ may define the ability of diacetylanthracenes to undergo deacylations and rearrangements according to Agranat-Gore rearrangement.

Another factor that may influence the tilting of the acetyl group and, as a consequence, the feasibility of acyl rearrangements, is the overcrowding. Its main source is the short contact distances between the carbonyl oxygen and the *peri*-hydrogen, or between the methyl group and *peri*-hydrogen. The intramolecular O···H distances in the crystal structures of the mono- and diacetylanthracenes under study are not particularly short, 221–246 pm, for the *Z*-acetyl groups, which corresponds to 0–9% penetration. There are no close contact distances caused by the *E*-acetyl groups.

2.1.2 Intermolecular interactions in monoacetylanthracenes and diacetylanthracenes

Aromatic–aromatic interactions are non-covalent intermolecular forces similar to hydrogen bonding [Janiak, 2000]. Aromatic systems may be arranged in three principal configurations:

- A stacked (**S**) configuration, or a π···π interaction, in which aromatic rings are face-to-face aligned, with the interplanar distances of about 3.3–3.8 Å [Janiak, 2000]. This configuration has the maximal overlap but it is rarely observed in real systems containing aromatic rings [Sinnokrot & Sherrill, 2006].
- The T-shaped configuration (**T**), or a C–H···π interaction, where one aromatic ring points at the center of another ring.
- The parallel displaced (**D**), or offset stacked, configuration; it is reached from the stacked configuration by the parallel shift of one aromatic ring relative to the other [Sinnokrot & Sherrill, 2006], and features both π-π and C–H···π interactions. The **T**- and **D**-type configurations are often observed in small aromatic compounds [Dahl, 1994] and proteins [Hunter et al., 1991].

The crystal structure of the parent anthracene (AN) has been studied [Brock & Dunitz, 1990; Sinclair et al., 1950; Murugan & Jha, 2009]. It crystallizes in the monoclinic space group $P2_1/a$. Within the unit cell, the anthracene molecules are packed in a "herringbone" pattern, similar to the parent PAH naphthalene [Desiraju & Gavezzotti, 1989]. In this motif, the C···C non-bonded interactions are between non-parallel nearest neighbor molecules. The herringbone packing is one of four basic structural types for PAH, which are defined depending on the shortest cell axis and the interplanar angle [Desiraju & Gavezzotti, 1989]. The structures with herringbone packing, "sandwich herringbone" packing and γ packing obtain crystal stabilization mainly from C···C interactions, but also from C···H interactions [Desiraju & Gavezzotti, 1989]. The "graphitic", or β, packing characterized by strong C···C interactions without much contribution from C···H contacts [Desiraju & Gavezzotti, 1989]. The selected geometric parameters of aromatic interactions in the mono- and diacetylanthracenes under study are presented in Table 4. $Cg1$ is the centroid for the C^1–C^2–C^3–C^4–C^{4a}–C^{9a} ring, $Cg2$ is the centroid for the C^{4a}–C^{10}–C^{10a}–C^{8a}–C^9–C^{9a} ring and $Cg3$ is the centroid for the C^5–C^6–C^7–C^8–C^{8a}–C^{10a} ring; $Cg4$–6 are the respective centroids of the second non-equivalent molecule in the unit cell, if it exists. Interplanar angle is the angle between the planes of adjacent molecules. Slippage distance is distance of one centroid from the projection of another centroid. Displacement angle is the angle between the ring normal and the centroid vector.

The molecules of 2-AcAN are packed in a "herringbone" pattern, with the interplanar angle of 51.0°. The anthracene moieties in the crystal structure of 2-AcAN adopt the **T**-configuration with the shortest centroid-centroid separation of 464.7 pm. The shortest distances between the centroids of one molecule and the carbon atoms of the other molecule are $Cg3'$···C^4=343.7 pm, $Cg2'$···C^8=351.2 pm, $Cg2'$···C^{10}=351.2 pm, $Cg3'$···C^9=357.6 pm and $Cg1'$···C^5=358.2 pm. The respective centroid-hydrogen distances are $Cg3'$···H^4=271.5 pm, $Cg2'$···H^8=283.3 pm, $Cg2'$···H^{10}=280.8 pm, $Cg3'$···H^9=288.2 pm and $Cg1'$···H^5=287.9 pm. The π···π interactions in 2-AcAN are very weak despite close lying parallel planes, as reflected in very long distances between the respective centroids (>584 pm). Thus, the aryl C–H···π interactions dominate in the crystal structure of 2-AcAN. The unit cell of 2-AcAN is shown in Figure 11.

The molecules of 1,5-Ac_2AN are packed in a "herringbone" pattern, with the interplanar angle of 56.2°. The anthracene moieties in the crystal structure of 1,5-Ac_2AN adopt the **T**-configuration with the shortest centroid-centroid separation of 462.9 and 470.5 pm. The shortest distances between the centroids of one molecule and the carbon atoms of the other molecule are $Cg1'$···C^4=341.9 pm, $Cg1'$···C^3=353.6 pm and $Cg2'$···C^4=376.3 pm. The respective centroid-hydrogen distances are $Cg1'$···H^4=264.3 pm, $Cg1'$···H^3=293.7 pm and $Cg2'$···H^4=342.9 pm. Thus, the aryl C–H···π interactions dominate in the crystal structure of 1,5-Ac_2AN, while the π···π interactions are not possible due to very long distances between molecules lying in the parallel planes (>600 pm). The unit cell of 1,5-Ac_2AN is shown in Figure 12.

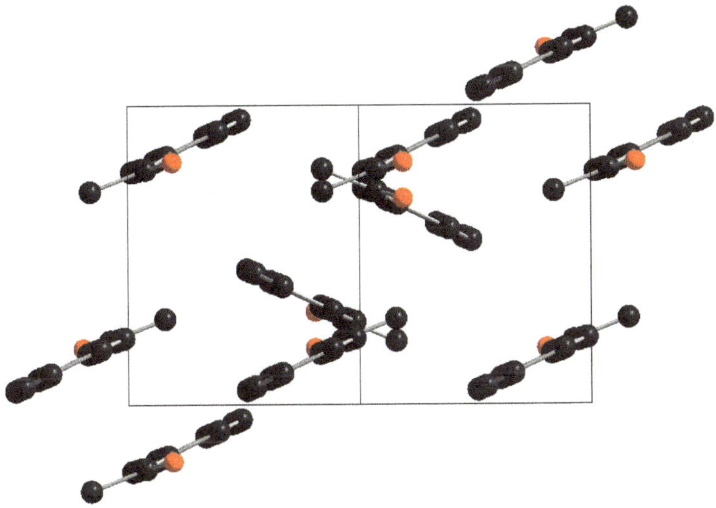

Fig. 11. The unit cell of 2-AcAN (view along *c* axis)

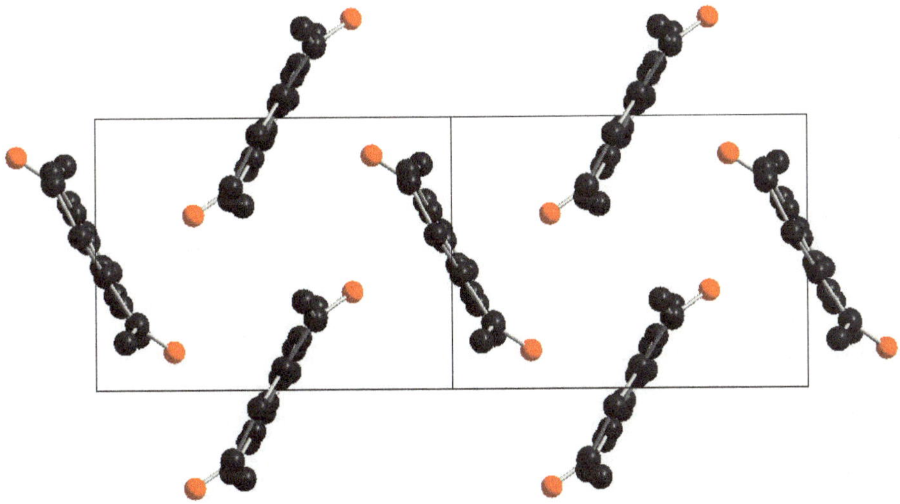

Fig. 12. The unit cell of 1,5-Ac_2AN (view along special axis *1,0,1*)

The molecules of 1,6-Ac_2AN are packed by β type, forming a layered structure made up of "graphitic" planes with zero interplanar angle. From the point of view of aromatic–aromatic interactions, the anthracene moieties in the crystal structure of 1,6-Ac_2AN are stacked by the **D**-type, with the centroid–centroid separation of 359.2 and 385.6 pm. The slippage distances of the centroids are relatively short, 94.0 and 107.1 pm. The shortest contact distances between the aromatic carbons in 1,6-Ac_2AN are $C^5...C^{7'}$=355.1 and $C^6...C^{8a'}$=358.5. The unit cell of 1,6-Ac_2AN is shown in Figure 13.

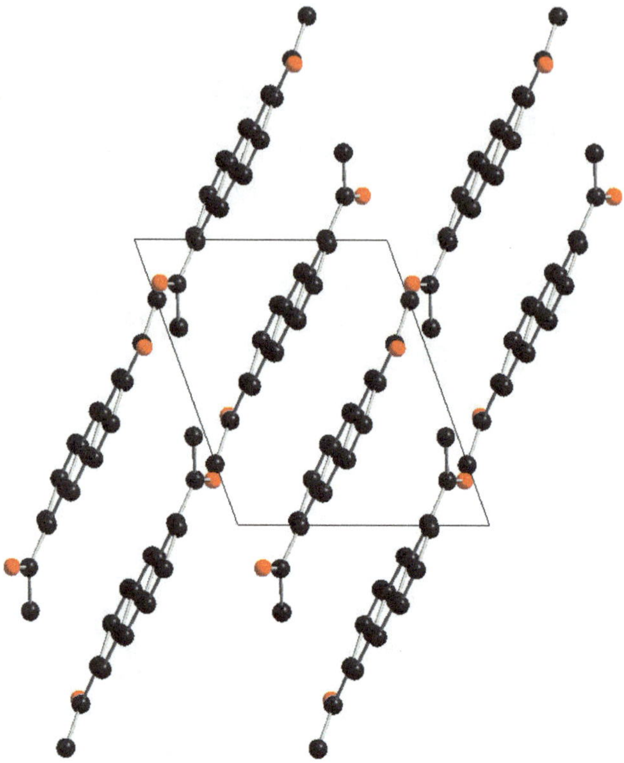

Fig. 13. The unit cell of 1,6-Ac_2AN (view along *c* axis)

The molecules of 1,7-Ac_2AN are also packed by β type. The anthracene moieties in the crystal structure of 1,7-Ac_2AN adopt the **D**-configuration, with the shortest centroid-centroid separation of 370 pm. Despite the longer slippage distance between centroids (154.4–154.8 pm), the contact distances in 1,7-Ac_2AN are shorter than those in 1,6-Ac_2AN: $C^3...C^{8'}$=333.3, $C^{4a}...C^{9'}$=336.4, $C^8...C^{9'}$=337.1 and $C^1...C^{10'}$=340.9. In both 1,6-Ac_2AN and 1,7-Ac_2AN the aromatic interactions are mainly those of the π···π type. The unit cell of 1,7-Ac_2AN is shown in Figure 14.

The molecules of 1,8-Ac_2AN are packed in a "herringbone" pattern, with the interplanar angle of 34.7°. The centroids of the anthracene molecules lying onto the parallel planes are separated by 580–581 pm. These distances together with the slippage distance of 493-494 pm render the aromatic interactions of either **S**- or **D**-type impossible. The **T**-type interactions in 1,8-Ac_2AN are too weak to be of any importance, due to the long distances between centroids (546–562 pm). However, the plane of the acetyl group (containing C^1, C^{11}, C^{13}, O^{15}) of molecule A forms the angle of 4.0° with the aromatic plane of molecule B. Analogously, the plane of the acetyl group (containing $C^{1'}$, $C^{11'}$, $C^{13'}$, $O^{15'}$) of molecule B is nearly parallel to the aromatic plane of molecule A, 3.8°. The distances between the anthracene systems and the carbonyl group are sufficiently small to consider the intermolecular π···π interactions: $Cg4'\cdots O^{1}$=353.8 pm, $Cg4'\cdots C^{11}$=384.3 pm, $Cg3\cdots O^{1'}$=363.3 pm and $Cg3\cdots C^{11'}$=398.2 pm. Thus, the crystal structure of 1,8-Ac_2AN features π–π-interactions not between two aromatic systems, but between the aromatic system and the carbonyl π-bond. The unit cell of 1,8-Ac_2AN is shown in Figure 15.

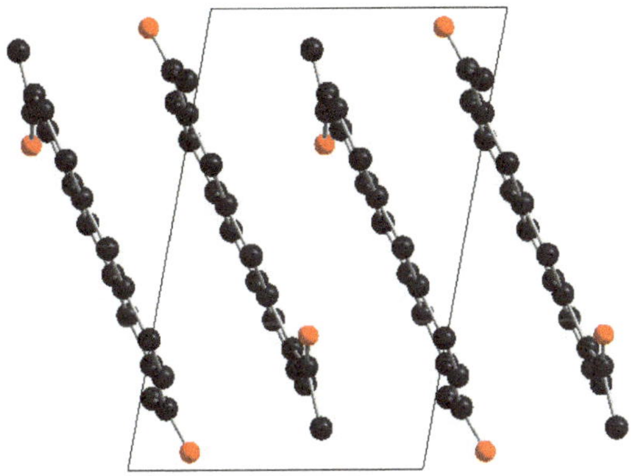

Fig. 14. The unit cell of 1,7-Ac_2AN (view along *b* axis)

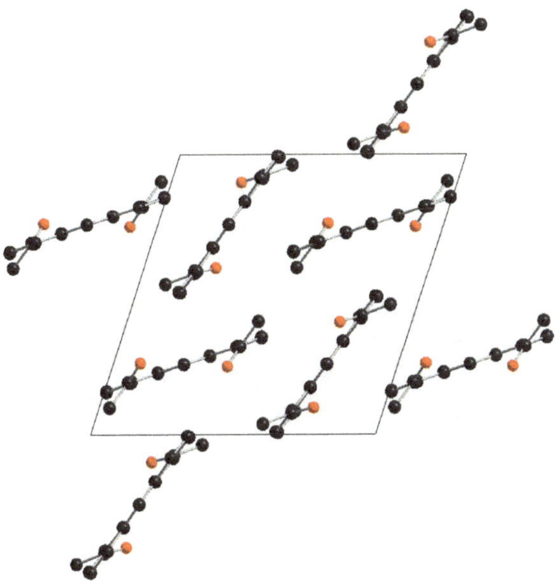

Fig. 15. The unit cell of 1,8-Ac_2AN (view along *b* axis)

The molecules of 2,7-Ac_2AN are packed in a "herringbone" pattern. The anthracene moieties in the crystal structure of 2,7-Ac_2AN adopt the **T**-configuration, similarly to 1,5-Ac_2AN. The planes of the adjacent molecules form the angle of 58.1°. The shortest distances between the centroids and the carbon atoms are $Cg3{\cdots}C^4$=358.4 pm and $Cg1{\cdots}C^5$=375.5 pm on the one side of the anthracene system, and $Cg3{\cdots}C^9$=374.0 pm, $Cg2{\cdots}C^8$=374.8 pm on the other side. The respective shortest centroid-aryl hydrogen distances are $Cg1{\cdots}H^5$=299.3 pm and $Cg3{\cdots}H^4$=283.5 pm. The **D**-type interactions in 2,7-Ac_2AN are very weak due to the large separation of centroids (420–433 pm) and large slippage distances (226–242 pm). The unit cell of 2,7-Ac_2AN is shown in Figure 16.

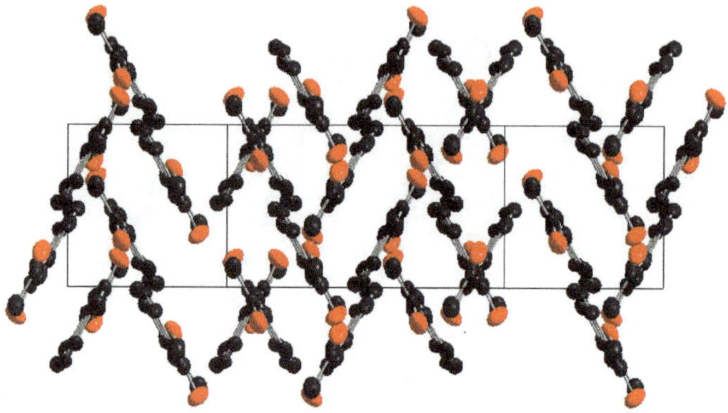

Fig. 16. The unit cell of 2,7-Ac_2AN (view towards plane *1,0,–5*)

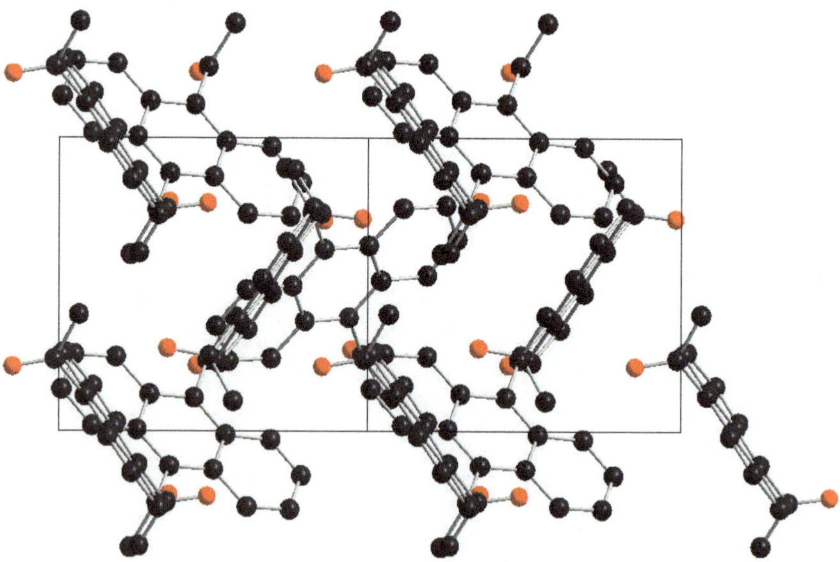

Fig. 17. The unit cell of 9,10-Ac_2AN (view along special axis *1,0,1*)

The anthracene moieties in the crystal structure of 9,10-Ac_2AN adopt the T-configuration, similarly to 1,5-Ac_2AN and 2,7-Ac_2AN. The planes of the adjacent molecules form the angle of 73.6°. The shortest distances between the centroids and the carbon atoms are $Cg2\cdots C^7$=356.1 pm, $Cg2\cdots C^8$=379.6 pm and $Cg1\cdots C^8$=384.7 pm. The respective shortest centroid–aryl hydrogen distances are $Cg2\cdots H^7$=287.9 pm, $Cg2\cdots H^8$=329.4 pm and $Cg1\cdots H^8$=294.4 pm. The **D**-type interactions in 9,10-Ac_2AN are non-existent. The molecules lying in the parallel planes are separated by >720 pm, probably due to the considerable twist angles of the acetyl group in 9,10-Ac_2AN (–85.0° and 87.0°), making the tighter arrangement impossible. The unit cell of 9,10-Ac_2AN is shown in Figure 17.

	Centroid	Centroid	Centroid centroid distance pm	Interplanar distance pm	Interplanar angle deg	Slippage distance pm	Displacement angle deg
2-AcAN	*Cg*1	*Cg*3'[a]	464.7	–	51.0	110.8	13.8
	*Cg*1	*Cg*3'[b]	477.2	–	51.0	103.6	12.5
	*Cg*1	*Cg*3'[c]	499.5	–	51.0	209.7	24.8
	*Cg*1	*Cg*3'[d]	511.1	–	51.0	205.9	23.8
	*Cg*1	*Cg*2'[e]	584.5	251.6	0.0	544.4	64.5
1,5-Ac_2AN	*Cg*2	*Cg*1'[f]	462.9	–	56.2		
	*Cg*1	*Cg*1'[g]	470.5	–	56.2		
	*Cg*2	*Cg*1'[h]	601.5	292.2	0.0	526.4	61.1
	*Cg*3	*Cg*2'[h]	601.5	292.2	0.0	525.8	60.9
1,6-Ac_2AN	*Cg*3	*Cg*3'[i]	359.2	346.1	0.0	94.0	15.2
	*Cg*1	*Cg*1'[j]	385.6	370.4	0.0	107.1	16.1
1,7-Ac_2AN	*Cg*1	*Cg*3'[i]	370.1	335.9	0.0	154.6	24.7
	*Cg*1	*Cg*2'[i]	370.4	335.9	0.0	154.8	24.7
	*Cg*2	*Cg*2'[i]	370.2	335.9	0.0	154.4	24.7
1,8-Ac_2AN	*Cg*1	*Cg*4'[k]	546.4	–	34.7		
	*Cg*1	*Cg*4'[l]	561.5	–	34.7		
	*Cg*1	*Cg*1'[i]	580.5	307.2	0.0	492.6	58.1
	*Cg*1	*Cg*1'[m]	580.9	305.9	0.0	493.9	58.2
2,7-Ac_2AN	*Cg*2	*Cg*3'[n]	419.8	354.6	0.0	226.2	32.6
	*Cg*3	*Cg*3'[n]	432.7	354.6	0.0	241.8	35.2
	*Cg*1	*Cg*3'[o]	481.0	–	58.1		
	*Cg*2	*Cg*2'[o]	486.6	–	58.1		
9,10-Ac_2AN	*Cg*2	*Cg*3'[p]	475.5	–	73.6		
	*Cg*2	*Cg*1'[q]	481.8	–	73.6		
	*Cg*3	*Cg*2'[r]	721.2	478.0	0.0	540.1	48.5
	*Cg*2	*Cg*1'[r]	722.3	478.0	0.0	541.5	48.6
	*Cg*3	*Cg*1'[r]	724.2	478.0	0.0	544.0	48.7

Symmetry codes: [a] 0.5–*x*, 0.5+*y*, 1.5–*z*; [b] 0.5–*x*, –0.5+*y*, 1.5–*z*; [c] 1.5–*x*, 0.5+*y*, 1.5–*z*; [d] 1.5–*x*, –0.5+*y*, 1.5–*z*; [e] -1+*x*, *y*, *z*; [f] *x*, 0.5–*y*, 0.5–*z*; [g] 1–*x*, 0.5+*y*, 0.5–*z*; [h] *x*, 1+*y*, *z*; [i] 1–*x*, 1–*y*, 1–*z*; [j] 1–*x*, –*y*, –*z*; [k] 1.5–*x*, 1+*y*, 1.5–*z*; [l] 1–*x*, –*y*, 1–*z*; [m] 0.5+*x*, –*y*, –0.5+*z*; [n] 0.5–*x*, 0.5–*y*, 0.5–*z*; [o] –x, 0.5+*y*, 0.5–*z*; [p] –0.5+*x*, 1.5–*y*, *z*; [q] 0.5+*x*, 0.5–*y*, *z*; [r] *x*, –1+*y*, *z*.

Table 4. Aromatic interactions in monoacetylanthracenes and diacetylanthracenes

Thus, the monoacetylanthracenes and diacetylanthracenes under study may be divided into two groups, based on the aromatic–aromatic interactions in their crystal structures. The anthracene units in 1,6-Ac_2AN and 1,7-Ac_2AN are offset stacked (the **D**-type arrangement) and feature aromatic–aromatic π···π interactions. The anthracene molecules in ketones 2-AcAN, 1,5-Ac_2AN, 2,7-Ac_2AN and 9,10-Ac_2AN adopt the **T**-type arrangement, and feature aryl C–H···π interactions. The analysis of the literature crystal structures of 1-AcAN and 9-AcAN shows that these ketones also adopt the **T**-type arrangement. In 1-AcAN, 9-AcAN, 1,5-Ac_2AN and 9,10-Ac_2AN the considerable twist angles of the acetyl groups prevents the molecules from being arranged in close lying parallel planes. The exception is the crystal structure of 1,8-Ac_2AN, which features π···π-interactions between the aromatic system and the carbonyl π-bond. Most likely the methyl groups are the reason for the lack of more examples of slipped-stacking and also in some cases the competing ketone–π system as well. It should be noted, however, that the centroid-centroid analysis can be misleading, and its limitations should not be overlooked.

Another kind of intermolecular interactions that could exist in acetylanthracenes is hydrogen bonds. No particular strong intermolecular aryl C–H···O bonds have been found in the diacetylanthracenes under study. The shortest contact distances between an oxygen and an aromatic hydrogen are $O^{15}\cdots H^{5}$=242.2 pm (9,10-Ac_2AN), $O^{15}\cdots H^{1}$=247.4 pm and $O^{16}\cdots H^{9}$=259.4 pm (2,7-Ac_2AN), $O^{15}\cdots H^{5}$=255.6 pm (1,6-Ac_2AN), $O^{16}\cdots H^{3}$=256.4 pm and $O^{15}\cdots H^{4}$=260.7 pm (1,7-Ac_2AN), $O^{15}\cdots H^{2'}$=260.5 pm (1,8-Ac_2AN), $O^{15}\cdots H^{2}$=284.6 pm (1,5-Ac_2AN). The shortest contact distances between an oxygen and a methyl hydrogen are of a similar magnitude: $O^{15}\cdots H^{14c}$=240.8 pm (2,7-Ac_2AN), $O^{16}\cdots H^{12c}$=254.7 pm (1,6-Ac_2AN), $O^{16}\cdots H^{12b}$=257.5 pm (1,7-Ac_2AN), $O^{15}\cdots H^{12c}$=259.4 pm (1,5-Ac_2AN), $O^{15}\cdots H^{12c}$=265.9 pm (9,10-Ac_2AN).

2.2 NMR Study of monoacetylanthracenes and diacetylanthracenes

The structure of a compound in crystal is not necessarily the same as that in solution. More often, in the case of substances that are not conformationally homogeneous, e.g. diacetylanthracenes, the crystal has a unique conformation and the conformational heterogeneity appears in fluid phases [Eliel & Wilen, 1994]. An insight into the conformations of mono- and diacetylanthracenes in solution may be gained from the chemical shifts of the aromatic protons adjacent to the carbonyl groups. The magnetic shielding (or deshielding) effect on the chemical shifts of protons that lie in or near the plane of the carbonyl group is well known. The McConnell equation [McConnel, 1957] predicts shielding for protons lying above the center of a carbon–oxygen double bond and deshielding for protons located within a cone aligned with the carbon–oxygen bond axis. The McConnell model, however, takes into account only the effect of magnetic anisotropy.

Recently, more detailed shielding model has been proposed [Martin et al., 2003]. According to this model, shielding is predicted for protons located in the region from over the center of the carbon–oxygen double bond to beyond the carbon atom; deshielding is predicted for protons located above and beyond the oxygen atom. Table 5 gives ^{1}H-NMR chemical shifts for the monoacetylanthracenes and diacetylanthracenes under study, together with the chemical shifts in parent anthracene (AN).

The data presented in Table 5 show that the protons at *ortho*-positions to an acetyl group are considerably deshielded as compared with the protons of unsubstituted anthracene. The magnitudes of the low field shifts of the *ortho*-protons are similar among monoacetylanthracenes and diacetylanthracenes: $\Delta\delta(H^1$, ppm)=0.70 (2-AcAN), 0.72 (2,7-Ac_2AN); $\Delta\delta(H^2$, ppm)=0.59 (1-AcAN), 0.63 (1,6-Ac_2AN), 0.67 (1,5-Ac_2AN), 0.67 (1,7-Ac_2AN), 0.73 (1,8-Ac_2AN); $\Delta\delta(H^3$, ppm)=0.61 (2-AcAN), 0.65 (2,7-Ac_2AN); $\Delta\delta(H^5$, ppm)=0.62 (1,6-Ac_2AN); $\Delta\delta(H^6$, ppm)=0.63 (1,7-Ac_2AN), 0.65 (2,7-Ac_2AN), 0.67 (1,5-Ac_2AN); $\Delta\delta(H^7$, ppm)=0.58 (1,6-Ac_2AN), 0.73 (1,8-Ac_2AN); $\Delta\delta(H^8$, ppm)=0.72 (2,7-Ac_2AN), 0.77 (1,7-Ac_2AN). The protons at *peri*-positions to an acetyl group are deshielded with even greater magnitudes: $\Delta\delta(H^9$, ppm)=1.06 (1,6-Ac_2AN), 1.08 (1-AcAN), 1.17 (1,5-Ac_2AN), 1.27 (1,7-Ac_2AN), 1.78 (1,8-Ac_2AN). The latter case is special because of the presence of two acetyl groups at *peri*-positions to H^9, which nearly double its low field chemical shift. Note that in 2-AcAN, 1,6-Ac_2AN, 1,7-Ac_2AN and 2,7-Ac_2AN both protons *ortho* to the acetyl groups demonstrate similar low field shifts, suggesting that these protons are located above the plane of the carbonyl group and near the oxygen atom [Martin et al., 2003]. Thus, the twist angles of the acetyl groups of mono- and diacetylanthracenes are small, in accordance with their respective X-ray crystal structures, and *E,Z*-diastereomerizations of the acetyl groups at both α (1, 5, 8) and β (2, 6, 7) positions are swift on the NMR time scale.

	H^1	H^2	H^3	H^4	H^5	H^6	H^7	H^8	H^9	H^{10}	CH_3	CH_3
AN	7.95	7.41	7.41	7.95	7.95	7.41	7.41	7.95	8.40	8.40		
1-AcAN		7.998	7.469	8.169	7.998	7.528–7.495	7.528–7.495	8.083	9.482	8.446	2.810	
2-AcAN	8.646		8.054–7.982	8.054–7.982	8.054–7.984	7.546	7.516	8.054–7.984	8.573	8.432	2.763	
9-AcAN	7.859	7.556–7.477	7.556–7.477	8.027	8.027	7.556–7.477	7.556–7.477	7.859		8.473	2.822	
1,5-Ac_2AN		8.083	7.530	8.262		8.083	7.530	8.262	9.570	9.570	2.818	2.818
1,6-Ac_2AN		8.040	7.490	8.153	8.570		7.994	8.064	9.457	8.523	2.796	2.730
1,7-Ac_2AN		8.080	7.559	8.199	8.036	8.036		8.719	9.673	8.460	2.836	2.773
1,8-Ac_2AN		8.140	7.514	7.964	7.964	7.514	8.140		10.175	8.471	2.840	2.840
2,7-Ac_2AN	8.670		8.063	8.063	8.063	8.063		8.670	8.718	8.449	2.775	2.775
9,10-Ac_2AN	7.881–7.845	7.571–7.537	7.571–7.537	7.881–7.845	7.881–7.845	7.571–7.537	7.571–7.537	7.881–7.845			2.816	2.816

Table 5. The ^{1}H-NMR chemical shifts (δ, ppm) of aromatic and methyl protons in anthracene (AN), monoacetylanthracenes and diacetylanthracenes under study.

Ketones 9-AcAN and 9,10-Ac_2AN differ from the rest of the mono- and diacetylanthracenes. The protons at *peri*-positions to the acetyl groups of 9-AcAN and 9,10-Ac_2AN are slightly shielded: $\Delta\delta(H^1$, ppm)= –0.09 (9-AcAN), –0.09 (9,10-Ac_2AN). This suggests that the carbonyl groups in 9-AcAN and 9,10-Ac_2AN are turned away of the protons at *peri*-positions, and these protons are located near the carbonyl carbon atoms, which implies high twist angles of the acetyl groups. It corresponds well to the respective X-ray crystal structures of 9-AcAN and 9,10-Ac_2AN.

2.3 DFT computational study of monoacetylanthracenes and diacetylanthracenes

DFT methods are capable of generating a variety of isolated molecular properties quite accurately, especially via the hybrid functional, and in a cost-effective way [deProft & Geerlings, 2001, Koch & Holthausen, 2000]. The B3LYP hybrid functional was successfully employed to treat overcrowded BAEs [Biedermann et al., 2001, Pogodin et al., 2006] and overcrowded naphthologues of BAEs-1, i.e. mono-bridged tetraarylethylenes [Assadi et al., 2009]. The monoacetylanthracenes and diacetylanthracenes under study were subjected to a systematic computational DFT study of their conformational spaces and of their relative stabilities. The B3LYP/6-31G(d) relative energies of the global minima conformations of certain diacetylanthracenes have been previously reported [Mala'bi et al., 2011]. The total and relative B3LYP/6-31G(d) energies (E_{Tot} and ΔE_{Tot}) and Gibbs free energies (ΔG_{298} and $\Delta\Delta G_{298}$) of the acetylanthracenes are presented in Table 6. Selected calculated geometrical parameters of the acetylanthracenes are also given in Table 6. The following geometrical parameters were considered: the twist angles τ_1, τ_2 and τ_9 and the respective twist angles υ around the anthracenyl–carbonyl bond; the dihedral angle θ between the least-square planes of the carbonyl group and the anthracene system; the dihedral angle φ between the least-square planes of two side rings of the anthracene system; the pyramidalization angles χ at C^1, C^2 and C^9.

2.3.1 Conformational space of monoacetylanthracenes and diacetylanthracenes

Monoacetylanthracenes may adopt two conformations, *Z* and *E*, defined by the twist angle of the carbonyl group. Diacetylanthracenes may adopt four conformations, i. e. *ZZ*, *ZE*, *EZ* and *EE*; in certain cases, *ZE* is identical to *EZ*. In addition, the oxygen atoms of two carbonyl groups may be located on the same side of the aromatic plane, or on the opposite sides, potentially resulting in *syn*- and *anti*-*ZZ*, *ZE*, *EZ* and *EE* conformations, respectively. Depending on the symmetry constraints and the twist angle τ, not all of the above-mentioned conformations exist for a given diacetylanthracene. The possible conformations of diacetylanthracene are shown in Fig. 18.

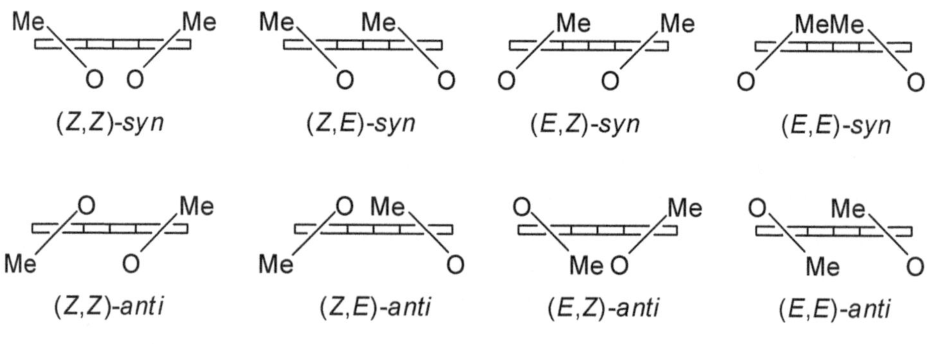

Fig. 18. Schematic representation of the eight possible conformations of a diacetylanthracene (view along the aromatic plane).

			E_{Tot}	ΔE_{Tot}	ΔG_{298}	$\Delta\Delta G_{298}$	τ[a]	υ[b]	θ	φ	C_{11}–C_{arom}[c]	χ
			Hartree	kJ/mol	kJ/mol	kJ/mol	deg	deg	deg	deg	pm	deg
1-AcAN	*Z*	C_s	–692.17301155	14.66	15.79	0.00	0.0	180.0	0.0	0.0	149.9	0.0
1-AcAN	[d] *Z*	C_1	–	–	–	–	27.1	–152.7	28.6	3.2	149.3	–0.1
1-AcAN	*E*	C_1	–692.16815672	27.41	28.80	13.01	150.8	–31.1	36.0	3.7	150.7	1.9
2-AcAN	*E*	C_s	–692.17859715	0.00	0.00	0.00	180.0	0.0	0.0	0.0	149.6	0.0
2-AcAN	[d] *E*		–	–	–	–	173.1	–5.3	5.9	0.4	149.6	–1.6
2-AcAN	*Z*	C_s	–692.17777414	2.16	2.24	2.24	0.0	180.0	0.0	0.0	149.9	0.0
9-AcAN	–	C_1	–692.16381815	38.80	36.94	0.00	–67.0	113.9	69.8	1.7	151.3	–1.0
9-AcAN	[d] –	C_1	–	–	–	–	87.9	–91.8	89.2	5.8	150.4	0.3
1,5-Ac_2AN	*ZZ*	C_{2h}	–844.81621983	24.25	27.69	0.00	0.0	180.0	0.0	0.0	149.8	0.0
1,5-Ac_2AN	[d] *ZZ*	C_i	–	–	–	–	20.0	–156.8	22.7	0.0	149.4	–3.2
1,5-Ac_2AN	*ZE*	C_1	–844.81074449	38.62	40.48	12.79	152.4	–29.3	34.1	3.8	150.6	1.6
							–1.1	178.9	2.3		150.0	0.0
1,5-Ac_2AN	*EEanti*	C_i	–844.80527143	52.99	55.06	27.36	150.6	–30.9	34.8	0.0	150.8	1.5
1,5-Ac_2AN	*EEsyn*	C_2	–844.80559961	52.13	55.74	28.05	151.9	–29.9	35.6	7.4	150.7	1.7
1,6-Ac_2AN	*ZE*	C_s	–844.82056764	12.83	13.45	0.00	0.0	180.0	0.0	0.0	150.0	0.0
							180.0	0.0	0.0		149.7	0.0
1,6-Ac_2AN	[d] *ZE*	C_1	–	–	–	–	30.0	–147.1	32.2		150.1	–2.9
							178.6	–0.7	1.9	1.3	149.3	–0.7
1,6-Ac_2AN	*ZZ*	C_s	–844.82005388	14.18	15.14	1.69	0.0	180.0	0.0	0.0	150.0	0.0
							0.0	180.0	0.0		150.0	0.0
1,6-Ac_2AN	*EE*	C_1	–844.81569542	25.63	27.03	13.57	150.6	–31.1	36.0	3.6	150.8	1.7
							179.9	–0.1	1.4		149.8	–0.1
1,6-Ac_2AN	*EZanti*	C_1	–844.81493981	27.61	28.88	15.43	150.6	–31.2	36.2		150.7	1.7
							–0.3	179.8	1.7		150.0	–0.1
1,7-Ac_2AN	*ZE*	C_s	–844.82110775	11.42	12.34	0.00	0.0	180.0	0.0	0.0	150.0	0.0
							180.0	0.0	0.0		149.7	0.0
1,7-Ac_2AN	[d] *ZE*	C_1	–	–	–	–	–15.2	162.9	16.0	2.3	149.8	1.9
							–176.6	3.7	4.5		149.0	0.3
1,7-Ac_2AN	*ZZ*	C_s	–844.81939574	15.91	15.83	3.50	0.0	180.0	0.0	0.0	150.1	0.0
							0.0	180.0	0.0		150.0	0.0
1,7-Ac_2AN	*EE*	C_1	–844.81562930	25.80	26.79	14.45	150.3	–31.4	36.3	3.7	150.8	1.7
							179.6	–0.5	1.7		149.8	0.0
1,7-Ac_2AN	*EZanti*	C_1	–844.81488173	27.76	28.96	16.62	150.8	–31.0	35.8	3.5	150.9	1.7
							0.2	–179.8	1.1		150.1	0.0
1,8-Ac_2AN	*ZZanti*	C_2	–844.81111292	37.66	38.89	0.00	–17.3	160.4	19.3	2.2	150.2	2.3
1,8-Ac_2AN	[d] *ZZ*	C_2	–	–	–	–	–34.0	145.4	36.0	0.3	149.3	0.6
							–32.4	144.9	35.4	3.4	148.9	2.7
1,8-Ac_2AN	*EZ*	C_1	–844.81126554	37.26	39.25	0.35	150.4	–31.2	36.1	3.5	151.1	1.6
							1.5	–178.3	2.6		150.0	0.2
1,8-Ac_2AN	*EEsyn*	C_s	–844.80423404	55.72	56.49	17.60	147.9	–33.8	40.2	7.0	150.7	1.7
1,8-Ac_2AN	*EEanti*	C_2	–844.80485619	54.08	56.56	17.66	148.1	–33.7	38.6	5.1	150.7	1.8
1,9-Ac_2AN	*ZZanti*	C_1	–844.79904569	69.34	70.32	0.00	–50.9	120.3	59.9	7.5	150.8	8.8
							–59.6	114.1	62.8		151.5	–6.3
1,9-Ac_2AN	*EZsyn*	C_1	–844.78990701	93.33	96.19	25.88	–141.2	45.3	56.7	10.7	151.2	–6.5
							44.8	–128.4	48.1		151.1	6.8
1,10-Ac_2AN	*ZE*	C_1	–844.80536578	52.75	49.45	0.00	0.2	–180.0	1.0	1.8	150.1	0.2
							–108.0	73.0	75.3		151.6	–1.0
1,10-Ac_2AN	*ZZ*	C_1	–844.80575464	51.73	50.43	0.98	1.8	–178.2	2.7	2.4	150.1	0.0
							–65.9	115.1	68.5		151.4	1.0

1,10-Ac_2AN	*EZanti*	C_1	-844.80066416	65.09	63.37	13.91	148.6	-33.3	38.0	2.9	150.7	1.9
							-70.6	110.9	72.5		151.6	-1.5
1,10-Ac_2AN	*EEanti*	C_1	-844.80064430	65.14	63.38	13.92	148.5	-33.6	37.9	2.8	150.8	2.1
							-106.6	75.1	76.7		151.6	1.7
1,10-Ac_2AN	*EEsyn*	C_1	-844.80035871	65.89	63.70	14.25	149.9	-31.7	37.8	5.4	150.8	1.6
							111.3	-68.9	71.9		151.6	0.2
9,10-Ac_2AN	*E*	C_i	-844.79648217	76.07	71.57	0.00	-72.6	108.5	74.7	0.0	151.6	-1.1
9,10-Ac_2AN [d]	*E*	C_1	-	-	-	-	-85.0	94.0	86.7	1.6	151.3	-1.0
							87.0	-93.7	86.5		151.5	-0.6
9,10-Ac_2AN	*Z*	C_s	-844.79616186	76.91	71.63	0.06	71.8	-108.9	74.2	2.9	151.6	-0.7
9,10-Ac_2AN	*E*	C_2	-844.79637404	76.35	72.55	0.99	75.4	-105.7	76.9	0.1	151.6	-1.1
9,10-Ac_2AN	*Z*	C_2	-844.79619082	76.84	73.69	2.13	-71.9	108.8	73.9	3.1	151.6	-0.7
2,6-Ac_2AN	*EE*	C_{2h}	-844.82603788	-1.53	0.40	0.00	180.0	0.0	0.0	0.0	149.8	0.0
2,6-Ac_2AN	*ZE*	C_s	-844.82517129	0.75	0.79	0.40	0.0	180.0	0.0	0.0	150.1	0.0
							180.0	0.0			149.8	0.0
2,6-Ac_2AN	*ZZ*	C_{2h}	-844.82448815	2.54	4.19	3.79	0.0	-180.0	0.0	0.0	150.1	0.0
2,7-Ac_2AN	*EZ*	C_s	-844.82545585	0.00	0.00	0.00	180.0	0.0	0.0	0.0	149.7	0.0
							0.0	180.0			150.0	0.0
2,7-Ac_2AN [d]	*EZ*	C_1	-	-	-	-	171.9	-3.3	9.8	2.7	149.0	-4.8
							0.9	-178.8	1.6		148.9	0.3
2,7-Ac_2AN	*EE*	C_{2v}	-844.82612845	-1.77	0.20	0.20	180.0	0.0	0.0	0.0	149.8	0.0
2,7-Ac_2AN	*ZZ*	C_{2v}	-844.82444406	2.66	4.29	4.29	0.0	180.0	0.0	0.0	150.1	0.0
2,9-Ac_2AN	*EE*	C_1	-844.81120818	37.41	32.29	0.00	-178.9	1.4	2.1	1.6	149.8	-0.4
				1.50			-106.9	73.9	75.8		151.6	0.8
2,9-Ac_2AN	*EZ*	C_1	-844.81178130	35.90	35.90	3.61	-178.9	1.3	1.8	2.5	149.9	-0.2
				0.00			-63.0	118.1	66.2		151.3	1.1
2,9-Ac_2AN	*ZE*	C_1	-844.81042981	39.45	37.37	5.08	-1.8	178.5	2.6	1.8	150.1	-0.3
				3.55			-114.4	66.0	69.1		151.5	0.4
2,10-Ac_2AN	*EE*	C_1	-844.81146291	36.74	34.49	0.00	179.9	-0.7	0.7	1.8	149.7	0.5
							-113.9	66.7	70.5		151.5	-0.6
2,10-Ac_2AN	*EZ*	C_1	-844.81128488	37.21	34.81	0.32	179.6	-0.5	0.3	1.7	149.7	-0.1
					0.32		-68.8	112.0	71.4		151.5	0.9
2,10-Ac_2AN	*ZE*	C_1	-844.81074527	38.62	36.64	2.15	0.9	-179.8	1.9	2.0	150.0	0.7
					2.15		-114.3	66.4	69.3		151.4	-0.8
2,10-Ac_2AN	*ZZ*	C_1	-844.81084444	38.36	38.36	3.87	1.0	-179.2	1.7	2.1	150.0	-0.2
					3.87		-65.7	115.3	68.6		151.4	1.0

[a] $\tau_1(C^{9a}–C^1–C^{11}–O^{15})$ for 1-AcAN, 1,5-Ac_2AN, 1,6-Ac_2AN, 1,7-Ac_2AN and 1,8-Ac_2AN, $\tau_2(C^1–C^2–C^{11}–O^{15})$ for 2-AcAN and 2,7-Ac_2AN, $\tau_2(C^5–C^6–C^{13}–O^{16})$ for 1,6-Ac_2AN, $\tau_2(C^8–C^7–C^{13}–O^{16})$ for 1,7-Ac_2AN, $\tau_9(C^{9a}–C^9–C^{11}–O^{15})$ for 9-AcAN and 9,10-Ac_2AN.

[b] $\upsilon_1(C^2–C^1–C^{11}–O^{15})$ for 1-AcAN, 1,5-Ac_2AN, 1,6-Ac_2AN, 1,7-Ac_2AN and 1,8-Ac_2AN, $\upsilon_2(C^3–C^2–C^{11}–O^{15})$ for 2-AcAN and 2,7-Ac_2AN, $\upsilon_2(C^7–C^6–C^{13}–O^{16})$ for 1,6-Ac_2AN, $\upsilon_2(C^6–C^7–C^{13}–O^{16})$ for 1,7-Ac_2AN, $\upsilon_9(C^{8a}–C^9–C^{11}–O^{15})$ for 9-AcAN and 9,10-Ac_2AN.

[c] $C^1–C^{11}$ for 1-AcAN, 1,5-Ac_2AN, 1,6-Ac_2AN, 1,7-Ac_2AN and 1,8-Ac_2AN, $C^2–C^{11}$ for 2-AcAN and 2,7-Ac_2AN, $C^6–C^{13}$ for 1,6-Ac_2AN, $C^7–C^{13}$ for 1,7-Ac_2AN, $C^9–C^{11}$ for 9-AcAN and 9,10-Ac_2AN.

[d] the selected geometrical parameters derived from the corresponding X-ray structures.

Table 6. Total energies (E_{Tot}), relative energies (ΔE_{Tot}) and Gibbs free energies (ΔG_{298}) and selected geometric parameters of mono- and diacetylanthracenes.

Ketone 1-AcAN adopts a C_s-*Z* conformation as its global minimum. The planar (excluding the methyl hydrogens) C_s-1*Z*-AcAN is overcrowded due to the short $O^{13}···H^9$ contact distance (the $O^{13}···H^9$ distance is 215 pm, 14% penetration, based on the sum of the wan-der-Vaals radii of oxygen and hydrogen, 244 pm [Zefirov, 1997]). The non-planar C_1-1*E*-AcAN conformation (the twist angle $\tau_1(C^{9a}–C^1–C^{11}–O^{13})$=150.8°) is higher in energy by 13.0 kJ/mol. The energy barrier for the *E*,*Z*-diastereomerization C_s-1*Z*-AcAN→C_1-1*E*-AcAN by the rotation of the acetyl group via a nearly orthogonal transition state is 19.5 kJ/mol. As mentioned above, 1-AcAN [Langer1993] crystallizes as the *Z*-diastereomer, which is correctly described by the calculated structure of C_s-1*Z*-AcAN. However, the carbonyl group in the crystal structure of 1-AcAN is considerably twisted out of the plane of the anthracene ring system, τ_1=27.1°. As a result, the calculated C_s-1*Z*-AcAN structure is more overcrowded than the X-ray structure (in the latter the $O^{13}...H^9$ distance is 223 pm).

Ketone 2-AcAN adopts a C_s-*E* conformation as its global minimum. Its local minimum C_s-2*Z*-AcAN conformation is 2.2 kJ/mol higher in energy. Both conformations are not overcrowded, lacking any *peri*-interactions. The energy barrier for the *E*,*Z*-diastereomerization C_s-2*E*-AcAN→C_s-2*Z*-AcAN by the rotation of the acetyl group via a nearly orthogonal transition state is 31.5 kJ/mol. The calculated C_s-2*E*-AcAN conformation corresponds well to the *E*-conformation of the crystal structure. The latter, however, features a small twist angle of $\tau_2(C^1–C^2–C^{11}–O^{13})$=173.1°, in contrast to the planar (excluding the methyl hydrogens) calculated structure.

In the global minimum conformation of 9-AcAN the twist angle $\tau_9(C^{9a}–C^9–C^{11}–O^{13})$ is -67.0°. This conformations cannot be defined as either *E* or *Z*, and no other minimum conformation was located. Comparing the calculated structure of 9-AcAN with the crystal structure of 9-AcAN reported in the literature [Zouev2011], the carbonyl group in the latter is almost orthogonal to the plane of the anthracene ring system: the twist angle $\tau_9(C^{9a}–C^9–C^{11}–O^{13})$=87.9° is considerably larger than the twist angle predicted by the DFT calculations. The energy barrier for the enantiomerization of 9-AcAN via the orthogonal [C_s-9-AcAN] transition state is only 3.6 kJ/mol. The low enantiomerization barrier as compared to the diastereomerization barriers in 1-AcAN and 2-AcAN is due to an already high twist angle in 9-AcAN.

Ketone 1,5-Ac_2AN adopt a C_{2h}-1*Z*,5*Z* conformation as its global minimum. The geometry optimizations under C_2 or C_i symmetry constraints converged to the C_{2h} symmetry structure. C_{2h}-1*Z*,5*Z*-Ac_2AN is considerably overcrowded due to the short $O^{15}...H^9/O^{16}...H^{10}$ contact distances (14% penetration). The C_{2h}-1*Z*,5*Z*-Ac_2AN conformation corresponds to the *Z*,*Z* X-ray structure of 1,5-Ac_2AN. However, the calculated structure is planar (excluding the methyl hydrogens), while the X-ray structure has the twist angle $\tau_1(C^{9a}–C^1–C^{11}–O^{15})$=20.0° and the dihedral angle θ=22.7°, and, as a result, is less overcrowded. In addition to the global minimum, there are three local minima conformations of 1,5-Ac_2AN: C_1-1*Z*,5*E*-Ac_2AN, C_i-1*E*,5*E*-*anti*-Ac_2AN and C_2-1*E*,5*E*-*syn*-Ac_2AN. The four conformations of 1,5-Ac_2AN undergo diastereomerizations by the rotation of one of the acetyl groups via "nearly orthogonal" transition states, in which the rotating acetyl group has the twist angle of τ=85-97°, and the other acetyl group retains its *E*- or *Z*-conformation. The rotation of an acetyl group of C_{2h}-1*Z*,5*Z*-Ac_2AN via [C_1-1*Z*,90-Ac_2AN] leads to the C_1-1*Z*,5*E*-Ac_2AN conformation, which is 12.8 kJ/mol higher in energy than the global minimum. The *E*-orientation of the acetyl group at the 5-position and the *peri*-interactions of its methyl hydrogens with H^{10} force the acetyl group out of the aromatic plane, thus decreasing the conjugation. Due to the twist angle $\tau_1(C^{10a}–C^5–C^{13}–O^{16})$=152.4° which differs from either 0° or 180°, rotation of the 1*Z*-acetyl group of C_1-1*Z*,5*E*-Ac_2AN may be realized in either *anti*-(via [C_1-90,5*E*-*anti*-Ac_2AN]) or in *syn*-direction (via [C_1-90,5*E*-*syn*-Ac_2AN]) relative to the 5*E*-acetyl group. These processes lead to the different local minima C_i-1*E*,5*E*-*anti*-Ac_2AN and C_2-1*E*,5*E*-*syn*-Ac_2AN conformations, respectively, which are 27.4 and 28.0 kJ/mol higher in energy than C_{2h}-1Z,5Z-Ac_2AN, due to both acetyl groups being forced out of the aromatic plane: $\tau_1(C^{9a}–C^1–C^{11}–O^{15})$=150.6° and 151.9°, respectively. In addition, the C_i-1*E*,5*E*-*anti*-Ac_2AN and C_2-1*E*,5*E*-*syn*-Ac_2AN conformations may undergo *syn,anti*-diastereomerization via the [C_1-1*E*,5E_{180}-Ac_2AN] transition state. It is a "nearly planar" transition state of a different type than the "nearly orthogonal" ones; the twist angle of the rotating acetyl group is close to zero, and the other acetyl group retains its *E*- or *Z*-conformation. The [C_1-1*E*,5E_{180}-Ac_2AN] transition state is considerably strained due to the short O^{16}...H^{10} distance (205.3 pm) and the distorted sp^2 angles $C^{13}–C^5–C^{10a}$ (127.9°) and $C^{13}–C^5–C^6$ (113.4°). The diastereomerization processes in 1,5-Ac_2AN are shown in Fig. 19.

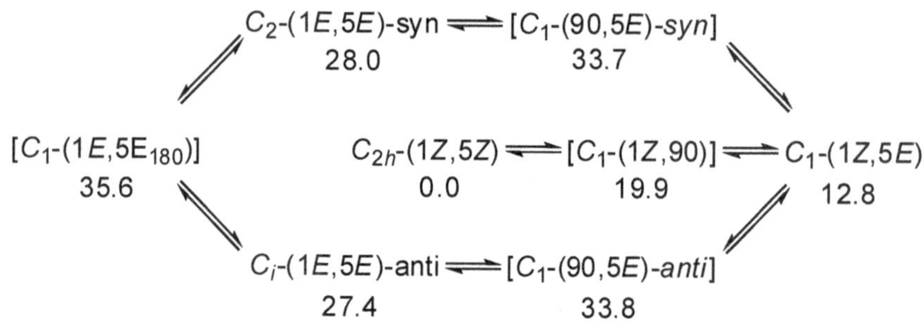

Fig. 19. The interconversion of conformations of 1,5-Ac_2AN and their relative Gibbs free energies (ΔG_{298}, kJ/mol)

Ketone 1,6-Ac_2AN adopts a C_s-1*Z*,6*E* conformation as its global minimum. Like C_{2h}-1Z,5Z-Ac_2AN, it is overcrowded due to the short O^{15}...H^9 contact distance (14% penetration). The C_s-1*Z*,6*E*-Ac_2AN conformation corresponds to the *Z,E* X-ray structure of 1,6-Ac_2AN. As in the case of 1,5-Ac_2AN, the DFT calculations predict a planar structure for 1,6-Ac_2AN, while the X-ray geometry features the twisted 1*Z*-acetyl group: the twist angle $\tau_1(C^{9a}–C^1–C^{11}–O^{15})$=30.0° and the dihedral angle θ=32.2°. The 6*E*-acetyl group remains in the aromatic plane in both calculated and the X-ray geometries. The rotation of the 1*Z*-acetyl group leads from C_s-1*Z*,6*E*-Ac_2AN via [C_1-90,6*E*-Ac_2AN] to the local minimum C_1-1*E*,6*E*-Ac_2AN, which is 13.6 kJ/mol higher in energy. The 6*E*-acetyl group, in contrast to the 1*E*-acetyl group, lies in the aromatic plane: $\tau_1(C^{9a}–C^1–C^{11}–O^{15})$=150.6° and $\tau_2(C^5–C^6–C^{13}–O^{16})$=179.9°. The rotation of the 6*E*-acetyl group of C_1-1*E*,6*E*-Ac_2AN may be realized either via [C_1-1*E*,90-*syn*-Ac_2AN] or via [C_1-1*E*,90-*anti*-Ac_2AN] transition states; both pathways lead to C_1-1*E*,6*Z*-*anti*-Ac_2AN, which is 15.4 kJ/mol higher in energy than the global minimum. The rotation of the 6*E*-acetyl group in C_s-1*Z*,6*E*-Ac_2AN via [C_1-1*Z*,90-Ac_2AN] leads to the local minimum C_s-1*Z*,6*Z*-Ac_2AN, which is only 1.7 kJ/mol higher in energy than the global minimum. The rotation of the 1*E*-acetyl group in C_1-1*E*,6*Z*-Ac_2AN via [C_1-90,6*Z*-Ac_2AN] also leads to C_s-1*Z*,6*Z*-Ac_2AN. The diastereomerization processes in 1,6-Ac_2AN are shown in Fig. 20.

Ketone 1,7-Ac_2AN, similarly to 1,6-Ac_2AN, adopts a C_s-1*Z*,7*E* conformation as its global minimum. It is overcrowded due to the short O^{15}...H^9 contact distance (15% penetration). The C_s-1*Z*,7*E*-Ac_2AN conformation corresponds to the *Z,E* X-ray structure of 1,7-Ac_2AN. The differences between the geometries of the planar DFT calculated structure of C_s-(1*Z*,7*E*)-Ac_2AN and the twisted X-ray structure of 1,7-Ac_2AN are smaller than in 1,5-Ac_2AN and 1,6-Ac_2AN. In the X-ray structure of 1,7-Ac_2AN the twist angles are $\tau_1(C^{9a}–C^1–C^{11}–O^{15})$=-15.2°

and $\tau_2(C^8–C^7–C^{13}–O^{16})$=-176.6. The relative stabilities of the conformations of 1,7-Ac_2AN and its conformational space are very similar to those of 1,6-Ac_2AN, both being α,β-diacetylanthracenes. The local minima conformations C_s-1*Z*,7*E*-Ac_2AN, C_1-1*E*,7*E*-Ac_2AN and C_1-1*E*,7*Z*-*anti*-Ac_2AN are higher in energy than the global minimum by 3.5, 14.5, and 16.6 kJ/mol, respectively. The diastereomerization processes in 1,7-Ac_2AN are shown in Fig. 21.

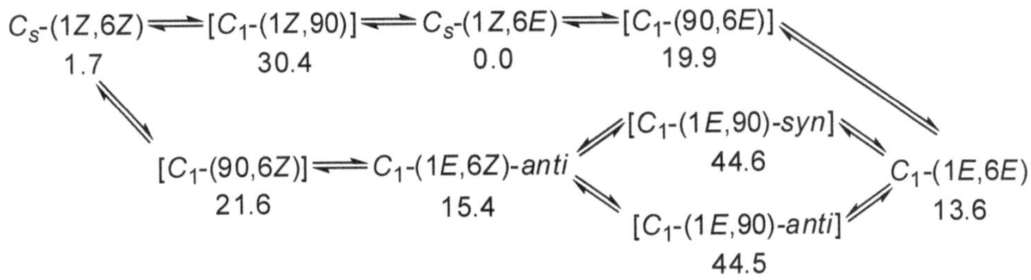

Fig. 20. The interconversion of conformations of 1,6-Ac_2AN and their relative Gibbs free energies (ΔG_{298}, kJ/mol)

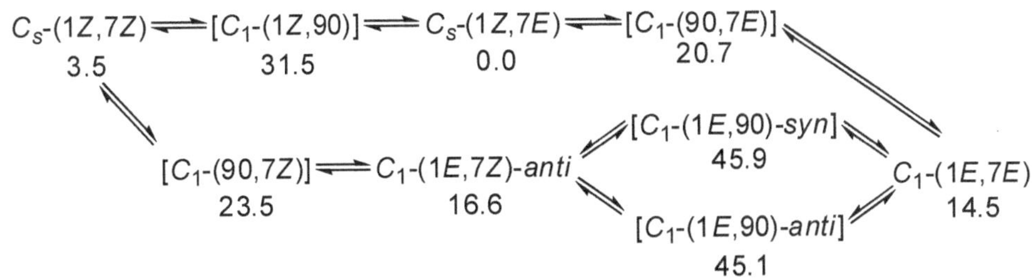

Fig. 21. The interconversion of conformations of 1,7-Ac_2AN and their relative Gibbs free energies (ΔG_{298}, kJ/mol)

The most interesting diacetylanthracene is 1,8-Ac_2AN. *Peri*-interactions $O^{15}...H^9$ and $O^{16}...H^9$ tilt both carbonyl groups out of the aromatic plane, rendering a planar conformation such as C_{2h}-1Z,5Z-Ac_2AN energetically highly unfavorable. Ketone 1,8-Ac_2AN adopts a C_2-1Z,8Z-*anti* conformation as its global minimum. It is overcrowded due to the short $O^{15}...H^9$ contact distance (12% penetration). The C_2-1Z,8Z-*anti*-Ac_2AN conformation corresponds to the Z,Z X-ray structure of 1,8-Ac_2AN. Both structures feature twisted carbonyl groups; however, in the X-ray structure the twist angles are more pronounced ($\tau_1(C^{9a}–C^1–C^{11}–O^{15})$=-32.4° and -34.0°, θ=35.4° and 36.0°) than in the calculated structure ($\tau_1(C^{9a}–C^1–C^{11}–O^{15})$=-17.3°, θ=19.3°). Although the conformational space of 1,8-Ac_2AN resembles that of another α,α-diacetylanthracene, 1,5-Ac_2AN, it is more complicated. There are three local minima conformations of 1,8-Ac_2AN: C_1-1Z,8*E*-Ac_2AN, C_s-1*E*,8*E*-*syn*-Ac_2AN and C_2-1*E*,8*E*-*anti*-Ac_2AN. Rotation of an acetyl group of C_2-1Z,8Z-Ac_2AN via [C_1-1Z,90-Ac_2AN] leads to the C_1-1Z,8*E*-Ac_2AN conformation, which is only 0.4 kJ/mol higher in energy. The tilting of the 8*E*-acetyl group ($\tau_2(C^{8a}–C^8–C^{13}–O^{16})$=150.4°) allows the 1Z-acetyl group to align itself with the aromatic plane ($\tau_1(C^{9a}–C^1–C^{11}–O^{15})$=1.5°), restoring the conjugation and thus stabilizing this conformation. The rotation of the 1Z-acetyl group of C_1-1Z,8*E*-Ac_2AN may be realized in either *syn*- (via [C_1-90,8*E*-*syn*-Ac_2AN]) or in *anti*-direction (via [C_1-90,8*E*-*anti*-Ac_2AN]) relative to the 8*E*-acetyl group. These pathways lead to the local minima C_s-1*E*,8*E*-*syn*-Ac_2AN and C_2-1*E*,8*E*-*anti*-Ac_2AN conformations, respectively, which are 17.6 and 17.7 kJ/mol higher in energy than the global minimum. These two conformations undergo interconversion via the [C_1-1*E*,8E_{180}-Ac_2AN] transition state. The diastereomerization processes in 1,8-Ac_2AN are shown in Fig. 22.

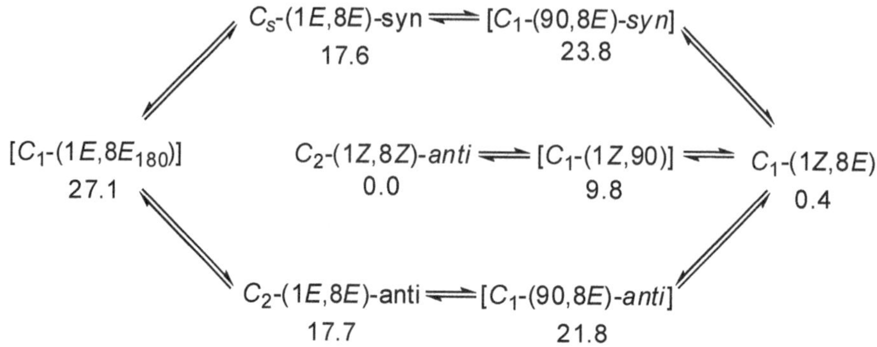

Fig. 22. The interconversion of conformations of 1,8-Ac_2AN and their relative Gibbs free energies (ΔG_{298}, kJ/mol)

Ketone 2,7-Ac_2AN adopts a C_s-2*E*,7*Z* conformation as its global minimum. It is not overcrowded, lacking *peri*-interactions. The C_s-(2*E*,7*Z*)-Ac_2AN conformation corresponds well to the *E*,*Z* X-ray structure of 2,7-Ac_2AN. The differences between the geometries of the planar DFT calculated structure of C_s-2*E*,7*Z*-Ac_2AN and the twisted X-ray structure of 2,7-Ac_2AN are not large: in the latter structure the twist angles are τ_2(C^1–C^2–C^{11}–O^{15})=171.9° and τ_2(C^8–C^7–C^{13}–O^{16})=0.9° (θ=9.8° and 1.6°, respectively). There are only two local minima conformations of 2,7-Ac_2AN, both are planar like the global minimum. Due to the twist angles τ_2 being either 0° or 180°, no *anti*-, *syn*-conformations are possible. The rotation of the 7*Z*-acetyl group in C_s-2*E*,7*Z*-Ac_2AN via [C_1-2*E*,90-Ac_2AN] leads to C_{2v}-2*E*,7*E*-Ac_2AN conformation, which is only 0.2 kJ/mol higher in energy. The rotation of the 2*E*-acetyl group of C_s-2*E*,7*Z*-Ac_2AN via [C_1-90,7*Z*-Ac_2AN] leads to the C_{2v}-2*Z*,7*Z*-Ac_2AN conformation, which is 4.3 kJ/mol higher in energy than the global minimum. The diastereomerization processes in 2,7-Ac_2AN are shown in Fig. 23.

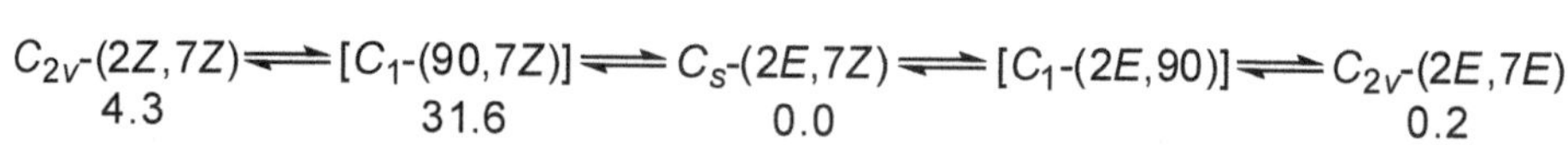

Fig. 23. The interconversion of conformations of 2,7-Ac_2AN and their relative Gibbs free energies (ΔG_{298}, kJ/mol)

The conformational space of 2,6-Ac_2AN is similar to that of 2,7-Ac_2AN. The global minimum is the C_{2h}-2*E*,6*E*-Ac_2AN conformation. Rotation of the 6*E*-acetyl group leads to C_s-2*E*,6*Z*-Ac_2AN conformation, which is only 0.4 kJ/mol higher in energy than the global minimum. The rotation of the 2*E*-acetyl group in C_s-2*E*,6*Z*-Ac_2AN leads to C_{2h}-2*Z*,6*Z*-Ac_2AN conformation, which is 3.8 kJ/mol higher in energy than the global minimum. The diastereomerization processes in 2,6-Ac_2AN are shown in Fig. 24.

$$C_{2h}\text{-}(2Z,6Z) \rightleftharpoons [C_1\text{-}(90,6Z)] \rightleftharpoons C_s\text{-}(2E,6Z) \rightleftharpoons [C_1\text{-}(2E,90)] \rightleftharpoons C_{2h}\text{-}(2E,6E)$$

3.8 0.4 0.0

Fig. 24. The interconversion of conformations of 2,6-Ac_2AN and their relative Gibbs free energies (ΔG_{298}, kJ/mol)

Ketone 9,10-Ac_2AN stands out of the other diacetylanthracenes by virtue of its acetyl groups being each flanked by two *peri*-hydrogens. In order to avoid short non-contact distances to $H^1/H^4/H^5/H^8$, the acetyl groups in all the conformations of 9,10-Ac_2AN are considerably twisted. Another mode for the relief of the steric strain in 9,10-Ac_2AN is elongation of the C^{11}–C^9 and C^{12}–C^{10} carbonyl bonds, 151.6 pm, as compared to 149.7 pm in planar C_s-(2*E*,7*Z*)-Ac_2AN and 149.8 pm in C_s-(2*E*,6*E*)-Ac_2AN. The global minimum of 9,10-Ac_2AN is a C_i-*E* conformation, with the twist angles $\tau_9(C^{9a}–C^9–C^{11}–O^{15})=-72.6°$, $\tau_9(C^{10a}–C^{10}–C^{13}–O^{16})=72.6°$ and the dihedral angle $\theta=74.7°$. It corresponds well to the X-ray structure, which features even higher twist angles $\tau_9=-85.0°$, 87.0° and the dihedral angles $\theta=86.7°$, 86.5°. The local minima conformations of 9,10-Ac_2AN are C_s-*Z* (0.1 kJ/mol), C_2-*E* (1.0 kJ/mol) and C_2-*Z* (2.1 kJ/mol). They all have high twist angles, ±71.8°, 75.4° and -71.9°, respectively. The similarity of the energies and the geometries of the four conformations of 9,10-Ac_2AN stems from the fact that in 9,10-Ac_2AN, each of the *Z* and *E* conformations is defined relative to the other acetyl group, and not by the twist angles of the carbonyl groups relative to the anthracene system, which are very similar for all four conformations of 9,10-Ac_2AN. The C_i-*E* global minimum undergoes diastereomerization to the C_2-*E* conformation via [C_1-(9*E*,10E_{180})] transition state, in which one of the carbonyl groups lies in the aromatic plane. The C_s-*Z* and C_2-*Z* conformations interconvert via the analogous [C_1-(9*Z*,10Z_0)] transition state. The C_i-*E* conformation diastereomerizes to the C_2-*Z* conformation and the C_2-*E* conformation diastereomerizes to the C_s-*Z* conformation via the pair of transition states [C_1-90-*syn*] and [C_1-90-*anti*], in which one of the carbonyl groups is orthogonal to the aromatic plane. The diastereomerization processes in 9,10-Ac_2AN are shown in Fig. 25.

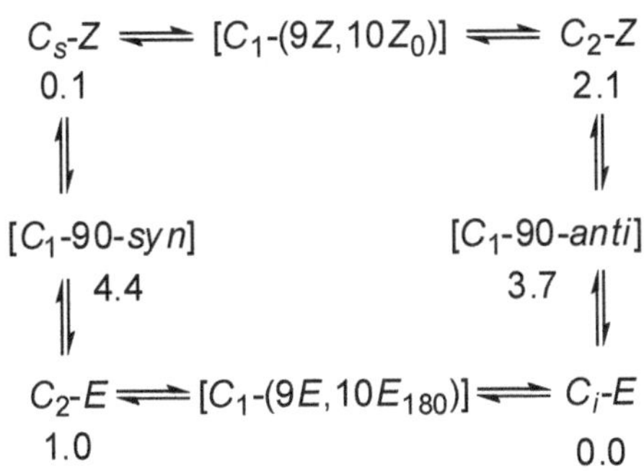

Fig. 25. The interconversion of conformations of 9,10-Ac_2AN and their relative Gibbs free energies (ΔG_{298}, kJ/mol)

Ketone 1,9-Ac_2AN has never been isolated. Recently 1,9-Ac_2AN has been claimed to be a putative intermediate in the Friedel–Crafts acyl rearrangements of 1,5-Ac_2AN, 1,8-Ac_2AN and 9,10-Ac_2AN in PPA to give 3-methylbenz[de]anthracen-1-one [Mala'bi et al., 2011]. Ketone 1,9-Ac_2AN adopts a C_1-1*Z*,9*Z*-*anti* conformation as its global minimum. Both acetyl groups are considerably twisted because of their mutual *peri*-positions: $\tau_1(C^{9a}–C^1–C^{11}–O^{15})$=-50.9°, $\tau_9(C^{9a}–C^9–C^{13}–O^{16})$=-59.6°. The local minimum conformation C_1-1*E*,9*Z*-*syn*-Ac_2AN is considerably higher in energy than the global minimum, 25.9 kJ/mol. Potentially, two more conformations may exist due to the twist angles τ_1 and τ_9 being different from 0° or 180°, i.e. C_1-1*Z*,9*Z*-*syn*-Ac_2AN and C_1-1*E*,9*Z*-*anti*-Ac_2AN. However, the search after these conformations has not resulted in any additional stationary points. The C_1-1*Z*,9*E*-Ac_2AN and C_1-1*E*,9*E*-Ac_2AN conformations have also not been found in the conformational space of 1,9-Ac_2AN, probably due to the considerable steric strain caused by the *peri*-interactions between the methyl of the 9*E*-acetyl group and the 1-acetyl group.
Ketone 1,10-Ac_2AN (which has never been synthesized [Mala'bi et al., 2011]) adopts a C_1-1*Z*,10*E* conformation as its global minimum. Contrary to 1,9-Ac_2AN, its acetyl groups do not affect directly each other. Hence, their twist angles, $\tau_1(C^{9a}–C^1–C^{11}–O^{15})$=0.2°, $\tau_9(C^{4a}–C^{10}–C^{13}–O^{16})$=-108.0°, are very close to the twist angles of the lone acetyl groups in C_s-1Z-AcAN (0.0°) and C_1-9-AcAN (-67.0°), respectively. Another consequence of the non-interacting acetyl groups in 1,10-Ac_2AN is the abundance of conformations - six minima conformations have been identified. The local minimum C_1-1Z,10Z-Ac_2AN conformation is only 1.0 kJ/mol less stable than the global minimum, and differs from it in the twist angle $\tau_9(C^{4a}–C^{10}–C^{13}–O^{16})$=-65.9°. There are four 1*E* conformations of 1,10-Ac_2AN, which have the twist angles $\tau_1(C^{9a}–C^1–C^{11}–O^{15})$ of 148–150° and the relative energy of 13.9–15.3 kJ/mol. The conformational behavior of 1,10-Ac_2AN is complicated. Depending on the rotational direction of the

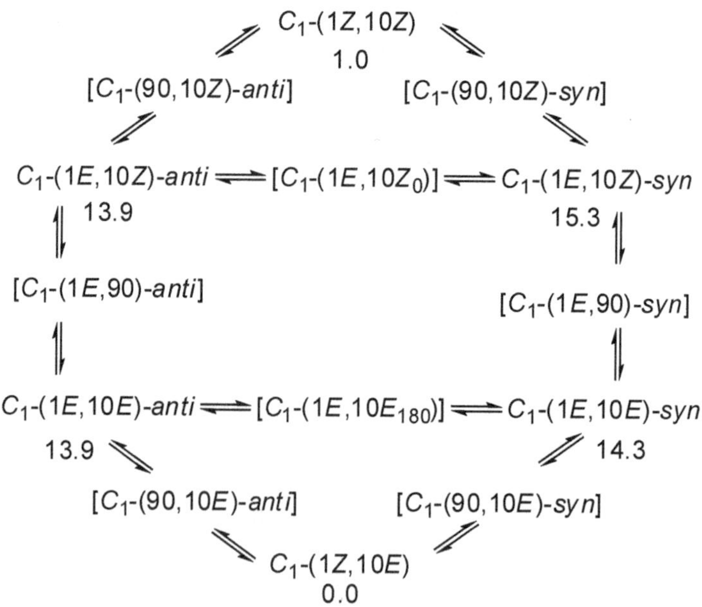

Fig. 26. The interconversion of conformations of 1,10-Ac_2AN and their relative Gibbs free energies (ΔG_{298}, kJ/mol)

1*Z*-acetyl group, the C_1-1*Z*,10*E*-Ac_2AN conformation may undergo diastereomerization to either C_1-1*E*,10*E*-*anti*-Ac_2AN or C_1-1*E*,10*E*-*syn*-Ac_2AN via the respective "nearly orthogonal" transition states. Analogously, C_1-1*Z*,10*Z*-Ac_2AN may undergo diastereomerization to either C_1-1*E*,10*Z*-*anti*-Ac_2AN or C_1-1*E*,10*Z*-*syn*-Ac_2AN. The C_1-1*E*,10*E*-*anti*-Ac_2AN and C_1-1*E*,10*Z*-*anti*-Ac_2AN conformations are interconnected via the [C_1-1*E*,90-*anti*-Ac_2AN] transition state, while C_1-1*E*,10*E*-*syn*-Ac_2AN and C_1-1*E*,10*Z*-*syn*-Ac_2AN are interconnected via the [C_1-1*E*,90-*syn*-Ac_2AN] transition state. Finally, C_1-1*E*,10*E*-*anti*-Ac_2AN is interconnected with C_1-1*E*,10*E*-*syn*-Ac_2AN and C_1-1*E*,10*Z*-*anti*-Ac_2AN is interconnected with C_1-1*E*,10*Z*-*syn*- Ac_2AN, via the "nearly planar" transition states [C_1-1*E*,10E_{180}-Ac_2AN] and [C_1-1*E*,10Z_0-Ac_2AN], respectively. The diastereomerization processes in 1,10-Ac_2AN are shown in Fig. 26.

Ketone 2,9-Ac_2AN (which has never been synthesized) adopts a C_1-2*E*,9*E* conformation as its global minimum. The acetyl groups in 2,9-Ac_2AN do not affect directly one another, and twist angles are similar to the respective twist angles in 2-AcAN and 9-AcAN: $\tau_2(C^1–C^2–C^{11}–O^{15})=-178.9°$ and $\tau_9(C^{4a}–C^{10}–C^{13}–O^{16})=-106.9°$. There are two local minima conformations of 2,9-Ac_2AN, C_1-2*E*,9*Z*-Ac_2AN (3.6 kJ/mol above the global minimum) and C_1-2*Z*,9*E*-Ac_2AN (5.1 kJ/mol). Surprisingly, the search after the C_1-2*Z*,9*Z*-Ac_2AN conformation was not successful. The acetyl groups in the putative C_1-2*Z*,9*Z*-Ac_2AN conformation are not expected to cause a steric hindrance more severe than in the C_1-1*Z*,9*Z*-*anti*-Ac_2AN conformation. Nevertheless, the latter conformation exists and even was found to be a global minimum, while the former does not seem to exist. The global minimum C_1-2*E*,9*E*-Ac_2AN conformation may diastereomerize either to the C_1-2*E*,9*Z*-Ac_2AN conformation via the [C_1-2*E*,90-Ac_2AN] transition state, or to the C_1-2*Z*,9*E*-Ac_2AN conformation via the [C_1-90,9*E*-Ac_2AN] transition state. The diastereomerization processes in 2,9-Ac_2AN are shown in Fig. 27.

$$\underset{3.6}{C_1\text{-}(2E,9Z)} \rightleftharpoons [C_1\text{-}(2E,90)] \rightleftharpoons \underset{0.0}{C_1\text{-}(2E,9E)} \rightleftharpoons [C_1\text{-}(90,9E)] \rightleftharpoons \underset{5.1}{C_1\text{-}(2Z,9E)}$$

Fig. 27. The interconversion of conformations of 2,9-Ac_2AN and their relative Gibbs free energies (ΔG_{298}, kJ/mol)

Ketone 2,10-Ac_2AN (which has never been synthesized) adopts a C_1-2*E*,10*E* conformation as its global minimum. The twist angles are $\tau_2(C^1–C^2–C^{11}–O^{15})=179.9°$ and $\tau_9(C^{4a}–C^{10}–C^{13}–O^{16})=-113.9°$. The global minimum C_1-2*E*,10*E*-Ac_2AN conformation may diastereomerize either to the C_1-2*Z*,10*E*-Ac_2AN conformation (the relative energy of 2.1 kJ/mol) via the [C_1-90,10*E*-Ac_2AN] transition state, or to the C_1-2*E*,10*Z*-Ac_2AN conformation (0.3 kJ/mol) via the [C_1-2*E*,90-Ac_2AN] transition state. Both these local minima configurations undergo diastereomerization to the C_1-2*Z*,10*Z*-Ac_2AN conformation (3.9 kJ/mol) via the [C_1-2*Z*,90-Ac_2AN] and [C_1-90,10*Z*-Ac_2AN] transition states, respectively. The diastereomerization processes in 2,10-Ac_2AN are shown in Fig. 28.

The comparison between the X-ray structures of mono- and diacetylanthracenes and their respective calculated geometries deserves a brief discussion. The absolute values of the twist angles of the B3LYP/6-31G(d) calculated conformations (including the local minima conformations) of mono- and diacetylanthracenes may be summarized as follows: $|\tau_1|=0–17.3°$[2] for the 1*Z*-acetyl groups, $|\tau_1|=141.2–152.4°$ for the 1*E*-acetyl groups, $|\tau_2|=0.0–1.8°$ for

---

[2] The 1,9-Ac_2AN is an outlier, having unusually high twist angle of the 1*Z*-acetyl group, $|\tau_1|=51.9°$, due to the steric strain caused by its interaction with the *peri* 9*Z*-acetyl group. The mutual *peri*-positions of the acetyl groups lead to an enhanced degree of overcrowding.

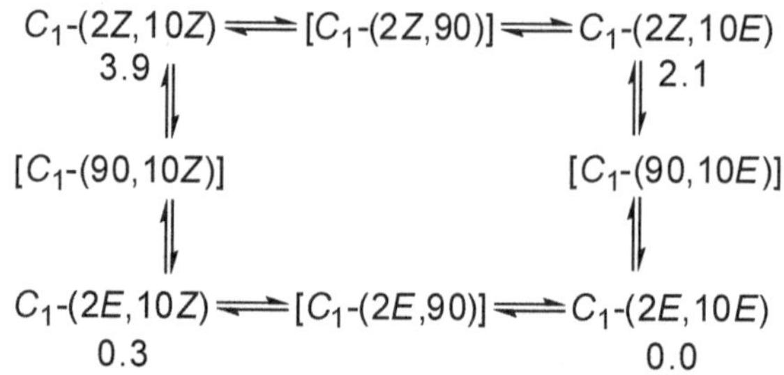

Fig. 28. The interconversion of conformations of 2,10-Ac_2AN and their relative Gibbs free energies (ΔG_{298}, kJ/mol)

the 2Z-acetyl groups, $|\tau_2|$=178.9–180.0° for the 2E-acetyl groups, and $|\tau_9|$=44.8–75.4° (180–$|\tau_9|$ values were taken for $|\tau_9|$>90°). The respective twist angles derived from the X-ray structures are $|\tau_1|$=15.2–34.0° for the 1Z-acetyl groups, $|\tau_2|$=0.9° for the 2Z-acetyl group, $|\tau_2|$=177.9–178.6° for the 2E-acetyl groups, and $|\tau_9|$=85.0–87.9°. There is no X-ray structure of acetylanthracenes featuring a 1E-acetyl group, and such conformations are always found to be local minima by the DFT calculations. The B3LYP/6-31G(d) calculations seem to underestimate the twist angles of the 1Z- and 9-acetyl groups in mono- and diacetylanthracenes. In a number of cases it leads to predicting planar and more overcrowded conformations than the respective twisted X-ray geometries. There is a limited number of reports in the literature that DFT methods, including B3LYP, could overstabilize planar conformations of biphenyl and related heteroaromatic compounds [Viruela et al., 1997; Karpfen et al., 1997]. As in the X-ray structures, the acetyl groups and the anthracene systems in the mono- and diacetylanthracenes under study are essentially planar. Thus, B3LYP/6-31G(d) satisfactorily predicts the overall conformations of mono- and diacetylanthracenes under study, i.e. the Z-conformation of 1-AcAN, the E-conformation of 2-AcAN, the twisted conformation of 9-AcAN, the Z,Z conformations of 1,5-Ac_2AN and 1,8-Ac_2AN, the Z,E conformations of 1,6-Ac_2AN, 1,7-Ac_2AN and 2,7-Ac_2AN, and the E,E conformation of 9,10-Ac_2AN. It has not escaped our mind, however, that the packing interactions in crystals can readily dominate and suppress any preference for one conformation or another, especially in the cases of low diastereomerization barriers and low energy differences. We also note the limitations of the DFT calculations in the gas phase and in the comparison of the computational results with the crystal structures.

The relative free Gibbs energies of the diacetylanthracenes under study are given in Table 6. It should be noted that 1,5-Ac_2AN is 11.2 kJ/mol more stable than its constitutional isomer 1,8-Ac_2AN. The acetyl groups of 1,5-Ac_2AN are attached to a starred and an unstarred aromatic carbons of alternant anthracene, while the acetyl groups of 1,8-Ac_2AN are both attached to starred aromatic carbons. Simple resonance considerations would favor the stabilization of 1,8-Ac_2AN over 1,5-Ac_2AN, due to the better delocalization of the partial positive charge in the dipolar Kekulé structures. However, the twist angle of the acetyl groups in 1,8-Ac_2AN are notably larger than that in 1,5-Ac_2AN, in both the crystal structures (32.4°/34° vs. 20.0°) and the DFT calculated geometries (17.3° vs. 0.0°). This increased twist angle decreases the conjugation between the acetyl group and the aromatic system, thus destabilizing 1,8-Ac_2AN relative to 1,5-Ac_2AN.

The order of stabilities of the global minima of the diacetylanthracenes is 2,7-Ac_2AN≈2,6-Ac_2AN>1,7-Ac_2AN≈1,6-Ac_2AN>1,5-Ac_2AN>2,9-Ac_2AN>2,10-Ac_2AN>1,8-Ac_2AN>1,10-Ac_2AN>>1,9-Ac_2AN>9,10-Ac_2AN. The acetyl groups at positions 1, 5 and 8 destabilize the diacetylanthracenes, while acetyl groups at positions 9 and 10 cause even greater destabilization. The destabilization of the 1, 5, 8, 9 and 10-substituted diacetylanthracenes relative to their 2, 6 and 7-substituted constitutional isomers stems from the overcrowding due to repulsive non-bonding interactions between the carbonyl oxygen/methyl group and the aromatic hydrogens in *peri*-positions, and from the decreased resonance stabilization between the carbonyl and the aromatic system. Thus, the acetyl groups in 9,10-Ac_2AN, 1,9-Ac_2AN, 1,10-Ac_2AN and 1,8-Ac_2AN, being considerably tilted out of the aromatic plane, reduce the relative stabilities of these diacetylanthracenes, potentially allowing deacylation, transacylation and acyl rearrangements. By contrast, 2,7-Ac_2AN and 2,6-Ac_2AN are not expected to undergo the Friedel-Crafts acyl rearrangements.

2.3.2 Activation barriers

As noted above, monoacetylanthracenes and diacetylanthracenes may undergo *E,Z*-diastereomerizations and *syn,anti*-diastereomerizations by rotation of their acetyl groups. The diastereomerization of the first type connects an *E*-diastereomer with a *Z*-diastereomer and proceeds via a "nearly orthogonal" transition state, in which the acetyl group, rotating around the C_{arom}–C_{carb} bond, has the twist angle of τ=85–97° (the other acetyl group in diacetylanthracenes retains its *E*- or *Z*-orientation). The diastereomerization of the second type occurs only in diacetylanthracenes and connects either an *E-syn*-diastereomer with an *E-anti*-diastereomer, or a *Z-syn*-diastereomer with a *Z-anti*-diastereomer. It proceeds via a "nearly planar" transition states, in which the twist angle of the rotating acetyl group is close to either 180° (*E*-diastereomer) or zero (*Z*-diastereomer), and the other acetyl group retains its *E*- or *Z*-orientation. Fig. 29 and Fig. 30 show the *E,Z*-diastereomerization and *syn,anti*-diastereomerization on the example of 1,8-Ac_2AN.

Table 7 gives the energy barriers for the *E,Z*-diastereomerization and *syn,anti*-diastereomerization in the monoacetylanthracenes and diacetylanthracenes under study by rotation of the acetyl groups via the respective nearly orthogonal or nearly planar transition states. The E,Z-diastereomerization energy barriers may be divided into three groups, depending on the position of the acetyl group that undergoes rotation and on its conformation. *E*-Acetyl groups at positions 1, 5 and 8 and acetyl groups at positions 9 and 10 have the lowest energy barriers, 3.6–9.5 kJ/mol, due to their already significant twist angles (τ=141–152° for *E*-acetyl groups at positions 1, 5 and 8 and τ=67–73° for acetyl groups at positions 9 and 10). *Z*-Acetyl groups at the same positions 1, 5 and 8 have medium energy barriers, 19.5–23.5 kJ/mol. The twist angles of these acetyl groups are smaller (τ=0–17°), but the *E,Z*-diastereomerization in this case is facilitated by the steric strain due to repulsive *peri*-interactions between the carbonyl oxygen and aromatic H^9/H^{10} hydrogens. Finally, both *E*- and *Z*-acetyl groups at positions 2, 6 and 7 have the highest energy barriers for diastereomerization, 27.3–31.6 kJ/mol, due to the lack of steric strain and negligible twist angles (less than 1°). For comparison, the experimental rotational energy barrier for methyl 1-(2-methylnaphthyl) ketone is 33.9 kJ/mol (–110 °C, [Wolf, 2008]). Table 8 gives the relative Gibbs free energies of the global minima and the most stable local minima of the acetylanthracenes whose crystal structures have been determined in this study or reported in the literature, i.e. 1-AcAN, 2-AcAN, 9-AcAN, 1,5-Ac_2AN, 1,6-Ac_2AN, 1,7-Ac_2AN, 1,8-Ac_2AN, 2,7-Ac_2AN and 9,10-Ac_2AN, as well the energy barriers for their *E,Z*-

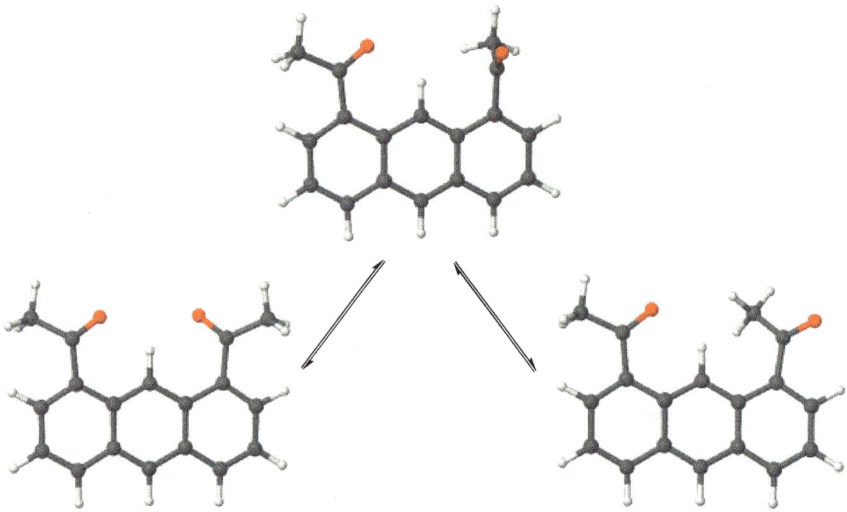

Fig. 29. *E*,*Z*-Diastereomerization of C_2-1Z,8Z-Ac$_2$AN to C_2-1Z,8*E*-Ac$_2$AN via [1Z,90-Ac$_2$AN] transition state

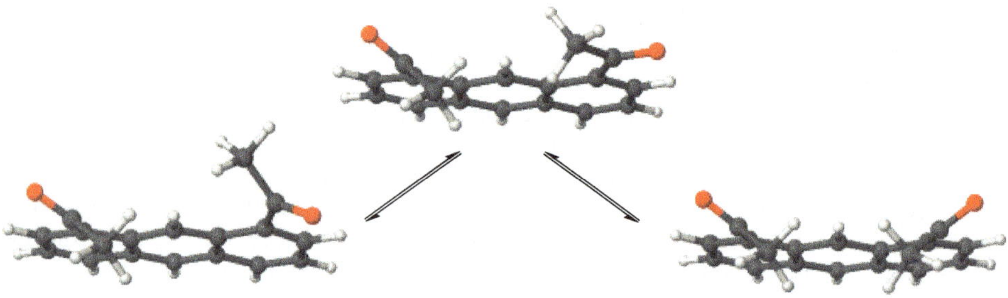

Fig. 30. *syn*,*anti*-Diastereomerization of C_2-1*E*,8*E*-*anti*-Ac$_2$AN to C_s-1*E*,8*E*-*syn*-Ac$_2$AN via [1*E*,8E_{180}-Ac$_2$AN] transition state

diastereomerizations. The energy barriers are in the range of 20–32 kJ/mol (relative to the respective global minima) for the rotation of the acetyl groups at 1, 2, 5, 6 and 7 positions. The lower energy barrier in 1,8-Ac$_2$AN (9.8) may be rationalized by destabilization of the global minimum due to the larger twist of the acetyl groups. This effect is even more pronounced in the case of 9-AcAN and 9,10-Ac$_2$AN, which have large twist values (67° and 73°, respectively) and remarkably low *E*,*Z*-diastereomerization barriers (less than 4 kJ/mol). All these barriers are sufficiently low to allow a swift *E*,*Z*-diastereomerizations on the NMR time scale (at room temperature), in accordance with the results of the NMR experiments (*vide supra*). The differences in the relative energies of the global minimum and the most stable local minimum of these acetylanthracenes are relatively small, 0.06-3.5 kJ/mol (with the exception of 1-AcAN and 1,5-Ac$_2$AN), which suggests the presence of both *E*- and *Z*-diastereomers in equilibrium mixture at ambient temperature.

E,*Z* diastereomerization or *syn*,*anti*-diastereomerization	Δ*G*‡ kJ/mol	Transition state	E_{Tot} Hartree
1*Z*-AcAN→1*E*-AcAN	19.52	[1-AcAN]	-692.16527547
1*E*-AcAN→1*Z*-AcAN	6.51		
2*Z*-AcAN→2*E*-AcAN	29.28	[2-AcAN]	-692.16644237
2*E*-AcAN→2*Z*-AcAN	31.52		
9-AcAN→9-AcAN*	3.64[a]	[9-AcAN]	-692.16355087
1*Z*,5*Z*-Ac_2AN→1*Z*,5*E*-Ac_2AN	19.93	[1*Z*,90-Ac_2AN]	-844.80776475
1*Z*,5*E*-Ac_2AN→1*Z*,5*Z*-Ac_2AN	7.14		
1*E*,5*E*-*syn*-Ac_2AN→1*E*,5*E*-*anti*-Ac_2AN	7.54	[1*E*,5E_{180}-Ac_2AN]	-844.80415197
1*E*,5*E*-*anti*-Ac_2AN→1*E*,5*E*-*syn*-Ac_2AN	8.22		
1*Z*,5*E*-Ac_2AN→1*E*,5*E*-*syn*-Ac_2AN	20.93	[90,5*E*-*syn*-Ac_2AN]	-844.80261144
1*E*,5*E*-syn-Ac_2AN→1*Z*,5*E*-Ac_2AN	5.67		
1*Z*,5*E*-Ac_2AN→1*E*,5*E*-*anti*-Ac_2AN	21.01	[90,5*E*-*anti*-Ac_2AN]	-844.80259178
1*E*,5*E*-*anti*-Ac_2AN→1*Z*,5*E*-Ac_2AN	6.44		
1*Z*,6*Z*-Ac_2AN→1*Z*,6*E*-Ac_2AN	28.66	[1*Z*,90-Ac_2AN]	-844.80884650
1*Z*,6*E*-Ac_2AN→1*Z*,6*Z*-Ac_2AN	30.35		
1*Z*,6*Z*-Ac_2AN→1*E*,6*Z*-Ac_2AN	19.86	[90,6*Z*-Ac_2AN]	-844.81218832
1*E*,6*Z*-Ac_2AN→1*Z*,6*Z*-Ac_2AN	6.13		
1*Z*,6*E*-Ac_2AN→1*E*,6*E*-Ac_2AN	19.86	[90,6*E*-Ac_2AN]	-844.81289093
1*E*,6*E*-Ac_2AN→1*Z*,6*E*-Ac_2AN	6.29		
1*E*,6*Z*-Ac_2AN→1*E*,6*E*-Ac_2AN	29.02	[1*E*,90-*anti*-Ac_2AN]	-844.80381925
1*E*,6*E*-Ac_2AN→1*E*,6*Z*-Ac_2AN	30.88		
1*E*,6*Z*-Ac_2AN→1*E*,6*E*-Ac_2AN	29.16	[1*E*,90-*syn*-Ac_2AN]	-844.80371345
1*E*,6*E*-Ac_2AN→1*E*,6*Z*-Ac_2AN	31.01		
1*Z*,7*Z*-Ac_2AN→1*Z*,7*E*-Ac_2AN	28.05	[1*Z*,90-Ac_2AN]	-844.80879074
1*Z*,7*E*-Ac_2AN→1*Z*,7*Z*-Ac_2AN	31.54		
1*Z*,7*Z*-Ac_2AN→1*E*,7*Z*-Ac_2AN	19.95	[90,7*Z*-Ac_2AN]	-844.81202172
1*E*,7*Z*-Ac_2AN→1*Z*,7*Z*-Ac_2AN	6.83		
1*Z*,7*E*-Ac_2AN→1*E*,7*E*-Ac_2AN	20.72	[90,7*E*-Ac_2AN]	-844.81303091
1*E*,7*E*-Ac_2AN→1*Z*,7*E*-Ac_2AN	6.26		
1*E*,7*Z*-Ac_2AN→1*E*,7*E*-Ac_2AN	28.53	[1*E*,90-*anti*-Ac_2AN]	-844.80384032
1*E*,7*E*-Ac_2AN→1*E*,7*Z*-Ac_2AN	30.69		
1*E*,7*Z*-Ac_2AN→1*E*,7*E*-Ac_2AN	29.29	[1*E*,90-*syn*-Ac_2AN]	-844.80370612
1*E*,7*E*-Ac_2AN→1*E*,7*Z*-Ac_2AN	31.45		
1*Z*,8*Z*-*anti*-Ac_2AN→1*Z*,8*E*-Ac_2AN	9.81	[1*Z*,90-Ac_2AN]	-844.80732674
1*Z*,8*E*-Ac_2AN→1*Z*,8*Z*-*anti*-Ac_2AN	9.46		

1*E*,8*E*-*anti*-Ac$_2$AN→1*Z*,8*E*-Ac$_2$AN	4.15	[90,8*E*-*anti*-Ac$_2$AN]	-844.80278079
1*Z*,8*E*-Ac$_2$AN→1*E*,8*E*-*anti*-Ac$_2$AN	21.46		
1*E*,8*E*-*syn*-Ac$_2$AN→1*Z*,8*E*-Ac$_2$AN	6.22	[90,8*E*-*syn*-Ac$_2$AN]	-844.80231227
1*Z*,8*E*-Ac$_2$AN→1*E*,8*E*-*syn*-Ac$_2$AN	23.46		
1*E*,8*E*-*syn*-Ac$_2$AN→1*E*,8*E*-*anti*-Ac$_2$AN	9.49	[1*E*,8E_{180}-Ac$_2$AN]	-844.80296243
1*E*,8*E*-*anti*-Ac$_2$AN→1*E*,8*E*-*syn*-Ac$_2$AN	9.42		
C_i-*E*-9,10-Ac$_2$AN→C_2-*Z*-9,10-Ac$_2$AN	3.73	[C_1-90-*anti*-9,10-Ac$_2$AN]	-844.79633239
C_2-*Z*-9,10-Ac$_2$AN→C_i-*E*-9,10-Ac$_2$AN	3.67		
C_2-*E*-9,10-Ac$_2$AN→C_s-*Z*-9,10-Ac$_2$AN	3.44	[C_1-90-*syn*-9,10-Ac$_2$AN]	-844.79605863
C_s-*Z*-9,10-Ac$_2$AN→C_2-*E*-9,10-Ac$_2$AN	4.36		
C_s-2*E*,7*Z*-Ac$_2$AN→C_s-2*Z*,7*Z*-Ac$_2$AN	31.63	[C_1-90,7*Z*-Ac$_2$AN]	-844.81338779
C_s-2*Z*,7*Z*-Ac$_2$AN→C_s-2*E*,7*Z*-Ac$_2$AN	27.34		

[a] enantiomerization barrier

Table 7. Energy barriers ($\Delta G^{\ddagger}$, kJ/mol) for diastereomerizations of monoacetylanthracenes and diacetylanthracenes

			ΔG_{298}	$\Delta\Delta G_{298}$	$\Delta G^{\ddagger}$
1-AcAN	*Z*	C_s	15.79	0.00	19.52
1-AcAN	*E*	C_1	28.80	13.01	6.51
2-AcAN	*E*	C_s	0.00	0.00	31.52
2-AcAN	*Z*	C_s	2.24	2.24	29.28
9-AcAN	-	C_1	36.94	0.00	3.64
1,5-Ac$_2$AN	*ZZ*	C_{2h}	27.69	0.00	19.93
1,5-Ac$_2$AN	*ZE*	C_1	40.48	12.79	7.14
1,6-Ac$_2$AN	*ZE*	C_s	13.45	0.00	30.35
1,6-Ac$_2$AN	*ZZ*	C_s	15.14	1.69	28.66
1,7-Ac$_2$AN	*ZE*	C_s	12.34	0.00	31.54
1,7-Ac$_2$AN	*ZZ*	C_s	15.83	3.50	28.05
1,8-Ac$_2$AN	*ZZ*	C_2	38.89	0.00	9.81
1,8-Ac$_2$AN	*EZ*	C_1	39.25	0.35	9.46
9,10-Ac$_2$AN	*E*	C_i	71.57	0.00	3.73
9,10-Ac$_2$AN	*Z*	C_s	71.63	0.06	3.67
2,7-Ac$_2$AN	*EZ*	C_s	0.00	0.00	31.63

Table 8. Relative energies (ΔG_{298} and $\Delta\Delta G_{298}$, kJ/mol) of selected monoacetylanthracenes and diacetylanthracenes and respective energy barriers for *E*,*Z*-diastereomerizations ($\Delta G^{\ddagger}$, kJ/mol)

3. Conclusions

The monoacetylanthracenes and diacetylanthracenes under study adopt non-planar conformations in their crystal structures. The twist angles are maximal for the 9-acetyl groups ($|\tau_9|$=85.0–87.9°) and significant for the 1Z-acetyl groups ($|\tau_1|$=15.2–34.0°), but very small for 2-acetyl groups. The conformations in solution are in agreement with the X-ray crystal structure conformations, according to the NMR data. The crystal structures are stabilized by intermolecular interactions: aromatic–aromatic π–π interactions (1,6-Ac_2AN and 1,7-Ac_2AN), C···H-π interactions (2-AcAN, 1,5-Ac_2AN, 2,7-Ac_2AN and 9,10-Ac_2AN), or π–π interactions between the anthracene unit and the carbonyl bond (1,8-Ac_2AN). The B3LYP/6-31G(d) calculated conformations of the monoacetylanthracenes and diacetylanthracenes are in good agreement with the X-ray crystal structures. The acetyl groups in the crystal structures and the B3LYP/6-31G(d) calculated global minima of the monoacetylanthracenes and diacetylanthracenes preferentially adopts 1Z and 2*E* conformations. The order of stabilities of the diacetylanthracenes under study is 2,7-Ac_2AN>1,7-Ac_2AN≈1,6-Ac_2AN>1,5-Ac_2AN>1,8-Ac_2AN>9,10-Ac_2AN. The acetyl groups at positions 1, 5 and 8 destabilize the diacetylanthracenes because of the repulsive interactions between the carbonyl oxygen/methyl group and the aromatic *peri*-hydrogens, and because of the decreased resonance stabilization. This effect is even more pronounced for the acetyl groups at positions 9 and 10. The B3LYP/6-31G(d) calculated energy barriers for the E,Z-diastereomerizations show that the *E,Z*-diastereomerizations is swift on the NMR time scale (at room temperature), in accordance with the results of the NMR experiments. The present results of the crystallographic and theoretical study of monoacetylanthracenes and diacetylanthracenes contribute to our understanding of the motifs of reversibility and thermodynamic control in the Friedel–Crafts acyl rearrangements of these representative PAKs.

4. Experimental section

Table 9 summarizes the applied methods of preparation of the monoacetylanthracenes and diacetylanthracenes. Melting points are uncorrected. All NMR spectra were recorded with Bruker DRX 500 MHz spectrometer. ^{1}H-NMR spectra were recorded at 500.13 MHz using $CDCl_3$ as solvent and as internal standard, δ($CDCl_3$)=7.263 ppm. ^{13}C-NMR spectra were recorded at 125.75 MHz using $CDCl_3$ as a solvent with internal standard, δ($CDCl_3$)=77.008 ppm. Complete assignments were made through 2-dimensional correlation spectroscopy (COSY, HSQC, HBMC and NOESY). Anthracene and nitrobenzene were obtained from Sigma-Aldrich; acetyl chloride and aluminum chloride were obtained from Acros. All the solvents were AR grade. Chloroform and dichloromethane were distilled before use. Single crystal X-ray diffraction was carried out on a Bruker SMART APEX CCD X-ray diffractometer, equipped with graphite monochromator and using MoKα radiation (λ=0.71073 Å). Low temperature was maintained with a Bruker KRYOFLEX nitrogen cryostat. The diffractometer was controlled by a Pentium-based PC running the SMART software package [Bruker AXS GmbH, 2002a]. Immediately after collection, the raw data frames were transferred to a second PC computer for integration and reduction by the SAINT program package [Bruker AXS GmbH, 2002b]. The structures were solved and refined by the SHELXTL software package [Bruker AXS GmbH, 2002c].

Compound	Method of preparation	Solvent	Melting point °C	Lit. melting point, °C	Solvent of recryst.	Reference
1-AcAN	Anthracene $AlCl_3$, Acetyl chloride	CH_2Cl_2	110	109	EtOH	Bassilios, 1962 Mala'bi et al.,2009
2-AcAN	Anthracene $AlCl_3$, Acetyl chloride	$C_6H_5NO_2$	177	174–178	$MeCO_2Et$	Mala'bi et al.,2009 Bassilios, 1966
9-AcAN	Anthracene $AlCl_3$, Acetyl chloride	CH_2Cl_2	78	75–76	$MeCO_2Et$	Bassilios, 1962 Bassilios, 1962
1,5-Ac_2AN	Anthracene $AlCl_3$, Acetyl chloride	CH_2Cl_2	215	213	$CHCl_3$	Bassilios, 1963
1,6-Ac_2AN	2-AcAN $AlCl_3$, Acetyl chloride	ClC_2H_4Cl	170–172	171–172	$MeCO_2Et$	Gore, 1966
1,7-Ac_2AN	2-AcAN $AlCl_3$, Acetyl chloride	ClC_2H_4Cl	102–103	-	EtOH	-
1,8-Ac_2AN	Anthracene $AlCl_3$, Acetyl chloride	CH_2Cl_2	179	174–176	$CHCl_3$	Sarobe & Jenneskens, 1997
2,7-Ac_2AN	2-AcAN $AlCl_3$, Acetyl chloride	$C_6H_5NO_2$	156–157		*i*PrOH	
9,10-Ac_2AN	9,10-dicarbomethoxy-anthracene, MeLi	Et_2O	247	249–250	CH_2Cl_2	Duerr, 1988

Table 9. Summary of methods of preparation of monoacetylanthracenes and diacetylanthracenes.

The quantum mechanical calculations were performed using the Gaussian03 [Frisch et al., 2004] package. Becke's three parameter hybrid density functional B3LYP [Becke, 1993], with the non-local correlation functional of Lee, Yang, and Parr [Lee et al., 1988] was used. The split valence 6-31G(d) basis set [Hariharan & Pople, 1973] was employed. All structures were fully optimized using symmetry constrains as indicated. Vibrational frequencies were computed at the same level of theory to verify the nature of the stationary points. For calculating the thermal corrections to Gibbs' free energy, the zero point energies were scaled by 0.9804 [Bauschlicher & Partridge, 1995].

5. References

Adams, C. J., Earle, M. J., Roberts, G. & Seddon, K. R. (1998) Friedel–Crafts Reactions in Room Temperature Ionic Liquids. *Chemical Communications*, No. 19, pp. 2097–2098, ISSN: 1364-548X

Agranat, I. & Shih, Y.-S. (1974) Haworth Synthesis as a Route to the Anthracene Ring System. *Synthesis*, No. 12, pp. 865–867, ISSN: 0039-7881

Agranat, I. & Shih, Y.-S. (1974) The Synthesis of Linearly Annelated Polycyclic Ketones by Friedel-Crafts Rearrangements of Their Angular Isomers. *Synthetic Communications*, Vol. 4, No. 2, pp. 119–126, ISSN: 0039-7911

Agranat, I., Shih, Y.-S. & Bentor, Y. (1974) Incursion of reversibility in Friedel-Crafts acylations. *Journal of American Chemical Society*, Vol. 96, No. 4, pp. 1259–1260, ISSN: 0002-7863

Agranat, I., Bentor, Y. & Shih, Y.-S. (1977) Remarkable Reversibility in Aromatic Friedel-Crafts Acylations. *Para↔Ortho* Acyl Rearrangements of Fluorofluorenones in Polyphosphoric Acid. *Journal of the American Chemical Society*, Vol. 99, No. 21, pp. 7068–7070, ISSN: 0002-7863

Anderson, K., Becker, H., Engelhardt, L. M., Hansen, L. & White, A. H. (1984) Molecular geometry and crystal structure of 9-acetylanthracene. *Australian Journal of Chemistry*, Vol. 37, No. 6, pp. 1337–1340, ISSN: 0004-9425

Andreou, A. D., Gore, P. H. & Morris, D. F. C. (1978) Acetyl exchange between acetylmesitylene and acetyl chloride under Friedel–Crafts conditions. *Journal of Chemical Society, Chemical Communications*, No. 6, pp. 271–272, ISSN: 0022-4936

Assadi, N., Pogodin, S., Cohen, S., Levy, A. & Agranat, I. (2009) Overcrowded naphthologs of mono-bridged tetraarylethylenes: analogs of bistricyclic aromatic enes. *Structural Chemistry*, Vol. 20, pp.541–556, ISSN: 1040-0400

Balaban, A. T. (1966) Deacylation of Non-conjugated Ketones and the Reversibility of C-acylations, In: *Omagiu Acad. Prof. Raluca Ripan*, Dragulesea, C., pp 103–109, Editura Academiei Republicii Socialiste Romania, Bucharest

Bassilios, H.F., Shawky, M. & Salem A.Y. (1963) Acetylation of anthracene by the Friedel-Crafts reaction, using ethylene chloride as the solvent. *Recueil des Travaux Chimiques des Pays-Bas*, Vol. 82, No. 3, pp. 298–301, ISSN: 0165-0513

Bassilios, H.F., Salem, A.Y. & Anous, M.T. (1966) Acetylation of anthracene by the Friedel-Crafts reaction using nitrobenzene as the solvent. *Bulletin des Societes Chimiques Belges*, Vol. 75, No. 9-10, pp. 582–588, ISSN: 0037-9646

Bauschlicher Jr., C. W. & Partridge, H. (1995) A modification of the Gaussian-2 approach using density functional theory. *Journal of Chemical Physics*, Vol. 103., No. 5, pp. 1788–1791, ISSN: 0021-9606

Becke, A. D. (1993) Density-functional thermochemistry. III. The role of exact exchange. *Journal of Chemical Physics*, Vol. 98, No. 7, pp. 5648–5652, ISSN: 0021-9606

Biedermann, P. U., Stezowski, J. J. & Agranat, I. (2001) Thermochromism of overcrowded bistricyclic aromatic enes (BAEs). A theoretical study. *Chemical Communications*, No. 11, pp. 954–955, ISSN: 1364-548X

Brock, C. P. & Dunitz, J. D. (1990) Temperature Dependence of Thermal Motion in Crystalline Anthracene. *Acta Crystallographica Section B*, Vol. 46, No. 6, pp. 795–806, ISSN: 0108-7681

Bruker AXS GmbH (2002) SAINT-NT V5.0, D-76181 Karlsruhe, Germany

Bruker AXS GmbH (2002) SHELXTL-NT V6.1, D-76181 Karlsruhe, Germany

Bruker AXS GmbH (2002) SMART-NT V5.6, D-76181 Karlsruhe, Germany

Buehler, C. A. & Pearson, D. E. (1970). *Survey of Organic Synthesis*, p. 652, Wiley-Interscience, ISBN: 978-0894644085, New York

Dahl, T. (1994) The Nature of Stacking Interactions between Organic Molecules Elucidated by Analysis of Crystal Structures. *Acta Chemica Scandinavica*, Vol. 48, pp. 95–106, ISSN:0904-213X

de Proft, F. & Geerlings, P. (2001) Conceptual and Computational DFT in the Study of Aromaticity. *Chemical Reviews*, Vol. 101, No. 5, pp. 1451–1464, ISSN: 0009-2665

Desiraju, G. R. & Gavezzotti, A. (1989) Crystal Structures of Polynuclear Aromatic Hydrocarbons. Classification, Rationalization and Prediction from Molecular Structure. *Acta Crystallographica Section B: Structural Science,* Vol. 45, No. 5, pp. 473–482, ISSN: 0108-7681

Dowdy, D., Gore, P. H. & Waters, D. N. (1991) The Friedel–Crafts acetylation of naphthalene in 1,2-dichloroethane solution. Kinetics and mechanism. *Journal of Chemical Society, Perkin Transactions 2,* No. 8, pp. 1149–1159, ISSN: 1472-779X

Duerr, B.F., Chung Y.-S. & Czarnik, A.W. (1988) Syntheses of 9,10-Disubstituted Anthracenes Derived from 9,10-Dilithioanthracene. *Journal of Organic Chemistry,* Vol. 53, No. 9, pp. 2120-2122, ISSN: 0022-3263

Effenberger, F., Klenk, H. & Reiter, P. L. (1973) New Aspects of the Fries Rearrangement. *Angewandte Chemie International Edition*, Vol. 12, No. 9, pp. 775–776, ISSN: 1433-7851

Eliel, E. L. & Wilen, S. H. (1994) *Stereochemistry of Organic Compounds,* pp. 24–31, Wiley, ISBN: 978-0471016700, New York

Frangopol, M., Genunche, A., Frangopol, P. T. & Balaban, A. T. (1964) Pyrylium salts formed by diacylation of olefins XV: ^{14}C-tracer study of the acetylation of 4-chloro-3,4-dimethyl-2-pentanone. *Tetrahedron,* Vol. 20, No. 8, pp. 1881–1888, ISSN: 0040-4020

Frisch, M. J., Trucks, G. W., Schlegel, H. B., Scuseria, G. E., Robb, M. A., Cheeseman, J. R., Montgomery Jr., J. A., Vreven, T., Kudin, K. N., Burant, J. C., Millam, J. M., Iyengar, S. S., Tomasi, J., Barone, V., Mennucci, B., Cossi, M., Scalmani, G., Rega, N., Petersson, G. A., Nakatsuji, H., Hada, M., Ehara, M., Toyota, K., Fukuda, R., Hasegawa, J., Ishida, M., Nakajima, T., Honda, Y., Kitao, O., Nakai, H., Klene, M., Li, X., Knox, J. E., Hratchian, H. P., Cross, J. B., Adamo, C., Jaramillo, J., Gomperts, R., Stratmann, R. E., Yazyev, O., Austin, A. J., Cammi, R., Pomelli, C., Ochterski, J. W. , Ayala, P. Y., Morokuma, K., Voth, G. A., Salvador, P., Dannenberg, J. J., Zakrzewski, V. G., Dapprich, S., Daniels, A. D., Strain, M. C., Farkas, O., Malick, D. K., Rabuck, A. D., Raghavachari, K., Foresman, J. B., Ortiz, J. V., Cui, Q., Baboul, A. G., Clifford, S., Cioslowski, J., Stefanov, B. B., Liu, G., Liashenko, A., Piskorz, P., Komaromi, I., Martin, R. L., Fox, D. J., Keith, T., Al-Laham, M. A., Peng, C. Y., Nanayakkara, A., Challacombe, M. Gill, P. M. W., Johnson, B., Chen, W., Wong, M. W., Gonzalez, C. & Pople, J. A. (2004) *Gaussian 03, Revision C.02*, Gaussian, Inc., Wallingford CT.

Gore, P. H. (1955) The Friedel-Crafts Acylation Reaction and its Application to Polycyclic Aromatic Hydrocarbons. *Chemical Review*, Vol. 55, No. 2, pp. 229–281, ISSN: 0009-2665

Gore, P. H. (1964) In: *Friedel–Crafts and Related Reactions*, Olah, G. A., Vol. 3 (Part 1), pp. 1-381, Wiley-Interscience, ISBN: 978-0470653241, New York

Gore, P. H. (1974). Friedel-Crafts acylations: some unusual aspects of selectivity. *Chemistry & Industry*, Vol. 18, pp. 727–731, ISSN: 0009-3068

Gore, P.H. & Thadani, C.K. (1966) The Formation of Mono- and Di-ketones in the Friedel-Crafts Acetylation of Anthracene. *Journal of the Chemical Society (C)*. No. pp. 1729–1733, ISSN: 0022-4952

Hariharan, P. C. & Pople, J. A. (1973) The Influence of Polarization Functions on Molecular Orbital Hydrogenation Energies. *Theoretica Chimica Acta*, Vol. 28, pp. 213–222, ISSN: 0040-5744

Heaney, H. (1991) The Intramolecular Aromatic Friedel–Crafts Reaction In: *Comprehensive Organic Synthesis*, Trost, B. M., Fleming, I., Heathcock, C. H., Vol. 2, pp. 753–768, Pergamon Press, ISBN: 978-0-08-052349-1, Oxford.

Hunter, C. A., Singh, J. & Thornton, J. M. (1991) π–π Interactions: the Geometry and Energetics of Phenylalanine-Phenylalanine Interactions in Proteins. *Journal of Molecular Biology*, Vol. 218, No. 4, pp. 837–846, ISSN: 00222836

Janiak, C. (2000) A critical account on π–π stacking in metal complexes with aromatic nitrogen-containing ligands. *Journal of the Chemical Society, Dalton Transactions*, No. 21, pp. 3885–3896, ISSN: 1477-9226

Karpfen, A., Choi, C. H. & Kertesz, M. (1997) Single-Bond Torsional Potentials in Conjugated Systems: A Comparison of ab Initio and Density Functional Results. *Journal of Physical Chemistry A*, Vol. 101, No. 40, pp. 7426–7433, ISSN: 1089-5639

Koch, W. & Holthausen, M.C. (2001) *A Chemist Guide to Density Functional Theory* (2nd Ed), Wiley-VCH, ISBN: 3-527-30372-3, Weinheim.

Langer, V. & Becker, H.-D. (1993) Crystal structure of 1-acetylanthracene, $C_{16}H_{12}O$. *Zeitschrift für Kristallographie*, Vol. 206, No. 1, pp. 155–157, ISSN: 0044-2968

Lee, C., Yang, W. & Parr, R. G. (1988) Development of the Colle-Salvetti correlation-energy formula into a functional of the electron density. *Physical Review B*, Vol. 37, No. 2, pp. 785–789, ISSN: 1098-0121

Levy, L., Pogodin, S., Cohen, S. & Agranat, I. (2007) Reversible Friedel–Crafts Acylations of Phenanthrene: Rearrangements of Acetylphenanthrenes. *Letters in Organic Chemistry*, Vol. 4, No. 5, pp. 314–318, ISSN: 1570-1786

Li, M.-Q. & Jing, L.-H. (2006) 1,5-Diacetylanthracene. *Acta Crystallographica Section E*, Vol. 62, No. 9, pp. o3852–o3853, ISSN: 1600-5368

Mala'bi, T., Pogodin, S. & Agranat, I. (2009) Reversible Friedel-Crafts Acylations of Anthracene: Rearrangements of Acetylanthracenes. *Letters in Organic Chemistry*, Vol. 6, No. 3, pp. 237–241, ISSN: 1570-1786

Mala'bi, T., Pogodin, S. & Agranat, I. (2011) Kinetic control wins out over thermodynamic control in Friedel–Crafts acyl rearrangements. *Tetrahedron Letters*, Vol. 52, No. 16, pp. 1854–1857, ISSN: 0040-4039

Martin, N. H., Allen III, N. W., Brown, J. D., Kmiec, D. M. Jr. & Vo, L. (2003) An NMR shielding model for protons above the plane of a carbonyl group. *Journal of Molecular Graphics and Modelling*, Vol. 22, No. 1, pp. 127–131, ISSN: 1093-3263

McConnell, H. M. (1957) Theory of Nuclear Magnetic Shielding in Molecules. I. Long-Range Dipolar Shielding of Protons. *Journal of Chemical Physics*, Vol. 27, No. 1, pp. 226-229, ISSN: 0021-9606

Moss, G. P. (1996) Basic Terminology of Stereochemistry (IUPAC recommendations 1996). *Pure & Applied Chemistry*, Vol. 68, No. 12, pp. 2193–2222, ISSN: 0033-4545; http://www.chem.qmul.ac.uk/iupac/stereo.

Muller, P. (1994) Glossary of Terms Used in Physical Organic Chemistry (IUPAC Recommendations 1994). *Pure & Applied Chemistry*, Vol. 66, No. 5, pp. 1077–1184, ISSN: 0033-4545

Murugan, N. A. & Jha, P. C. (2009) Pressure dependence of crystal structure and molecular packing in anthracene. *Molecular Physics*, Vol. 107, No. 16, 1689–1695, ISSN: 0026-8976

Nenitzescu, C. D. & Balaban, A. T. (1964) In: *Friedel–Crafts and Related Reactions*; Olah, G. A., Vol. 3 (part 2), pp. 1033–1152, Wiley-Interscience, ISBN: 978-0470653265, New York.

Norman, R. O. C. & Taylor, R. (1965). *Electrophilic Substitution in Benzenoid Compounds*, p. 174–182, Elsevier, ISBN: 978-0444404275, London

Okamoto, A. & Yonezawa, N. (2009) Reversible ArS_E Aroylation of Naphthalene Derivatives. *Chemistry Letters*, Vol. 38, No. 9, pp. 914–915, ISSN: 0366-7022

Olah, G. A. (1973). *Friedel–Crafts Chemistry*, p. 102, Wiley-Interscience, ISBN: 0471653152, New York

Pearson, D. E. & Buehler, C. A. (1971). Unusual Electrophilic Aromatic Substitution in Synthesis. *Synthesis*, No. 9, pp. 455–477, ISSN: 0039-7881

Pogodin, S., Rae, I. D. & Agranat, I. (2006) The Effects of Fluorine and Chlorine Substituents across the Fjords of Bifluorenylidenes: Overcrowding and Stereochemistry. *European Journal of Organic Chemistry*, No. 22, pp. 5059–5068, ISSN: 1434-193X

Sarobe, M. & Jenneskens, L.W. (1997) High-Temperature Behavior of 1,8-Diethynylanthracene. Benz[*mno*]aceanthrylene, a Transient Intermediate for Cyclopenta[*cd*]pyrene. *Journal of Organic Chemistry*, Vol. 62, pp. 8247–8250, ISSN: 0022-3263

Sinclair, V. C., Robertson, J. M. & Mathieson, A. McL. (1950) The Crystal and Molecular Structure of Anthracene. II. Structure Investigation by the Triple Fourier Series Method. *Acta Crystallographica*, Vol. 3, No. 4, pp. 251–256, ISSN: 0365-110X

Sinnokrot, M. O. & Sherrill, C. D. (2006) High-Accuracy Quantum Mechanical Studies of π–π Interactions in Benzene Dimers. *Journal of Physical Chemistry A*, Vol. 110, No. 37, pp. 10656–10668, ISSN: 1089-5639

Titinchi, S. J. J., Kamounah, F. S., Abbo, H. S. & Hammerich, O. (2008) The synthesis of mono- and diacetyl-9*H*-fluorenes. Reactivity and selectivity in the Lewis acid catalyzed Friedel-Crafts acetylation of 9*H*-fluorene. *ARKIVOC*, No. xiii, pp. 91–105, ISSN: 1551-7004

Viruela, P. M., Viruela, R., Ortí, E. & Brédas, J.-L. (1997) Geometric Structure and Torsional Potential of Biisothianaphthene. A Comparative DFT and ab Initio Study. *Journal of the American Chemical Society*, Vol. 119, No. 6, pp. 1360–1369, ISSN: 0002-7863

Wang, Z. (2009). Friedel-Crafts Acylation, In: *Comprehensive Organic Name Reactions and Reagents*, Vol. 1, Chapt. 248, pp. 1126–1130, Wiley, ISBN: 978-0471704508, New York

Wolf, C. (2008) *Dynamic Stereochemistry of Chiral Compounds. Principles and Applications*, pp. 95-96, RSC Publishing, ISBN: 978-0-85404-246-3, Washington.

Zefirov, Y. V. (1997) Comparative Analysis of Systems of van der Waals Radii. *Crystallography Reports (Kristallografiya)*, Vol. 42, No. 1, pp. 111–116, ISSN: 1063-7745

Zouev, I., Cao, D.-K., Sreevidya, T. V., Telzhensky, M., Botoshansky, M. & Kaftory, M. (2011) Photodimerization of anthracene derivatives in their neat solid state and in solid molecular compounds. *CrystEngComm*, Vol. 13, No. 13, pp. 4376–4381, ISSN: 1466-8033

3

Novel Challenges in Crystal Engineering: Polymorphs and New Crystal Forms of Active Pharmaceutical Ingredients

Vânia André and M. Teresa Duarte
Centro de Química Estrutural, Dept. of Chemical and Biological Engineering, Instituto Superior Técnico, Universidade Técnica de Lisboa, Lisbon Portugal

1. Introduction

Crystal engineering and co-crystallization have evolved in recent years and gained a special interest not only in academia but also in the pharmaceutical field as it has been shown that the physical and pharmacokinetic properties of new crystal forms (solvates, salts, molecular salts, co-crystals, polymorphs) are different when compared to pure APIs[1-16]. Actually, producing co-crystals of pharmaceuticals has been reported to change their melting points[3], solubility and dissolution rates[2, 4], moisture uptake[17], physical and chemical stability[18] and *in vivo* exposure[9, 19-21]. The leading idea is that the potentiality of new different forms may open to innovation and new drug discoveries as well as to intellectual property protection via patenting of new forms of "old drugs"[5, 7, 22]. The diversity of forms that crystalline solids may attain is mainly due to non-covalent interactions resulting in different molecular assemblies that imply an energetic interplay between enthalpy and entropy.

Although organic salts are traditionally the preferred crystal form of APIs because of their higher solubility and/or increased degree of crystallinity, the potential number of suitable organic salts is limited to the counterions specified by the Food and Drug Administration (FDA) as generally regarded as safe (GRAS). This limitation stimulates the development of other suitable forms and recently co-crystals have been gaining relevance in studies and some of them have already shown to improve therapeutic utility as well as reducing the side effects even when compared with marketed drugs. Consequently, APIs represent a particular great challenge to crystal engineers, because they are inherently predisposed for self-assembly since their utility is usually the result of the presence of one or more exofunctional supramolecular moieties. However, the crystal packing of APIs is even less predictable than that of other organics due to their multiple avenues for self-assembly. Additionally, APIs are commonly valuable chemical entities and therefore the diversity of the crystal forms of those molecules is of great importance for the variability of properties and potential intellectual property.

Co-crystals are most commonly thought of as structural homogeneous crystalline materials that contain two or more neutral building blocks that are present in definite stoichiometric amounts and are obtained through the establishment of strong hydrogen bonds and other non-covalent interactions such as halogen bonds, π-π and coulombic interactions. However, if the groups involved in these bonds have the tendency to transfer protons between acids and bases, then the result may be a molecular salt instead of a co-crystal. In principle, this event replaces the X-H···Y interaction by a charge assisted $X^{-}\cdots H\text{-}Y^{+}$ hydrogen bond[22].

The formation of multicomponent crystal forms relies mainly on the hydrogen-bond synthons that are possible to form and their relative robustness. Hence a thorough datamining based on the Cambridge Structural Database (CSD)[23], is required for a successful design of the new crystallines. One notable obstacle in the path of rational co-crystal design is the phenomenon of polymorphism, to which organic molecules are predisposed. Polymorphic co-crystals are also not uncommon and a few systems have already been reported to date[24].

Even though co-crystals are traditionally obtained by solution techniques, often limited by differences in solubility of co-crystal components and/or solvent/solute interactions[25], the best strategies to attain the desired forms consist on a judicious choice of synthetic and crystallization conditions, which also contemplate the environment-friendly techniques of mechanochemistry (neat (NG), liquid-assisted (LAG) and ion- and liquid-assisted grinding (ILAG)) that have demonstrated to be an efficient method in co-crystallization screening and synthesis: solid-state grinding allows the formation of multicomponent forms even with low-solubility components that would be difficult to use with the traditional solution techniques; the addition of catalytic amounts of a liquid to the grinding mixture further improves the efficiency of grinding co-crystallization, as already proven[26-30].

Novel crystallines are usually fully characterized using powder (XRPD) and single crystal (SCXRD) X-ray diffraction techniques, differential scanning calorimetry (DSC), thermogravimetric analysis (TGA), hot-stage microscopy (HSM) and spectroscopic methods, such as Fourier transform infrared (FTIR) and Raman.

A wide range of studies have been performed in the last few years, here we will focus on some of those we have been recently engaged. Several studies with gabapentin, a neuroleptic drug, have been reported[31-34] and a few multicomponent crystal forms with carboxylic acids have been exploited and will be deeply discussed in this chapter. With the antibiotic 4-aminosalicylic acid (4ASA) some new crystal forms were disclosed, solvates, and molecular salt[35], showing a clear tendency of this API to form multicomponent crystallines with lone-pair containing heterocycles, such as dioxane, morpholine and piperazine, disrupting the typical $R_2^2(8)$ homosynthon in 4ASA (Figure 1).

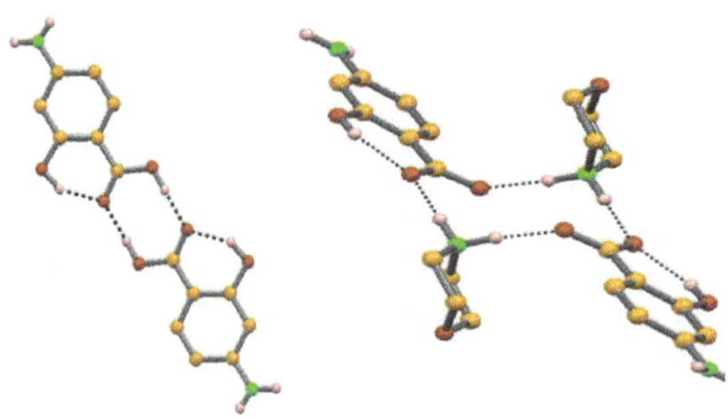

a b

Fig. 1. (a) $R_2^2(8)$ synthon in the 4ASA crystal packing; (b) disrupted $\mathbf{R_2^2(8)}$ synthon in the 4ASA:morpholine molecular salt.[36]

Another example is the widely used paracetamol, which also presents multicomponent crystal forms that were identified and characterized[37-39]. Perindopril erbumine is an antihypertensive drug existent in quite a few polymorphic forms and several hydrates were also disclosed[40-48]. A detailed discussion of some of these compounds will be addressed in this chapter.

But as important as the synthesis of these new compounds is the injunction of their structure-properties relationships. Solubility is a major factor that is known to strongly affect API's performance and therefore its correlation with structural and thermal data is of upmost importance.

API coordination complexes are another related topic that has recently been disclosed as a new pathway for the development of improved crystal forms. In this area, as example, two complexes of 4ASA with silver and one complex coordinating piracetam, a nootropic API, to nickel were prepared and characterized[49].

2. Gabapentin: Polymorphs and multicomponent crystal forms[1,2]

Gabapentin (1-(aminomethyl)cyclohexane acetic acid, GBP) is an analogue of gamma-aminobutyric acid (GABA) and exhibits anticonvulsant properties. It is a neuroleptic drug prescribed for the prevention of seizure, for the treatment of mood disorders, anxiety and tardive diskinesia[50-57], as well as for the treatment of neuropathic pain[58]. More recently, GBP has also been applied in the treatment of limb tremor[59, 60]. This API is highly soluble but has limited and variable bioavailability, probably due to its dependence on a low-capacity aminoacid transporter expressed in a limited region of the upper small intestine. Changes in solid state structure can have marked influence on the physiological absorption characteristics supporting the search for multicomponents crystal forms as means of improving the limited bioavailability of the drug[7, 33, 61, 62].

Gabapentin is known to exist in three anhydrous polymorphic forms[32, 63, 64], which have been the object of many patent applications and issued patents. The nomenclature is not uniform and different publications refer to different forms making use of the same name. One hydrate form, labelled form I, is known[65], while form II is the anhydrous commercial form[66]. The crystal structures of these forms are present in the Cambridge Structural Database (CSD) with the refcodes QIMKOM for form I[67] at -120°C and QIMKIG and QIMKIG01 for form II at -120°C[67] and at RT[58], respectively. Form III as labelled by Braga and co-workers[68], has been described by Pesachovich *et al* [69], and its crystal structure has been reported by Reece and Levendis as form α[70]. A patent by Chen et al[71]. describes a new form of gabapentin, dehydrated A, which is consistent with form β reported by Reece and Levendis[70], labelled form IV by Braga and co-workers[68]. Another crystalline form of gabapentin was described by Lladò *et al*[72], but its powder pattern is very similar to that of

[1] Adapted with permission from On the Track of New Multicomponent Gabapentin Crystal Forms: Synthon Competition and pH Stability,Vânia André, Auguste Fernandes, Pedro Paulo Santos, and M. Teresa Duarte, *Crystal Growth & Design*, 2011, *11* (6), pp 2325–2334,DOI: 10.1021/cg200008z. Copyright (2011) American Chemical Society.

[2] Adapted with permission of the Royal Society of Chemistry (RSC) from Polymorphic gabapentin: thermal behaviour, reactivity and interconversion of forms in solution and solid-state Dario Braga, Fabrizia Grepioni, Lucia Maini, Katia Rubini, Marco Polito, Roberto Brescello, Livius Cotarca, M. Teresa Duarte, Vânia André and M. Fátima M. Piedade *New J. Chem.*, 2008, 32, 1788-1795, DOI: 10.1039/B809662G.

form IV. Kumar *et al*[73] reported a form of gabapentin in an international patent application that is most likely a mixture of forms III and IV (Figure 2).

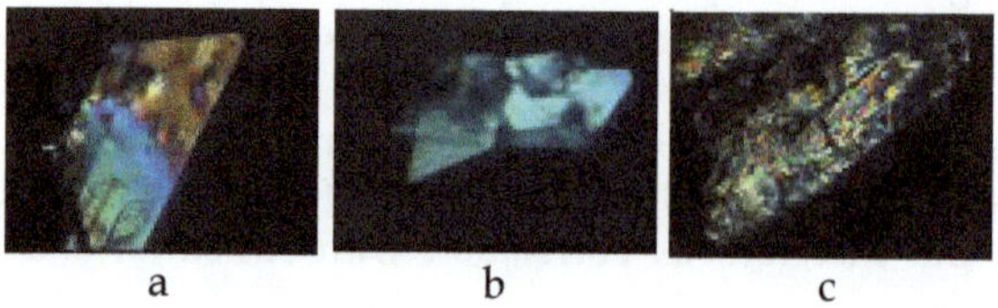

Fig. 2. Crystals' morphology of gabapentin polymorphs (a) II, (b) III and (c) IV.

In all characterized forms of gabapentin the molecule crystallizes in its zwitterionic form and hydrogen-bonds consist of charge-assisted N^{+}-H•••O^{-} interactions all the supramolecular arrangements relying on chain motifs.

Polymorph II forms double chains that can be seen along b, in which the molecules are oriented so that the substituent groups of the cyclohexane are turned to each other. This orientation of the NH_3^{+} and COO^{-} results in no interactions with neighboring chains along *b* and only one hydrogen bond is used to connect this chain to the parallel chain formed exactly in the same orientation (Figure 3).

In the crystal structure of polymorph III molecules form a 2D sheet along *b*, in which the gabapentin molecules are organized in chains. In the same chain, molecules orient the substituent groups of the cyclohexane ring in the same direction, although the cyclohexane ring is 70.52° alternately rotated. In consecutive chains, the substituent groups are anti-parallel oriented and there are interactions between these two chains (Figure 3).

Several similarities between packing of both forms III and IV of gabapentin can be detected and, indeed, the main difference is that in the latter an intramolecular interaction is established. Form IV also forms chains of gabapentin molecules, in which the cyclohexane structures are rotated alternately rotated by 65.83°. Just as in form III, in each chain the substituent groups are oriented in the same direction and in the chain below have an anti-parallel orientation. The chains are connected in pairs because of the anti-parallel orientation of the substituent groups, which leaves no opportunity to establish hydrogen-bonds with a third chain. However in this form no direct interactions between molecules in the same chain are observed: all of them connect with molecules in the chain above/below and that same molecule that interacts with "initial" molecule will also be connected to the molecule just besides the first one in the other chain. Consecutive parallel chains are formed just ones behind the others, exactly with the same orientation (Figure 3).

Overlapping the structures of the three gabapentin's forms (II, III and IV) it is possible to see that these are conformational polymorphs (Figure 4).

In the three known polymorphs of gabapentin, the carboxylate C-O bond lengths differ by 0.020 Å, 0.017 Å and 0.016 Å in forms II, III and V, respectively. For all these polymorphs, the longer bond involves the O atom that forms two NH···O contacts and the shorter bond is involved only in one NH···O interaction. In all the three structures, the cyclohexane ring adopts an almost perfect chair conformation.

Thermal characterization of the three polymorphs is also reported[68]. In the first DSC heating cycle form II shows an endothermic peak at 158°C (Figure 5) while in the second heating cycle the endothermic peak is at 87°C, thus representing the melting point of gabapentin-lactam[74], confirmed by recrystallization and second heating cycle from the melt on HSM

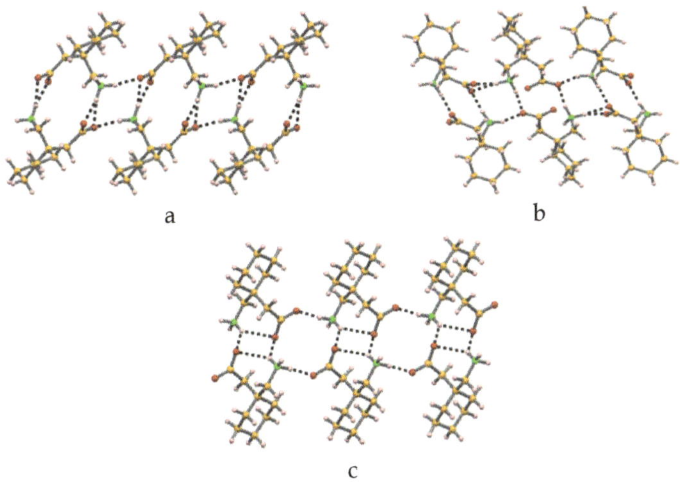

Fig. 3. Supramolecular arrangements of gabapentin forms (a) II, showing the double chains along b, with the substituent groups aligned, (b) III, showing the double chains along b, with the substituent groups rotated and (c) IV showing the double chains assisted by the intramolecular hydrogen bond.

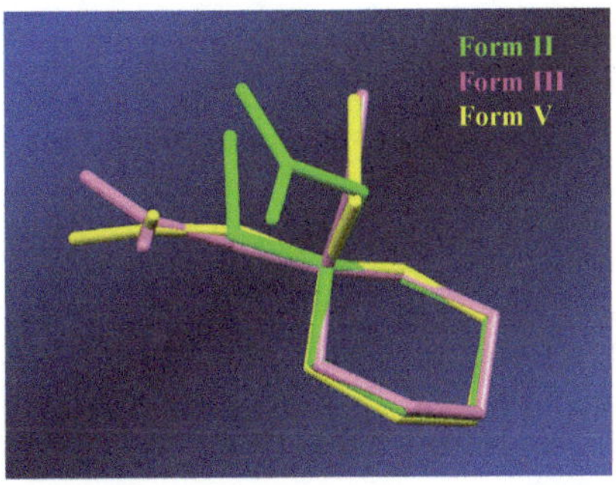

Fig. 4. Gabapentin forms II, III and IV structures overlapped. Hydrogen atoms were omitted for a better visualization.

(Figure 6). The formation of gabapentin-lactam is not surprising, and it is known that gabapentin is unstable in aqueous solutions and undergoes an intramolecular dehydration reaction yielding the lactam[75]; the formation of gabapentin-lactam has also been observed in the solid state[74]. Therefore the endothermic peak observed in the first heating cycle does not correspond to the melting of gabapentin, but covers several events: the cyclization, the release of water and the melting of gabapentin-lactam; these events could not be separated even with a slow scanning rate. Form III shows a similar thermal behaviour, with a broad endothermic peak at 165°C, which is again due to the cyclization process with formation of gabapentin-lactam, water release and melting of gabapentin-lactam[68].
As noted above, the thermodynamic form II undergoes the reaction at a slightly lower temperature than the metastable form III. The different behaviour might be explained by the slightly different pattern of hydrogen bonds in their crystal structures[70].

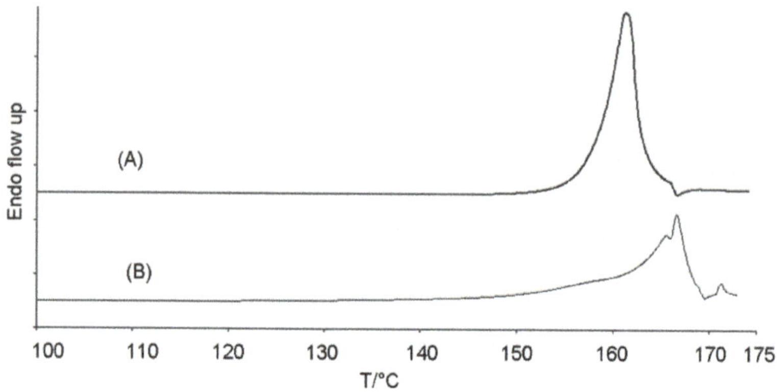

Fig. 5. DSC traces (open pan) of gabapentin Form II (a), with onset temperature = 158.0°C and peak temperature = 161.1°C, and of gabapentin Form III (b), with onset temperature = 165.0°C and peak temperature = 166.6°C[68].

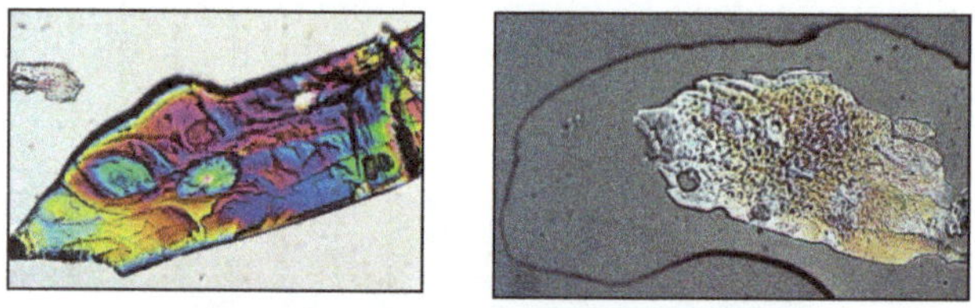

a b

Fig. 6. Hot-stage microscopy of crystalline gabapentin-lactam (a) as obtained in the first DSC heating cycle of gabapentin, showing melting at 89°C (b)[68].

Several attempts to produce pure form IV in reasonable quantity to be used in DSC measurements were not successful; still a HSM experiment on single crystals of form IV isolated within an oil drop was possible. Figure 7 clearly shows the release of water as gas bubbles in the temperature range 152-155°C, immediately followed by melting of the gabapentin-lactam thus formed[68].

a b c

Fig. 7. Hot-stage microscopy on gabapentin form IV crystals (preserved in Fomblin oil): (a) single crystal at 32°C, (b) evolution of water bubbles at 153°C and (c) complete melting of gabapentin-lactam at 157°C (amplification 100x)[68].

The unique presence in crystals of form IV of an intramolecular N-H··O hydrogen-bond, associated with a smaller number of intermolecular hydrogen bonds with respect to the other two forms, must be responsible for the lower reaction temperature observed[70].

Furthermore, a monohydrate[76-78], two polymorphic chloride hemihydrates[58, 63, 79, 80] an hemisulfate hemihydrate[79, 80] and an heptahydrate under high pressure[81] forms are also known. Coordination complexes of this API with Cu and Zn were isolated and characterized[82]. An extensive pH stability of gabapentin has been disclosed where an ester derivative obtained at low pH was reported[83].

Multicomponent crystal forms (co-crystals and molecular salts) involving GBP with different carboxylic acids were also recently disclosed[1, 32-34].

A search in the Cambridge Structural Database (CSD)[23] (July 2011) has shown that the $R_4^2(8)$ synthon is the most common between cationic amine and carboxylate moieties. The expected synthons to be formed between the API and the coformers should be based on carboxyl···carboxylate and amine···carboxylate interactions. Accordingly to a CSD[23] survey (July 2011), the preferred interactions should be amine···carboxylate followed by the carboxyl···carboxylate.

Therefore, carboxylic acids were chosen as potential coformers of multicomponent crystal forms of gabapentin. Mono and di-carboxylic acids bearing one or more hydroxyl moieties have previously been exploited by Reddy et al[33] revealing an important role of the OH group in the supramolecular arrangements of the new forms. A series of mono-, di- and tricarboxylic acids, without further hydroxyl moieties, were considered by André *et al*[84] to exploit the use of one or more equivalents of carboxylic moieties avoiding the hydroxyl competition.

Five new multicomponent crystal forms of the neuroleptic drug gabapentin with isophthalic acid (pKa1=3.5; pKa2=4.5[85]), phthalic acid (pKa1=3.0; pKa2=5.3[85]), L-glutamine (pKa1=2.1; pKa2=4.3[85]), terephthalic (pKa1=3.5; pKa2=4.5[85]) and trimesic (pKa1=3.1; pKa2=3.9; pKa3=4.7[85]) acids have been reported and are characterized by XRPD. Despite all the crystallization attempts, single crystals suitable for SCXRD were only grown for the compounds with terephthalic (**4**) and trimesic (**5**) acids, which are further characterized by SCXRD, DSC, TGA, HSM and IR. The strong homomeric $R_2^2(8)$ and heteromeric $R_4^2(8)$ synthons observed in the carboxylic acids and gabapentin, respectively, were disrupted and competing synthons based on carboxyl···carboxylate and amine···carboxylate interactions were formed in the new crystallines with trimesic and terephthalic acids[84].

With L-glutamine, a new crystal form **1** characterized by XRPD is obtained by solution techniques. Both solution and LAG experiments with phthalic acid resulted in a mixture of the coformer and a new crystalline **2**. With isophthalic acid, a mixture of a new crystal form **3**, isophthalic acid and gabapentin polymorph III was identified. In this case, the yield of the supramolecular reaction is low and both reagents are also detected, though gabapentin is detected in a different polymorphic form from the starting material. The full conversion into the new form with terephthalic acid, **4**, was only attained by LAG, as terephthalic acid displays solubility problems and showing not only that the formation of this salt is independent of reactional pH but also the advantage of this method when using highly insoluble compounds. The multicomponent crystal form comprising gabapentin and trimesic acid, **5**, was obtained as a single phase both by solution and LAG techniques (Figure 8).

The asymmetric unit of **4** consists on one gabapentin cation and half a terephthalic acid anion residing on an inversion centre. In this structure there is clear evidence of proton transfer between both compounds within the structure and thus this form a molecular salt.

Gabapentin cations are connected to terephthalic anions through three different charge-assisted interactions (Scheme I.a), two of which involve the protonated amine moiety of the

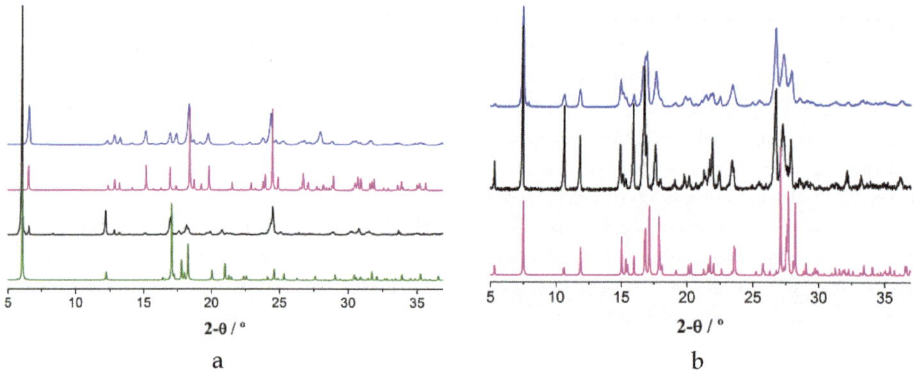

Fig. 8. (a) Experimental XRPD patterns obtained from mechanochemistry (blue); a mixture of gabapentin polymorph III and **4** obtained by solution techniques (black); and gabapentin polymorph III (green); theoretical XRPD patterns obtained from SCXRD data of **4**, at 150K (pink); (b) Experimental XRPD pattern obtained from **5** obtained by LAG (blue) and solution (black) techniques; theoretical powder diffraction pattern obtained from single-crystal data, at 150K (pink).

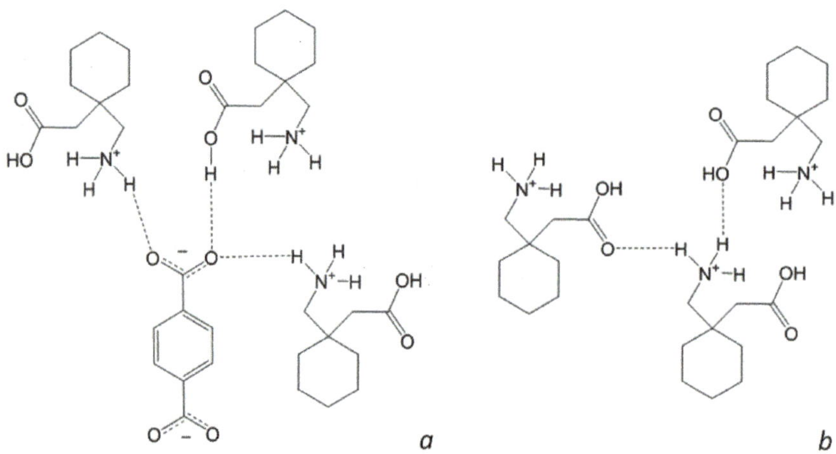

Scheme I. Main hydrogen bond interactions present in molecular salt **4**

API and the carboxylate of the anion, $N^+H_{GBP}\cdots O^-_{TA}$, and a third one concerning the carboxylic group of GBP and again the carboxylate of terephthalic acid, $OH_{GBP}\cdots O^-_{TA}$ (Figure 9.a). The donor group - either the amine and/or the carboxylic moiety - belong to GBP, while the carboxylate groups of terephthalic acid always act as acceptors, with O3 behaving as a bifurcated acceptor.

Gabapentin cations interact among them by the $N^+H_{GBP}\cdots O_{GBP}$ interactions (Scheme I.b), one of them being slightly longer. It is possible to see that these interactions among four gabapentin cations form a tape supported by $R_4^4(12)$ synthons growing along *c* (Figure 9.b). The interplay of these hydrogen bonds gives rise to terephthalic acid anions acting as spacers between GBP cationic tapes, very clear in a view along the *b* axis (Figure 9.c). Within the terephthalic acid row, the anionic spacers alternate with a rotation of 27°; it can also be seen that in the gabapentin tape cations are rotated by 39°.

The formation of GBP tapes is a common pattern both in the structure of the three polymorphs of gabapentin and in 4, in the latter the coformer links consecutive tapes. The formation of 4 disrupts the $R_2^2(8)$ synthons typical of the terephthalic acid while increasing the number of hydrogen-bond interactions in which gabapentin is involved when compared with any of the three known polymorphic forms.

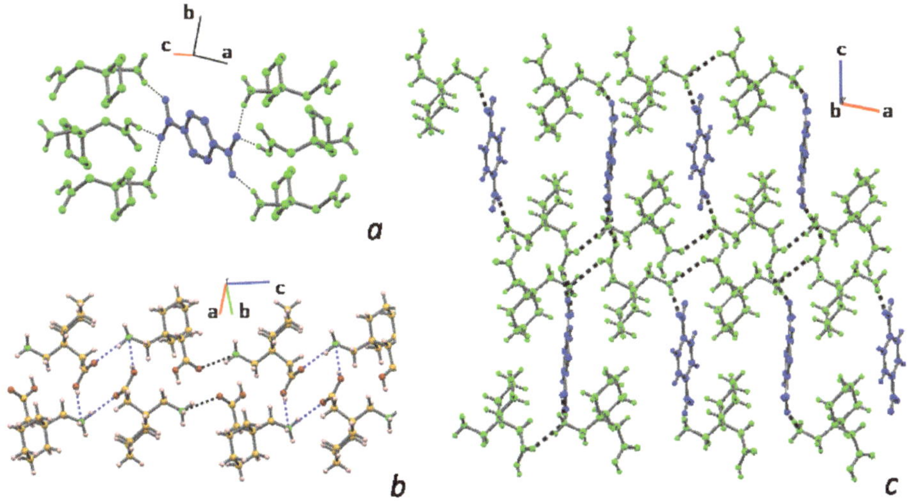

Fig. 9. Packing diagrams obtained from **4**; (a) detailed hydrogen-bonding system in **4**, (b) view showing the cationic GBP tape and depicting $R_4^4(12)$ synthons represented in blue, (c) view along *b* of the supramolecular packing where the spacer function of the coformer is clear. Color code: green - GBP; blue - coformer.

Asymmetric unit of crystalline **5** consists of one gabapentin zwitterion, one trimesic acid molecule and one water molecule. In this structure there is clear evidence that the proton transfer occurs within gabapentin, resulting in its zwitterionic form; therefore all the molecules involved in this structure are globally neutral and thus we have a hydrated co-crystal.

An intramolecular hydrogen bond is established in each gabapentin molecule [$N^+H_{GBP}\cdots O^-_{GBP}$] and gabapentin zwitterions interact among them using the amine and the carboxylate moieties, $N^+H_{GBP}\cdots O^-_{GBP}$. Both interactions are responsible for the formation of dimers based on $R_4^2(8)$ synthons (Scheme II.a).

Two of the carboxylic acid groups of trimesic acid are used to form the usual $R_2^2(8)$ synthon through $OH_{TA}\cdots O_{TA}$. The third COOH no longer maintains this typical pattern but interacts with three independent gabapentin zwitterions (Scheme II.b), two of which are involved in the previously mentioned GBP $R_4^2(8)$ dimer. In these GBP$\cdots$TA interactions, C=O acts as an acceptor for one NH of gabapentin [$N^+H_{GBP}\cdots O_{TA}$]; OH works both as acceptor, from another gabapentin's amine moiety [$N^+H_{GBP}\cdots O_{TA}$], and as donor to a CO of a third gabapentin molecule [$OH_{TA}\cdots O^-_{GBP}$] (Figure 10.a). A tape of GBP zwitterionic dimers assisted by trimesic acid moieties is formed (Figure 10.b). Actually these tapes are further reinforced by water molecules as each gabapentin zwitterion interacts with three water molecules *via* $N^+H_{GBP}\cdots O_W$ and two $OH_W\cdots O^-_{GBP}$ (Scheme II.a).

The supramolecular arrangement of **5** can be described as alternated gabapentin zwitterionic ondulated chains and trimesic acid zigzag chains, with water molecules lying in the space between them (Figure 10.c). Trimesic acid besides supporting the gabapentin tapes also acts as spacer between them, similarly to compound **4**.
Comparing this structure with the three known GBP polymorphs, the intramolecular bond is similar to the one formed in polymorph IV and the $R^2_4(8)$ synthons are observed also in polymorph III. The typical $R^2_2(8)$ synthon between trimesic acid molecules is maintained in 2/3 of its interactions and it is only disrupted to establish connections with GBP zwitterions, increasing the number of hydrogen-bonds in which they both are involved.
The presence of the intramolecular bond in gabapentin zwitterions could suggest an analogue conformation of GBP molecules in this co-crystal and in polymorph IV, but this is not observed and, in fact GBP adopts different conformations.

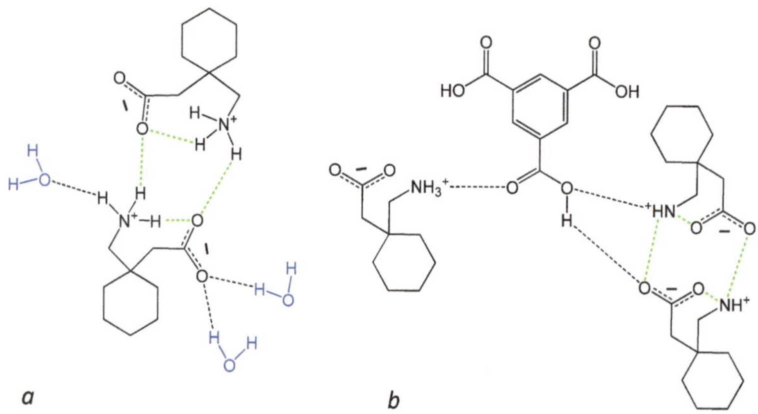

a *b*

Scheme II. Main hydrogen bond interactions present in **5**.

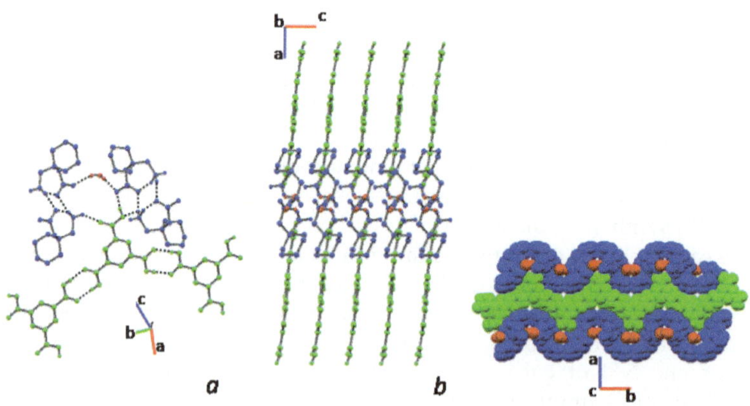

a *b*

Fig. 10. Packing diagrams for co-crystal **5** (a) detailed hydrogen-bonding system in GBP:trimesic acid hydrate; (b) view along b showing both the tape made of GBP dimers assisted by water and trimesic acid spacers; (c) space filling diagram viewed along the *c*-axis. Color code: green - GBP; blue - coformer; red- water.

In both gabapentin multicomponent crystals' structures **4** and **5**, gabapentin's cyclohexane ring adopts a chair conformation in which the aminomethyl group is in an equatorial position, with the carboxymethyl group in the axial position. The relative positioning of the substituent groups is similar to the one observed in gabapentin polymorph IV; in the other two polymorphic forms of gabapentin the aminomethyl and carboxymethyl groups occupy the inverted positions (Figure 11).

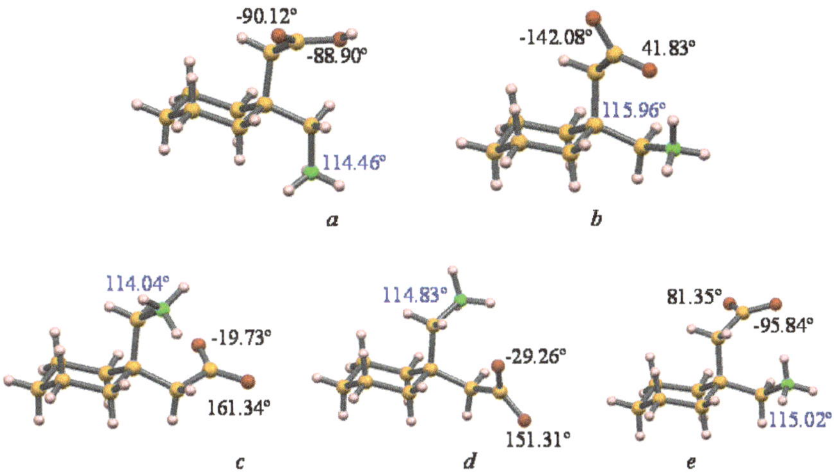

Fig. 11. A comparison of the GBP conformation in: (a) GBP:terephthalic acid molecular salt; (b) GBP:trimesic acid co-crystal; (c) GBP polymorph II; (d) GBP polymorph III; (e) GBP polymorph IV. C-C-N bond angles are given in blue and both C-C-C-O dihedral angles in black[84].

Analyzing all the unveiled multicomponent forms of gabapentin[32-34, 58, 63, 76, 79], there was no systematic behavior concerning the relative positioning of the aminomethyl and carboxymethyl substituent groups, what can lead us to conclude that this is governed by the supramolecular interactions.

As expected the carboxylate···amine interactions in GBP and the $R_2^2(8)$ in the carboxylic acids are partially disrupted and new hydrogen-bonding patterns were induced by the introduction of the coformer. Although there is proton transfer in **4** and not in **5**, in both forms GBP interacts with the acid coformer through carboxyl···carboxylate and amine···carboxyl/carboxylate synthons represented in Scheme III. The interactions *via* synthons I and II are in agreement with the results previously presented by Reddy et al[33] and Kavuru et al[86].

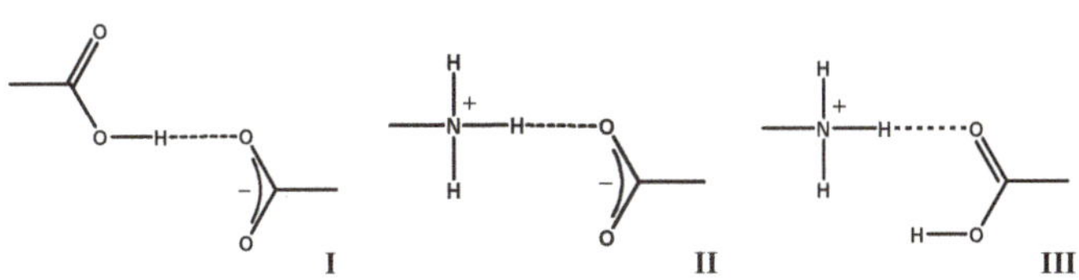

Scheme III. Main hydrogen bonded synthons observed in **4** and **5**.

Thermal studies were performed on the new crystal forms **4** and **5** and a combination of DSC, TGA and HSM data allowed some conclusions on the thermal stability of these compounds. The thermogram of **4** (Figure 12.a) is characterized by an endothermic peak at 150ºC, corresponding to the melting of the compound. The melting peak is found at a lower temperature than any of the reported polymorphic forms of gabapentin[32] and within the range obtained for other multicomponent forms[33, 34]. This peak encloses the cyclisation/ lactamization of gabapentin[32] implying water release that is observed on HSM experiments (Figure 13) and detected in TGA.

The thermogram obtained from **5** (Figure 12.b) is characterized by a wide bump between 70 and 120ºC and one broad endothermic peak at 159ºC. The first peak is due to the slow release of crystallization water and the second peak encloses lactamization of gabapentin and melting as seen in **4**. Both these phenomena are supported by TGA and HSM (Figures 12 and 14).

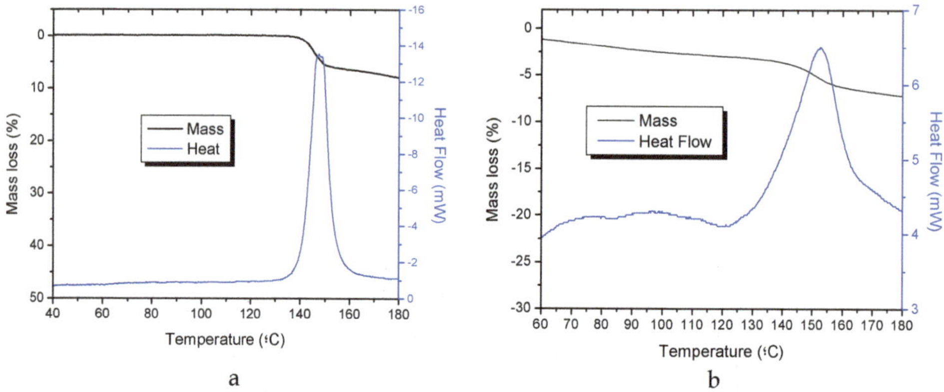

Fig. 12. DSC and TGA obtained for (a) molecular salt **4**, and (b) co-crystal **5**.

As previously mentioned, HSM experiments with compounds **4** and **5** were also performed and are in agreement with what was observed in the DSC and TGA experiments and were used in the interpretation of these results.

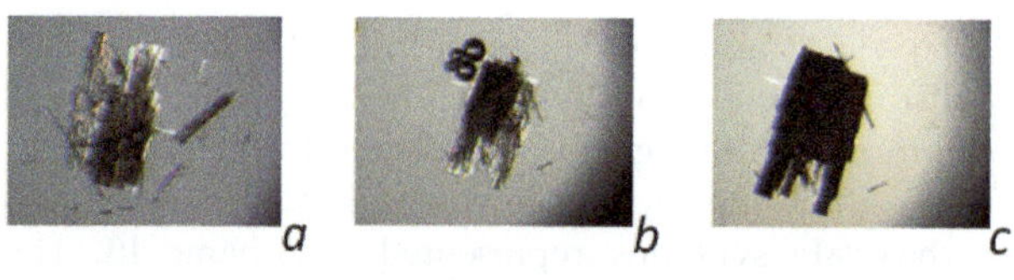

Fig. 13. HSM images for **4** at a) 25 ºC; b) 140ºC - water being released in the lactamization process; c) 144.5ºC - crystal appearance just before melting[84].

Fig. 14. HSM images for **5** at a) 25 ºC; b) 90ºC - slow release of crystallization water; c) 160ºC - lactamization and melting[84].

IR spectroscopy (Figure 15) complemented the characterization of the new crystal forms **4** and **5**. In both spectra, the presence of the NH_3^+ group is evidenced by the peaks corresponding to the symmetric and antisymmetric bending frequencies (1500 and 1610 cm^{-1}) and by the peak corresponding to the stretching frequency (2650 cm^{-1}). In **4**, the carboxylate group of the acid and the carboxylic group of the API are also well distinguished: the carbonyl band is exhibited at frequency > 1700 cm^{-1} typical of a aliphatic carboxylic group; proton transfer between the coformer and the API is confirmed by the presence of coformer carboxylate bands together with the absence of the carbonyl band typical (1680 cm^{-1}) of the terephthalic group. In **5**, although a clear identification of the carboxylate of the API and the carboxylic group of the acid is not so ascertained, it is possible to note the absence of the carboxylic moiety of gabapentin and identify, by comparison with the spectra of the pure coformer, the peak of the carboxylic moieties of trimesic acid; therefore the existence of the carboxylate in gabapentin is inferred.

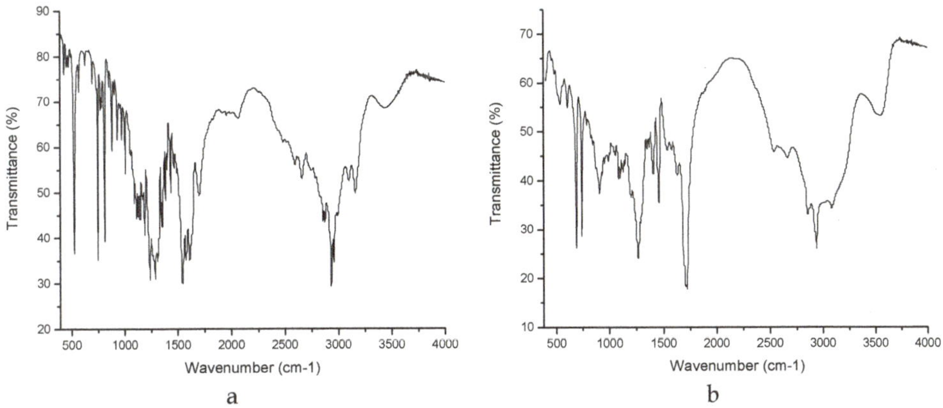

Fig. 15. IR spectra for **4** (a) and **5** (b) obtained by liquid-assisted grinding[84].

The solubility of the new multicomponent forms is lower than that for gabapentin, as desired. As previous studies on gabapentin indicate that this API is especially dependent on the pH of the environment[31], pH dependent stability of these two new forms was also studied and significant differences were found for **4** and **5**, the first being stable in quite a narrower pH range (Figures 16 and 17).

Coformer	☑	☑	☑	☑	☑	☑
Molecular salt					☑	☑
pH	**1.8**	**2.4**	**3.7**	**4.8**	**6.1**	**7.9**

Fig. 16. pH dependent stability of **4**

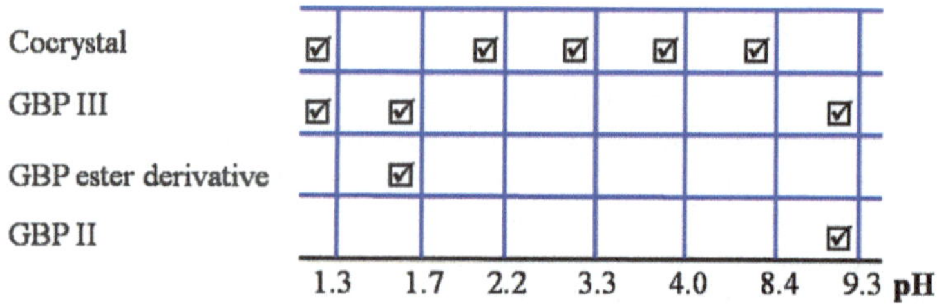

Fig. 17. pH dependent stability of **5**. [Ester derivative reported in[83]]

3. Perindopril: polymorphs and hydrates[3]

Perindopril, 2-methylpropane-2-amine-(2S,3aS,7aS)-1-[(2S)-2-[[(1S)-1-ethoxy-carbonyl-butyl] amino]propanoyl]octahydro-1H-indole-2-carboxylic acid, is an antihypertensive drug that acts through the inhibition of angiotensin converting enzyme (ACE), a zinc metalloenzyme involved in the control of blood pressure. It is effective in the treatment and prevention of several medical conditions, such as reducing blood pressure, reversing abnormalities of vascular structure and function in patients with essential hypertension, congestive heart failure, post-myocardial infarction and diabetic nephropathy[87-91]. Perindopril along with ramipril were associated with lower mortality than most other ACE inhibitors[92]. Besides the antihypertensive properties, it also comprises vasculoprotective and antithrombotic effects, playing a favourable role in terms of cardiovascular morbidity[93-99].

This API is, in fact, an acid-ester prodrug that is converted into the active diacid perindoprilat by hydrolysis promoted by the liver esterases after administration[93, 100]. It is orally administered in the form of tablets containing its 1:1 salts with erbumine (*tert*-butylamine) (Aceon®) or L-arginine (Coversyl®)[43, 101]. The perindopril L-arginine salt is equivalent to perindopril erbumine (Figure 18) but it is more stable and therefore it can be distributed to all the climatic zones without the need for specific packaging[101].

Over the last years, several forms of perindopril erbumine have been disclosed and several patents have been filed mainly based on their typical powder XRPD patterns[44, 45, 102-105]. Perindopril erbumine is known to exist in several polymorphic forms[46, 48, 102, 103, 105-107], as well as mono-, di- and sesqui-hydrated forms, characterized by XRPD, vibrational spectroscopy and thermal analysis methods[47, 108]. Also amorphous compositions have been patented[42] as well as a perindopril tosylate form[109].

Some of the different pharmacological and adverse effects exerted by ACE inhibitors may depend on the different phisicochemical (solubility, lipophilicity, acidity) and pharmacokinetic (absorption, protein binding, half-life and metabolic disposition) properties but also on their ability to penetrate and bind tissue sites[110]. Theoretical studies on p*K*a, lipophilicity, solubility, absorption and polar surface of ACE inhibitors, including perindopril, and its active metabolite, perindoprilat, have been reported[111]. In 2009, Remko presented theoretical calculations of molecular structure and stability of the arginine and erbumine salts of perindopril[43].

---

[3] Adapted with permission from First Crystal Structures of the Antihypertensive Drug Perindopril Erbumine: A Novel Hydrated Form and Polymorphs α and β, Vânia André, Luis Cunha-Silva, M. Teresa Duarte, and Pedro Paulo Santos, *Crystal Growth & Design*, 2011, *11* (9), pp 3703–3706, DOI: 10.1021/cg200430z. Copyright (2011) American Chemical Society".

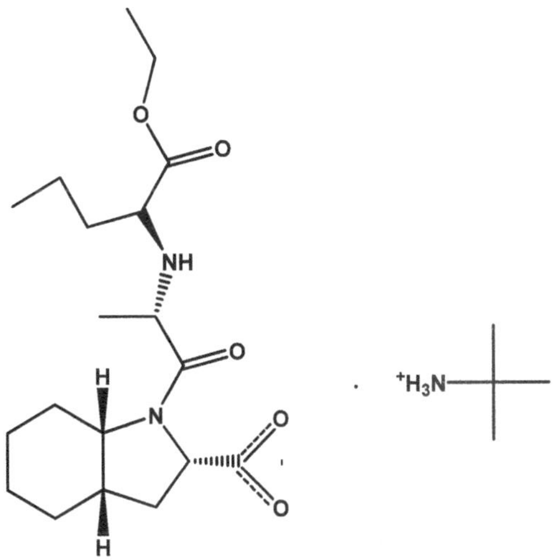

Fig. 18. Chemical diagram of perindopril erbumine salt.

Careful searches in the literature and in the Cambridge Structural Database[112] revealed that, although this API is known since 1981, until very recently only the crystal structure of perindoprilat, the pharmacologically active compound, had been determined in 1991[93]. In 2011, Remko and co-workers[41] unveiled the crystal structure of perindopril erbumine dehydrate.

Also in 2011, during a polymorphic screening of perindopril erbumine, the molecular structures of its α and β polymorphs[45, 113] have been determined by SCXRD as well as an unprecedented hydrated form of formula $(C_4H_{12}N)(C_{19}H_{31}N_2O_5)\cdot1.25H_2O$[40, 114]. Elemental and Karl-Fischer analyses confirmed the water contents of the three forms, that were were fully characterized by XRPD, vibrational spectroscopy (ATR-FT-IR and FT-Raman) and thermal analysis methods (TGA, DSC and HSM)[40]. Furthermore, stability, solubility and dissolution profile studies were performed.

The crystal packing of polymorphic forms α and β show similar hydrogen bonding interactions involving the perindopril and the erbumine ions. Perindopril anions interact with erbumine cations in an extended NH···O hydrogen bonding network leading to a supramolecular structure with the moieties organized in a double-chain arrangement. Each erbumine cation connects with three perindopril anions *via* the amine moiety: two of them are in the same chain whereas the other perindopril belongs to the opposite chain where the positioning of the anions in their respective chains, it is possible to notice that they assume antiparallel orientations i.e., perindopril anions of one chain are rotated of 180° relatively to the anions in the adjacent chain. Consequently two related types of $C_2^2(6)$ synthons are formed in both chains that are connected among them by $D_1^1(2)$ motifs.

The NH···O hydrogen bond distances are within the ranges of 2.707 - 2.803 Å and 2.738 - 2.788 Å in α and β forms, respectively. These double-chains do not establish classical hydrogen bonds among them neither in α nor β forms.

The new 1:1:1.25 hydrated form crystallizes with a triclinic symmetry, in the P1 chiral space group. This hydrated form was obtained both by solution and by LAG, which, as previously said, has several advantages not only in the preparation process, where equally yield and purity are obtained, but also in an environmental context[115-119]. Its asymmetric unit consists of two crystallographic independent perindopril anions, two erbumine cations and 2.5 water molecules. The CO distances in the carboxylate moiety and the location of the three hydrogen atoms in the amine moieties from the electron density map confirmed the presence of the salt.

The chiral centers in both perindopril crystallographic independent anions of the hydrated form exhibit the (S) configuration, corresponding exactly to the same configuration of the starting form α as well as of form β, what is important to assure the pharmacological activity of the API (Figure 18). The main conformational differences between these crystallographic independent anions are noted in the $-CH_2CH_2CH_3$ terminal groups (torsion angles of -58.2(4)° *vs* 175.1(9)°). The crystal packing of this hydrated form is very similar to the one described for polymorphic forms α and β, involving similar hydrogen bonding interactions between the perindopril and the erbumine ions (Figure 19). The NH···O hydrogen bond distances are within the ranges of 2.75 - 2.781 Å. The main difference between this hydrate and the polymorphs previously described is that while the double-chains do not establish classical hydrogen bonds among them neither in α nor β forms, in the hydrated form water molecules play an important role by linking adjacent chains through interactions between two crystallographically independent perindopril anions *via* the carbonyl group of one [$O_W \cdots O_{C=O}$ distance of 2.717Å] and the amine moiety of the other [$N_{N\text{-}H} \cdots O_W$ distance of 2.430Å]. Water molecules lie in the free spaces arising from the supramolecular arrangement described (Figure 19) and interact through cooperative O_W-H···O_W hydrogen bonds forming trimeric water clusters [O···O distances in the cluster: 2.644, 2.687 and 2.932 Å] (Figures 19 and 20).

Vibrational spectroscopy (FT-IR and FT-Raman) studies support the structural features unveiled by SCXRD data which are reflected in the spectra through a number of diagnostic

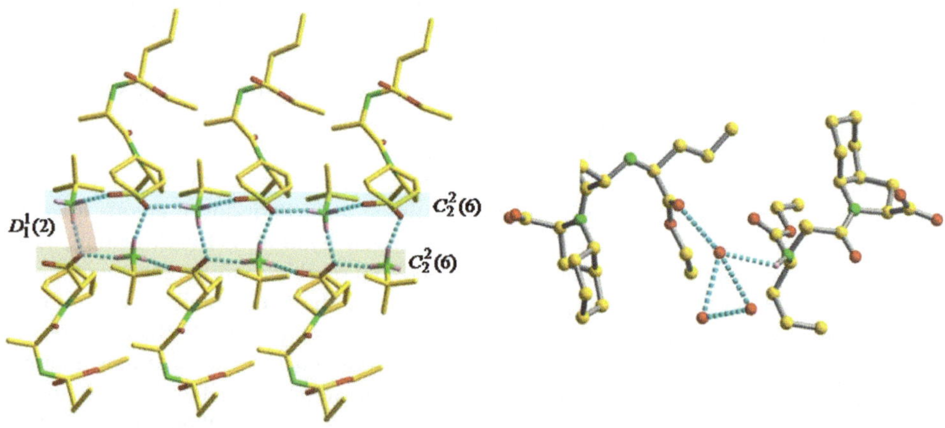

a b

Fig. 19. Crystal packing of the novel hydrated form of perindopril erbumine (1:1:1.25): (a) supramolecular arrangement with the perindopril anions and erbumine cations organized in double-chains; H bonds represented as blue dashed lines; water molecules were omitted for clarity; (b) detailed hydrogen bonding within the water cluster. Only hydrogen atoms involved in hydrogen bonding are shown, with exception of water molecules for which no hydrogen atoms are displayed[40].

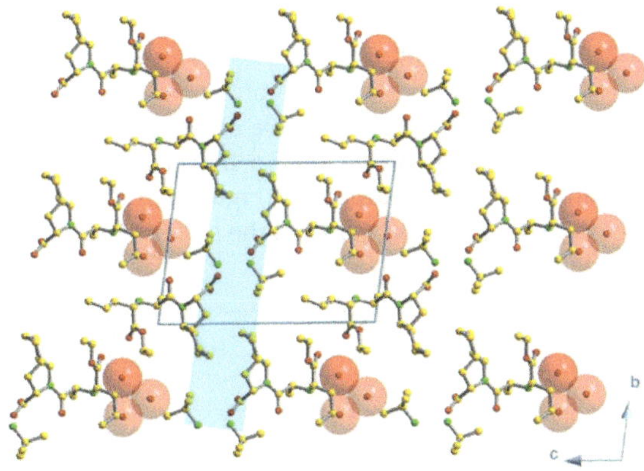

Fig. 20. Crystalline packing of the novel hydrated form of perindopril erbumine (1:1:1.25): Double-chain array formed by $C_2^2(6)$ and $D_1^1(2)$ motifs is highlighted in blue[40].

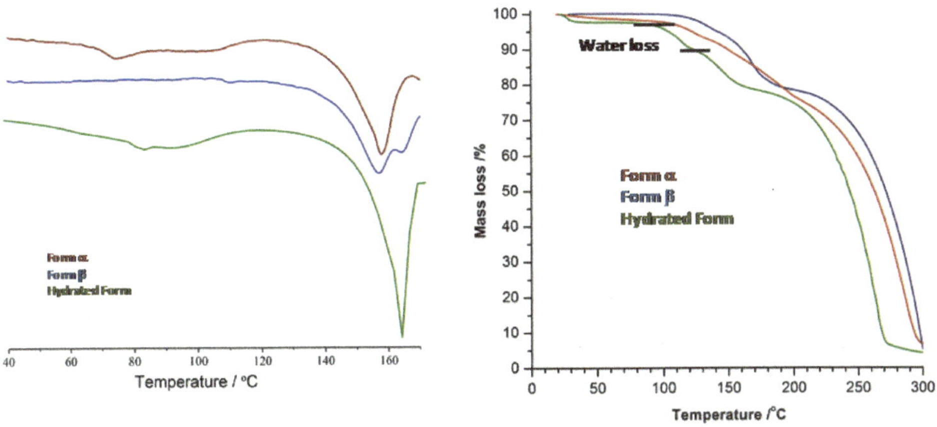

a b

Fig. 21. (a) DSC and (b) TGA pattern for all the forms of perindopril erbumine discussed.

bands (Figure 21). In particular the strong bands in the range of 3200-2600 cm^{-1} of the FT-Raman spectra are attributed to the υ_s(C–H) and υ_s(N–H) stretching vibrational modes diagnosing the presence of NH and NH_3^+ groups in the perindopril and erbumine cation, respectively. The strong bands around 1642, 1569 cm^{-1} and 1387 cm^{-1} (observed in both the FT-IR and FT-Raman spectra) are assigned to the $\upsilon_s(COO^-)$ and $\upsilon_{as}(COO^-)$ respectively, confirming the deprotonation of the carboxylic acid group. Contrasting with the FT-IR spectra of forms α and β, the spectrum of the hydrated form in the 3200-2600 cm^{-1} range reflects the presence of crystallization water molecules involved in well defined hydrogen bonds, by the presence of resolved peaks.

The combination of data obtained from DSC, TGA and HSM indicates that the novel hydrated form is stable until approximately 80°C, temperature at which a peak is observed in the DSC (Figure 21.a), a smooth mass loss is detected in the TGA (Figure 21.b) and bubbles start to appear in the HSM. The water loss occurs from this temperature until approximately 120°C. At 164°C melting and decomposition take place. TGA for forms α and β reveals that there is no mass loss before 120°C, confirming the absence of water in both these forms.

The new 1:1:1.25 hydrate has shown to be as stable on shelf as form α for eighteen months and water slurry experiments revealed that it as a thermodynamically stable form. It has also shown to have a similar dissolution profile (Figure 22) as the commercially available drug and to be slightly more soluble in water than the α form[40].

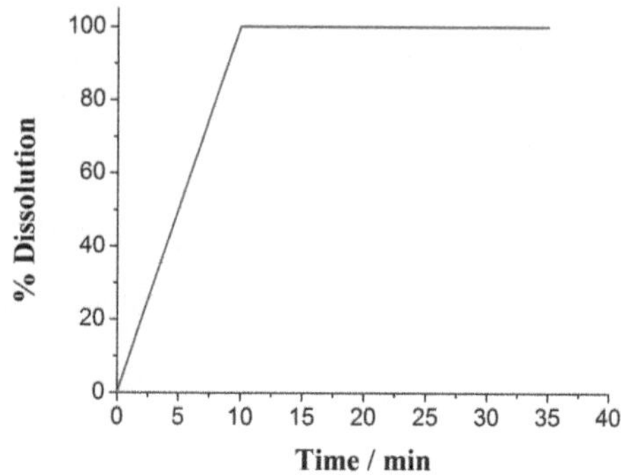

Fig. 22. Dissolution profile for the 1:1:1.25 hydrated form.

A probable reason for this is the enhanced stability provided by the presence of the water molecules linking the erbumine-perindopril double chains. Analysis of crystal structure has again proven to be quite important for the establishment of the intermolecular interactions responsible for the supramolecular arrangement and thus the physicochemical properties of APIs.

4. Concluding remarks

Over the last two decades crystal engineering, a key tool for the design of new crystal forms, has made possible the synthesis of novel pharmaceutical materials as well as molecular level control of crystallization and phase transformations. Advances in crystal engineering and supramolecular chemistry invite us to consider new perspectives and perhaps definitions of the various solid-state forms that the same and/or different molecules may adopt in terms of molecular assemblies and architectures.

Pharmaceutical co-crystals have proven to offer potential benefits of superior efficacy, solubility and stability in drug formulation. It seems reasonable to assert that co-crystal approaches should be considered routinely as part of a broader set of form and formulation explorations to achieve the best possible drug products. Although the interest in co-crystals and polymorphs and their utility is obvious, identifying and implementing an efficient discovery and control method remains a challenge.

5. Acknowledgements

The authors acknowledge Fundação para a Ciência e a Tecnologia, MCTES, Portugal, for funding the Project POCI/QUI/58791/2004, PEst-OE/QUI/UI0100 /2011, and the Ph.D. Grant SFRH/40474/2007 (V.A.).

6. References

[1] H. T. Wu, G. Kaur, and G. L. Griffiths, An improved synthesis of a fluorescent gabapentin-choline conjugate for single molecule detection. *Tetrahedron Letters*, 50, (2009):2100.

[2] R. D. B. Walsh, M. W. Bradner, S. Fleischman, L. A. Morales, B. Moulton, N. Rodriguez-Hornedo, and M. J. Zaworotko, Crystal engineering of the composition of pharmaceutical phases. *Chemical Communications*, (2003):186.

[3] S. L. Childs, L. J. Chyall, J. T. Dunlap, V. N. Smolenskaya, B. C. Stahly, and G. P. Stahly, Crystal engineering approach to forming co-crystals of amine hydrochlorides with organic acids. Molecular complexes of fluoxetine hydrochloride with benzoic, succinic, and fumaric acids. *Journal of the American Chemical Society*, 126, (2004):13335.

[4] O. Almarsson and M. J. Zaworotko, Crystal engineering of the composition of pharmaceutical phases. Do pharmaceutical co-crystals represent a new path to improved medicines? *Chemical Communications*, (2004):1889.

[5] N. Shan and M. J. Zaworotko, The role of co-crystals in pharmaceutical science. *Drug Discovery Today*, 13, (2008):440.

[6] P. Vishweshwar, J. A. McMahon, J. A. Bis, and M. J. Zaworotko, Pharmaceutical co-crystal. *Journal of Pharmaceutical Sciences*, 95, (2006):499.

[7] J. F. Remenar, S. L. Morissette, M. L. Peterson, B. Moulton, J. M. MacPhee, H. R. Guzman, and O. Almarsson, Crystal engineering of novel co-crystals of a triazole drug with 1,4-dicarboxylic acids. *Journal of the American Chemical Society*, 125, (2003):8456.

[8] D. P. McNamara, S. L. Childs, J. Giordano, A. Iarriccio, J. Cassidy, M. S. Shet, R. Mannion, E. O'Donnell, and A. Park, Use of a glutaric acid co-crystal to improve oral bioavailability of a low solubility API. *Pharmaceutical Research*, 23, (2006):1888.

[9] T. Friscic, A. V. Trask, W. D. S. Motherwell, and W. Jones, Guest-directed assembly of caffeine and succinic acid into topologically different heteromolecular host networks upon grinding. *Crystal Growth & Design*, 8, (2008):1605.

[10] M. B. Hickey, M. L. Peterson, L. A. Scoppettuolo, S. L. Morrisette, A. Vetter, H. Guzman, J. F. Remenar, Z. Zhang, M. D. Tawa, S. Haley, M. J. Zaworotko, and O. Almarsson, Performance comparison of a co-crystal of carbamazepine with marketed product. *European Journal of Pharmaceutics and Biopharmaceutics*, 67, (2007):112.

[11] J. F. Remenar, M. L. Peterson, P. W. Stephens, Z. Zhang, Y. Zimenkov, and M. B. Hickey, Celecoxib : Nicotinamide dissociation: Using excipients to capture the co-crystal's potential. *Molecular Pharmaceutics*, 4, (2007):386.

[12] S. G. Fleischman, S. S. Kuduva, J. A. McMahon, B. Moulton, R. D. B. Walsh, N. Rodriguez-Hornedo, and M. J. Zaworotko, Crystal engineering of the composition of pharmaceutical phases: Multiple-component crystalline solids involving carbamazepine. *Crystal Growth & Design*, 3, (2003):909.

[13] J. A. McMahon, J. A. Bis, P. Vishweshwar, T. R. Shattock, O. L. McLaughlin, and M. J. Zaworotko, Crystal engineering of the composition of pharmaceutical phases. 3. Primary amide supramolecular heterosynthons and their role in the design of pharmaceutical co-crystals. *Zeitschrift fur Kristallographie*, 220, (2005):340.

[14] K. Shiraki, N. Takata, R. Takano, Y. Hayashi, and K. Terada, Dissolution improvement and the mechanism of the improvement from co-crystallization of poorly water-soluble compounds. *Pharmaceutical Research*, 25, (2008):2581.
[15] M. K. Stanton, S. Tufekcic, C. Morgan, and A. Bak, Drug Substance and Former Structure Property Relationships in 15 Diverse Pharmaceutical Co-Crystals. *Crystal Growth & Design*, 9, (2009):1344.
[16] A. V. Trask, W. D. S. Motherwell, and W. Jones, Pharmaceutical co-crystallization: Engineering a remedy for caffeine hydration. *Crystal Growth & Design*, 5, (2005):1013.
[17] A. V. Trask, W. D. S. Motherwell, and W. Jones, Physical stability enhancement of theophylline via co-crystallization. *International Journal of Pharmaceutics*, 320, (2006):114.
[18] A. Bak, A. Gore, E. Yanez, M. Stanton, S. Tufekcic, R. Syed, A. Akrami, M. Rose, S. Surapaneni, T. Bostick, A. King, S. Neervannan, D. Ostovic, and A. Koparkar, The co-crystal approach to improve the exposure of a water-insoluble compound: AMG 517 sorbic acid co-crystal characterization and pharmacokinetics. *Journal of Pharmaceutical Sciences*, 97, (2008):3942.
[19] A. M. Chen, M. E. Ellison, A. Peresypkin, R. M. Wenslow, N. Variankaval, C. G. Savarin, T. K. Natishan, D. J. Mathre, P. G. Dormer, D. H. Euler, R. G. Ball, Z. X. Ye, Y. L. Wang, and I. Santos, Development of a pharmaceutical co-crystal of a monophosphate salt with phosphoric acid. *Chemical Communications*, (2007):419.
[20] N. Variankaval, R. Wenslow, J. Murry, R. Hartman, R. Helmy, E. Kwong, S. D. Clas, C. Dalton, and I. Santos, Preparation and solid-state characterization of nonstoichiometric co-crystals off a phosphodiesterase-IV inhibitor annul L-tartaric acid. *Crystal Growth & Design*, 6, (2006):690.
[21] S. L. Childs and K. I. Hardcastle, Co-crystals of chlorzoxazone with carboxylic acids. *Crystengcomm*, 9, (2007):363.
[22] C. B. Aakeroy, M. E. Fasulo, and J. Desper, Co-crystal or salt: Does it really matter? *Molecular Pharmaceutics*, 4, (2007):317.
[23] F. H. Allen, The Cambridge Structural Database: a quarter of a million crystal structures and rising. *Acta Crystallographica Section B-Structural Science*, 58, (2002):380.
[24] W. W. Porter, S. C. Elie, and A. J. Matzger, Polymorphism in carbamazepine co-crystals. *Crystal Growth & Design*, 8, (2008):14.
[25] S. Karki, T. Friscic, and W. Jones, Control and interconversion of co-crystal stoichiometry in grinding: stepwise mechanism for the formation of a hydrogen-bonded co-crystal. *Crystengcomm*, 11, (2009):470.
[26] T. Friscic, S. L. Childs, S. A. A. Rizvi, and W. Jones, The role of solvent in mechanochemical and sonochemical co-crystal formation: a solubility-based approach for predicting co-crystallisation outcome. *Crystengcomm*, 11, (2009):418.
[27] T. Friscic, L. Fabian, J. C. Burley, W. Jones, and W. D. S. Motherwell, Exploring co-crystal - co-crystal reactivity via liquid-assisted grinding: the assembling of racemic and dismantling of enantiomeric co-crystals. *Chemical Communications*, (2006):5009.
[28] S. Karki, T. Friscic, W. Jones, and W. D. S. Motherwell, Screening for pharmaceutical co-crystal hydrates via neat and liquid-assisted grinding. *Molecular Pharmaceutics*, 4, (2007):347.

[29] D. Braga, L. Maini, G. de Sanctis, K. Rubini, F. Grepioni, M. R. Chierotti, and R. Gobetto, Mechanochemical preparation of hydrogen-bonded adducts between the diamine 1,4-diazabicyclo [2.2.2] octane and dicarboxylic acids of variable chain length: An x-ray diffraction and solid-state NMR study. *Chemistry-A European Journal*, 9, (2003):5538.

[30] D. Braga and F. Grepioni, Making crystals from crystals: a green route to crystal engineering and polymorphism. *Chemical Communications*, (2005):3635.

[31] V. Andre, M. M. Marques, M. F. M. da Piedade, and M. T. Duarte, An ester derivative of the drug gabapentin: pH dependent crystal stability. *Journal of Molecular Structure*, 973, (2010):173.

[32] D. Braga, F. Grepioni, L. Maini, K. Rubini, M. Polito, R. Brescello, L. Cotarca, M. T. Duarte, V. Andre, and M. F. M. Piedade, Polymorphic gabapentin: thermal behaviour, reactivity and interconversion of forms in solution and solid-state. *New Journal of Chemistry*, 32, (2008):1788.

[33] L. S. Reddy, S. J. Bethune, J. W. Kampf, and N. Rodriguez-Hornedo, Co-crystals and Salts of Gabapentin: pH Dependent Co-crystal Stability and Solubility. *Crystal Growth & Design*, 9, (2009):378.

[34] M. Wenger and J. Bernstein, An alternate crystal form of gabapentin: A co-crystal with oxalic acid. *Crystal Growth & Design*, 8, (2008):1595.

[35] V. Andre, D. Braga, F. Grepioni, and M. T. Duarte, Crystal Forms of the Antibiotic 4-Aminosalicylic Acid: Solvates and Molecular Salts with Dioxane, Morpholine, and Piperazine. *Crystal Growth & Design*, 9, (2009):5108.

[36] V. Andre, D. Braga, F. Grepioni, and M. Duarte, Crystal Forms of the Antibiotic 4-Aminosalicylic Acid: Solvates and Molecular Salts with Dioxane, Morpholine, and Piperazine. *Crystal Growth & Design*, 9, (2009):5108.

[37] F. P. A. Fabbiani, D. R. Allan, A. Dawson, W. I. F. David, P. A. McGregor, I. D. H. Oswald, S. Parsons, and C. R. Pulham, Pressure-induced formation of a solvate of paracetamol. *Chemical Communications*, (2003):3004.

[38] I. D. H. Oswald, D. R. Allan, P. A. McGregor, W. D. S. Motherwell, S. Parsons, and C. R. Pulham, The formation of paracetamol (acetaminophen) adducts with hydrogen-bond acceptors. *Acta Crystallographica Section B-Structural Science*, 58, (2002):1057.

[39] I. D. H. Oswald, W. D. S. Motherwell, S. Parsons, and C. R. Pulham, A paracetamol-morpholine adduct. *Acta Crystallographica Section E-Structure Reports Online*, 58, (2002):O1290.

[40] V. n. André, L. s. Cunha-Silva, M. T. Duarte, and P. P. Santos, First Crystal Structures of the Antihypertensive Drug Perindopril Erbumine: A Novel Hydrated Form and Polymorphs α and β. *Crystal Growth & Design*, 11, (2011):3703.

[41] M. Remko, J. Bojarska, P. Jezko, L. Sieron, A. Olczak, and W. Maniukiewicz, Crystal and molecular structure of perindopril erbumine salt. *Journal of Molecular Structure*, 997, (2011):103.

[42] P. Gupta, N. Bhandari, S. Gore, M. K. Pananchukunnath, and I. Bhushan, IN200802491-I4, 2010.

[43] M. Remko, Molecular structure and stability of perindopril erbumine and perindopril L-arginine complexes. *European Journal of Medicinal Chemistry*, 44, (2009):101.

[44] A. Kumar, S. R. Soudagar, A. Mathur, S. T. Gunjal, N. B. Panda, and D. U. Jadhav, EP1987828-A1, 2008.

[45] A. Kumar, S. R. Soudagar, A. Mathur, and N. B. Panda, US2008183011-A1; EP1964836-A2; IN200601843-I3; EP1964836-A3, 2008.

[46] A. Ujagare, D. A. Kochrekar, and P. Sarjekar, WO2007017894-A2, 2007.

[47] A. Ujagare, D. A. Kochrekar, and P. Sarjekar, WO2007017893-A2, 2007.

[48] C. Straessler, V. Lellek, R. Faessler, R. Fassler, C. Strassler, C. Strossler, R. Fossler, S. Christoph, L. Vit, and F. Roger, WO2004113293-A1; NO200600256-A; EP1636185-A1; AU2004249345-A1; BR200411966-A; KR2006035636-A; MX2005013811-A1; CN1812971-A; JP2007507418-W; US2007135512-A1; ZA200600655-A; CN100395235-C; CN101333181-A; NZ544160-A; AU2004249345-B2; AU2004249345-B8; MX268989-B; US7705046-B2; US2010160404-A1, 2005.

[49] D. Braga, F. Grepioni, V. Andre, and M. T. Duarte, Drug-containing coordination and hydrogen bonding networks obtained mechanochemically. *Crystengcomm*, 11, (2009):2618.

[50] A. A. Jensen, J. Mosbacher, S. Elg, K. Lingenhoehl, T. Lohmann, T. N. Johansen, B. Abrahamsen, J. P. Mattsson, A. Lehmann, B. Bettler, and H. Brauner-Osborne, The anticonvulsant gabapentin (Neurontin) does not act through gamma-aminobutyric acid-B receptors. *Molecular Pharmacology*, 61, (2002):1377.

[51] C. P. Taylor, Emerging Perspectives on the Mechanism of Action of Gabapentin. *Neurology*, 44, (1994):10.

[52] C. P. Taylor, N. S. Gee, T. Z. Su, J. D. Kocsis, D. F. Welty, J. P. Brown, D. J. Dooley, P. Boden, and L. Singh, A summary of mechanistic hypotheses of gabapentin pharmacology. *Epilepsy Research*, 29, (1998):233.

[53] C. M. Santi, F. S. Cayabyab, K. G. Sutton, J. E. McRory, J. Mezeyova, K. S. Hamming, D. Parker, A. Stea, and T. P. Snutch, Differential inhibition of T-type calcium channels by neuroleptics. *Journal of Neuroscience*, 22, (2002):396.

[54] A. C. Errington, T. Stohr, and G. Lees, Voltage gated ion channels: Targets for anticonvulsant drugs. *Current Topics in Medicinal Chemistry*, 5, (2005):15.

[55] A. B. Ettinger and C. E. Argoff, Use of Antiepileptic drugs for nonepileptic conditions: psychiatric disorders and chronic pain. *Neurotherapeutics*, 4, (2007):75.

[56] A. S. Kato and D. S. Bredt, Pharmacological regulation of ion channels by auxiliary subunits. *Current Opinion in Drug Discovery & Development*, 10, (2007):565.

[57] E. Eisenberg, Y. River, A. Shifrin, and N. Krivoy, Antiepileptic drugs in the treatment of neuropathic pain. *Drugs*, 67, (2007):1265.

[58] K. Ananda, S. Aravinda, P. G. Vasudev, K. M. P. Raja, H. Sivaramakrishnan, K. Nagarajan, N. Shamala, and P. Balaram, Stereochemistry of gabapentin and several derivatives: Solid state conformations and solution equilibria. *Current Science*, 85, (2003):1002.

[59] F. Henle, J. Leemhuis, C. Schmid, T. J. Feuerstein, and D. K. Meyer, Gabapentin-lactam induces the formation of dendritic branches in cultured hippocampal neurons. *Naunyn-Schmiedebergs Archives of Pharmacology*, 369, (2004):342.

[60] F. Henle, J. Leemhuis, C. Fischer, H. H. Bock, K. Lindemeyer, T. J. Feuerstein, and D. K. Meyer, Gabapentin-lactam induces dendritic filopodia and motility in cultured hippocampal neurons. *Journal of Pharmacology and Experimental Therapeutics*, 319, (2006):181.

[61] P. Revill, J. Bolos, N. Serradell, and M. Bayes, Gabapentin enacarbil. Treatment of restless legs syndrome, treatment of postherpetic neuralgia, treatment of neuropathic pain. *Drugs of the Future*, 31, (2006):771.

[62] K. C. Cundy, T. Annamalai, L. Bu, J. De Vera, J. Estrela, W. Luo, P. Shirsat, A. Torneros, F. M. Yao, J. Zou, R. W. Barrett, and M. A. Gallop, XP13512 [(+/-)-1-([(alpha-isobutanoyloxyethoxy)carbonyl] aminomethyl)-1-cyclohexane acetic acid], a novel gabapentin prodrug: II. Improved oral bioavailability, dose proportionality, and colonic absorption compared with gabapentin in rats and monkeys. *Journal of Pharmacology and Experimental Therapeutics*, 311, (2004):324.

[63] J. A. Ibers, Gabapentin and gabapentin monohydrate. *Acta Crystallographica Section C-Crystal Structure Communications*, 57, (2001):641.

[64] H. A. Reece and D. C. Levendis, Polymorphs of gabapenitin. *Acta Crystallographica Section C-Crystal Structure Communications*, 64, (2008):O105.

[65] D. E. Butter and B. J. Greenman, US Patent, No. 4 894 476, 1990.

[66] G. Satzinger, J. Hartenstein, M. Herrmann, and W. Heldt, US Patent, No. 4 024 175, 1977.

[67] J. A. Ibers, Gabapentin and gabapentin monohydrate. *Acta Crystallographica, Section C: Crystal Structure Communications*, C57, (2001):641.

[68] D. Braga, F. Grepioni, L. Maini, K. Rubini, M. Polito, R. Brescello, L. Cotarca, M. Duarte, V. Andre, and M. Piedade, Polymorphic gabapentin: thermal behaviour, reactivity and interconversion of forms in solution and solid-state. *New Journal of Chemistry*, 32, (2008):1788.

[69] P. Pesachovich, C. Singer, and G. Pilarski, US Patent, No. 6 255 526 B1, 2001.

[70] H. A. Reece and D. C. Levendis, Polymorphs of gabapentin. *Acta Crystallographica Section C*, 64, (2008):o105.

[71] L. R. Chen, S. R. Babu, C. J. Calvitt, and B. Tobias, US Patent, No. 6 800 782 B2, 2004.

[72] J. Bosh Lladò, R. G. Cruz, E. M. Grau, and M. d. C. O. Miguel, US Patent, No. 6 521 787 B1, 2003.

[73] Y. Kumar, C. H. Khanduri, K. K. Ganagakhedkar, R. Chakraborty, h. N. Dorwal, A. Rohtagi, and A. K. Panda, International Publication Number Patent, No. WO 2004/106281 A1, 2004.

[74] A. Cutrignelli, N. Denora, A. Lopedota, A. Trapani, V. Laquintana, A. Latrofa, G. Trapani, and G. Liso, Comparative effects of some hydrophilic excipients on the rate of gabapentin and baclofen lactamization in lyophilized formulations. *International Journal of Pharmaceutics*, 332, (2007):98.

[75] A. S. Kearney, S. C. Mehta, and G. W. Radebaugh, The effect of cyclodextrins on the rate of intramolecular lactamization of gabapentin in aqueous solution. *International Journal of Pharmaceutics*, 78, (1992):25.

[76] P. G. Vasudev, S. Aravinda, K. Ananda, S. D. Veena, K. Nagarajan, N. Shamala, and P. Balaram, Crystal Structures of a New Polymorphic Form of Gabapentin Monohydrate and the E and Z Isomers of 4-Tertiarybutylgabapentin. *Chemical Biology & Drug Design*, 73, (2009):83.

[77] D. E. Butler and B. J. Greenman, EP340677-A2; AU8932678-A; DK8902126-A; FI8902067-A; JP2011546-A; US4894476-A; US4960931-A; CA1306755-C; EP340677-B1; DE68912819-E; ES2061774-T3; IE62958-B; JP2619951-B2; DK200200056-A; DK175127-B; DK175400-B.

[78] L. J. Bosch, M. D. C. Onrubia Miguel, L. E. Pagans, M. C. Onrubia Miguel, L. J. Bosch, M. M. C. Onrubia, and L. E. Pagans, WO200064857-A; EP1174418-A; WO200064857-A1; AU200035592-A; EP1174418-A1; ES2164527-A1; HU200200741-A2; US6528682-B1; ES2164527-B1; EP1174418-B1; DE60007734-E; ES2213571-T3; IN200101052-P2; IL145920-A.

[79] J. P. Jasinski, R. J. Butcher, H. S. Yathirajan, L. Mallesha, K. N. Mohana, and B. Narayana, Crystal Structure of a Second Polymorph of Gabapentin Hydrochloride Hemihydrate with a Three-Center Bifurcated Hydrogen Bond. *JOURNAL OF CHEMICAL CRYSTALLOGRAPHY*, 39, (2009):777.

[80] S. Chava, S. R. Gorantla, V. S. K. Indukuri, V. S. K. Indukurl, C. Satyanarayana, G. S. Ramanjaneyulu, and I. V. Kumar, WO2004093780-A2; EP1615875-A2; US2006235079-A1; US7439387-B2.

[81] F. P. A. Fabbiani, D. C. Levendis, G. Buth, W. F. Kuhs, N. Shankland, and H. Sowa, Searching for novel crystal forms by in situ high-pressure crystallisation: the example of gabapentin heptahydrate. *Crystengcomm*, 12, (2010):2354.

[82] D. Braga, F. Grepioni, L. Maini, R. Brescello, and L. Cotarca, Simple and quantitative mechanochemical preparation of the first zinc and copper complexes of the neuroleptic drug gabapentin. *Crystengcomm*, 10, (2008):469.

[83] V. Andre, M. Marques, M. da Piedade, and M. Duarte, An ester derivative of the drug gabapentin: pH dependent crystal stability. *Journal of Molecular Structure*, 973, (2010):173.

[84] V. Andre, A. Fernandes, P. P. Santos, and M. T. Duarte, On the Track of New Multicomponent Gabapentin Crystal Forms: Synthon Competition and pH Stability. *Crystal Growth & Design*, 11, (2011):2325.

[85] R. Williams, pKa Data Compiled by R. Williams. http://research.chem.psu.edu/brpgroup/pKa_compilation.pdf.

[86] P. Kavuru, D. Aboarayes, K. K. Arora, H. D. Clarke, A. Kennedy, L. Marshall, T. T. Ong, J. Perman, T. Pujari, L. Wojtas, and M. J. Zaworotko, Hierarchy of Supramolecular Synthons: Persistent Hydrogen Bonds Between Carboxylates and Weakly Acidic Hydroxyl Moieties in Co-crystals of Zwitterions. *Crystal Growth & Design*, 10, (2010):3568.

[87] L. H. Opie, Inhibition of the Cerebral Renin-Angiotensin System to Limit Cognitive Decline in Elderly Hypertensive Persons. *Cardiovascular Drugs and Therapy*, 25, (2011):277.

[88] L. Opie, T. Dalby, and D. P. Naidoo, Perindopril (Coversyl) prevents cardiovascular death and MI in coronary disease patients regardless of their cardiovascular risk. *Cardiovascular journal of South Africa : official journal for Southern Africa Cardiac Society [and] South African Society of Cardiac Practitioners*, 14, (2003):277.

[89] J. Wong, R. A. Patel, and P. R. Kowey, The clinical use of angiotensin-converting enzyme inhibitors. *Progress in Cardiovascular Diseases*, 47, (2004):116.

[90] M. E. Bertrand, Provision of cardiovascular protection by ACE inhibitors: a review of recent trials. *Current Medical Research and Opinion*, 20, (2004):1559.

[91] S. Laurent, Clinical benefit of very-low-dose perindopril-indapamide combination in hypertension. *Journal of Hypertension*, 19, (2001):S9.

[92] L. Pilote, M. Abrahamowicz, E. Rodrigues, M. J. Eisenberg, and E. Rahme, Mortality rates in elderly patients who take different angiotensin-converting enzyme

inhibitors after acute myocardial infarction: A class effect? *Annals of Internal Medicine*, 141, (2004):102.

[93] C. Pascard, J. Guilhem, M. Vincent, G. Remond, B. Portevin, and M. Laubie, CONFIGURATION AND PREFERENTIAL SOLID-STATE CONFORMATIONS OF PERINDOPRILAT (S-9780) - COMPARISON WITH THE CRYSTAL-STRUCTURES OF OTHER ACE INHIBITORS AND CONCLUSIONS RELATED TO STRUCTURE-ACTIVITY-RELATIONSHIPS. *Journal of Medicinal Chemistry*, 34, (1991):663.

[94] A. Remkova and H. Kratochvil'ova, Effect of the angiotensin-converting enzyme inhibitor perindopril on haemostasis in essential hypertension. *Blood Coagulation & Fibrinolysis*, 11, (2000):641.

[95] A. Okrucka, J. Pechan, and H. Kratochvilova, Effects of the angiotensin-converting enzyme (ACE) inhibitor perindopril on endothelial and platelet functions in essential hypertension. *Platelets*, 9, (1998):63.

[96] A. Remkova, H. Kratochvil'ova, and J. Durina, Impact of the therapy by renin-angiotensin system targeting antihypertensive agents perindopril versus telmisartan on prothrombotic state in essential hypertension. *Journal of Human Hypertension*, 22, (2008):338.

[97] D. Neglia, E. Fommei, A. Varela-Carver, M. Mancini, S. Ghione, M. Lombardi, P. Pisani, H. Parker, G. D'Amati, L. Donato, and P. G. Camici, Perindopril and indapamide reverse coronary microvascular remodelling and improve flow in arterial hypertension. *Journal of Hypertension*, 29, (2011):364.

[98] S. Laurent, Predictors of cardiovascular mortality and morbidity in hypertension. *Current medical research and opinion*, 21 Suppl 5, (2005):S7.

[99] A. Remkova and M. Remko, The Role of Renin-Angiotensin System in Prothrombotic State in Essential Hypertension. *Physiological Research*, 59, (2010):13.

[100] B. Lecocq, C. Funckbrentano, V. Lecocq, A. Ferry, M. E. Gardin, M. Devissaguet, and P. Jaillon, INFLUENCE OF FOOD ON THE PHARMACOKINETICS OF PERINDOPRIL AND THE TIME COURSE OF ANGIOTENSIN-CONVERTING ENZYME-INHIBITION IN SERUM. *Clinical Pharmacology & Therapeutics*, 47, (1990):397.

[101] E. Telejko, Perindopril arginine: benefits of a new salt of the ACE inhibitor perindopril. *Current Medical Research and Opinion*, 23, (2007):953.

[102] R. Rucman, WO2005068490-A1; EP1709066-A1; EP1709066-B1; DE602005009319-E; EP2003142-A1; RU2372353-C2, 2005.

[103] R. Rucman, WO2005068425-A1; EP1713771-A1; RU2387641-C2, 2005.

[104] R. Rucman and P. Zupet, WO2007058634-A1; EP1948224-A1; ZA200803889-A, 2007.

[105] R. Rucman and P. Zupet, WO2008150245-A2; WO2008150245-A3; EP2164469-A2; CN101742986-A, 2009.

[106] D. Churchley and A. Amberkhane, WO2008050185-A2, 2008.

[107] S. N. Devarakonda, M. Asnani, S. R. Bonnareddy, P. R. Padi, U. K. Chandramohan, S. S. Chitre, V. Nalivella, S. K. Vasamsetti, A. Minakshi, B. S. Reddy, and P. P. Reddy, WO2007092758-A2; WO2007092758-A3; IN200804031-P4, 2008.

[108] R. Rucman and P. Zupet, EP1647547-A1, 2006.

[109] M. Piran, M. Cohen, S. Costi, M. Khan, M. Gharpure, A. K. Pandey, R. S. Yadav, P. V. Patil, S. Lad, and G. Patle, WO2010148396-A1, 2010.

[110] N. J. Brown and D. E. Vaughan, Angiotensin-converting enzyme inhibitors. *Circulation*, 97, (1998):1411.

[111] M. Remko, Acidity, lipophilicity, solubility, absorption, and polar surface area of some ACE inhibitors. *Chemical Papers*, 61, (2007):133.

[112] L. Fabian, Cambridge Structural Database Analysis of Molecular Complementarity in Co-crystals. *Crystal Growth & Design*, 9, (2009):1436.

[113] N. S. Joshi, S. B. Bhirud, and K. E. Rao, US2005250706-A1; WO2005108365-A1; IN200400531-I3; IN220637-B, 2005.

[114] V. Andre, L. Cunha-Silva, P. P. Santos, and M. T. Duarte, Portugal Patent, No., 2010.

[115] E. H. H. Chow, F. C. Strobridge, and T. Friscic, Mechanochemistry of magnesium oxide revisited: facile derivatisation of pharmaceuticals using coordination and supramolecular chemistry. *Chemical Communications*, 46, (2010):6368.

[116] T. Friscic, New opportunities for materials synthesis using mechanochemistry. *Journal of Materials Chemistry*, 20, (2010):7599.

[117] D. R. Weyna, T. Shattock, P. Vishweshwar, and M. J. Zaworotko, Synthesis and Structural Characterization of Co-crystals and Pharmaceutical Co-crystals: Mechanochemistry vs Slow Evaporation A from Solution. *Crystal Growth & Design*, 9, (2009):1106.

[118] A. V. Trask, D. A. Haynes, W. D. S. Motherwell, and W. Jones, Screening for crystalline salts via mechanochemistry. *Chemical Communications*, (2006):51.

[119] T. Friscic and W. Jones, Recent Advances in Understanding the Mechanism of Co-crystal Formation via Grinding. *Crystal Growth & Design*, 9, (2009):1621.

4

Supramolecular Arrangements in Organotellurium Compounds *via* Te···Halogen Contacts

Angel Alvarez-Larena, Joan Farran and Joan F. Piniella
Universitat Autònoma de Barcelona
Spain

In memory of Professor Gabriel Germain, for his guidance and his friendship

1. Introduction

The understanding of the atomic interactions involved in crystal structures is fundamental in the crystal engineering field. In a first instance, this knowledge can be correlated with the crystal properties and, in a second instance, it can be applied to the design of crystalline materials for specific applications.

In molecular crystals, the crystal cohesion is attributed to weak attractive forces, unlike strong covalent interactions that hold the atoms bonded in a molecule. Some of these weak interactions, for example the hydrogen bonds between electronegative atoms, have been recognized and studied for a long time (Pimentel & McClellan, 1960; Jeffrey & Saenger, 1991; Steiner, 2002). Nonconventional and weaker hydrogen bonds between AH (AH = OH, NH, etc.) and soft bases (π systems) or between CH and B (B = O, N, etc.), π···π interactions, halogen bonds, cation···π interactions, have become the focus of interest in the last decades due to their potential role in supramolecular chemistry and in biochemical processes (Desiraju & Steiner, 2001; Metrangolo et al., 2008; Nishio et al., 2009; Schneider, 2009; Salonen et al., 2011).

Soft interactions between heavy p-block elements and electronegative atoms are frequent and have been shown to play a significant role in supramolecular chemistry. Tellurium is a chemical element showing this kind of soft interactions, also known as secondary bonding (Alcock, 1972). Organotellurium compounds have been investigated mainly in organic synthesis (Singh & Sharma, 2000; Petragnani & Stefani, 2005), but also in medicine (Ba et al., 2010), in materials science (Steigerwald & Sprinkle, 1987; Hails et al., 2001) and recently in polymerization processes (organotellurium-mediated living radical polymerization (TERP)) (Kitayama et al., 2010) and to protect materials (lubricants, polymers) from oxidation (Shanks et al., 2006).

The type of the Te···X interaction (and the secondary bonding interaction) and its relevance in the tellurium coordination polyhedra have been discussed (Alcock, 1972; Zukerman-Schpector & Haiduc, 2001). This chapter deals with the weak Te···halogen (Te···X) interactions found in organotellurium crystal structures and with the supramolecular

arrangements derived from them. Some supramolecular self-assemblies based on Te···halogen secondary bonds have been described (Haiduc & Zukerman-Schpector, 2001; Zukerman-Schpector et al., 2002; Srivastava et al., 2004; Cozzolino et al., 2011).
This chapter presents a systematic update including quantitative aspects. It is performed an analysis of the influence of several factors on the type of supramolecular pattern such as the polymorphism or the nature of halogen.

2. The Te···X contact in organotellurium compounds

The study of Te···X contacts was based on analysis of data from Version 5.32 (last update of May 2011) of the Cambridge Structural Database (CSD) (Allen, 2002), where organic and organometallic crystal structures determined from X-ray (or neutron) diffraction data are deposited. A search was performed with the aid of the ConQuest program (Version 1.13) (Bruno et al., 2002) in order to retrieve crystal structures containing organotellurium compounds and halogen atoms. Only organic structures with available atomic coordinates were considered and several additional exclusion conditions were applied. Structures with R factor > 0.1, and structures where tellurium or halogen atoms are disordered, were omitted. Moreover, structures with charged fragments were also eliminated. After an additional checking to remove multiple entries of the same structure, a set of 481 structures was accepted. A search of intermolecular distances between tellurium and halogen atoms was performed and the results were analyzed using the Vista program (Version 2.1) (CCDC, 1994).
If a random distribution of halogens around tellurium atom is supposed, a distance histogram showing only an exponential growth would be expected. However this was not the case and a maximum was observed before the exponential growth (Figure 1). This maximum can be attributed to the existence of a Te···X interaction. A similar behaviour has been reported, for example, in O···O distance histograms where a H-bonding maximum has been observed (Rowland & Taylor, 1996). In the case of Te···F distances, most compounds are perfluorinated, introducing a bias in their distribution. When these compounds are removed, the remaining population size is so small that the resulting histogram is not statistically significant. When Te···X secondary and Te-X covalent bond distance ranges are compared, the former is broader (Table 1), according to the weaker nature of Te···X interaction. The maximum of the Te···X peak was located at a distance lower than the sum of van der Waals radii (Bondi, 1964) eventhough the peak spreads beyond this reference value. The sum of van der Waals radii is highly used in order to determine whether an interaction is present or not. However, this value is rather arbitrary (it is an approximation) and it can not be considered as a cut-off. In fact, in the case of H-bonding the use of such cut-off criterium has been discouraged (Jeffrey & Saenger, 1991). So, in the case of Te···X, it seems that the interaction can be present beyond the sum of van der Waals radii although its force decreases, according with the electrostatic character of the interaction. The upper limit for the peak analysis has been situated at $1.10 \cdot \Sigma r_{vdW}$, corresponding approximately to the local minimum in the histogram (for X = Cl, Br, and I).
Two more considerations about Te···X distance ranges should be taken into account. The first one is the temperature effect on the distance. Half the structures used in the present study have been determined at low temperature. It is known that weak interaction distances are more temperature dependent than covalent bond distances (Forni et al., 2004). Two CSD structures, DMTEII and QIXZAY, showing Te···I contacts and measured both at 130K and at

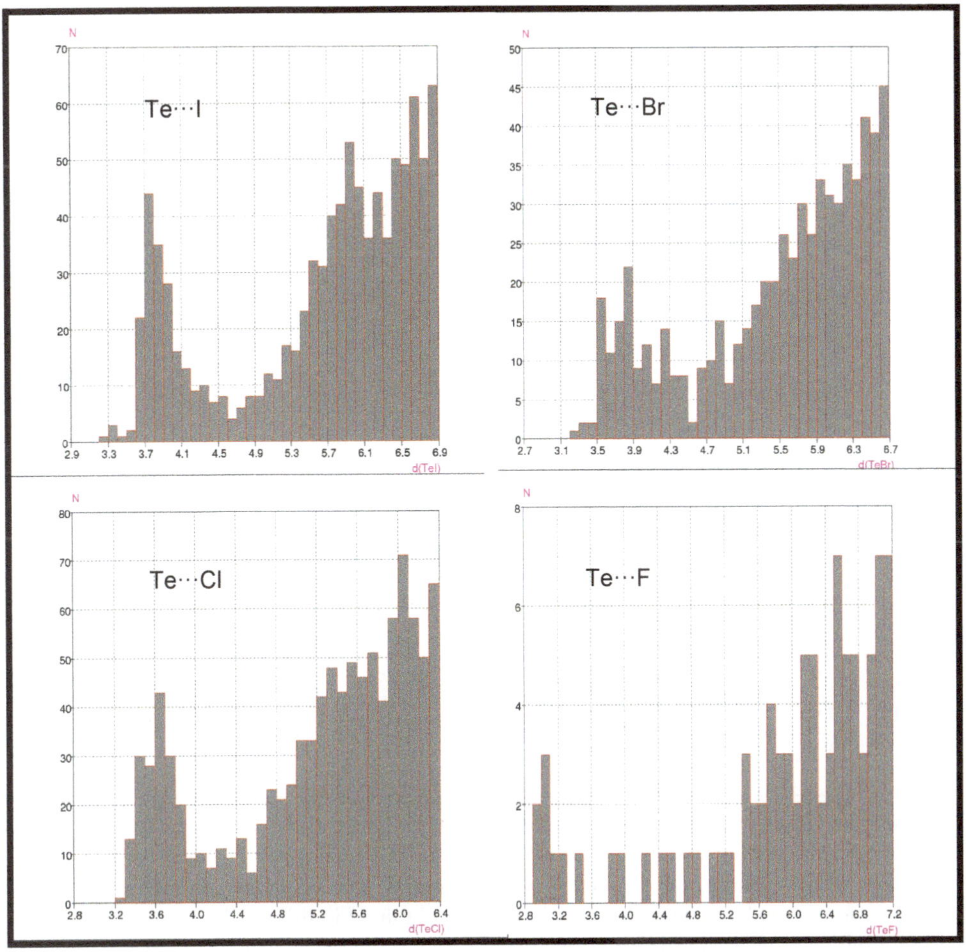

Fig. 1. Histograms for the distribution of intermolecular Te···X distances

room temperature (see section 3.8), are well suited to carry out a distance comparison. In the case of QIXZAY the 130K experiment reveals a mean Te···I shortening of 0.050 Å while the mean Te-I bond shortening is just 0.005 Å, close to the experimental error (0.002 Å). DMTEII behaves in a similar way being the Te···I shortening even greater: 0.082 Å. So, every peak in the distance histograms is, in fact, the superposition of two peaks. However, the estimated difference in maximum positions (0.05 Å) is small compared with the peak width. Moreover, the number of structures defining every peak is not big enough to undertake a study of the temperature effect. The second consideration is the potential relation of Te···X distances vs the tellurium coordination. Eventhough this dependence should be expected, no differences were found. A greater number of structures may help to establish this dependence. In this context, a study carried out on tellurium supramolecular synthons established that no correlation exists between secondary bond distance and the coordination number of tellurium (Cozzolino et al., 2011).

The existence of intermolecular Te···X interactions is frequent in crystal structures of organotellurium compounds containing halogen atoms. Using the values showed in Table 1, the estimation of structures with Te···X interactions is about 60%. Of the total number of structures showing Te···X contacts, 13% of them exhibit Te···F interaction, 38%, Te···Cl, 20%, Te··· Br and 29%, Te···I.

	d(σ)	q_{10}, q_{90}	n
Te···F	3.26(11)	2.96, 3.80	10
Te-F	1.989(4)	1.95, 2.02	43
Σr_{vdW}	3.53		
Te···Cl	3.672(15)	3.42, 3.97	191
Te-Cl	2.521 (3)	2.47, 2.61	546
Σr_{vdW}	3.81		
Te···Br	3.88 (2)	3.56, 4.25	121
Te-Br	2.693(5)	2.63, 2.79	234
Σr_{vdW}	3.91		
Te···I	3.95 (2)	3.67, 4.37	203
Te-I	2.944(4)	2.87, 3.04	383
Σr_{vdW}	4.04		

Table 1. Te···X secondary and Te-X covalent bond distances (Å) in organotellurium crystal structures. The sum of van der Waals radii (Bondi, 1964), Σr_{vdW}, is also included. d: mean value; σ: mean standard deviation; q_{10}, q_{90}: 10% and 90% percentiles; n: number of observations; covalent bond distances have been obtained in this work from organotellurium(IV) crystal structures from CSD

3. Supramolecular arrangements in organotellurium compounds

3.1 Method

A CSD search considering organic structures with deposited atomic coordinates, with R factor < 0.1, and without disorder involving Te and X was undertaken. After removing multiple entries of the same structure, a new search was carried out in order to retrieve structures showing intermolecular Te···X contacts. According to the results presented in section 2, a cut-off of Σr_{vdW} + 10% Å was considered. In these structures, supramolecular patterns were investigated using the program Mercury (Version 2.4) (Macrae et al., 2008). A set of 616 organotellurium structures having halogen atoms was collected. The greater part of them, 490 (80%), contain the Te-X unit, *i.e.* at least one tellurium-halogen covalent bond. Four groups can be considered: (i) structures containing tellurium atoms with only one halogen atom bonded to it, *i.e.* having the C-Te(-X)$_1$ unit, (ii) structures containing tellurium atoms bonded to only two halogen atoms, *i.e.* having the C-Te(-X)$_2$ unit, (iii) structures containing tellurium atoms with only three halogen atoms bonded to it, *i.e.* having the C-Te(-X)$_3$ unit, and (iv) structures containing tellurium atoms bonded to four halogen atoms, *i.e.* having the C-Te(-X)$_4$ unit. A fifth group (v) includes the remaining 126 structures of organotellurium compounds having halogen atoms but not Te-X units: structures with C-Te(-X)$_0$.

3.2 Structures of compounds containing the Te(-X)$_1$ unit

Te(-X)$_1$ unit (tellurium atom bonded to one halogen only), is a very simple unit and it is a good starting point to study supramolecular arrangements *via* Te···X contacts. Two arrangements are the more habitual in this group: (a) dimeric assembly, and (b) simple chain (Figure 2, Table 2).

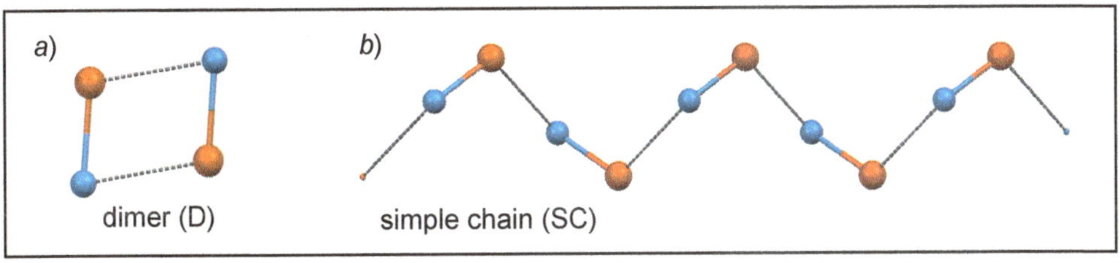

Fig. 2. Main supramolecular arrangements of compounds containing the Te(-X)$_1$ unit

	F	Cl	Br	I	Total
Dimer	0	10	12	9	31
Simple chain	2	6	3	3	14
Total	2	16	15	12	45

Table 2. Summary of structures containing the Te(-X)$_1$ unit

a. In the dimer, the two Te-X rods are bonded by two Te···X secondary bonds. The majority is centrosymmetric, with X···Te-X angles around 90°.
   In C-Te(II)-X (X = Cl, Br, I) compounds, dimers were observed These compounds are not stable and the secondary bond stabilizes them. A new pattern is observed when weaker secondary bonds are considered (contact distances < Σ r_{vdW} + 20%): a chain of dimers like a zigzag ladder (Figure 3) where two neighbour dimers are related by a symmetry centre. In this way, three TeX distances are present, being the Te-X rod length the shortest one. A great dispersion of distances was observed, not only in secondary bonds but in the "primary" bond as well.
   Dimeric arrays were also observed in some molecules containing several Te-X units. In these cases the covalent skeleton increases the dimensionality of the whole arrangement. In this way, if two Te-X units are present, the structure contains chains, if there are four Te-X units by molecule, a sheet of dimers is formed (Figure 4).
b. In the simple chain, every rod is bonded to its neighbour using only one secondary bond. In some cases the chain is planar and neighbour rods are equivalent by translation. In the other cases, chains are generated by a screw binary axis or by a glide plane.

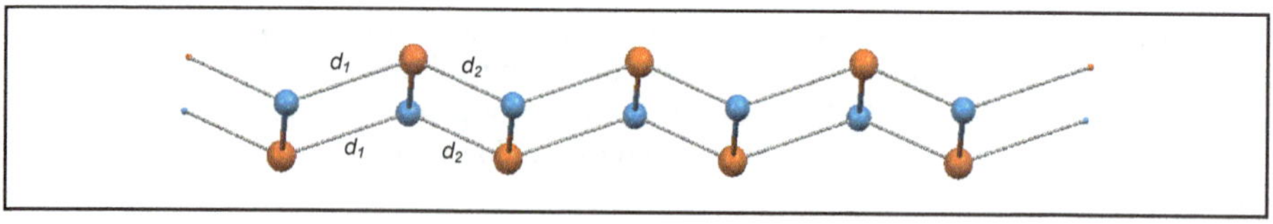

Fig. 3. Chain of dimers like a zigzag ladder ($d_1 > d_2$)

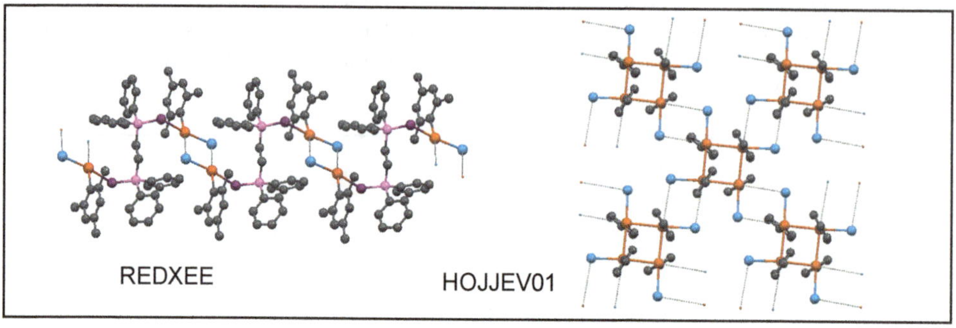

Fig. 4. Examples of chain and sheet resulting from the combination of dimers and a covalent skeleton

3.3 Structures of compounds containing the Te(-X)$_2$ unit

The most populated group of organotellurium compounds contains the C-Te(-X)$_2$ unit (only two halogen atoms bonded to tellurium). A lot of them have a C_2TeX_2 core showing a pseudotrigonal bipyramidal disposition with the halogen atoms in axial positions and a lone electron pair in an equatorial site (Figure 5).

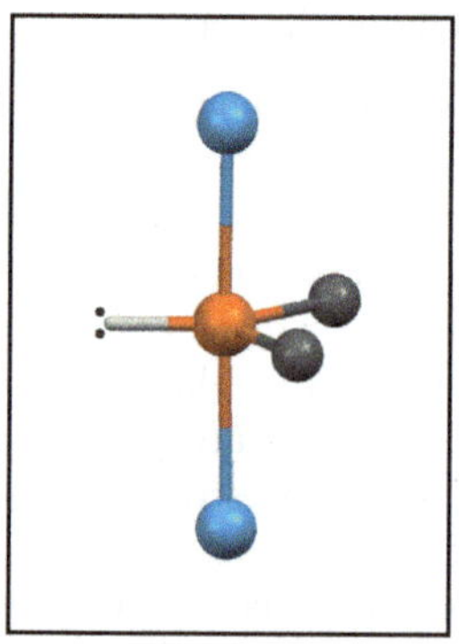

Fig. 5. Pseudotrigonal bipyramidal coordination frequent in R_2TeX_2 compounds

A great diversity of arrangements is present in this group. Three of such arrangements are clearly the most usual: (a) dimer, (b) simple chain, and (c) chain of four-sided rings sharing opposite vertices. Other three arrangements, with more complicated patterns of secondary bonds, have also significant populations: (d) tetramer, (e) sheet, and (f) 3D-network (Figure 6, Table 3).

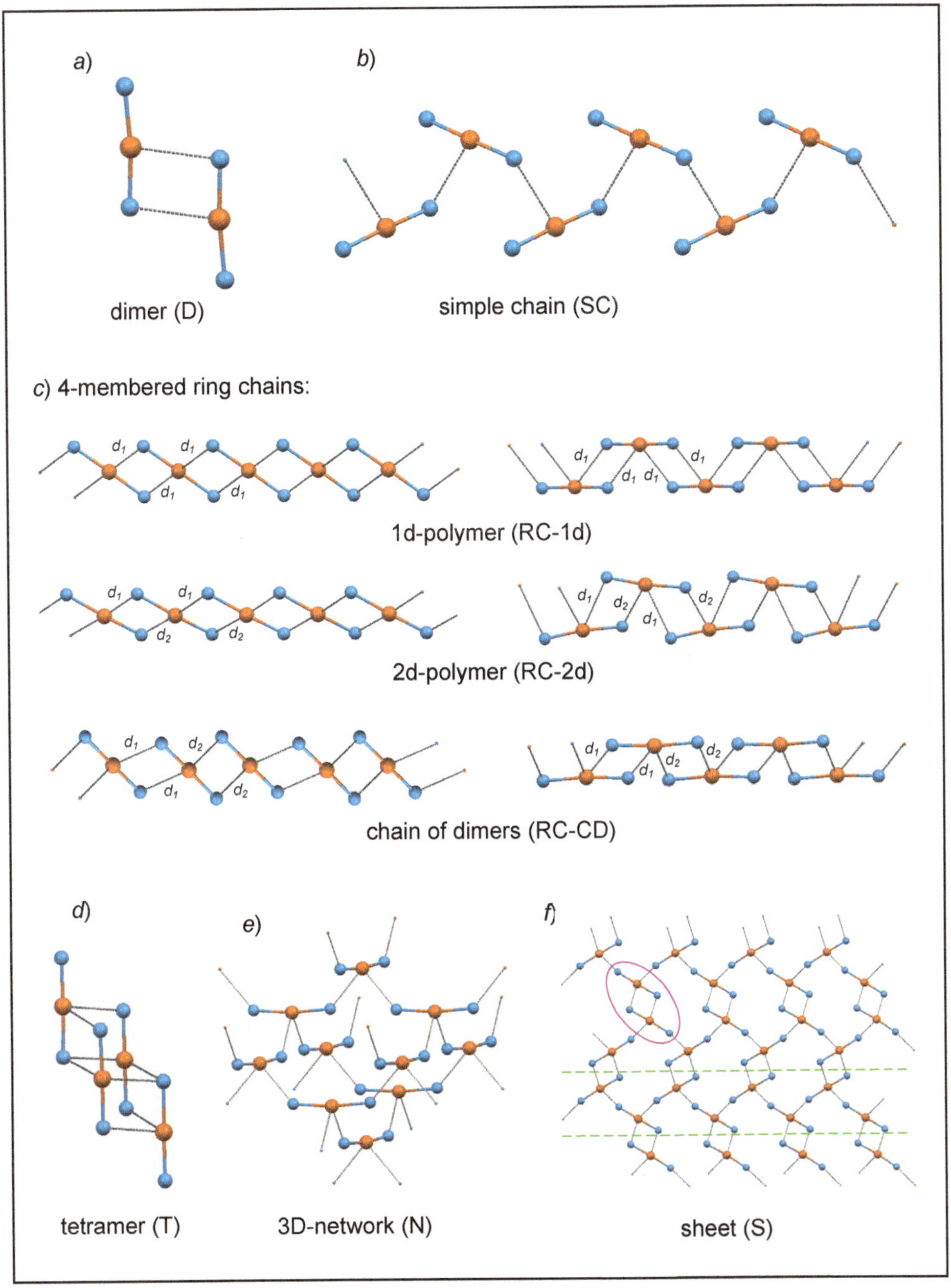

Fig. 6. Main supramolecular arrangements of compounds containing the Te(-X)$_2$ unit

	F	Cl	Br	I	Total
Dimer	2	18	7	18	45
Simple chain	0	17	10	4	31
Chain of 4-membered rings	2	11	5	6	24
Tetramer	0	5	1	3	9
Sheet	0	2	0	2	4
3D-Network	0	1	2	2	5
Less frequent arrangements	2	1	2	3	8
Total	6	55	27	38	126

Table 3. Summary of structures containing the $Te(-X)_2$ unit

a. Dimers formed by $Te(-X)_2$ have the same connectivity as those described for $Te(-X)_1$, *i.e.* two Te···X secondary bonds, and the majority are also centrosymmetric. Non centrosymmetric dimers are almost planar in absence of other interactions.
Most X-Te···X angles are lesser than 90°, *i.e.* two rods are "moved away" relative to the rectangular disposition. This is a small difference respect to the $Te(-X)_1$ dimers, where deviation from 90° were present in both directions. Only three dimers of angular $Te(-X)_2$ were found, centrosymmetric all of them.

b. The $Te(-X)_2$ simple chain is also referable to $Te(-X)_1$ simple chains but here, planar chains are absent, the reason being that planar chains have the translation as unique symmetry element. In the case of $Te(-X)_2$, this would imply the presence of an additional Te···X contact leading to a different kind of chain. So, rods in $Te(-X)_2$ simple chains are equivalent by screw binary axes or by glide planes.

c. The third most common arrangement is a di-bridged chain made by 4-membered rings sharing opposite vertices (Te atoms). Three types of chain (polymer) can be considered when Te···X distances are analyzed. So, if 4-membered rings are not equal (6 structures), a dimer is present and this type of polymer will be named chain of dimers. If all 4-membered rings are equal, the basic unit in the polymer is an X-Te-X rod and two cases are possible. In the more symmetric one (7 structures), the two secondary bonds between neighbour rods are equal while in the less symmetric one (11 structures) are not. These two types of polymer will be named 1d- and 2d-polymer respectively.
The three types of chain have also different internal symmetry. In the chains of dimers, 4-membered rings are centrosymmetric (rhomboids) and are disposed in zigzag. In six of the more symmetric polymers, there are symmetry centres in the middle of all rings, and moreover, binary axes, perpendicular to the chain direction, through the Te atoms, and glide planes. One symmetric polymer is helical, generated by a screw 4-fold axis, vs the zigzag disposition found in the other 6 structures. At last, polymers with two different secondary bonds between neighbours are generated by glide planes and adopt a zigzag conformation.

d. Another finite arrangement of $Te(-X)_2$ is a cyclic tetramer, where a $Te(-X)_2$ unit is bonded to every neighbour with two secondary bonds in a step-like manner. The tetramer has two different types of tellurium atoms: the two "middle" Te atoms defining a Te_2X_2 ring are different from the two "terminal" Te atoms placed out of the Te_2X_2 ring.

This arrangement implies the existence of two non-equivalent rods, thus forcing the structure to have more than one molecule in the asymmetric unit ($Z' > 1$). All the tetramers found are centrosymmetric.

e. Five structures with infrequent spatial groups (*Fdd*2 and $I4_1/acd$) adopt a symmetric polar 3D-network where all secondary bonds are equal. Every Te(-X)$_2$ rod is bonded to four neighbours, with two bonds from Te and one from every halogen.

f. A sheet with two secondary bond distances and where every rod is bonded to three neighbours is relatively usual. This arrangement can be described in two ways, depending on the distance ratio. In two structures, the best description is to consider them as centrosymmetric dimers (type (a)) bonded to four neighbours, involving the two Te and the two distant halogens of the dimer. For another two structures it is better to think in simple chains (type (b)) where every rod is bonded to another rod of a neighbouring chain by means of two Te···X bonds. Anyway, 4- and 12-membered rings are present.

g. Finally, four pairs of structures were found, each pair with its own arrangement: two chains and two sheets (Figure 7).

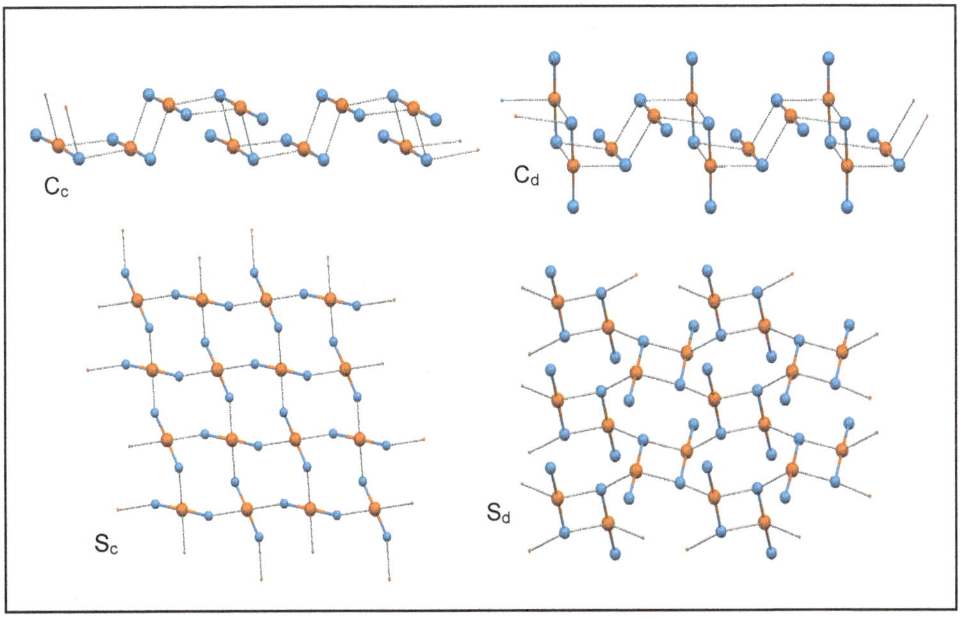

Fig. 7. Less frequent supramolecular arrangements of compounds containing the Te(-X)$_2$ unit: C_c and C_d chains; S_c and S_d sheets

3.4 Structures of compounds containing the Te(-X)$_3$ unit

This unit shows a T disposition derived of the tendency of halogen atoms to occupy axial positions (angle X-Te-X near to 180°). The dimeric arrangement (a) is adopted by most structures. The simple chain (b) and the chain of 4-membered rings (c) are present but are much less usual (Figure 8, Table 4).

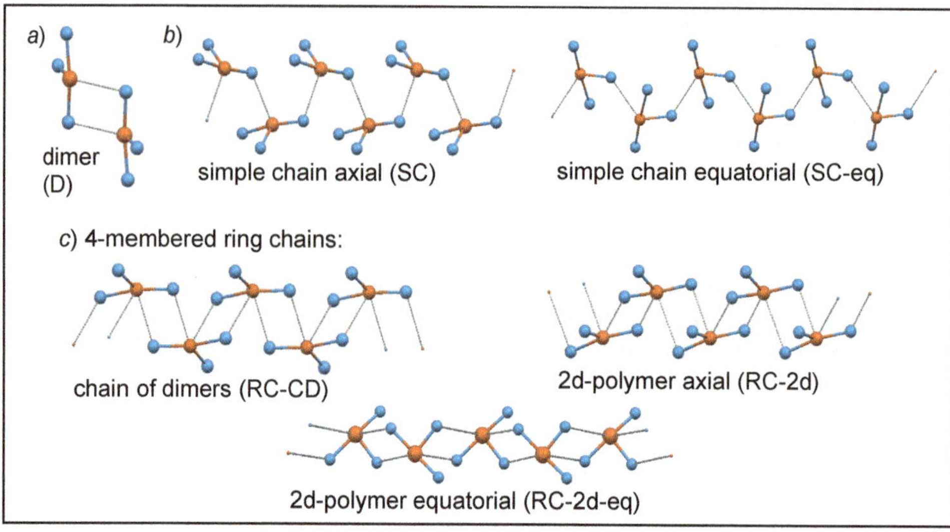

Fig. 8. Main supramolecular arrangements of compounds containing the Te(-X)$_3$ unit: dimer (D); simple chain involving only axial halogen atoms (SC) or equatorial halogen atoms (SC-eq); 4-membered ring chain: chain of dimers (RC-CD), 2d-polymer involving only axial halogen atoms (RC-2d), and 2d-polymer involving equatorial (RC-2d-eq) halogen atoms

	F	Cl	Br	I	Total
Dimer	0	10	4	3	17
Simple chain	0	3	1	1	5
Chain of 4-membered rings	1	1	2	1	5
Total	1	14	7	5	27

Table 4. Summary of structures containing the Te(-X)$_3$ unit

a. Dimer. As in Te(-X)$_1$ and Te(-X)$_2$ dimers, two units are linked by two secondary bonds. Axial halogen atoms are involved in them. Dimers are centrosymmetric and the equatorial halogen atom can be in different orientations. In most cases, the angle between the Te-X$_{eq}$ bond and the normal to planar core is less than 20° but in three cases this angle is significantly higher, 37-53°.
b. Simple chain. As in Te(-X)$_1$ and Te(-X)$_2$ simple chains, two units Te(-X)$_3$ are linked by one Te···X secondary bond. The link involves either axial (3 structures) or equatorial halogen (2 structures). As in Te(-X)$_2$ chains, units in Te(-X)$_3$ simple chains are equivalent by screw axes or by glide planes. It is remarkable that in one case, the screw axis is ternary and the chain turns to be helical, vs the zigzag disposition of the chains where binary screw axes are present.
c. In chains of 4-membered rings two cases have been found. In one of them, rings are defined by one axial halogen and the equatorial one. Chains are polymers with two Te···X distances (screw binary axis). In the other case, analogous to the Te(-X)$_2$ 4-membered ring chains, rings are defined only by axial halogen atoms and two

dispositions can be adopted: chain of dimers (centrosymmetric rings) and polymer with two Te···X distances (glide planes).

3.5 Structures of compounds containing the Te(-X)$_4$ unit

These compounds are ionic (or zwitterionic) and the coordination around the Te can be described as a tetragonal pyramid with the halogens and the tellurium in the basal plane and the carbon in the apical position of the pyramid. Two supramolecular patterns are relevant: (a) dimers and (b) simple chains (Figure 9, Table 5).

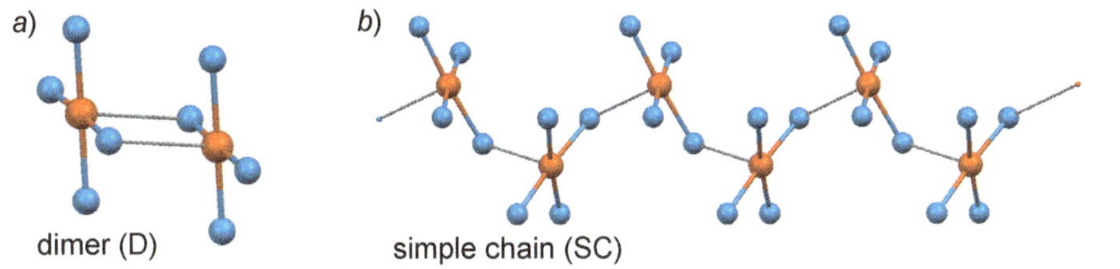

Fig. 9. Main supramolecular arrangements of compounds containing the Te(-X)$_4$ unit

	F	Cl	Br	I	Total
Dimer	0	6	4	6	16
Simple chain	0	2	3	2	7
Total	0	8	7	8	23

Table 5. Summary of structures containing the Te(-X)$_4$ unit

a. Dimers. As in Te(-X)$_1$, Te(-X)$_2$ and Te(-X)$_3$, dimers contain two secondary bonds. These bonds complete an octahedral coordination for every Te. All dimers are centrosymmetric except one with a binary axis instead.
b. Simple chains. Again, the secondary bond completes the octahedral coordination of tellurium atom. As in Te(-X)$_2$ and Te(-X)$_3$ chains, units in Te(-X)$_4$ simple chains are equivalent by screw axes or by glide planes.

3.6 Structures of compounds without Te-X units [Te(-X)$_0$]

This group is very heterogeneous from a chemical point of view, nevertheless some subsets can be established (Figure 10). A first subset (12 structures) was considered where neutral molecules contain one Te and halogen atoms.

This subset can be considered a general case of the compounds studied in the previous sections: now the tellurium atom and the halogens are separated by more than one covalent bond, hereinafter Te(---X)$_n$. Therefore some above described arrangements were also found here: dimer (4 structures), simple chain (2 structures), and chain of dimers (2 structures).

In a second subset of 11 neutral ditellurides, known patterns were also found in more than half the cases, considering them as X---Te-Te---X. Every Te---X unit affords its own arrangement of secondary bonds: dimers (3 structures), and simple chains (3 structures).

When Te-Te bridges are considered, arrangements of higher dimensionality can exist. For example, a Te-Te unit can be bridging dimers to afford a chain of dimers.
Analogously, the combination of chains of secondary bonds and Te-Te bridges can give three more complex arrangements: (i) double chains, (ii) sheets composed by parallel chains, and (iii) double sheets linked by perpendicular Te-Te bridges in a grid-like array.
In other neutral molecules having two tellurium atoms separated by other atoms, single chains of secondary bonds linked by the Te---Te core were observed.
The rest of structures (50) are ionic. In 19 of them, tellurium atoms are in the cationic moiety, halogen atoms are in the anion, and one secondary bond links both ions to give an 1:1 adduct. More complex adducts exist when more than two ions or solvent molecules are bonded by Te···X bonds.

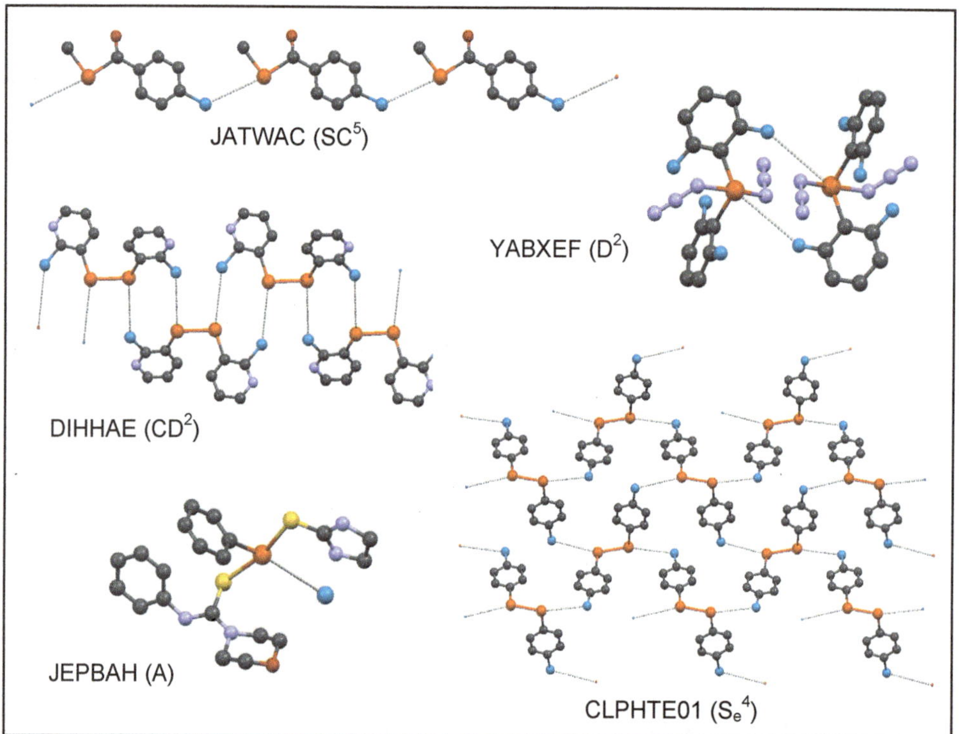

Fig. 10. Examples of supramolecular arrangements in the Te(-X)$_0$ group: dimer (D); simple chain (SC); chain of dimers (CD); specific sheet (S$_e$); adduct (A). Superscripts stand for the number of atoms between Te and X atoms

3.7 Additional structural considerations

3.7.1 Polymorphism

A search for polymorphism was performed in order to study its relation with the arrangement. Eleven organotellurium compounds containing halogen atoms showed two polymorphic structures each one: two in the Te(-X)$_1$ group, seven in the Te(-X)$_2$ group, one in the Te(-X)$_3$ group, and one in the Te(-X)$_0$ group (Figure 11, Table 6). In ten cases, different arrangements were observed for the polymorphic pair while in one case polymorphism is associated to a different packing of the same supramolecular entity (tetramers). Moreover, it is usual to observe a change of dimensionality between polymorphs.

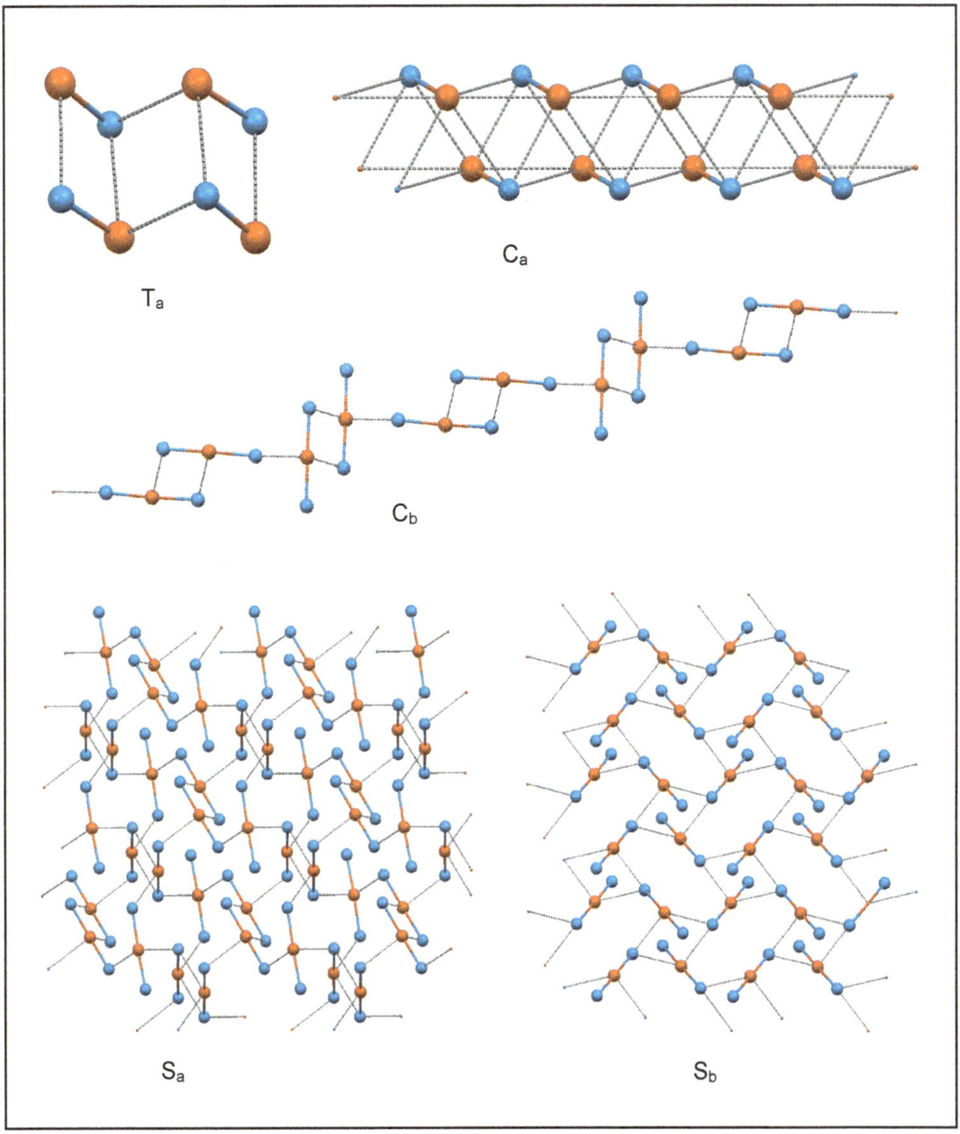

Fig. 11. Some specific arrangements present in polymorphic structures: T$_a$ tetramer; C$_a$ and C$_b$ chains; S$_a$ and S$_b$ sheets

	Refcode	S. G.	Arrang.	D_c (g cm^{-3})
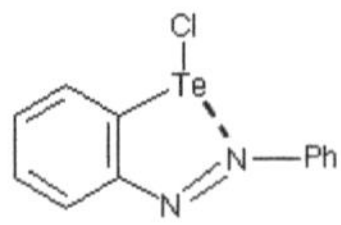	PAZPTE PAZPTE01	P-1 $I2/a$	T_a C_a	1.942 1.966
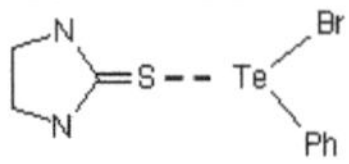	BTUPTE BTUPTE01	$C2/c$ $P2_1/c$	0 D	2.106 2.017
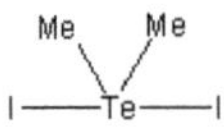	DIDMTE DIDMTE01	$P2_1/c$ $C2/c$	S_a RC	3.393 3.410
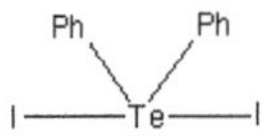	CIFLEI CIFLEI01	$I4_1$ $P2_1/c$	N C_b	2.556 2.534
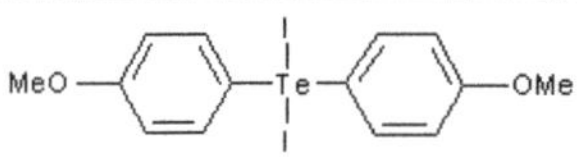	NUNHUZ NUNHUZ01	P-1 P-1	T T	2.256 2.287
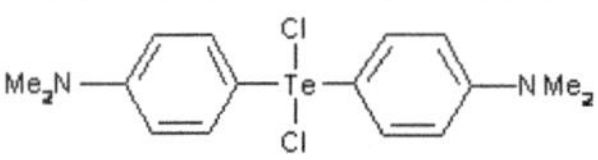	SABCII SABCII01	Pn $Ibca$	0 D	1.669 1.666
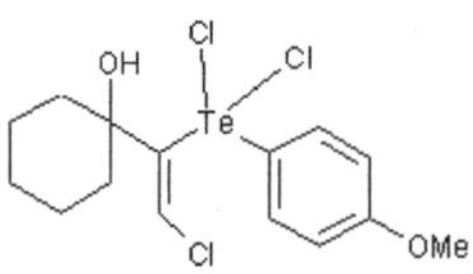	ASEHUB ASEHUB01	$P2_1/c$ $P2_1/a$	0 SC	1.759 1.737
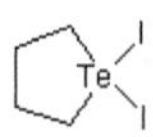	QIXZAY QIXZAY03	$P2_1/n$ $P2_1/c$	C_c C_d	3.138 3.263
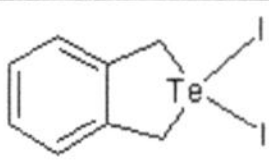	DIBTEP02 DIBTEP10	$Fdd2$ $P2_1/c$	N S_b	2.843 2.887
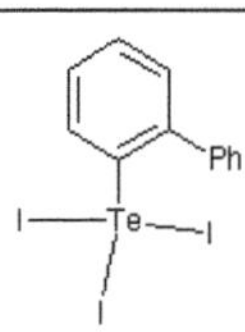	BIPTEI BIPTEI01	$P2_1/n$ $P2_1/c$	SC D	2.861 2.777
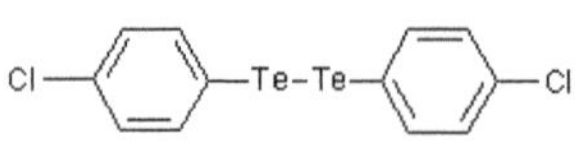	CLPHTE CLPHTE01	$P2_1/n$ $P2_12_12_1$	N_a S_e^4	2.316 2.365

Table 6. Supramolecular arrangements found in polymorphic structures. 0 = no Te···X contacts; D = dimer; SC = simple chain; RC = chain of 4-membered rings; T = tetramer; N = 3D-network; C_c, C_d: specific chains (Fig. 7); S_e^4: specific chain (Fig. 10); C_a, C_b, T_a, S_a, S_b: specific chains, tetramer, sheets (Fig. 11); N_a specific 3D-network

3.7.2 Halogen series

The set of organotellurium structures containing halogen atoms includes three complete series of halo compounds with the same molecular structure but different halogen, X = F, Cl, Br, I (Table 7). In the three series, bromo and iodo members are isostructural. In the second series, the chloro compound is also isostructural. When isostructurality is not present, arrangements are different.

Series	F	Cl	Br	I	
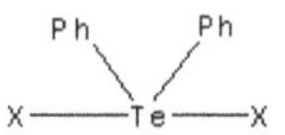	KEHBIJ S_c	BIBBOE S	BIBCEV C_c	QIXZAY C_c	QIXZAY03 C_d
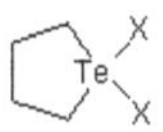	GETHOC S_c	BIBCAR N	BIBBUK N	DIBTEP02 N	DIBTEP10 S_b
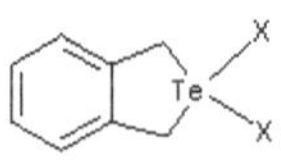	FPHTEL RC	OPNTED10 SC	DPHTEB01 N	CIFLEI N	CIFLEI01 C_b

Table 7. Supramolecular arrangements in complete halogen series. SC = simple chain; RC = chain of 4-membered rings; S = sheet; N = 3D-network; C_c, C_d, S_c: specific chains and sheet (Fig. 7); C_b, S_b: specific chain and sheet (Fig. 11)

Partial series of three and two members are also present in the set of compounds studied, and its abundance is significant: 20 and 47 series respectively. In these partial series, isostructurality is also frequent but examples of non-isostructural members with the same kind of supramolecular arrangement are present (Table 8).

Series	F	Cl	Br	I
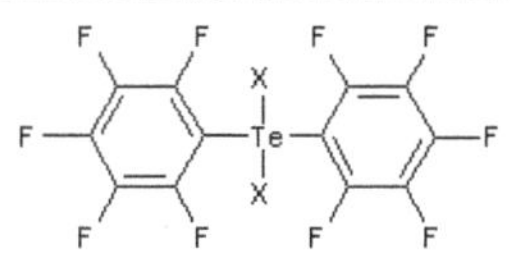		YASVAP RC	YASVET RC	
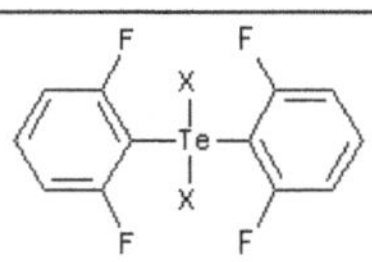	IDEXIY RC	IDEXOZ RC		
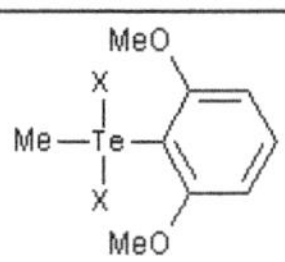			TUWPAC RC	TUWPIK RC

Table 8. Supramolecular arrangements in partial halogen series. RC = chain of 4-membered rings

3.8 Selection of structures

The following list collects relevant structures of this section (Table 9).

Refcode	Journal	Vol.	Page	Year
Structures of compounds containing the Te(-X)$_1$ unit				
a) Dimers				
BSEUTE	*Acta Chem.Scand.A*	29	763	1975
BTUPTE01	*Acta Chem.Scand.A*	29	738	1975
BZETBS	*Acta Chem.Scand.*	20	132	1966
BZETCS	*Acta Chem.Scand.*	20	132	1966
CEHFAW	*Organometallics*	18	803	1999
CETUTE	*Acta Chem.Scand.A*	29	763	1975
DONQEC	*Inorg.Chem.*	38	3994	1999
DONQIG	*Inorg.Chem.*	38	3994	1999
DONQOM	*Inorg.Chem.*	38	3994	1999
ERIBIQ	*Organometallics*	22	5473	2003
GEDSEO	*Z.Kristallogr.*	221	166	2006
HIFVOH	*Acta Crystallogr.,Sect.C*	52	424	1996
OCICIN	*Chem.-Eur.J.*	12	2515	2006
OLODUO	*Acta Crystallogr.,Sect.C*	59	o571	2003
OLOFAW	*Acta Crystallogr.,Sect.C*	59	o571	2003
PETBIA	*Z.Anorg.Allg.Chem.*	633	127	2007
PETBOG	*Z.Anorg.Allg.Chem.*	633	127	2007
POYGEP	*J.Organomet.Chem.*	549	257	1997
POYGIT	*J.Organomet.Chem.*	549	257	1997
QANWEH	*J.Organomet.Chem.*	605	39	2000
QIYBAB	*J.Organomet.Chem.*	623	74	2001
QOKZIZ	*Polyhedron*	20	203	2001
ROCTAE	*Inorg.Chem.*	36	1890	1997
TEGZIP	*Acta Crystallogr.,Sect.C*	52	2022	1996
WEXBIK	*Khim.Get.Soedin.,SSSR*		417	1994
WOMDIL	*Phosphorus,Sulfur,Silicon*	123	313	1997
YEXWIH	*Inorg.Chem.*	33	6154	1994
YEXWON	*Inorg.Chem.*	33	6154	1994
YEXWUT	*Inorg.Chem.*	33	6154	1994
Dimers connected by covalent skeleton				
HOJJEV01	*Angew.Chem.,Int.Ed.*	39	1796	2000
REDXEE	*Eur.J.Inorg.Chem.*		1294	2006
b) Simple chains				
BRTEBA	*Acta Crystallogr.,Sect.B*	35	849	1979
CEHKEF	*Organometallics*	18	803	1999
CIDKOQ	*Inorg.Chem.*	46	3275	2007
JABWUE	*J.Am.Chem.Soc.*	110	6762	1988

JEKDAE	*Can.J.Chem.*	67	1735	1989
KEZMEI	*Acta Crystallogr.,Sect.E*	63	o1674	2007
MIYPAL	*Z.Anorg.Allg.Chem.*	628	833	2002
OFUWER	*J.Fluorine Chem.*	112	207	2001
RIFTAB	*Can.J.Chem.*	74	1968	1996
RIGKIB	*Tetrahedron:Asymm.*	7	2797	1996
ROCTEI	*Inorg.Chem.*	36	1890	1997
TATZAP	*Inorg.Chem.*	35	2831	1996
ZISWON	*Organometallics*	14	4755	1995

Structures of compounds containing the Te(-X)$_2$ unit

a) Dimers

CIDKEG	*Inorg.Chem.*	46	3275	2007
CLPTEI	*Acta Crystallogr.*	15	887	1962
DMTEII	*Inorg.Chem.*	18	311	1979
GACXUD01	*Acta Crystallogr.,Sect.C*	47	703	1991
GIPRED	*Angew.Chem.,Int.Ed.*	46	8277	2007
GOKDUG	*Z.Kristallogr.*	223	536	2008
HEXBOB	*Can.J.Chem.*	72	1844	1994
HOFKES	*Acta Crystallogr.,Sect.C*	55	1339	1999
JATSIG	*J.Chem.Soc.,Dalton Trans.*		1697	1989
KOBYAB	*Acta Crystallogr.,Sect.C*	47	1858	1991
LAFNUB	*Z.Naturforsch.,B:Chem.Sci.*	48	199	1993
OCICUZ	*Chem.-Eur.J.*	12	2515	2006
OFUTOZ	*Dalton Trans.*		4023	2008
PEKLUM	*J.Crystallogr.Spectrosc.Res.*	23	181	1993
SABCII01	*Acta Crystallogr.,Sect.E*	61	o986	2005
SEPMAB	*J.Chem.Soc.,Dalton Trans.*		907	1990
TUWNOO	*Dalton Trans.*		973	2003
TUWNUU	*Dalton Trans.*		973	2003
TUWPEG	*Dalton Trans.*		973	2003
UJAMIC	*J.Organomet.Chem.*	695	2532	2010
UJAMUO	*J.Organomet.Chem.*	695	2532	2010
UJORUH	*J.Organomet.Chem.*	695	2118	2010
UKAFAO	*Inorg.Chem.*	49	7577	2010
UKAFES	*Inorg.Chem.*	49	7577	2010
UKAFIW	*Inorg.Chem.*	49	7577	2010
UKAFOC	*Inorg.Chem.*	49	7577	2010
UKAGET	*Inorg.Chem.*	49	7577	2010
UKAGIX	*Inorg.Chem.*	49	7577	2010
VOHVET	*J.Organomet.Chem.*	418	339	1991
WECQIF	*Z.Anorg.Allg.Chem.*	632	659	2006
WECQOL	*Z.Anorg.Allg.Chem.*	632	659	2006
WIDZOZ	*Z.Anorg.Allg.Chem.*	633	1261	2007

WIDZUF	*Z.Anorg.Allg.Chem.*	633	1261	2007
XAWNUF	*Z.Anorg.Allg.Chem.*	631	2677	2005
XEXJAM	*J.Organomet.Chem.*	691	5867	2006
XEXKUH	*J.Organomet.Chem.*	691	5887	2006
XOTBIS	*J.Organomet.Chem.*	694	1557	2009
YEZXEH	*Organometallics*	26	1955	2007
ZARCIE	*Can.J.Chem.*	73	255	1995
b) Simple chains				
ASEHUB01	*Z.Kristallogr.*	218	636	2003
BOQJIA	*Acta Crystallogr.,Sect.C*	55	1930	1999
CEDDAQ	*J.Org.Chem.*	48	5149	1983
CEJPAJ	*Heteroat.Chem.*	16	316	2005
CLCHTE	*Cryst.Struct.Commun.*	9	533	1980
GOKFAO	*Z.Kristallogr.*	223	536	2008
IDEXEU	*Inorg.Chem.*	40	5169	2001
JILXAD	*Acta Crystallogr.,Sect.C*	47	960	1991
KAKCOP	*Aust.J.Chem.*	58	119	2005
KEQCIT	*Inorg.Chim.Acta*	359	4619	2006
MIDWAY	*Struct.Chem.*	18	181	2007
NIJQOM	*Polyhedron*	16	2441	1997
OFUVER	*Dalton Trans.*		4023	2008
OFUVIV	*Dalton Trans.*		4023	2008
OPNTED10	*J.Chem.Soc.,Dalton Trans.*		251	1982
QOGPAD	*Acta Crystallogr.,Sect.C*	57	749	2001
SEJPIG	*Acta Crystallogr.,Sect.C*	46	251	1990
STECHB	*Inorg.Chem.*	11	3026	1972
TLCLTE	*Cryst.Struct.Commun.*	9	539	1980
UJOSAO	*J.Organomet.Chem.*	695	2118	2010
UJOSOC	*J.Organomet.Chem.*	695	2118	2010
VUZJIJ	*Acta Crystallogr.,Sect.C*	44	2182	1988
XUCNIS	*Chin.J.Chem.*	19	457	2001
XUCNOY	*Chin.J.Chem.*	19	457	2001
XURRAE	*Dalton Trans.*	39	2637	2010
XURSEJ	*Dalton Trans.*	39	2637	2010
YOWMEC	*Acta Crystallogr.,Sect.C*	51	861	1995
YUFNUI	*Z.Kristallogr.*	210	306	1995
c) Chains of 4-membered rings				
BZTELI10	*Inorg.Chem.*	14	1142	1975
CEGNOR	*J.Chem.Soc.,Dalton Trans.*		23	1984
HIQSUV	*Acta Crystallogr.,Sect.C*	55	648	1999
IDEXIY	*Inorg.Chem.*	40	5169	2001
PTELDI10	*Inorg.Chem.*	12	2669	1973
YASVAP	*Z.Anorg.Allg.Chem.*	619	1269	1993
YASVET	*Z.Anorg.Allg.Chem.*	619	1269	1993

IDEXOE	*Inorg.Chem.*	40	5169	2001
IDEXUK	*Inorg.Chem.*	40	5169	2001
LEVKED	*J.Am.Chem.Soc.*	128	14949	2006
MIHFIS	*J.Fluorine Chem.*	102	301	2000
TUWNII	*Dalton Trans.*		973	2003
TUWPAC	*Dalton Trans.*		973	2003
TUWPIK	*Dalton Trans.*		973	2003
UKAFUI	*Inorg.Chem.*	49	7577	2010
UKAGAP	*Inorg.Chem.*	49	7577	2010
VOTSIG	*Organometallics*	11	954	1992
XUPXAI	*Heterocycles*	80	1339	2010
CERWIF	*J.Organomet.Chem.*	262	151	1984
COBPOY	*J.Organomet.Chem.*	268	141	1984
DIDMTE01	*Dalton Trans.*		6274	2008
FPHTEL	*J.Chem.Soc.,Dalton Trans.*		2306	1980
MIHFEO	*J.Fluorine Chem.*	102	301	2000
YUPHIB	*Z.Anorg.Allg.Chem.*	635	862	2009
d) Tetramers				
BOHNAN	*Acta Crystallogr.,Sect.C*	39	45	1983
BUNDIY	*Inorg.Chem.*	48	10330	2009
CIMFOT	*Acta Crystallogr.,Sect.C*	40	1349	1984
IHOFIU	*J.Organomet.Chem.*	669	149	2003
LIMZEM	*Acta Crystallogr.,Sect.C*	51	639	1995
NUNHUZ	*Acta Crystallogr.,Sect.C*	54	995	1998
NUNHUZ01	*Acta Crystallogr.,Sect.C*	54	995	1998
PHTECL	*Inorg.Chem.*	19	2556	1980
PIXPES	*J.Organomet.Chem.*	693	957	2008
e) Sheets				
BIBBOE	*Polyhedron*	23	1629	2004
LUTBUX	*Can.J.Chem.*	80	1530	2002
OTEHEX	*Inorg.Chem.*	12	2665	1973
VIVGEM	*J.Organomet.Chem.*	410	309	1991
f) 3D-Networks				
BIBBUK	*Polyhedron*	23	1629	2004
BIBCAR	*Polyhedron*	23	1629	2004
CIFLEI	*J.Chem.Soc.,Dalton Trans.*		869	1984
DIBTEP02	*Inorg.Chem.*	24	1814	1985
DPHTEB01	*Acta Crystallogr.,Sect.E*	60	o2511	2004
g) Less frequent arrangements				
BIBCEV	*Polyhedron*	23	1629	2004
QIXZAY	*J.Organomet.Chem.*	623	74	2001
QIXZAY03	*Inorg.Chem.*	43	3742	2004
SIDGIW	*Struct.Chem.*	18	223	2007
COQNOL01	*Inorg.Chem.*	38	3994	1999

DMTECL01	*J.Am.Chem.Soc.*	105	229	1983
GETHOC	*J.Organomet.Chem.*	349	95	1988
KEHBIJ	*Z.Naturforsch.,B:Chem.Sci.*	61	528	2006

Structures of compounds containing the Te(-X)$_3$ unit

a) Dimers

AZOFAW	*J.Organomet.Chem.*	689	2092	2004
BIPTEI01	*Inorg.Chem.*	16	2318	1977
BPYTEB	*Inorg.Chem.*	16	612	1977
CIDKAC	*Inorg.Chem.*	46	3275	2007
CUBFIP	*Inorg.Chem.*	48	3239	2009
DEFHAX	*Acta Crystallogr.,Sect.C*	41	1545	1985
EOCTEC	*J.Organomet.Chem.*	181	335	1979
GEHDIH	*Z.Anorg.Allg.Chem.*	632	893	2006
GEHDON	*Z.Anorg.Allg.Chem.*	632	893	2006
GIJZEE	*J.Chem.Soc.,Dalton Trans.*		2363	1988
GOPLEC	*Polyhedron*	17	2153	1998
IMOZUF	*Organometallics*	22	5069	2003
SEPLUU	*J.Chem.Soc.,Dalton Trans.*		907	1990
WEQVAP	*J.Organomet.Chem.*	598	49	2000
XOPROK	*Angew.Chem.,Int.Ed.*	47	9982	2008
ZATROB	*Z.Kristallogr.*	210	552	1995

b) Simple chains

BONHUH	*Cryst.Struct.Commun.*	11	1593	1982
DOTLUT	*J.Org.Chem.*	51	1692	1986
PAFLIR	*Inorg.Chem.*	31	1431	1992
BIPTEI	*Inorg.Chem.*	15	2728	1976
UDUHUW	*Organometallics*	21	526	2002

c) Chains of 4-membered rings

WEQVET	*J.Organomet.Chem.*	598	49	2000
MUWBIP	*J.Organomet.Chem.*	664	306	2002
OLODOI	*Acta Crystallogr.,Sect.C*	59	o571	2003
WERWUM	*J.Organomet.Chem.*	691	4807	2006
YIHRAJ	*Z.Anorg.Allg.Chem.*	633	1618	2007

Structures of compounds containing the Te(-X)$_4$ unit

a) Dimers

CIFVOD	*J.Organomet.Chem.*	692	3081	2007
CIFVUJ	*J.Organomet.Chem.*	692	3081	2007
CIFWAQ	*J.Organomet.Chem.*	692	3081	2007
CITMAT	*Tetrahedron*	40	1607	1984
CUCGOW	*Can.J.Chem.*	62	32	1984
DEBXOY	*Appl.Organomet.Chem.*	19	1196	2005
DEBXUE	*Appl.Organomet.Chem.*	19	1196	2005

ELAVAP	*Inorg.Chim.Acta*	365	492	2011
IHIMAO	*J.Organomet.Chem.*	694	2463	2009
IHIMES	*J.Organomet.Chem.*	694	2463	2009
JASCAI	*Z.Anorg.Allg.Chem.*	631	1524	2005
MECGEH	*Eur.J.Inorg.Chem.*		958	2006
MECGOR	*Eur.J.Inorg.Chem.*		958	2006
MTEMTE	*J.Chem.Soc.A*		2018	1967
WUGVUQ	*J.Braz.Chem.Soc.*	17	1566	2006
WUTJOK	*Z.Anorg.Allg.Chem.*	629	215	2003
b) Simple chains				
IJOHUK	*Inorg.Chem.Commun.*	6	869	2003
IJOJAS	*Inorg.Chem.Commun.*	6	869	2003
JASBUB	*Z.Anorg.Allg.Chem.*	631	1524	2005
KEQCAL	*Inorg.Chim.Acta*	359	4619	2006
MECGUX	*Eur.J.Inorg.Chem.*		958	2006
WUGVIE	*J.Braz.Chem.Soc.*	17	1566	2006
WUGVOK	*J.Braz.Chem.Soc.*	17	1566	2006

Structures of compounds without Te-X units [Te(-X)$_0$] in Figure

CLPHTE01	*Acta Crystallogr.,Sect.C*	40	1011	1984
DIHHAE	*Bioinorg.Chem.Appl.*		69263	2007
JATWAC	*Inorg.Chem.*	38	519	1999
JEPBAH	*Acta Chem.Scand.*	44	464	1990
YABXEF	*J.Fluorine Chem.*	125	997	2004

Polymorphs

ASEHUB	*Z.Kristallogr.*	218	636	2003
ASEHUB01	*Z.Kristallogr.*	218	636	2003
BIPTEI	*Inorg.Chem.*	15	2728	1976
BIPTEI01	*Inorg.Chem.*	16	2318	1977
BTUPTE	*Acta Chem.Scand.A*	29	738	1975
BTUPTE01	*Acta Chem.Scand.A*	29	738	1975
CIFLEI	*J.Chem.Soc.,Dalton Trans.*		869	1984
CIFLEI01	*J.Chem.Soc.,Dalton Trans.*		869	1984
CLPHTE	*Acta Crystallogr*	10	201	1957
CLPHTE01	*Acta Crystallogr.,Sect.C*	40	1011	1984
DIBTEP10	*J.Organomet.Chem.*	178	423	1979
DIDMTE	*J.Chem.Soc.,Dalton Trans.*		316	1972
DIDMTE01	*Dalton Trans.*		6274	2008
NUNHUZ	*Acta Crystallogr.,Sect.C*	54	995	1998
NUNHUZ01	*Acta Crystallogr.,Sect.C*	54	995	1998
PAZPTE	*J.Chem.Res.*	145	1901	1979
PAZPTE01	*J.Organomet.Chem.*	549	257	1997
QIXZAY	*J.Organomet.Chem.*	623	74	2001

QIXZAY03	*Inorg.Chem.*	43	3742	2004
SABCII	*Inorg.Chem.*	43	4106	2004
SABCII01	*Acta Crystallogr.,Sect.E*	61	o986	2005
Halogen series				
KEHBIJ	*Z.Naturforsch.,B:Chem.Sci.*	61	528	2006
BIBBOE	*Polyhedron*	23	1629	2004
BIBCEV	*Polyhedron*	23	1629	2004
QIXZAY	*J.Organomet.Chem.*	623	74	2001
QIXZAY03	*Inorg.Chem.*	43	3742	2004
GETHOC	*J.Organomet.Chem.*	349	95	1988
BIBCAR	*Polyhedron*	23	1629	2004
BIBBUK	*Polyhedron*	23	1629	2004
DIBTEP02	*Inorg.Chem.*	24	1814	1985
DIBTEP10	*J.Organomet.Chem.*	178	423	1979
FPHTEL	*J.Chem.Soc.,Dalton Trans.*		2306	1980
OPNTED10	*J.Chem.Soc.,Dalton Trans.*		251	1982
DPHTEB01	*Acta Crystallogr.,Sect.E*	60	o2511	2004
CIFLEI	*J.Chem.Soc.,Dalton Trans.*		869	1984
CIFLEI01	*J.Chem.Soc.,Dalton Trans.*		869	1984
YASVAP	*Z.Anorg.Allg.Chem.*	619	1269	1993
YASVET	*Z.Anorg.Allg.Chem.*	619	1269	1993
IDEXOE	*Inorg.Chem.*	40	5169	2001
IDEXIY	*Inorg.Chem.*	40	5169	2001
TUWPAC	*Dalton Trans.*		973	2003
TUWPIK	*Dalton Trans.*		973	2003

Table 9. Short-form references to individual CSD entries. In blue, structures measured at low temperature

4. Conclusion

In the majority of structures having tellurium and halogens, intermolecular Te···X contacts are observed.

Despite the great chemical diversity found in these compounds, the majority of them can be grouped considering only a few supramolecular patterns, *i.e.*, dimer, simple chain and chain of 4-membered rings.

Isostructurality is frequent in series of compounds with the same molecular structure but different halogen and, moreover, in some non isostructural compounds the supramolecular arrangement is retained.

No relations have been found between supramolecular arrangements of polymorphs although due to the reduced number of cases can not ruled them out.

5. Acknowledgment

The authors gratefully acknowledge Dr D. Cendón for his valuable participation in the discussion and correction of the text.

6. References

Alcock, N. W. (1972). Secondary bonding to nonmetallic elements. *Adv. Inorg. Chem. Radiochem.*, Vol. 15(1), pp. 1-58, ISSN 0065-2792

Allen, F. H. (2002). The Cambridge Structural Database: a quarter of a million crystal structures and rising. *Acta Cryst.*, Vol. B58(3), pp. 380-388, ISSN 0108-7681

Ba, L. A.; Doering, M.; Jamier, V. & Jacob, C. (2010). Tellurium: an element with great biological potency and potential. *Org. Biomol. Chem.*, Vol. 8(19), pp. 4203-4216, ISSN 1477-0520

Bondi, A. (1964). Van der Waals Volumes and Radii. *J. Phys. Chem.*, Vol. 68(3), pp. 441-451, ISSN 1089-5639

Bruno, I. J.; Cole, J. C.; Edgington, P. R.; Kessler, M.; Macrae, C. F.; McCabe, P.; Pearson, J. & Taylor, R. (2002). New software for searching the Cambridge Structural Database and visualising crystal structures. *Acta Cryst.*, Vol. B58(3), pp. 389-397, ISSN 0108-7681

CCDC - Cambridge Crystallographic Data Centre (1994). Vista - A Program for the Analysis and Display of Data Retrieved from the CSD

Cozzolino, A. F.; Elder, P. J. W. & Vargas-Baca, I. (2011). A survey of tellurium-centered secondary-bonding supramolecular synthons. *Coord. Chem. Rev.*, Vol. 255(11-12), pp. 1426-1438, ISSN 0010-8545

Desiraju, G. R. & Steiner, T. (2001). *The weak hydrogen bond in structural chemistry and biology*, Oxford University Press, ISBN 0-19-850970-7, Oxford, UK

Forni, A.; Metrangolo, P.; Pilati, T. & Resnati, G. (2004). Halogen Bond Distance as a Function of Temperature. *Cryst. Growth Des.*, Vol. 4(2), pp. 291-295, ISSN 0022-0248

Haiduc, I. & Zukerman-Schpector, J. (2001). Supramolecular Self-Assembly Through Secondary Bonds in Organotellurium Chemistry. *Phosphorus, Sulfur, Silicon, Relat. Elem.*, Vol. 171(1), pp. 171-185, ISSN 1042-6507

Hails, J. E.; Cole-Hamilton, D. J.; Stevenson, J.; Bell, W.; Foster, D. F. & Ellis, D. (2001). A chemical assessment of the suitability of allyl iso-propyltelluride as a Te precursor for metal organic vapour phase epitaxy. *J. Cryst. Growth*, Vol. 224(1-2), pp. 21-31, ISSN 0022-0248

Jeffrey, G. A. & Saenger, W. (1991). *Hydrogen Bonding in Biological Structures*, Springer Verlag, ISBN 3-540-50839-2, Berlin, Germany

Kitayama, Y.; Chaiyasat, A.; Minami, H. & Okubo, M. (2010). Emulsifier-Free Organotellurium-Mediated Living Radical Emulsion Polymerization of Styrene; Polymerization Loci. *Macromolecules*, Vol. 43(18), pp. 7465-7471, ISSN 0024-9297

Macrae, C. F.; Bruno, I. J.; Chisholm, J. A.; Edgington, P. R.; McCabe, P.; Pidcock, E.; Rodriguez-Monge, L.; Taylor, R.; van de Streek, J. & Wood, P. A. (2008). Mercury CSD 2.0 - New Features for the Visualization and Investigation of Crystal Structures. *J. Appl. Cryst.*, Vol. 41(2), pp. 466-470, ISSN 0021-8898

Metrangolo, P.; Meyer, F.; Pilati, T.; Resnati, G. & Terraneo, G. (2008). Halogen bonding in supramolecular chemistry. *Angew. Chem. Int. Ed.*, Vol. 47(33), pp. 6114-6127, ISSN 1433-7581

Nishio, M.; Umezawa, Y.; Honda, K.; Tsuboyama, S. & Suezawa, H. (2009). CH/π hydrogen bonds in organic and organometallic chemistry. *Cryst. Eng. Comm.*, Vol. 11(9), pp. 1757-1788, ISSN 1466-8033

Petragnani, N. & Stefani, H. A. (2005). Advances in organic tellurium chemistry. *Tetrahedron*, Vol. 61(7), pp. 1613-1679, ISSN 0040-4020

Pimentel, G. C. & McClellan, A. L. (1960). *The Hydrogen Bond*, W. H. Freeman and Company, ISBN 978-0198558835, San Francisco, USA

Rowland, R. S. & Taylor, R. (1996). Intermolecular Nonbonded Contact Distances in Organic Crystal Structures: Comparison with Distances Expected from van der Waals Radii. *J. Phys. Chem.*, Vol. 100(18), pp. 7384-7391, ISSN 1089-5639

Salonen, L. M.; Ellermann, M. & Diederich, F. (2011). Aromatic rings in chemical and biological recognition: energetics and structures. *Angew. Chem. Int. Ed.*, Vol. 50(21), pp. 4808-4842, ISSN 1433-7581

Schneider, H. J. (2009). Binding Mechanisms in Supramolecular Complexes. *Angew. Chem. Int. Ed.*, Vol. 48(22), pp. 3924-3977, ISSN 1433-7581

Shanks, D.; Engman, L.; Mayte Pach, M. & Thomas Norrby, T. (2006). Probing the antioxidative properties of combinations of an organotellurium compound, BHT and thiol in oil. *Lubr. Sci.*, Vol. 18(2), pp. 87-94, ISSN 0954-0075

Singh, A. K. & Sharma, S. (2000). Recent developments in the ligand chemistry of tellurium. *Coord. Chem. Rev.*, Vol. 209(1), pp. 49-98, ISSN 0010-8545

Srivastava, P. C.; Bajpai, S.; Lath, R.; Bajpai, S. M.; Kumar, R. & Butcher, R. J. (2004). Molecular aggregates, zig-zag 2D-stairs, -ribbons and 3D-supramolecular networks of cyclic telluranes assisted by intermolecular Te···Cl and Te···Br secondary bonding. *Polyhedron*, Vol. 23(9), pp. 1629-1639, ISSN 0277-5387

Steigerwald, M. L. & Sprinkle, C. R. (1987). Organometallic synthesis of II-VI semiconductors. 1. Formation and decomposition of bis(organotelluro)mercury and bis(organotelluro)cadmium compounds. *J. Am. Chem. Soc.*, Vol. 109(23), pp. 7200-7201, ISSN 0002-7863

Steiner, T. (2002). The Hydrogen Bond in the Solid State. *Angew. Chem. Int. Ed.*, Vol. 41(1), pp. 48-76, ISSN 1433-7581

Zukerman-Schpector, J. & Haiduc, I. (2001). Diorganotellurium(IV) Dihalides and Secondary Bonding: Revisiting the Coordination Polyhedra. *Phosphorus, Sulfur, Silicon, Relat. Elem.*, Vol. 171(1), pp. 73-112, ISSN 1042-6507

Zukerman-Schpector, J.; Haiduc, I.; Camillo, R. L.; Comasseto, J. V., Cunha, R. L. O. R. & Jorge, A. (2002). Supramolecular self-assembly through tellurium···halogen secondary bonds: A hexagonal grid of Te_2Cl_2 and Te_6Cl_6 rings in the solid state structure of 1,1,3-trichloro-2,4,5,6-tetrahydro-1*H*-1λ^4-benzo[b]tellurophene. *Can. J. Chem.*, Vol. 80(11), pp. 1530-1537, ISSN 0008-4042

5

σ-Bonded *p*-Dioxolene Transition Metal Complexes

Anastasios D. Keramidas[1], Chryssoula Drouza[2] and Marios Stylianou[1]
[1]*University of Cyprus*
[2]*Cyprus University of Technology*
Cyprus

1. Introduction

Hydroquinones(HQ) are molecules of great importance in chemistry and biology. They undergo proton-coupled electron transfer to afford neutral *p*-semiquinone(SQ) and *p*-quinone(Q) species as illustrated in figure 1.

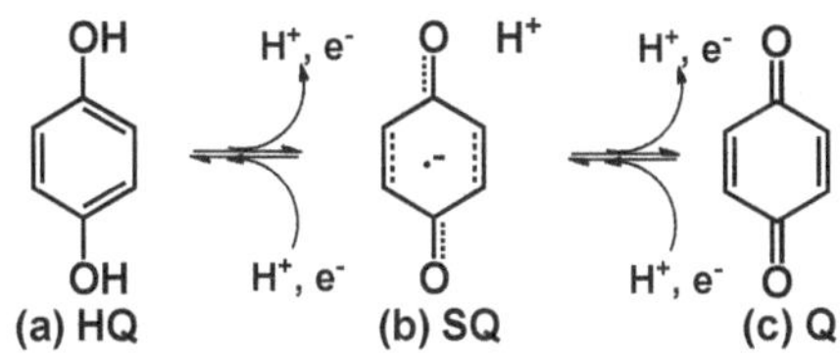

Fig. 1. Proton-coupled electron transfer in hydroquinone molecules

Metal ions are known to lie in close proximity with these species in biological systems, thus resulting in immediate interaction. The two coupled, metal and organic redox centers have been found to participate in several biological processes such as, the oxidative maintenance of biological amine levels, (Klinman, 1996) tissue (collagen and elastin) formation, (Klinman, 1996) photosynthesis (Calvo, et al., 2000) and respiration (Iwata, et al., 1998). Although the crystal structures of many of these enzymes have been solved, the role of the metal ions in these reactions is still controversial. From another point of view, quinonoid metal complexes exhibit rich redox, magnetic and photochemical properties and thus can underpin key technological advances in the areas of energy storage, sensors, catalysis and "smart materials" (Evangelio & Ruiz-Molina, 2005; Stylianou, et al., 2008).
Metal ions interact with hydroquinone systems, through σ-bonding to the oxygen atoms and/or through π-bonding to the carbocyclic ring. The structurally characterized σ-bonded hydroquinone metal complexes are surprisingly limited. Structures of metal ions with *p*-semiquinones and quinones are even rarer, mainly due to the absence of a chelate coordination site in simple *p*-(hydro/semi)quinone and the low p*K* values of the semiquinone and quinone oxygen atoms. A strategy to synthesize stable metal complexes with hydroquinone species is to use substituted hydroquinones in *o*-position with substituents containing one or more donor atoms, enabling in this way the metal atom to

form chelate rings. In addition to the stabilization of the metal complexes, hydroquinones substitution offers a direct control of the redox properties of the metal ion and increases the number of new possible structural motives by changing the number and the type of the donor atoms of the chelating group. One of the problems that someone has to face working with "non-innocent" ligands, such as hydroquinones, is the determination of their formal charge in the complex. Sometimes, physicochemical properties of the complexes, such as strong magnetic coupling between the metal ion and the organic radical, may give misleading results regarding the oxidation states. It has been shown that X-ray crystallography can be used for the determination of the oxidation states of the non innocent ligands in the complexes. For example, the $C\text{-}O_{hydroquinonate}$ and the C-C bond lengths of the *p*-dioxolene ligands are strongly dependent on the formal charge of the ligands.

In this chapter we demonstrate that the rich structural chemistry of hydroquinonate complexes is predicated on a) the ability of the metal ions to reversibly deprotonate the -OH groups, b) the remote and adjacent bridge ligating modes of hydroquinone and c) the reversible metal ion - hydroquinone electron transfer which results in stabilization of the *p*-semiquinone oxidation state. The determination of the oxidation state of the *p*-dioxolene ligand based of C-O and intraring bond distances is also analyzed. The application of a statistical approach for the determination of the ligand formal charges is being discussed. In addition, a graphical method for the assignment of the oxidation states has been included in this chapter. Finally, the factors that promote the stabilization of the semiquinone radical versus the hydroquinone are discussed based on the structural data. Here, we will mainly focus on the $V^{IV/V}$ complexes with the 2,5- bisubstituted hydroquinone with iminodiacetic acid or bis(2-methylpyridyl)amine in *o*-position. These are the only universally structurally characterized *p*-semiquinone examples in the literature up to today and the structure of the hydroquinone complexes can be directly compared with that of the *p*-semiquinone analogues. These compounds are oxidized from the atmospheric oxygen to form stable semiquinone radicals, trapping intermediates of dioxygen reduction that have been identified by X-ray crystallography. This is an important development towards the better understanding of the catalytic reduction mechanisms of dioxygen from metal ions in biological systems as well as in the catalytic oxidation of organic substrates from metal complexes.

It is clear that σ-bonded hydroquinone/*p*-semiquinone-metal complexes have many interesting properties that have only begun to be explored or exploited (*vide infra*). X-ray crystallography represents a basic and irreplaceable tool in this exploration. This chapter will provide a glimpse of the fascinating structural chemistry exhibited by hydroquinones/*p*-semiquinones metal complexes and the utilization of X-ray crystallography into the exploration of the chemistry and the development of hydroquinones/*p*-semiquinones based functional bioinorganic models.

2. Structural studies of hydroquinonate/p-semiquinonate/p-quinone transition metal complexes

Structural investigation has proven to be an essential tool for the characterization of *p*-dioxolene complexes. Metal-oxygen bond lengths are often characteristic of a particular oxidation state of the metal, and the *p*-dioxolene carbon-oxygen lengths are sensitive to the charge of the ligand. Apart from providing indirect information on the charge distribution within the complex, crystallographic studies have revealed the donor-acceptor tendency for complexation. The first hydroquinone complex characterized by crystallography has been reported 30 years ago (Heistand, et al., 1982). However only the last 10 years the number of the characterized by crystallography *p*-dioxolene complexes has increased significantly, including the first *p*-semiquinone complex in 2002 (Drouza, et al., 2002). This is in marked contrast to the extensive structural chemistry of chelate stabilized *o*-dioxolene metal complexes reported in the literature(Pierpont, 2001). This is mainly due to the absence of a chelate coordination site in simple *p*-dioxolenes and their low pKa values. The oxygen atoms of *p*-dioxolenes act as unidentate donor atoms, as shown in figure 2 for hydroquinones. Hydroquinone may ligate one metal ion or two metal ions bridged from two different or from the same oxygen donor atoms (figure 2).

Substituted in *o*-position *p*-dioxolenes, with substituents containing one or more donor atoms, stabilize metal ion ligation through the formation of chelate rings. A systematic collection of the substituents reported in the literature including their transition metal complexes characterized by X-ray crystallography is illustrated in figure 3. The transition metal complexes of these ligands together with some important crystallographic data are summarized in table 1. The type of the substituent is very important because it may control the stabilization of certain metal ions defining the oxidation states of the metal ions and of the *p*-dioxolenes, as well the structure of the molecule.

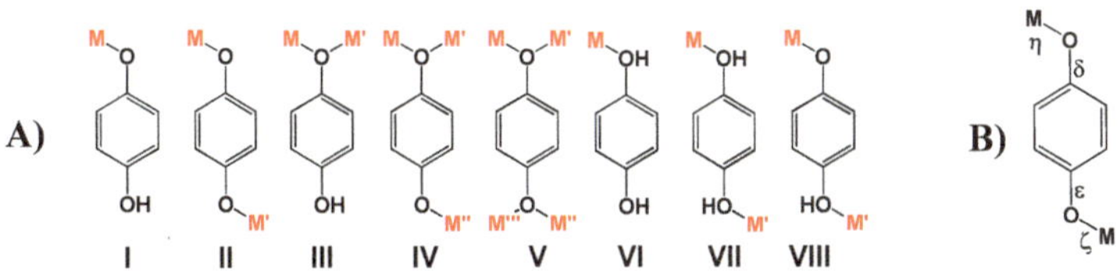

Fig. 2. A) Coordination modes of hydroquinones **I**) monodentate, **II**) remote bridged, **III**) adjacent bridged, **IV**) remote and one adjacent bridges, **V**) remote and two adjacent bridges **VI**) protonated monodentate, **VII**) protonated bridged, **VIII**) monoprotonated bridged, B) Labeling of the M-O and C-O bonds

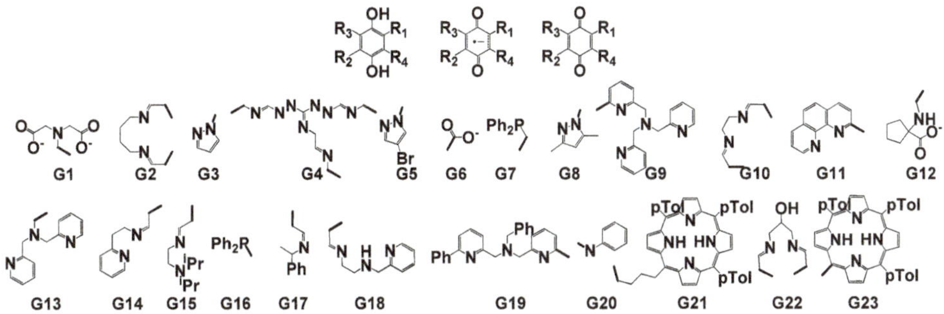

Fig. 3. Substituents of hydroquinone / *p*-semiquinone / *p*-quinones used for transition metal ion ligation

Comp	Metal	η /Å	δ /Å	ε /Å	ζ /Å	C-O-M (°)	Arom. Substitution	Ref.
1[a]	Fe^{III}	1.903(4) 1.864(4)	1.374(8) 1.372(8)	1.316(6) 1.330(7)	---	126.7(3) 129.6(3)	R1=G2 R2=R3=R4=H	(Sanmartin, et al., 2004)
2[a]	Ti^{III}	1.828(8) 1.775(8)	1.38(3) 1.38(2)	1.34 (1) 1.35(2)	---	146.5(7) 138.5(8)	R1=R2=R3=R4=H	(Errington, et al., 2007)

3[a]	Pt^{II}	2.030(3)	1.380(6)	1.334(5)	---	130.3 (3)	R1=G7 R2=R3=R4=H	(Sembiring, et al., 1999)
4[a]	Cu^{II}	1.900(4)	1.386(7)	1.327(7)	---	122.0(3)	R1=G9 R2=R3=R4=H	(He, et al., 2003)
5[a]	$Cu^{II,I}$	1.924(1)	1.381(3)	1.329(2)	---	126.4(1) 127.2(1)	R1=G10 R2=R3=R4=H	(Margraf, et al., 2006)
6[a]	Ni^{II}	1.854(2) 1.860(3)	1.397(5) 1.386(5)	1.321(5) 1.322(4)	---	127.8 (2) 127.6(2)	R1=G10 R3= tBu R2=R4=H	(Margraf, et al., 2006)
7[a]	Pd^{II}	1.940(7)	1.39(1)	1.34(1)	---	124.5(5)	R1=G11 R2=R3=R4=H	(Berthon, et al., 1992)
8[a]	Cr^{III}	1.924(2)	1.391(3)	1.362(3)	---	120.9(1)	R1=G1 R2=R3=R4=H	(Huang, et al., 2008)
9[a]	Fe^{III}	1.927(2), 1.920(2)	1.379(4)	1.343(4)	---	127.7(2), 126.4(2)	R1= G17, R2=R3=R4=H	(Becker, et al., 2010)
10[a]	Ni^{II}	1.827(4) 1.827(4)	1.386(9) 1.393(8)	1.332(7) 1.325(7)	---	127.3(3), 127.4(3)	R1= G10, R2=R3=R4=H	(Kondo, et al., 2003)
11[a]	Cu^{II}	1.870(4)	1.384(9)	1.321(7)	---	126.0(4)	R1= G18, R2=R3=R4=H	(Li, et al., 2000)
12[a]	Ni^{II}	1.90(1) 1.87(1)	1.38(2) 1.36(2)	1.37(2) 1.35(5)	---	119.8(8), 118.9(8)	R1= G7, R2=R3=R4=H	(Sembiring, et al., 1992)
13[a]	Mo^{VI}	1.945(2) 1.955(2)	1.381(6) 1.359(5)	1.353(5) 1.348(5)	---	137.8(2), 133.5(2)	R1= G6, R2=R3=R4=H	(Litos, et al., 2006)
14[a]	W^{VI}	1.88(3)	1.36(4)	1.36(4)	---	166(2), 157(3),	R1=R2=R3=R4=H	(Vaid, et al., 2001)
15[b]	V^{IV}	1.887(4) 1.887(4)	1.322(6)	1.322(6)	---	137.0(1), 137.0(1)	R1=R2=G1 R3=R4=H	(Drouza, et al., 2002)
16[b]	V^{V}	1.878(3) 1.865(3)	1.353(6) 1.353(5)	1.352(6) 1.353(5)	1.878(3) 1.865(3)	131.9(3) 131.5(3)	R1=R2=G1 R3=R4=H	(Drouza, et al., 2002)
17[b]	W^{V}	1.948(6)	1.362(9)	1.362(9)	1.948(6)	133.3(5), 133.3(5)	R1=R2=R3=R4=H	(Stobie, et al., 2003)
18[b]	Cu^{II}	1.803(3)	1.300(2)	1.300(2)	1.803(3)	126.8(1), 126.8(1)	R1=R2=G3 R3=R4=H	(Dinnebier, et al., 2002)
19[b]	Fe^{III}	1.862(1)	1.3492(4)	1.3492(4)	1.8616(1)	132.13(3), 132.13(3)	R1=R2=R3=R4=H	(Heistand, et al., 1982)
20[b]	Cu^{I}	1.91(2) 1.916(2)	1.322(3) 1.327(3)	1.322(3) 1.327(3)	1.91(2) 1.916(2)	123.9(2) 122.6(2)	R1=R2=G5 R3=R4=H	(Margraf, et al., 2009)
21[b]	Ru^{III}	1.975(7)	1.38(1)	1.34 (1)	1.966(5)	115.9 (6)	R1=R2=G8 R3=R4=H	(Kumbhakar, et al., 2008)
22[b]	Ru^{III}	1.983(2)	1.346(4)	1.346(4)	1.976(2)	118.4(2), 115.9(2)	R1=R2=G8 R3=R4=H	(Kumbhakar, et al., 2008)
23[b]	Cu^{II}	1.915(1)	1.322(2)	1.322(2)	1.915(1)	121.4(1), 121.4(1)	R1=R2=G3 R3=R4=H	(Margraf, et al., 2006)
24[b]	Ti^{II}	1.785(5)	1.360(8)	1.360(8)	1.785(5)	165.1 (4) 165.1 (4)	R1=R4=Me R2=R3=H	(Arévalo, et al., 2003)
25[b]	Mn^{III}	2.193(4)	1.253(7)	1.253(7)	2.193(4)	180.000	R1=R2= R3=R4=Cl	(Brandon, et al., 1998)
26[b]	V^{IV}	1.951(3) 1.952(3)	1.364(5) 1.352(5)	1.364(5) 1.352(5)	1.951(3) 1.952(3)	128.2(3), 128.2(3)	R1=R2=G1 R3=R4=H	(Drouza & Keramidas, 2008)
27[b]	V^{V}	1.866(2) 1.824(2)	1.346(3) 1.346(3)	1.338(3) 1.338(3)	1.824(2) 1.866(2)	132.3(1), 137.0(1)	R1=R2=G1 R3=R4=H	(Drouza & Keramidas,

						137.0(1), 132.3(1)		2008)
28[b]	V^{V}	1.827(2) 1.823(2)	1.346(3) 1.325(3)	1.335(3) 1.327(3)	1.865(2) 1.878(2)	137.8(2), 132.3(2) 137.2(2), 135.9(2)	R1=R2=G1 R3=R4=H	(Drouza & Keramidas, 2008)
29[b]	$V^{IV/V}$	1.937(2) 1.879(2)	1.314(3) 1.350(4)	1.308(3) 1.345(4)	1.880(2) 1.879(2)	138.1(2), 136.4(2) 134.8(2), 131.8(2)	R1=R2=G1 R3=R4=H	(Drouza & Keramidas, 2008)
30[b]	$V^{IV/V}$	1.884(2) 1.898(3)	1.338(4) 1.302(4)	1.350(4) 1.303(4)	1.8512(2) 1.913(2)	134.3(2), 134.8(2)	R1=R2=G1 R3=R4=H	(Drouza & Keramidas, 2008)
31[b]	V^{III}	1.877(9)	1.38(2)	1.34(2)	1.886(7)	134.1(9), 129.9(8)	R1=R2=R3=R4=H	(Tanski & Wolczanski, 2001)
32[b]	Ti^{IV}	1.882(4) 1.915(6)	1.373(8) 1.37(1)	1.373(8) 1.36(1)	1.882(4) 1.874(6)	137.4(4), 137.4(4)	R1=R2=R3=R4=H	(Vaid, et al., 1997)
33[b]	Zr^{IV}	1.978(2)	1.357(3)	1.357(3)	1.978(2)	144.8(1), 144.8(1)	R1=R2=R3=R4=H	(Evans, et al., 1998)
34[b]	Ti^{III}	1.870(3),	1.360(7)	1.369(7)	1.898(4)	148.5(3), 1.429(3)	R1=R2=R3=R4=H	(Tanski, et al., 2000)
35[b]	Mo^{IV}	1.924(8) 1.974(8) 1.924(8)	1.34(2) 1.37 (2) 1.33 (2)	1.33(2) 1.38 (2) 1.34(2)	1.937(8) 1.935(8) 1.92(1)	137.3(9), 142.3(9) 127.5(9),	R1=R2=R3=R4=H	(McQuillan, et al., 1998)
36[b]	Cu^{II}	1.880(3)	1.337(5)	1.337(5)	1.88(3)	127.2(2), 127.2(2)	R1=G15 R2=R3=R4=H	(Kretz, et al., 2006)
37[b]	Ti^{III}	1.865(2), 1.867(2)	1.349(4)	1.348(4)	1.867(2)	165.3(2), 169.6(2)	R1=R2=R3=R4=H	(Horacek, et al., 2010)
38[b]	Ti^{III}	1.864(4), 1.86(4)	1.353(3)	1.353(3)	1.864(4)	155.2(2), 155.2(2)	R1=R2=R3=R4=H	(Kunzel, et al., 1996)
39[b]	Cu^{II}	2.370(3), 2.464(2)	1.386(4)	1.380(4)	2.370(3), 2.464(2)	111.7(2), 112.1(2)	R1=R2= G1, R3=R4=H	(Stylianou, et al., 2008)
40[b]	Pd^{II}	1.981(2)	1.341(4)	1.341(4)	1.981(2)	118.3(2)	R1=R2= G7, R3=R4=H	(Caldwell, et al., 2008)
41[b]	Mo^{VI}	1.922(8)	1.35(1)	1.35(1)	1.922(8)	136.1(7)	R1=R2=R3=R4=H	(Ung, et al., 1996)
42[b]	W^{VI}	1.93(1)	1.36(2)	1.36(2)	1.927(1)	137(1)	R1=R2=R3=R4=H	(McQuillan, et al., 1996)
43[b]	Cu^{II}	1.880(3)	1.337(5)	1.337(5)	1.880(3)	127.2(2)	R1=R2= G15, R3=R4=H	(Kretz, et al., 2006)
45[b]	Fe^{III}	1.874(8)	1.27(1)	1.27(1)	1.874(8)	169.4(7)	R1=R2=R3=R4=Cl	(Rheingold & Miller, 2003)
46[b]	Mo^{V}	1.948(9) 1.954(8)	1.36 (2) 1.36 (2)	1.38(2) 1.35(2)	1.914(8) 1.953(8)	140.9(8), 135.6(8)	R1=R2=R3=R4=H	(Ung, et al., 1999)
47[c]	Cu^{II}	1.889(7), 2.326(7)	1.40(2) 1.36(2)	1.32(1) 1.31(1)	---	121.1(5), 127.6(5),	R1= G4 R2=R3=R4=H	(Zharkouskaya, et al., 2005)
48[c]	Zn^{II}	2..021(3), 2.030 (3)	1.347(5)	1.347(5)	2..021(3), 2.030 (3)	123.2(2), 1233(2)	R1=R2=G6 R3=R4=H	(Rosi, et al., 2005)
49[c]	Cu^{II}	1.924(2), 1.980(2)	1.527(6)	1.347(4)	---	132.8(2), 121.7(2)	R1=G22 R2=R3=R4=H	(Song, et al., 2007)
50[c]	Cu^{II}	1.934(2), 1.930(3)	1.364(3)	1.371(4)	---	134.4(2), 1203(2)	R1=G12 R2-R3=R4=H	(Sreenivasulu, et al., 2006)
51[c]	Ti^{IV}	2.048(3),	1.373(7)	1.371(6)	---	126.2(3),	R1=G12	(Vaid, et al.,

		2.043(3)				126.8(3)	R2-R3=R4=H	1997)
g		1.782(3)	1.387(6)	1.361(6)	2.208(4)	173.0(3), 125.9(3)	R1=R2=R3=R4=H	(Vaid, et al., 1997)
b		1.794(4)	1.376(6)	1.351(6)	1.923(3)	164.1(3), 129.3(3)	R1=R2=R3=R4=H	(Vaid, et al., 1997)
52[c]	Cu^{II}	1.971(6), 1.955(7) 1.922(7), 1.929(7)	1.38(1)	1.32(2)	---	130.1(5), 130.1(6)	R1=G14 R2=R3=R4=H	(Gelling, et al., 1990)
53[a]	Zn^{II}	1.946(5) 1.962(6)	1.37(1) 1.39(1)	1.33(1) 1.35(1)	---	126.8(5) 123.8(5)	R1=G2 R2=R3=R4=H	(Matalobos, et al., 2004)
c		2.046(5) 2.043(5)	1.377(9) 1.368(8)	1.343(9) 1.327(8)	---	126.1(5), 127.4(5) 124.0(5), 128.8(5)	R1=G2 R2=R3=R4=H	(Matalobos, et al., 2004)
54[c]	Cu^{II}	1.893(2), 2.463(2)	1.374(4)	1.345(3)	---	118.0(2), 126.8(2)	R1= G19, R2= R3=R4=H	(Matalobos, et al., 2004)
55[c]	Cu^{II}	1.936(6), 1.978(6), 2.358(6), 1.945(6)	1.39(1), 1.37(1)	1.36(1), 1.35(1)	---	129.9(6), 119.2(5), 108(5), 128.8(5)	R1= G1, R2= R3=R4=H	(Stylianou, et al., 2008)
56[c]	Cu^{II}	1.971(6), 1.934(6)	1.389(9)	1.350(8)	---	128.1(5), 132.1(6)	R1= G18, R2=R3=R4=H	(Li, et al., 2000)
57[c]	Pd^{II}	2.02(2), 2.13(1)	1.40(4)	1.36(3)	---	118(1), 122(1)	R1= G7, R2=R3=R4=H	(Sembiring, et al., 1995)
58[c]	Cu^{II}	1.971(6), 1.955(7)	1.38(1)	1.32(1)	---	130.1(5), 130.1(6)	R1=R3= G14, R2=R4=H	(Gelling, et al., 1990)
59[d]	Ti^{IV}	2.07(1), 2.014(7)	1.38(2)	1.37(2)	1.78(1)	126.6(7), 126(1), 153(1)	R1=R2=R3=R4=H	(Vaid, et al., 1999)
d		2.014(7), 2.07(1)	1.382)	1.37(2)	1.78(1)	126.6(7), 126(1), 153(1)		
b		1.836(8),	1.35(2)	1.37(2)	1.814(7)			
d		2.019(8), 2.08(1)	1.38(2)	1.39(2)	1.80(1)	147.0(7), 140.5(7)		
b		1.854(7), 1.80(1)	1.33(2)	1.33(2)	1.854(7)	134.2(8),		
d		2.08(1), 2.019(8)	1.38(2)	1.39(2)	1.80(1)	145(1)		
b		1.814(7)	1.37(1)	1.35(1)	1.836(8)	127.8(7), 124.5(9), 145(1)		
b		1.807(9)	1.38(2)	1.38(2)	1.807(9)	146.8(2)		
60[e]	Cu^{II}	2.375(1)	1.382(2)	1.376(2)	---	109.22(9)	R1=G1	(Zhang, et al., 2009)
61[e]	Cu^{II}	2.653(3), 2.547(3)	1.216(5) 1.219(4)	1.217(4) 1.218(5)	---	106.7(2), 112.1(2)	R1=tBu R3=G9 G2=G4=H	(Philibert, et al., 2003)
62[f]	Cu^{II}	2.359(2)	1.382(3)	1.382(3)	---	108.6(1)	R1= R2= G1, R3=R4=H	(Stylianou, et al., 2008)
63[c]	Ti^{IV}	2.048(3)	1.373(7)	1.371(6)	---	126.8(3)	R1=G12 R2-R3=R4=H	(Vaid, et al., 1999)
g		1.782(3)	1.387(6)	1.361(6)	2.208(4)	173.0(3),	R1=R2=R3=R4=H	(Vaid, et al.,

		1.782(3)	1.387(6)	1.361(6)	2.208(4)	125.9(3) 173.0(3), 125.9(3)		1999)
[b]		1.794(4)	1.376(7)	1.351(6)	1.923(3)	164.1(5), 129.3(3)	R1=R2=R3=R4=H	(Vaid, et al., 1999)
64[h]	Cu^{II}	1.921(2), 1.908(2)	1.403(8)	1.4025(8)	1.921(2), 1.908(2)	128.76(6), 125.70(6), 128.76(6), 125.70(6),	R1= R2= G14, R3= R4= H	(Phan, et al., 2011)
66[h]	Cu^{II}	1.931(7), 2.338(6)	1.36(1)	1.34(1)	1.920(7), 2.343(6)	116.2(5), 124.1(5),	R1= R2= G14, R3= R4= H	(Phan, et al., 2011)
67[h]	Zn^{II}	2.00(2)	1.36(3)	1.36(3)	2.00(2)	128(1), 130(1)	R1= R2= G6, R3=R4=H	(Dietzel, et al., 2008)
68[h]	Zn^{II}	1.96(1), 1.99(2)	1.38(3)	1.38(3)	1.96(1), 1.99(2)	125(1), 129(1)	R1= R2= G6, R3=R4=H	(Dietzel, et al., 2008)
69[h]	Cu^{II}	1.922(4), 2.514(4)	1.350(6)	1.350(6)	1.922(4), 2.514(4)	124.6(3), 113.6(3)	R1= R2= G14, R3=R4=H	(Kretz, et al., 2006)
70[a]	Zn^{II}	2.222(6)	1.233(9)	1.242(9)	---	135.7(5)	R_1= G23, R_2= R_3= CH_3, R_4= H	(Senge, et al., 1999)
71[b]	Rh^{II}	2.25(1)	1.24(2)	1.24(2)	2.25(1)	136.8(8)	R_1= R_2= CH_3, R_3= R_4= H	(Handa, et al., 1996)
72[b]	Mo^{II}	2.619(9)	1.28()	1.28(2)	2.594(9)	152.7(8), 141.7(8)	R_1= R_3= CH_3, R_2= R_4= H	(Handa, et al., 1995)
73[a]	Mo^{II}	2.569(6)	1.21(1)	1.24(1)	---	140.6(5)	R_1= R_3= tBu, R_2= R_4= H	(Handa, et al., 1995)

Table 1. Summary of structurally characterized hydroquinone/*p*-semiquinone/*p*-quinone transition metal complexes. Some important crystallographic data are also included. Abbreviations are according to figures 2 and 3. [a] Mode I, [b] mode II, [c] mode III, [d] mode IV, [e] mode V, [f] mode VI, [g] mode VII, [h] mode VIII according to figure 2

2.1 Simple hydroquinones

The first crystallographic report on a transition metal hydroquinone appeared in 1982 (Heistand, et al., 1982). Heistand et.al reported the structure of a binuclear iron(III) complex containing two iron-salen units bridged together with a simple deprotonated hydroquinone (coordination mode **II**, figure 4). Although the $C\text{-}O_{hydroquinonate}$ bond length [1.349(3) Å] is shorter than the C-O bond of free hydroquinone (1.398 Å), it is within the limits expected for this *p*-dioxolene's oxidation state. It is worth noticing here that the respective catecholate complex found in the crystal structure is ligated with Fe^{III} in a monodentate fashion, in contrast to the hydroquinone complex which even in 50 fold excess of $[Fe(salen)]^+$ crystallizes as dimer. Heistand et al. have assigned the formation of the dimer to the crystallization process. However, the fist coordination of Fe^{III} to *p*-hydroquinone enhances the acidity of the second OH, and this may account for the stabilization of the dinuclear complex. In contrast, the intra molecular H-bond stabilization in catechols reduces the acidity of the second OH favoring the formation of the mononuclear complex. Nevertheless, this is the first example showing that hydroquinone can function as bridging ligand.

Other examples of dinuclear complexes following a mode II coordination have been reported with simple hydroquinone to bridge two $Zr(acac)_3^+$ (**33**) (Evans, et al., 1998), or Fe^{III}(5,10,15,20-tetraphenylporphyrinato)$^+$ (**45**) (Rheingold & Miller, 2003), or $Ti^{IV}Cl(CP^*)_2^+$ (**38**) (Kunzel, et al., 1996), or $Ti^{III}(CP^*)_2$ (**37**) (Horacek, et al., 2010) or W^VOCl[hydrogen tris(3,5-dimethylpyrazolyl)borate]$^{+}$ (**17**) (Stobie, et al., 2003) or Mo^{V}OCl[hydrogen tris(3,5-dimethylpyrazolyl)borate]$^{+}$ (**41**) (Ung, et al., 1996). The last two moieties present additional interest because they form trinuclear complexes with two hydroquinones bridging three of these groups in an open structure (**46**) (Ung, et al., 1999) or three hydroquinones bridging three groups in a close triangular structure (**42**) (**35**) (McQuillan, et al., 1998; McQuillan, et al., 1996) (Figure 5). The C-$O_{hydroquinonate}$ bond distances range from 1.320 up to 1.394 Å indicative of the hydroquinone oxidation state of the ligand.

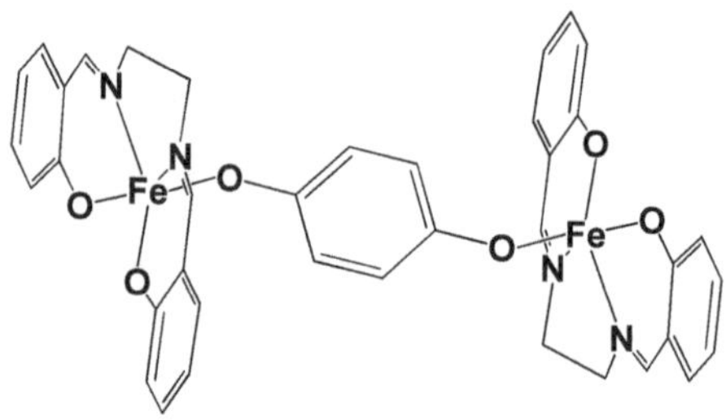

Fig. 4. Drawing of the molecular structure of first bridged hydroquinone complex **19**. The numbering of complexes is according to table 1

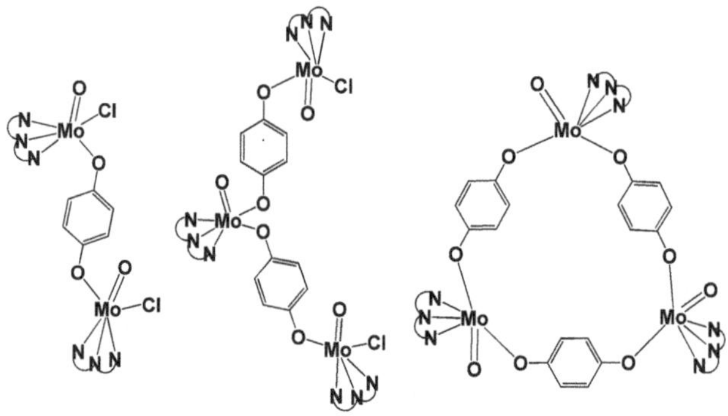

41 **46** **35**

Fig. 5. Drawings of the structure of the **41** monuclear, **46** trinuclear open structure, and **35** trinuclear trigonal structure, complexes of hydroquinone with Mo^{V}OCl[hydrogen tris(3,5-dimethylpyrazolyl)borate]$^{+}$ group. The numbering of complexes is according to table 1

In exploiting the bridging properties of hydroquinone, polymeric complexes of $Ti^{III/IV}$ and $V^{III/IV}$ have been synthesized and structurally characterized. The successful synthesis of polymeric architectures is based on the use of non chelating monodentate co-ligands, such as pyridine, which leave several positions open for coordination of the metal ion from hydroquinone. The crystal structures of these compounds show the hydroquinone to ligate metal ions in various bridging modes including, mode **II** (**32**), (**34**), (**31**) (Tanski, et al., 2000; Tanski & Wolczanski, 2001; Vaid, et al., 1997), mode **IV** (**63**) (Vaid, et al., 1997) and mode **VIII** (**51**) (Vaid, et al., 1997). A noticeable feature of these structures is the dependence of the M-$O_{hydroquinonate}$ bond distances on the coordination mode. For example, for the coordination

mode **II** the Ti-$O_{hydroquinonate}$ bond distance ranges from 1.782 up to 1.923 Å, which are shorter than the distances of the bridged Ti - μ-$O_{hydroquinonate}$ - Ti (2.042, 2.048 Å) and the Ti-$OH_{hydroquinonate}$ distance (2.207 Å). The crystal packing of these structures reveal the formation of 1D (**31**) (Tanski & Wolczanski, 2001), 2D (**34**) (Tanski, et al., 2000) and 3D polymers (**32**), (**63**) (Vaid, et al., 1997; Vaid, et al., 1999). The formation of lower dimension 1D polymers for the V^{IV} complex compared with the titanium ones, is mainly due to the less available positions for hydroquinone coordination (two for the vanadium and four for the titanium complexes) because of the presence of the V=O_{oxo} group and the smaller coordination number (5).
Structural characterized complexes with simple hydroquinone to ligate metal ions in monodentate manner (mode **I**) are very rare (table 1). Very interesting examples are the 3D hydrogen bonded structures of the homoleptic six coordinated tungsten complex [W(4-hydrophenoxy)$_6$] (**14**). (Vaid, et al., 2001)

2.2 Substituted hydroquinones

The substituted hydroquinones with chelate groups can be separated into two main categories, the 2- monosubstituted and the 2,5- bisubstituted.
Monosubstituted hydroquinones result in the formation of type **I**, **III**, or **VI** complexes (figure 2). The structure of the complex seems to be controlled from the metal and the substitution. In general, larger substitutions with groups that can form more chelate rings favor the monodentate coordination mode **I** and smaller groups tend to form binuclear M - μ-$O_{hydroquinonate}$ - M bridged type **III** complexes. An example, are the Cu^{II} complexes of the Schiff pyridine complexes **56** and **11** shown in figure 6. (Li, et al., 2000)

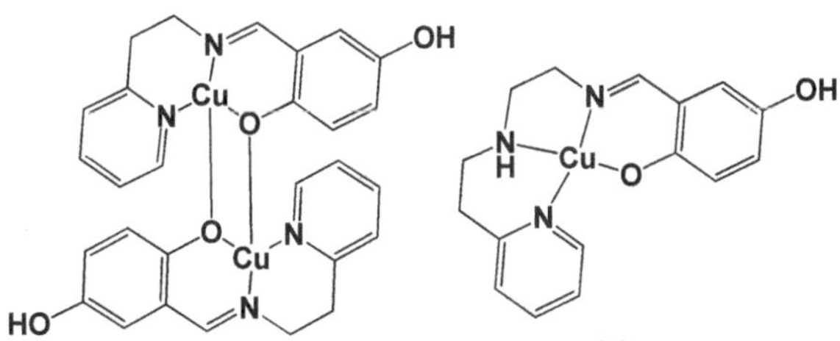

56 **11**

Fig. 6. Drawings of the structures of complexes **56** and **11** showing the effect of the size of the chelate group in the preference to type **III** and type **I** coordination modes respectively. The numbering of complexes is according to table 1

The Cu-$O_{hydroquinonate}$ bond distance of **11** (1.870 Å) is sufficiently shorter than the respective distances of **56** (1.932 - 1.976 Å) as expected. The C-$O_{hydroquinonate}$ bond distance of the ligated to the metal oxygen (1.321 and 1.350 Å) is also shorter than the free C-$OH_{hydroquinonate}$ bond (1.384 and 1.389 Å).
The protonated **VI** and the depproronated **I**, **III** of a complex are possible to be present in the solution of the reaction mixture and are dependent on the acidity-basicity of the solution. The speciation of water soluble complexes is controlled by pH and thus, various structural different complexes can be isolated. (Stylianou, et al., 2008) An example of the structures of the iminodiacetate monosubstituted hydroquinones isolated at different pHs are shown in figure 7.

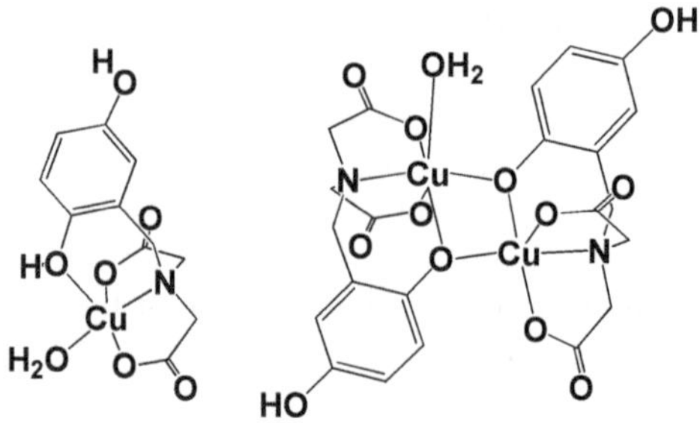

60 55

Fig. 7. Drawings of the structures of complexes **60** and **55** isolated from acidic and alkaline pHs respectively. (Stylianou, et al., 2008; Zhang, et al., 2009) The numbering of complexes is according to table 1

The $Cu^{II}_2O_2$ phenolate is a rare example of an asymmetric bridging core containing two copper atoms, one having a square pyramidal and the other an octahedral coordination sphere. This is the second example reported in the literature of the asymmetric $Cu^{II}_2O_2$ phenolate bridged complexes with the two copper ions exhibiting different coordination geometry (octahedral and square pyramidal). It is worth noticing here that although the type **III** dinuclear always crystallizes out of the alkaline solution, speciation studies have shown that the major species in solution is the type **I** mononuclear species. (Stylianou, et al., 2010)

Bisubstituted hydroquinones with chelate groups have been found to form with metal ions type **II, V** and **VII** structures. The same principles that we discussed for monosubstituted complexes are valid here too. However, the bisubstituted hydroquinone is a bridging ligand and chelate groups that leave the metal ion unsaturated lead to polymerization through additional coordination (for example complex **18**) (Dinnebier, et al., 2002) or through M-$O_{hydroquinonate}$-M bridging (for example complex **69**) (Kretz, et al., 2006)(figure 8).

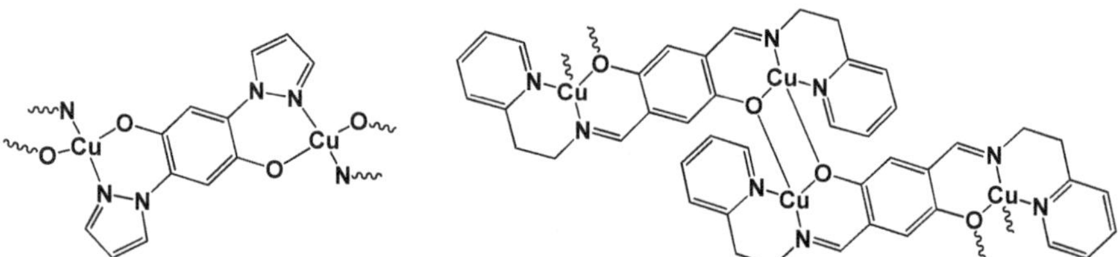

18 69

Fig. 8. Drawings of the polymeric structures of **18** and **69**. The numbering of complexes is according to table 1

Two very interesting examples of coordination modes **VII** and **V** are the structures of the complexes **62** and **55** (figure 9). In both complexes Cu^{II} ions are ligated to the same ligand (H_6bicah, figure 10), but they have been isolated from acidic and basic pHs respectively. Despite the fact that the hydroquinonate oxygen atom is deprotonated, the bond distances from Cu^{II} are very long [2.370(2) Å, 2.464(4) Å]. Deprotonation of the distant Y495 tyrosine in galactose oxidase accompanied with the strengthening of the interaction with copper ion has been proposed to be an important step in the catalytic activity of the enzyme. In this structure, despite the deprotonation of the hydroquinone, the oxygen atom remains in axial position and therefore could be considered as an intermediate of this process.

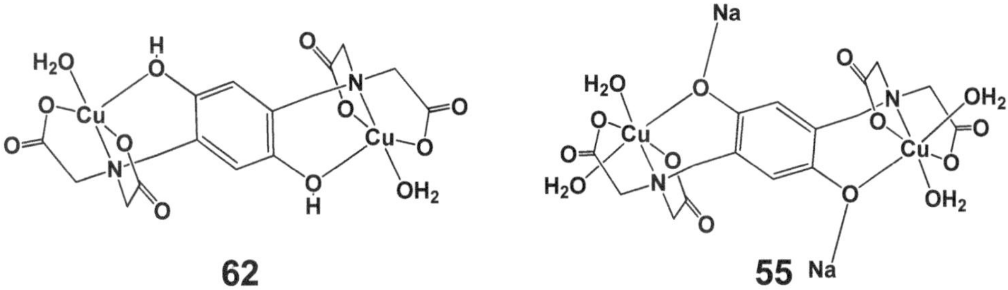

62 **55**

Fig. 9. Drawings of the structure of **62** and **55**. The numbering of complexes is according to table 1

2.3 Semiquinonates

There are only three *p*-semiquinonate complexes characterized by X-ray crystallography, and in all cases the semiquinone ligand bridges two V^{IV} ions. The ligands which are used for the stabilization of the semiquinone radicals are shown in figure 10.

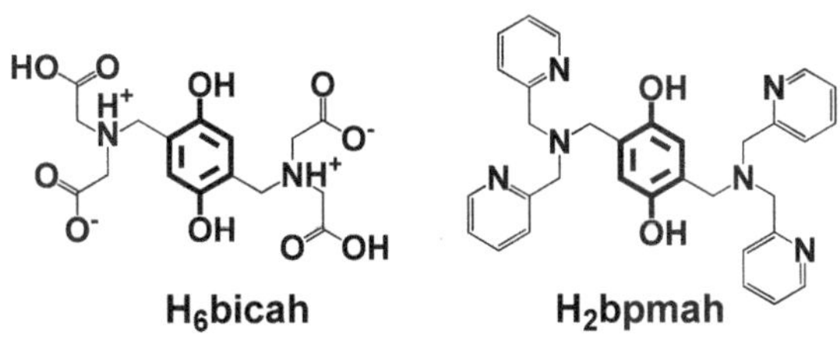

H_6**bicah** H_2**bpmah**

Fig. 10. 2,5- substituted iminodiacetic (H_6**bicah**) and bis(2-methylpyridyl)amine (H_2**bpmah**) hydroquinone ligands

The two of the three complexes have a rectangular shape structure, similarly to complexes **27, 28**: one has all vanadium atoms in oxidation state IV and both ligands in semiquinonate oxidation state (**15, 78**) and the other has two V^{IV} and two V^{V} ions ligated to a semiquinone and a hydroquinone ligand respectively (**29, 30**). The third complex is a V^{IV} dinuclear semiquinonate complex (**75, 79**) (figure 11). In addition to these compounds, the structures of the complexes with the ligands in hydroquinone oxidation state and the vanadium ions in IV or V or IV/V have been solved (figure 11). Because the structures of the complexes remain the same independently of the oxidation state of the ligand or the metal ion, this study gives us the opportunity for direct comparison between the compounds X-ray data.
The $C\text{-}O_{semiquinonate}$ bond distance ranges between 1.294 and 1.322 Å which is significantly shorter than the respective $C\text{-}O_{hydroquinonate}$ bond distance (1.328-1.366 Å) and designates the oxidation state of the ligand. The $V^{IV}\text{-}O_{semiquinonate}$ bond distances are significantly shorter than the respective $V^{IV}\text{-}O_{hydroquinonate}$ when complexes of the same ligand are compared. The mean values of the $V^{IV}\text{-}O_{semiquinonate}$ bond distances (1.89 ± 0.03 Å) calculated for all

complexes are also smaller than the mean values of V^{IV}-$O_{hydroquinonate}$ (1.94 ± 0.03 Å). Apparently, V^{IV} ions have higher affinity to bind the semiquinone oxygen atom than the hydroquinone and thus the ligation of V^{IV} is the main factor for stabilization of semiquinone radical in these complexes. More information about the affinity of the V^{IV} ion for the semiquinonate oxygen is being obtained from the C-$O_{hydroquinonate/semiquinonate}$-V angle. The C-$O_{semiquinonate}$-$V^{IV}$ angle (~137 °) is similar to the angle of C-$O_{hydroquinonate}$-V^{V} and indicates that the V^{IV}-$O_{semiquinonate}$ bond has significant π character. In contrast, the C-$O_{hydroquinonate}$-V^{IV} angle (~127 °) is significantly smaller, reflecting the weaker bonding between V^{IV} and $O_{hydroquinonate}$. However, the most interesting differences appear in the mixed oxidation ligand tetranuclear complexes **29** and **30** where the differences between the C-$O_{semiquinonate}$-V^{IV} and the C-$O_{hydroquinonate}$-V^{V} angles can be compared directly in the same molecule (Table 2). In these complexes the C-$O_{semiquinonate}$-V^{IV} angles (136.4 - 138.1 °) are significantly larger than the C-$O_{hydroquinonate}$-V^{V} (131.8-134.8 °), revealing that V^{IV}-$O_{semiquinonate}$ π bonding is stronger even than V^{V}-$O_{hydroquinonate}$.

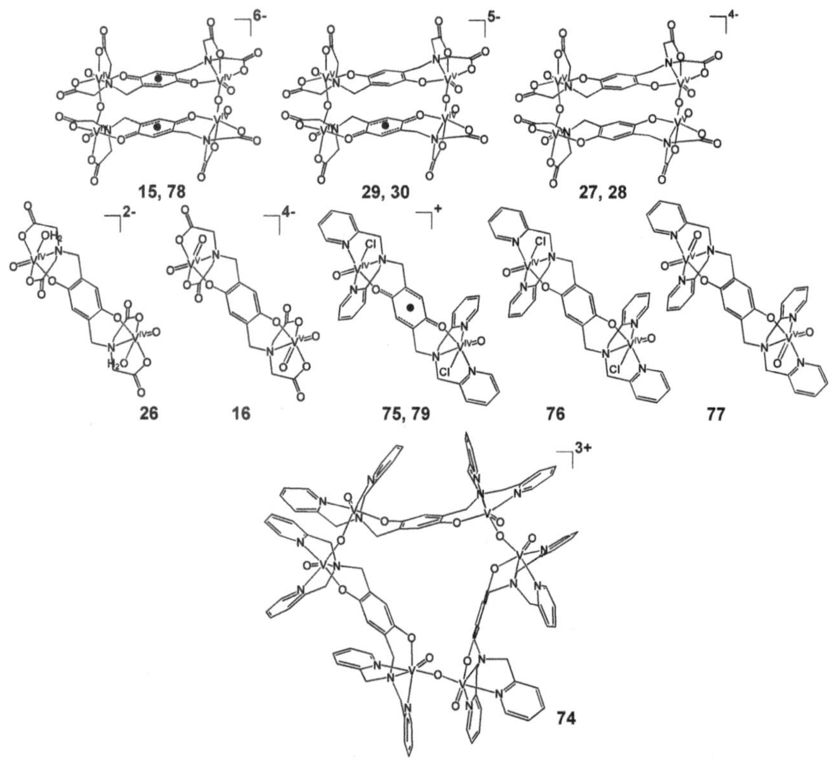

Fig. 11. Structures of vanadium complexes: 15, 16, 26-30, 74-79

2.4 *p*-Quinones

Complexes of *p*-quinones with transition metal ions are also rare. The quinone oxygen is a poor electron donor and thus, does not form easily σ- bonded complexes with transition metal ions. A few examples that have been included in Table 1 show quinone to form complexes with soft metal centers. The most peculiar feature of these structures is that coordination of the $O_{quinone}$ causes elongation of the bond in comparison with the free quinone which is opposite to what is observed for the hydroquinones. For example, in the type **I** complex **70** the $C{=}O_{quinone}$ of the ligated oxygen is 1.242 Å, which is 0.09 Å larger than the length of $C{=}O_{quinone}$ of the free oxygen atom (Senge, et al., 1999). This lengthening shows that there is a significant π-back donation.

3. Charge distribution

One of the main perspective for developing the transition metal coordination chemistry of ligands derived from hydroquinone/*p*-semiquinone/quinone is associated with their rich redox activity and the potential of forming compounds that may exist in a number of electronic states due to the combined electrochemical activity of the metal ion and one or more quinone ligands (Drouza & Keramidas, 2007). Structural studies on X-ray crystallography has proven to be a valuable tool in the determination of charge distribution and delocalization in the coupling redox active metal ions and ligands. The *p*-dioxolene transition metal complexes have shown that the delocalization is not important.

3.1 Charge calculations based on the C-O and the intraring bond lengths

The C-O and the C-C intraring bond lengths of the *p*-dioxolene ligands are strongly dependent on the formal charge of the ligands (Drouza & Keramidas, 2007). The C-O distance exhibits the largest differences in the three oxidation states, ~1.35 Å in hydroquinone, ~1.32 Å in *p*-semiquinone and 1.24 Å in *p*-quinone and therefore, can be used for the assignment of the oxidation state of the ligand (table 1). In addition, the C-C interatomic bond distances are almost indistinguishable in hydroquinone ring (~1.40 Å), but they exhibit a long-short-long pattern in *p*-semiquinones (~ 1.42, ~ 1.37 and ~ 1.42 Å) which becomes more distinct in *p*-quinones (~ 1.46, ~ 1.33 and ~ 1.46 Å).

A quantitative approach in the evaluation of quinones charge is the linear relationship of the charge of dioxolenes (D) versus the C-O and all the six intraring C/C distances of the ligand, given by the equations 1 and 2; where d_i is the experimental *i*th bond length, and d_{1i} and d_{2i} are the *i*th bond lengths of the pure forms of catecholate and *o*-benzoquinone respectively.

$$\Delta_i = -2(d_i - d_{2i})/(d_{1i} - d_{2i}) \qquad (1)$$

$$\Delta = (\Sigma w_i \Delta_i)/(\Sigma w_i) \qquad (2)$$

This equation has been primarily used for the *o*-dioxolenes by Carugo et.al (Carugo, et al., 1992). Recently, these equations have been applied on the determination of the charge of *p*-dioxolenes using the bond lengths of a large number of ligands ligated to various metal ions (Drouza & Keramidas, 2007). The values of d_{1i} and d_{2i} for hydroquinone and *p*-quinones respectively are considered as the experimental bond lengths of the uncomplexed organic molecules. The ideal values of Δ are 0, -1 and -2 for Q, SQ and HQ respectively. The application of equations 1 and 2 on these bond lengths for all the complexes afforded Δ values that were statistical significant, matching the expected hydroquinonate, semiquinonate or quinone nature of *p*-dioxolenes ligated either to one or to two metal ions.

Important crystallographic data, calculation of the oxidation state of the metal ions using bond valence sums (BVS) and application of Δ calculations on the vanadium complexes of figure 11 are summarized in table 2. The results from these calculations confirm the oxidation states of the metal ions and the organic redox centers. Δ values deviate from the ideal values, with the largest deviation to be observed in one of the hydroquinones of complex **5**, (Δ = -1.56 instead of -2). These deviations are probably due to the coordination of the metal ion to the ligand and to the ionic or dipolar interactions of the complex with the counter ions and solvent molecules, which transfer charge from the organic ligands, thus distorting further the structures of *p*-dioxolenes. Despite these deviations the prediction of the organic ligand oxidation state has been successful and is in agreement with the oxidation state found from other spectroscopic techniques.

Comp.	V=O/Å	$V-O_{HQ/SQ}$ / Å	$V-O_{bridge}$ / Å	$C-O_{HQ/SQ}$/ Å	$C-O_{HQ/SQ}-V$ / deg	Δ	BVS ($V^{IV/V}$)
H_2bpmah				1.366(1)		-1.88, HQ	
75	1.598(2)	1.913(2)		1.294(3)	135.3(2)	-0.97, SQ	+4.14 (V^{IV})
76	1.606(2)	1.941(1)		1.329(2)	126.8(1)	-1.67, HQ	+3.92 (V^{IV})
	1.613(2)	1.938(2)		1.345(4)	127.5(2)	-1.77, HQ	+3.90 (V^{IV})
79	1.605(2)	1.914(2)		1.300(3)	134.8(1)	-1.13, SQ	+4.08 (V^{IV})
77	1.645(3)	1.903(2)	1.630(3)	1.345(4)	131.6(2)	-1.61, HQ	+4.98 (V^{V})
74	1.597(4)	1.854(3)	1.720(3)	1.360(5)	136.7(6)	-1.71, HQ	+5.11 (V^{V})
	1.593(3)	1.905(3)	1.843(3)	1.329(5)	128.4(7)		+4.35 (V^{IV})
	1.599(3)	1.940(3)	1.899(3)	1.330(5)	127.2(6)	-1.84, HQ	+4.19 (V^{IV})
	1.598(3)	1.844(3)	1.736(3)	1.343(6)	134.0(6)		+5.04 (V^{V})
	1.589(4)	1.925(4)	1.886(3)	1.369(6)	126.3(6)	-1.71, HQ	+4.28 (V^{IV})
	1.599(3)	1.848(4)	1.774(3)	1.352(6)	136.5(6)		+4.99 (V^{V})
26	1.601(3)	1.950(3)	2.002(3)	1.366(5)	124.8(3)	-1.94, HQ	+3.93 (V^{IV})
	1.606(3)	1.952(3)	2.009(3)	1.353(5)	128.2(2)	-1.85, HQ	+3.92 (V^{IV})
27	1.603(2)	1.824(1)	1.840(2)	1.338(2)	137.0(1)	-1.67, HQ	+5.03 (V^{V})
	1.610(2)	1.866(1)	1.748(2)	1.346(2)	132.3(1)		+5.05 (V^{V})
28	1.599(2)	1.827(2)	1.818(2)	1.335(3)	133.8(2)	-1.75, HQ	+5.10 (V^{V})
	1.620(2)	1.865(2)	1.729(2)	1.346(3)	132.3(2)		+5.07 (V^{V})
	1.606(2)	1.825(2)	1.875(2)	1.327(3)	137.2(2)	-1.56, HQ	+4.97 (V^{V})
	1.613(2)	1.878(2)	1.759(2)	1.325(3)	135.9(2)		+5.07 (V^{V})
29	1.622(2)	1.878(2)	1.697(2)	1.346(3)	134.8(2)	-1.72, HQ	+5.12 (V^{V})
	1.625(2)	1.879(2)	1.699(2)	1.352(3)	131.8(2)		+5.06 (V^{V})
	1.597(2)	1.880(2)	1.914(2)	1.312(3)	138.1(2)	-1.30, SQ	+4.24 (V^{IV})
	1.604(2)	1.936(2)	1.934(2)	1.309(3)	136.4(2)		+4.07 (V^{IV})
30	1.611(3)	1.885(2)	1.707(3)	1.348(4)	134.2(2)	-1.70, HQ	+5.11 (V^{V})
	1.630(2)	1.851(2)	1.699(2)	1.335(4)	134.8(2)		+5.11 (V^{V})
	1.609(2)	1.915(2)	1.926(2)	1.299(4)	137.2(2)	-1.24, SQ	+4.18 (V^{IV})
	1.607(3)	1.903(2)	1.915(3)	1.293(4)	137.1(2)		+4.20 (V^{IV})
78	1.616(2)	1.867(2)	1.813(2)	1.321(3)	135.2(2)	-1.38, SQ	+4.40 (V^{IV})
	1.618(2)	1.879(2)	1.809(2)	1.313(3)	136.4(2)		+4.42 (V^{IV})
15	1.620(3)	1.886(3)	1.807(1)	1.322(5)	137.0(3)	-1.39, SQ	+4.35 (V^{IV})
16	1.652(3)	1.864(3)	1.620(4)	1.354(5)	131.6(3)	-2.02, HQ	+5.22 (V^{V})
	1.641(4)	1.878(3)	1.626(4)	1.353(6)	131.9(3)	-1.95, HQ	+5.22 (V^{V})

Table 2. Comparison of Selected Chemical Bonds, and Δ Values for ligand, Binuclear V^{IV}/V and Hexanuclear V^{IV} Complexes Respectively

3.2 Graphical discrimination of *p*-dioxolenes oxidation state

A nice graphical overview of the dependence of the C-$O_{Hydroquinonate/Semiquinonate}$ and V-$O_{Hydroquinonate/Semiquinonate}$ bond lengths, C-$O_{Hydroquinonate/Semiquinonate}$-V angle and Δ on the oxidation state of vanadium ion and the *p*-dioxolene ligand for the vanadium complexes contained in Table 2 is shown in figure 12.

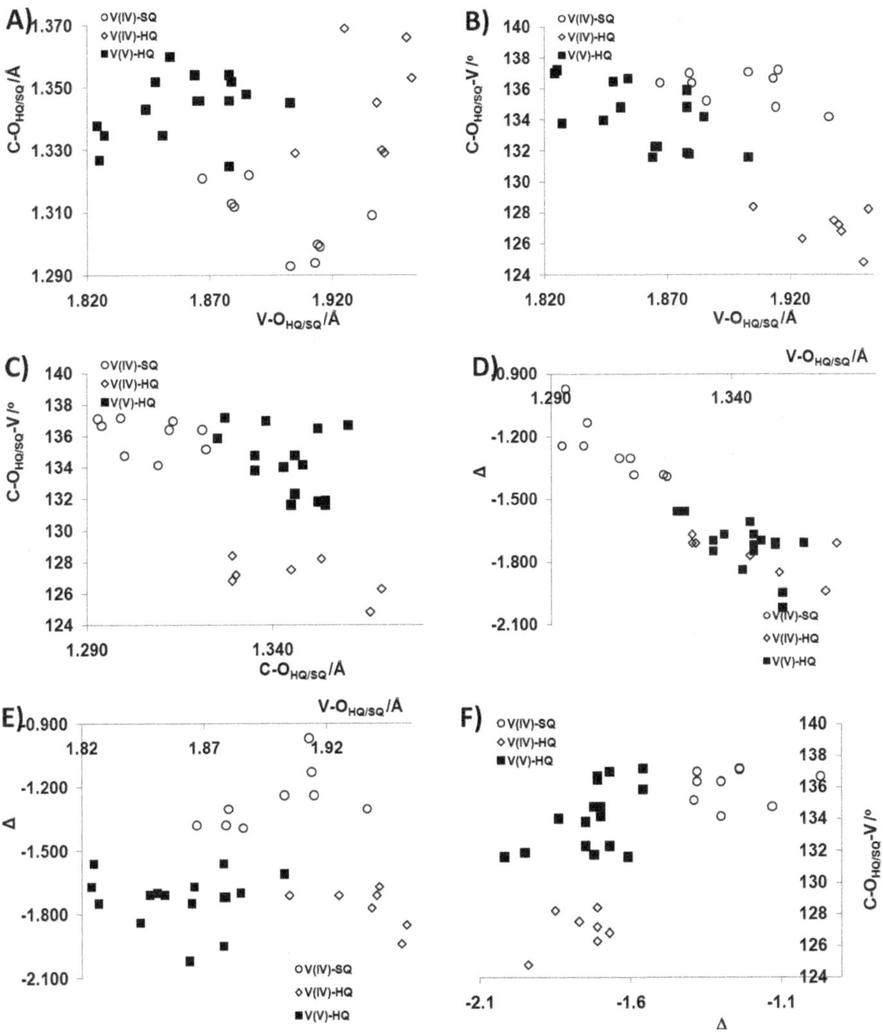

Fig. 12. Graphical presentations of the dependence of the C-$O_{HQ/SQ}$ / V-$O_{HQ/SQ}$ bond lengths, C-$O_{HQ/SQ}$-V and Δ/V-$O_{HQ/SQ}$ over the oxidation state of vanadium ion and the *p*-dioxolene ligand for the vanadium complexes in figure 11. HQ=hydroquinone, SQ=Semiquinone

The graph of C-$O_{Hydroquinonate/Semiquinonate}$ / V-$O_{Hydroquinonate/Semiquinonate}$ (figure 12A) gives separate regions for the different metal ligand oxidation states, however some of the values are at the border lines, mainly because the respective V-$O_{Hydroquinonate/Semiquinonate}$ deviate from the expected average bond lengths. The graphs of the C-$O_{Hydroquinonate/Semiquinonate}$-V angle versus V-$O_{Hydroquinonate/Semiquinonate}$ and C-$O_{Hydroquinonate/Semiquinonate}$ (Figure 12B, 12C) separate better the ligands oxidation state. There is a poor separation between the V^{V}-Hydroquinonate and V^{IV}-Semiquinonate complexes in C-$O_{Hydroquinonate/Semiquinonate}$ -V/V-$O_{Hydroquinonate/Semiquinonate}$ graphs because of the similar lengths between V^{V}-$O_{Hydroquinonate}$ and V^{IV}-$O_{Semiquinonate}$ bond lenghts. On the other hand C-$O_{Hydroquinonate/Semiquinonate}$-V/C-$O_{Hydroquinonate/Semiquinonate}$ graph gives a better separation between the different oxidation states. Among the graphs of C-$O_{Hydroquinonate/Semiquinonate}$, V-$O_{Hydroquinonate/Semiquinonate}$ and C-$O_{Hydroquinonate/Semiquinonate}$-V versus Δ (Figure 12D, 12E, 12F) the latter gives the best redox discrimination. Concluding, Δ and C-$O_{Hydroquinonate/Semiquinonate}$-V angle are the best parameters for the differentiation of the ligands oxidation state and of the strong V-O bonded V^{V}-Hydroquinonate and V^{IV}-Semiquinonate compounds than the weaker V^{IV}-Hydroquinonate respectively.

4. Utilization of X-ray crystallography as a characterization tool for the development of functional bioinorganic models based on hydroquinone/*p*-semiquinone ligands

The study of the bioinorganic model compounds significantly contributes to: a) the investigation of metal ions mechanisms and reactivity in biological systems and b) the development of efficient and selective stable catalysts that will mimic the activity of the enzymes. For example, the development of metal-quinone complexes and stabilization of the radicals has been inspired mainly from two redox center metal-enzymes that use modified *p*-semiquinone and phenolic radicals to facilitate two electron reduction of O_2 to H_2O_2 with subsequent oxidation of an organic substrate. The O_2/H_2O_2 oxidative C-H activation by the transition metal bioinorganic-based complexes constitutes a very important industrial process. X-ray crystallography has proven to be an essential tool for the successful implementation of the targets of bioinorganic chemistry, in particular, in deciphering the complicated multivariate aqueous systems. Some applications of crystallography may include the gathering of information about the speciation of the complexes in aqueous solutions and the providing of evidence for the molecules reactivity. In this chapter, we will discuss our most recent work in the development of vanadium based hydroquinone/semiquinone model systems for the investigation of reduction of O_2 induced by the metal ions in acidic aqueous solutions.

4.1 $V^{V/IV}$-hydroquinone/semiquinone models compounds

The V^{IV} complexes with the modified-hydroquinones H_6bicah, H_2bpmah (figure 10), have some very interesting features: they are water soluble, they stabilize *p*-semiquinone radicals in acidic conditions and they reduce dioxygen. Apparently, these compounds function both as activators and as radical traps and thus provide important information for the investigation of O_2 activation reactions by the isolation of reactive intermediates.

The tetranuclear V^{IV} bicah^{6-} complexes (figure 11) exhibit a pH induced electron transfer from the metal center to the ligand and subsequent fast oxidation from the atmospheric oxygen, in acidic pHs (4.5 - 3.0). (Drouza & Keramidas, 2008) As described above for this solution, the three tetranuclear complexes **15**, **29** and **27** (figure 11) were isolated and characterized by crystallography at pHs 4.5, 3.4 and 3.0 respectively. Based on these data a possible mechanism of a two step reaction of **15** has been proposed and is shown in figure 13. At the first step, an aqueous solution of **15** is acidified (pH=3.4), resulting in one electron transfer from one of the attached to semiquinones V^{IV} to the semiquinonate radical. The formed hydroquinone is immediately oxidized to semiquinone from the atmospheric oxygen and then the second V^{IV} is oxidized from the semiquinone to give **29**. At the second step, the semiquinone of the second dinuclear unit is also reduced, oxidizing one of the attached V^{IV} ions at pH=3.0 and proceeds to the formation of **27** following a similar mechanism. The trigger for this electron transfer, which finally results in oxidation of the complex from dioxygen, is the semiquinone protonation at acidic conditions. The mechanism is supported mainly by the isolation and the structural characterization of the initial and final products **15, 29, 27**, however none of the intermediates **id1-4** has been isolated or characterized. The reasons for not isolating the intermediates are a) the facile oxidation of the formed hydroquinone from the atmospheric oxygen and b) the almost overlapping redox potentials of the one electron $V^{V} \rightarrow V^{IV}$ and semiquinone→hydroquinone redox processes, and the structural similarity between the redox species which permits the fast electron transfer from the V^{IV} to the semiquinone radical.

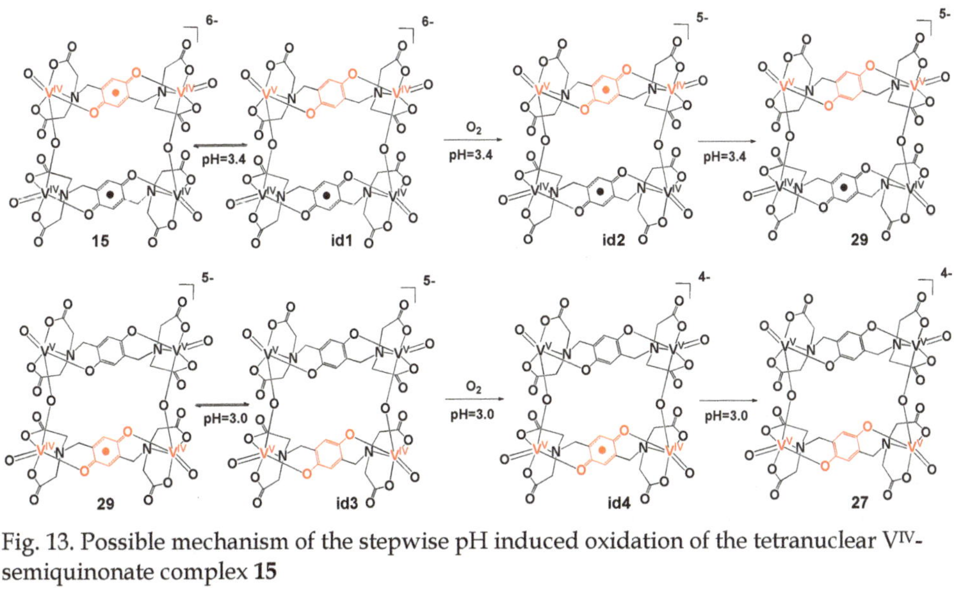

Fig. 13. Possible mechanism of the stepwise pH induced oxidation of the tetranuclear V^{IV}-semiquinonate complex **15**

Isolation and structural characterization of the intermediates will confirm our hypothesis, however the stabilization of these species is required. The strategy we followed (Stylianou, et al., 2011) was the stabilization of vanadium in oxidation state IV. This was achieved by the increase of the $V^{V} \rightarrow V^{IV}$ redox potential with the replacement of the hard oxygen carboxylate donor atoms ($bicah^{6-}$) with the softer pyridine nitrogen donor atoms ($bpmah^{2-}$) (figure 10). Thus, the respective $bpmah^{2-}$-intermediate complexes will be trapped because this semiquinone is not strong enough to oxidize the stabilized V^{IV} to V^{V}.

Indeed, complex **76** is being oxidized in aqueous solutions in two steps, first oxidation of the organic ligand to semiquinone (complex **75**) at pH 2.8 and then transfer of an electron from V^{IV} to the semiquinone at pH 2.2 resulting in the formation of the hexanuclear complex **74** (figure 14).

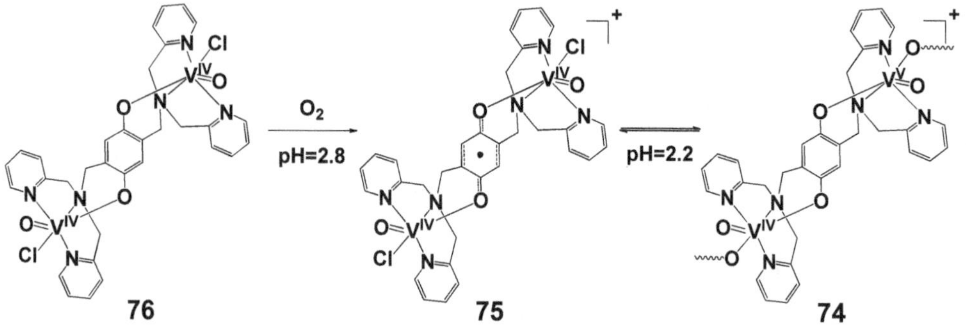

Fig. 14. Mechanism of the two step oxidation of **76** in aqueous solution

Although the $bpmah^{2-}$ complexes (two dinuclear oxo-bridged units) are structurally different from the respective compounds of $bicah^{6-}$ (one dinuclear unit), from the comparison of the reaction mechanism in figure 13 with this in figure 14, it is apparent that complex **76** is analogous to **id1**, **id3**, complex **75** to **id2**, **id4** and complex **74** to **29**, **27**. Complex **76** is stable in solution at pH 3.4, however, it is very rapidly oxidized to semiquinone radical at pH=2.8 from dioxygen. As we have predicted, the pyridine nitrogen donor atoms stabilize the oxidation state IV of vanadium. At pH 2.2 the oxophilicity of the metal increases, resulting in replacement of the chlorine with an oxo bridging group. This change of vanadium coordination environment is expected to reduce the $V^{IV} \rightarrow V^{V}$ redox potential and in combination with the stronger oxidative ability of semiquinone at lower pH (Baruah, et al., 2002; Drouza & Keramidas, 2008; Rath, et al., 1999) it results in electron transfer from V^{IV} to semiquinone and the slow formation of the V^{IV}/V^{V} mixed valent hydroquinone complex **74**.

Hydroquinone is not a strong reducing agent and thus cannot directly reduce O_2. It is certain that the interaction of the metal ion with O_2 increases the one electron reduction potential of dioxygen to superoxide radical oxidizing hydroquinone to semiquinone. Superoxide is further reduced to peroxide. One of the most surprising features of the semiquinone complex **75** is its low oxophilicity that increases at lower pH, as shown by the formation of **74**. Examples in the literature show that V^{IV} complexes with pyridine type donors are highly oxophilic, thus, it is probable that the bonding with the semiquinone radical softens the metal ion. This is important for further understanding the reactivity of vanadium at the oxidation reactions of organic substrates from the metal activated O_2. However, full understanding of the properties of these radical complexes requires more *p*-semiquinone complexes to be synthesized and crystallographically characterized. In detailed investigation of these mechanisms, the involvement of additional electrochemical, magnetic and spectroscopic techniques are required. However, crystallography is always the best way to start. It constitutes the first priority in the development of new bioinorganic model compounds.

5. Conclusions

The difficulties in the synthesis of stable transition metal - hydroquinone/semiquinone/quinone complexes have delayed the development of this chemistry relative to the phenol and catechol ones. The notion that *p*-dioxolene chemistry resembles that of phenols or *o*-dioxolenes and thus the study of those molecules also covers hydroquinones is mistaken. *p*-Dioxolenes have different reactivity. They are more reactive than phenols and less reactive than *o*-dioxolenes. In addition, as it has been shown, it is a bridging ligand which provides functional polymeric materials with novel optical, redox and magnetic properties. The last ten years, the importance of these ligands in the synthesis of bioinorganic models, in the development of bioinspiring "green" catalysts and of functional materials has been recognized, resulting in an increase of the structurally characterized *p*-dioxolene transition metal complexes.

Here, we have reviewed the rich coordination chemistry of *p*-dioxolenes with transition metals found in their crystal structures, examined the structural data that can be applied for the calculation of ligand charge and understood the factors in the metal induced stabilization of *p*-semiquinone radicals. V^{IV} is the only ion found up to now to stabilize the σ-bonded *p*-semiquinone radical. The stabilization is a result of the very strong bond between the metal and the oxygen of the dioxolene ligand. These binary metal-organic redox bioinorganic models have rich pH induced redox chemistry in aqueous solution as it has been proven from the detailed crystallographic study of the species produced at various pHs. Particular impetus for future research aimed at these molecules is provided by the established significance in O_2 activation reactions.

The structural chemistry of *p*-dioxolenes transition metal complexes is to a large extent unexplored. However, the unique redox properties and structural diversity of these ligands in combination with the recent advances in novel syntheses for the stabilization of the complexes have attracted the scientific curiosity, and thus, a prosperous future for this chemistry is waiting to be seen.

6. Acknowledgment

We thank the Research Promotion Foundation of Cyprus and the European Structural Funds for the financial support of this work with the proposals ΔΙΔΑΚΤΩΡ/ΔΙΣΕΚ/0308/49.

7. References

Arévalo, S., Bonillo, M.R., de Jesús, E., de la Mata, F.J., Flores, J.C., Gómez, R., Gómez-Sal, P. & Ortega, P., (2003). Synthesis of polymetallic Group 4 complexes bridged by benzenediolate and triolate ligands. X-ray crystal structure of [{Ti(C5Me5)Cl2}2{[mu]-1,4-O(2,3-C6H2Me2)O---}]. *Journal of Organometallic Chemistry*, Vol.681, No.1-2, pp. 228-236, ISSN 0022-328X

Baruah, B., Das, S. & Chakravorty, A., (2002). A family of vanadate esters of monoionized and diionized aromatic 1,2-diols: Synthesis, structure, and redox activity. *Inorganic Chemistry*, Vol.41, No.17, pp. 4502-4508, ISSN 0020-1669

Becker, J.M., Barker, J., Clarkson, G.J., van Gorkum, R., Johal, G.K., Walton, R.I. & Scott, P., (2010). Chirality and diastereoselection in the mu-oxo diiron complexes L(2)Fe-O-

FeL(2) (L = bidentate salicylaldiminato). *Dalton Transactions*, Vol.39, No.9, pp. 2309-2326, ISSN 1477-9226

Berthon, R.A., Colbran, S.B. & Craig, D.C., (1992). Palladium(II) complexes of 2-(2,5-dimethoxyphenyl)-1,10-phenathrolinr (phenhqme2) and 2-(2,5-hydroquinonyl)-1,10-phenathroline (phenhqh2) - the x-ray crystal-structure of PDCL (phenhqh).H2O.(CH3)2SO. *Polyhedron*, Vol.11, No.2, (Jan), pp. 243-250, ISSN 0277-5387

Brandon, E.J., Rogers, R.D., Burkhart, B.M. & Miller, J.S., (1998). The structure and ferrimagnetic behavior of meso-tetraphenyl-porphinatomanganese(III) tetrachloro-1,4-benzoquinonide, (MnTPP)-T-III (+) QCl(4) (center dot-)center dot PhMe: Evidence of a quinoidal structure for QCl(4) (center dot-). *Chemistry-a European Journal*, Vol.4, No.10, (Oct), pp. 1938-1943, ISSN 0947-6539

Caldwell, S.L., Gilroy, J.B., Jain, R., Crawford, E., Patrick, B.O. & Hicks, R.G., (2008). Synthesis and redox properties of a phosphine-subsituted para-dioxolene and its bimetallic palladium complex. *Canadian Journal of Chemistry-Revue Canadienne De Chimie*, Vol.86, No.10, (Oct), pp. 976-981, ISSN 0008-4042

Calvo, R., Abresch, E.C., Bittl, R., Feher, G., Hofbauer, W., Isaacson, R.A., Lubitz, W., Okamura, M.Y. & Paddock, M.L., (2000). EPR study of the molecular and electronic structure of the semiquinone biradical Q(A)(-center dot) Q(B)(-center dot) in photosynthetic reaction centers from Rhodobacter sphaeroides. *Journal of the American Chemical Society*, Vol.122, No.30, (Aug), pp. 7327-7341, ISSN 0002-7863

Carugo, O., Castellani, C.B., Djinović, K. & Rizzi, M., (1992). Ligands derived from o-benzoquinone: Statistical correlation between oxidation state and structural features. *Journal of the Chemical Society, Dalton Transactions*, No.5, pp. 837-841, ISSN 0300-9246

Dietzel, P.D.C., Johnsen, R.E., Blom, R. & Fjellvag, H., (2008). Structural changes and coordinatively unsaturated metal atoms on dehydration of honeycomb analogous microporous metal-organic frameworks. *Chemistry-a European Journal*, Vol.14, No.8, pp. 2389-2397, ISSN 0947-6539

Dinnebier, R., Lerner, H.W., Ding, L., Shankland, K., David, W.I.F., Stephens, P.W. & Wagner, M., (2002). One-dimensional spin chains from Cu-II ions and 2,5-bis(pyrazol-1-yl)-1,4-dihydroxybenzene. *Zeitschrift Fur Anorganische Und Allgemeine Chemie*, Vol.628, No.1, (Jan), pp. 310-314, ISSN 0044-2313

Drouza, C. & Keramidas, A.D. (2007). Charge distribution in vanadium p-(hydro/semi)quinonate complexes. In: *Vanadium: The Versatile Metal*, Kustin, K., et al (Ed.), pp.352-363, American Chemical Society, ISBN 9780841274464, Washington, DC

Drouza, C. & Keramidas, A.D., (2008). Solid state and aqueous solution characterization of rectangular tetranuclear V-IV/V-p-semiquinonate/hydroquinonate complexes exhibiting a proton induced electron transfer. *Inorganic Chemistry*, Vol.47, No.16, (Aug), pp. 7211-7224, ISSN 0020-1669

Drouza, C., Tolis, V., Gramlich, V., Raptopoulou, C., Terzis, A., Sigalas, M.P., Kabanos, T.A. & Keramidas, A.D., (2002). p-hydroquinone-metal compounds: synthesis and crystal structure of two novel V-V-p-hydroquinonate and V-IV-p-semiquinonate species. *Chemical Communications*, No.23, pp. 2786-2787, ISSN 1359-7345

Errington, R.J., Petkar, S.S., Middleton, P.S., McFarlane, W., Clegg, W., Coxall, R.A. & Harrington, R.W., (2007). Non-aqueous synthetic methodology for TiW5 polyoxometalates: protonolysis of (MeO) TiW5O18 (3-) with alcohols, water and phenols. *Dalton Transactions*, No.44, pp. 5211-5222, ISSN 1477-9226

Evangelio, E. & Ruiz-Molina, D., (2005). Valence tautomerism: New challenges for electroactive ligands. *European Journal of Inorganic Chemistry*, No.15, (Aug), pp. 2957-2971, ISSN 1434-1948

Evans, W.J., Ansari, M.A. & Ziller, J.W., (1998). The reactivity of zirconium acetylacetonate with phenols. *Polyhedron*, Vol.17, No.2-3, pp. 299-304, ISSN 0277-5387

Gelling, O.J., Meetsma, A. & Feringa, B.L., (1990). Bimetallic oxidation catalysts-synthesis, x-ray-analysis, and reactivity of a binuclear para-hydroquinone-containing copper(II) complex. *Inorganic Chemistry*, Vol.29, No.15, (Jul), pp. 2816-2822, ISSN 0020-1669

Handa, M., Matsumoto, H., Namura, T., Nagaoka, T., Kasuga, K., Mikuriya, M., Kotera, T. & Nukada, R., (1995). Chain and discrete dimer complexes of molybdenum(II) trifluoroacetate axially coordinated by p-quinones, Mo-2(O2CCF3)(4)(2,6-Me-BQ) (N) (2,6-Me-BQ=2,6-dimethyl-p-benzoquinone) and Mo-2(O2CCF3)(4)(2,6-t-Bu-BQ)(2) (2,6-t-Bu-BQ=2,6-di-t-Butyl-p-benzoquinone). *Chemistry Letters*, No.10, (Oct), pp. 903-904, ISSN 0366-7022

Handa, M., Mikuriya, M., Sato, Y., Kotera, T., Nukada, R., Yoshioka, D. & Kasuga, K., (1996). Chain compounds of rhodium(II) trifluoroacetate linked by p-quinone Rh-2(O2CCF3)(4)(p-Q) (n), p-Q=1,4-benzoquinone, 1,4-naphthoquinone, and 2,3-dimethyl-1,4-benzoquinone. *Bulletin of the Chemical Society of Japan*, Vol.69, No.12, (Dec), pp. 3483-3488, ISSN 0009-2673

He, Z.C., Colbran, S.B. & Craig, D.C., (2003). Could redox-switched binding of a redox-active ligand to a copper(II) centie drive a conformational proton pump gale? A synthetic model study. *Chemistry-a European Journal*, Vol.9, No.1, (Jan), pp. 116-129, ISSN 0947-6539

Heistand, R.H., Roe, A.L. & Que, L., (1982). Dioxygenase models - crystal - structures of N,N'-(1,2-phenylene)bis(salycyldeniminato)(catecholato-O)iron(III) and mu-(1,4-benzenediolato-O,O')-bis N,N'-ethylnebis(salicylideniminato)iron(III). *Inorganic Chemistry*, Vol.21, No.2, pp. 676-681, ISSN 0020-1669

Horacek, M., Gyepes, R., Cisarova, I., Kubista, J., Pinkas, J. & Mach, K., (2010). Synthesis and structure of dinuclear dimethylene- or 1,4-phenylene-linked bis(decamethyltitanoceneoxide) (Ti(III)) complexes. *Journal of Organometallic Chemistry*, Vol.695, No.21, (Oct), pp. 2338-2344, ISSN 0022-328X

Huang, F.P., Wang, Y.X., Zhao, J., Bian, H.D., Yu, Q. & Liang, H., (2008). p-hydroquinone-metal complex: Synthesis, crystal structure and electrochemical property. *Chinese Journal of Inorganic Chemistry*, Vol.24, No.9, (Sep), pp. 1523-1526, ISSN 1001-4861

Iwata, S., Lee, J.W., Okada, K., Lee, J.K., Iwata, M., Rasmussen, B., Link, T.A., Ramaswamy, S. & Jap, B.K., (1998). Complete structure of the 11-subunit bovine mitochondrial cytochrome bc(1) complex. *Science*, Vol.281, No.5373, (Jul), pp. 64-71, ISSN 0036-8075

Klinman, J.P., (1996). Mechanisms whereby mononuclear copper proteins functionalize organic substrates. *Chemical Reviews*, Vol.96, No.7, (Nov), pp. 2541-2561, ISSN 0009-2665

Kondo, M., Nabari, K., Horiba, T., Irie, Y., Kabir, M.K., Sarker, R.P., Shimizu, E., Shimizu, Y. & Fuwa, Y., (2003). Synthesis and crystal structure of Ni{bis(2,5-dihydroxysalicylidene)ethylenediaminato} : a hydrogen bonded assembly of Ni(II)-salen complex. *Inorganic Chemistry Communications*, Vol.6, No.2, (Feb), pp. 154-156, ISSN 1387-7003

Kretz, T., Bats, J.W., Losi, S., Wolf, B., Lerner, H.W., Lang, M., Zanello, P. & Wagner, M., (2006). Hydroquinone-bridged dinuclear CuII complexes and single-crystalline Cu-II coordination polymers. *Dalton Transactions*, No.41, pp. 4914-4921, ISSN 1477-9226

Kumbhakar, D., Sarkar, B., Maji, S., Mobin, S.M., Fiedler, J., Urbanos, F.A., Jimenez-Aparicio, R., Kaim, W. & Lahiri, G.K., (2008). Intramolecular Valence and Spin Interaction in meso and rac Diastereomers of a p-Quinonoid-Bridged Diruthenium Complex. *Journal of the American Chemical Society*, Vol.130, No.51, (Dec), pp. 17575-17583, ISSN 0002-7863

Kunzel, A., Sokolow, M., Liu, F.Q., Roesky, H.W., Noltemeyer, M., Schmidt, H.G. & Uson, I., (1996). Synthesis and characterisation of quinonide bridged dinuclear complexes of titanium and zirconium. *Journal of the Chemical Society-Dalton Transactions*, No.6, (Mar), pp. 913-919, ISSN 0300-9246

Li, P.Y., Solanki, N.K.I., Ehrenberg, H., Feeder, N., Davies, J.E., Rawson, J.M. & Halcrow, M.A., (2000). Copper(II) complexes of hydroquinone-containing Schiff bases. Towards a structural model for copper amine oxidases. *Journal of the Chemical Society-Dalton Transactions*, No.10, pp. 1559-1565, ISSN 0300-9246

Litos, C., Terzis, A., Raptopoulou, C., Rontoylanni, A. & Karaliota, A., (2006). Polynuclear oxomolybdenum(VI) complexes of dihydroxybenzoic acids: Synthesis, spectroscopic and structure characterization of a tetranuclear catecholato-type coordinated 2,3-dihydroxybenzoate and a novel tridentate salicylato-type coordinated 2,5-dihydroxybenzoate trinuclear complex. *Polyhedron*, Vol.25, No.6, (Apr), pp. 1337-1347, ISSN 0277-5387

Margraf, G., Bats, J.W. & Wagner, M., (2009). *Cambridge Structural Database* pp. ISSN

Margraf, G., Kretz, T., de Biani, F.F., Laschi, F., Losi, S., Zanello, P., Bats, J.W., Wolf, B., Removic-Langer, K., Lang, M., Prokofiev, A., Assmus, W., Lerner, H.W. & Wagner, M., (2006). Mono-, di-, and oligonuclear complexes of Cu-II ions and p-hydroquinone ligands: Syntheses, electrochemical properties, and magnetic Behavior. *Inorganic Chemistry*, Vol.45, No.3, (Feb), pp. 1277-1288, ISSN 0020-1669

Matalobos, J.S., Garcia-Deibe, A.M., Fondo, M., Navarro, D. & Bermejo, M.R., (2004). A di-mu-phenoxo bridged zinc dimer with unfamiliar spatial arrangement. *Inorganic Chemistry Communications*, Vol.7, No.2, (Feb), pp. 311-314, ISSN 1387-7003

McQuillan, F.S., Berridge, T.E., Chen, H.L., Hamor, T.A. & Jones, C.J., (1998). Bi-, tri-, and tetranuclear metallomacrocycles constructed by metal-directed reactions involving resorcinol or hydroquinone or by addition of octahedral metal centers to tetrahexylcalix 4 resorcinarene: X-ray crystal structure of syn,syn- Mo{HB(3,5-Me2C3HN2)(3)}(NO)(1,4-O2C6H4) (3). *Inorganic Chemistry*, Vol.37, No.19, (Sep), pp. 4959-4970, ISSN 0020-1669

McQuillan, F.S., Chen, H.G., Hamor, T.A. & Jones, C.J., (1996). Metal directed syntheses of metallocyclophanes containing triangular arrays of interacting redox active octahedral metal centres, X-ray crystal structure of W(Tp*)(NO)(OC6H4O) (3). *Polyhedron*, Vol.15, No.21, pp. 3909-3913, ISSN 0277-5387

Phan, N.H., Halasz, I., Opahle, I., Alig, E., Fink, L., Bats, J.W., Cong, P.T., Lerner, H.W., Sarkar, B., Wolf, B., Jeschke, H.O., Lang, M., Valenti, R., Dinnebier, R. & Wagner, M., (2011). Thermally induced crystal-to-crystal transformations accompanied by changes in the magnetic properties of a Cu(II)-p-hydroquinonate polymer. *Crystengcomm*, Vol.13, No.2, pp. 391-395, ISSN 1466-8033

Philibert, A., Thomas, F., Philouze, C., Hamman, S., Saint-Aman, E. & Pierre, J.L., (2003). Galactose oxidase models: Tuning the properties of Cu-II-phenoxyl radicals. *Chemistry-a European Journal*, Vol.9, No.16, (Aug), pp. 3803-3812, ISSN 0947-6539

Pierpont, C.G., (2001). Redox isomerism for quinone complexes of chromium and chromium oxidation state assignment from X-ray absorption spectroscopy. *Inorganic Chemistry*, Vol.40, No.22, (Oct), pp. 5727-+, ISSN 0020-1669

Rath, S.P., Rajak, K.K. & Chakravorty, A., (1999). Synthesis, structure, and catecholase reaction of a vanadate ester system incorporating monoionized catechol chelation. *Inorganic Chemistry*, Vol.38, No.20, pp. 4376-4377, ISSN 0020-1669

Rheingold, A.L. & Miller, J., (2003). *Cambridge Structural Database*, pp. ISSN

Rosi, N.L., Kim, J., Eddaoudi, M., Chen, B.L., O'Keeffe, M. & Yaghi, O.M., (2005). Rod packings and metal-organic frameworks constructed from rod-shaped secondary building units. *Journal of the American Chemical Society*, Vol.127, No.5, (Feb), pp. 1504-1518, ISSN 0002-7863

Sanmartin, J., Garcia-Deibe, A.M., Fondo, M., Navarro, D. & Bermejo, M.R., (2004). Synthesis and crystal structure of a mononuclear iron(III) (eta(2)-acetato) complex of a beta-cis folded salen type ligand. *Polyhedron*, Vol.23, No.6, (Mar), pp. 963-967, ISSN 0277-5387

Sembiring, S.B., Colbran, S.B. & Craig, D.C., (1999). Synthesis and electrochemistry of platinum complexes of hydroquinon-2-ylmethyl- and p-benzoquinon-2-ylmethyl-diphenylphosphine. *Journal of the Chemical Society-Dalton Transactions*, No.10, (May), pp. 1543-1554, ISSN 0300-9246

Sembiring, S.B., Colbran, S.B., Craig, D.C. & Scudder, M.L., (1995). Palladium(II) 2-diphenylphosphinohydroquinone (H(2)PPHQ) complexes - preparation and structures of a novel cluster, (PDBR(HPPHQ))(4) CENTER-DOT-2H2O, and a phosphine-phosphite complex, cis- PDBR2(C6H3(OH)-1,PPH(2)-3,PPH(2)O-4) CENTER-DOT-2H(2)O. *Journal of the Chemical Society-Dalton Transactions*, No.22, (Nov), pp. 3731-3741, ISSN 0300-9246

Sembiring, S.B., Colbran, S.B. & Hanton, L.R., (1992). The preparation, properties and x-ray crystal-structure of the nickel(II) hydroquinonylphosphine complex cis NI(PPHQH)2 .H2O.2(CH3)2NCHO. *Inorganica Chimica Acta*, Vol.202, No.1, (Dec), pp. 67-72, ISSN 0020-1693

Senge, M.O., Speck, M., Wiehe, A., Dieks, H., Aguirre, S. & Kurreck, H., (1999). Structure and conformation of photosynthetic pigments and related compounds. 12. A crystallographic analysis of porphyrin-quinones and their precursors. *Photochemistry and Photobiology*, Vol.70, No.2, (Aug), pp. 206-216, ISSN 0031-8655

Song, Y.F., van Albada, G.A., Tang, J., Mutikainen, I., Turpeinen, U., Massera, C., Roubeau, O., Costa, J.S., Gamez, P. & Reedijk, J., (2007). Controlled copper-mediated chlorination of phenol rings under mild conditions. *Inorganic Chemistry*, Vol.46, No.12, (Jun), pp. 4944-4950, ISSN 0020-1669

Sreenivasulu, B., Zhao, F., Gao, S. & Vittal, J.J., (2006). Synthesis, structures and catecholase activity of a new series of dicopper(II) complexes of reduced Schiff base ligands. *European Journal of Inorganic Chemistry*, No.13, (Jul), pp. 2656-2670, ISSN 1434-1948

Stobie, K.M., Bell, Z.R., Munhoven, T.W., Maher, J.P., McCleverty, J.A., Ward, M.D., McInnes, E.J.L., Totti, F. & Gatteschi, D., (2003). Mono- and di-nuclear tris(pyrazolyl)borato-oxo-tungsten(v) complexes with phenolate ligands: syntheses and structures, and magnetic, electrochemical and UV/Vis/NIR spectroscopic properties. *Dalton Transactions*, No.1, pp. 36-45, ISSN 1477-9226

Stylianou, M., Drouza, C., Athanasopoulos, G.I., Giapintzakis, J. & Keramidas, A.D., (2011). Vanadium(IV)-hydroquinonate radical traps towards the investigation of dioxygen activation from vanadium complexes at acidic pHs. *Submitted*, pp. ISSN

Stylianou, M., Drouza, C., Viskadourakis, Z., Giapintzakis, J. & Keramidas, A.D., (2008). Synthesis, structure, magnetic properties and aqueous solution characterization of p-hydroquinone and phenol iminodiacetate copper(II) complexes. *Dalton Transactions*, No.44, pp. 6188-6204, ISSN 1477-9226

Stylianou, M., Keramidas, A.D. & Drouza, C., (2010). PH-potentiometric investigation towards chelating tendencies of p -hydroquinone and phenol iminodiacetate copper(II) complexes. *Bioinorganic Chemistry and Applications*, Vol.2010, pp. ISSN 1565-3633

Tanski, J.M., Vaid, T.P., Lobkovsky, E.B. & Wolczanski, P.T., (2000). Covalent metal-organic networks: Pyridines induce 2-dimensional oligomerization of (mu-OC6H4O)(2)Mpy(2) (M = Ti, V, Zr). *Inorganic Chemistry*, Vol.39, No.21, (Oct), pp. 4756-4765, ISSN 0020-1669

Tanski, J.M. & Wolczanski, P.T., (2001). A covalent vanadium(III) 2-dimensional network and vanadyl chains linked by aryldioxides. *Inorganic Chemistry*, Vol.40, No.2, (Jan), pp. 346-353, ISSN 0020-1669

Ung, V.A., Bardwell, D.A., Jeffery, J.C., Maher, J.P., McCleverty, J.A., Ward, M.D. & Williamson, A., (1996). Dinuclear oxomolybdenum(V) complexes showing strong interactions across diphenol bridging ligands: Syntheses, structures, electrochemical properties, and EPR spectroscopic properties. *Inorganic Chemistry*, Vol.35, No.18, (Aug), pp. 5290-5299, ISSN 0020-1669

Ung, V.A., Couchman, S.M., Jeffery, J.C., McCleverty, J.A., Ward, M.D., Totti, F. & Gatteschi, D., (1999). Electrochemical and magnetic exchange interactions in trinuclear chain complexes containing Oxo-Mo(V) fragments as a function of the topology of the bridging ligand. *Inorganic Chemistry*, Vol.38, No.2, (Jan), pp. 365-369, ISSN 0020-1669

Vaid, T.P., Lobkovsky, E.B. & Wolczanski, P.T., (1997). Covalent 3- and 2-dimensional titanium-quinone networks. *Journal of the American Chemical Society*, Vol.119, No.37, (Sep), pp. 8742-8743, ISSN 0002-7863

Vaid, T.P., Sydora, O.L., Douthwaite, R.E., Wolczanski, P.T. & Lobkovsky, E.B., (2001). Hydrogen bonds between polyphenol (p-HOC6H4O)(6)W and bipyridines: (4,4 '-bipy center dot HOC6H4O)(6)W and 3-D networks {4,4 '-(nC(5)H(4))(2)(CH2CH2)}(n){(HOC6H4O)(6)W} (infinity) (n=2, 3). *Chemical Communications*, No.14, pp. 1300-1301, ISSN 1359-7345

Vaid, T.P., Tanski, J.M., Pette, J.M., Lobkovsky, E.B. & Wolczanski, P.T., (1999). Covalent three-dimensional titanium(IV)-aryloxide networks. *Inorganic Chemistry*, Vol.38, No.14, (Jul), pp. 3394-3405, ISSN 0020-1669

Zhang, X.Q., Huang, F.P., Bian, H.D., Yu, Q. & Liang, H., (2009). Aqua N-(2,5-dihydroxybenzyl)iminodiacetato copper(II). *Acta Crystallographica Section E-Structure Reports Online*, Vol.65, (Nov), pp. M1393-U1393, ISSN 1600-5368

Zharkouskaya, A., Buchholz, A. & Plass, W., (2005). A new coordination polymer architecture with (10,3)-a network containing chiral hydrophilic 3-D channels. *European Journal of Inorganic Chemistry*, No.24, (Dec), pp. 4875-4879, ISSN 1434-1948

Structural Diversity on Copper(I) Schiff Base Complexes

Aliakbar Dehno Khalaji
Department of Chemistry, Faculty of Science, Golestan University, Gorgan
Iran

1. Introduction

Azomethines (known as Schiff bases), are perspective materials for wide spectrum of applications, particularly for anion sensor [1], antimicrobial agents [2-4] and nonlinear optical materials [5,6]. There has been considerable interest in some Schiff bases derived from salicylaldehyde because they show photochromism and thermochromism in the solid state [7]. The preparation of these compounds is simple and elegant. Since their discovery by Hugo Schiff in 1864 [8], they are prepared by condensing an active carbonyl compounds (ketone or aldehyde) with an amine, generally in refluxing alcohol [9-15]. Schiff bases are often used as ligands in inorganic chemistry [16-22].
In recent years, there has been a growing interest in the synthesis, characterization and crystal structures of copper(I) Schiff base complexes, not only because they have interesting properties and structural diversity [23-25] but also because they have found important application in catalysis for the coupling of phenylacetylene with halobenzene [26], preparation of supramolecular assemblies [27,28], the design of single and double-stranded architectures [29,30] and the grid complexes [31,32]. Then, Many efforts have been devoted to the design and synthesis of new Schiff base ligands that would be able to control the crystal structure of copper(I) complexes [33-40]. The purpose of this chapter is to present the current status of chemistry of copper(I) Schiff base complexes.

2. Schiff base ligands

Schiff bases are functional groups that contain a carbon-nitrogen double bond (C=N) with the nitrogen atom connected to an aryl or alkyl group, but not hydrogen. They are of the general formula R1R2C=N-R3, where R3 is an aryl or alkyl group that makes the Schiff base a stable imine. Schiff base compounds can be synthesized from an amine and a carbonyl compound by nucleophilic addition, followed by a dehydration to generate an imine [9-15], and are broadly classified as bidnetate and bis-bidentate Schiff bases.

2.1 Bidentate schiff-bases

2.1.1 Symmetric bidentate schiff bases

The basic symmetric bidentate Schiff base ligands (Scheme 1) have different R1 and R2 substituents [41-62]. Schiff bases based on aldehydes have hydrogen atom as one of the substituents (R1) at carbon atom of azomethine group, while second substituent R2 may be an alkyl, an aryl or a heterocyclic group. Schiff bases based on ketones have an alkyl, an aryl or a heterocyclic group in both the substituents at carbon atom of azomethine group, which may be same or different.

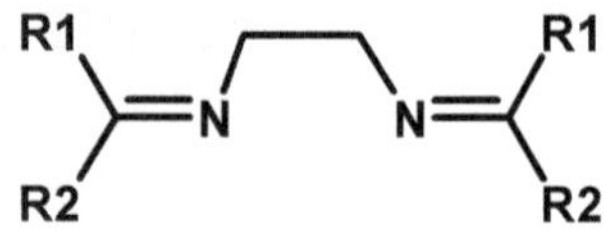

Scheme 1.

The symmetric bidentate Schiff base ligands have two arms connected via a ring, or C-C bond, as for example, shown in Schemes 2-8.

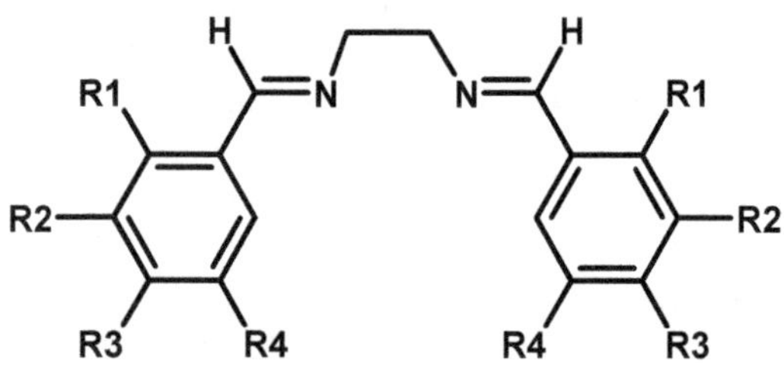

R1 = R2 = R4 = H	R3 = CH_3	L^1 [41]
R1 = H	R2 = R3 = R4 = CH_3O	L^2 [41,42]
R1 = R3 = R4 = H	R2 = CH_3O	L^3 [43]
R1 = R2 = R4 = H	R3 = CH_3O	L^4 [44]
R2 = R3 = R4 = H	R1 = Cl	L^5 [45]
R1 = R2 = R3 = R4 = H		L^6 [46]

Scheme 2.

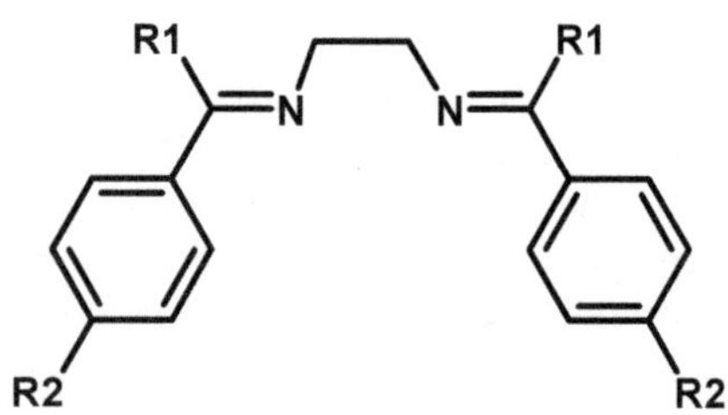

R1 = Ph	R2 = H	L^7 [47]
R1 = CH_3	R2 = Cl	L^8 [48]
R1 = CH_3	R2 = NO_2	L^9 [48]

Scheme 3.

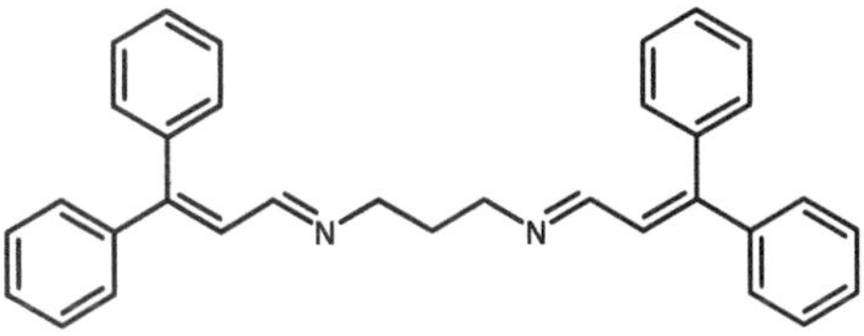

L^{10} [49]

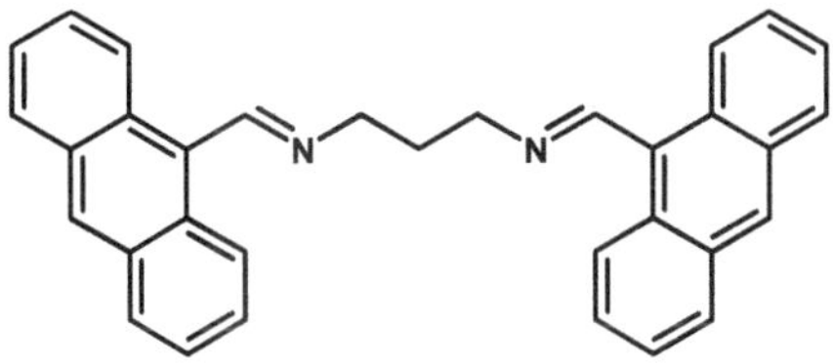

L^{11} [50]

Scheme 4.

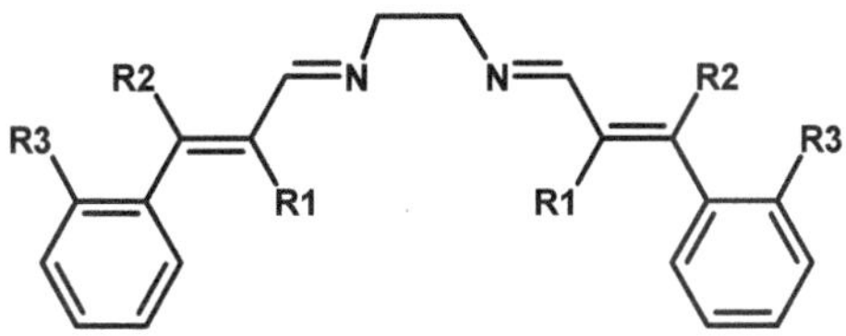

R1 = R2 = R3 = H		L^{12} [51,52]
R1 = R2 = H	R3 = NO_2	L^{13} [53]
R1 = R3 = H	R2 = Ph	L^{14} [54,55]
R2 = R3 = H	R1 = CH_3	L^{15} [56]

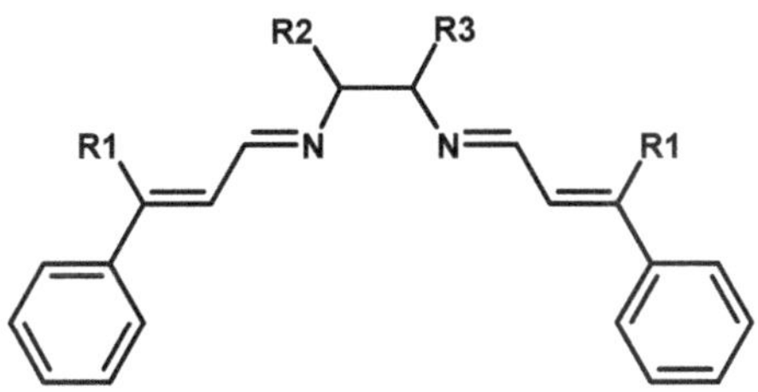

R1 = R2 = H	R3 = CH_3	L^{16} [57]
R1 = R2 = R3 = Ph		L^{17} [57]

Scheme 5.

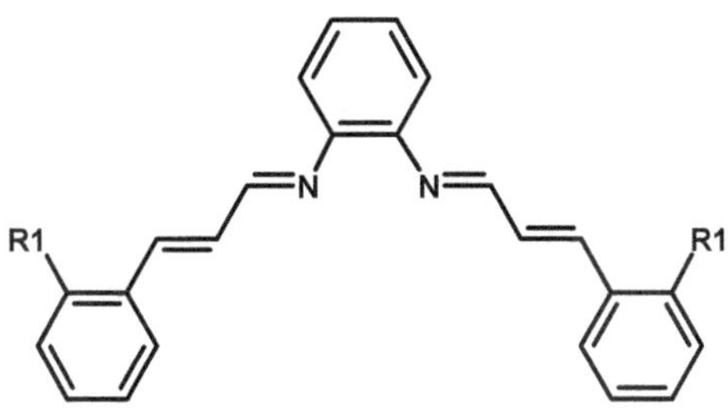

R1 = H	L^{18} [58]
R1 = NO_2	L^{19} [59]

Scheme 6.

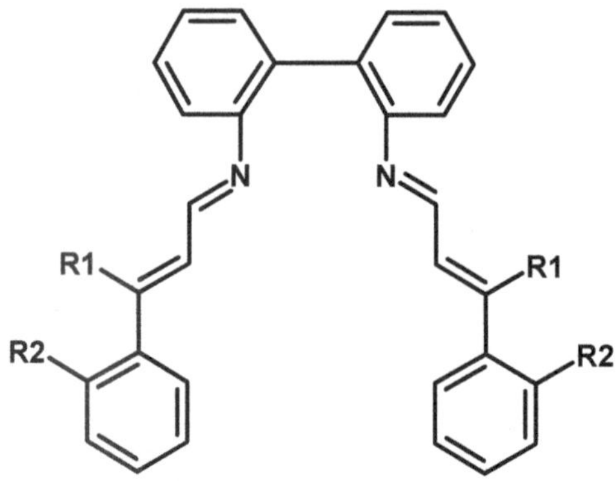

R1 = Ph , R2 = H L^{20} [60,61] R1 = H, R2 = NO_2L^{21} [62]

Scheme 7.

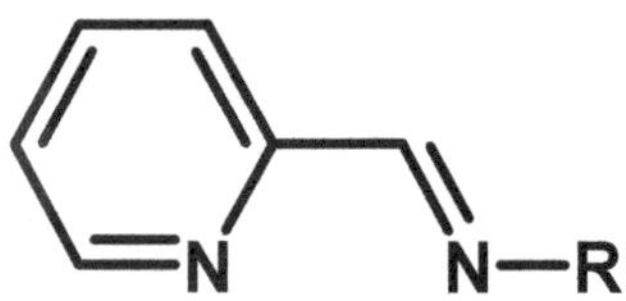

Scheme 8.

2.1.2 Asymmetric bidentate schiff bases

Asymmetric Schiff base ligands (Scheme 9) have been synthesized from an amine and pyridinecarboxaldehyde. They can be classified by different R substituent [6365].

R	L [Ref]	R	L [Ref]	R	L [Ref]
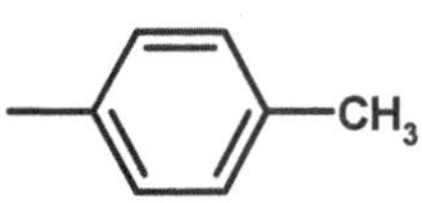	L^{22} [63]	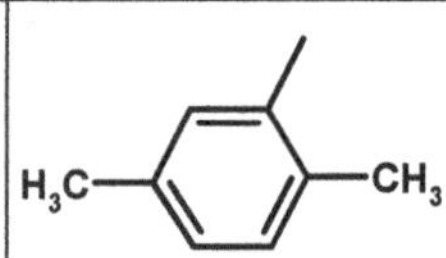	L^{25} [63]	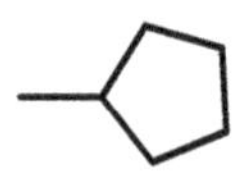	L^{28} [65]
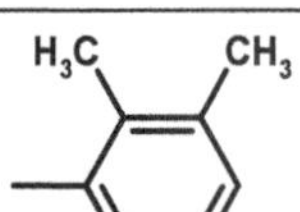	L^{23} [63]	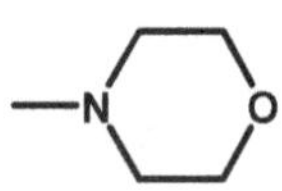	L^{26} [64]	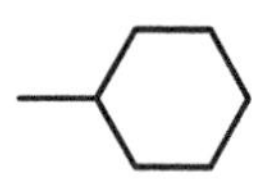	L^{29} [65]
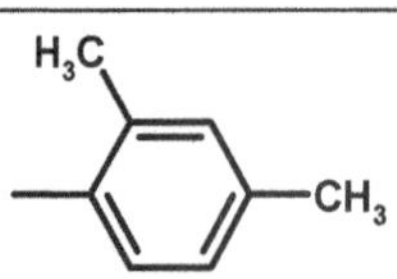	L^{24} [63]	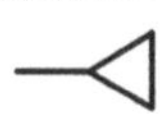	L^{27} [65]	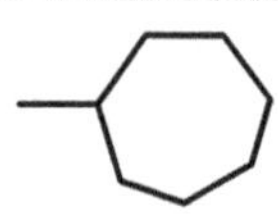	L^{30} [65]

Scheme 9.

2.2 Bis-bidentate schiff-bases

The basic bis-bidentate Schiff base ligands have two arms connected via a ring, or C-C bond, as for example, shown in Schemes 10 – 14 [27-30,32,35-40,66-72].

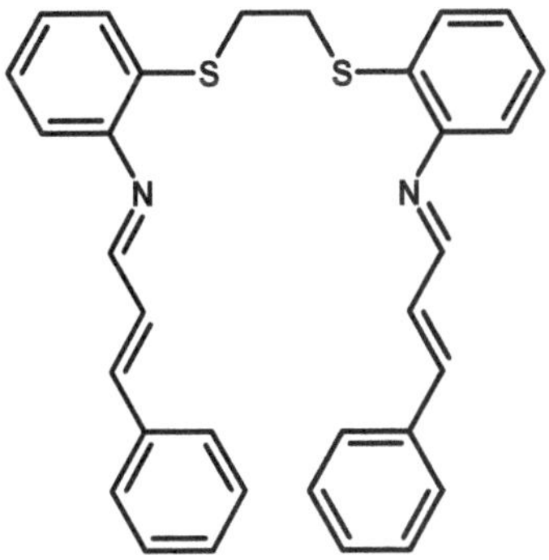

L^{31} [35,66]

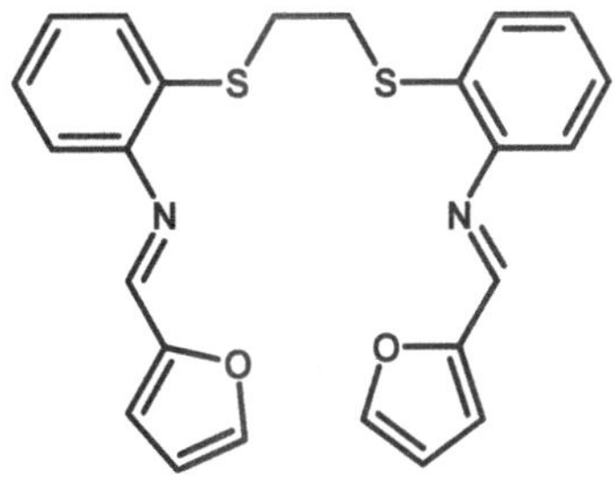

L^{32} [36]

Scheme 10.

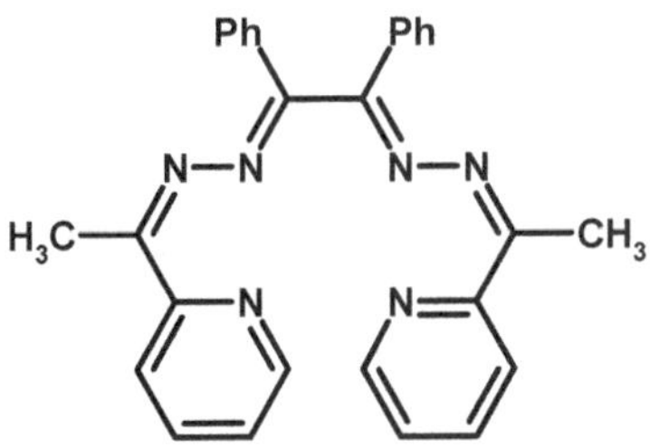

L^{33} [29, 67]

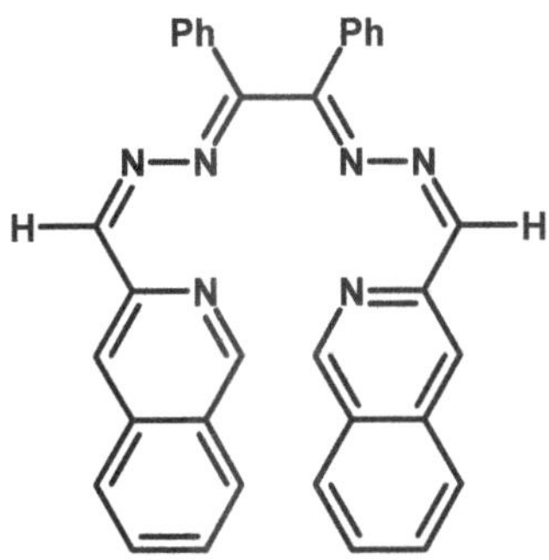

L^{34} [68]

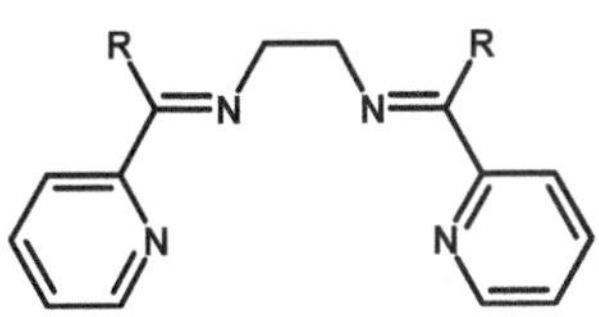

R = H L^{35} [30, 38]

R = CH_3

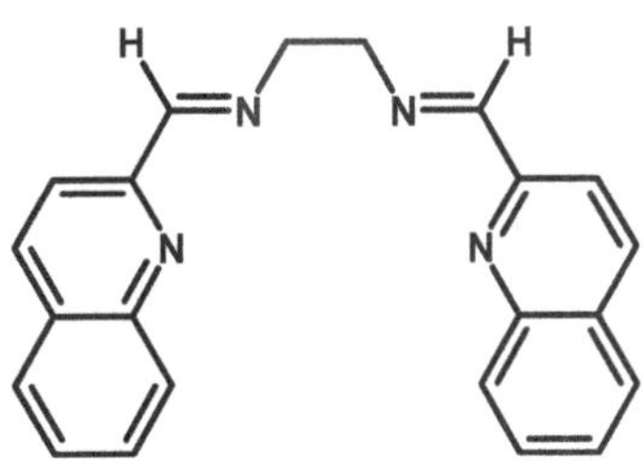

L^{37} [67]

L^{36} [39,69]

Scheme 11.

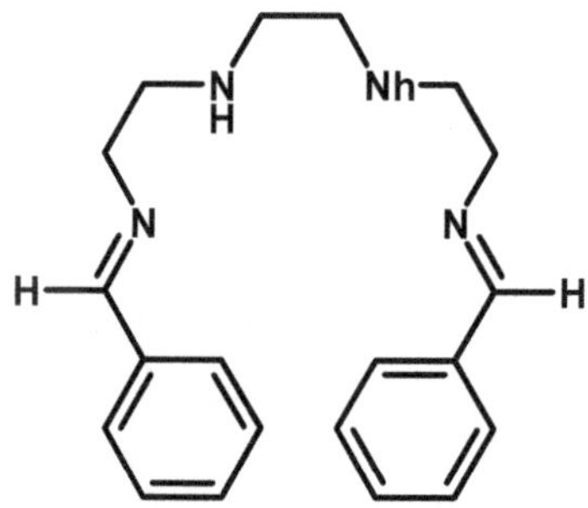

L^{38} [37]

Scheme 12.

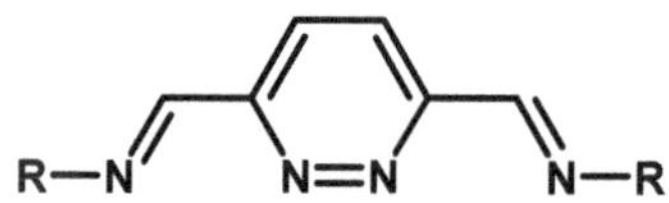

Scheme 13 [32].

R	L	R	L
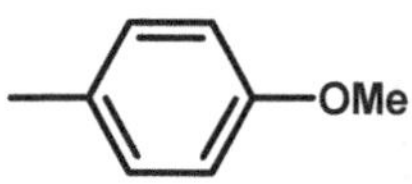	L^{39}	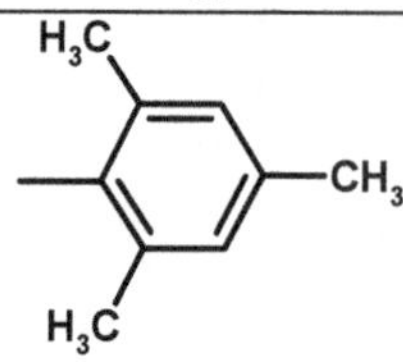	L^{43}
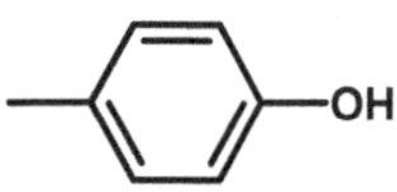	L^{40}	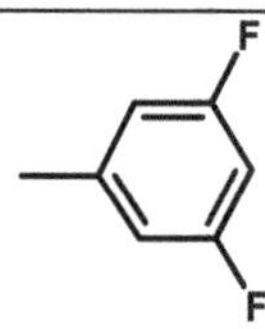	L^{44}
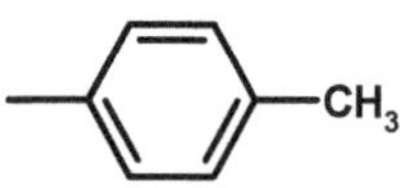	L^{41}	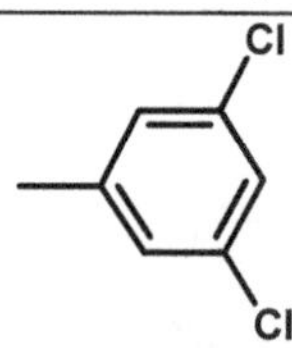	L^{45}
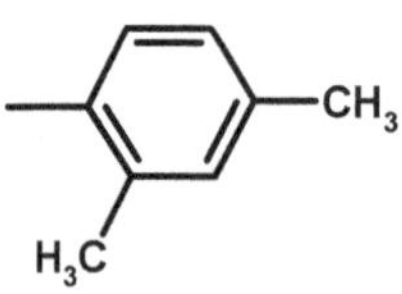	L^{42}	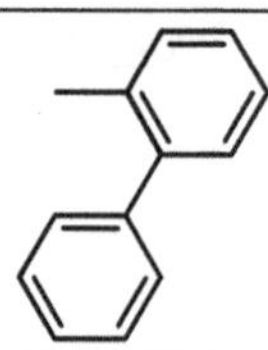	L^{46}

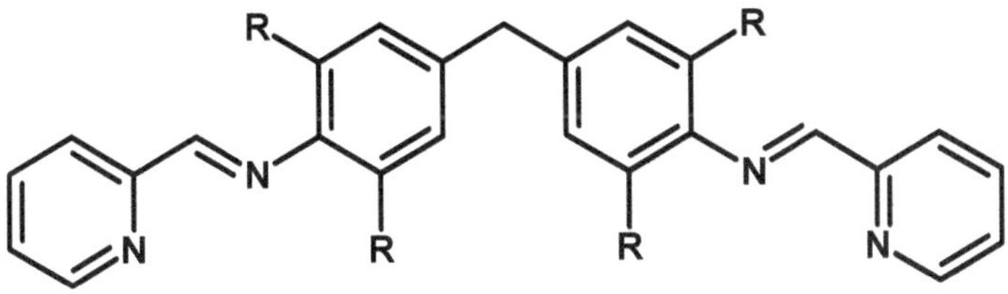

R = H L^{47} [28,70,71]

R = CH_3 L^{48} [72]

R = C_2H_5 L^{49} [27]

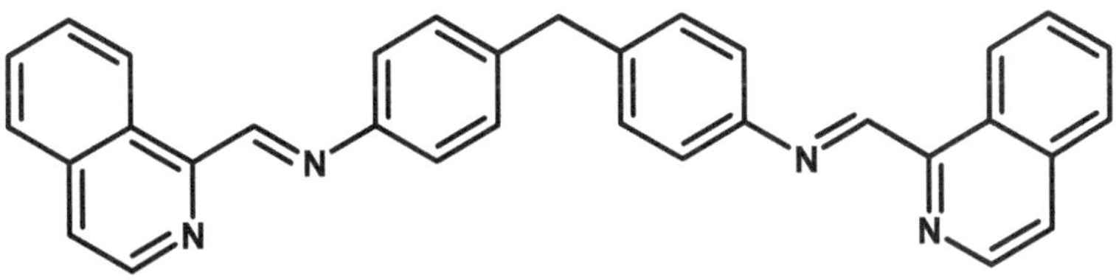

L^{50} [40]

Scheme 14.

3. Bonding modes of schiff base ligands

A number of bonding modes have been observed for the Schiff bases in their neutral form. The binding occurs via nitrogen atom of azomethine group in chelating and bridging modes, I - VIII (Scheme 15).

4. Copper(I) schiff base complexes

Copper(I) Schiff base complexes were generally prepared by the reaction of a CuX, where X = halide (Cl^-, Br^-, I^-), pseudohalide (NCS^-, N_3^-), or $[Cu(CH_3CN)_4]Y$ (Y = the non-coordinating anions (ClO_4^-, BPh_4^-) with a suitable Schiff base ligands in an acetonitrile solvent, followed by room temperature stirring under an N_2 or air atmosphere [41-62]. In these complexes, Schiff-bases act as chelating ligands and cause the geometry around the copper(I) atom will be distorted tetrahedral.

Several different crystallization techniques have been used to grow crystals of copper(I) complexes with Schiff base ligands. Some of them are:

1. Slow diffusion of Et_2O into the concentrated solution of complex at room temperature
2. Very slow evaporation of the solvent at ambient temperatures for several days
3. Layering technique
4. Precipitation and re-crystallization from a mixture of solvents

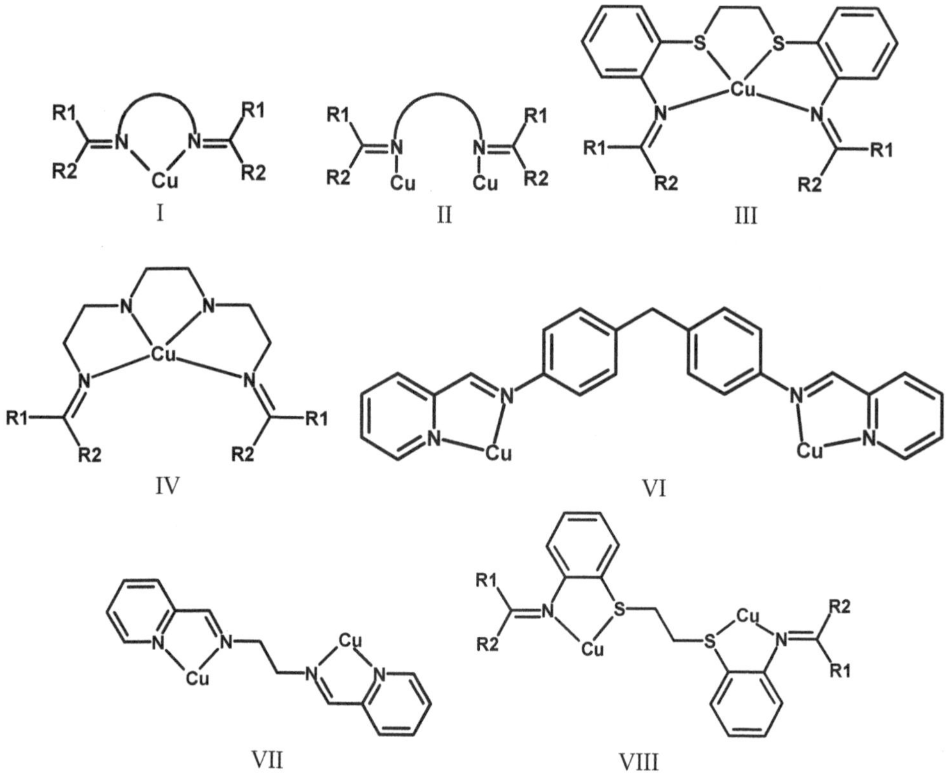

Scheme 15.

4.1 Copper(I) complexes with symmetric bidentate Schiff base ligand

4.1.1 Copper(I) iodide complexes

The reaction of equimolar amounts of Meca$_2$en (L[15]) at the presence of copper(I) iodide in acetonitrile at room temperature yielded yellow microcrystalline powder of mononuclear copper(I) complex Cu(Meca$_2$en)(CH_3CN)I (1) (Scheme 16) [56]. Slow evaporation of the solvent, 1:1 *v/v* chloroform:acetonitrile, at 273 K gave suitable yellow crystals. The coordination geometry around the copper ion is distorted tetrahedron formed by two N atoms from Meca$_2$en, one N atom from acetonitrile and one iodine atom. The Meca$_2$en acts as a bidentate ligand coordinating *via* two N atoms to the copper.

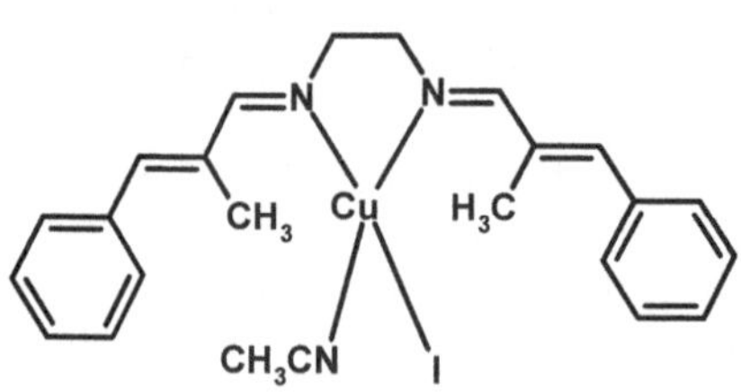

Scheme 16.

By changing the Schiff base ligand from Meca$_2$en to Phca$_2$pn (L[10]), ca$_2$en (L[12]) and ca$_2$ph (L[18]), dinuclear copper(I) complexes [Cu(Phca$_2$pn)I]$_2$ (2) [49], [Cu(ca$_2$en)I]$_2$ (3) [51] and [Cu(ca$_2$ph)I]$_2$ (4) [58] (Scheme 17) have been prepared. Slow diffusion of diethylether vapour into the concentrated acetonitrile solution at 298 K gave suitable orange crystals (2 and 3), while dark red crystals of complex 4 were obtained at 273 K. The coordination geometry around the copper ion in 2 and 3 is distorted tetrahedron formed by two N atoms from the Schiff base ligand and two iodine substituents. The $Cu^IN_2I_2$ tetrahedra share iodine atoms which therefore double-bridge the neighboring copper(I) ions.

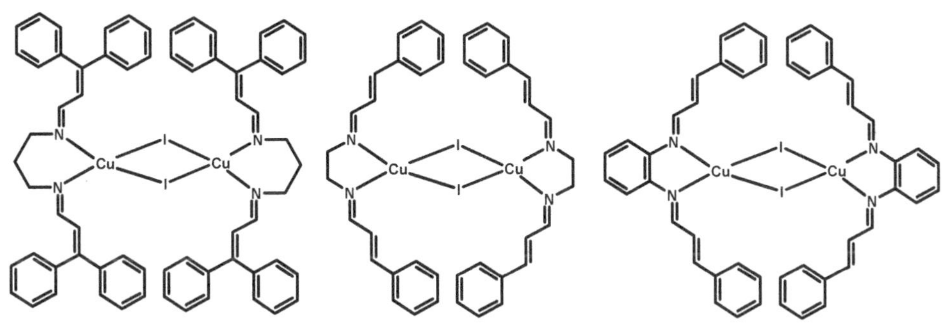

2 3 4

Scheme 17.

However, the chelate N-Cu-N angle in 2 (100.17(12)°) is larger than the corresponding angle in 3 (82.7(3)°), while the I-Cu-I angle in 2 (104.517(17)°) is larger thanthe corresponding one in 3 (119.65(4)°), making the distortion of the tetrahedron considerably larger in (3) because of restricting bite angle of the chelating ligand. The N-Cu-I bond angles range from 100.53(9) to 122.48(9)° in 2 are similar to the corresponding angle in 3 (103.8(2) - 121.8(2)°). The bond distances Cu-I (2.7264(6) Å) and Cu-N (2.055(3) and 2.065(3) Å) in 2 are similar to the corresponding bond distances in 3 (2.6903(14), 2.069(7) and 2.085(7) Å, respectively). The dinuclear unit in these complexes is centrosymmetric with crystallographic center of symmetry located between the two copper(I) ion. The distances between the two copper(I) ions are 3.372(2) Å in 2 and 2.635(2) Å in 3.

The reaction of equimolar amounts of bz$_2$en (L[7]), phca$_2$en (L[14]) and Phca$_2$-dab (L[17]) at the presence of copper(I) iodide in acetonitrile at room temperature yielded mononuclear copper(I) complexes [Cu(bz$_2$en)][Cu$_2$I$_4$] (6) [47], [Cu(Phca$_2$en)$_2$][I$_3^-$] (6) [54] and [Cu(Phca$_2$-dab)$_2$][[CuI$_2$] (7) [61] (Scheme 18). Slow diffusion of diethylether vapour into the concentrated acetonitrile-dichloromethane solution for 5, acetonitrile solution for 6 and chloroform solution for 7 at 298 K gave suitable dark red crystals.

The Schiff base ligand Phca$_2$-dab (L[17]) have two different phenyl rings (the inner and outer phenyl rings) [61]. They play different role in the arrangement of the cations and anions of [Cu(Phca$_2$-dab)$_2$][CuI] (7). As a result, a one-dimensional net of 7 molecules is generated by 4-PE (fourfold phenyl embrace) including Ri rings (Fig. 1a). The Ro rings are not involved in the 4-PE local interactions. The [CuI$_2$]$^-$ anions in the crystal structure are located in the space between Ro rings of the two adjacent cations (Fig. 1b) [61].

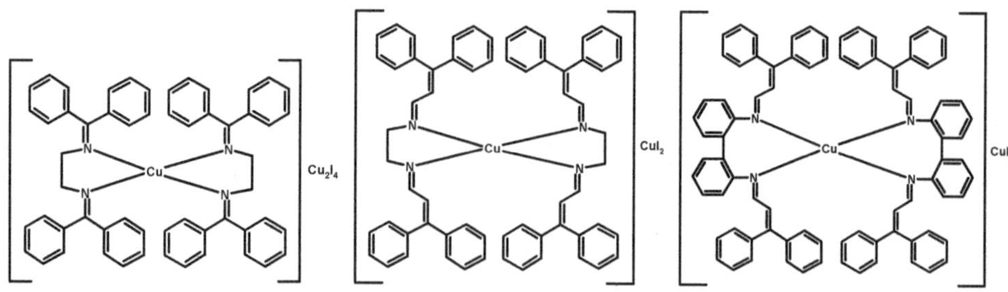

5 6 7

Scheme 18.

The solid-state structure of complexes 5-7 reveals ionic complexes containing a cation of copper(I) ion coordinated to four nitrogen atoms of two Schiff base ligands. The coordination geometry around the copper ion is distorted tetrahedron. The complexes contain anions di-μ-iodo-diiododicuprate(I) for 5, diiodocuprate(I) for 6 and I_3^- for 7.

While a tetrahedral geometry might be expected for copper(I) complexes 6 and 7, the coordination sphere around copper(I) in these complexes is distorted by the restricting bite angle of the chelating Schiff base ligand. The average Cu-N bond distance of 2.023 Å in 6 is similar to the corresponding bond distance in 7 (2.035 Å). The chelate N-Cu-N angle in 6 (83.67(11)°) is shorter than similar to the corresponding bond angle in 7 (102.59(15)°).

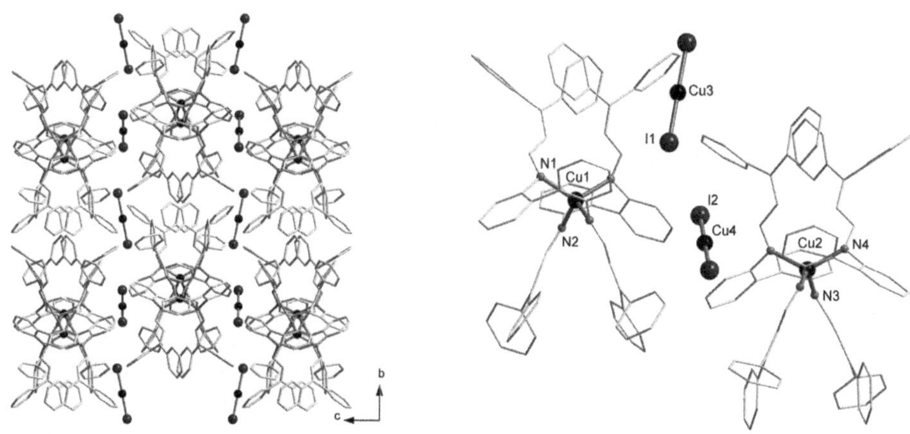

a b

Fig. 1. a) One-dimensional net of fourfold phenyl embrace (4-PE) in 7. b) Representation of anions (CuI_2) associated with phenyl rings Ro.

The reaction of equimolar amounts of (3,4,5-MeO-ba)$_2$bn (L^{43}) in the presences of copper(I) iodide yields polymeric copper(I) chain complex $[Cu_2(\mu\text{-}(3,4,5\text{-MeO-ba})_2bn)(\mu\text{-}I)_2]_n$ (8) (Scheme 19) [33]. Slow evaporation of the solvent at 273 K gave suitable yellow crystals.

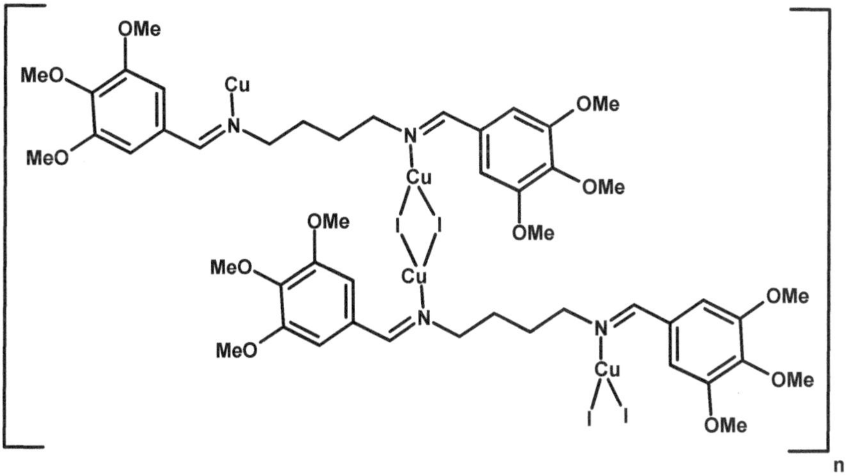

Scheme 19.

The copper(I) atom in 8 is coordinated by two iodine ions and one nitrogen atom from Schiff-base ligand in distorted trigonal planar coordination geometry. In the crystal structure of 8 the (3,4,5-MeO-ba)$_2$bn acts as a bridging ligand with the nitrogen atoms of the two imine functions. This leads to formation of dinuclear [Cu_2(μ-(3,4,5-MeO-ba)$_2$bn)] groups which are bridged by two iodine anions [(μ-I)$_2$] into a neutral 1D-chain copper(I) iodide coordination polymer (Fig. 2).

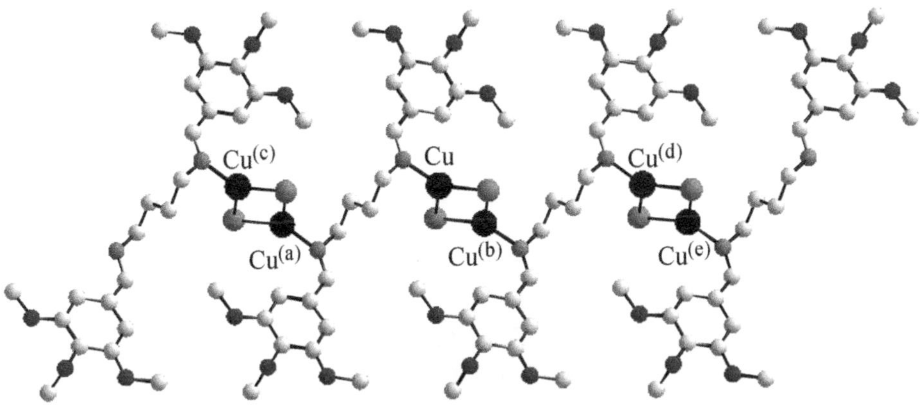

Fig. 2. D-chain of complex 8 running along the [001] direction. Symmetry codes: (a) -x+1,-y,-z+1; (b) -x+1,-y,-z+2

The Cu-N distance 1.9733(19) Å is similar like in other copper(I) complexes [41,43]. The Cu-I distances of 2.5491(4) and 2.5648(3) Å are slightly shorter than those of 2.6650(5) and 2.7393(5) Å in the 1D-chain copper(I) complex [Cu_2(μ-I)$_2$(PPh_3)$_2$(4,4'-bpy)]$_n$ [73]. The Cu···Cu distance in the Cu-(μ-I)$_2$-Cu fragment of 2 is 2.5583(6) Å, i.e. shorter than in other similar Cu-(μ-I)$_2$-Cu complexes [73].

The reaction of dinuclear copper(I) complex $[Cu(ca_2en)(I)]_2$ (3) [51] or mononuclear copper(I) complex $[Cu(Phca_2en)_2][I_3^-]$ (6) [54] with PPh_3 in an acetonitrile solution mononuclear copper(I) complex $[Cu(ca_2en)(PPh_3)I]$ (9) [52] and $[Cu(Phca_2en)(PPh_3)I]$ (10) [55] have been prepared (Scheme 20). Slow diffusion of diethylether vapour into the concentrated acetonitrile solution for 9 and 10 at 298 K gave suitable orange crystals.

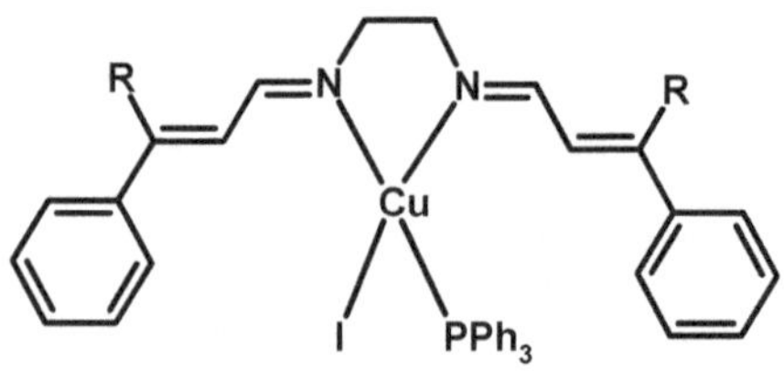

R = H (9) R = Ph (10)

Scheme 20.

The copper(I) ion in 9 and 10 is coordinated to two N atoms of one Schiff base ligand, one P from PPh_3 and one iodine group in geometry distorted by the restricting bite angle of the chelating ligands. The chelate N-Cu-N angle (81.33(7)°) in 9 is similar to the corresponding bond distance and angle in 10 (2.104 Å and 81.74(5)°). The P-Cu-I angle is 120.964(15) and 115.582(11)° in 9 and 10, respectively, being larger than the tetrahedral values. The Cu-N (2.0915(17), 2.0918(17) Å), Cu-I (2.6469(4) Å) and Cu-P (2.2327(6) Å) bond distances in 9 agree well with the corresponding distances in 10 (2.1129(11), 2.0961(12), 2.6501(2) and 2.2342(4) Å, respectively).
Compounds containing M-PPh_3 moieties are recognized to form multiple phenyl embrance (MPE) as predominant supramolecular motifs [74,75]. In the complex 10, phenyl rings of PPh_3 groups exhibit different conformations that are illustrated in Fig. 3. The Cu-PPh_3 moiety of roughly threefold symmetry forms dimers that are separated from each other 7.270(7) Å, with Cu-P···P angle of 173.61(1)° [55].

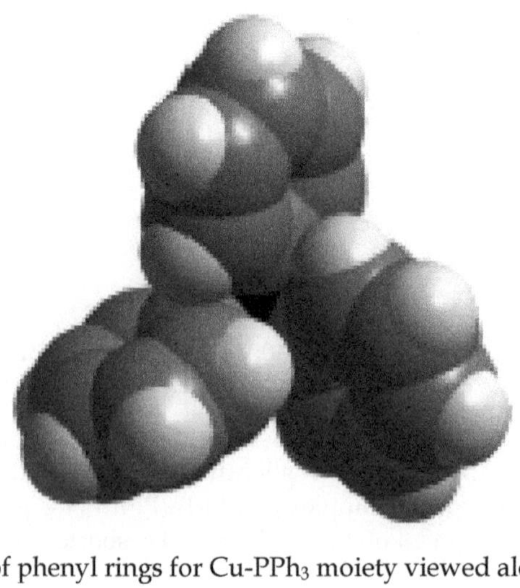

Fig. 3. Conformation of phenyl rings for Cu-PPh_3 moiety viewed along P-M in 10.

4.1.2 Copper(I) thiocyanate complexes

The pseudohalide NCS^- is known to coordinate to metal ions in both terminal and bridging modes [76,77]. Thiocyanate ion can afford a variety of coordination bridging modes (Scheme 21), such as μ-1,3-NCS, μ-1,1-NCS, μ-1,1-SCN, μ-1,1,3-NCS and μ-1,1,3-SCN, and link one or more transition metal ions to form zero, one-, two- and three-dimensional complexes [76,77].

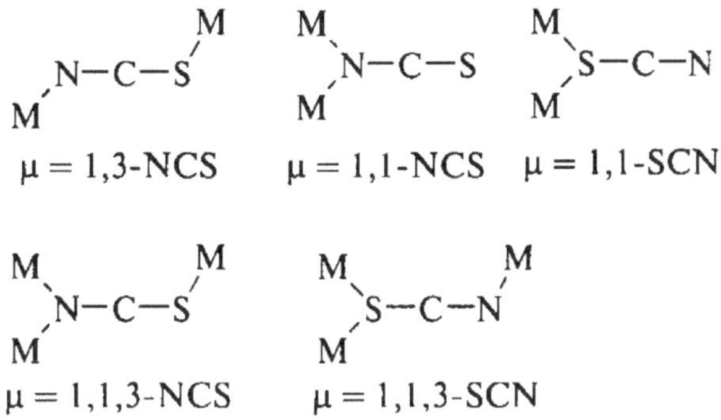

Scheme 21.

The reaction of bidentate Schiff base ligand (3-Me-ba)$_2$en (L^1), (3,4,5-Meo-ba)$_2$en (L^2) and (2-MeO-ba)$_2$en (L^3) with CuSCN in an acetonitrile, three copper(I) thiocyanate coordination polymers, [Cu((3-Me-ba)$_2$en)(SCN)]$_n$ (11), [Cu((3,4,5-MeO-ba)$_2$en)(SCN)]$_n$ (12) and [Cu((2-MeO-ba)$_2$en)(SCN)]$_n$ (13), with one-dimensional coordination polymer chains (Scheme 22), have been prepared [41,43]. Slow evaporation of the solvent at 273 K gave suitable yellow crystals of 11-13.

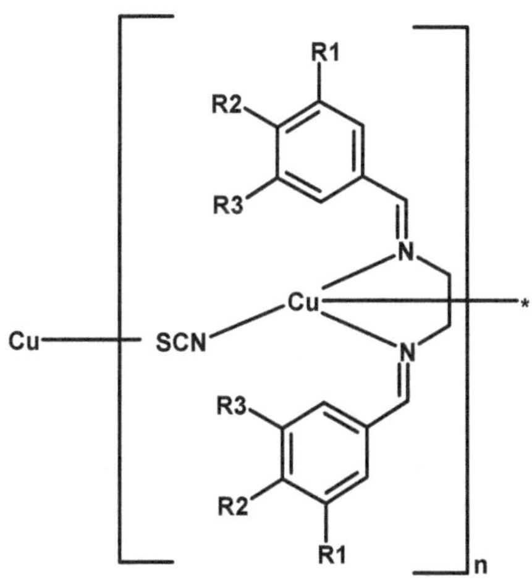

R1 = R3 = H, R2 = CH_3 (11) R1 = R2 = R3 = CH_3O (12)
R2 = R3 = H, R1 = CH_3O (13)

Scheme 22.

The crystal structures of these complexes have common features. In these complexes distorted tetrahedral coordination at Cu(I) is defined by the two N donor atoms of the Schiff base ligands and the N and S donor atoms of two thiocyanate ligands (Scheme 23). The Schiff base ligands act as chelating ligands while the SCN^- anion acts as a bridging ligand.

In 12, the Cu...Cu separation through the bridging thiocyanate ligand is 5.604 Å. The Cu-S bond length 2.3468(7) Å is greater compared with those of the Cu-N bonds (mean distance 2.06 Å), which causes a significant distortion of the coordination polyhedron around the Cu(I) ion. The values observed for Cu-N and Cu-S distances are comparable with those found in other tetrahedral Cu(I) complexes. Moreover, the Cu-N bond to the bridging SCN group is shorter than Cu-N bonds to the Schiff base ligands. The bridging SCN anion is almost linear with S-C-N bond angle of 179.4(2)°.

4.1.3 Copper(I) perchlorate complexes

$Cu^{I}N_4$ chromophores which exhibit photoluminiscence in solution are known only in copper(I) complexes of substituted 1,10-phenanthrolines and Schiff bases. Many advances in the chemistry of copper(I) complexes have been achieved on four coordinated tetrahedral complexes of the type $[Cu(NN)_2]^+$ or $[Cu(NN)(P)_2]^+$ (Scheme 23), where NN is Schiff base ligand and P is triphenylphosphine [78-80]. These complexes have relatively stable framework and exhibit metal-to-ligand charge transfer transition in the visible region.

The reaction of L^6 and L^7 in the presences of $[Cu(CH_3CN)_4]ClO_4$ (2:1 molar ratio) in degassed methanol under dry N_2 atmosphere, yielded yellow and red precipitates of $[Cu(ba_2en)_2]ClO_4$ (13) and $[Cu(bz_2en)_2]ClO_4$ (14), respectively (Scheme 24). Single crystals of 13 and 14, were grown by slow diffusion of n-hexane into a dilute dichloromethane solution of the complexes [46].

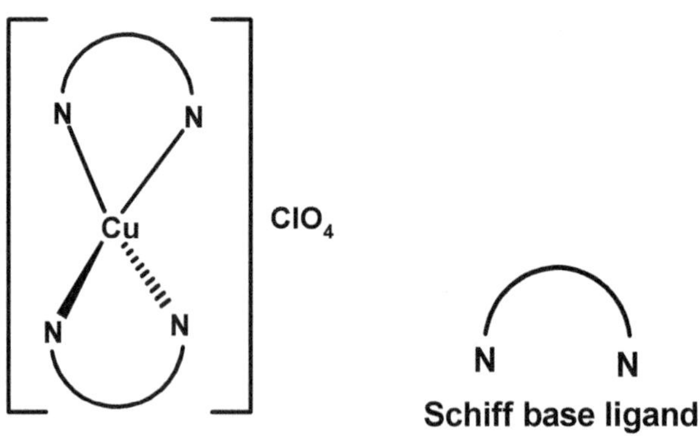

Scheme 23.

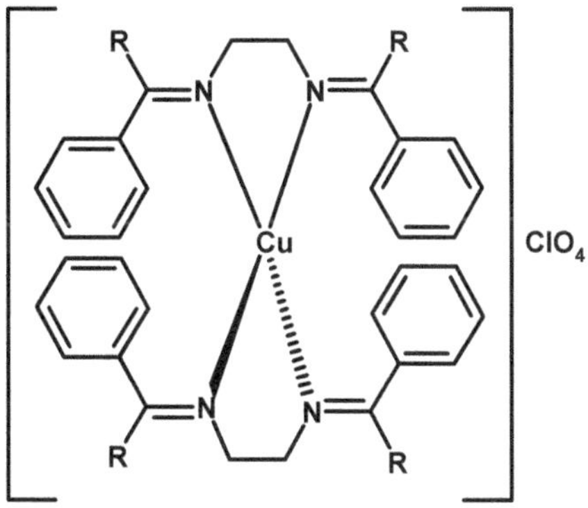

R = H (13) R = Ph (14)

Scheme 24.

The average bite angle of the ligand (N-Cu-N) in both complexes is 85.3° [46]. In 13, the average Cu-N bond length is 2.073 Å, with an average dihedral angle of 73.1° between the two CuN_2 coordination planes, while the Cu-N bond lengths in 14 are in range from 2.014(4) to 2.183(4) Å, with dihedral angle of 63.9(2)° [46]. Complex 13 shows a quasireversible $Cu^{II/I}$ couple in cyclic voltametery with an $E_{1/2}$ of 0.81 V *vs* SCE, while this couple is much more reversible in complex 14 with an $E_{1/2}$ of 0.66 V *vs* SCE [46]. Interestingly, the emission spectrum of complex 14 shows an additionalband around 550 nm in methanol albeit very weak which is absent in dichloromethane. Both the emissions in 14 in methanol originate from the MLCT state [46].

In recent years, Dehghanpour *et al.* have systematically studied on copper(I) complexes of the type $[Cu(NN)_2]^+$ or $[Cu(NN)(P)^2]^+$ with unconjugated Schiff base ligands [44,48,53,57,59,62]. For example, the reaction of nca²en (L^{13}) in the presences of $[Cu(CH_3CN)_4]ClO_4$ (2:1 molar ratio) in acetonitrile, yielded orange-red precipitates of $[Cu(nca_2en)_2]ClO_4$ (15) (Scheme 25). Single crystals of 15, were grown by slow diffusion of Et_2O into a concentrated acetonitrile solution of the complex [53]. Complex 15 shows a quasireversible $Cu^{II/I}$ couple in cyclic voltametery with an $E_{1/2}$ of 0.55 V *vs* SCE.

The solid-state structures of complexes 15-17 (Scheme 25) reveal copper(I) ion coordinated to four N atoms of two Schiff base ligands. The coordination geometry around the copper ion is distorted by the restricting bite angle of the chelating ligands. The chelate N-Cu-N angle (84.2(2)°) in 15 is similar to the corresponding bond angle in 16 (82.27(5) and 82.29(9)°) and is shorter than the corresponding bond angle in 17 (99.3(18) and 100.7(18)°). The Cu-N (2.044(3) and 2.020(3) Å) bond distances in 15 agree well with the same distances in 16 (2.035(2), 2.046(2), 2.062(2) and 2.038(2) Å) and in 17 (2.019(4), 2.022(4), 2.041(5) and 2.046(4) Å).

The structure of Schiff base ligands L^2 and L^{11}, are very similarto the ligands reported previously, but the side substitutes are changed to more bulky groups (3 methoxy groups in L^2 and one anthryl group in L^{11}), which is hence expected to exhibit unusual coordination mode with copper(I) ion. The reaction of L^2 in the presence of $[Cu(CH_3CN)_4]ClO_4$ (2:1 molar ratio) in acetonitrile, yielded yellow precipitate of $[Cu((3,4,5\text{-MeO-ba})_2en)(CH_3CN)_2]ClO_4$ (18) (Scheme 26). Single crystals of 18 were grown by slow diffusion of Et_2O into an acetonitrile solution of the complex [42]. The copper(I) ion is coordinated by two N atoms from one bidentate Schiff base ligand and two N atoms from two acetonitrile groups. The copper atom adopts a tetrahedral geometry. The Cu-N(ligand) distances are 2.076(3) and 2.076(3) Å, and the Cu-N(acetonitrile) distances are 1.964(4) and 1.975(4) Å.

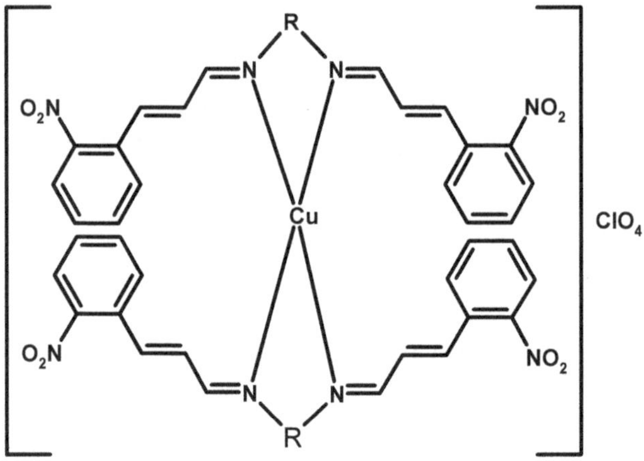

R = $-CH_2-CH_2-$ (15) R = -Ph- (16) R = -Ph-Ph- (17)

Scheme 25.

The reaction of L^{11}, at the presence of $[Cu(CH_3CN)_4]BF_4$ (2:1 molar ratio) in methanol, gave yellow-green precipitates of $[Cu(Anca_2en)_2]BF_4$ (19) (Fig. 4) [50]. Single crystals suitable for X-ray diffraction analysis were grown from a dichloromethane-ethanol (*v/v*, 1/1) solution of complex. The structure of 19 is shown in Fig. 23. In 19 the copper(I) ion is coordinated to three N atoms from two bidentate Schiff base ligands in a distorted trigonal planar geometry. The remaining imine N atom remains uncoordinated. The Cu-N distances are 2.111(3), 1.946(4) and 1.930(3) Å, respectively.

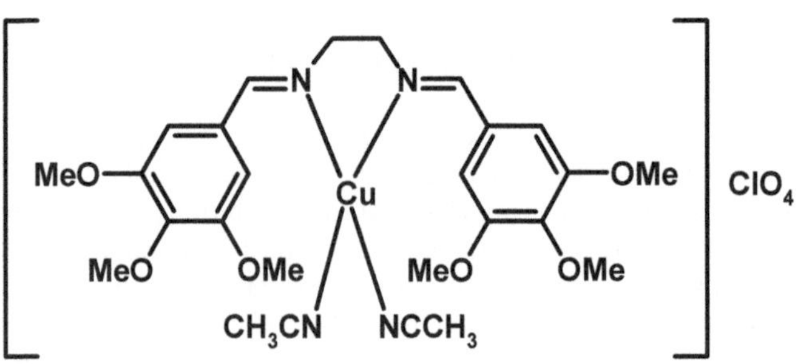

Scheme 26.

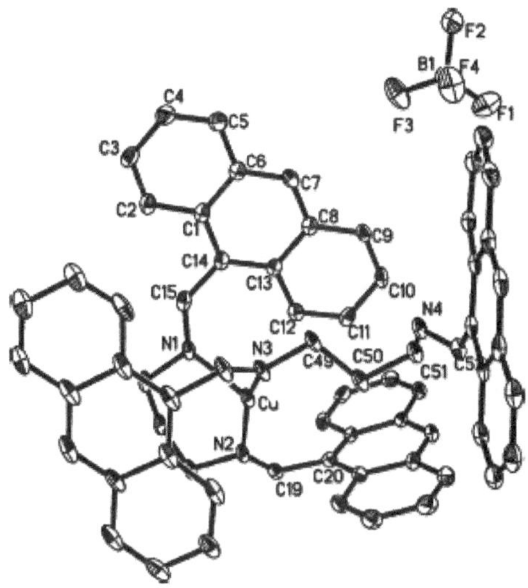

Fig. 4. An ORTEP view of 19, showing the atomic numbering scheme. Displacement ellipsoids are drawn at the 30% probability level. H atoms have been omitted for clarity.

Apparently, the coordination pattern of 19 is very different from the copper(I) complexes reported so far [46,47]. In this complex, the bulky anthryl group has larger steric hindrance and may inhibit another imine-N of the second ligand to coordinated with copper(I), resulting in more stable three-coordinated conformation.

4.2 Copper(I) complexes with asymmetric bidentate Schiff base ligand

In recent years, Dehghanpour *et al.* have systematically studied on copper(I) complexes of the type $[Cu(NN)_2]^+$ or $[Cu(NN)(P)_2]^+$ with asymmetric Schiff base ligands [60, 63-65]. For example, the reaction of 4-ampc (L^{26}) (Scheme 9) and dpa-qa (L^{51}) (Scheme 27) in the presences of $[Cu(CH_3CN)_4]BPh_4$ (2:1 molar ratio) in acetonitrile, yielded orange and dark-red precipitates of $[Cu(4\text{-ampc})_2]BPh_4$ (20) [64] (Scheme 27) and $[Cu(dpa\text{-}qa)_2]BPh_4$ (21) [60] (Scheme 27), respectively. Single crystals of 20 and 21, were grown by slow diffusion of Et_2O into a concentrated acetonitrile solution of the complex.

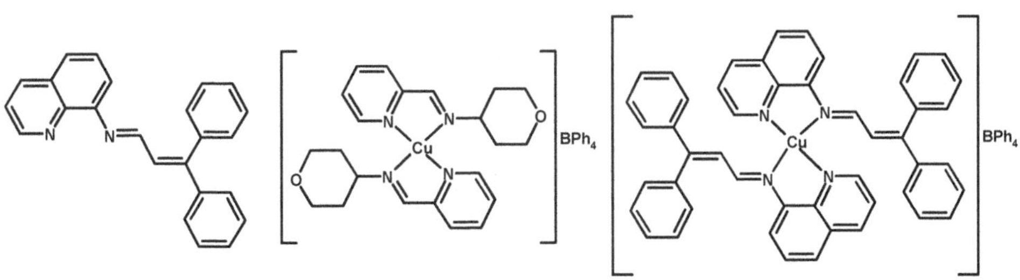

L^{51} [60] 20 21

Scheme 27.

Since no d-d transitions are expected for a d^{10} complex, the UV-Vis bands are assigned to MLCT or ligand-centered $\pi \rightarrow \pi^*$ transitions [60]. The absorption spectrum of 21, in $CHCl_3$ reveals a band with a true maximum at 503 nm. This observation is in ag5reement with the higher conjugation in the coordinated dpa-qa [60]. Complex 20 shows a quasireversible $Cu^{II/I}$ couple in cyclic voltametery with an $E_{1/2}$ of 0.47 V *vs* SCE [60].

4.3 Copper(I) complexes with symmetric tetradentate Schiff base ligands

Flexible N/S or NN donor ligands have received much attention due to their use in constructing coordination frameworks by self-assembly, and also in investigating the mechanism of supramolecular interactions [81-86]. With flexible ligand, the competition between bridging and chelating coordination modes is an important factor in producing mono, di, and polynuclear metal complexes [81-86].

4.3.1 $(NS)_2$ Schiff bases with a flexible spacer

In recent years, Morshedi *et al.* have systematically studied on copper(I) complexes with flexible N/S donor Schiff base ligands [35,36,66]. The tetradentate $(NS)_2$ Schiff bases L^{31} and L^{32} were prepared as reported elsewhere [35,36,66]. The reaction of $(thio)_2$dapte (L^{32}) in the presences of CuI (1:1 molar ratio) in acetonitrile, yielded dark-red precipitates of new one-dimensional copper(I) coordination polymer $[Cu_2(\mu\text{-}I)_2(\mu\text{-}(thio)_2dapte)]_n$ (22) (Scheme 28). Single crystals of 22 were grown by slow evaporation of solvent at room temperature for several days [36]. The solubility of 22 in common organic solvents is very low. Complex 22 crystallizes with the asymmetric unit formed by one copper(I) and one iodine ion located in general positions, and one centrosymmetric organic ligand $(thio)_2$dapte placed in the inversion center (1/2, 1, 1/2). In this complex, the flexible Schiff base ligand $(thio)_2$dapte acts as a bis-chelating ligand through its two iminic nitrogens and two sulfur atoms of ethanedithiolate group, creating the $[Cu_2(\mu\text{-}(thio)_2dapte)]$ dinuclear fragment. Such dinuclear entities are linked to each other by the two iodine anions acting as a doubly μ_2-bridging ligand $[(\mu\text{-}I)_2]$ and forming one-dimensional copper(I) coordination polymer with the general formula $[Cu_2(\mu\text{-}I)_2(\mu\text{-}(thio)_2dapte)]_n$ (22). In this complex, copper(I) ion has a distorted tetrahedral coordination geometry formed by one nitrogen and one sulfur atom from the Schiff base ligand and two iodine substituents, with significantly different bond distances [Cu1-N1 2.073(4), Cu1-S1 2.342(2), Cu1-I1 2.6008(7) and Cu1-$I1^b$ 2.6575(7) Å]. The highly distorted tetrahedral geometry around the copper(I) ion is also due to the small bite angle of the $(thio)_2$dapte bis-chelating ligand. The Cu1···$Cu1^b$ distance of copper atoms connected through the bridging iodine ligands (2.6745(11) Å) is shorter than the corresponding distance of copper atoms connected through the bis-chelating ligand (Cu1···$Cu1^a$ 6.942(1) Å).

By using the bis-chelating Schiff base ligand ca_2dapte [35,66] and at the similar condition to 22, the reaction of ca_2dapte (L^{31}) in the presences of CuSCN (1:1 molar ratio) in acetonitrile under N_2 yielded orange-red precipitates of mononuclear copper(I) complex $[Cu(ca_2dapte)(NCS)]$ (23) (Scheme 29). Single crystals of 23, were grown by slow diffusion of Et_2O vapor into a concentrated solution of complex in acetonitrile at room temperature [66]. In the ^1H-NMR spectra of 23, the iminic protons corresponding to the coordinated and non coordinated imine groups appear at 8.95 and 8.97 ppm, respectively (free ligand has 8.14 ppm).

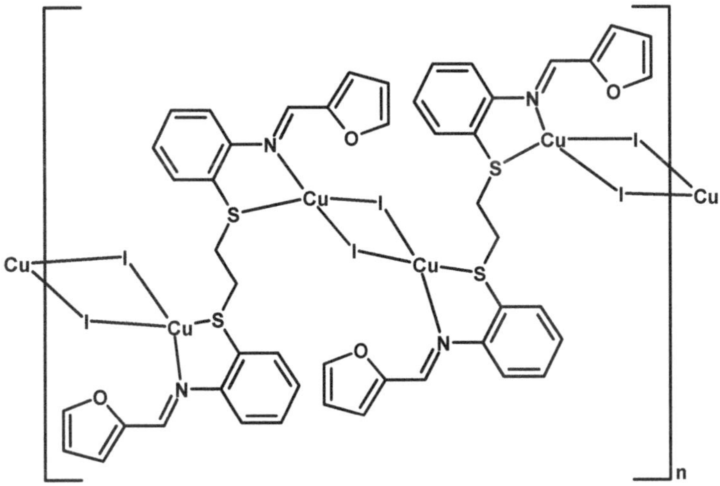

Scheme 28.

Scheme 29.

The asymmetric unit of complex 23 contains one copper atom, one NCS^- ligand and a ca_2dapte Schiff base ligand. In this complex, the NCS^- pseudohalide anion acts as a terminal ligand through its nitrogen atom, and ca_2dapte acts as a tridentate ligand coordinated through one N-imino, and two sulfur atoms. The second imino arm is directed away of the copper(I) center. Copper(I) ion has a distorted tetrahedral coordination geometry formed by one nitrogen and two sulfur atom from the Schiff base ligand and one nitrogen from thiocyanate, with significantly different bond distances [Cu-N1 2.064(2), Cu-S1 2.2886(8), Cu-S2 2.3824(7) and Cu-N3 1.903(2) Å] [66]. The high distortion of the coordination tetrahedron is essentially due to the steric requirements of the ca_2dapte bis-chelating ligand leading to strong deviation of the bond angles around the copper(I) ion from ideal tetrahedric values [N-Cu-S, N-Cu-N and S-Cu-S bond angles are in the range 86.2-130.0°] [66].

In the mononuclear copper(I) complex $[Cu(ca_2dapte)(NCS)]$ (23), the N_2S_2 Schiff base ligand ca_2dapte, acts as a tridentate ligand surrounding the copper(I) ion [66]. In $[Cu_2(ca_2dapte)(PPh_3)_2X_2]$ (X = I (24) and Br (25)) complexes [35], ca_2dapte acts as two independent bidentate ligands connected by a flexible bridge (Scheme 30), and in [Cu(ca$_2$dapte)]ClO_4 (26) the ligand is tetradentate surrounding the copper(I) ion [35] (Scheme 30).

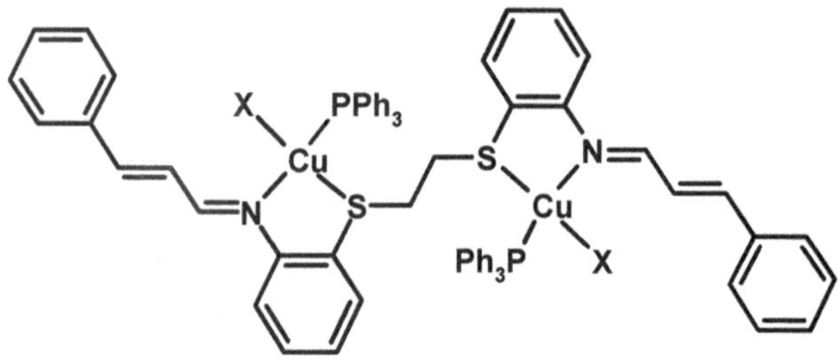

X = I (24) X = Br (25)

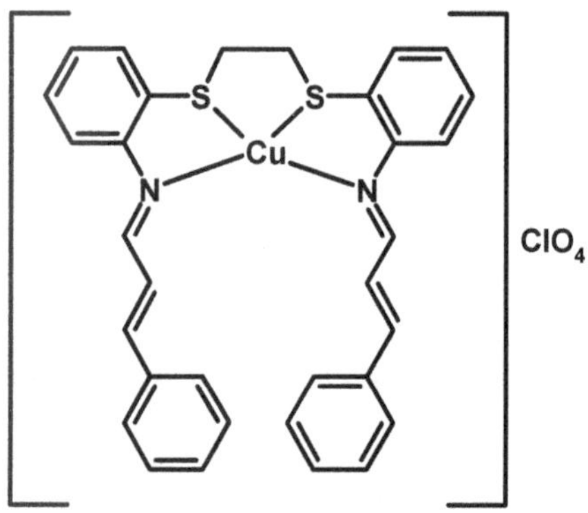

26

Scheme 30.

Dinuclear complexes 24 and 25 were prepared by a similar procedure [35]. To a solution of PPh_3 in acetonitrile was added a solution of CuX (X = I (24) and Br (25)) in acetonitrile and the mixture was stirred at room temperature for about 10 min to give a clear solution. Then, the ligand ca$_2$dapte was added and a clear orange-red solution was obtained. Single crystals of 24 and 25, were grown by slow diffusion of Et_2O vapor into a concentrated solution of complex [35]. These complexes are centrosymmetric dimers with pairs of Cu(PPh_3)X units bridged by ca$_2$dapte. The solid-state structure of complexes 24 and 25 reveals copper(I) ion coordinated to one N atoms and one sulfur atom of one Schiff base ligand, one P from PPh_3 and one halide group. The central Cu(I)/Cu(Br) ion has a common irregular pseudo-tetrahedral geometry arising from the low intraligand N-Cu-S chelate angles 80.72(8)° in 24 and 80.54(19)° in 25. The P-Cu-X angle is 126.17(3) and 121.41(7)° in 24 and 10, respectively, being larger than the tetrahedral values. In the ^{1}H-NMR spectra of 23 and 24, the iminic protons corresponding to the coordinated imine groups appear as a doublet at 8.41 and 8.33 ppm, respectively (free ligand has 8.14 ppm).

To a solution of [Cu(CH_3CN)]ClO_4 in acetonitrile was added, with continuous stirring, a solution of ca$_2$dapte (L^{31}) in the minimum amount of chloroform at room temperature and

then stirred for 10 min to give a clear dark-red solution. Single crystals of 26, were grown by slow diffusion of Et_2O vapor into a concentrated solution of the complex [35]. There are four similar but independent complex units in the asymmetric unit of 26. In the absence of other coordinating ligands the ca_2dapte acts as a tetradentate ligand. However, [Cu(ca_2dapte)NCS] complex is formed in the presence of NCS [66] causing that the ca_2dapte acts as a tridentate ligand leaving a free site for the NCS to coordinate.
The solid-state structure of complex 26 reveals copper(I) ion coordinated to two N atoms and two sulfur atoms of one Schiff base ligand. The central Cu(I) ion has the common irregular pseudo-tetrahedral geometry arising from the low intraligand N-Cu-S chelate angles 86.55(12) and 86.93(11)° and S-Cu-S chelate angle 93.11(5)°. In the ^{1}H-NMR spectrum of 26, the iminic protons corresponding to the coordinated imine groups appear as a doublet at 8.23 ppm (free ligand has 8.14 ppm).

4.3.2 (NN')$_2$ Schiff bases with a flexible spacer

In this type of complexes, N-donor ligands are mainly focused on Schiff bases containing pyridine group (see L^{35} and L^{36}).
Dinuclear copper(I) complexes were obtained when the ligands with the -C=N- group in *ortho*-position of pyridine ring were used in the synthesis. The reaction of P_2en (L^{35}) [30,38] or (Mepk)$_2$en (L^{36}) [39,69] in the presences of CuI and PPh_3 (1:2:2 molar ratio) in acetonitrile, yielded dark-red or orange precipitates of dinuclear copper(I) complexes [Cu_2I_2(μ-P_2en)(PPh_3)$_2$] (27) or [Cu_2I_2(μ-(Mepk)$_2$en)(PPh_3)$_2$]·2CH_3CN (28). Single crystals of 27 and 28, were grown by slow evaporation of solvent at room temperature for several days [38, 69]. These complexes are centrosymmetric dimers with pairs of Cu(PPh_3)I units bridged by P_2en [38] and (Mepk)$_2$en [69]. The solid-state structure of complexes 27 and 28 reveals copper(I) ion coordinated to two N atoms of one Schiff base ligand, one P from PPh_3 and one iodine group (Scheme 31). The central Cu(I) ion has the common irregular pseudo-tetrahedral coordination geometry arising from the low intraligand N-Cu-N chelate angles 79.40(15)° in 27 and 78.86(10)° in 28. The P-Cu-I angle is 123.11(4) and 119.05(3)° in 27 and 28, respectively, being larger than the ideal tetrahedral values.

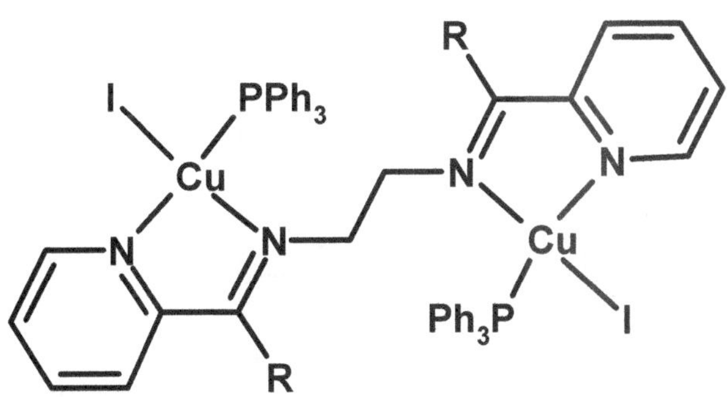

R = H (27) R = CH_3 (28)

Scheme 31.

The reaction of (Mepk)$_2$en (L^{36}) in the presence of [Cu(CH_3CN)$_4$]ClO_4 and PPh_3 (1:2:4 molar ratio) in acetonitrile yielded yellow precipitates of dinuclear copper(I) complexes [Cu_2(μ-(Mepk)$_2$en)(PPh_3)$_4$](ClO_4)$_2$·2$CHCl_3$ (29) [39]. Single crystals of 29, were grown by slow diffusion of Et_2O into the concentrated solution at 273 K for several days [39]. These complexes are centrosymmetric dimers with pairs of $Cu(PPh_3)_2$ units bridged $(Mepk)_2en$. The solid-state structure of complex 29 reveals copper(I) ion coordinated to two N atoms of one Schiff base ligand and two P from two PPh_3 (Scheme 32). The central Cu(I) ion has the common irregular pseudo-tetrahedral geometry arising from the low intraligands N-Cu-N chelate angles 79.4(2)° and the P-Cu-P angle is 118.24(7)°. The N-Cu-P bond angles range from 111 to 120°, i.e. slightly more than in a regular tetrahedron (109.5°). The P-Cu-P angle has opened up due to the steric effect from the bulky PPh_3 ligands. The Cu···Cu' separation distance 7.06(1) Å in 29 is longer than that observed in 27 (6.2577(6) Å) which is also caused by steric effect of PPh_3 ligand and also ClO_4 anions.

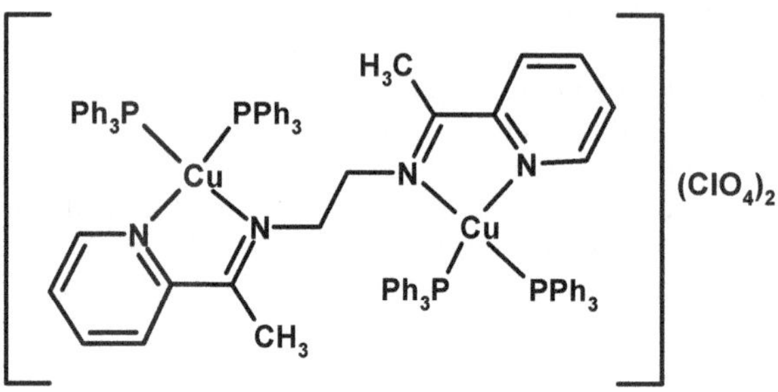

Scheme 32.

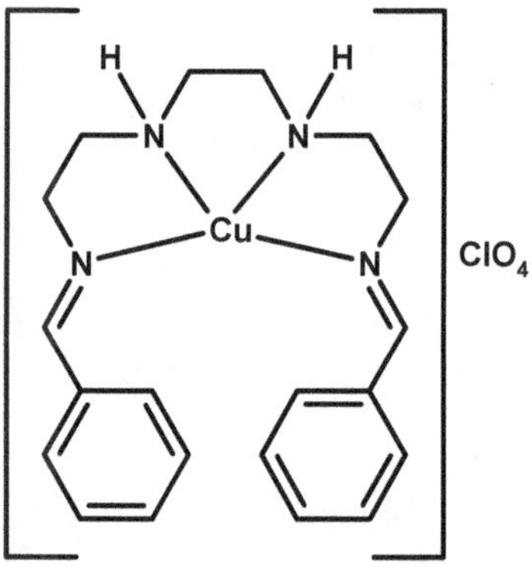

Scheme 33.

The ligand L[38] is a 2:1 condensate of benzaldehyde and triethylenetetramine (trien). The reaction of ba_2trien in the presences of $[Cu(CH_3CN)_4]ClO_4$ (1:1 molar ratio) in degassed methanol under N_2 atmosphere yielded reddish yellow precipitates of mononuclear copper(I) complexe $[Cu(ba_2trien)]ClO_4$ (30) [37] (Scheme 33). Single crystals of 30, were grown by slow evaporation of dichloromethane-hexane mixture. The N_4 coordination sphere of the copper(I) ion is significantly distorted from tetrahedral due to the steric constraints of the ligand with the $Cu-N_{imino}$ bond lengths shorter than the $Cu-N_{amino}$ ones.

The cation has a crystallographic C_2 axis. This complex displays a quasireversible $Cu^{II/I}$ couple with a half-wave potential of 0.12 V *vs.* SCE [37].
Though 30 does not show any emission in methanol at room temperature, at 77 K in methanol glass it displays two very distinct emission bands at 470 and 495 nm together with a faint one at 545 nm. It should be noted that in the rigid matrix at 77 K, methanol coordination to the copper(II) center in the excited state is not operative.

4.4 Copper(I) grid complexes

In recent years, grid complexes have been targeted by many researchers as a template to arrange two bidentate ligand binding pockets orthogonal to each other. Brooker *et al.* systematically studied on grid complexes of first transition metal ions that exhibit interesting structural, electrochemical and magnetic properties [31,32, 87-90]. The flat bis-bidentate ligand dppn [3,6-bis(2-pyridiyl)pyridazine] has been used for preparation of copper(I) gridlike complexes in 1992 (Scheme 34) [91]. In recent years, the incorporation of such grid-forming bis-bidentate Schiff base ligands into dppn has been explored (Scheme 13). They have been prepared from the reaction of 3,6-diformylpyridazine and substituted amino-benzenes. It was suggested that these ligands are preorganised to form a tetranuclear [2 × 2] grid complexes with a tetrahedrally coordinated metal ion such as copper(I) and silver(I) [32].

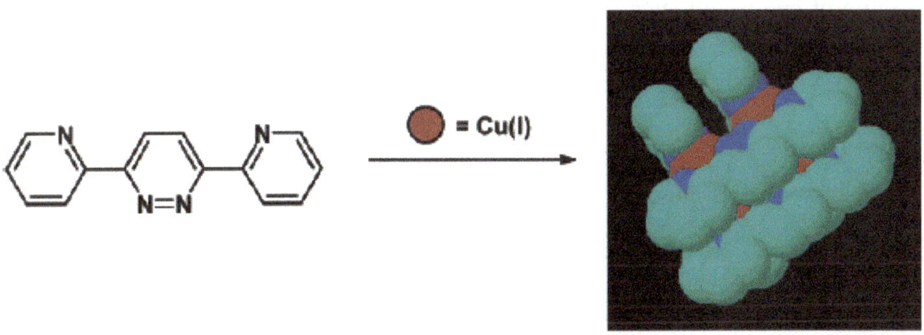

Scheme 34.

For preparation of grid copper(I) complexes, the bis-bidentate Schiff base ligands (Scheme 25) were suspended in pure dry acetone and degassed with nitrogen for 15 min. An equimolar amount of tetrakis(acetonitrile)copper(I) hexaflurophosphate was added and the reaction mixture refluxed under nitrogen for 2h. The diffusion of Et_2O vapour into an acetone solution yielded single crystals of gridelike copper(I) complexes $[Cu_4(L^{45})_4](PF_6)_4$ (31) and $[Cu_4(L^{41})_4](PF_6)_4$ (32) (Fig. 5), while the layering of an acetone solution of the complex on benzene gave single crystals of $[Cu_4(L^{42})_4](PF_6)_4$ (33) (Fig. 5) [32].
While the tetranuclear cations $[Cu_4(L^{45})_4](PF_6)_4$ and $[Cu_4(L^{41})_4](PF_6)_4$ has no crystallographic imposed symmetry, the cation $[Cu_4(L^{45})_4](PF_6)_4$ contains a two-fold rotation axis relating two halves of the grid from the two ligand strands and two copper(I) ions (Fig. 42). In all of the [2 × 2] copper(I) grid complexes, the coordination geometry of the copper(I) centers is a

considerably flattened tetrahedron, formed by four nitrogens from the two Schiff base ligands, with significantly different Cu-N bond distances.

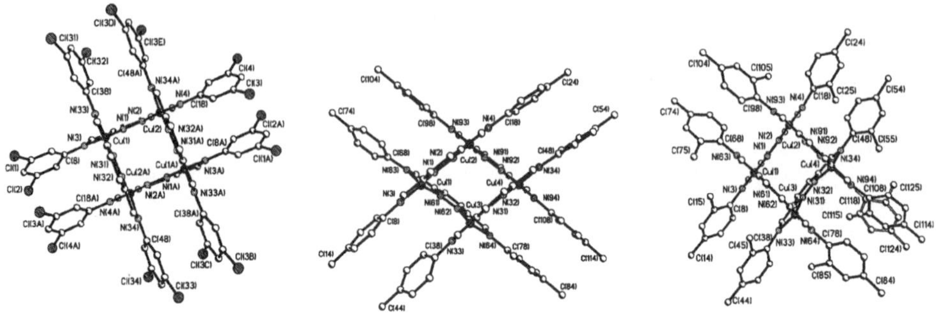

31 32 33

Fig. 5. An ORTEP views of 31-33, showing the atomic numbering scheme. H atoms are omitted for clarity.

These complexes show multiple reversible redox couples. As anticipated, the $E_{1/2}$ values are observed to vary depending on the electron donating/withdrawing nature of the substituents. Complex 31, with chloro withdrawing sustituents, exhibits a total of four reversible one-electron reduction processes. It is likely that each of these processes involves reduction of one ligand strand which can also be taken as a supporting evidence for the existence of the tetranuclear grid structure. For the two complexes 32 and 33 all of the processes are shifted anodically, and hence the initial portion of the reversible couple is swamped by a strong stripping peak, preventing this fourth process from being resolved [32].

4.5 Double-stranded dinuclear copper(I) helicate complexes

Helicity continues to receive considerable attention as it allows for better understanding of the self-assemby processes involved in supramolecular chemistry [92]. Until now, many examples of both single and double-stranded architectures have now been reported [92]. Studies on inorganic helical complexes containing transition metal ions have attracted much attention recently [27-29,40,67-72]. In these supramolecular self-assembleies, the formation of the helicates can be described as the result of reading molecular information stored in the ligands by metal ions following the coordination algorithm such as tetrahedral. It seems that the self-assembly process of supramolecular helicates is not only controlled by the coordination geometry of metal ions and flexibility of the spacer groups in the ligand, but also influenced by the linkage mode of the spacer group [27-29,40,67-72].

Recently Pal *et al.* have reported a double-stranded, dinuclear, homotopic and neutral copper(I) helicate $[Cu_2(p_2en)_2](ClO_4)_2$ (34) with L^4 [30] (Scheme 35). Copper(I) complex 34 has been obtained by reaction of p_2en with $[Cu(CH_3CN)_4]ClO_4$ in equimolar proportion in anhydrous methanol under N_2 atmosphere. This complex is quite stable in solid state as well as in methanol or dichloromethane solution towards aerial oxidation. It is found to be a double-stranded helicate. The N4 coordination sphere around each copper center is distorted tetrahedral, with angles in the range 80.2 - 81.3° (chelate angles) and 116.2 - 136.6° (interligand ones). The Cu-N bond distances are in the range 2.007 - 2.110 Å. For each copper, one Cu-N bond is significantly longer than others. The longer bonds are formed by one of the two terminal pyridyl N's.

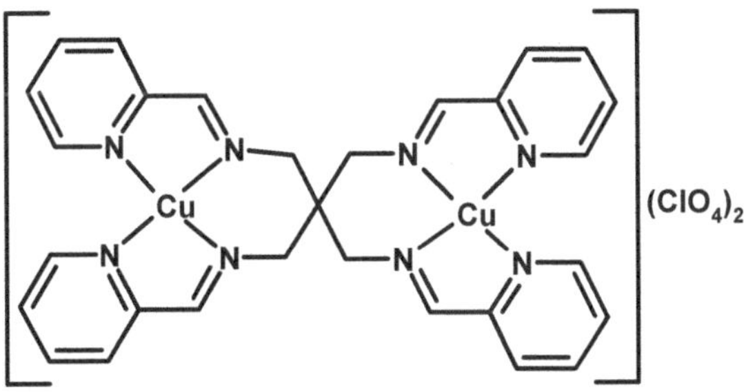

Scheme 35.

Complex 34 does not exhibit any emission in methanol and dichloromethane at ambient temperature, because coordination of the anion ClO_4 to the copper(II) center generated in the photoexcited state brings about a special type of quenching [30], while copper(I) complex $[Cu_2(p_2en)_2](PF_6)_2$ (35) is as stable as 34 towards aerial oxidation. Upon excitation at 360 or 470 nm, 35 display a single broad emission band with the maxima at 540 nm in methanol at room temperature [30].

In recent years, Hannon *et al.* have systematically studied on helical metallo-supramolecular arrays with multidentate Schiff bas ligands L^{47}–L^{50} [27,28,40,71,72]. For this kind of compounds, the ligand must offer sufficient flexibility for multiple strands to wrap around two or more metal centers, and it should also be sufficiently rigid to impose the same stereochemistry as both metals [27,28]. Non-helical isomers arise when the ligand is too flexible to impose the same stereochemistry at both metal centers [27,28]. For example, while the Schiff base ligand L^{47} is optimal for octahedral metal triple-helix formation [27,28], with tetrahedral metal ions a mixture of double-stranded helices (*rac* isomers) and boxes (*meso* isomers) are formed [27,28] (Scheme 36). These two isomers are in equilibrium in solution and the box is favored enthalpically, while the helix is favored by entropy. Thus at low temperature the box conformation dominates but as the temperature is raised the proportion of helix conformation grows.

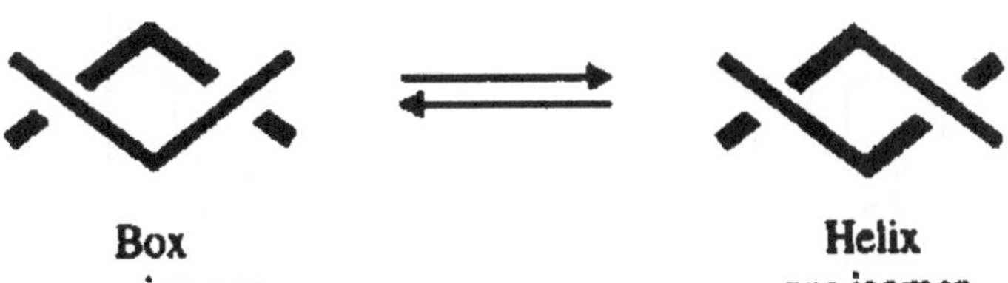

Scheme 36.

Reaction of L^{47} with monocations, such as copper(I), leads to dinuclear double-stranded complexes with 2:2 stoichiometry $[Cu_2L^{47}_2]^{2+}$ (36). In the ^{1}H-NMR spectrum of 36, the central

CH_2 protons in the helix can be used to identify the two isomers. The CH_2 protons in the helix are equivalent and thus appear as a singlet resonance, while in the box they are non-equivalent and thus appear as two doublets [28].

Addition of ethyl groups to the central spacer destabilizes the cyclophane configuration so that only $[Cu_2L^{49}{}_2]^{2+}$ (36) helicate isomer is present in solution [70]. This is evident from the ^{1}H-NMR spectrum that reveals single specie at room temperature and low temperature, and the central CH_2 resonance as a single confirming the helical conformation [70]. The double-helicate structure of complex 36 has been further confirmed by crystallography. The double helicate $[Cu_2L^{49}{}_2]^{2+}$ cation represents a dicationic cylinder (Scheme 38) where each copper(I) center bound to two pyridylimine units, one from each ligand. Each ligand bridges two copper(I) centers giving rise to the dinuclear-stranded helicate architecture. In $[Cu_2L^{50}{}_2](BF_4)_2$ (37) each ligand bridges two copper(I) centers giving rise to the dinuclear-stranded helicate architecture (Scheme 37) [70].

^{1}H-NMR spectra of copper(I) complex 37 have been recorded in CD_3NO_2 and CD_3CN solution. In CD_3NO_2 solution at room temperature, two sets of resonance signals are observed consistent with the prersence of two solution species, the *meso*- and *rac*-isomers, and while in CD_3CN solution at room temperature, a singlet set of resonance signals is observed because the ligands exchange is more rapid in this solvent [40].

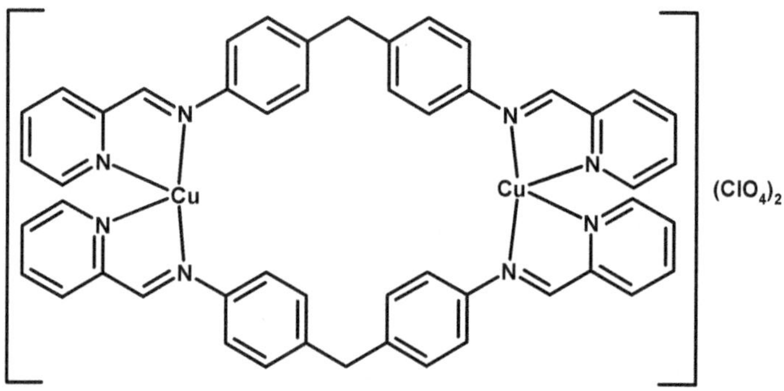

Scheme 37.

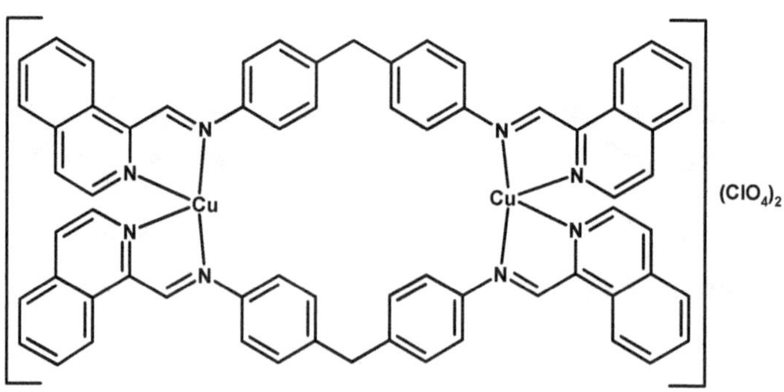

Scheme 38.

5. References

[1] Y.M. Hijji, B. Barare, A.P. Kennedy, R. Butcher, Sensors and Actators B136 (2009) 297-302.
[2] M. Aslantas, E. Kendi, N. Demir, A.E. Sabik, M. Tumer, M. Kertmen, Spectrochim. Acta A74 (2009) 617-624.
[3] H. Unver, M. Tildiz, B. Dulger, O. Ozgen, E. Kendi, T.N. Durlu, J. Mol. Struct. 737 (2005) 159-164.
[4] M. Yildiz, H. Unver, B. Dulger, D. Erdener, N. Ocak, A. Erdonmez, T.N. Durlu, J. Mol. Struct. 738 (2005) 253-260.
[5] A. Karakas, H. Unver, A. Elmali, J. Mol. Struct. 877 (2008) 152-157.
[6] A. Karakas, A. Elmali, U. Unver, I. Svoboda, Spectrochim. Acta A61 (2005) 2979-2987.
[7] E. Hadjoudis, A. Rontoyianni, K. Ambroziak, T. Dziembowska, I.M. Mavridis, J. Photochem. Photobiol. A162 (2004) 521-530.
[8] F.A. Carey, Organic Chemistry, 5th edn. McGraw-Hill, New York, (2003) 724.
[9] C.G. Hamaker, O.S. Maryashina, D.K. Daley, A.L. Walder, J. Chem. Crystallogr. 40 (2010) 34-39.
[10] B. Jarzabek, B. Kaczmarczyk, D. Sek, Spectrochim. Acta A74 (2009) 949-954.
[11] A.D. Khalaji, K. Fejfarova, M. Dusek, Acta Chim. Slov. 57 (2010) 257-261.
[12] V. Jeseentharani, J. Selvakumar, A. Dayalan, B. Varghese, K.S. Nagaraja, J. Mol. Struct. 966 (2010) 122-128.
[13] H. Petek, C. Albayrak, M. Odabasoglu, I. Senel, O. Buyukgungor, Struct. Chem. 21 (2010) 681-690.
[14] C. Albayrak, B. Kosar, S. Demir, M. Odabasoglu, O. Buyukgungor, J. Mol. Struct. 963 (2010) 211-218.
[15] T.R. van den Ancker, G.W.V. Cave, C.L. Raston, Green Chem. 8 (2006) 50-53.
[16] A.D. Khalaji, S. Triki, Russ. J. Coord. Chem. 37 (2011) 518-522.
[17] S. Jalali Akerdi, G. Grivani, H. Stoeckli-Evans, A.D. khalaji, Russ. J. Coord. Chem. 37 (2011) 394-398.
[18] A.D. Khalaji, S. Maghsodlou Rad, G. Grivani, D. Das, J. therm. Anal. Calorim. 103 (2011) 747-751.
[19] A. Ray, S. Mitra, A.D. Khalaji, C. Atmani, N. Cosquer, S. Triki, J. M. Clemente-Juan, S. Cardona-Serra, C.J. Gomez-Garcia, R.J. Butcher, E. Garribba, D. Xu, Inorg. Chim. Acta 363 (2010) 3580-3588.
[20] A.D. Khalaji, M. Weil, G. grivani, S. Jalali Akerdi, Monatsh. Chem. 141 (2010) 539-543.
[21] A.D. Khalaji, H. hadadzadeh, K. Fejfarova, M. Dusek, Polyhedron 29 (2010) 807-812.
[22] A.D. Khalaji, H. Stoeckli-Evans, Polyhedron 28 (2009) 3769-3773.
[23] A. Mukherjee, R. Chakrabarty, S.W. Ng, G.K. Patra, Inorg. Chim. Acta 363 (2010) 1707-1712.
[24] D. Petrovic, L.M.R. Hill, P.G. Jones, W.B. Tolman, M. Tamm, Dalton Trans. (2008) 887-894.
[25] X. Li, J. Ding, W. Jin, Y. Cheng, Inorg. Chim. Acta 362 (2009) 233-237.
[26] A. Fazel, S. Al-Fayez, L.H. Abdel-Rahman, Z.S. Seddigi, A.R. Al-Arfaj, B.E. Ali, M.A. Dastageer, M.A. Gondal, M. Fettouhi, Polyhedron 28 (2009) 4072-4076.
[27] L.J. Childs, J. Malina, B.E. Rolfsnes, M. Pascu, M.J. Prieto, M.J. Broome, P.M. Rodger, E. Sletten, V. Moreno, A. Rodger, M.J. Hannon, Chem. Eur. J. 12 (2006) 4919-4927.

[28] L.J. Childs, M. Pascu, A.J. Clarke, N.W. Alcock, M.J. Hannon, Chem. Eur. J. 10 (2004) 4291-4300.
[29] S. Chowdhury, P.B. Iveson, M.G.B. Drew, D.A. Tocher, D. Datta, New J. Chem. 27 (2003) 193-196.
[30] P.K. Pal, S. Chowdhury, P. Purkayastha, D.A. Tocher, D. Datta, Inorg. Chem. Commun. 3 (2000) 585-589.
[31] S. Brooker, S.J. Hay, P.G. Plieger, Angew. Chem. Int. Ed. 39 (2000) 1968-1970.
[32] J.R. Price, Y. Lan, S. Brooker, Dalton Trans. (2007) 1807-1820.
[33] A.D. Khalaji, S. Triki, D. Das, J. Therm. Anal. Calorim. 103 (2010) 779-783.
[34] A.D. Khalaji, R. Welter, Inorg. Chim. Acta 359 (2006) 4403-4406.
[35] M. Morshedi, M. Amirnasr, A.M.Z. Slawin, J.D. Woollins, A.D. Khalaji, Polyhedron 28 (2009) 167-171.
[36] M. Morshedi, M. Amirnasr, S. Triki, A.D. Khalaji, Inorg. Chim. Acta 362 (2009) 1637-1640.
[37] S. Panja, S. Chowdhury, M.G.B. drew, D. Datta, Inorg. Chem. Commun. 5 (2002) 304-307.
[38] X.-H. Zhou, T. Wu, D. Li, Inorg. Chim. Acta 359 (2006) 1442-1448.
[39] A.D. Khalaji, M. Amirnasr, R. Welter, Russ. J. Coord. Chem. 36 (2010) 835-837.
[40] M. Pascu, G.J. Clarkson, B.M. Mariuki, M.J. Hannon, Dalton Trans. (2006) 2635-2642.
[41] A.D. Khalaji, M. Weil, H. hadadzadeh, M. Daryanavard, Inorg. Chim Acta 362 (2009) 4837-4842.
[42] A.D. Khalaji, K. Brad, Y. Zhang, Acta Crystallogr. E64 (2009) m189.
[43] A.D. Khalaji, H. hadadzadeh, K. Gotoh, H. Ishida, Acta Crystallogr. E64 (2009) m70.
[44] S. Dehghanpour, L. Forouhi, M.M. Amini, H.R. Khavasi, K. Jahani, F. Nouroozi, E. Zamanifar, J. Coord. Chem. 61 (2008) 455-463.
[45] M.H. Habibi, M. Montazerozohori, K. Barati, R.W. Harrington, W. Clegg, Anal. Sci. 23 (2007) x45-x46.
[46] S. Chowdhury, G.K. Patra, M.G.B. drew, N. Chattopadhyay, D. Datta, J. Chem. Soc. Dalton Trans. (2000) 235-237.
[47] R. Kia, V. Mirkhani, S. Harkema, G.J. van Hummel, Inorg. Chim. Acta 360 (2007) 3369-3375.
[48] S. Dehghanpour, M. Khalaj, A. Mahmoudi, J. Coord. Chem. 62 (2009) 2957-2965.
[49] A.D. Khalaji, M. Amirnasr, J.-C. Daran, Acta Crystallogr. E62 (2006) m2200-m2201.
[50] G. Zhang, G. Yang, J.S. Ma, J. Chem. Crystallogr. 36 (2006) 631-635.
[51] G. kickelbick, M. Amirnasr, A.D. Khalaji, S. Dehghanpour, Acta Crystallogr. E58 (2002) m381-m382.
[52] G. kickelbick, M. Amirnasr, A.D. Khalaji, S. Dehghanpour, Aust. J. Chem. 56 (2003) 323-328.
[53] S. Dehghanpour, R. Kempe, S. Balireddi, L. Fotouhi, F. Tabasi, F. Mojahed, S. Salek, Z. Anorg. Allg. Chem. 632 (2006) 2321-2325.
[54] A.D. Khalaji, M. Amirnasr, J.-C. Daran, Acta Crystallogr. E62 (2006) m3183-m3184.
[55] S. Dehghanpour, A.H. Mahmoudkhani, M. Amirnasr, Struct. Chem. 17 (2006) 255-262.
[56] A.D. Khalaji, R. Welter, M. Amirnasr, A.H. Barry, Anal. Sci. 24 (2008) x137-x138.

[57] S. Dehghanpour, F. Rominger, K. Aoki, Synt. React. Inorg. Metal-Org & Nano-Met Chem. 39 (2009) 230-235.
[58] A.D. Khalaji, M. Amirnasr, R. Welter, Anal. Sci. 22 (2006) x49-x50.
[59] R. Hadjikhani, S. Dehghanpour, A. Mahmoudi, F. Mojahed, Z. Anorg. Allg. Chem. 632 (2006) 723-725.
[60] M. Amirnasr, G. Kickelbick, S. Dehghanpour, Helv. Chim. Acta 89 (2006) 274-284.
[61] M. Amirnasr, A.D. khalaji, L.R. Falvello, Inorg. Chim. Acta 359 (2006) 713-717.
[62] S. Dehghanpour, F. Rominger, Russ. J. Coord. Chem. 35 (2009) 259-263.
[63] S. Dehghanpour, N. Bouslimani, R. Welter, F. Mojahed, Polyhedron 26 (2007) 154-162.
[64] S. Dehghanpour, F. Rominger, Synt. React. Inorg. Metal-Org & Nano-Met Chem. 38 (2008) 598-603.
[65] W. Massa, S. Dehghanpour, K. Jahani, Inorg. Chim. Acta 362 (2009) 2872-2878.
[66] M. Morshedi, M. Amirnasr, S. Triki, A.D. Khalaji, J. Chem. Crystallogr. 41 (2011) 39-43.
[67] S. De, S. Chowdhury, M.G.B. Drew, D. Datta, Ind. J. Chem. A50 (2011) 171-175.
[68] G.K. Patra, I. Goldberg, New J. Chem. 27 (2003) 1124-1131.
[69] A.D. Khalaji, M. Amirnasr, R. Welter, Acta Crystallogr. E62 (2006) m2950-m2951.
[70] J. Keegan, P.E. Kruger, M. Nieuwenhuyzen, N. Martin, Cryst. Growth Des. 2 (2002) 329-332.
[71] J.M.C.A. Kerckhoffs, J.C. Peberdy, I. Meistermann, L.J. Childs, C.J. Isaac, C.R. Pearmund, V. Reudegger, S. Khalid, N.W. Alcock, M.J. Hannon, A. rodger, Dalton Trans. (2007) 734-742.
[72] L.J. Childs, N.W. Alcock, M.J. Hannon, Angew. Chem. Int. Ed. 41 (2002) 4244-4247.
[73] R.-Z. Li, D. Li, X.-C. Huang, Z.-Y. Qi, X.-M. Chen, Inorg. Chem. Commun. 6 (2003) 1017-1019.
[74] I. Dance, M. Scudder, J. Chem. Soc. Dalton Trans. (2000) 1579-1586.
[75] I. Dance, M. Scudder, J. Chem. Soc. Dalton Trans. (2000) 1587- 1594.
[76] Q.M. Wang, G.-C. Guo, T.C.W. Mak, Chem. Commun. (1999) 1849-1850.
[77] Y. Li, Z.X. Zhang, T. Li, K.C. Li, Russ. J. Coord. Chem. 36 (2010) 48-52.
[78] D.R. McMillin, J.R. Kirchhoff, K.V. Goodwin, Coord. Chem. Rev. 64 (1985) 83-92.
[79] M.K. Eggleston, P.E. Fanwick, A.J. Pallenberg, D.R. McMillin, Inorg. Chem. 36 (1997) 4007-4010.
[80] M.K. Eggleston, D.R. McMillin, K.S. Koenig, A.J. Pallenberg, Inorg. Chem. 36 (1997) 172-176.
[81] R.F. Song, Y.B. Xie, X.H. Bu, J.Mol.Struct. 657 (2003) 311- 316.
[82] Y.B. Xie, C.H. Cui, X.H. Bu, J. Mol. Struct. 705 (2004) 11- 14.
[83] Y. Zhang, M. Du, J.R. li, R.H. Zhang, X.H. Bu, Dalton Trans. (2003) 1509- 1514.
[84] Y.B. Xie, C. Zhang, J.R. Li, X.H. Bu, Dalton Trans. (2003) 562- Dalton Trans. (2003) 1509.
[85] T.K. Ranson, H. Adams, M.D. Ward, CrystEng Commun. 8 (2006) 497- 501.
[86] S. Tavacoli, T.A. Miller, R.L. Paul, J.C. Jeffery, M.D. ward, Polyhedron 22 (2003) 507-514.
[87] J. Hausmann, G.B. Jameson, S. Brooker, Chem. Commun. (2003) 2992-2993.
[88] J.R. Price, Y. Lan, G.B. Jameson, S. Brooker, Dalton Trans. (2006) 1491-1494.
[89] J. Hausmann, S. Brooker, Chem. Commun. (2004) 1530-1531.

[90] J.R. Price, N.G. White, A. Perez-Velasco, G.B. Jameson, C.A. Hunter, S. Brooker, Inorg. Chem. 47 (2008) 10729-10739.

[91] M.-T. Youinou, N. Rahmouni, J. Fischer, J.A. Osborn, Angew. Chem. Int. Ed. Engl. 31 (1992) 733- 735.

[92] G. Dong, P. Ke-Liang, D. Chun-Ying, H. Cheng, M. Aing-Jin, Inorg. Chem. 41 (2002) 5978-5985.

7

Intramolecular N–H···X (X = F, Cl, Br, I, and S) Hydrogen Bonding in Aromatic Amide Derivatives - The X-Ray Crystallographic Investigation

Dan-Wei Zhang and Zhan-Ting Li
Department of Chemistry, Fudan University, Shanghai
China

1. Introduction

Hydrogen bonding (H-bonding) has recently been defined by IUPAC as "an attractive interaction between a hydrogen atom from a molecule or a molecular fragment X-H in which X is more electronegative than H, and an atom or a group of atoms in the same or a different molecule, in which there is evidence of bond formation". In most cases, the strength of an H-bond increases with the increase of the electronegativity value of the acceptor atom (Pauling, 1960). This is exactly the case for oxygen and nitrogen atoms. The H-bonds formed between them and the NH and OH groups are usually strong, which play essential roles in studies in supramolecular, crystal engineering, materials, and life sciences (Scheiner, 1997; Jeffrey,1997). As s result of their growing applications in supramolecular chemistry and crystal engineering, in the past two decades, the critical assessment of the weaker H-bonds has also become an important topic (Desiraju & Steiner, 2001). In this context, organic halogen and sulfur atoms, C-X (X = F, Cl, Br, I, S), have all been demonstrated to be weak H-bonding acceptors (Dunitz & Taylor, 1997), although their electronegativities (Pauling scale: 3.98, 3.16, 2.96, 2.66, and 2.58, respectively) are all higher than that of hydrogen (2.20). Indeed, over years it has been accepted that organic fluorine "hardly ever accepts hydrogen bonds (Dunitz, 2004)," presumably due to its low polarizability and tightly contracted lone pairs. For other organic heteroatoms, the increased van der Waals radius and decreased electronegativities may also weaken their capacity of forming the intramolecular electrostatic interaction, i.e., H-bonding, with the amide hydrogen and lose the competition with the amide oxygen of another molecule which forms the intermolecular N–H···O=C H-bonding. In contrast, the halogen anions are capable of forming strong intermolecular H-bonding with NH, OH or even CH protons (Harrell & McDaniel, 1964; Simonov et al., 1996; Del Bene & Jordan, 2001).

This chapter summarizes recent progresses in the assessment of the weak intramolecular six- and five-membered H-bonding patterns formed by aromatic amides bearing the above five atoms. Theoretical investigations show that similar intermolecular H-bonding patterns can be formed by fluorine in DNA or RNA base analogues (Frey et al., 2006; Koller et al., 2010; Manjunatha et al., 2010), although they are difficult to be confirmed in solution experimentally. The crystal structures of many organic halogen or sulfur (ether) compounds exhibit such intermolecular short contacts, which may be mainly driven by the intrinsic preference of these atoms in forming the H-bonding or formed due to the assistance of the intermolecular stacking and van der Waals force (Toth et al., 2007) and other intra- and intermolecular interactions.

Due to the increased conformational flexibility of the backbones and the decreased acidity of the amide proton, the H-bonding in aliphatic amide derivatives is expected to be even weaker. However, five-membered intramolecular N–H···F (F: Hughes & Small, 1972; O'Hagan et al., 2006), N–H···Cl (de Sousa et al., 2007; Kalyanaraman et al., 1978) and N–H···I (Savinkina et al., 2008) H-bonding patterns have been observed in aliphatic amides. To the best of our knowledge, the six-membered one is not available yet in simple organic molecules.

One consideration for exploiting the intramolecular N–H···X (X = F, Cl, Br, I, S) H-bonding of the aromatic amides is that the new patterns may find applications in designing new preorganized building blocks for crystal and supramolecular engineering (Biradha, 2003; Desiraju, 2005). Furthermore, new H-bonding motifs may also be useful in building foldamers (Zhu et al., 2011; Zhao & Li, 2010; Saraogi & Hamilton, 2009; Li et al,, 2008; Li et al., 2006; Huc, 2004; Sanford & Gong, 2003), the artificial secondary structures, and for designing biologically or medicinally useful structures (Tew et al., 2010; Li et al., 2008; Bautista et al., 2007). For doing this, the more competitive intermolecular N–H···O=C H-bonding of the amide unit has to be suppressed. There are two approaches for realizing this purpose. The first one concerns the introduction of a strong intramolecular H-bond to "lock" the amide proton. The second one is to introduce one or more bulky groups to impede the contact of the amides. In these ways, the very weak intramolecular N–H···I hydrogen bonding can be observed. There are several techniques for investigating the formation of the weak intramolecular H-bonding. The NMR spectroscopy is promising for studies in solution (Manjunatha et al., 2010), and the infrared spectroscopy can be used to detect samples in both the solution and solid state (Legon, 1990), while the computational modeling can provide useful information about the effects of discrete factors on the stability of the H-bonds (Dunitz, 2004; Liu et al., 2009), which are particularly valuable when experimental evidences are not available. In view of the feature of this book, we will focus on the investigations by the X-ray crystallography.

The crystal structure of an aromatic amide molecule is affected by many factors, including the stacking pattern, van der Waals force, intra- and intermolecular hydrogen and halogen bonding, and shape matching of the molecule. The entrapped solvent molecules, particularly those containing heteroatoms, may also play an important role because they are able to form hydrogen or halogen bonding with the molecule and thus affect the stacking pattern to suppress or promote the formation of the intramolecular H-bonding. Concerning the criterion for the formation of the weak intramolecular H-bonding, we simply check the distance between the heteroatom and the amide hydrogen in the crystal structure. If it is shorter than the sum of the radius of the two atoms, we consider that an H-bonding is formed (Desiraju & Steiner, 2001). Although in X-ray structures the proton/hydrogen is not located accurately and may bend toward or away from the acceptor, for clarity we simply use the reported distances between the two concerned atoms as the criteria.

2. N–H···F Hydrogen bonding

Fluorine atom has the highest electronegativity. In 1996, Howard et al. carried out a review on the short F···H contacts from all of the organofluorine compounds deposited in the

Cambridge Structural Database System (CSDS) and concluded that organic fluorine is at best only a weak H-bonding acceptor (Howard et al., 1996). In 1997, Dunitz and Taylor also executed an intensive search of the CSDS and confirmed that organic fluorine accepts hydrogen bonds only in the absence of a better acceptor (Dunitz & Taylor, 1997). They also examined the evidence for H-bonding to organic fluorine in protein–ligand complexes and found that it is unconvincing. They thus proposed that, due to its low polarizability and tightly contracted lone pairs, organic fluorine does not compete with stronger H-bond acceptors such as oxygen or nitrogen, and only when other better acceptor atoms are sterically hindered that the O-H···F or N-H···F H-bonding can be formed (Barbarich et al., 1999).

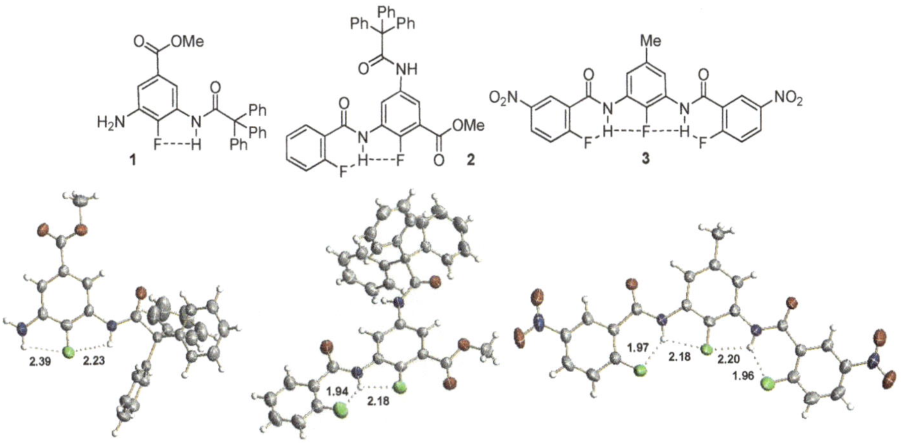

Fig. 1. Compounds 1-3 and their crystal structures.

In 1982, Kato et al. reported the crystal structure of 2-fluorobenzamide (Kato & Sakurai, 1982). Although the positions of hydrogen atoms were not determined, the N···F distance is 2.80 Å, which corresponded to an *N*H···F distance of 2.15 Å by molecular modeling. Clearly, an intramolecular six-membered N-H···F hydrogen bond exists in the crystal. In 2003, Li et al. found that 2-fluorobenzamide derivatives might promote the stability of hydrazide-based quadruply hydrogen-bonded heterodimers by forming six-membered intramolecular N–H···F hydrogen bonding (Zhao et al, 2003). A number of model compounds were then designed and prepared (Li et al., 2005). The crystal structures of compounds **1-3**, which bear one triphenylmethyl or two nitro groups to increase their cystallinity (Corbin et al, 2003; Yin et al., 2003), were obtained (Figure 1). All the three compounds adopt a well-defined planar conformation rigidified by the intramolecular N–H···F H-bonds. The F···H (amide) distance of compound **1** is 2.23 Å, and the N–H···F angle is 106°. The fluorine atoms of both **2** and **3** are located to the proximity of the amide hydrogen due to the formation of the three-centered H-bonds, which is common for similar alkoxyl-substituted aromatic amide (Gong, 2001). The F···H (amide) distance of the six- and five-membered H-bonds is 1.94 and 2.18 Å in **2**, and 1.97 and 2.18 Å in **3**, respectively. The corresponding F···H–N angle is 136 and 108° for **2**, and 136 and 111° for **3**. All these values fall into the range of the criterion for the judgment of a F···H–N H-bond—the F···*H*N distance < 2.3 Å and the N–H···F angle > 90°

proposed by Dunitz and Taylor (Dunitz & Taylor, 1997). The N*H*···F distance of the amino group of **1** is 2.39 Å, which is larger than that of the amide, also reflecting the preference of the amide proton to form the intramolecular hydrogen bond. ^{1}H NMR experiments also support that the five- and six-membered and three-center H-bonds are formed in solution.

Recently, the crystal structures of more *N*-aryl 2-fluorobenzamides have been reported, most of which display the six-membered N–H···F H-bonding motif. The structures of compounds 4 and 5 are shown in Figure 2 as examples (Chopra & Row, 2008, 2005). The crystal structures of many *N*-(2-fluorophenyl)amides are also available, which usually exhibit the intramolecular five-membered N–H···F H-bonding. As examples, the structures of 6 and 7 are provided in Figure 2 (Chopra & Row, 2005; Buyukgungor & Odabasoglu, 2008). It is worthy to note that no intramolecular N–H···F H-bonding is generated by the 2-fluorobenzenamine cation of 7. Its three ammonium protons only form intermolecular H-bonding with the oxygen atoms of the anion, reflecting that organic fluorine is weaker than oxygen as proton acceptor.

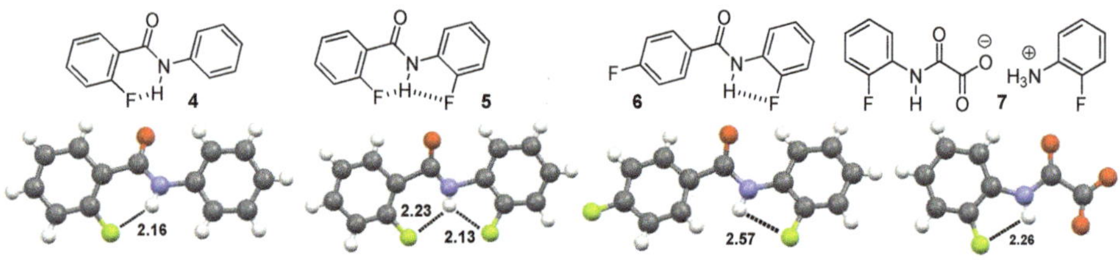

Fig. 2. Compounds 4-7 and their crystal structures.

Generally, 2-fluorobenzamides have a large preference of forming the six-membered N–H···F H-bonding. When there exist other strong competitive interactions, this H-bonding may be suppressed. This occurs for selenourea derivative **8** (Kampf et al., 2004). This compound forms a dimeric pattern in the crystal stabilized by two strong N–H···Se=C H-bonds (Figure 3), which causes a large torsion (51°) of the amide unit from the benzene plane. As a result, the intramolecular N–H···F H-bonding is not formed. For *N*-(2-fluorophenyl)amides, the five-membered N–H···F H-bonding may also be suppressed, as revealed in the crystal structure of **9** (Lewis et al., R. J.; 1991). This compound exists in two conformations in the crystal structure. One of them forms the intramolecular five-membered H-bonding, while another one displays an intramolecular F···O=C contact (Figure 3). These observations indicate that, although fluorine is quite strong to form the five- and six-membered H-bonds, in the presence of other strong interactions, they may still be inhibited.

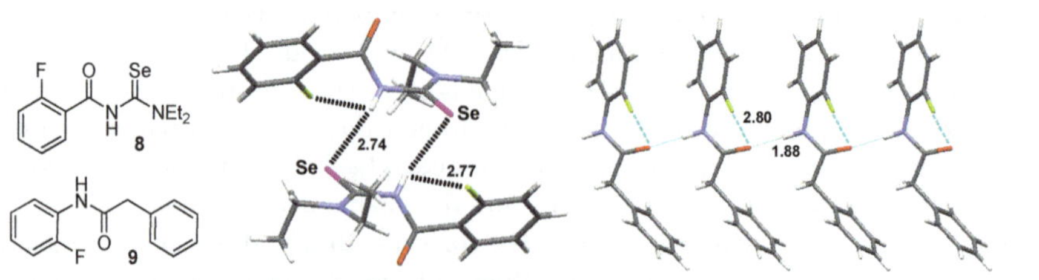

Fig. 3. Compounds 8 and 9 and their crystal structures.

To compare with their methoxyl-bearing analogues that form the N-H···OMe H-bonding, Li et al. also prepared compounds 10-12 (Zhu et al., 2007). The fluorine atoms in these compounds all form the intramolecular six- and five-membered N-H···F H-bonds (Figure 4). This three-center H-bonding pattern has been revealed for many alkoxyl-substituted linear aromatic amides (Li et al., 2006). The amide protons in these fluorine-bearing compounds further form intermolecular N-H···O=C H-bonding with the carbonyl oxygen. Similar intermolecular H-bonding is not displayed for the methoxyl-substituted analogues. Two factors are proposed to cause this difference. The first is that the intramolecular N-H···OMe H-bonding is strong and reduces the ability of the amide proton to form other H-bonding. The second is that methoxyl group is larger than fluorine and thus has a larger steric hindrance to suppress the formation of additional intermolecular H-bonding. The three-center H-bonding pattern is also observed for 13 (Chisholm et al., 2002), the 3-oxobutanamide unit of which forms a strong six-membered N-H···O=C H-bond. We may expect that the two H-bonds stabilize each other by co-inhibiting the intermolecular N-H···O=C H-bond.

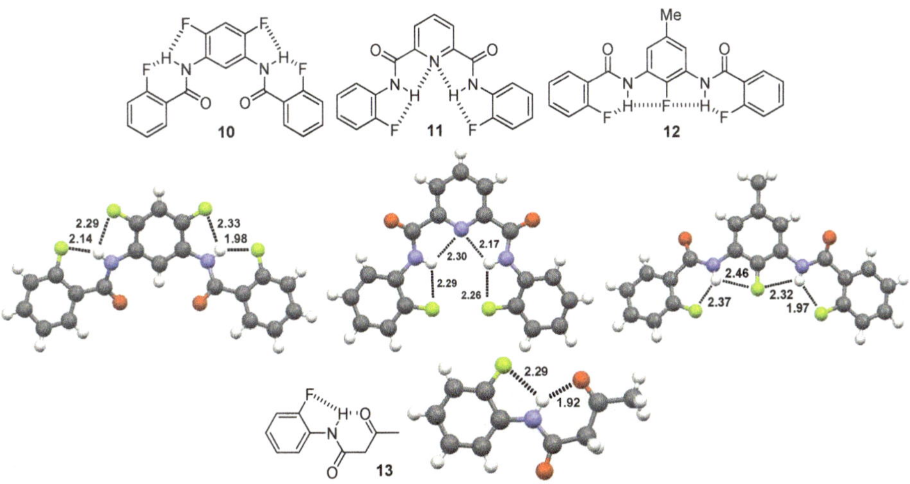

Fig. 4. Compounds 10-13 and their crystal structures.

The three-center H-bonding pattern does not always survive. For example, obviously due to the rigidity of the macrocyclic skeleton, the 2-fluoroisophthalamide units of macrocycle **14** form only one six-membered N-H···F H-bond (Figure 5) (Zhu et al., 2009). In the crystal structure of compound 15 (Guo et al., 2009), the molecules form a dimer which is stabilized by two N-H···O=C H-bonds between the carboxylic and amide units (Figure 5). The two H-bonds strengthen the torsion of the amide units from the two benzene planes. As a result, it only exhibits one weak six-membered N-H···F H-bond, and the 2-F of the aniline does not form the expected five-membered N-H···F H-bond. Instead, it displays an intermolecular F···OH contact (the distance: 2.81 Å). This compound contains several fluorine atoms and a carboxylic acid group and thus can produce discrete weak interactions. This result again reflects that the molecular conformation formed in the crystal is the outcome of the competition of different intra- and intermolecular interactions.

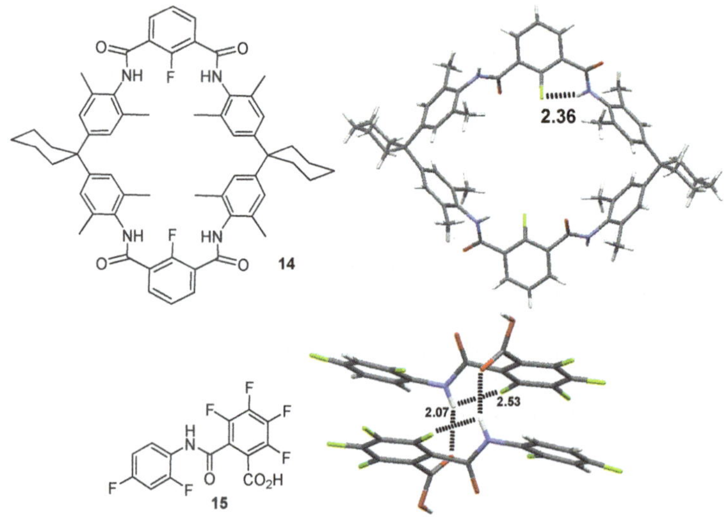

14

15

Fig. 5. Compounds 14and 15 and their crystal structures.

3. N–H···Cl hydrogen bonding

In 1974, Kato et al. reported the crystal structure of 2-chlorobenzamide (Kato et al., 1974), which exhibits a dimeric structure stabilized by the intermolecular eight-membered N-H···O=C H-bonding (Etter, 1990). The amide units further form a chain of the N-H···O=C H-bonding, which is typical for benzamides, but no six-membered N-H···Cl H-bonding is displayed. In recent years, the crystal structures of many 2-chloro-*N*-phenylbenzamide derivatives have been reported. Most of them do not form the intramolecular six-membered N-H···Cl H-bonding. However, compounds 16a-c (Arslan et al., 2007; Binzet et al., 2006; Binzet et al., 2004), 17 (Caleta et al., 2008) and 18 (Zhu et al., 2008) do form this weak H-bonding (Figure 6). The thiourea unit in 16a-c and the benzothiazol unit in 17 should increase the acidity of the amide protons, which, together with the possible steric effect, may facilitate the formation of the intramolecular six-membered N-H···Cl H-bonding. For 18, the large trityl group suppresses the intermolecular N-H···O=C H-bonding. Thus, the weak N-H···Cl H-bonding can be formed. When the trityl group is replaced with an adamantyl group, the resulting amide do not give rise to the N-H···Cl H-bonding in the crystal structure. Although ^{1}H NMR experiments in chloroform-d reveal that the intramolecular six-membered N-H···Cl H-bonding is generated in solution (Zhu et al., 2008), in the crystal structure, the amide units are only engaged in intermolecular N-H···O=C H-bonding.

N-(2-chlorophenyl)acetamide 19a forms a five-membered N-H···Cl H-bond (Figure 7) (Wan et al., 2006). The crystal structures of a number of its analogues are also available. In most cases, for example, for 19b (Gowda et al., 2007d), 19c (Zhu et al., 2008), 19d (Gowda et al., 2007a), 19e (Gowda et al., 2001), 19f (Gowda et al., 2000), 19g (Gowda et al., 2009), and 19h (Gowda et al., 2010), the intramolecular N-H···Cl H-bonding is formed (Figure 7). The intermolecular N-H···O=C H-bonding is also formed for all the compounds. Thus, we may consider that their strengths are comparable. The Cl and Br atoms on the methyl groups of

19d-h do not form the similar five-membered H-bonding. The crystal structure of methacrylamide derivative 19i does not display the five-membered H-bonding (Figure 7) (Kashino et al., 1994). Instead, a weak intermolecular C=C–H···π contact, together with the strong intermolecular N–H···O=C· H-bonding, is observed, indicating that the existence of other additional intermolecular interaction may also be able to suppress this intramolecular five-membered N–H···Cl H-bonding.

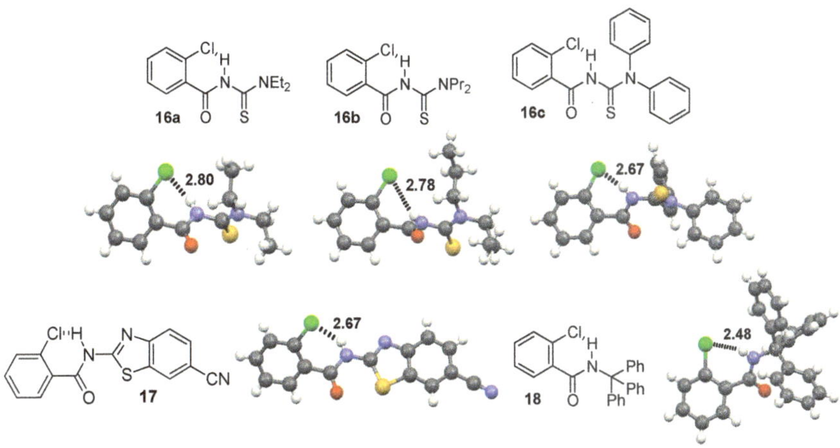

Fig. 6. Compounds 16a-c, 17 and 18 and their crystal structures.

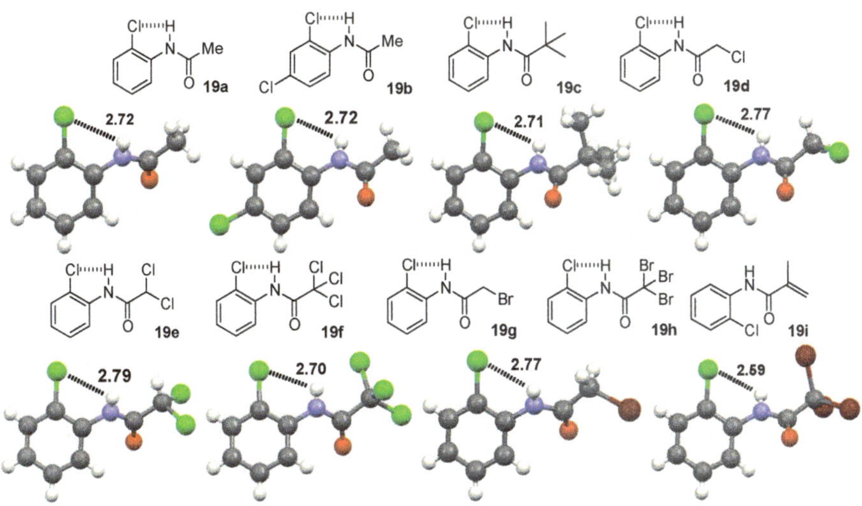

Fig. 7. Compounds 19a-i and the crystal structures of 19a-h.

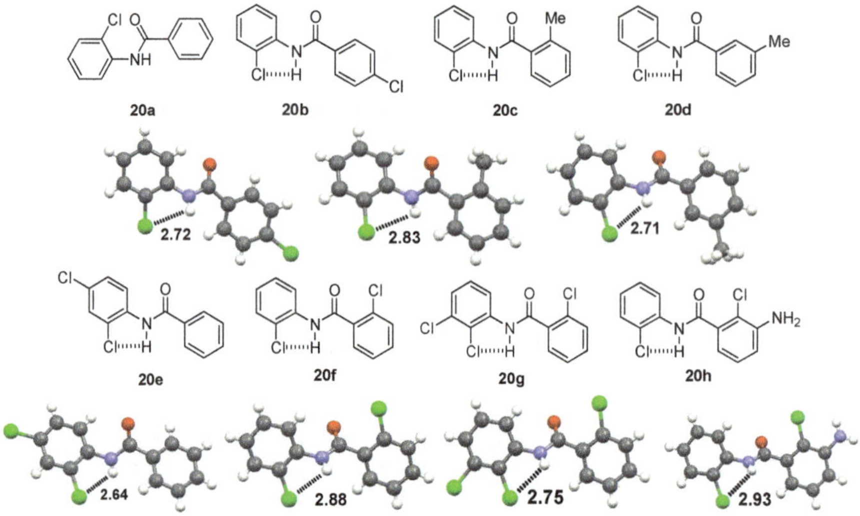

Fig. 8. Compounds 20a-h and the crystal structures of 20b-h.

Concerning the *N*-phenyl benzamide backbone, the crystal structure of *N*-(2-chlorophenyl)-benzamide 20a does not form the five-membered N-H···Cl H-bonding (Gowda et al., 2007c), because its amide unit is distorted too much (65°) from the 2-chlorobenzene plane due to the competition of the strong intermolecular N-H···O=C H-bonding. However, the five-membered H-bonding is observed in the crystal structures of its derivatives 20b (Saeed et al., 2008), 20c (Gowda et al., 2008b), 20d (Rodrigues et al., 2010), 20e (Gowda, B. T. Et al, 2008c), 20f (Gowda, B. T. et al., 2007b), 20g (Gowda, B. T. et al., 2008a), and 20h (Zhu et al., 2008) (Figure 8). The increase of the molecular size might be one of the factors that favor the formation of the bonding. Different from the fluorine-bearing analogues, compounds 20g-i do not form the six-membered N-H···Cl H-bonding on the benzoyl side because of a large torsion of the amide unit from the benzene ring which favors the intermolecular N-H···O=C H-bonding.

Although the intramolecular six-membered N-H···Cl H-bonding is not observed in the crystal structures of compound 20e-h, it occurs in the crystal structures of compounds 21 and 22 (Figure 9) (Sindt & Mackay, 1979; Khan et al., 2007). Compound 21 forms the shortest N*H*···Cl contact (2.23 Å) (Sindt & Mackay, 1979). Another three intramolecular H-bonds formed by the two hydroxyl groups should remarkably facilitate its formation because they not only inhibit the intermolecular N-H···O=C H-bonding, but also promote the co-planarity of the benzamide unit. The intramolecular six-membered N-H···O=C H-bonding in 22 should also help the formation of its six-membered N-H···Cl H-bonding, because it prevents the amide proton from forming the intermolecular N-H···O=C H-bonding (Khan et al., 2007). The hydroxyl group also forms a strong intermolecular H-bond (the O*H*···O=C distance: 1.79 Å) with the amide oxygen of another molecule, which might further facilitates the formation of the N-H···Cl H-bonding by enhancing the planarity of the benzamide unit.

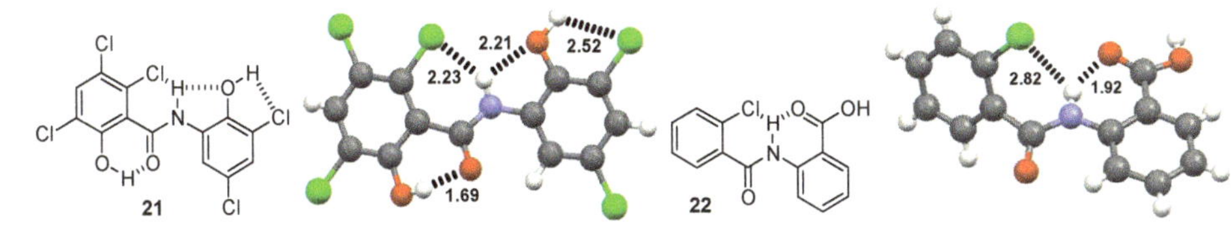

Fig. 9. Compounds 21 and 22 and their crystal structures.

4. N–H···Br hydrogen bonding

In 1972, Izumi reported the crystal structure of 2-bromobenzamide (Izumi & Okamoto, 1972), which displays a stacking pattern similar to that of 2-chlorobenzamide, with no six-membered N-H···Br H-bonding being formed (Kato et al., 1974). *N*-substituted derivatives 23a-c also do not form this H-bonding in the crystal structures (Figure 10) (Zhu et al., 2008, 2009), even though they bear the bulky trityl group, which helps to promote the formation of the intramolecular H-bonding for 18 (Zhu et al., 2008). The amide unit of 23a is distorted to be nearly perpendicular (89°) to the benzamide plane to form the continued intermolecular N-H···O=C H-bonding (Zhu et al., 2008). This continued intermolecular H-bonding is not observed in the crystal structures of 23b and 23c (Zhu et al., 2009). The Br atom of 23b chooses to form weak intermolecular trifurcate Br···O_2N contacts (the distance: 2.80Å), while the benzamide carbonyl oxygen forms an H-bond with the proton of another amide of the neighboring molecule. These results confirm that the 2-Br atom of benzamide is even weaker than Cl at the same position as the H-bonding acceptor. To verify if this weak H-bonding occurs, Zhu et al. prepared compounds 24a and 24b (Zhu et al., 2009). The crystal structures of both compounds show the formation of the intramolecular six-membered N-H···Br H-bonding (Figure 10). Compound 24a displays a dimeric motif stabilized by two intermolecular N-H···O=C (amino) H-bonds, which also prevent the amide from forming the intermolecular N-H···O=C H-bonding. This dimeric structure should also promote the co-planarity of the benzamide unit and thus facilitate the bromine atom to approach the amide proton to form the N-H···Br H-bonding. The Br···*H*N distance is 2.70 Å, which is pronouncedly shorter than the sum of the van der Waals radii of bromine and hydrogen (3.05 Å). The two trityl groups of 24b provide large enough steric hindrance to suppress the intermolecular N-H···O=C H-bonding. They also create a cavity seized by an ethyl acetate molecule, the C=O oxygen of which is H-bonded to acetamide proton, which may also facilitate the formation of the N-H···Br H-bonding by weakening the capacity of the molecule to interact intermolecularly. The Br···*H*N distance is 2.84 Å, indicating that this H-bonding is weaker than that in 24a. Compound 24c adopts two conformations in the crystal structure (Narayana et al., 2007). One of them forms the intramolecular N-H···Br H-bonding. Both the amide O and H are H-bonded to water trapped in the crystal and thus no intermolecular N-H···O=C H-bonding is formed, which may play a key role in promoting the formation of the N-H···Br H-bonding. Another conformation does not form the N-H···Br H-bonding. Its Br atom is engaged in a very weak intermolecular C=*N*···Br contact (the distance: 3.39 Å). The amino, triphenylacetamido and methoxyl groups in 24a-c are all electron donors. This feature may also make a contribution for the formation of the six-membered H-bonding.

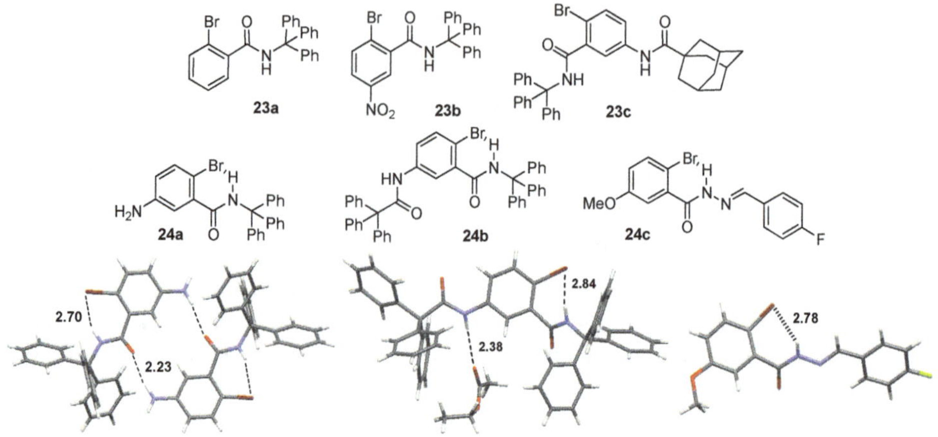

Fig. 10. Compounds 23a-c and 24a-c and the crystal structures of 24a-c.

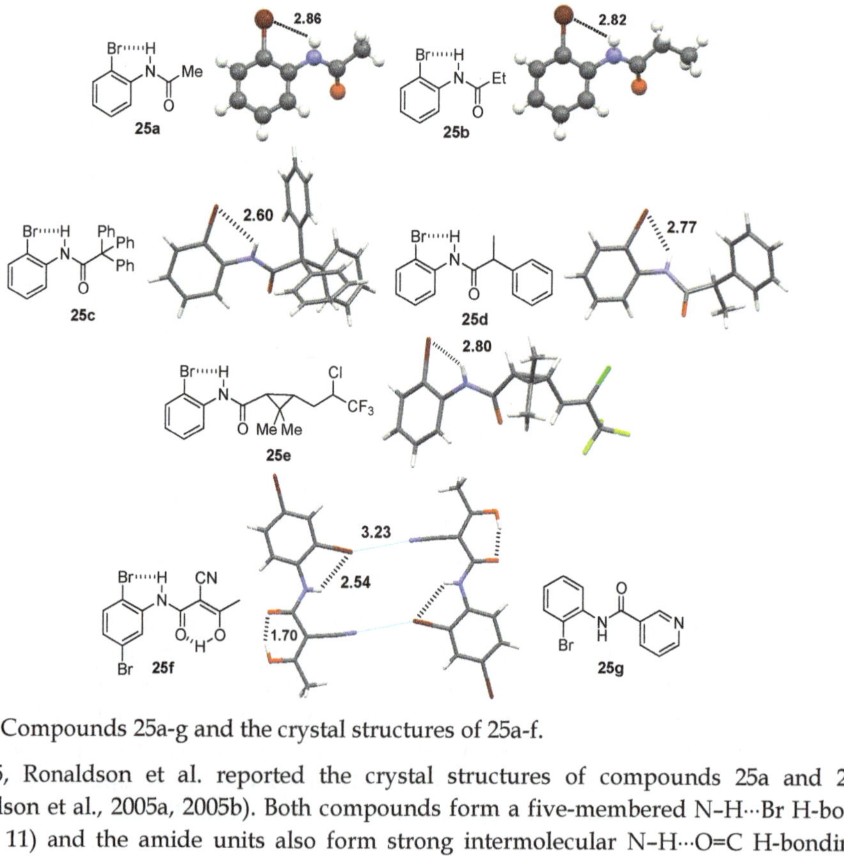

Fig. 11. Compounds 25a-g and the crystal structures of 25a-f.

In 2005, Ronaldson et al. reported the crystal structures of compounds 25a and 25b (Ronaldson et al., 2005a, 2005b). Both compounds form a five-membered N-H···Br H-bond (Figure 11) and the amide units also form strong intermolecular N-H···O=C H-bonding, which is similar to that observed for 19a. The crystal structures of compounds 25c (Zhu et al., 2008), 25d (Huang & Xu, 2006), 25e (Liu & Yan, 2007), 25f (Venkatachalam et al., 2005), 25g (Percival et al., 2007) are also available (Figure 11). Only compound 25g does not form the intramolecular N–H···Br H-bonding. The Br···*H*N distance of 25c is pronouncedly shorter than that of 25d and 25e, which may be attributed to the bulky trityl group in 25c which prevents it from forming the intermolecular N–H···O=C H-bonding. By contrast, both 25d and 25e form the intermolecular N–H···O=C H-bonding, which causes a large torsion of the amide unit from the attached benzene ring. Compound 25f displays the shortest Br···*H*N contact (2.54 Å), together with a strong intramolecular OH···O=C H-bond. It also exists as a dimer in the crystal stabilized by two intermolecular Br···N≡C bonds. These two bonds should remarkably promote the co-planarity of the benzamide unit and consequently the formation of the N–H···Br H-bonding because the benzamide unit displays a very small torsion (2°). Compound 25g does not form the N–H···Br H-bonding, because the strong intermolecular N–H···O=C H-bonding induces a large torsion (64°) of the amide unit from the benzene ring.

5. N–H···I hydrogen bonding

The crystal structures of 2-iodobenazamide 26a (Nakata et al., 1976) and its derivatives 26b (Balavoine et al., 1999), 26c (Wardell et al., 2005), 26d (Garden et al., 2005), and 26e (Zhu et al., 2008), have been reported. All these compounds do not form the intramolecular six-membered N–H···I H-bonding, but give rise to the intermolecular N–H···O=C H-bonding, which causes the amide unit to distort greatly from the attached benzene ring. For compound 26e which bears a bulky trityl group, the torsion angle is as high as 80° for the formation of the intermolecular N–H···O=C H-bonding (the *N*H···O distance: 2.39 Å). Even so, the intramolecular N–H···I H-bonding can not compete with this intermolecular H-bonding. However, the crystal structure of compound 26f does display this weak six-membered N–H···I H-bonding (Figure 12) (Zhu et al., 2009). Similar to 24a, 26f also exists as a dimer stabilized by two intermolecular C=O···H(NH) H-bonds. Such a dimeric stacking pattern remarkably enhances the co-planarity of the benzamide unit. As a result, the benzamide has a relatively small torsion of 39°, enabling the formation of the N–H···I H-bonding.

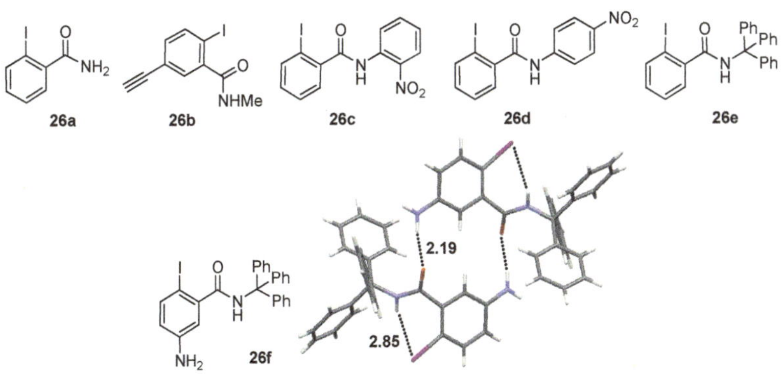

Fig. 12. Compounds 26a-f and the crystal structures of 26f.

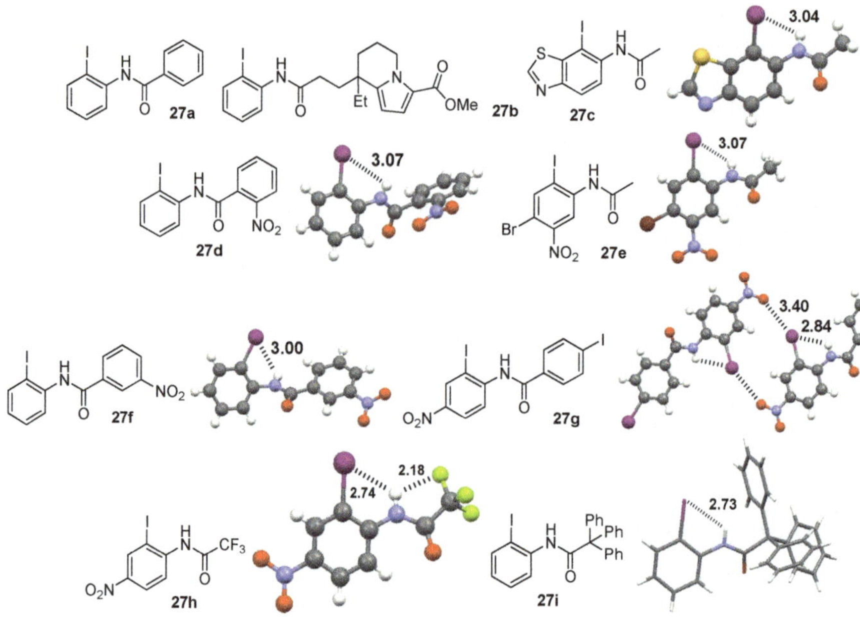

Fig. 13. Compounds 27a-i and the crystal structures of 27c-i.

The crystal structures of compounds 27a (Zhu et al., 2008), 27b (Bowie et al., 2005), 27c (Cicak et al., 2010), 27d (Wardell et al., 2005), 27e (Demartin et al., 2004), 27f (Wardell et al., 2006), 27g (Glidewell et al., 2003), 27h (Garden et al., 2006) and 27i (Zhu et al., 2008), are also available (Figure 13). No intramolecular five-membered N-H···I H-bonding is observed for 27a and 27b. However, this weak H-bonding is formed in the crystals of 27c-i. The intermolecular C=O···H-N H-bonding is also observed for 27c-h. However, this H-bonding is very weak for 27h (the O···H distance: 2.67 Å). As a result, its N*H*···I distance is shorter than that of 27d-g. The amide proton of 27h is also H-bonded to one of the F atom on the CF_3 group, which may also help to strengthen the N-H···I H-bonding by weakening the intermolecular N-H···O=C H-bonding. All the I atoms in the nitro derivatives also produce the intermolecular I···O(NO) contact. However, only 27g forms a dimeric structure. Compound 27i exhibits the shortest I···*H*N contact. As observed for 25c, its bulky trityl group completely suppresses the intermolecular N-H···O=C H-bonding. Clearly, without the competition of this stronger H-bonding, the weak N-H···I H-bonding can be formed more easily.

6. N–H···S hydrogen bonding

In 2009, Du et al. reported the crystal structures of compounds 28a-c (Figure 14) (Du et al., 2009). Compound 28a does not give rise to the intramolecular six-membered N-H···S H-bonding. However, compounds 28b and 28c do. The introduction of the bulky trityl group inhibits the formation of the intermolecular N-H···O=C H-bonding for both compounds. Compound 28b also forms two intermolecular (3,4)C-H···O=C contacts (2.58 and 2.64 Å) with one benzene of another molecule, which might also strengthen the torsion of the amide unit from the attached benzene ring and weaken its intramolecular N–H···S H-bonding. Thus, its N*H*···S distance is larger than that of 28c.

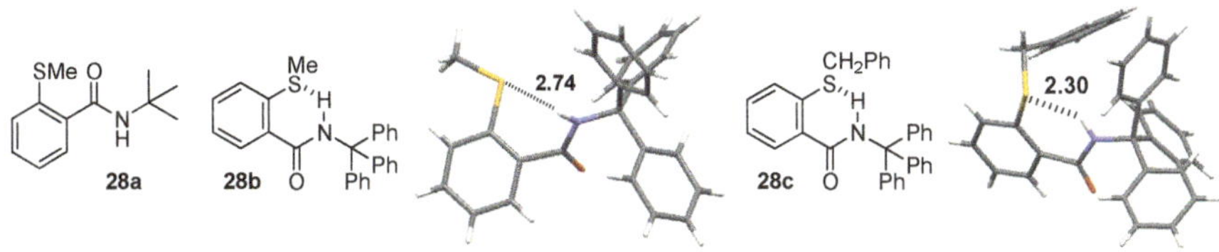

Fig. 14. Compounds 28a-c and the crystal structures of 28b and 28c.

In 1980, Rahman et al. reported that compound 29 formed a weak intramolecular five-membered N–H···S H-bonding (the S···H distance: 2.92 Å) in the crystal structure (Figure 15) (Rahman & van der Helm, 1980). The two amide units form two intermolecular N–H···O=C H-bonds, which lead to a dimeric binding pattern. Because the two molecules do not come from one stacking layer, the formation of the dimeric pattern causes a large torsion of the amide unit from the MeS-substituted benzene ring. In contrast, compound 30 did not form the intramolecular N–H···S H-bonding (Du et al., 2009). Its amide unit is greatly distorted from the benzene ring to form the intermolecular N–H···O=C H-bonding. In 2004, Liu et al. reported that 31 formed stronger intramolecular five-membered N–H···S H-bonds (Figure 15) (Liu et al., 2004). The two amide units are also engaged in five-membered (Py)N···H–N H-bonding, which inhibits the intermolecular C=O···H–N H-bonding and thus facilitates the formation of the neighboring N–H···S H-bonding. The similar three-center H-bonding is also generated in the crystal structure of compound 32 (Figure 15) (Choi et al., 2009). In this compound, two ether O atoms act as H-bonding acceptors to form two additional intramolecular six-membered N–H···O H-bonds, which prevent the amides from forming the intermolecular C=O···H–N H-bonding. Similar three-center hydrogen bonding pattern is not observed in the crystal structure of 33 (Du et al., 2009). (Figure 15), reflecting the weakness of the six-membered S···H–N H-bonding.

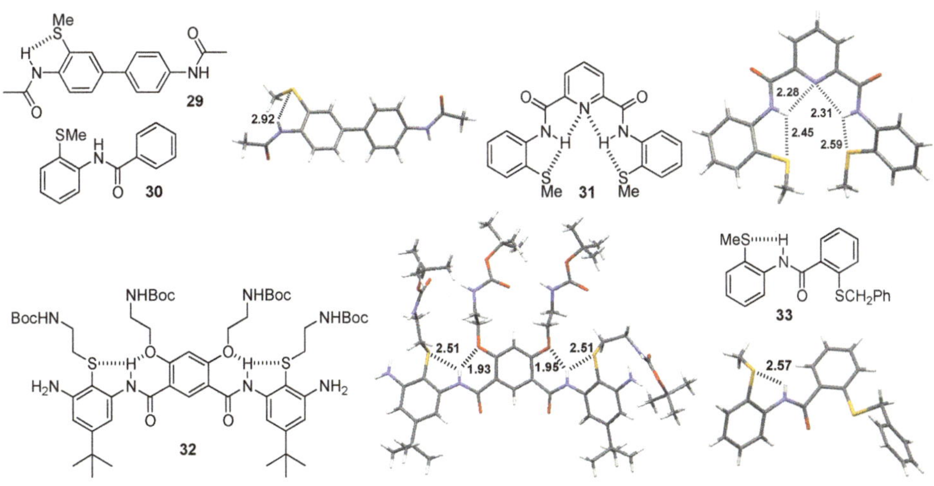

Fig. 15. Compounds 29-33 and the crystal structures of 29 and 31-33.

7. Conclusion

This chapter summarizes the crystal structures of aromatic amide derivatives that bear one halogen or sulfur atom at the ortho position of the amide unit to observe the possibility of forming the weak intramolecular N–H···X (X = F, Cl, Br, I, S) H-bonds. Generally, the five-membered H-bonds are easier to form than the six-membered ones for the identical aromatic backbone. Considering the difference of the chemical, steric and electronic environments of the heteroatom at the two different positions, this observation does not lead to conclusion that the former H-bonds are stronger than the latter ones. One straightforward reason for this difference is that the formation of the six-membered H-bonding requires to confine the rotation of three single bonds, while the five-membered one just needs to confine two. Since many complicated factors may affect the formation and stability of these intramolecular H-bonds, it is still difficult to predict whether or not a compound forms such H-bonding in the crystal. However, in most cases, if the competition of the intermolecular C=O···H–N H-bonding is suppressed, there will be a good chance of observing them, and for the halogen derivatives of the same backbone, it is obvious that their capacity of accepting the amide proton is in the order of F>Cl>Br>I, which is in the same order as electronegativity but reverse of atomic size and polarizability.

The fact that a compound does not form the above intramolecular H-bonding in crystal does not mean that it does not form the intramolecular H-bonding in solution. In crystal, the compound usually has one or two conformations, while in solution it generally exists as a dynamic mixture of several conformers of low energy and their distribution ratios will be affected by strong and weak interactions. In crystal, the structure and conformation of a compound is affected remarkably by intermolecular interactions, while in solution, the intermolecular interactions are highly concentration- and solvent-dependent. In a solvent of high polarity, the intermolecular interactions are broken by the solvent molecules, and the intramolecular H-bonding may also be weakened by the solvent to the extent that it is difficult to be detected. However, in a solvent of low polarity, a compound of low concentration should have a good chance to form the above intramolecular H-bonding.

In the past decades, the conventional, strong N–H···O and N–H···N H-bonds of amide derivatives have been the "protagonists" in studies in molecular recognition, crystal engineering, materials and biological sciences. In recent years, the above relatively weak H-bonding patterns have been used in discrete research areas. For example, the N–H···F and N–H···Cl H-bonds have been utilized to construct artificial secondary structures (Li et al., 2005; Gan et al., 2010, 2011a, 2011b), and the N–H···S H-bond has been used to create antimicrobial agents by restraining their conformations (Tew et al., 2010; Choi et al., 2009). We may expect that they will find more applications in the future, in particular in crystal engineering and supramolecular chemistry.

8. Acknowledgement

Works in the authors' laboratory are financially supported by the National Natural Science Foundation (20732007, 20921091, 20872167, 20974118), and the Ministry of Science and

Technology of China (2007CB808001), the Science and Technology Committee of Shanghai Municipality, and the Chinese Academy of Sciences.

9. References

Arslan, H., Florke, U., Kulcu, N., & Binzet, G. (1997) *Spectrochim. Acta A*, 68, 1347.

Balavoine, F., Madec, D. , & Mioskowski, C. (1999) *Tetrahedron Lett.*, 40, 8351.

Barbarich, T. J., Rithner, C. D., Miller, S. M., Anderson, O. P., & Strauss, S. S. (1999) *J. Am. Chem. Soc.*, 121, 4280.

Bautista, A. D., Craig, C. J., Harker, E. A., & Schepartz, A. (2007) *Curr. Opin. Chem. Biol.*, 11, 685.

Binzet, G., Arslan, H., Florke, U., Kulcu, N., & Duran, N. (2006) *J. Coord. Chem.*, 59, 1395.

Binzet, G., Florke, U., Kulcu, N., & Arslan, H. (2004) *Z. Kristallogr.*, 219, 395.

Biradha, K. (2003) *CrystEngComm*, *5*, 374.

Bowie, A. L., Jr., Hughes, C. C., & Trauner, D. (2005) *Org. Lett.*, 7, 5207.

Buyukgungor, O., & Odabasoglu, M. (2008) *Acta Crystallogr. E*, 64, o808.

Caleta, I., Cincic, D., Karminski-Zamola, G., & Kaitner, B. (2008) *J. Chem. Cryst.*, 38, 775.

Chisholm, G., Kennedy, A. R., Beaton, L., & Brook, E. (2002) *Acta Crystallogr. C*, 58, o645.

Choi, S., Isaacs, A., Clements, D., Liu, D., Kim, H., Scott, R. W., Winkler, J. D., & DeGrado, W. F. (2009) *Proc. Nat. Acad. Sci. U.S.A.*, 106, 6968.

Chopra, D., & Row, T. N. G. (2008) *CrystEngComm*, 10, 54.

Chopra, D., & Row, T. N. G. (2005) *Cryst. Growth Des.*, 5, 1679.

Cicak, H., Dakovic, M., Mihalic, Z., Pavlovic, G., Racane, L., & Tralic-Kulenovic, V. (2010) *J. Mol. Struct.*, 975, 115.

Corbin, P. S., Lawless, L. J., Li, Z.-T., Ma, Y., Witmer, M. J., & Zimmerman, S. C. (2002) *Proc. Natl. Acad. Sci. U.S.A.*, 99, 5099.

Del Bene, J. E., & Jordan, M. J. T. (2001) *J. Mol. Struct.*, 573, 11.

Demartin, F., Filippini, G., Gavezzotti, A., & Rizzato, S. (2004) *Acta Crystallogr. B*, 60, 609.

Desiraju, G. R. (2005) *Chem. Commun.*, 2995.

Desiraju, G. R., & Steiner, T. (2001) *The Weak Hydrogen Bond: In Structural Chemistry and Biology*, Oxford University Press, New York.

de Sousa, A. S., Marques, H. M., Fernandes, M. A. (2007) *Acta Crystallogr. E*, 63, o1642.

Du, P., Jiang, X.-K., & Li, Z.-T. (2009) *Tetrahedron Lett.*, 50, 320.

Dunitz, J. D. (2004) *ChemBioChem*, 5, 614.

Dunitz, J. D., & Taylor, R. (1997) *Chem. Eur. J.*, 3, 89.

Etter, M. C. (1990) *Acc. Chem. Res.*, 23, 120.

Frey, J. A., Leist, R., & Leutwyler, S. (2006) *J. Phys. Chem. A*, 110, 4188.

Gan, Q., Li, F., Li, G., Kauffmann, B., Xiang, J., Huc, I., & Jiang, H. (2010) *Chem. Commun.*, 46, 297.

Gan, Q., Ferrand, Y., Bao, C., Kauffmann, B., Grélard, A., Jiang, H., & Huc, I. (2011) *Science*, 331, 1172.

Gan, Q., Wang, Y., & Jiang, H. (2011) *Curr. Org. Chem.*, 15, 1293.

Garden, S. J., Glidewell, C., Low, J. N., Skakle, J. M. S., & Wardell, J. L. (2005) *Acta Crystallogr. C*, 61, o450.

Garden, S. J., Wardell, J. L., Low, J. N., Skakle, J. M. S., & Glidewell, C. (2006) *Acta Crystallogr. E*, 62, o3762.

Glidewell, C., Low, J. N., Skakle, J. M. S., & Wardell, J. L. (2003) *Acta Crystallogr. C*, 59, o636.

Gong, B. (2001) *Chem. Eur. J.*, 7, 4337.

Gowda, B. T., Paulus, H., & Fuess, H. (2000) *Z.Naturforsch. A*, 55, 711.

Gowda, B. T., Paulus, H., & Fuess, H. (2001) *Z. Naturforsch. A*, 56, 386.

Gowda, B. T., Foro, S., & Fuess, H. (2007) *Acta Crystallogr. E*, 63, o4611.

Gowda, B. T., Foro, S., Sowmya, B. P., & Fuess, H. (2007) *Acta Crystallogr. E*, 63, o3789.

Gowda, B. T., Sowmya, B. P., Kozisek, J., Tokarcik, M., & Fuess, H. (2007) *Acta Crystallogr. E*, 63, o2906.

Gowda, B. T., Svoboda, I., & Fuess, H. (2007) *Acta Crystallogr. E*, 63, o3308.

Gowda, B. T., Foro, S., Sowmya, B. P., & Fuess, H. (2008) *Acta Crystallogr. E*, 64, o1342.

Gowda, B. T., Foro, S., Sowmya, B. P., & Fuess, H. (2008) *Acta Crystallogr. E*, 64, o1421.

Gowda, B. T., Tokarcik, M., Kozisek, J., Sowmya, B. P., & Fuess, H. (2008) *Acta Crystallogr. E*, 64, o950.

Gowda, B. T., Foro, S., Suchetan, P. A., Fuess, H., & Terao, H. (2009) *Acta Crystallogr. E*, 65, o1998.

Gowda, B. T., Foro, S., Suchetan, P. A., & Fuess, H. (2010) *Acta Crystallogr. E*, 66, o386.

Guo, D., Nichol, G. S., Cain, J. P., & Yalkowsky, S. H. (2009) *Acta Crystallogr. E*, 65, o2644.

Harrell, S. A., & McDaniel, D. H. (1964) *J. Am. Chem. Soc.*, 86, 4497.

Howard, J. A. K., Roy, V. J., O'Hagan, D., & Smith, G. T. (2006) *Tetrahedron*, 52, 12613.

Huang, J.-Y., & Xu, W. (2006) *Acta Crystallogr. E*, 62, o2651.

Huc, I. (2004) *Eur. J. Org. Chem.*, 17.

Hughes, D. O., Small, R. W. H. (1972) *Acta Crystallogr. B*, 28, 2520.

Izumi, T., & Okamoto, N. (1972) *Mem. Chubu Inst. Technol.*, 8, 139.

Jeffrey, G. A. (1997) *An Introduction to Hydrogen Bonding*, Oxford University Press, Oxford.

Kalyanaraman, B., Kispert, L. D., Atwood, J. L. (1978) *J. Cryst. Mol. Struct.*, 8, 175.

Kampf, M., Richter, R., Hennig, L., Eidner, A., Baldamus, J., & Kirmse, R. (2004) *Z. Anorg. Allg. Chem.*, 630, 2677.

Kashino, S., Iwamoto, T., Yamamoto, E., & Shiraga, T. (1994) *Bull. Chem. Soc. Jpn.*, 67, 1226.

Kato, Y., & Sakurai, K. (1982) *Bull. Chem. Soc. Jpn.*, 55, 1643.

Kato, Y., Takaki, Y., & Sakurai, K. (1974) *Acta Crystallogr. B*, 30, 2683.

Khan, Z. A., Khan, K. M., & Anjum, S. (2007) *Acta Crystallogr. E*, 63, o4103.

Koller, A. N., Bozilovic, J., Engels, J. W., & Gohlke, H. (2010) *Nucleic Acids Res.*, 38, 3133.

Legon, A. C. (1990) *Chem. Soc. Rev.*, 19, 197.

Lewis, R. J., Camilleri, P., Kirby, A. J., Marby, C. A., Slawin, A. A., & Williams, D. J. (1991) *J. Chem. Soc. Perkin Trans. 2*, 1625.

Li, C., Ren, S.-F., Hou, J.-L., Yi, H.-P., Zhu, S.-Z., Jiang, X.-K., & Li, Z.-T. (2005) *Angew. Chem. Int. Ed.*, 44, 5725.

Li, X., Wu, Y.-D., & Yang, D. (2008) *Acc. Chem. Res.*, 41, 1428.

Li, Z.-T., Hou, J.-L., & Li, C. (2008) *Acc. Chem. Res.*, 41, 1343.

Li, Z.-T., Hou, J.-L., Li, C., & Yi, H.-P. (2006) *Chem. Asian J.*, 1, 766.

Liu, D.-Q., & Yan, F.-Y. (2007) *Acta Crystallogr. E*, 63, o4202.

Liu, S.-G., Li, Y.-Z., Zuo, J.-L., & You, X.-Z. (2004) *Acta Crystallogr. E*, 60, o1527.

Liu, Z., Remsing, R. C., Liu, D., Moyna, G., & Pophristic, V. (2009) *J. Phys. Chem. B*, 113, 7041.

Manjunatha, R. G. N., Vasantha, K. M. V., Row, G. T. N., & Suryaprakash, N. (2010) *Phys. Chem. Chem. Phys.*, 12, 13232.

Nakata, K., Tateno, T., & Sakurai, K. (1976) *Mem. Osaka Kyoiku Univ.*, 25, 61.

Narayana, B., Sunil, K., Yathirajan, H. S., Sarojini, B. K., & Bolte, M. (2007) *Acta Crystallogr. E*, 63, o2948.

O'Hagan, D., Rzepa, H. S., Schuler, M., Slawin, A. M. Z. (2006) *Beilst. J. Org.Chem.*, 2, 19.

Pauling, L. (1960) *The Nature of the Chemical Bond*, 3rd ed. Cornell University Press: Ithaca, New York.

Percival, D., Storey, J. M. D., & Harrison, W. T. A. (2007) *Acta Crystallogr. E*, 63, o1851.

Rahman, A., & van der Helm, D. (1980) *Acta Crystallogr. B*, 36, 2444.

Rodrigues, V. Z., Tokarcik, M., Gowda, B. T., & Kozisek, J. (2010) *Acta Crystallogr. E*, 66, o891.

Ronaldson, V., Storey, J. M. D., & Harrison, W. T. A. (2005) *Acta Crystallogr. E*, 61, o315.

Ronaldson, V., Storey, J. M. D., & Harrison, W. T. A. (2005) *Acta Crystallogr. E*, 61, o3200.

Saeed, A., Khera, R. A., Gotoh, K., & Ishida, H. (2008) *Acta Crystallogr. E*, 64, o1934.

Sanford, A. R., & Gong, B. (2003) *Curr. Org. Chem.*, 7, 1649.

Saraogi, I., & Hamilton, A. D. (2009) *Chem. Soc. Rev.*, 38, 1726.

Savinkina, E. V., Zamilatskov, I. A., Buravlev, E. A., Albov, D. V. (2008) *Mendeleev Commun.*, 18, 131.

Scheiner, S. (1997) *Hydrogen Bonding: A Theoretical Perspective*, Oxford University Press, New York.

Simonov, Y. A., Fonari, M. S., Lipkowski, J., Gelmboldt, V. O., & Ganin, E. V. (1996) *J. Inclus. Phenom. Mol. Recogn. Chem.*, 24, 149.

Sindt, A. C., & Mackay, M. F. (1979) *Acta Crystallogr. B*, 35, 2103.

Tew, G. N., Scott, R. W., Klein, M. L., & DeGrado, W. F. (2010) *Acc. Chem. Res.*, 43, 30.

Toth, G., Bowers, S. G., Truong, A. P., & Probst, G. (2007) *Curr. Pharm. Des.*, 13, 3476.

Venkatachalam, T. K., Zheng, Y., Ghosh, S., & Uckun, F. M. (2005) *J. Mol. Struct.*, 743, 103.

Wan, X., Ma, Z., Li, B., Zhang, K., Cao, S., Zhang, S., & Shi, Z. (2006) *J. Am. Chem. Soc.*, 128, 7416.

Wardell, J. L., Low, J. N., Skakle, J. M. S., & Glidewell, C. (2006) *Acta Crystallogr. B*, 62, 931.

Wardell, J. L., Skakle, J. M. S., Low, J. N., & Glidewell, C. (2005) *Acta Crystallogr. C*, 61, o634.

Yin, Z., Jiang, L., He, J., & Cheng, J.-P. (2003) *Chem. Commun.*, 2326.

Zhao, X., & Li, Z.-T. (2010) *Chem. Commun.*, 46, 1601.

Zhao, X., Wang, X.-Z., Jiang, X.-K., Chen, Y.-Q., Li, Z.-T., & Chen, G.-J. (2003) *J. Am. Chem. Soc.*, 125, 15128.

Zhu, S., Nieger, M., Daniels, J., Felder, T., Kossev, I., Schmidt, T., Sokolowski, M., Vögtle, F., & Schalley, C. A. (2009) *Chem. Eur. J.*, 15, 5040.

Zhu, Y.-Y., Jiang, L., & Li, Z.-T. (2009) *CrystEngComm*, 11, 235.

Zhu, Y.-Y., Wang, G.-T., & Li, Z.-T. (2011) *Curr. Org. Chem.*, 15, 1266.

Zhu, Y.-Y., Wu, J., Li, C., Zhu, J., Hou, J.-L., Li, C.-Z., Jiang, X.-K., & Li, Z.-T. (2007) *Cryst. Growth Des.*, 7, 1490.
Zhu, Y.-Y., Yi, H.-P., Li, C., Jiang, X.-K., & Li, Z.-T. (2008) *Cryst. Growth Des.*, 8, 1294.

8

Features of Structure, Geometrical, and Spectral Characteristics of the $(HL)_2[CuX_4]$ and $(HL)_2[Cu_2X_6]$ (X = Cl, Br) Complexes

Olga V. Kovalchukova
Peoples' Friendship University of Russia
Russian Federation

1. Introduction

Coordinate compounds of copper(II) are widely spread both as biological objects (metallo-proteins and metallo-enzymes), and in engineering. Among the functions of the copper proteins are: the electron transfer involving the Cu(I)/Cu(II) couple; mono- terminal-oxidases, which form either water or hydrogen peroxide from dioxygen; oxygenases, which incorporate an oxygen atom into a substrate; superoxide degradation to form dioxygen and peroxide; and the oxygen transport. From a structural point of view, there are three main types of biologically active copper centres found in the copper proteins (Cowan, 1993). These are "blue" copper centres, where copper atoms are normally coordinated to two nitrogens and two sulphurs, "non-blue" copper centres, where copper atoms are coordinated to two or three nitrogens as well as oxygens, and copper dimers. The nitrogens come from histidine groups, the sulfur from methionine and cysteine, the oxygens from the carboxyllic acid in the protein. So called "non-blue" and dimeric copper-containing proteins are of a a great similarity with complex halo (chloro-, bromo-) cuprates (Abolmaali et al., 1998). Thus, studies of structural and spectral characteristics of anionic complex halides of copper(II) can help to explain electronic structures as well as high reactional abilities and selectivities of active sites of copper-containing biopolymers in catalytic processes.

It is also evident that anionic halocuprate(II) complexes are catalytically active species responsible for the increased reactivity in a lot of organic reactions (oxidation and polymerization of phenols, reactions of tertiary ammines, dimerization of primary alkyl groups et al.). Various investigations show that catalytic activities of complex copper(II) halides depend upon structures of their coordination polyhedra (Allen et al., 2009).

And finally, cupric halo-complexes relate to classic magneto-active systems containing 3d-metals, magnetic properties of which significantly depend on features of the spatial structure of complex anions (Rakitin & Kalinnikov, 1994).

d^9-Electronic subshell of Cu(II) is responsible for distortions of symmetry of the coordination polyhedron (Gerloch & Constable, 1994). This deals with the Jahn-Teller effect (as a result of electron-vibrational interactions), and a large spin-orbital interaction constant. These two effects are of comparable values, and this fact complicates the prediction of structures of complexes of such types as well as physico-chemical properties and biological activity of Cu(II) complexes are in many respects determined by features of their structures.

The X-Ray analysis on single crystals can unequivocally determine structures of substances but isolation of single crystals is a complicated process which may not be achieved successively. Thus, a great role should belong to site-methods of structure determination (spectral, magnetic et al). Such correlations like "structure - spectral parameters - magnetic characteristics" help to describe features of the structure of complexes and as a result to predict their possible physical properties and areas of application.

2. Electronic structure and coordination geometry of Cu(II)-compounds

The ground state for the Cu^{2+} cation in its octahedral coordination is $^2E_g(t_{2g})^6(e_g)^3$, and the one for the square coordination is $^2E_g\, d_{xz}{}^2 d_{yz}{}^2 d_{z^2}{}^2 d_{xy}{}^2 d_{x^2-y^2}{}^1$. The only excited state might relate to $^2T_{2g}(t_{2g})^5(e_g)^4$ with the regular and distorted tetrahedral coordination (energy difference 10Dq) (Cotton & Wilkinson, 1966). The maximal coordination number of copper(II) is 6 which relates to octahedral complexes of the $(t_{2g})^6$ $(e_g)^3$ - $d_{xz}{}^2 d_{yz}{}^2 d_{xy}{}^2 d_{z^2}{}^2 d_{x^2-y^2}{}^1$ or $d_{xz}{}^2 d_{yz}{}^2 d_{xy}{}^2 d_{x^2-y^2}{}^2 d_{z^2}{}^1$ configurations. As the $d_{x^2-y^2}$ orbital, when compared to the d_{z^2} one, contains only one electron, it forms stronger Cu-to-ligand bonds. Thus, four planar ligands are stronger joint to the central ion than the two axial ligands which are bounded to Cu(II) along the *z*-axis. Sometimes the difference is so great that Cu(II) complexes may be considered as square. Most often, the coordination numbers of Cu(II) are 4 (square) or 6 (distorted octahedron). Tetrahedral coordination (T_d symmetry) and distorted tetrahedral coordinations (D_{2d}) also exist. The distorted octahedron is characterized by four short Cu–L bonds at one plane and two longer axial bonds in the trans-position (4+2 coordination) or 2 short and 4 long bonds (2+4 coordination). The square environment of Cu(II) appears as a limiting case of a tetragonal distortion of an octahedron.
The copper(II) also forms a lot of complexes with the coordinate number 5. This can be explained by blocking of one of the tops of the octahedron by an e_g-lone electron pair. As a result, a square pyramid is formed (4+1 configuration). The square-pyramidal configuration exists for example in copper(II) pycrate diaquaacetylccetonate and $K[Cu(NH_3)_5](PF_6)_3$ (Adman, 1991). It usually appears in copper(II) complexes with pyridine and other bases, and is also typical for a Cu–β-alanyl-L-hystidine compound which is used as a model of M–protein interactions (Gerloch & Constable, 1994). In the case that ligands are stereochemically movable, a more symmetric structure of a trigonal bipyramid is formed as in $[CuBr_5]^{3-}$ or $[CuDipy_2I]^+$ (Gillard & Wilkinson, 1963).
The copper(II) forms both cationic and anionic complexes. The anionic ones - cuprates(II) are most often formed in excess of hydrohalic (HCl, HBr), cyanic (HCN) or thiocyanic (HSCN) acids. HI usually reduces Cu(II) into Cu(I). The complex anions of the type of $M^{+1}[CuX_3]$ and $M_2{}^{+1}[CuX_4]$ (X = Cl^-, Br^-, CN^-, SCN^-) are stabilized by counter-ions the most simple of which are cations of alkaline metals (Willett & Geiser, 1984). For example, the structure of a red $CsCuCl_3$ (Schluetes et. al., 1966) contains Cu(II) ions octahedrically surrounded by six chloride anions. The $CuCl_4{}^{2-}$ in a yellow Cs_2CuCl_4 (Helmholz & Kruh, 1952) is in tetragonal distortion. Brownish-red compounds of $LiCuCl_3{\cdot}2H_2O$ (Vossos et al., 1963) and $KCuBr_3$ (Geiser et al., 1986a) consist of planar $Cu_2X_6{}^{2-}$ anions (X = Cl, Br) with symmetrical Cu–X–Cu bridges.
Actually, $M_2[CuX_4]$, X = Cl, Br compounds with inorganic monovalent cations in the outer sphere can be mostly considered as double salts of a very low stability. For example, for

$[CuCl_4]^{2-}$ the formation constants were found to be log β_1 4.0; log β_2 4.7; log β_3 1.96; log β_4 0.23 (Khan & Schwing-Weill, 1976). The complexes are easily destroyed in polar solvents. Halogen-ions in the inner sphere are easily replaced by other ligands (ammonia, water at al). Majority of complex halides are not stable in air and destroyed by absorption of water vapors.
Much more interesting are anionic halocuprates(II) containing protonated organic molecules as counter-ions. They are stabilized with the help of formation of a system of intra- and intermolecular hydrogen bonds (H-bonds) or extra coordinate bonds via lone electron pairs of donating atoms (N, O, S) or vacant molecular orbitals of organic molecules (so-called dative bonds with the metal-to-ligand charge transfer M→L). The present chapter belongs to development of features of the structures and properties of such type of compounds.

3. Structure characteristics and properties of $CuCl_4^{2-}$ species

Stereochemistry of copper(II) halides is rather rich (Smith, 1976). The latest results are summarized in the review (Murphy & Hathaway, 2003). Two types of coordination exists: the "common" one with ionic radii of about 0.5 Å and semi-coordinated where the bond lengths are 0.3 - 1.0 Å longer (Hathaway, 1982). The shape of coordination polyhedra changes from square planar (Harlow et al., 1975) to distorted tetrahedral (Diaz et al., 1999).
The degree of distortion of CuX_4^{2-} coordination polyhedra is determined by the mean value of the flattering or trans-angle θ (Fig.1). It is evident that the non-distorted tetrahedral configuration (T_d) of CuX_4^{2-} corresponds to mean θ-values up to 109 deg. as the planar distortion increases its value up to 180 deg.

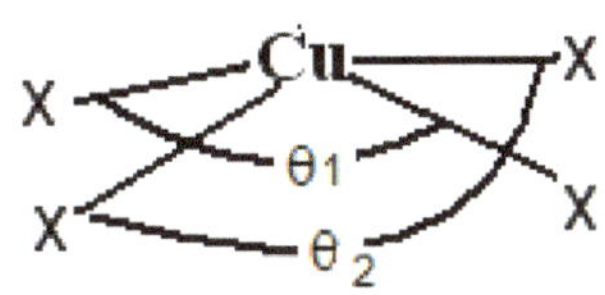

Fig. 1. Determination of the degree of distortion of coordination polyhedra of Cu(II) complexes

3.1 Structure description

Since the last century, a lot of structures of common formulae $(HL)_2CuCl_4$ and $(HL)_2CuBr_4$ where L is an organic base were determined by X-Ray crystallography. It was stated that the geometry of CuX_4^{2-} depends upon stability of H-bonds and other electrostatic and steric interactions between counter-ions and halide-ions of tetrahalocuprate(II) fragments (Murphy & Hathaway, 2003). H-bonds flatten the structure towards D_{4h} configuration. The decrease in the abilities of organic cations to form H-bonds with the inorganic anion leads to the decrease in distortion of its tetrahedral geometry. Really, in $Cs_2[CuX_4]$ the mean values of trans-angle θ is determined to be 124 deg. for X = Cl (Helmholz & Kruh, 1952) and 128.4 deg. for X = Br (Morosin & Lingafelter, 1960) which corresponds to slightly distorted tetrahedron, as well as tetrahalocuprates containing diprotonated 3-amminopyridine as counter-ions are characterized by planar structures of inorganic anions (θ 170.60 deg. for tetrachlorocuprate and 170.56 deg. for its bromo-analogue (Willet et al., 1988). More complicated organic compounds provoke intermediate characters of inorganic anions (Table 1).

L	$\theta(CuCl_4^{2-})$, deg	Ref.	$\theta(CuBr_4^{2-})$, deg	Ref.
Cs+	124	Helmholz & Kruh, 1952	128.4	Morosin & Lingafelter, 1960
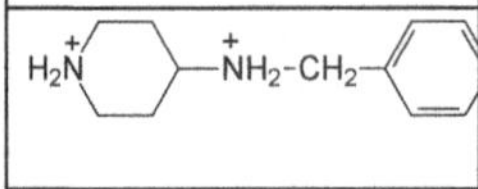	133	Antolini et al., 1981	125.6	Place & Willett, 1988
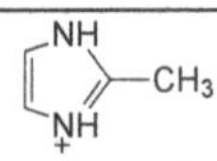	134.4	Kovalchukova et al., 2008a	131.64	Koval'chukova et al., 2009b
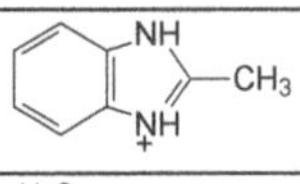	134.69	Bhattacharya et al., 2004	134.2	Bhattacharya et al., 2004
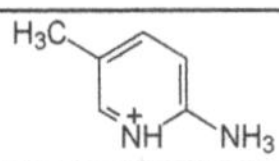	140	Place & Willett, 1987b	137	Place & Willett, 1987b
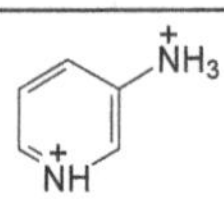	170.56	Willet et al, 1988	170.6	Willet et al, 1988
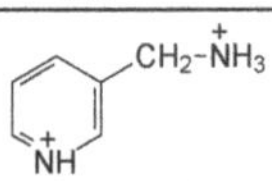	177.4	Long et al., 1997	178	Long et al., 1997
$H_3\overset{+}{N}-CH_2-CH_2-\overset{+}{N}H_3$	180	Tichý et al., 1978	180	Rubenacker et al., 1984

Table 1. Some characteristics of coordination polyhedra of some $(HL)_2[CuX_4]$ structures.

Plotting of degrees of distortion of the CuX_4^{2-} coordination polyhedron *vs.* the type of the halogenide-anion (Fig 2) gives a straight-line dependence with closely related values of $\theta(CuCl_4^{2-})$ and $\theta(CuBr_4^{2-})$ for the same organic cations (Koval'chukova et al., 2009b). Thus, the degree of distortion of coordination polyhedron slightly depends upon the nature of X (Cl^-, Br^-), and is mainly determined by the nature of the organic cation.

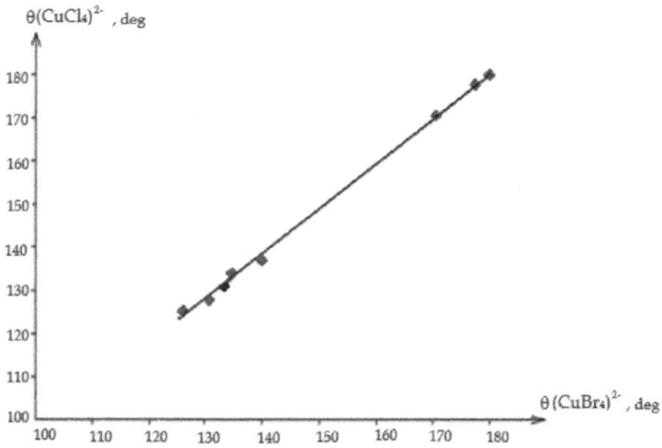

Fig. 2. The dependence of flattening angle of CuX_4^{2-} coordination polyhedron on the nature of inorganic anion (Koval'chukova et al., 2009b).

Among the features of crystal structures of tetrabromocuprates(II) in the D_{4h} configuration stabilized by cations of organic aliphatic amines (such as cyclopentylamine (Luque et al., 2001), phenylethylamine (Arend et al., 1978), methyl(2-phenylethylamine (Willett, 1990)) there are tight interactions of tetrabromocuprate(II) anions. This leads to the appearance of two semi-coordinate bonds with Cu–Br distances 3.0 – 3.1 Å as compared with common 2.41 - 2.45 Å. As a result, the coordination number of copper becomes 6 (4+2 coordination) (Fig. 3).
The same type of (4+2) coordination of Cu^{2+} was found in $(HL)_2[CuCl_4]$, L = 2-methyl-l,3-di-2-pyridyl-2-propanol (Schlemper et al., 1989). Cu^{2+} ions are in distorted square-bipyramidal coordination. The bridging angles Cu'ClCu are 125.1 and 129.5 deg, and Cu-CI distances 2.457 and 2.868 Å.

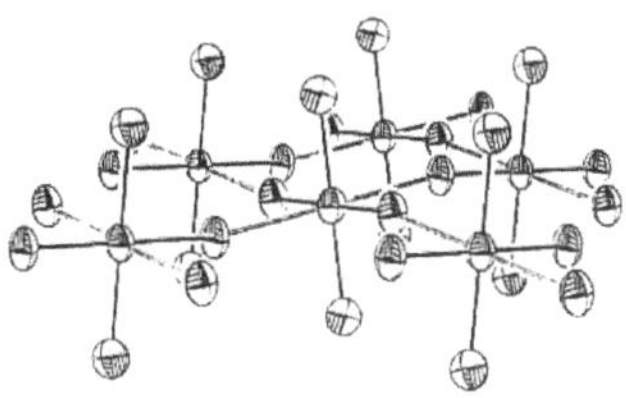

Fig. 3. The formation of a layer of $(CuBr_4)_n$ in $(PhC_2H_4NH_3)_2CuBr_4$ (Arend et al., 1978)

Tetrachlorocuprate(II) ions may also coordinate other d-metals. Paramagnetic metallic ions form bridging anti-spin magnetic chained systems which are of a great interest for physics (Landee et al., 1988). The Cu-Cl distances of bridging fragments may reach 2.325 Å, and those of terminate character shorten till 2.181 Å.
The possibility of bromination of organic cations with using for neutralization of anionic bromocuprates may also take place. That was described for example by (Place & Willett, 1987a). While the formation of anionic tetrabromocuprates(II) containing protonated 2-amino-3-methylpyrydine as a counter-ion, partial bromination of an organic specie was observed. Both 2-amino-3-methylpyrydinium and 2-amino-5-bromo-3-methylpyrydinium cations neutralize the negative charge of tetrabromocuprate anion in the crystal lattice.

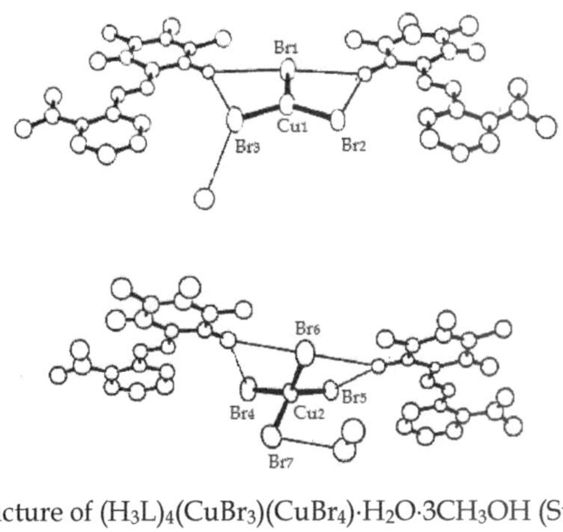

Fig. 4. The crystal structure of $(H_3L)_4(CuBr_3)(CuBr_4)\cdot H_2O\cdot 3CH_3OH$ (Sundberg et al., 1992).

Sundberg and co-workers (Sundberg et al., 1992) showed that while complex formation of $CuBr_2$ with 6-amino-1,3-dimethyl-5-((2-carboxyphenyl)azo)uracyl (HL) in the presence of HBr, partial reduction of Cu(II) into Cu(I) is observed. As a result, a complex $(H_3L)_4(CuBr_3)(CuBr_4)\cdot H_2O\cdot 3CH_3OH$ contains copper atoms in oxidation states +2 and +1. The coordination number for Cu(I) is 3 ($CuBr_3^{2-}$ as a planar anion), and 4 for Cu(II) ($CuBr_4^{2-}$ as a distorted tetrahedron). In lattice, both $(H_3L)_2(CuBr_3)$ and $(H_3L)_2(CuBr_4)$ exist as independent fragments. So the above complex may be considered as co-crystallization of two complexes containing copper in two different oxidation states. Water and methanol molecules have a lattice character (Fig. 4).

As it was intimated (Place & Willett, 1987a), the H-bonding between organic protonated molecules and inorganic tetrahalocuprate(II) anions is not released for structures with $\theta <$ 132 deg. Such types of structures are described, for example, for bis(4-azafluorene-9-onium) tetrabromocuprate(II) monohydrate (Kovalchukova et al., 2009a) (θ 130.98 deg.) crystal package of which is present on Fig. 5. It is evident that cations and anions form altermating layers with no H-bonds between them. The only one intermolecular interaction existing in the structure belongs to the hydrated water molecule.

On the other hand, H-bonds are found in the $(H_2L)[CuBr_4]$ structure with the flattening angle θ 120 deg (L - 2,4,6-tris(N,N-dimethylamino)methylphenol) (Kovalchukova ett al., 2010). The Cu-Br distances in the distorted $CuBr_4^{2-}$ anion range from 2.469(2) to 2.515(1) Å. The minimal Cu–N distance is 3.65 Å, and its value rejects the possibility of formation of a ligand-to-metal coordinate bond. Despite a small value of the flattening angle θ, a hydrogen N-H...Br bond exists between an H-atom of a protonated tertiary amino-group of the organic dication and one of a Br-atom of the inorganic anion (r N-H 1.05; H…Br 2.42; N…Br 3.46 Å; ∠ NHBr 169.5 deg.).

Flattening of the CuX_4^{2-} structure increases the number and strength of the H-bonds. As an example, for $(HL)_2[CuX_4]$ (L = 2-methylimidazolium, X = Cl, θ 134.4 deg (Kovalchukova et al., 2008a); X = Br, θ 132.64 deg (Koval'chukova et al., 2009b)), multiple H-bonds connect counter-ions in crystals (Fig. 6). Similar structures were found for N-benzylpiperazinium tetrachlorocuprate(II) (Antolini et al., 1981) containing an unequally flattened $[CuCl_4]^{2-}$ tetrahedron (θ 166.4 deg.) (fig. 7) as well as for bis(ethylenediammoniummonobromide) tetrabromocuprate(II), θ 159.2 deg (Anderson & Willett, 1971), and a square-planar limit (θ 180 deg) for creatinium tetrachlorocuprate(II) (Battaglia et al., 1979). Probably H-bondings play the crucial role in these limiting planar structures as compared with crystal packing forces or the cation bulk. Nearly linear strong H-bonds (Fig. 7) produce such a reduction of the ligand-to-ligand electrostatic repulsion to allow a nearly square-planar geometry, favored by the crystal field stabilization (Smith, 1976).

H-bonds change not only bond angles but also Cu-X bond lengths. In (Antolini et al., 1981) the shortest Cu-Cl distances (~2.260 Å) were found for chlorine atoms which do not form H-bonds. Abnormally long Cu-Cl bonds (~2.334 Å) were determined for Cl-atoms engaged in a strong bifurcate H-bounding (Halvorson et al., 1990). Graphical correlations of parameters of H-bonds, i.e. r(H-Br) on r(N-Br) (Koval'chukova et al., 2009b), $N-\hat{H}...Cl$ angle on r(Cu-Cl) (Kovalchukova et al., 2008a) (Fig. 8, 9) are linear but the dependencies of the NHBr angles on θ (Fig. 10) for tetrachlorocuprates(II) and tetrabromocuprates(II) have opposite derivative signs.

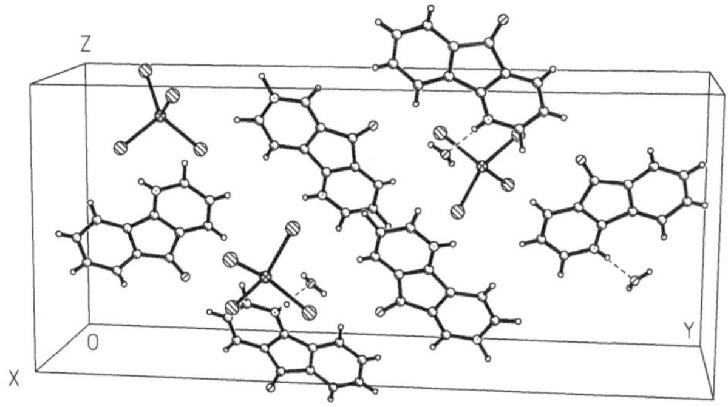

Fig. 5. The crystal package of bis(4-azafluorene-9-onium) tetrabromocuprate(II) monohydrate (Kovalchukova et al., 2009a).

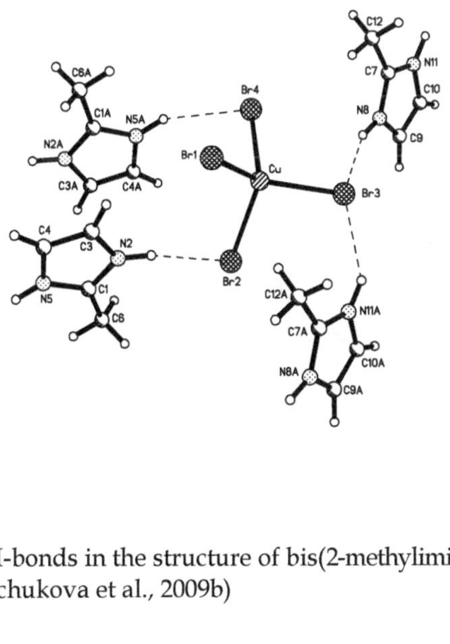

Fig. 6. The formation of H-bonds in the structure of bis(2-methylimidazolium) tetrabromocuprate (Koval'chukova et al., 2009b)

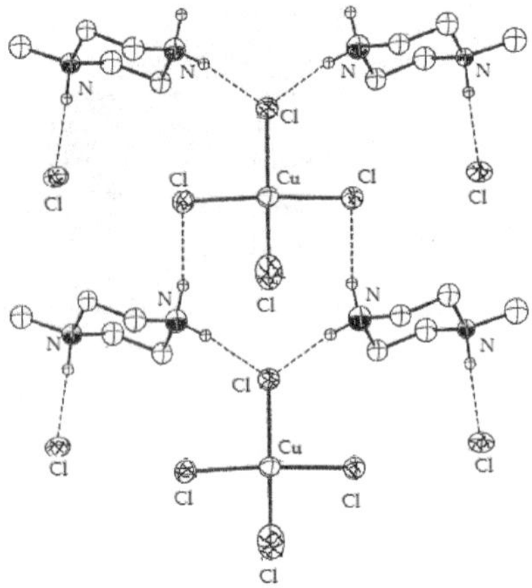

Fig. 7. Hydrogen bonds (dashed lines) in N-benzylpiperazinium tetrachlorocuprate(II). C-bonded H atoms and phenyl rings are omitted for clarity (Halvorson et al., 1990).

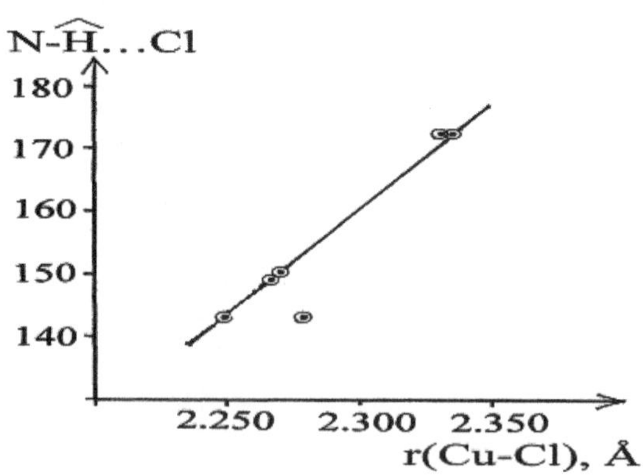

Fig. 8. The dependence of values of bond angles at a bridging H-atom $N-\hat{H}...Cl$ on the Cu-Cl distance for anionic tetrachlorocuprates(II) containing N-protonated organic bases as counter-ions (Kovalchukova et al., 2008a).

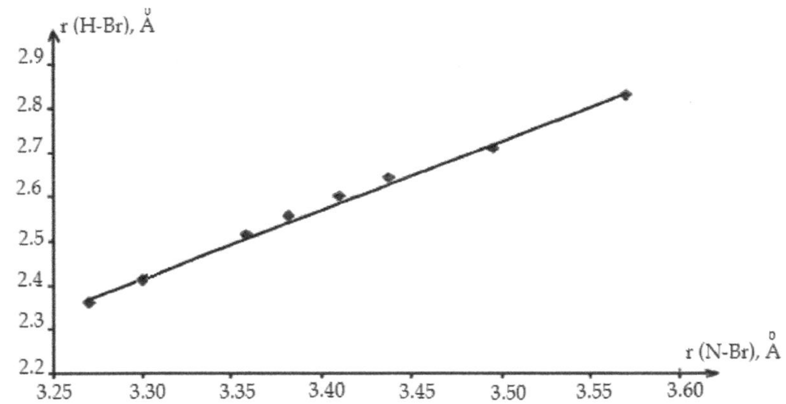

Fig. 9. The dependence of values of H–Br on N–Br distances for anionic tetrabromocuprates(II) containing N-protonated organic bases as counter-ions (Koval'chukova et al., 2009b).

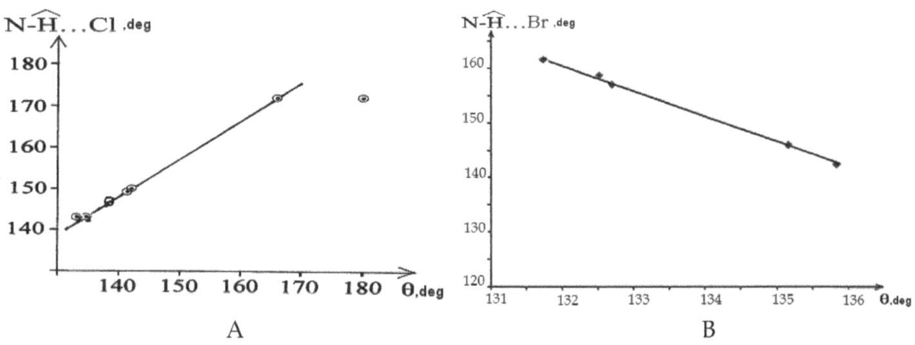

Fig. 10. The dependence of the degree of distortion of coordination polyhedra of anionic tetrachlorocuprates(II) (A) and tetrabromocuprates(II) (B) on the value of bond angles at a bridging H-atom $N-\hat{H}...X$ (X = Cl, Br) (Kovalchukova et al., 2008a; Koval'chukova et al., 2009b).

3.2 Spectral-structural correlations

It is evident that isolation of single crystals for determination of the type of a CuX_4^{2-} coordination polyhedron and degree of its distortion is not always possible. On the other hand, the above parameters mostly affect physico-chemical properties of halocuprates(II). Correlations of spectroscopic characteristics of substances with the features of their crystal structures are useful for these purposes. The major role belongs to electronic spectroscopy.

Electronic absorption spectra of $(HL_2)[CuX_4]$, X = Cl, Br are characterized by 3 types of absorption bands. The first ones are d-d transitions of Cu^{2+} cations which lie at 16000 - 5500 cm^{-1} for tetrachlorocuprates(II) and at 9090 - 6000 cm^{-1} for tetrabromocuprates(II). One wide band is present in the spectra at a room temperature as below 77 K it is split into three sharp bands which relate to $^2B_2 \rightarrow ^2A_1$, $^2B_2 \rightarrow ^2B_1$, and $^2B_2 \rightarrow ^2E$ electron transitions (Halvorson et al., 1990). The correlation between the type and degree of distortion of $[CuCl_4]^{2-}$ polyhedra and maxima of d-d transition bands for D_{2d} and D_{4h} symmetries are described with the help of semi-empiric (McDonald et al., (1988) and empiric (Wasson et al., 1977) formulae and are presented on Fig. 11, 12.

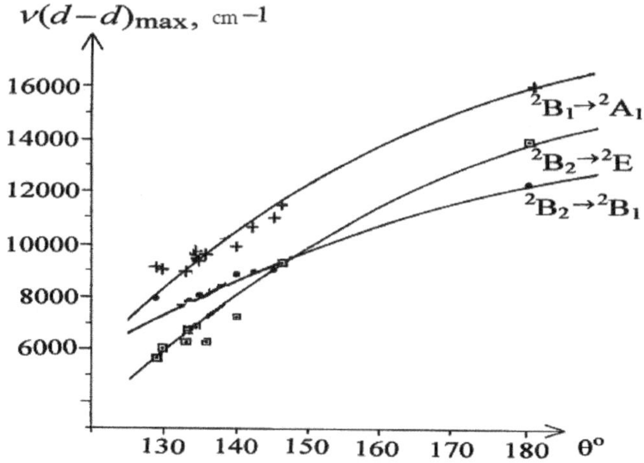

Fig. 11. The dependence of maxima of d-d transition bands in electronic absorption spectra of compounds containing $[CuCl_4]^{2-}$ anions on the flattening angle of the coordination polyhedron below 77 K (the lines present calculated data (McDonald et al., (1988)).

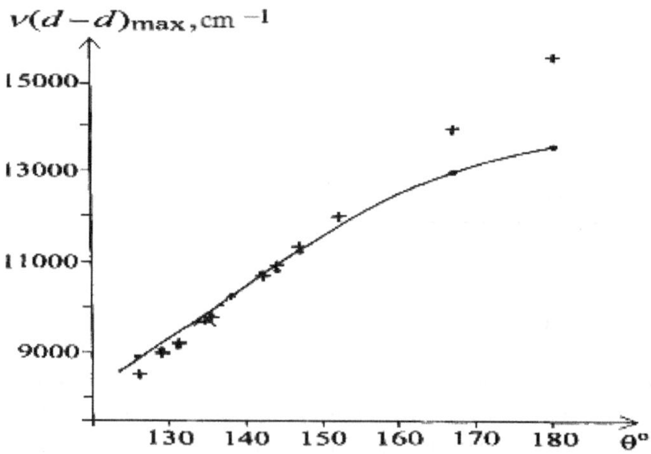

Fig. 12. The dependence of maximum of d-d transition bands in electronic absorption spectra of compounds containing $[CuCl_4]^{2-}$ anions on the flattening angle of the coordination polyhedron at a room temperature (the lines present calculated data (Wasson et al., 1977)).

The second type of absorption bands in UV and visible parts of electronic spectra of halocuprates(II) relates to $X \rightarrow Cu^{2+}$ (X = Cl, Br) charge transfer (CT). Their high intensities can be explained by the ability of Cu^{2+} to be reduced into Cu^+. The transition frequencies are determined both by the nature of X (Cl, Br) and by the degree of distortion of the CuX_4^{2-} polyhedron (Fig. 13) (Koval'chukova et al., 2009b).

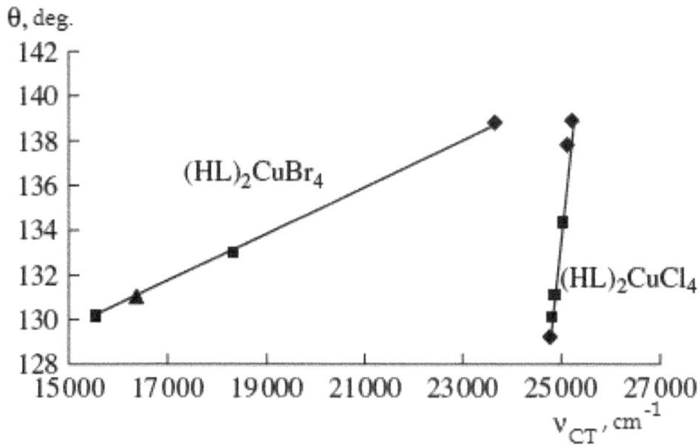

Fig. 13. The dependence of the position of the $X \rightarrow Cu^{2+}$ (X = Cl, Br) charge transfer bands in the electronic absorption spectra of tetrahalocuprates on the degree of distortion of the anionic polyhedron (Koval'chukova et al., 2009b).

From the data of Fig. 13, empiric formulae for calculation of the degree of distortion of CuX_4^{2-} (X = Cl, Br) polyhedra from the position of charge transfer bands in the electronic absorption spectra were obtained:

$$\theta = 0.0272 \cdot \nu_{CT} - 544.56 \text{ (for } (HL)_2CuCl_4 \text{ at } R^2=0.9718 \quad (1)$$

$$\theta = 0.001 \cdot \nu_{CT} + 113.97 \text{ (for } (HL)_2CuBr_4 \text{ at } R^2=0.9963 \quad (2)$$

$$\nu_{CT}(CuBr_4^{2-}) = 272 \cdot \nu_{CT}(CuCl_4^{2-}) - 658 \quad (3)$$

The 3rd group of bands in the electronic absorption spectra of halocuprates relates to electron transitions in organic cations. They are of the highest intensities and may overlap with the charge transfer transitions. Unfortunately, the dependence of positions of ligand bonds in the electronic absorption spectra on the nature of H-bonds is not so evident, and no sufficient correlations were found.

The same conclusion was made by Marcotrigiano and co-authors (Grigereit et al., 1987) with using IR spectroscopy data. The attempt to correlate the character of H-bonds in polycrystalline $(H_2L)CuBr_4$ (L = 1-methylpyperazine; 2-methylpyperazine), as well as $(HL)_2CuBr_4$ (L = 1-methylpyperazine) with the shift in ligand absorption bands in IR spectra of complexes with respect to those in corresponding hydrobromides also failed. This might deal with the existence of strong H-bonds of different nature both in complex halocuprates and in initial organic hydrobromides.

4. Structure characteristics and properties of $Cu_2Cl_6^{2-}$ structures

As already noted, halide-ions may take place in coordination as bridging ligands. This leads to the formation of dimeric, oligomeric, and even polymeric $Cu_nX_{2n+2}^{2-}$ (X = Cl, Br) anions

(Grigereit et al., 1987). Studies of structures and magnetic properties (O'Bannon & Willett, 1986) of $KCuCl_3$ and NH_4CuCl_3 showed the existence of characteristic antiferromagnetically-joint dimeric systems, i.e. the formulae of the above substances should be presented as $K_2Cu_2Cl_6$ и $(NH_4)_2Cu_2Cl_6$.
Symmetric dibridged structures of dimeric $A_2Cu_2X_4$ (X = Cl, Br) structures are based on $Cu_2X_6^{2-}$ dimers in one of the three mostly possible geometric configurations (Fig. 14). The planar dimer A is described by the φ-angle and usually has the (4+2) coordination mode (Landee et al., 1988; Bencini & Gatteschi, 1986). The structures present anionic dimers containing two four-coordinated coppers(II) with the D_{2d} geometry. Four coordinate bonds of each metallic atom are formed by Cl-atoms two of which are bridging ligands. Counter-ions are protonated tetramethylene sulfoxide (Scott & Willett, 1991) or tetrapropylammonium (Landee et al., 1988) cations. The structures are stabilized by axial semi-coordinate Cu-Cl bonds involving the terminate halides in $Cu_2Cl_6^{2-}$ anions.

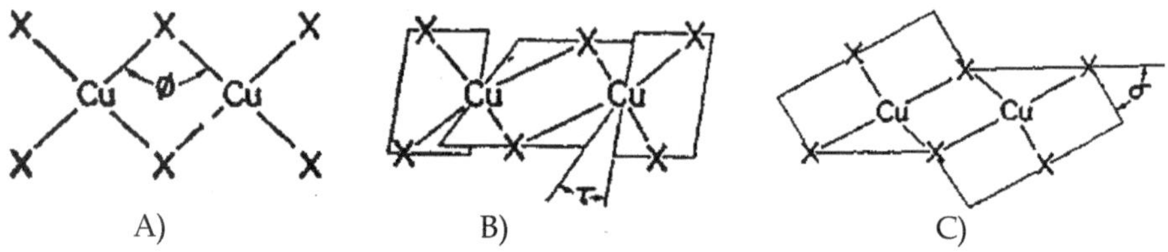

Fig. 14. The types of distortion of $Cu_2X_6^{2-}$ anions: A) planar; B) twisted (screwed); C) folded (Willett & Geiser, 1984).

4.1 Planar $Cu_2X_6^{2-}$ structures

One of the common types of planar anionic halocuprates(II) is a pseudo-planar dibridged structure (Harlow et al., 1975). Each of Cu(II) atom forms a primary planar square configuration. Variations of the crystal package may exist. For example, this may be presented in the graphical form for $KCuCl_3$ (Fig. 15a) (Willett et al., 1963), $(H_2mel)[Cu_2Cl_6]$, H_2mel - melaninium cation (Fig. 15b) (Colombo et al., 1985), $(CH_3)_2CHNH_3CuCl_3$ (Fig. 15c) (Roberts et al., 1981) or $(CH_3)_2NH_2CuCl_3$ (Fig. 15d) (Willett, 1966).
Authors (Scott & Willett, 1991) described a crystal structure of bis(tetrapropylammonium) hexabromodicuprate(II) with 6-coordinated copper ((4+2) geometry). The axial coordination involves terminal Br-atoms of neighboring $Cu_2Br_6^{2-}$ anions (r Cu–Br_{axial} 2.876 Å) and N-atoms of organic cations. The coordination polyhedron of Cu^{2+} lies between square-pyramidal and distorted octahedral. The same type of structure was determined for dimorpholinium hexahalodicuprate (II) salts (Scott et al., 1988).
The (4+2) coordination of Cu(II) in planar $Cu_2Br_6^{2-}$ anions of dibenzotetrathiafulvalenium hexabromodicuprate(II) (Honda et al., 1985) was achieved with the help of the formation of two Cu…S semi-coordination bonds (r Cu–S from 3.425 to 4.580 Å).
In bis(4-aminopyridinium) hexabromodicuprate(II) diaquatetrabromodicopper(II) (Place & Willett, 1994) the coordination polyhedron in a planar $Cu_2Br_6^{2-}$ is completed to a square pyramidal (coordination number 5) by Br-atoms of a $Cu_2Br_2(H_2O)_2$ fragments as 4-aminopyridinium molecules are not involved in the coordination (Fig. 16). The Cu-Br distances inside the dimer are 2.401 Å (Cu-Br terminal) and 2.443 Å (Cu-Br bridging); the Cu…Br bond lengths of semi-coordinate bonds lie in the range 2.904 to 3.200 Å.

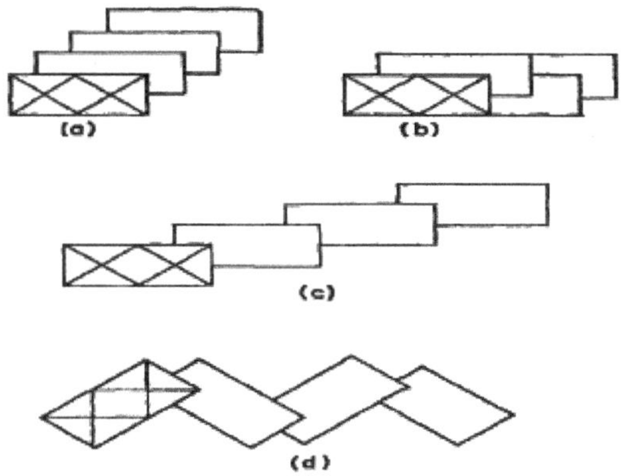

Fig. 15. The crystal package of $Cu_2X_6^{2-}$ dimmers: (a) double chain; (b) "zig-zag" double chain; (c) alternative chain; (d) "zig-zag" alternative chain (Willett & Geiser, 1984).

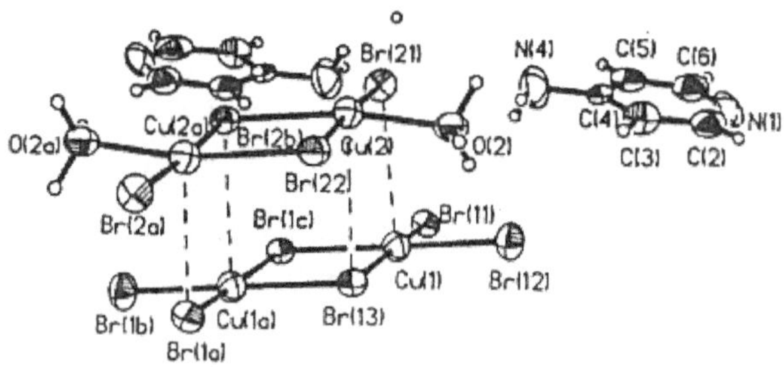

Fig. 16. The crystal structure of bis(4-aminopyridinium) hexabromodicuprate(II) diaquatetrabromodicopper(II) (Place & Willett, 1994) .

4.2 Twisted $Cu_2X_6^{2-}$ structures

The distortion of planar $Cu_2X_6^{2-}$ anions may take place in two directions. The first brings to twisted (or screwed) structures (Fig. 14 B) which are characterized by the twisting angle τ between the bridging Cu_2X_2 plane and two terminate CuX_2 planes. Twisted dimers are most often formed in complexes with large organic cations with a little tendency in formation of the H-bonds. The possibility to form twisted structures relates to the electrostatic repulsion between halide anions which provides the possibility of tetrahedral distortion to the coordination sphere of Cu^{2+} cations.

Such types of twisted $Cu_2X_6^{2-}$ structures were found for example for $(Ph_4As)_2[Cu_2Cl_6]$ (Willett & Chow, 1974), $(Ph_4Sb)_2[Cu_2Cl_6]$ (Bencini et al., 1985), and $(Ph_4P)_2[Cu_2Cl_6]$ (Textor et al., 1974). In the above structures, the coordination geometry of Cu(II) is intermediate between square-planar and tetrahedral. The τ-angle is close to 50 deg. for both the structures.

The authors (Kovalchukova et al., 2008b) report the crystal structure of bis(4-pyperidyl-1)-2-phenylpyrido[2,3-a]anthraquinonium-7,12) hexachlorodicuprate(II) which consists of separated organic cations and twisted $Cu_2Cl_6^{2-}$ anions (Fig. 17). The terminal C1(2)-Cu-C1(3) plane is rotated around the central Cu_2Cl_2 plane through 50.5 deg. The coordinate polyhedron of each Cu(II) atom is a distorted tetrahedron with the bond angles 100.5 (C1(2)-Cu-C1(3)), 96.6 (C1(2)-Cu-C1(1)), 87.5 (Cl(l)-Cu-C1(1*), and 92.5 deg. (CI(3)-Cu-CI(I*)). In the crystal, the inorganic anion and planar organic cations form a skeleton structure of H-bonds involving H-atoms at C(24) and C(25) of the pyridine fragment of the cation and Cl(2) and Cl(2*) atoms of the hexachlorodicuprate(II) anion.

Two crystalline phases of $(TTM\text{-}TTF)_2[Cu_2Cl_6]$ (TTM-TTF = tetra(methylthio)tetrathiafulvalene) were detected by H. Enders (Endres, 1987). They differ in the arrangement of the $(TTM\text{-}TTF)^+$ radical cations and the geometry of the $[Cu_2Cl_6]^{2-}$ anions. Both of them are of the twisted geometry with the τ-angles of 10.0 (almost planar) and 31.5 deg. This fact reflects the "plasticity" of the coordination sphere of Cu(II).

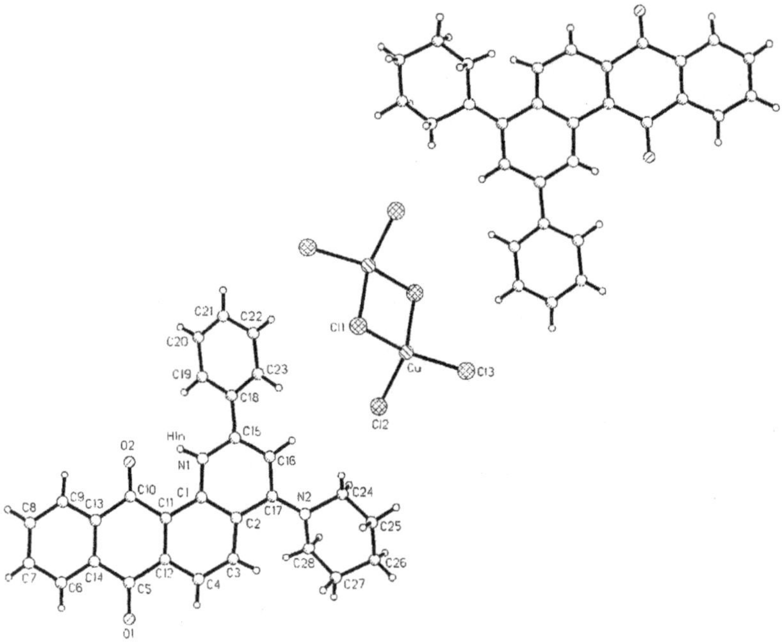

Fig. 17. The crystal structure of bis(4-pyperidyl-1)-2-phenylpyrido[2,3-a]anthraquinonium-7,12) hexachlorodicuprate(II) (Kovalchukova et al., 2008b).

4.3 Bifolded $Cu_2X_6^{2-}$ structures

The second type of the distortion of a planar $Cu_2X_6^{2-}$ anion is known as a folded or "*sedia*" structure (Fig. 14C) where two terminal halide-ions (each one for every side of the dimer) go out of the conjunction plane. This type of distortion is characterized by the σ-angle between the central Cu_2X_2, and the terminal CuX_3 planes.

The bifolded structures usually have (4+1) type of coordination of Cu(II) ions with the σ-angle inside the interval from 19 to 32.5 deg (Geiser et al., 1986a). The degree of folding of the dimer increases in case if all the five ligands are approaching the copper atom. This leads to the transformation of the coordination polyhedron from square-pyramidal (SP) to the trigonal-bipyramidal (TBP) one (Fig. 18). Addition of one more coordination bond ((4+2) coordination) increases the Cu–L distances and leads to formation of square-bipyramidal (SBP) structures.

The change in the type of the coordination polyhedron in the folded $Cu_2X_6^{2-}$ anions is described by the change in degrees of distortion which are calculated as a mathematical difference of two θ-angles (fig. 18) (Blanchette & Willett, 1988). In case of the square-pyramidal structure (SP), $\theta_1 \approx \theta_2$, as for trigonal-bipyramidal (TBP) configurations the limits $\theta_1 \rightarrow 120$ deg.; $\theta_2 \rightarrow 180$ deg., $\Delta = \theta_2 - \theta_1 = 60$ deg. It is evident that the majority of the determined structures are characterized by Cu–Cl distances in the range 2.65 to 2.75 Å, and the Δ range 15 to 30 deg.

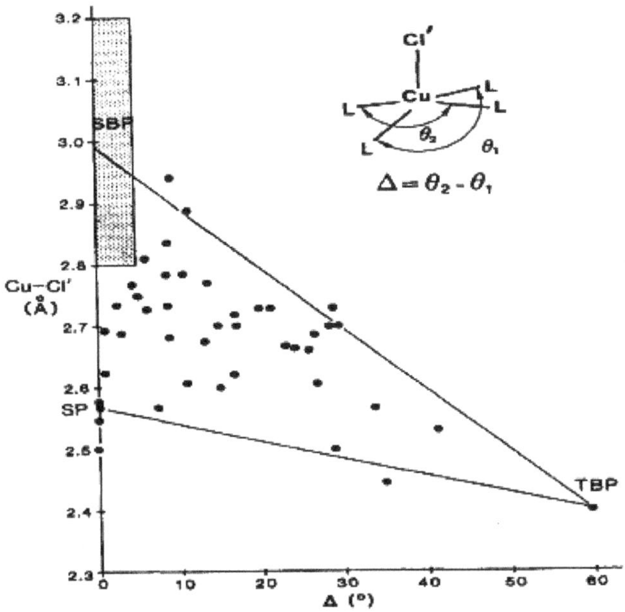

Fig. 18. The dependence of Cu–Cl distances in hexachlorodicuprates(II) on the degree of distortion Δ and the type of coordination polyhedron of Cu(II) (Blanchette & Willett, 1988).

A lot of bifolded five-coordinated halocuprates(II) were reported by various authors and cited in (O'Brien et al., 1988), and three different types of the formation of coordination polyhedra were found. In the first case, all the five coordination sites of Cu(II) are occupied by halide ions. That was described for example for bis(benzimidazolium) hexachlorodicuprate(II) (Bukowska-Strzyzewska & Tosik, 1985). As it was shown (Fig. 19), polymeric slightly bifolded $[Cu_2Cl_6^{2-}]_\infty$ chains elongated along the y axis are formed by stacking of $Cu_2Cl_6^{2-}$ dimers involving one of the terminal Cl-atom from each side of the monomeric unit. The coordination polyhedra around the Cu atoms of each crystallography independent chain may be described as a distorted square pyramid. The Cu–Cl bond lengths and angles in both crystallography independent anionic chains are not identical. The bridging Cu(1)-Cl(3) and Cu(2)-Cl(6) bonds (2.298 and 2.320 Å, respectively) correlate well to those for other described bonds of such a type (Murray-Rust, 1975). The terminal

Cu–Cl bonds, shorter than the bridging ones, are not equal. The Cu(1)–Cl(1) and Cu(2)–Cl(4) bonds (2.291 and 2.284 Å, respectively) linking the adjacent dimmers, are significantly longer than the terminate Cu(1)–Cl(2) and Cu(2)–Cl(5) bonds (2.245 and 2.256 Å, respectively) which are not involved into the interchain stacking. The Cu...Cu distances inside the $Cu_2Cl_6^{2-}$ dimmers are 3.464 and 3.470 Å, as the Cu...Cu distances between the adjanced dimmers are 3.716 and 3.777 Å. The inorganic anions and organic cations are joint by bifurcated H-bonds between NH^+ fragments of benzimidazolium cations and Cl(4) Cl(5) atoms of inorganic anions (r N-H 1.00 Å; r H...Cl 2.33 - 2.64 Å; r N...Cl 3.175 - 3.268 Å; ∠ N-H...Cl 119 - 144 deg.). Another type of H-bonds involves the NH-fragments of the organic cation and Cl(1) atoms (r N-H 1.00 Å; r H...Cl 2.24 and 2.73 Å for two unequivalent chains; r N...Cl 3.216 and 3.396 Å; ∠ N-H...Cl 164 and 124 deg.).

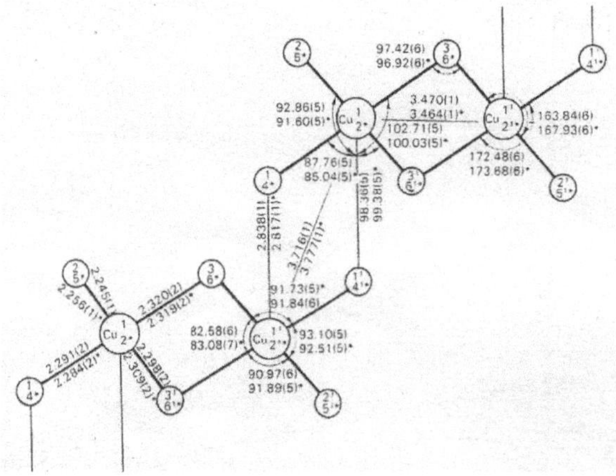

Fig. 19. Bond lengths (Å) and angles (deg.) in the two crystallographically independent $[Cu_2Cl_6^{2-}]_\infty$ chains of bis(benzimidazolium) hexachlorodicuprate(II) (Bukowska-Strzyzewska & Tosik, 1985).

Two similar structures of polymeric chlorocuprate(II) containing piperidinium and piperazinium counter-ions (Battaglia et al., 1988) also incorporate unequivalent $[Cu_2Cl_6^{2-}]_\infty$ chains joint together by Cu–Cl axial bonds. The $Cu_2Cl_6^{2-}$ monomers are bifolded with the σ-angles 29.6 and 23.2 deg. for piperidinium and piperazinium salts respectively. The bifolded distortion gives each Cu(II) ion a (4+1) coordination geometry but the chains differ in their configurations. In the piperazinium hexachlorodicuprate, adjacent dimers are related by unit-cell translation as illustrated on Fig. 15c. From the other hand, in the piperidinium salt the $[Cu_2Cl_6^{2-}]_\infty$ fragments are related by a *c*-glide operation (Fig. 15d). The smallest trans Cl–Cu-Cl angle is 150.43 deg. for the piperidinium salt and 156.8 in the piperazinium one. The equatorial Cu-Cl bond lengths are considerably shorter for the piperidinium complex (average 2.267 Å *vs.* 2.298 Å in the piperazinium one)but the axial semi-coordinate distances are identical (2.612 *vs.* 2.622 Å). The intradimer bridging Cu–Cl–Cu φ-angles are 95.5 *vs.* 95.8, and the interdimer bridging Cu–Cl–Cu φ′-angles are 87.1 *vs.* 89.1 deg., respectively.

The folded structure of $Cu_2Cl_6^{2-}$ anions is also observed in bis(4-azfluorene-9-onium) hexachlorodiaquadicuprate(II) dehydrate (Koval'chukova et al., 2009a) but the coordinate sphere of Cu^{2+} ions differs from previous cases. Both the two central atoms of the dimer are coordinated by four Cl atoms and one water molecule (Fig. 20). The polyhedron $CuCl_4H_2O$ is a distorted square pyramid with one chlorine atom in the axial position. Three Cl atoms and one water molecule form the pyramid basis, and the Cu atom is shifted towards the pyramid top by 0.27 Å. The bridging Cu–Cl distances are 2.274 and 2.307 Å, and the terminal Cu–Cl distances are 2.265 and 2.593 (axial bond) Å. The bifolded $Cu_2Cl_6^{2-}$ dimers (σ 20 deg.) are isolated. The organic cations are adjusted to the inorganic anion via an extra water molecule.

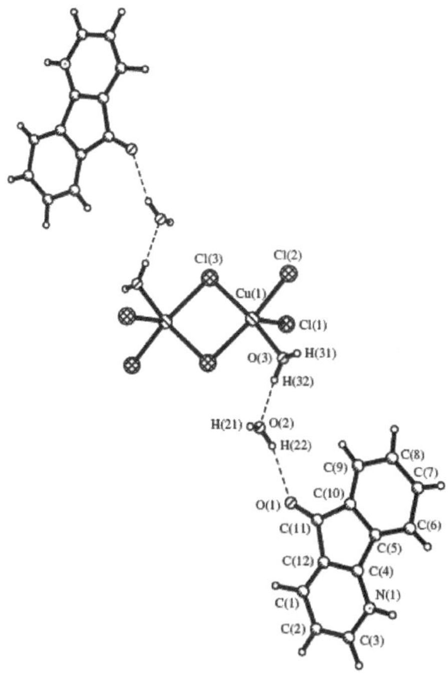

Fig. 20. The crystal structure of $(HL)_2[\{CuCl_2(H_2O)_2\}_2(\mu\text{-}Cl)_2 \cdot 2H_2O$ (L - 4-azfluorene-9-one) (Koval'chukova et al., 2009a)

At the same time, in bis(1-amino-4-azfluorene-9-olium) hexachlorodicuprate(II) (Kuz'mina et al., 2002) the fifth coordinate bond of Cu(II) cations is realized at the expense of the interaction of the lone electron pair of the O-atom of the hydroxo-group of the organic cation which is protonated by its N-heterocyclic atom with a vacant Cu(II) d-orbital (Fig. 21). The $CuCl_4O$ polyhedron represents the form of distorted square pyramid where the O-atom occupies the axial position. All the four Cl atoms lie at one plane (the mean deviation from planarity is 0.055 Å) and the Cu atom is 0.174 Å shifted towards the pyramid top. The hexachlorodicuprate(II) anion has a bifolded configuration (σ = 26.6 deg.). The bridging Cu–Cl distances (2.299 and 2.324 Å) are a little bit longer than the terminal ones (2.226 and 2.262 Å). In the lattice, the inorganic dimers are adjacent in chains along the *x*-axis with the help of O-H...Cl hydrogen bonds. The shortest interdimer Cl...Cl contacts in chains are 4.584 Å.

The chains are adjacent in three dimensional skeleton structure via N–H...Cl hydrogen bonds (both N atoms of the pyridine ring and the amino-group are involved in the H-bounding).

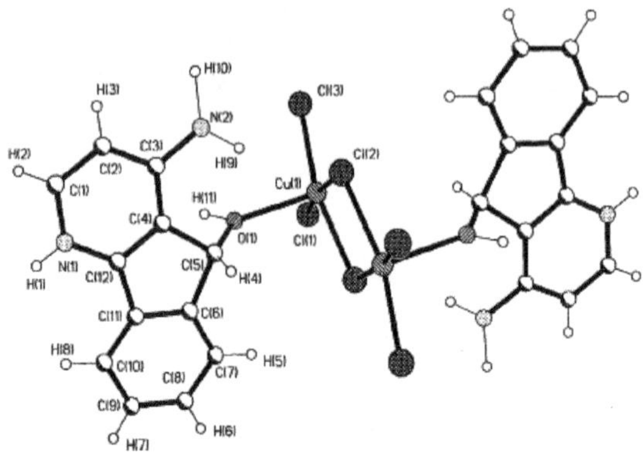

Fig. 21. The crystal structure of bis(1-amino-4-azfluorene-9-olium) hexachlorodicuprate(II) (Kuz'mina et al., 2002).

In the similar structure of bis(3-aminopyridinium) hexabromodicuprate(II) monohydrate (Blanchette & Willett, 1988), one of the terminal coordinate bond is released at the expense of the formation of a Cu–N bond involving the amino-group of the bifolded organic cation (Fig. 22). One of the Br atoms of the inorganic anion forms the axial bond.

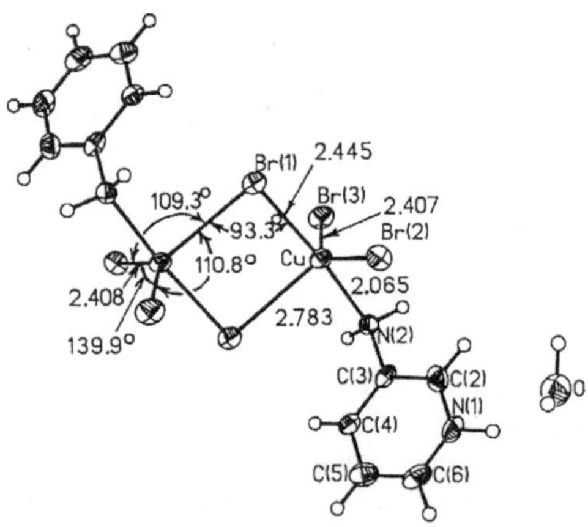

Fig. 22. The crystal structure of bis(3-aminopyridinium) hexabromodicuprate(II) monohydrate (Blanchette & Willett, 1988).

The analogous coordination of Cu(II) atoms via N-atoms of amino-groups of protonated organic bases was described for the co-crystallized bis(3-amino-2-methyl-4H-pyrido[1,2-a]pyrimidine-4-onium) hexachlorodicuprate(II) and 3-amino-2-methyl-4H-pyrido[1,2-a]pyrimidine-4-onium chloride (Kuz'mina et al., 2003) (Fig. 23).

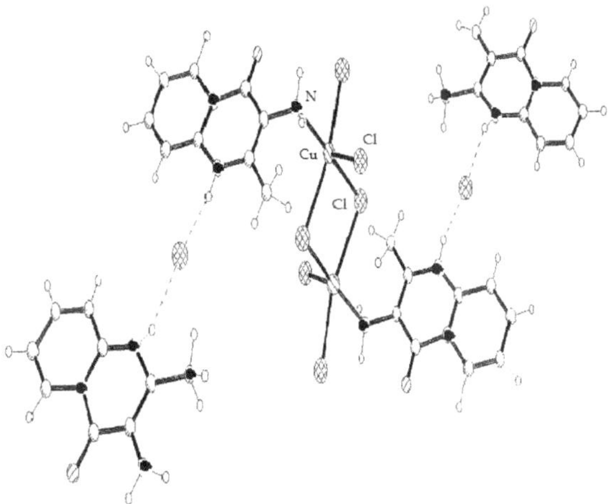

Fig. 23. The crystal structure of co-crystallized bis(3-amino-2-methyl-4H-pyrido[1,2-a]pyrimidine-4-onium) hexachlorodicuprate(II) and 3-amino-2-methyl-4H-pyrido[1,2-a]pyrimidine-4-onium chloride (Kuz'mina et al., 2003).

The fifth coordinate bond is formed via a lone electron pair of an amino-group of the organic catrion which is protonted by an N-atom of a heterocycle. The coordination polyhedron $CuCl_4N$ of each Cu atoms in the dimer is a trigonal bipyramid with the N-atom as an axial ligand (*r* Cu–N 2.080 Å). The Cu–Cl axial distance (2.316 Å) does not differ from the three others. The $Cu_2Cl_6^{2-}$ anion is isolated from other inorganic species and exists as a bifolded centro-symmetric dimer with the folding angle σ 13.3 deg. The structure is stabilized by a set of H-bonds between the H-atoms of the organic molecule and Cl^- ions of the inorganic anion.

4.4 Spectral-magneto-structural correlations

Analysis of spectral and structural data of hexachlorodicuprates(II) show linear dependences of positions of the absorption bands relating to the Cl→Cu charge transfer (CT) on the degree of distortion of the anionic polyhedron ($\Delta = \theta_1 - \theta_2$) (Koval'chukova et al., 2009a). From the Fig. 24 it is evident, that the Δ value changes from 0 to 6 deg. in non-distorted or slightly distorted square-pyramidal structures (ν_{CT} 19610 - 19900 cm^{-1}), and from 15 to 50 deg. in trigonal-bipyramidal structures (ν_{CT} 23880 - 24390 cm^{-1}). Thus, the position of the charge transfer bands in the electronic absorption spectra of the compounds of the general formula $(HL)_2[Cu_2Cl_6]$ may characterize the degree of distortion of dimeric coordination polyhedra. The attempt to deduce an analogous dependence for hexabromodicuprate(II) failed because of poor information in the literature.

It is well known that a slight exchange interaction exists in bridging dimeric hexachlorodicuprates and the magnetic behavior depends upon features of the structure of $Cu_2Cl_6^{2-}$ complex anions (Hay et al., 1975; Willett et al., 1983). For the folded structures, the value of the exchange interaction (J/k) is determined by the folding σ-angle and the bridging φ-angle. It was determined (O'Brien et al., 1988; Battaglia et al., 1988) that if φ = 95 - 96 deg., the absence of the exchange interaction (J/k = 0) is observed at σ = 23 deg. In the case when σ < 23 deg., the antiferromagnetic exchange interaction occurs, and the ferromagnetic exchange interaction is observed at σ > 23 deg. (Fig. 25).

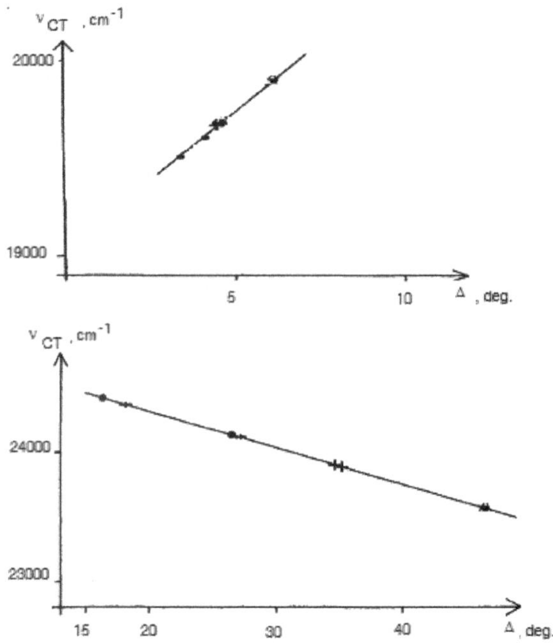

Fig. 24. The dependences of the positions of the absorption bands relating to the Cl→Cu charge transfer (CT) on the degree of distortion of the anionic polyhedron ($\Delta = \theta_1 - \theta_2$) (Koval'chukova et al., 2009a).

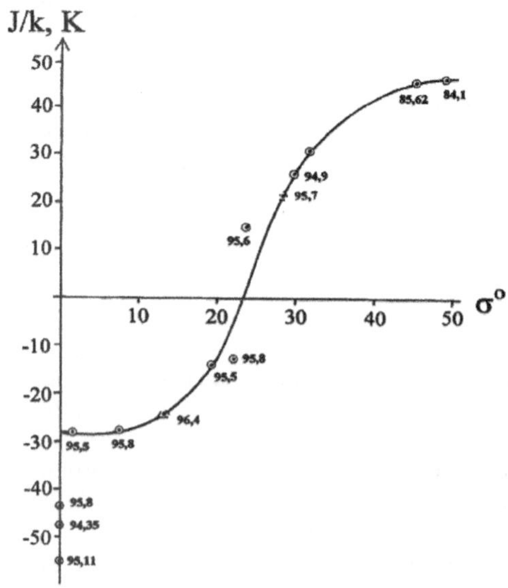

Fig. 25. The dependence of magnetic properties of bifolded hexchlorodicuprates(II) on the folding σ-angle and the bridging φ-angle (O'Brien et al., 1988).

The value of antiferromagnetic depends upon the shape of the coordination polyhedron of Cu(II) atoms: in the bifolded dimeric hexachlorodicuprates(II) the minimal antiferromagnetic exchange is observed for square-pyramidal structures, as well as the maximal one is observed for trigonal-bipyramidal complexes (Hay et al., 1975). The antiferromagnetic exchange increases with shortening of the bridging Cu–Cl bonds and with an increase of the electron density at the bridging atoms. In layered structures $[Cu_2Cl_6^{2-}]_\infty$ the existence of short intralayer Cl...Cl contacts (less than 3,94 Å) provoke the intralayer antiterromagnetic interaction which increases with the shortening of Cl...Cl distances (Scott et al., 1988).

5. Conclusion

Anionic halocuprate(II) complexes are of a great interest for scientists because of the areas of their application. They show a large variety in composition and coordinational geometry, and may be presented by mononuclear CuX_2^{2-} species (X = Cl, Br) or form $Cu_2X_6^{2-}$ dimers which can exist separately or be arranged in polymeric $[Cu_2Cl_6^{2-}]_\infty$ chains. Dimeric structures may exist as planar, twisted, or folded fragments. The coordination number of Cu(II) can change from 4 (tetrahedrons with different degrees of tetrahonal distortion or planar square configuration) to 5 (from the square pyramidal to the trigonal-bipyramidal configurations) and even 6 (more or less distorted octahedra). More complicated structures can also exist despite the event they were out of our interest.

Among the factors controlling on the type and finer details of coordination mode of Cu(II) there is the nature of the cation which neutralizes the negative charge of halocuprate(II) species, i.e. cation size, shape, flexibility, as well as the ability of formation of H-bonds. The nature of the halide atom (Cl or Br) affects the type of the structure much less. Unfortunately the influence of the organic cation on the features of the structure of halocuprates(II) is not unequivocal, and the determination of the crystalline structures not available at every instant. The relationship between the structure and spectral or magnetic properties appeared to be helpful by the prediction of the features of coordination modes, and physical properties as well as possible areas of the application of newly synthesized halocuprates(II).

6. Acknowledgments

This work was supported by the Russian Foundation for Basic Research (Grant 10-03-00003-a)

7. References

Abolmaali, B., Taylor, H.V. & Weser, U. (1998). Evolutionary aspects of copper binding centers in copper proteins. *Structure and Bonding*, Vol. 91, pp. 91-190, ISSN 0081-5993

Adman, E.T. (1991) Copper protein structures. *Adv. Protein Chemistry*, Vol. 42, pp. 145-197, ISBN 9780120342426

Allen, C.P., Benkovics, T., Turek, A.K. & Yoon, T.P. (2009). Oxaziridine-Mediated Intramolecular Amination of sp^3-Hybridized C−H Bonds. *J. Amer. Chem. Soc.* Vol. 131, No. 35 (September 2009), pp. 12560-12561, ISSN 0002-7863

Anderson, D.N. & Willett, R.D. (1971). The Crystal Structure of Bis(Ethylenediammoniummonobromide) tetrabromocuprate(II). *Inorg. Chim. Acta Rev.*, Vol. 5, No 1, (January 1971), pp. 41-45, ISSN 0073-8085

Antolini, L., Menabue, L., Pellacani, G.C., Saladini, M., Marcotrigiano, G. & Porzio, W. (1981). Spectroscopic and structural investigation of two N-benzylpiperazinium tetrachlorocuprates(II), one hemihydrate and one anhydrous: two compounds containing unequally flattened $[CuCl_4]^{2-}$ tetrahedra. *J. Chem. Soc., Dalton Trans.*, pp. 1753 - 1759

Arend, H., Huber, W., Mischgofsky, F.H. & Richter-Van Leeuwen, O.K. (1978). Layer Perovskites of the $(C_nH_{2n+1}NH_3)_2MX_4$ and $NH_3(CH_2)_mNH_3MX_4$ Families with M = Cd, Cu, Fe, Mn or Pd and X = Cl or Br: Importance, Solubilities and Simple Growth Techniques. *J. Cryst Growth*, Vol. 43, No 1, pp. 213-223, ISSN 0022-0248

Battaglia, L.P., Bonamartini-Corrad,i A., Geiser, U., Willett, R.D., Motori, A., Sandrolini, F., Antolini, L., Manfredini, T., Menabue, L. & Pellacani G.C. (1988). The Crystal Structures, Magnetic and Electrical Properties of Two Polymeric Chlorocuprate(II) Compounds. *J. Chem. Soc, Dalton Trans.*, No 2, (February 1988), pp. 265-271

Battaglia, L.P., Bonamartini-Corradi, A., Marcotrigiano, G., Menabue, L. & Pellacani, G.C. (1979). Halocuprates(II) of the N-Phenyipiperazinium Mono- and Dications: Crystal and Molecular Structure of N-Phenylpiperazinium Tetrachlorocuprate(II). *Inorg. Chem.*, Vol. 18, No 1, (January 1979), pp. 148-152, ISSN 0020-1669

Bencini, A. & Gatteschi, D. (1986). Calculations of the Electronic Structure and Magnetic Properties of Weakly Coupled Transition-Metal Clusters. The $[Cu_2Cl_6]^{2-}$ Dimers. *J. Amer. Chem. Soc.*, Vol. 108, No 19, (September 1986), pp. 5763-5771, ISSN 0002-7863

Bencini, A., Gatteschi, D. & Zanchini, C. (1985). Anisotropic Exchange in Transition-Metal Dinuclear Complexes. *Inorg. Chem.*, Vol. 24, No 5, (February 1985), pp. 704-708, ISSN 0020-1669

Bhattacharya, R., Chandra, S., Bocelli, G., Cantoni, A. & Ghosh, A. (2004). Synthesis and crystal structure of bis(2-methylbenzimidazolium) tetrahalocuprate(II). *J. Chem. Crystallogr.*, Vol. 34, No 6, (June 2004), pp. 393-400, ISSN 1074-1542

Blanchette, J.T. & Willett, R.D. (1988). Magnetic and Structural Correlations in $[(C_5H_5N)NH)]_2Cu_2Cl_6$ and $[(C_5H_5N)NH_2]_2Cu_2Br_6{\cdot}H_2O$. *Inorg. Chem.*, Vol. 27, No 5, (March 1988), pp. 843-849, ISSN 0020-1669

Bukowska-Strzyzewska, M. & Tosik, A. (1985). Structure of Polimeric Bis(benzimidazolium) Hexachlorodicuprate(II), $2C_7H_7N_2^+{\cdot}Cu_2Cl_6{}^{2-}$. *Acta Crystallogr., Sect. C.*, Vol. C41, No 8, (August 1985), pp. 1184-1186, ISSN 0108-7673

Colombo, A., Menabue, L., Motori, A., Pellacani, G.C., Porzio, W. & Sandrolini, F. (1985). Crystal Structure and Spectroscopic, Magnetic, and Electrical Properties of a Copper(II) Dimmer, Melaminium Hexachlorodicuprate(II), Exhibiting a New Stacking Interaction. *Inorg. Chem.*, Vol. 24, No 19, (September 1985), pp. 2900-2905, ISSN 0020-1669

Cotton, F.A. & Wilkinson, G. (1966). *Advanced inorganic chemistry* (2nd edition), Interscience publishers, N.Y. - London - Sydney

Cowan, J.A. (1993). *Inorganic Biochemistry: An Introduction,* VCH Publishers, ISBN 0-471-18895-6, New York

Diaz, I., Fernandes, V., Belsky, V.K. & Martinez J.L. (1999). Synthesis and Structural Study of the Thermochromic Compounds Bis(2-amino-4-oxo-6-methylpyrimidinium) Tetrachlorocuprate(II) and Bis(2-amino-4-chloro-6-methylpyrimidinium) Hexachlorodicuprate(II). *Z. Naturforsch. B,* Vol. 54, No 6, (June 1999), pp. 718-724, ISSN 0932-0776

Endres, H. (1987). Two Modifications of Di[tetra(methylthio)tetrathiafulvalenium] Hexachlorodicuprate(II), $[(TTMTTF)^+]_2[Cu_2Cl_6]^{2-}$. *Z. Naturforsch. B,* Vol. 42, No 1, (January 1987), pp. 5-11, ISSN 0932-0776

Geiser, U., Willett, R.D., Lindbeck, M. & Emerson K. (1986a). Crystal Structure of Bis(trimethylammonium) Decabromotetracuprate(II): A Review of Stacking Patterns in Pseudo planar $Cu_nX_{2n+2}^{2-}$ and $Cu_nX_{2n}L_2$ Oligomers. *J. Amer. Chem. Soc.,* Vol. 108, No 6, (March 1986), pp. 1173-1179, ISSN 0002-7863

Geiser, U., Willett, R.D., Lindbeck, M. & Emerson, K. (1986b). Crystal Structure of Bis(trimethylammonium) Decabromotetracuprate(II): A Review of Stacking Patterns in Pseudoplanar Cu_nX_{2n+2} and $Cu_nX_{2n}L_2$ Oligomers. *J. Amer. Chem. Soc,* Vol. 108, No 6, (March 1986), pp. 1173-1179, ISSN 0002-7863

Gerloch, M. & Constable, E.C. (1994). *Transition Metal Chemistry.* VCH Verlagsgesellschafft, Weinheim & VCH Publishers, ISBN 3-527-29218-7, Federal Republic of Germany & New York, NY (USA)

Gillard, R.D. & Wilkinson, G. (1963). Some Quinqueco-ordinated Copper(II) Complexes. *J. Chem. Soc.,* pp. 5399-5406

Grigereit, T.E., Ramakrishna, B.L., Place, H., Willett, R.D., Pellacani, G.C., Manfredini, T., Menabue, L., Bonamartini-Corradi, A. & Battaglia L.P. (1987). Structures and Magnetic Properties of Trinuclear Copper(II) Halide Salts. *Inorg. Chem.,* Vol. 26, No 14, (July 1987), pp. 2235-2243, ISSN 0020-1669

Halvorson, K.E., Patterson, C. & Willett, R.D. (1990). Structures of Bis(4-aminopyridinium) Tetrachlorocuprate(II) Monohydrate, $[C_5H_7N_2]_2[CuCl_4]\cdot H_2O$, and Bis(2-amino-3-hydroxypyridinium) Tetrachlorocuprate(II), $[C_5H_7N_2O]_2[CuCl_4]$: Correlation of $CuCl_4^{2-}$ Geometry with Hydrogen Bonding and Electronic Structure. *Acta Crystallogr., Sect. B,* Vol. B46, No 4, (August 1990), pp. 508-519, ISSN 0108-7673

Harlow, R.L., Wells, W.J., Watt, G.W. & Simonsen, S.H. (1975). Crystal and Molecular Structure of Bis(N,N-dimethylphenetylammonium) Tetrachlorocuprate(II). Relationship between the Electronic Spectrum and Distortion of the $CuCl_4^{2-}$ Chromophore from Tetrahedral symmetry. *Inorg. Chem.,* Vol. 14, No 8, (August 1975), pp. 1786-1772, ISSN 0020-1669

Hathaway, B. (1982). Copper. *Coord. Chem. Rev.,* Vol. 41, pp. 423-487, ISSN 0010-8545

Hay, P.J., Thibeault, J.C. & Hoffmann, R. (1975). Orbital Interactions in Metal Dimmer Complexes. *J. Amer. Chem. Soc,* Vol. 97, No 17, (August 1975), pp. 4884-4899, ISSN 0002-7863

Helmholz, L. & Kruh, R.F. (1952). The Crystal Structure of Cesium Chlorocuprate, Cs_2CuCl_4, and the Spectrum of the Chlorocuprate Ion. *J. Amer. Chem. Soc.*, Vol. 74, No 5, (March 1952), pp. 1176-1181, ISSN 0002-7863

Honda, M., Katayama, C., Tanaka, J. & Tanaka, M. (1985). Structure of the 2/1 Complex Dibenzotetrathiafulvalenium Hexabromodicuprate(II), $2C_{14}H_8S_4^{+}\cdot Cu_2Br_6^{2-}$. *Acta Crystallogr., Sect. C.*, Vol. C41, No 5, (May, 1985), pp. 688-690, ISSN 0108-7673

Khan, M.A. & Schwing-Weill, M.J. (1976). Stability and Electronic Spectra of the Copper(II) Chloro Complexes in Aqueous Solutions. *Inorg. Chem.*, Vol. 15, No 9, (September 1976), pp. 2202-2205, ISSN 0020-1669

Koval'chukova, O.V., Stash, A.I., Strashnova, S.B. & Zaitsev, B.E. (2009a). Syntheses, Structures, and Geometric and Spectral Characteristics of the Hexachlorodicuprate(II) Complexes with the Protonated 4-Azafluoren-9-one Derivatives: The Crystal and Molecular Structures of Bis(4-Azafluoren-9-onium) Hexachlorodiaquadicuprate(II) Dihydrate. *Russian Journal of Coordination Chemistry.* Vo. 35, No 1 (January 2009), pp. 6-12, ISSN 1070-3284

Koval'chukova, O.V., Strashnova, S.B., Stash, A.I., Bel'skii, V.K. & Zaitsev, B.E. (2009b). Geometric and Spectral Characteristic of the Tetrahalocuprate(II) Complexes $(HL)_2CuX_4$ (X = Cl, Br). Crystal and Molecular Structures of Bis(2-Methylimidazolium) Tetrabromocuprate(II). *Russian Journal of Coordination Chemistry.* Vo. 35, No 7 (July 2009), pp. 505-512, ISSN 1070-3284

Kovalchukova, O.V., Palkina, K.K., Strashnova, S.B. & Zaitsev, B.E. (2008a). Synthesis, Structure, Geometrical, and Spectral Characteristics of the $(HL^{n})_2[CuCl_4]$ Complexes. Crystal and Molecular Structure of Bis(2-methylimidazolium) Tetrachlorocuprate(II). *Russian Journal of Coordination Chemistry.* Vo. 34, No 11 (November 2008), pp. 830-836, ISSN 1070-3284

Kovalchukova, O.V., Palkina, K.K., Tung, T.T., Strashnova, S.B., Zaitsev, B.E., Polyakova, N.V., & Levov, A.N. (2007). Interaction of Copper(II) Halides with 4-Azafluorene Derivatives in Neutral and Acid Media. Crystal and Molecular Structure of 4-Aza-9-oxofluorenium Tetrabromocuprate Hydrate $(HL^4)_2CuB_4\cdot H_2O$. *Russian Journal of Inorganic Chemistry,* Vol. 52, No. 5, (May 2007), pp. 733-742, ISSN 0036-0236

Kovalchukova, O.V., Stash, A.I., Strashnova, S.B., Belsky, V.K., Tung, T.T. & Zaitsev, B.E. (2008b). Interaction of Copper(II) Halides with 4-(Piperidyl-1)-2-Phenylpyrido[2,3-*a*]anthraquinone-7,12 (*L*) in Acidic Media: Crystal Structure and Spectral Characteristics of $(HL)_2[Cu_2Cl_6]$ and $(HL)[CuBr_2]$. *Crystallography Reports,* Vol. 53, No 3, (March 2008), pp. 451-455, ISSN 1063-7745

Kovalchukova, O.V., Stash, A.I., Strashnova, S.B., Romashkina, E.P. & Zaitsev, B.E. (2010). Synthesis and the Crystal and Molecular Structures of $(H_3L\cdot Cl)[CoCl_4]$ and $H_2L[CuBr_4]$ (*L* Is 2,4,6-Tri(*N,N*-Dimethylamino) methylphenol). *Crystallography Reports,* Vol. 55, No 3, (March 2010), pp. 428-432, ISSN 1063-7745

Kuz'mina, N.E., Palkina, K.K., Polyakova, N.V., Golubev, I.N., Medvedev, A.N., Koval'chukova, O.V., Strashnova, S.B., Zaitsev, B.E., Levov, A.N., & Toze, F. (2002). Synthesis, Structure, and Magnetic Properties of Bis(1-amino-9-hydroxy-4-azafluorenium) Hexachlorocuprate(II) $Cu_2Cl_6(C_{12}H_{11}N_2O)_2$. *Russian Journal of Inorganic Chemistry*, Vol. 47, No. 11, (Nowember 2002), pp. 1693-1698, ISSN 0036-0236

Kuz'mina, N.E., Palkina, K.K., Polyakova, N.V., Golubev, I.N., Medvedev, A.N., Kovalchukova, O.V., Strashnova, S.B., Mordovina, N.I., Nikitin, S.V. & Zaitsev, B.E. (2003). Synthesis, Structure, and Magnetic Properties of a Copper(II) Chloride Complex with 3-Amino-2-methyl-4*H*-pyrido[1,2-*a*]pyrimidin-4-one: The Crystal Structure of $Cu_2Cl_6(C_9H_{10}N_3O)_2 \cdot (C_9H_{10}N_3O)Cl$. *Russian Journal of Inorganic Chemistry*, Vol. 48, No. 2, (February 2003), pp. 205-211, ISSN 0036-0236

Landee, C.P., Djili, A., Mudgett, D.F., Newhall, M., Place, H., Scott, B. & Willett, R.D. (1988). Alternating Exchange in Homonuclear Ferrimagnetic Linear Chains. Tetrakis(tetramethylene sulfoxide) copper(II) Hexahalodicuprate(II) (Halo = Chloro, Bromo): Crystal Structures and Magnetic Susceptibilities. *Inorg. Chem.*, Vol. 27, No 4, (February 1988), pp. 620-627, ISSN 0020-1669

Long, G.S., Wei, M. & Willett, R.D. (1997). Crystal Structures and Magnetic Properties of a Novel Layer Perovskite System: (3-Picoliniumylammonium) CuX_4 (X= Cl, Br). *Inorg. Chem.*, Vol. 36, No 14, (July 1997), pp. 3102-3107, ISSN 0020-1669

Luque, A., Sertucha, J., Castillo, O. & Roman, P. (2001). Crystal packing and physical properties of pyridinium tetrabromocuprate(II) complexes assembled via hydrogen bonds and aromatic stacking interactions. *New J. Chem.*, Vol. 25, No 11, (November 2001), pp. 1208-1214, ISSN 1144-0546

Marcotrigiano, G., Menabue, L. & Pellacani, G.C. (1979). Tetrahalo and (mixed-tetrahalo) cuprates of the 1-methyl- and 2-methylpiperazinium dications. Hydrogen bonding effects on the coordination geometry of the CuX_4^{2-} anions. *J. Coord. Chem.*, Vol. 9, No 3, pp. 141-149, ISSN 0095-8972

McDonald, R.G., Riley, M.J. & Hitchman MA. (1988). Angular Overlap Treatment of the Variation of the Intensities and Energies of the d-d Transitions of the $CuCl_4^{2-}$ Ion on Distortion from a Planar toward a Tetrahedral Geometry: Interpretation of the Electronic Spectra of Bis(N-benzylpiperazinium) Tetrachlorocuprate(II) bis(hydrochloride) and N-(2-Ammonioethyl)morpholinium Tetrachlorocuprate(II). *Inorg. Chem.*, Vol. 27, No 5, (March 1988), pp. 895-900, ISSN 0020-1669

Morosin, B. & Lingafelter, E.C. (1960). The crystal structure of cesium tetrabromocuprate(II). *Acta Crystallogr.*, Vol. 13, No 10, (October 1960), pp. 807-809, ISSN 0108-7673

Murphy, B. & Hathaway B. (2003). The Stereochemistry of the Copper(II) Ion in the Solid-State - Some Recent Perspectives Linking the Jahn-Teller Effect, Vibronic Coupling, Structure Correlation Analysis Structural Pathways and Comparative X-ray Crystallography. *Coord. Chem. Rev.*, Vol. 243, No 1/2, (August 2003), pp. 237-262, ISSN 0010-8545

Murray-Rust, P. (1975). N,N'-Dimethyl-4,4'-bipyridylium (Paraquat) Hexachlorodicuprate(II). *Acta Crystallogr. Sect. B.*, Vol.B31, No 6, (June 1975), pp. 1771-1772, ISSN 0108-7673

O'Bannon, G. & Willett R.D. (1986). A Redetermination of the Crystal Structure of NH_4CuCl_3 and a Magnetic Study of NH_4CuX_3, (X = Cl, Br). *Inorg. Chim. Acta,* Vol. 53, No 3, (March 1981), pp. L131-L132, ISSN 0020-1693

O'Brien, S., Gaura ,R.M., Landee, CP., Ramakhrishna, B.L. & Willett, R.D. (1988). Magneto-Structural Correlations in Chains of Bifolded $Cu_2Cl_6^{2-}$ Dimers. *Inorg. Chim. Acta,* Vol. 141, No 1, (January 1988), pp. 83-89, ISSN 0020-1693

Place, H. & Willett, R.D. (1994). Bis(4-aminopyridinium) Hexabromodicuprate(II) Diaquatetrabromodicopper(II). *Acta Crystallogr., Sect. C,* Vol. C50, No 6, (June 1994), pp. 862-864, ISSN 0108-7673

Place, H. & Willett, R. (1988). Structure of N-Benzylpiperazinium Tetrabromocuprate(II) Hydrate and Bis[methyl(2-phenylethyl)ammonium] Tetrabromocuprate(II). *Acta Crystallogr., Sect. C.,* Vol. C44, No 1, (January 1988), pp. 34-38, ISSN 0108-7673

Place, H. & Willett, R.D. (1987a). Structure of Catalytically Related Species Involving Copper (II) Halides. III. 2-Amino-5-bromo-3-methylpyridinium 2-Amino-3-methylpyridinium Tetrabromocuprate(II). *Acta Crystallogr., Sect. C,* Vol. C43, No 8, (August 1987), pp. 1497-1500, ISSN 0108-7673

Place, H. & Willett, R.D. (1987b). Structure of Bis(2-amino-5-methylpyridinium) Tetrachlorocuprate (II) and Bis(2-amino-5-methylpyridinium) Tetrabromocuprate (II). *Acta Crystallogr., Sect. C,* Vol. C43, No 6, (June 1987), pp. 1050-1053, ISSN 0108-7673

Rakitin, Yu.V. & Kalinnikov, V.T. (1994). *Modern magnetochemistry.* Publishing House "Nauka", ISBN 5-02-024719-7, St-Petersburg

Roberts, S.A., Bloomquist, D.R., Willett, R.D. & Dodgen, H.W. (1981). Thermochromic Phase Transitions in Copper(II) Halide Salts. 1. Crystal Structure and Magnetic Resonance Studies of $(CH_3)_2CHNH_3CuCl_3$. *J. Amer. Chem. Soc.,* Vol. 103, No 10, (May 1981), pp. 2603-2610, ISSN 0002-7863

Rubenacker, G.V., Haines, D.N. & Drum Heller, J.E. (1984). Magnetic properties of the alkanediammonium copper halides. *J. Magn. Magn. Mat.,* Vol. 43, No 3, (March 1984), p. 238-242, ISSN 0304-8853

Schlemper, E.O., Patterson, C, & Stunkel J.A. (1989). Structure Containing Binuclear Cationic Cu^{II} Complexes of the 2-Methyl-l,3-di-2-pyridyl-2-propanol Monoanion as the Tetrachlorocuprate(II) Salt. *Acta Crystallogr., Sect. C,* Vol. C45, No 2, (February 1989), pp. 199-201, ISSN 0108-7673

Schluetes, A.W., Jacobson, R.A. & Rundle, R.E. (1966). A Redetermination of the Crystal Structure of $CsCuCl_3$. *Inorg. Chem.,* Vol. 5, No 2, (February 1966), pp. 277-280, ISSN 0020-1669

Scott, B. & Willett, R.D. (1991). A Copper(II) Bromide Dimmer System Exhibiting Piezochromic and Thermochromic Properties: The Crystal Structure and Electronic Spectroscopy of the Two Room-Temperature Phases of Bis(tetrapropylammonium)

Hexabromodicuprate(II). *J. Amer. Chem. Soc.*, Vol. 113, No 14, (July 1991), pp. 5253-5258, ISSN 0002-7863

Scott, B., Geiser, U., Willett, R.D., Patya,l B., Landee, C.P., Greeney, R., Manfredini, T., Pellacani, G.C., Bonamartini, C.A. & Battaglia, L.P. (1988). Structural and Magnetic Properties of Dimorpholinium Hexahalodicuprate(II) Salts: Study of Two Planar $Cu_2X_6^{2-}$ Dimmers. *Inorg. Chem.*, Vol. 27, No 14, (July 1988), pp. 2454-2460, ISSN 0020-1669

Smith, D.W. (1976). Chlorocuprates(II). *Coord. Chem. Rev.*, Vol. 21, No 2/3, (December 1976), pp. 93-157, ISSN 0010-8545

Sundberg, M.R., Kivekas, R., Ruiz, J., Moreno, J. M. & Colacio, E. (1992). Crystal Structure and Spectroscopic Properties of a Novel Mixed-Valence Compound Containing Tetrabromocuprate(II) and Tribromocuprate(I) Anions. *Inorg. Chem.*, Vol. 31, No 9, (April 1992), pp. 1062-1066, ISSN 0020-1669

Textor, M., Dubier, E. & Oswald, H.R. (1974). Crystal and Molecular Structure of Tetraphenylphosphonium Trichlorocuprate(II), $(C_6H_5)_4PCuC1_3$, a Complex Containing Nonplanar Dimeric $[Cu_2C1_6]^{2-}$ Units. *Inorg. Chem.*, Vol. 13, No 6, (June 1974), pp. 1361-1365, ISSN 0020-1669

Tichý, K., Benes, J., Hälg, W. & Arend, H. (1978). Neutron diffraction study of twinned crystals of ethylenediammonium copper tetrachloride and ethylenediammonium manganese tetrachloride. *Acta Crystallogr., Sect. B.,* Vol. B34, No 10, (October 1978), pp. 2970-2981, ISSN 0108-7673

Vossos, P.H, Fitzwater, D.R. & Rundle, R.E. (1963). The Crystal Structure of $LiCuCl_3{\cdot}2H_2O$. *Acta Crystallogr.*, Vol. 16, No 10, (October 1963), pp. 1037-1045, ISSN 0108-7673

Wasson, J.R., Hall, J.W, Richardson, W. & Hatfield, W.E. (1977). Imidazolium, Nicotinium, and Quinidinium Tetrachlorocuprates(II). *Inorg. Chem.*, Vol. 16, No 2, (February 1977), pp. 458-461, ISSN 0020-1669

Willet, R.D, Place, H. & Middleton M. (1988). Crystal structures of three new copper(II) halide layered perovskites: structural, crystallographic, and magnetic correlations. *J. Amer. Chem. Soc.*, Vol. 110, No 26, (December 1988), pp. 8639-8650, ISSN 0002-7863

Willett, R.D. & Chow, C. (1974). The Crystal Structure of $(Ph_4As)CuC1_3$ and the Molecular Structure and Electronic Spectrum of a Non-planar $Cu_2Cl_6^{2-}$ Ion. *Acta Crystallogr., Sect. B.*, Vol. B30, No 1, (January 1974), pp. 207-214, ISSN 0108-7673

Willett, R.D. & Geiser, U. (1984). Structural Characteristics of $ACuCl_3$ Salts. *Croat. Chem. Acta*, Vol. 57, No 3, pp. 737-747, ISSN 0011-1643

Willett, R.D. (1966). Crystal Structure and Optical Properties of $(CH_3)_2NH_2CuC1_3$. *J. Chem. Phys.*, Vol. 44, No 1, (January 1966), pp. 39-42, ISSN 0021-9606

Willett, R.D. (1990). Structures of the Antiferrodistortive Layer Perovskites Bis(phenethylammonium) Tetrahalocuprate (II), Halo = Cl^-, Br^-. *Acta Crystallogr., Sect. C,* Vol. C46, No 4, (April 1990), pp. 565-568, ISSN 0108-7673

Willett, R.D., Dwiggens, C.J., Kruh, R.F. & Rundle, R.E. (1963). Crystal Structures of $KCuCl_3$ and NH_4CuCl_3. *J. Chem. Phys.*, Vol. 38, No 10, (October 1963), pp. 2429-2436, ISSN 0021-9606

Willett, R.D., Gatteschi, D. & Kahn, O. (1983). Magneto-Structural Correlation in Exchange Coupled Systems. *Proceedings of Proc. of the NATO Advanced Study Inst, on Magneto-Structural Correlation in Exchange Coupled Systems*, Castiglione della Pescaia, Italy, June, 1983

9

Ruthenium(II)-Pyridylamine Complexes Having Functional Groups via Amide Linkages

Soushi Miyazaki[1] and Takahiko Kojima[2]
[1]*Research Center for Materials Science, Nagoya University*
[2]*Department of Chemistry, University of Tsukuba*
Japan

1. Introduction

Functionalization of metal complexes by introduction of functional groups has been recognized to be important toward the development of further functionality of metal complexes, including ion sensing, molecular recognition, and selective catalysis.

Convergence of functional groups into certain direction and appropriate spatial arrangement can be achieved by coordination of metal ions to ligands with those groups to perform novel functions that cannot be achieved by organic ligand molecules for themselves. This strategy allows us to access multifunctional molecules more easily than that with well-designed organic molecules in terms of synthetic availability.

Ruthenium complexes bearing chelating pyridylamine ligands are robust enough to hold those ligands in the coordination spheres for the convergence of functional groups attached to the ligands and to maintain their appropriate spatial geometry. We have used tris(2-pyridylmethyl)amine (TPA) and its derivatives which coordinate to the ruthenium ion as tetradentate ligands. Introduction of functional groups to the 6-position of pyridine rings of TPA can provide additional functionality for ruthenium-TPA complexes (Figure 1). The concept, i.e., the introduction of amide groups at the 6-positions of pyridine rings in TPA, has been originally introduced by Masuda and coworkers to construct a hydrophobic and sterically protected environment in copper complexes by using pivaloylamide groups (Harata *et al.*, 1994, 1995, 1998; Wada *et al.*, 1998). They have succeeded in a number of important metal complexes in bioinorganic chemistry. Inspired by their works, we have developed our concept to functionalize ruthenium-TPA complexes by introducing various functional groups via amide linkages. In our case, the ruthenium complexes bearing tri-substituted TPA is not suitable for functionalization due to its large steric crowding. Therefore we have applied bisamide and monoamide-TPA as ligands. In this chapter, we will present an overview of a chemistry of ruthenium complexes bearing bisamide-TPA and monoamide-TPA as ligands and their characteristics.

2. Convergence of hydrophobic functional groups in the coordination sphere of ruthenium complexes

According to the strategy mentioned above, we introduced hydrophobic groups to TPA toward molecular recognition based on van der Waals interactions in the coordination sphere of Ru(II) complexes. The TPA derivatives synthesized were those having 1-naphtyl, 2-naphtyl, and isobutyl groups via amide linkages (*N,N*-bis(6-(1-naphthoylamide)-2-pyridylmethyl)-*N*-(2-pyridylmethyl)amine = $(1\text{-Naph})_2$-TPA, *N,N*-bis(6-(2-naphthoyl-amide)-2-pyridylmethyl)-*N*-(2-pyridylmethyl)amine = $(2\text{-Naph})_2$-TPA, *N,N*-bis(6-(iso-butyrylamide)-2-pyridylmethyl)-*N*-(2-pyridyl-methyl)amine = $(\text{Isob})_2$-TPA, *N*-(6-(1-naphthoylamide)-2-pyridylmethyl)-*N,N*-bis(2-pyridylmethyl)amine = 1-Naph-TPA) and their Ru(II) complexes ($[\text{RuCl}((1\text{-Naph})_2\text{-TPA})]\text{PF}_6$ (**1**), $[\text{RuCl}((2\text{-naph})_2\text{TPA})]\text{PF}_6$ (**2**), $[\text{RuCl}((\text{Isob})_2\text{-TPA})]\text{PF}_6$, (**3**), $[\text{RuCl}((1\text{-Naph})\text{-TPA})(\text{DMSO})]\text{PF}_6$ (**4**)) were also prepared, respectively (Figure 1) (Kojima *et al.*, 2000, 2004a, 2005).

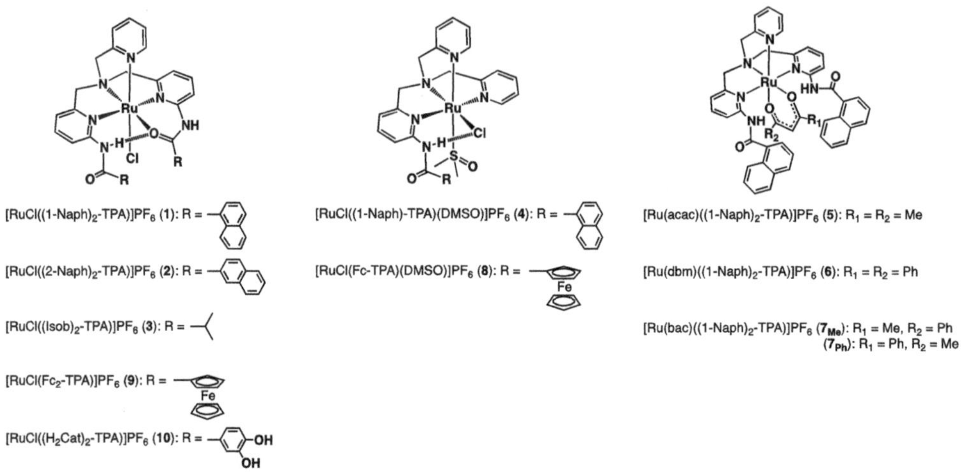

Fig. 1. Structures of Ru-bisamide-TPA and Ru-monoamide-TPA complexes

2.1 Crystal structures of ruthenium(II) complexes bearing TPA with 1-naphthyl, 2-naphthyl, and isopropyl groups via amide linkages

The crystal structures of **1-3** are shown in Figure 2. Those Ru(II)-bisamide-TPA complexes have basically the same coordination environment around the Ru(II) centers. The bisamide-TPA ligands in those complexes coordinate to the Ru(II) ions as pentadentate ligands by the chelation involving three pyridine nitrogen atoms, one tertiary amino nitrogen, and one amide oxygen. Besides those, one chloride ion binds to the Ru(II) center at the *trans* position to the unsubstituted pyridine ring, providing distorted octahedral environments of the Ru(II) complexes. The amide oxygen coordinates to the *trans* site to the tertiary amino nitrogen to afford an asymmetric geometry of the complex. The substituted pyridine rings are located at *trans* positions to each other, and the N-H hydrogen atoms of the uncoordinated amide moiety forms intramolecular hydrogen bonding with the coordinated oxygen atom. The hydrophobic groups were converged and fixed into one direction due to the intramolecular hydrogen bond between two amide arms.

In the structure of **1** (Figure 2a), the bond lengths between the Ru(II) center and the nitrogen atoms of the substituted pyridines are different from the Ru(II) complexes with unsubstituted TPA, reflecting the asymmetric coordination environment due to the coordination of amide oxygen (1.992(5) Å for Ru1-N3 and 2.125(5) Å for Ru1-N4). The dihedral angles between the amide moieties and the amide-linked pyridine rings are 13.4° for the coordinated amide group and 4.3° for the uncoordinated group, respectively. Compared to those dihedral angles, the naphthalene rings show larger dihedral angels with respect to the amide moieties (39.3° for the coordinated amide and 45.6° for the uncoordinated amide). The coordinated amide oxygen and the uncoordinated amide N-H hydrogen forms the intramolecular hydrogen bond with the distance of 3.036(6) Å for O1···N6 to converge the two naphthyl groups. The two naphthalene rings are close enough to form intramolecular π-π interactions as represented by the shortest contact of 3.36(1) Å for C29-C34 and other interatomic distance are listed in Table 1, and are almost parallel to each other with the dihedral angle of 5.9°, indicating the formation of a face-to-face π-π interaction (Figure 3a) (Hunter & Sanders, 1990; Janiak, 2000). Complex **2** bearing two 2-naphthyl groups exhibits almost the same structural features as those of **1** (Figure 2b). Dihedral angles between the amide-linked pyridine ring and the amide plane were calculated to be 10.4 and 20.7° for the coordinated and the uncoordinated ones, respectively. The distortion between the amide planes and the naphthalene rings (20.6° for the coordinated amide and 20.0° for the uncoordinated amide) are smaller than those observed for **1**, probably due to the lack of steric hindrance derived from the peri hydrogen atoms in **1**. The distance of the hydrogen bond between the coordinated amide oxygen and the

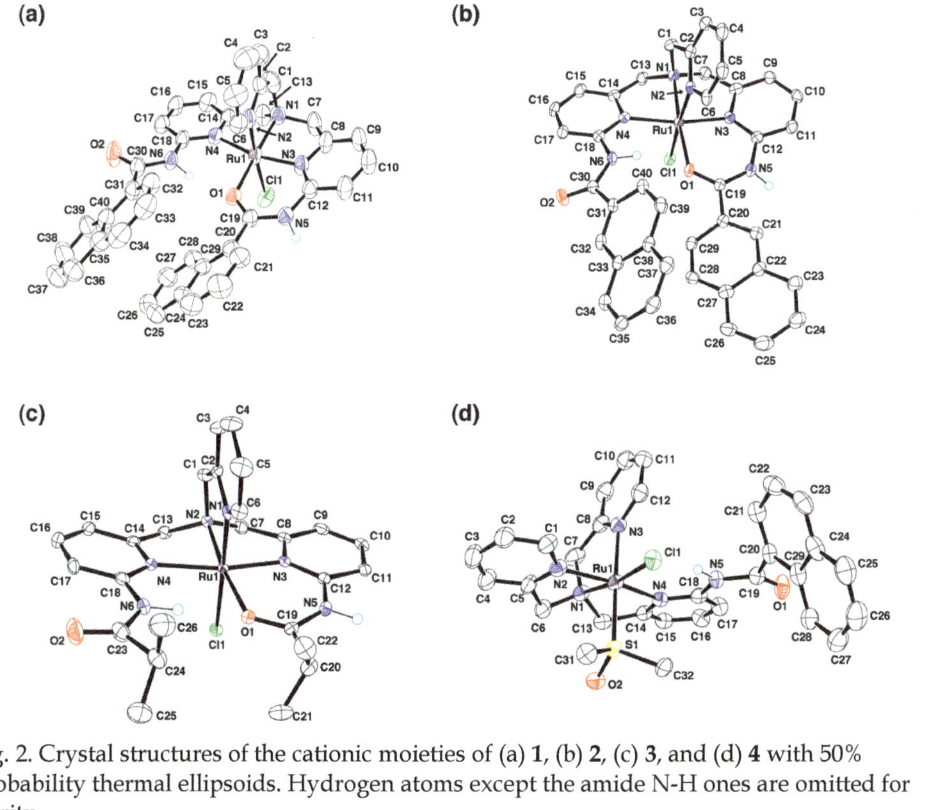

Fig. 2. Crystal structures of the cationic moieties of (a) **1**, (b) **2**, (c) **3**, and (d) **4** with 50% probability thermal ellipsoids. Hydrogen atoms except the amide N-H ones are omitted for clarity.

	1		2
C20···C33	3.68(1)	C26···C32	3.699(4)
C24···C34	3.65(1)	C26···C39	3.528(4)
C24···C35	3.50(1)	C26···C40	3.496(4)
C24···C40	3.43(1)	C27···C31	3.466(4)
C25···C35	3.63(1)	C27···C37	3.596(4)
C25···C40	3.59(1)	C27···C38	3.398(4)
C26···C38	3.69(1)		
C26···C39	3.57(1)		
C27···C31	3.61(1)		
C28···C32	3.64(1)		
C28···C33	3.563(9)		
C29···C33	3.65(1)		
C29···C34	3.36(1)		

Table 1. Interatomic Distances (Å) for Intramolecular π–π Interactions in **1** and **2**.

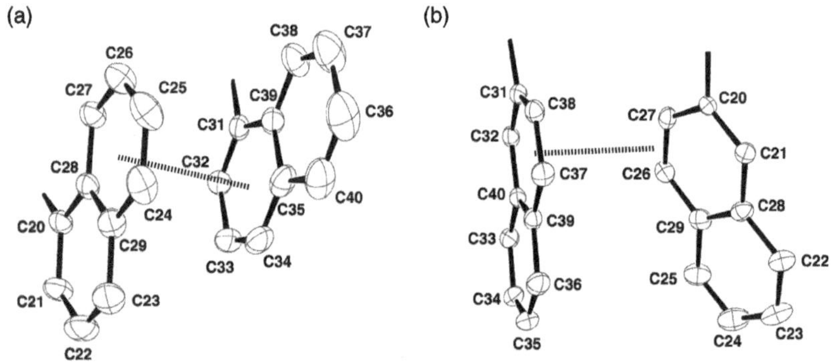

Fig. 3. Intramolecular π-π interactions between two naphthalene rings of (a) **1** and (b) **2**.

uncoordinated amide N-H moiety is 2.977(3) Å (O1···N6). Two naphthalene rings were converged with the shortest contact of 3.398(4) Å for C27-C38. However, two naphthalene rings of **2** show a larger dihedral angle of 27.1° than that (5.9°) of **1**, indicating a sort of edge-to-face π-π interaction (Figure 3b and Table 1). In this complex, the 2-naphthyl groups exhibit fluxional behavior and the thermodynamic parameters of the π-π interaction has been estimated to be $\Delta H° = -2.3$ kJ mol^{-1}; thus, $\Delta G° = -0.9$ kJ mol^{-1} and $\Delta S° = -7.7$ J mol^{-1} K^{-1} at 233 K in CD_3CN.

Concerning the crystal structure of **3** having two isopropyl groups (Figure 2c), two substituted pyridine rings occupy the *trans* sites to each other with the intramolecular hydrogen bond between the coordinated amide oxygen and the N-H hydrogen of the uncoordinated amide moiety (3.041(6) Å) in a similar manner to those in **1** and **2**. This observation suggests that the formation of the intramolecular hydrogen bond provides the asymmetric coordination environments of those complexes and convergence of the naphthalene rings to form intramolecular van der Waals interactions.

A Ru(II) complex having a monoamide-TPA ligand has been also synthesized employing (1-Naph)-TPA as a ligand. The crystal structure of the cation moiety of **4** is presented in Figure 2d. The (1-Naph)-TPA ligand coordinates to the Ru(II) center as a tetradentate ligand without the coordination of its amide oxygen as observed in Ru(II)-bisamide-TPA complexes mentioned above. The amide-substituted pyridine ring occupies the *trans* position to one of the unsubstituted pyridine rings. The chloride anion binds to the Ru(II) ion at the *trans* site to the tertiary amino group; the DMSO molecule binds to Ru(II) ion with the S atom at the *trans* site to the unsubstituted pyridine ring. The intramolecular hydrogen bond can be observed between the N-H group of the amide moiety and the chloride ligand (Cl1···N5, 3.135(3) Å). Complex **4** shows a larger dihedral angle between the amide plane and the 1-naphthyl group of 57.65(4)° than those (39.3° and 45.6°) of **1**.

As an important characteristics of the Ru(II)-bisamide-TPA complexes **1** - **3**, the coordinated amide moiety can exhibit reversible deprotonation and protonation, as depicted in Scheme 1, to regulate the redox potential of the Ru^{II}/Ru^{III} couple of the ruthenium center: The redox potential can be reversibly controlled in the range of ~500 mV (500 mV for **1**, 450 mV for **2**, and 480 mV for **3**) in CH_3CN.

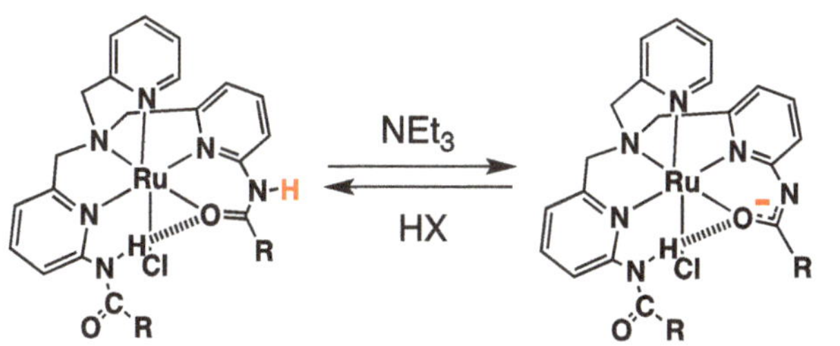

R = 1-naphthyl, 2-naphthyl, and isopropyl

Scheme 1. Reversible deprotonation and protonation of the coordinated amide moiety.

2.2 Crystal structures of ruthenium(II)-TPA-β-diketonato complexes and intramolecular rearrangement of coordination geometry of β-diketonato ligands

The convergence of functional groups to form hydrophobic environments in the coordination sphere of Ru(II) complexes has been achieved by introduction of naphthoylamide groups to the TPA ligand. Then, the ability to form π-π and CH/π (Nishio & Hirota, 1989) interactions in coordination spheres to regulate the stereochemistry of Ru centers has been examined using the Ru(II)-$(1\text{-Naph})_2$-TPA complex (**1**) and β-diketones, such as acetyl acetone (Hacac), dibenzoylmethane (Hdbm), and benzoyl acetone (Hbac). The reactions between **1** and β-diketones have been performed in ethylene glycol at 100°C in the presence of 2,6-lutidine as a base to afford corresponding β-diketonato complexes, [Ru(acac)($(1\text{-Naph})_2$-TPA)]PF_6 (**5**), [Ru(dbm)($(1\text{-Naph})_2$-TPA)]PF_6 (**6**), and [Ru(bac)($(1\text{-Naph})_2$-TPA)]PF_6 ($\mathbf{7_{Me}}$ and $\mathbf{7_{Ph}}$) as depicted in Figure 1 (Kojima *et al.*, 2004b). In those complexes, in contrast to the case of **1**, the $(1\text{-Naph})_2$-TPA ligand coordinates to the Ru(II) center as a tetradentate ligand by the TPA moiety, and the β-diketonato ligands bind to it as bidentate monoanionic ligands. Both of the amide oxygens in the $(1\text{-Naph})_2$-TPA ligand direct to the opposite sides from the metal center and both of the amide N-H hydrogens form intramolecular hydrogen bonds with one of the oxygen atoms of the β-diketonato ligand.

In the structure of **5** (Figure 4a), one of the methyl group of the acac ligand is sandwiched between the two naphthalene rings of the (1-Naph)$_2$-TPA ligand and forms CH/π interactions with the shortest distance of 3.34 Å for C41···C32 (Table 2 and Figure 5a). The formation of the CH/π interactions is also demonstrated in solution by ^{1}H NMR measurements in CD_2Cl_2. The singlet assigned to the included methyl protons of the acac ligand shows a large upfield shift to δ= -0.27 ppm compared to that of the nonincluded methyl group (δ= 1.86 ppm) due to the shielding by the π electrons of the naphthalene rings.

Crystal structure of **6** shows inclusion of one of the two phenyl groups of the dbm ligand in the hydrophobic cavity made of the two 1-naphthyl groups (Figure 4b). Close contact can be observed between the included phenyl ring of dbm and the naphthalene rings to form intramolecular π-π interactions (Table 2). The dihedral angles between the included phenyl ring of the dbm ligand and the two naphthalene rings are estimated to be 11.3(7)° and 62.4(8)°, indicating that one of the π-π interactions between them is a face-to-face type and the other isedge-to face (Figure 5b). In ^{1}H NMR measurements in CD_2Cl_2, large upfield shifts of signals assigned to the included phenyl protons compared to those of the nonincluded phenyl group are observed at δ = 5.54, 6.12, 6.50 ppm. This observation indicates the insertion of the phenyl ring into the two naphthalene rings even in solution.

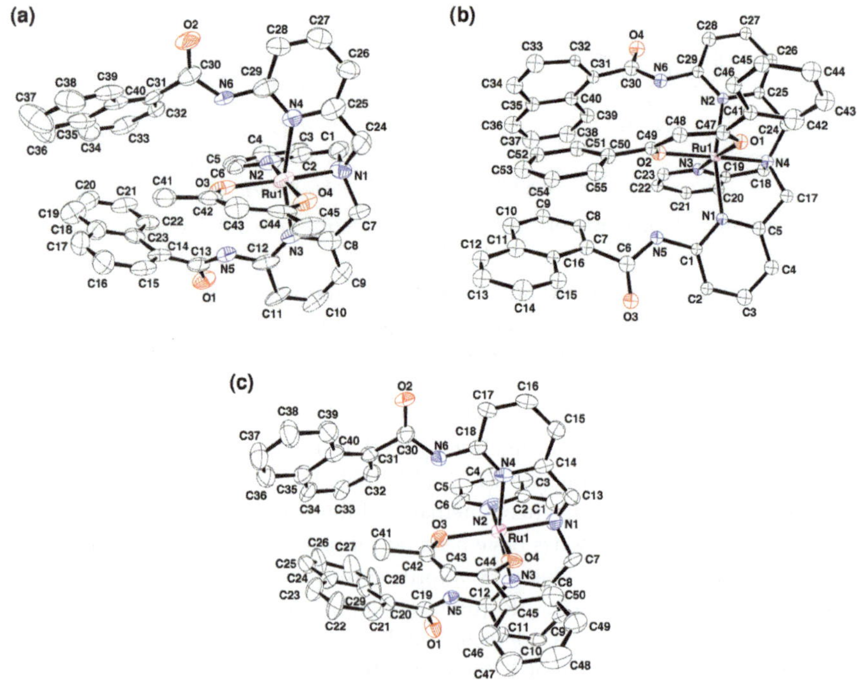

Fig. 4. ORTEP drawings of (a) **5** and (b) **6** with 50 % probability thermal ellipsoids and (c) $\mathbf{7_{Me}}$ with 30 % probability thermal ellipsoid.

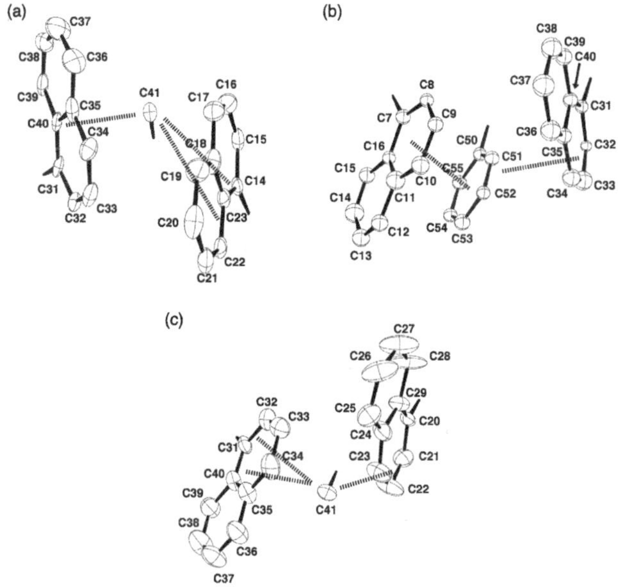

Fig. 5. Intramolecular CH/π and π-π interactions of (a) **5**, (b) **6**, and (c) **7**$_{Me}$.

	5		6		7_{Me}
CH/π interactions		π-π interactions		CH/π interactions	
C41···C14	3.59	C51···C7	3.46	C41···C21	3.38
C41···C22	3.62	C51···C8	3.45	C41···C22	3.41
C41···C23	3.45	C51···C9	3.66	C41···C31	3.51
C41···C32	3.34	C52···C9	3.66	C41···C32	3.66
C41···C33	3.52	C52···C10	3.31	C41···C40	3.64
		C52···C16	3.42		
		C53···C11	3.35		
		C53···C12	3.43		
		C53···C13	3.62		
		C53···C15	3.69		
		C54···C14	3.59		
		C51···C32	3.36		
		C51···C33	3.48		
		C52···C33	3.47		
hydrogen bonds					
O3···N5	2.79	O2···N5	2.87	O3···N5	2.87
O3···N6	2.85	O2···N6	2.92	O3···N6	2.92

Table 2. Interatomic distances (Å) for intramolecular π-π interactions in **5**, **6**, and **7**$_{Me}$.

The structure of **7**$_{Me}$ has been determined as shown in Figure 4c. The methyl group of the bac ligand is also included between the two naphthyl groups as observed in the structure of

5 (Table 2 and Figure 5c). Although the single crystal of $\mathbf{7_{Ph}}$ could not be obtained, the van der Waals interaction between the substituents of bac and naphthalene rings has been indicated by large upfield shift of the methyl signal of $\mathbf{7_{Me}}$ and those of the phenyl signals of $\mathbf{7_{Ph}}$ in their ^{1}H NMR spectra.

The formation of CH/π and π-π interactions in the hydrophobic cavity made of the two naphthyl groups in the Ru(II) coordination sphere has been demonstrated for **5** with acac having two methyl groups and **6** with dbm having two phenyl groups, respectively. Then, the selectivity between π-π and CH/π interactions has been examined by using an asymmetric β-diketone, Hbac, which has both methyl and phenyl groups. After the reaction of **1** with Hbac for 3 h at 100 °C in ethylene glycol, all the starting material was consumed and a mixture of $\mathbf{7_{Me}}$ and $\mathbf{7_{Ph}}$ was obtained at the ratio of 1.8 to 1, accompanying the dissociation of the amide oxygen in **1**. This observation indicates no kinetic selectivity in the coordination mode of the bac ligand to the Ru(II) ion. However, continuous heating of this mixture of the two isomers gave only $\mathbf{7_{Me}}$ due to the selective conversion of $\mathbf{7_{Me}}$ to $\mathbf{7_{Ph}}$, indicating that $\mathbf{7_{Me}}$ is thermodynamically more favored relative to $\mathbf{7_{Ph}}$. The reaction rate of the conversion from $\mathbf{7_{Me}}$ to $\mathbf{7_{Ph}}$ has been revealed to be independent of the Hbac concentration. This result indicates that the isomerization reaction proceeds as a one-directional intramolecular rearrangement. Kinetic analysis of the rearrangement in light of Arrhenius equation has been made to determine the activation energy to be 52 kJ mol^{-1}. A proposed reaction mechanism involving a putative transition state is described in Figure 6 (Kojima *et al.*, 2004b). This intramolecular rearrangement is assumed to proceed without bond rupture and the transition state should involve weakened coordination bonds for the bac ligand.

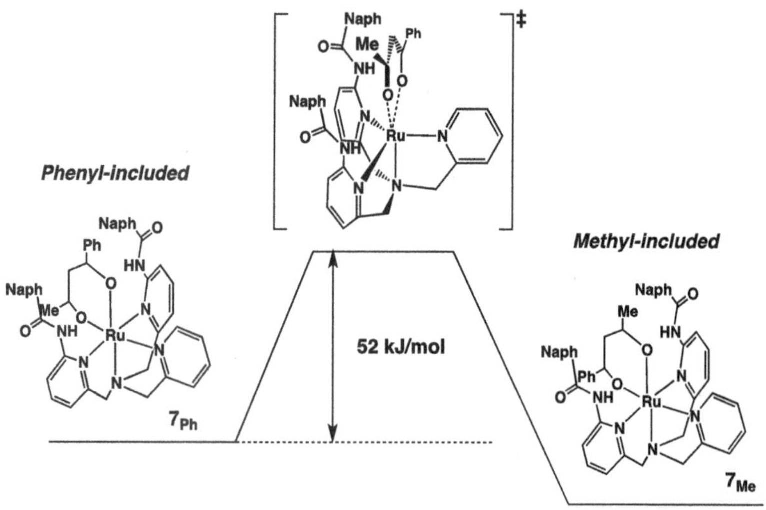

Fig. 6. A proposed reaction mechanism of the selective intramolecular rearrangement from $\mathbf{7_{Me}}$ to $\mathbf{7_{Ph}}$ in ethylene glycol.

We have also examined whether the selectivity of $\mathbf{7_{Me}}$ is derived from electronic or steric effect. According to PM3 calculation, the two oxygen atoms of bac exhibits no significant difference in its negative charges (-0.466 for Ph-C*O*- and -0.476 for Me-C*O*-), suggesting the electronic effect is not a determining factor of the regioselectivity of 7_{Me}. The reaction of **1** with an equimolar mixture of Hacac and Hdbm has been examined to evaluate the steric effect of methyl and phenyl groups on the selective conversion to 7_{Me}. After the reaction in ethylene glycol at 100 °C for 24 h, complexes **5** and **6** were obtained in the ratio of 4.8:1. This accessibility of the methyl group in the hydrophobic cavity of the complex suggests that the steric effect may contribute to the regioselectivity. However, the complete conversion of **6** to **5** has not been observed unlike the intramolecular rearrangement from 7_{Ph} to 7_{Me}. Therefore, the conversion from 7_{Ph} to 7_{Me} is supposed to be consequences of both the steric effect and the thermodynamic preference of CH/π interactions over π-π interactions in the hydrophobic cavity of the Ru(II)-(1-Naph)$_2$-TPA complex, even though superior thermodynamic stability gained by π-π interactions to that obtained by CH/π interactions is generally accepted (Steed & Atwood, 2000). The introduction of the 1-naphtyl groups attached to the TPA ligand via amide linkage has allowed us to regulate the stereochemistry of the ruthenium complex by selective inclusion of the methyl group of the bac ligand by virtue of the intramolecular CH/π interaction as well as the steric effect. The preference of CH/π interaction observed here can be attributed to the polarization of the C-H bond of the methyl group of the bac ligand bound to the positively charged Ru(II) center, making the C-H bond more acidic to strengthen the interaction between the positively polarized hydrogen atom and the negatively charged π-electron cloud of the naphthyl group.

3. Ferrocene-appended TPA derivatives and their ruthenium complexes: Intramolecular magnetic interactions between ruthenium and iron centers induced by oxidations

Ferrocene is a well-known redox-active molecule and has been introduced to many kinds of functional molecules. Among those, ferrocene-containing ligands have been reported on their multiple redox behavior and mixed-valence states toward a construction of redox-responsive functional molecules. Thus, we have prepared ferrocene-appended TPA derivatives (*N*-(6-ferrocenoylamide-2-pyridylmethyl)-*N*,*N*-bis(2-pyridylmethyl)amine = Fc-TPA, and *N*,*N*-bis(6-ferrocenoylamide-2-pyridylmethyl)-*N*-(2-pyridylmethyl)amine = Fc$_2$-TPA) and their Ru(II) complexes. We have focused on intramolecular magnetic interactions between the Fe(III) center in the ferrocenium moiety and the Ru(III) center upon oxidation. (Kojima *et al.*, 2008)

3.1 Crystal structures of a ferrocene-appended TPA derivative and its ruthenium(II) complex

In the crystal of Fc-TPA (Figure 7a), two adjacent molecules form intermolecular hydrogen bonds with each other between the amide N-H and one of unsubstituted pyridine rings (3.042(2) Å for N2···N5′, 160.5° for N2–H1′–N5′). An intermolecular C-H···O hydrogen bond is also found between the amide oxygen and one of unsubstituted pyridine ring (3.334(2) Å for C21···O1; Figure 8a). Intermolecular π-π interactions are observed between a pyridine ring and a substituted cyclopentadienyl (Cp) ring, a nonsubstituted Cp ring and adjacent pyridine rings, and a substitute pyridine ring and nonsubstituted pyridine ring (3.325(4) Å for C14···C7, 3.638(3) Å for C19···C1, 3.618(4) Å for C22···C4, 3.257(3) Å for C12-C28; Figure 8b). The π-planes of the substituted pyridine ring, the amide plane, and the amide-linked Cp ring show a high coplanarity with the dihedral angle of 18.1° between the pyridine plane and the amide plane and 7.4° between the amide plane and the Cp ring due to the extended conjugation of π-systems.

The coordination geometry of [RuCl(Fc-TPA)(DMSO)]PF_6 **(8)** has been revealed to be similar to that of **4** (Figure 7b). The six coordination sites of the Ru(II) center of **8** are occupied by Fc-TPA as a tetradentate ligand, one chloride ion, and one DMSO molecule coordinating via the S atom. The high coplanarity of the substituted pyridine ring–the amide plane-the linked Cp ring observed in the free ligand is retained after the formation of the Ru(II) complex: The dihedral angle between the substituted pyridine ring and the amide plane is 22.8° and that between the amide plane and the Cp ring is 13.7°. An intramolecular hydrogen bond is observed between the chloride and the amide N-H hydrogen. The separation between the Ru(II) center and the Fe(II) center has been determined to be 6.656Å.

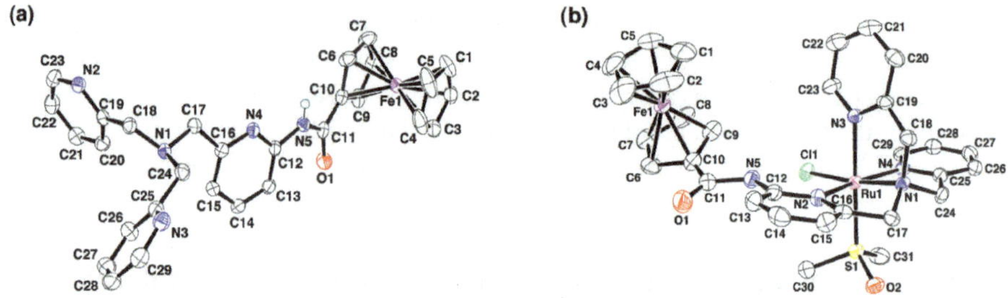

Fig. 7. ORTEP drawings of (a) Fc-TPA and (b) the cationic moiety of **8** with 50% probability thermal ellipsoids. Hydrogen atoms are omitted for clarity.

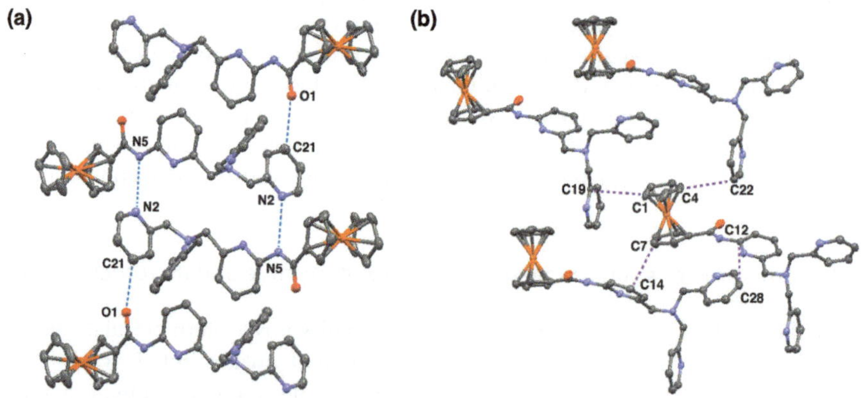

Fig. 8. Intermolecular interactions found in the crystal structure of Fc-TPA: (a) Intermolecular hydrogen bonds and (b) Intermolecular π-π interactions.

3.2 Redox behavior of ferrocene-appended ruthenium(II)-TPA complexes

In order to examine the redox properties of Fc-TPA, Fc_2-TPA, **8**, and [RuCl(Fc_2-TPA)](PF_6) **(9)** that has been proposed to possess the same geometry as those of **1–3** with one coordinated amide oxygen, cyclic voltammetry (CV) has been performed in CH_3CN containing 0.1 M TBAP as an electrolyte at room temperature. Redox potentials were determined relative to the ferrocene/ferrocenium ion couple as 0 V.

The cyclic voltammograms of Fc-TPA and Fc_2-TPA showed reversible redox waves at $E_{1/2}$ = 0.21 V and 0.23 V assigned to the Fc/Fc^+ couples, respectively.

The CV of **8** showed two redox waves at 0.23 V (ΔE = 86 mV) and 0.77 V (ΔE = 159 mV) at 100 mv/s. The first redox wave has been assigned to the Fc/Fc^+ couple of the coordinated Fc-TPA based on the redox potential of the free Fc-TPA. The redox potential of the second one is similar to that of **4** (0.79 V for Ru^{II}/Ru^{III}), therefore the second redox process has been assigned to the Ru^{II}/Ru^{III} couple. The Ru^{II}/Ru^{III} couple has exhibited a change of peak current with altering the scan rate, suggesting that this process is quasi-reversible.

Complex **9** shows reversible redox waves at 0.27 V and 0.46 V at the scan rate of 20 mV/s. The peak current of the first wave is twice as large as the second one. Thus, the first peak can be assigned to an overlapped wave derived from the Fc/Fc^+ couple of the uncoordinated Fc-amide moiety and the Ru^{II}/Ru^{III} couple based on the redox potential of Fc_2-TPA and **3** (0.27 V for Ru^{II}/Ru^{III}) and Ru(II)-bisamide-TPA (0.34 V for **1**, 0.30 V for **2**, and 0.27 V for **3**). The second redox couple was assigned to the Fc/Fc^+ couple of the coordinated Fc-amide moiety. The redox potentials of the two Fc moieties have been separated due to the coordination of the amide oxygen of one of the Fc-amide arms in **9**.

3.3 Magnetic interaction between the Ru(III) center and the Fe(III) center in ferrocene-appended ruthenium(II)-TPA complexes

We examined intramolecular magnetic interactions between the Ru center and the Fe center in **8** and **9** upon oxidation by $[Ru(bpy)_3]^{3+}$ in CH_3CN. Variable-temperature EPR spectra of two-electron oxidized species of **8**, $[Ru^{III}Cl(Fc^+\text{-TPA})^+(DMSO)]^{3+}$, showed a signal derived from a forbidden transition (Δm_s = 2) in a triplet state at g = 4.28 and a signal due to an allowed transition (Δm_s = 1) at g = 2.0038. The obtained EPR spectrum is different from reported EPR signal of Fc^+ with $g_{//}$ = 4.36 and $g_{\perp}$ = 1.30 in acetone at 20 K (Prins & Reinders, 1969) and also the EPR signal of the one-electron oxidized species of Fc-TPA with an intense signal at g = 2.0186 and a weak signal at g = 4.22 in CH_3CN at 2.6 K. The intensity of the signal at g = 4.28 of $[Ru^{III}Cl(Fc^+\text{-TPA})^+(DMSO)]^{3+}$ increased as lowering temperature as shown in Figure 9a. Therefore, the ground state of the two-electron oxidized species should be a triplet state (S = 1). A ferromagnetic coupling constant has been estimated to be J = 13.7 cm^{-1} based on the Bleaney-Bowers equation (Bleaney & Bowers, 1952) using the temperature-dependent change of the intensity of the signal at g = 4.28 (see Figure 9b) (Kojima *et al.*, 2008).

In order to gain structural insights into the triplet state, we have applied DFT calculations to optimize the structure of the two-electron oxidized species of **8** in the S = 1 state. The optimized structure has longer interatomic distance between Ru(III) center and Fe(III) center (7.523 Å) than that observed in the crystal structure of **8** (6.656 Å). The dihedral angle between the amide plane and the Cp ring (42.8°) has enlarged compared to that in the crystal structure of **8**. The calculated spin densities at the Ru(III) center (+0.925) and Fe(III) center (+1.397) suggests the localization of unpaired electrons at those metal centers. We assumed that the ferromagnetic interaction between Ru(III) and Fe(III) centers was induced by a superexchage interaction through the amide linkage (Kojima *et al.*, 2008).

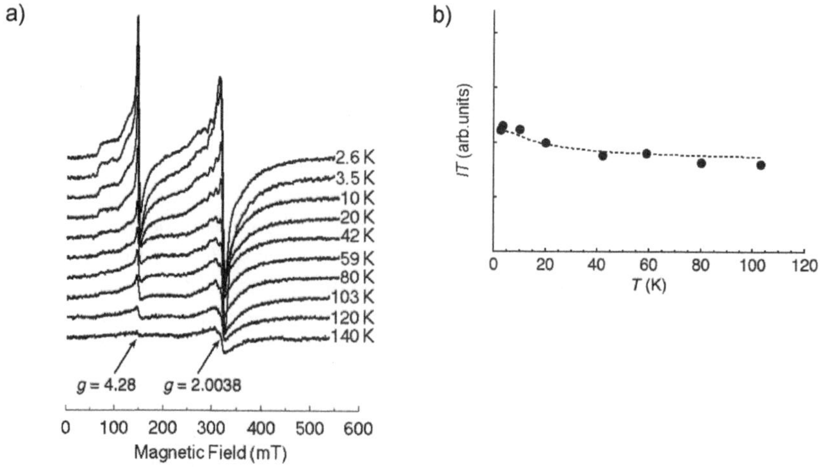

Fig. 9. (a) Variable-temperature EPR spectra of the two-electron oxidized species of **8** in the range of 2.6 - 140 K in CH_3CN. (b) Temperature dependence of *IT* (*I* reporesents the integration intensity of the signal at $g = 4.28$) of the two-electron oxidized species of **8**, together with the curve-fit of the data to the Bleaney-Bowers equation.

EPR measurements of **9** have been also made in CH_3CN at –150°C. In the case of its two-electron oxidation, the uncoordinated Fc and the Ru center of **9** can be oxidized according to the CV measurement, and the product ($[Ru^{III}Cl(Fc_2\text{-TPA})^+]^{3+}$) exhibited EPR signals at g = 2.01 ($\Delta m_s = 1$) and $g = 4.11$ ($\Delta m_s = 2$) as observed in $[Ru^{III}Cl(Fc\text{-TPA})^+]^{3+}$, indicating the formation of a resemble triplet state to that of $[Ru^{III}Cl(Fc\text{-TPA})^+]^{3+}$.

4. Ruthenium(II)-TPA complex having catechol pendants via amide linkages

Catechol and its derivatives have been known to form stable complexes with metal ions. Introduction of catechols to TPA enables us to create a ruthenium-TPA complex having a metal binding site formed by convergence of the catechol moiety. Such compound may allow us to readily create multinuclear metal complexes with multistep redox systems involving electronic interactions among those redox centers. Thus, we have synthesized a catechol-containing TPA (*N,N*-bis[6-{3,4-(dihydroxy)benzamide}-2-pyridylmethyl]-*N*-(2-pyridyl-methyl)amine = $(H_2Cat)_2$-TPA) and its ruthenium complex, $[RuCl((H_2Cat)_2\text{-TPA})]Cl$ (**10**) (Kojima *et al.*, 2010).

4.1 Crystal structure of a ruthenium(II) complex with two catechol pendants via amide linkages

The coordination environment of the Ru(II) center in the structure of **10** is similar to those of other related Ru(II)-bisamide-TPA complexes in terms of the coordination of Cat_2-TPA as a pentadendate ligand involving the coordination of one amide oxygen, as depicted in Figure 10. The chloride ion binds to the Ru(II) center at the *trans* position of the unsubstituted pyridine ring as can be seen in other related complexes (see Figure 2). The coordinated amide oxygen is located at the *trans* site to the tertiary amine, and the pyridine ring linked to the coordinated amide moiety is located at the *trans* position to the other amide

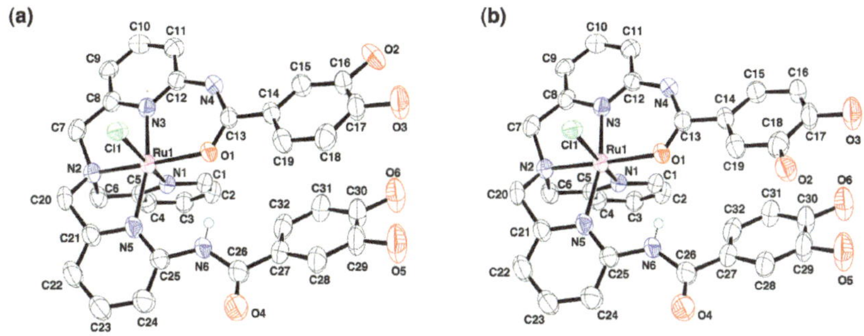

Fig. 10. ORTEP drawings of the cation moieties of two isomers A (a) and B (b) of **10**.

substituted pyridine ring to form intramolecular hydrogen bond (2.938 (7) Å for O1···N6, 148.9(4)° for O1···H-N6). Disorder of the O2 atom can be found in the catechol moiety connected to the coordinated amide arm, giving rise to two conformational (rotational) isomers in the crystal structure (Form A and B in Figure 10). The intramolecular π-π interactions between the two catechol moieties are observed in both isomers with the shortest distance of 3.36 Å. The O3 atom in the catechol rings forms a hydrogen bond with a water molecule of crystallization (2.90 Å for O3···O8). One of the oxygen atoms of the catechol moiety (O2) in the form A binds with O3 of an adjacent molecule through an intermolecular hydrogen bond with the distance of 2.637 Å (Figure 11).

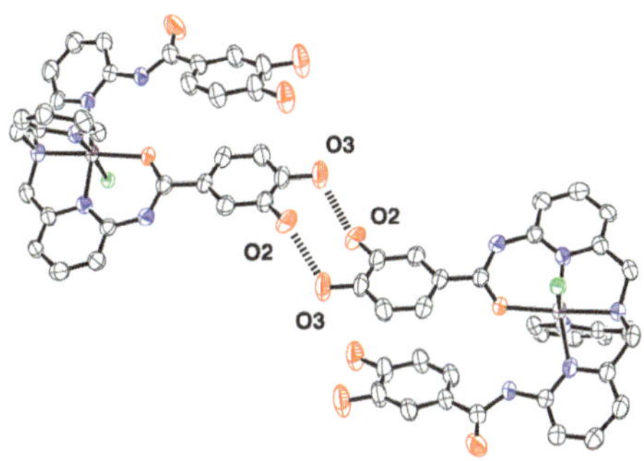

Fig. 11. Intermolecular hydrogen bonds between catechol moieties of form A of **10**.

4.2 Cu(II) binding to the converged catechol moieties to form a unique bimetallic complex

As observed in the crystal structure of **10**, the two independent catechol moieties have been successfully converged into one direction as a metal-binding site. Prior to the examination of metal binding to the catechol moieties, we have shed some lights on the deprotonation processes of **10**. The UV-vis titration of **10** with tetramethylammonium hydroxide (TMAOH) in DMF shows three-step spectral change. We assigned the first-step spectral change observed in the course of the addition of 0-2 equiv of TMAOH to the deprotonation of the coordinated amide N-H as observed for other Ru(II)-bisamide-TPA complexes (see Scheme 1 in Section 2.1) and one of the catechol O-H having lowest pK_a value as a consequence of convergence of catechol moieties (Scheme 2). The second and third deprotonation steps observed in the course of the addition of 2-3 and 3-5 equiv of TMAOH have been assigned to deprotonation of one catechol O-H and remaining two catechol O-H protons, respectively, as shown in Scheme 2.

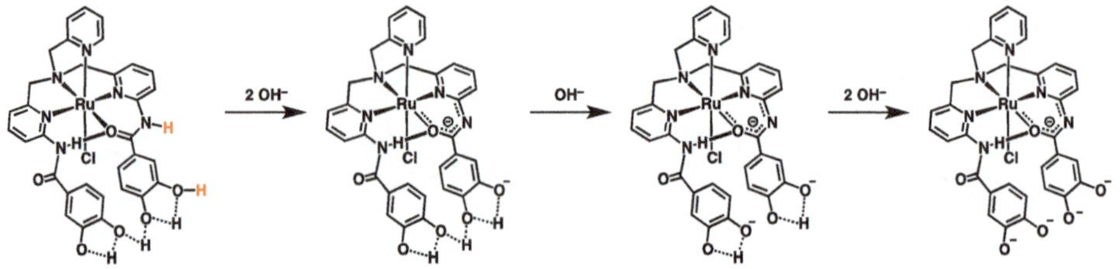

Scheme 2. Deprotonation processes of **10**.

As a precursor for the complexation of the Cu(II) ion to **10**, we have employed $[Cu^{II}(NO_3)_2(TMEDA)]$ (TMEDA = *N,N,N′,N′*-tetramethylethylenediamine) (Pavkovic *et al.* 1977), since the Cu(II) ion can form the most stable complex in light of the Irwing-Williams series. The UV-vis titration of **10** with $[Cu^{II}(NO_3)_2(TMEDA)]$ in MeOH allows us to reveal that the catechol moieties of **10** is capable of binding to only one Cu(II)-TMEDA unit. The reaction between **10** and $[Cu^{II}(NO_3)_2(TMEDA)]$ in MeOH proceeds in the presence of NEt_3. The addition of $TBAPF_6$ to a mixture of **10** and $[Cu^{II}(NO_3)_2(TMEDA)]$ gives a powder of $[RuCl((HCat)_2\text{-}TPA)\{Cu(TMEDA)\}]PF_6$ (**11**). ESI-MS measurements on **11** afforded a peak cluster at m/Z = 906.1 assigned to $[RuCl((HCat)_2\text{-}TPA)\{Cu(TMEDA)\}]^+$, indicating that each catechol moiety releases one proton and the converged two catechol moieties can accept one Cu(II)-TMEDA unit as demonstrated by the spectroscopic titration.

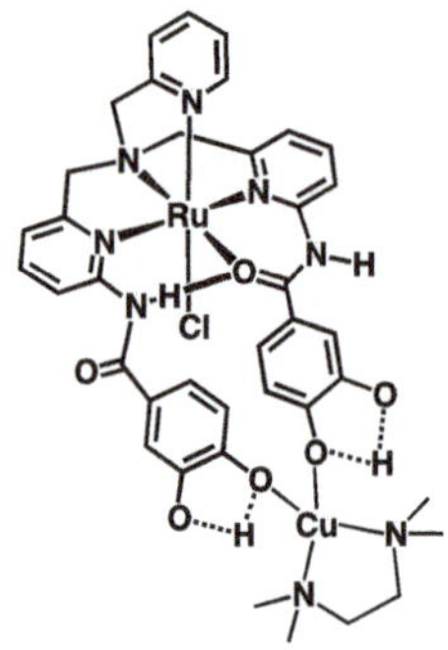

Fig. 12. Putative structure of **11**.

The EPR spectrum of **11** in MeOH at 77 K shows an anisotropic signal at $g_{//}$ = 2.239 and $g_{\perp}$ = 2.060, together with hyperfine coupling constant (hfc) of $A_{//}$ = 184 G due to Cu(II). These parameters are different from those of [Cu(catecholato)(TMEDA)] ($g_{//}$ = 2.241, $g_{\perp}$ = 2.071, and $A_{//}$ = 184 G) (Kodera *et al.*, 2003), even though the signals indicate that both Cu(II) complexes are with a monomeric tetragonal geometry in the $(d_{x^2-y^2})^1$ ground state (Wojciechowski *et al.*, 2009; Nielsen *et al.*, 2008). Considering the deprotonation behavior of catechol moieties, the complex **11** is supposed to have a unique coordination environment of a Cu(II)-catecholate complex as described in Figure 12. DFT calculations at the B3LYP/LANL2DZ level of theory suggest that the optimized structure of **11** shown in Figure 12 is more stable in 3.5 kcal/mol than that of a bimetallic complex holding the Cu-TMEDA unit at one catecholate moiety in a bidentate fashion.

5. Summary

In this chapter, we have presented an overview of ruthenium complexes having TPA derivatives with functional groups via amide linkages. The strategy has allowed us to construct unique second coordination spheres of ruthenium complexes gaining novel functionality. The ruthenium coordination to bisamide-TPA derivatives affords the convergence of functional groups attached by the amide linkage into one direction to form specific environments, such as a hydrophobic cavity. The hydrophobic cavity made of the two naphthyl groups forms noncovalent van der Waals interactions with inserted substituents. Under this situation, we can observe the uni-directional intramolecular rearrangement based on the theromodynamic preference of CH/π interactions to π-π interactions. The introduction of ferrocene moiety to the TPA ligand makes possible intramolecular ferromagnetic interaction between Ru(III) and Fe(III) centers in the ruthenium complex with ferrocene units as pendants. A catechol-appended TPA ligand can provide a preorganized metal binding site consisting of the two catechol moieties on the ruthenium complex to accept one Cu(II) complex in a unique coordination mode. Recently, we have also prepared a TPA derivative having two 1,10-phenanthroline (phen) moieties via amide linkages and its Ru(II) complex: The complex can form two appended Ru(II)-phen complexes and perform cooperative transfer hydrogenation of ketones (Kojima *et al.*, 2011). In those complexes, the intramolecular hydrogen bond formed between the amide moieties can facilitate convergence of the functional groups and stabilize the complexes. This can be a significant advantage for the development of transition-metal-based functional molecules. Thus, the design of Ru-TPA complexes having functional groups via amide linkage will provide a nice opportunity to create cooperative molecular functionality on the basis of characteristics of both the ruthenium center and functional groups to provide a unique second coordination sphere as well as other metal ions attached to the functional groups.

6. References

Bleaney, B., Bowers, K. D. (1952). Anomalous Paramagnetism of Copper Acetate. *Proc. R. Soc. London, Ser. A*, Vol.214, No.1119, pp. 451-465.

Harata, M., Jitsukawa, K., Masuda, H., Einaga, H. (1994). A Structurally Characterized Mononuclear Copper(II)-Superoxo Complex. *J. Am. Chem. Soc.*, Vol.116, No.23, pp. 10817-10818.

Harata, M., Jitsukawa, K., Masuda, H., Einaga, H. (1995). Synthesis and Structure of a New Tripodal Polypyridine Copper(II) Complex That Enables to Recognize a Small Molecule. *Chem. Lett.*, Vol.24, No.1, pp.61-62.

Harata, M., Hasegawa, K.; Jitsukawa, K., Masuda, H., Einaga, H. (1998). Preparations, Structures, and Properties of Copper(II) Complexes with New Tripodal Tetradentate Ligand. *N*-(2-Pyridylmethyl)bis(6-pyvalamido-2-pyridylmethyl)amine, and Reactivities of the Cu(I) Complex with Dioxygen. *Bull. Chem. Soc. Jpn.*, Vol. 71, No.5, pp. 1031-1038.

Hunter, C. A., Sanders, J. K. M. (1990). The Nature of π-π Interactions. *J. Am. Chem. Soc.*, Vol.112, No.14, pp. 5525-5534.

Janiak, C. (2000). A Critical Account on π-π Stacking in Metal Complexes with Aromatic Nitrogen-containing Ligands. *Inorg. Chem.*, Vol.49 No.8, pp. 3737-3745.

Kodera, M., Kawata, T., Kano, K., Tachi, Y., Itoh, S., Kojo, S. (2003). Mechanism for Aerobic Oxidation of 3,5-Di-*tert*-butylcatechol to 3,5-Di-*tert*-butyl-*o*-benzoquinone Catalyzed by Di-μ-hydroxo-dicopper(II) Complexes of Peralkylated Ethylelnediamine Ligands. *Bull. Chem. Soc. Jpn.*, Vol.76, No.10, pp. 1957-1964.

Kojima, T., Hayashi, K., Matsuda, Y. (2000). A Ruthenium(II)-Pyridylamine Complex Showing a Fluxional Intramolecular π-π Interaction. *Chem. Lett.*, Vol.29, No.9, 1008-1009.

Kojima, T., Hayashi, K., Matsuda, Y. (2004a). Structures and Properties of Ruthenium(II) Complexes of Pyridylamine Ligands with Oxygen-Bound Amide Moieties: Regulation of Structures and Proton-Coupled Electron Transfer. *Inorg. Chem.*, Vol.43, No.21, pp. 6793-6804.

Kojima, T., Miyazaki, S., Hayashi, K., Shimazaki, Y., Tani, F., Naruta, Y., Matsuda, Y. (2004b). Intramolecular Rearrangement for Regioselective Complexation by Intramolecular CH/π Interaction in a Hydrophobic Cavity of a Ruthenium Coordination Sphere. *Chem.-Eur. J.*, Vol.10, No.24, pp. 6402-6410.

Kojima, T., Matsuda, Y. (2005). A Novel Ru(II)-DMSO Complex Having Non-coordinating 1-Naphthoylamide Arm: Effects of Intramolecular Hydrogen Bonding on Redox Potential of the Ruthenium Center. *Chem. Lett.*, Vol.34, No.2, pp. 258-259.

Kojima, T., Noguchi, D., Nakayama, T., Inagaki, Y., Shiota, Y., Yoshizawa, K., Ohkubo, K., Fukuzumi, S. (2008). Synthesis and Characterization of Novel Ferrocene-Containing Pyridylamine Ligands and Their Ruthenium(II) Complexes: Electronic Communication through Hydrogen-Bonded Amide Linkage. *Inorg. Chem.*, Vol.47, No.3, pp. 886-895.

Kojima, T., Hirasa, N., Noguchi, D., Ishizuka, T., Miyazaki, S., Shiota, Y., Yoshizawa, K., Fukuzumi, S. (2010). Synthesis and Characterization of Ruthenium(II)-Pyridylamine Complexes with Catechol Pendants as Metal Binding Sites. *Inorg. Chem.*, Vol.49, No.8, pp. 3737-3745.

Nielsen, P., Toftlund, H., Boas, J. F., Pilbrow, J. R., Moubaraki, B., Murray, K. S., Neville, S. M. (2008). Electron Paramagnetic Resonance and Crystallographic Resolution of an Axially Compressed Octahedral Complex $[Cu(terpyR8)_2](ClO_4)_2$ Containing the C_8 Alkyl-chain Ligand TerpyR8; 4′-octoxy-2,2′:6′,2″-terpyridine. *Inorg. Chim. Acta*, Vol.361, No.12-13, pp. 3453-3461.

Nishio, M., Hirota, M. (1989). CH/π Interaction: Implications in Organic Chemistry. *Tetrahedron*, Vol.45, No.23, pp. 7201-7245.

Pavkovic, S. F., Miller, D., Brown, J. N. (1977). Dinitrato(*N,N,N',N'*-tetramethylethylene-diamine)copper(II). *Acta Crystallogr.*, Vol.B33, No.9, pp. 2894-2896.

Prins, R., Reinders, F. J., (1969). Electron Spin Resonance of the Cation of Ferrocene. *J. Am. Chem. Soc.*, Vol.91, No.17, pp. 4929-4931.

Steed, J., Atwood, J. (2000). *Supramolecular Chemistry*, Wiley, ISBN 978-0-470-51234-0, Chichester, UK.

Wada, A., Harata, M., Hasegawa, K., Jitsukawa, K., Masuda, H., Mukai, M., Kitagawa, T., Einaga, H. (1998) Structural and Spectroscopic Characterization of a Mononuclear Hydroperoxo-Copper(II) Complex with Tripodal Pyridylamine Ligands. *Angew. Chem., Int. Ed.*, Vol.37, No.6, 798-799.

Wojciechowski, K., Bitner, A., Bernardinelli, G., Brynda, M. (2009). Azacrown Ether-Copper(II)-hexanoate Complexes. From Monomer to 1-D metal Organic Polymer. *Dalton Trans.*, No.7, pp. 1114-1122.

Yano, Y., Kojima, T., Fukuzumi, S. (2011). Cooperative Catalysis of a Trinuclear Ruthenium(II) Complex in Transfer Hydrogenation of Ketones by Formic Acid. *Inorg. Chim. Acta*, Vol. 374, No. 1, pp. 104-111.

10

Role of X-Ray Crystallography in Structural Studies of Pyridyl-Ruthenium Complexes

Dai Oyama
Fukushima University
Japan

1. Introduction

It is extremely important for chemists to establish the structure of new compounds. A more accurate understanding of the structure leads to the construction of appropriate reaction systems. For example, although the well-known platinum complex, $[PtCl_2(NH_3)_2]$, has two geometrical isomers (*cis* and *trans* in Fig. 1), only the *cis*-isomer (so-called cisplatin) exhibits prominent antitumor activity (Reedijk, 1996; Rosenberg et al., 1969).

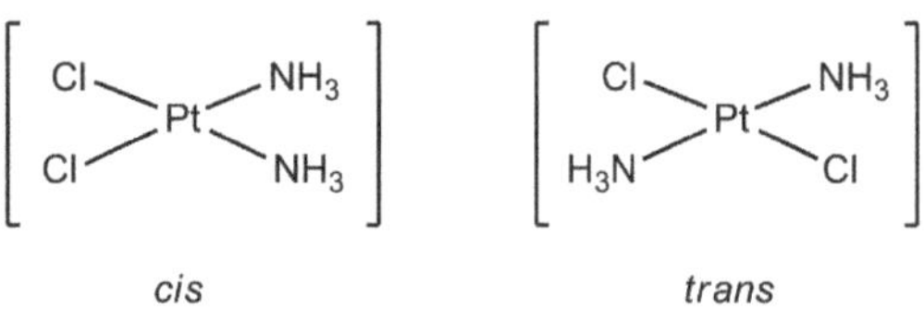

Fig. 1. Two geometrical isomers in $[PtCl_2(NH_3)_2]$.

Characterization of most structures of simple organic compounds is usually carried out by various spectroscopic measurements including NMR, IR and MS. On the other hand, detailed structural characterization of coordination compounds is achieved by X-ray diffraction methods, which can provide not only information on the geometrical structures, but also full bond parameters. In particular, single crystal X-ray diffraction is the most powerful tool for the detailed structural analysis of crystalline coordination compounds.

The aim of this chapter is to demonstrate particular advantages of X-ray structural analysis when compared with other techniques on coordination compounds. Herein, X-ray analysis of a variety of mononuclear ruthenium complexes containing pyridyl substituents is mainly described.

2. Coordination geometries and coordination modes

Some isomeric pairs of $[Ru(tpy)(L)Cl]^{n+}$ type complexes (tpy = 2,2′:6′,2′′-terpyridine, L = asymmetrical pyridyl-based bidentate ligands in Fig. 2) have been prepared and structurally characterized as precatalysts to investigate the effect of isomeric structural features on the catalytic epoxidation process (Chowdhury et al., 2011; Dakkach et al., 2010). In the complex $[Ru(tpy)(1)(OH_2)]^{2+}$, the distal isomer exhibits better activity because it contains a pyridine

C-H bond nearly parallel to the Ru-O bond, whereas for the proximal isomer this position is occupied by a C-CH_3 group and thus exerts a much stronger steric effect (Dakkach et al., 2010). In the complex $[Ru(tpy)(2)(OH_2)]^+$, on the other hand, the proximal isomer has been established to be an excellent catalyst for the chemoselective epoxidation though limited differences in electronic structural features exist between the isomeric pair (Chowdhury et al., 2011). These examples indicate that it is important to clearly distinguish the molecular structures of the compounds. In this section, representative examples of ruthenium complexes which possess asymmetrical multidentate ligands or multiple donor atoms are surveyed.

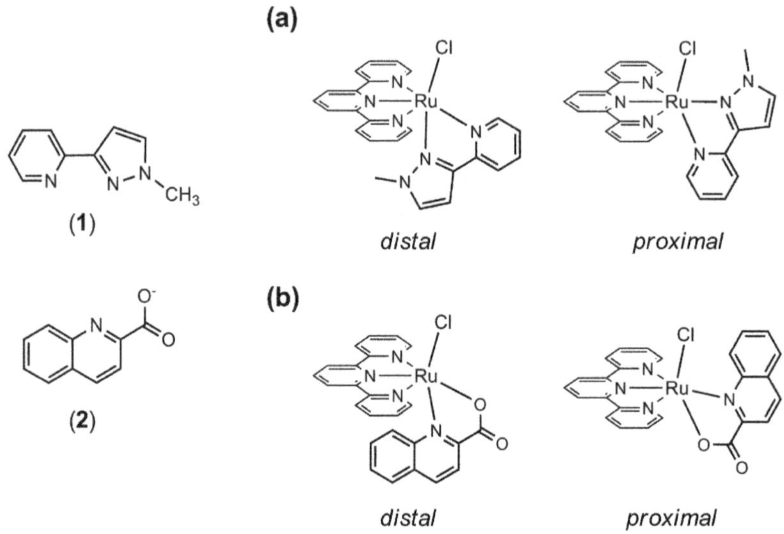

Fig. 2. Asymmetrical pyridyl-based ligands (1 and 2) and their ruthenium complexes. (a): $[Ru(tpy)(1)Cl]^+$, (b): [Ru(tpy)(2)Cl].

2.1 Coordination geometries of azopyridyl complexes

2-Azopyridyl derivatives (Fig. 3) behave as mono-, bi- and tridentate ligands, thus a wide variety of mononuclear complexes can be prepared. For example, 2-phenylazopyridine (3), which represents the most fundamental azopyridyl compound, is a bidentate ligand that can coordinate to a metal ion through the lone pairs on the pyridine and the azo nitrogen atoms, thereby forming a stable chelating 5-membered ring.

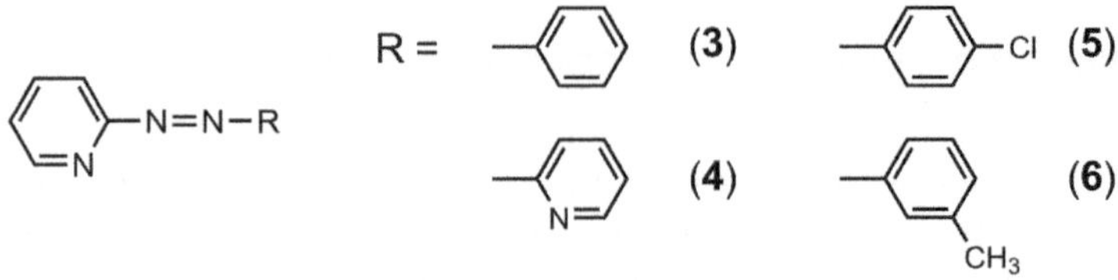

Fig. 3. Bidentate 2-azopyridyl ligands described herein.

Since the bidentate ligand (3) lacks a two-fold symmetry axis, there are five possible isomers in $[Ru(3)_2Cl_2]$ (Fig. 4) (Bao, K. Krause & R. A. Krause, 1988; Goswami, Chakravarty & Chakravorty, 1981; R. A. Krause & K. Krause, 1982 & 1980; Velders et al., 2004). Among them, α, β and ε adopt the *cis*-geometry with respect to the two chlorido ligands, whereas γ and δ are the *trans*-geometry. Besides the ε isomer, the structures have been determined by X-ray crystallography: the molecular structures of two $[Ru(3)_2Cl_2]$ isomers (α- and β-isomers) were published in 1984 (Seal & Ray, 1984), the third (γ) and the fourth (δ) isomers were reported in 2000 and 2004, respectively (Velders et al., 2004 & 2000).

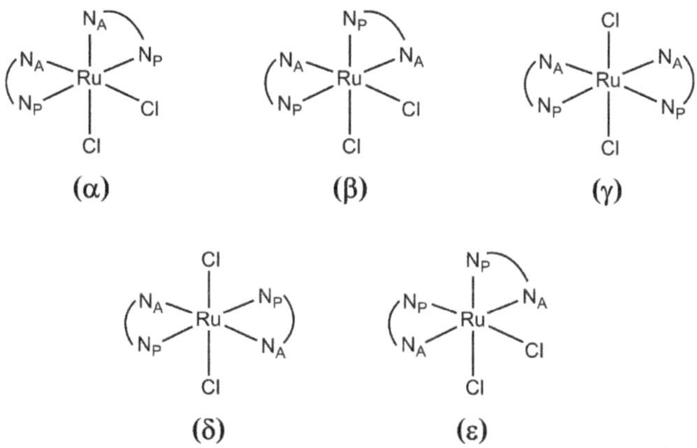

Fig. 4. Chemical structures of the five stereoisomers of $[Ru(3)_2Cl_2]$. N_A and N_P represent the azo and pyridine nitrogen atoms, respectively.

Ruthenium(II) complexes containing both Ru-C bonds and 3 as supporting ligands were synthesized by substituting chlorido ligand(s) in $[Ru(3)_2Cl_2]$ with CO or CN^- (Oyama, Takatsuki & Fujita, 2010). The molecular structure of $[Ru(3)_2(CO)Cl]^+$ determined by X-ray crystallography is shown in Fig. 5(a). The geometry of the complex corresponds to the α-isomer shown in Fig. 4. The carbonyl C-O and Ru-Cl bond distances, and the Ru-C-O bond angle are consistent with those observed for closely related complexes ($[RuL_2(CO)Cl]^+$: L = bidentate pyridine-based ligands), whereas the Ru-C bond distance is longer than those in similar complexes (Clear et al., 1980; Kepert et al., 2004). This is because of the presence of two 3 ligands with azo moieties. In general, the Ru-N(azo) bond distances are shorter than those of the Ru-N(pyridine) bonds in 3 in the ruthenium complexes (Velders et al., 2004). However, one of the Ru-N(azo) bonds is very long; even longer than the distances of the Ru-N(pyridine) bonds. This is caused by the *trans* influence of the CO ligand.

The crystal structure of $[Ru(3)_2(CN)_2]$ is shown in Fig. 5(b). Unexpectedly, the two CN groups are *trans* to each other, and the δ-form (shown in Fig. 4) is observed. A δ arrangement is very rare, and was only recently reported for δ-$[Ru(3)_2Cl_2]$ (Velders et al., 2004). In the δ-isomer, the two azo bonds (in mutual *trans* positions) should compete with each other for the Ru(II) 4d electron density, resulting in relatively long Ru-N(azo) bonds in $[Ru(3)_2(CN)_2]$ with respect to the equivalent bonds in γ-$[Ru(3)_2Cl_2]$ (Velders et al., 2004). This confirms that the Ru-N(azo) bond order decreases in $[Ru(3)_2(CN)_2]$.

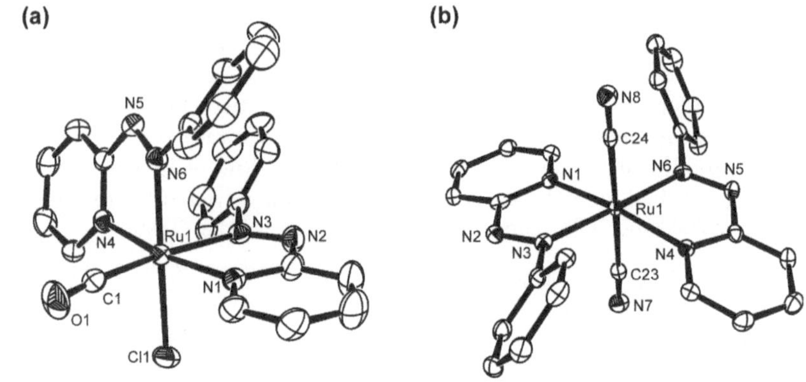

Fig. 5. Molecular structures of (a): $[Ru(3)_2(CO)Cl]^+$, (b): $[Ru(3)_2(CN)_2]$.

Although only the *cis*-isomer of $[Ru(3)_2(CO)Cl]^+$ is formed, the *trans*-isomer of $[Ru(3)_2(CN)_2]$ formed preferentially despite the use of a *cis*-$[Ru(3)_2Cl_2]$ precursor. In $[RuL_2(CN)_2]$ (L = bidentate pyridine-based ligands) complexes, the *cis* form is the preferred geometry for 2,2′-bipyridine and its analogues (Abe, Suzuki & Shinozaki, 2010; Cheng, Cheung & Che, 1997; Chow et al., 2005; Zhou et al., 2009), whereas the *trans* geometry is more stable in $[Ru(3)_2(CN)_2]$.

Many analogous bis-azopyridine complexes, $[Ru(azopyridine)_2L_2]^{n+}$, have also been reported (Das et al., 2009; Hotze et al., 2000 & 2005; Samanta et al., 2001; Shivakumar et al., 2000). In these complexes, the $[Ru(4)_2(CO)(PPh_3)]^{+/0}$ complexes form the unique ε-isomer examples.

In analogy with the bis-azopyridine complexes, the geometry of the complexes should also be consistent for the mono-azopyridine complexes because of the asymmetrical units of the azopyridines. For example, ruthenium complexes containing the ligand 3 and its analogue 5 ($[Ru(azopyridine)(H)(CO)(PPh_3)_2]$) exhibit two different geometries (Fig. 6) (Shivakumar et al., 2000). Therefore, it is important to control the orientation of the azopyridine ligands in preparing complexes that rely on the steric and/or electronic nature of the ligands. Such examples have been demonstrated in $[Ru(3)(bpy)Cl(L)]^+$ (bpy = 2,2′-bipyridine, L = CO or CH_3CN) complexes. Here, the configuration of the incoming azopyridine ligand 3 can be regulated by the strength of the π-bonding ability of the co-ligands L situated in the *trans* position: the pyridine nitrogen (weaker π-bonding character) of the incoming ligand 3 is selectively situated *trans* to the L ligand when L is a carbonyl, whereas the azo nitrogen (stronger π-bonding character) is situated *trans* to the acetonitrile ligand. Thus, the combination of CO-pyridine nitrogen (strong-weak π-bonding character) and CH_3CN-azo nitrogen (weak-strong π-bonding character) are quite reasonable (Oyama, Asuma & Takase, 2008). The molecular structures of $[Ru(3)(bpy)Cl(L)]^+$ (L = CO or CH_3CN) are shown in Fig. 7. The azo group of 3 in $[Ru(3)(bpy)Cl(L)]^+$ is clearly situated *cis* (L = CO) and *trans* (L = CH_3CN) to the L ligand. The corresponding complex with 4 ($[Ru(4)(bpy)(CO)Cl]^+$) is also selectively synthesized by using the above methodology (Oyama et al., 2009).

Molecular structures of various ruthenium mono-terpyridine complexes of the type $[Ru(tpy)(azopyridine)L]^{n+}$ (azopyridine = 3, 4, 6; L = Cl, OH_2, CH_3CN, NO_2) have been determined by X-ray analysis (Corral et al., 2006; Mondal, Walawalkar & Lahili, 2000; Mondal et al., 2001; Ooyama & Asaha, 2005; Oyama, Fujita & Yui, 2008). The asymmetrical nature of azopyridines lead to the possibility of two isomers as shown in Fig. 8; however, only one isomer (a in Fig. 8) has been characterized in all complexes.

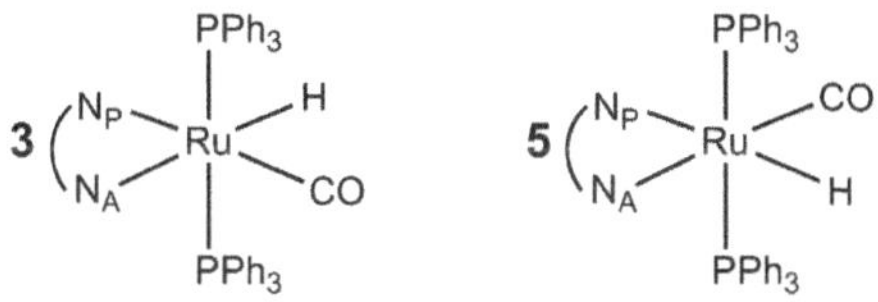

Fig. 6. Two different geometries in *trans*(PPh_3)-[Ru(azopyridine)(H)(CO)(PPh_3)$_2$] (azopyridine = (3) or (5)).

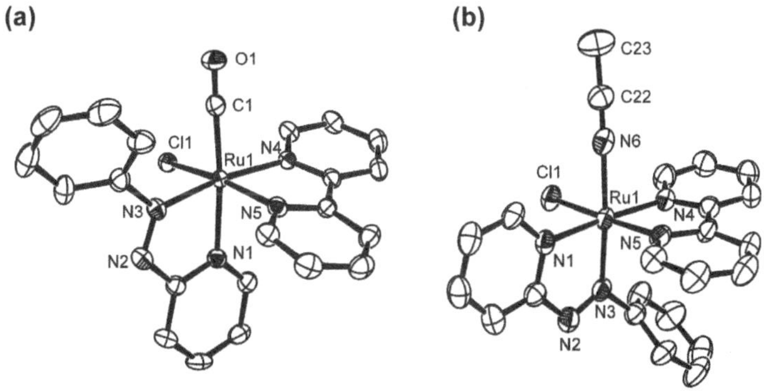

Fig. 7. Molecular structures of (a): [Ru(3)(bpy)(CO)Cl]$^+$, (b): [Ru(3)(bpy)(CH_3CN)Cl]$^+$.

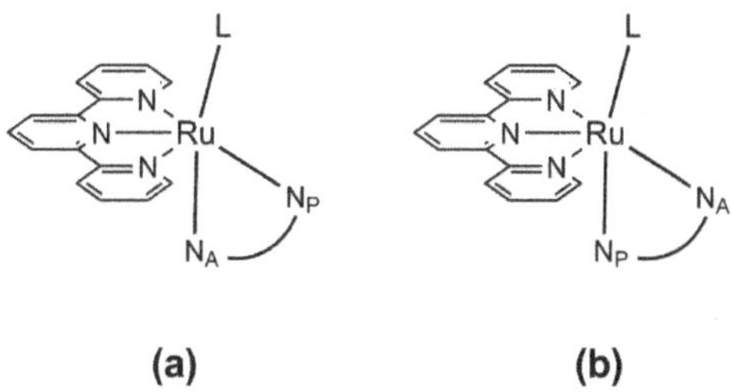

Fig. 8. Possible isomers in [Ru(tpy)(azopyridine)L]$^{n+}$ type complexes.

Finally, tris-bidentate and bis-tridentate azopyridine complexes are presented. The compound [Ru(azopyridine)$_3$]$^{n+}$ theoretically exists in two different isomeric forms (disregarding Δ, Λ enantiomerism), i.e., the meridional (*mer*) and facial (*fac*) isomers (Fig. 9). Only *mer*-[Ru(azopyridine)$_3$]$^{2+}$ were obtained and crystallographically characterized due to the facial isomer of [Ru(azopyridine)$_3$]$^{2+}$ being sterically unfavorable because of the crowding of the three phenyl groups (Das et al., 2003; Hotze et al., 2005). Ruthenium complexes with tridentate azopyridine ligands in Fig. 10 have been reported (Samanta et al., 2008). The configuration is bis-meridional, and the relative orientations of the pairs of coordinated atoms are *cis* (pyridyl-N), *trans* (azo-N) and *cis* (arylanilido-N), respectively.

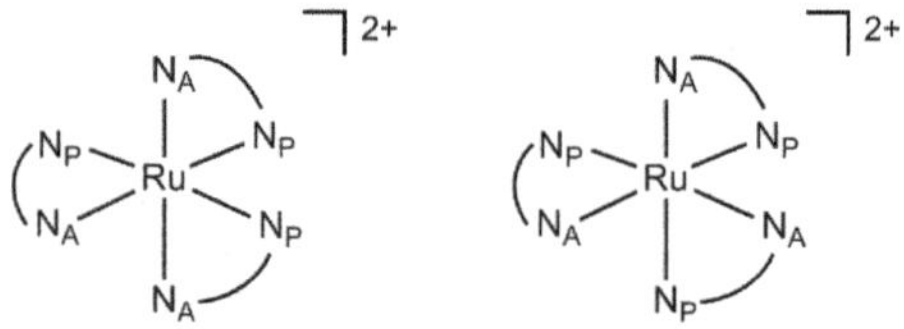

(*mer*) (*fac*)

Fig. 9. Possible isomers in $[Ru(azopyridine)_3]^{n+}$ type complexes.

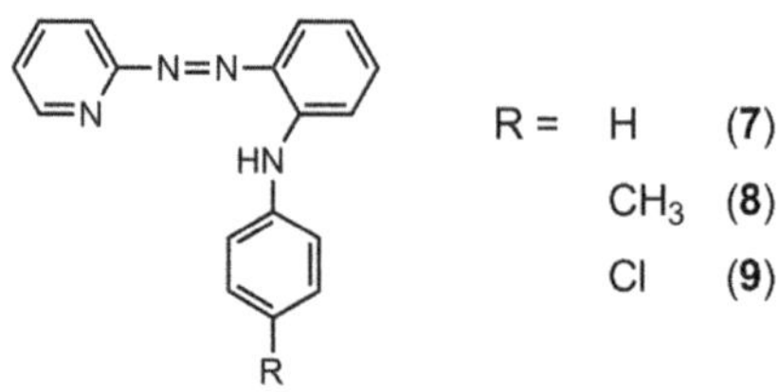

Fig. 10. Tridentate 2-azopyridyl ligands described herein.

2.2 Coordination geometries of other pyridyl-based complexes

The naphthyridines consist of a group of diazanaphthalenes with one nitrogen in each ring. In particular, the bidentate naphthyridine, 2-(2-pyridyl)-1,8-naphthyridine (pynp; 10), is a useful ligand for mononuclear systems (Fig. 11). For complexes with ligand 10; however, there is a stereochemical question with regard to the binding of ligand 10 because it has an asymmetrical structure. Consequently, it would be very interesting to study complexes involving 10 in an effort to understand the relationship between the coordination geometries of these complexes and their reactivities.

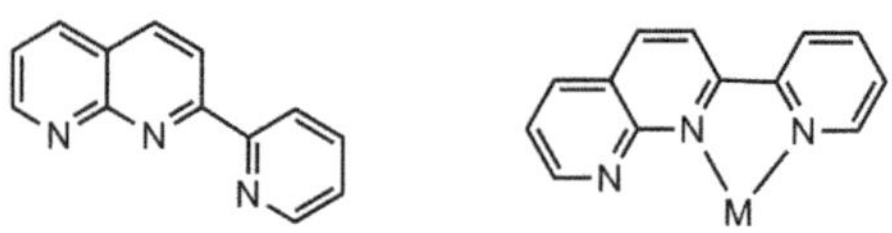

(**10**)

Fig. 11. Chemical structure of 2-(2-pyridyl)-1,8-naphthyridine (pynp) and its coordination to a metal center.

The $[Ru(tpy)(10)Cl]^+$ complex acts as a catalyst for water oxidation: it shows excellent catalytic activity with turnover numbers (TNs) of 1,170 (Tseng et al., 2008). This value is the highest in the analogous $[Ru(tpy)(L)Cl]^+$ type complexes (L = bidentate polypyridyl ligands) whose TNs are in the range of 0–570. The ligand 10 differs from all of the other bidentate (L) because it is asymmetric. Although there are two possible isomers in $[Ru(tpy)(10)Cl]^+$ (Fig. 12), only one isomer (Fig. 12(b)) is isolated and its structure has been determined by X-ray crystallography (Tseng et al., 2008). Two isomeric mixtures of $[Ru(tpy)(10)Cl]^+$ exist in solution (approximately 1:1), both of which can be separated by column chromatography. Alternatively, irradiation of visible light to isomeric mixtures of $[Ru(tpy)(10)L]^{2+}$ (L = OH_2 or CH_3CN) leads the formation of only one isomer in particular organic solvents (Oyama,

Yuzuriya & Takase, 2011). The resulting complex ($[Ru(tpy)(10)(OH_2)]^{2+}$) is the proximal configuration (Fig. 12(a)), which differs from the corresponding chlorido complex.

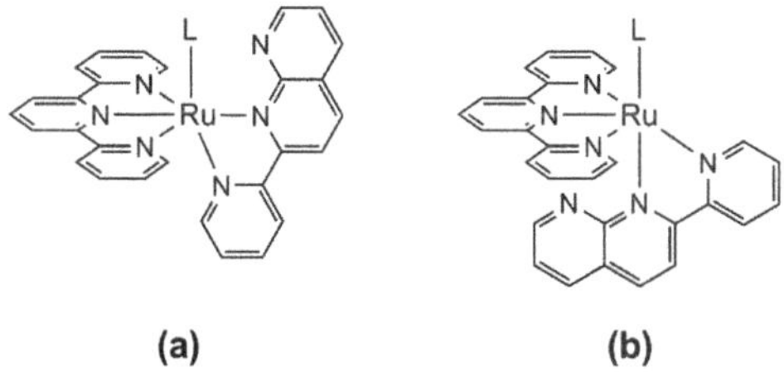

Fig. 12. Possible isomers in $[Ru(tpy)(10)L]^{n+}$ type complexes.

In contrast, particular stereoisomers could be selectively prepared in the *cis*-$[Ru(bpy)(10)(CO)Cl]^+$ system (Oyama, Hamada & Takase, 2011). As shown in Fig. 13, although there are four possible geometries in the complex, only two types of the complex could be selectively synthesized. Although the prepared complexes are all single species as determined from the spectroscopic measurements, their structures cannot be assigned because of their spectral resemblance. Consequently, X-ray analysis was required to determine their detailed structures. The molecular structures of two $[Ru(bpy)(10)(CO)Cl]^+$ isomers are shown in Fig. 14. One isomer corresponds to (a) in Fig. 13 (the pyridyl ring of 10 is situated *trans* to the CO ligand), the another isomer corresponds to (b) in Fig. 13 (the pyridyl ring of 10 is situated *cis* to the CO ligand).

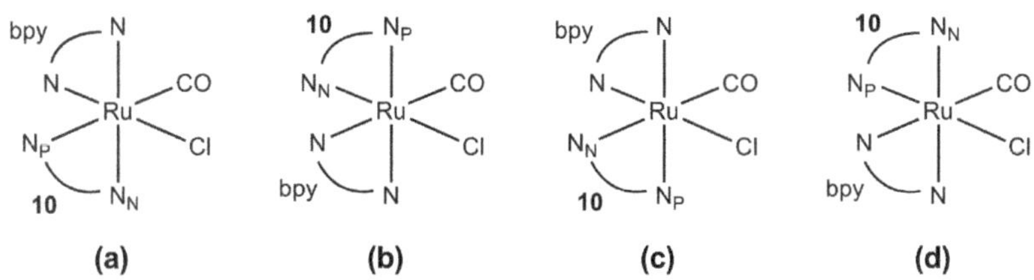

Fig. 13. Chemical structures of four stereoisomers in *cis*-$[Ru(bpy)(10)(CO)Cl]^+$. N_N and N_P denote the naphthyridine nitrogen and the pyridine nitrogen, respectively.

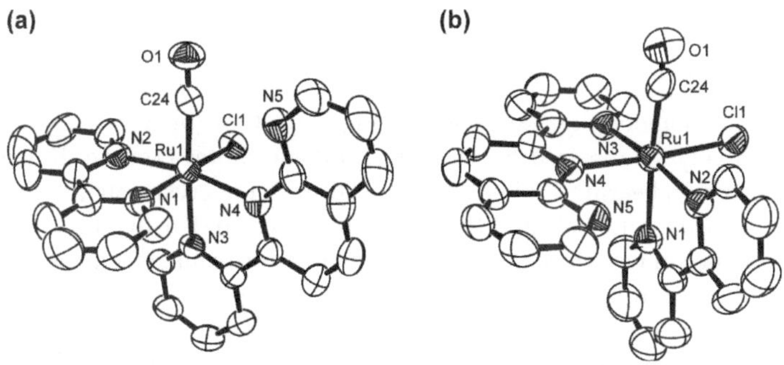

Fig. 14. Molecular structures of $[Ru(bpy)(10)(CO)Cl]^+$. (a): isomer "a" in Fig. 13, (b): isomer "b" in Fig. 13.

Picolinato (pic; 11) can also coordinate to a metal ion as an asymmetrical bidentate ligand (Fig. 15). The mono-pic complex, $[Ru(11)(CO)_2Cl_2]^-$, has been prepared and fully characterized, including X-ray analysis (Ooyama et al., 2003). Although three geometrical isomers are expected for $[Ru(11)(CO)_2Cl_2]^-$ (Fig. 16(a)), only form (i) in Fig. 16(a) was confirmed, and neither form (ii) nor (iii) was detected in the solid state and in solutions. The asymmetrical chelate of 11 gives substantially different effects to the two Ru–CO bonds of $[Ru(11)(CO)_2Cl_2]^-$. The Ru–C bond distance *trans* to the oxygen atom is shorter than another Ru–C bond *trans* to the nitrogen atom (Fig. 17(a)). The complex containing mono-11, *cis*-$[Ru(bpy)(11)(CO)_2]^+$ was also prepared by the reaction of Hpic with $[Ru(bpy)(CO)_2Cl_2]$, and only the type (iv) of the two possible isomers (iv) and (v) was obtained (Fig. 16(b) and Fig. 17(b)) (Ooyama et al., 2003).

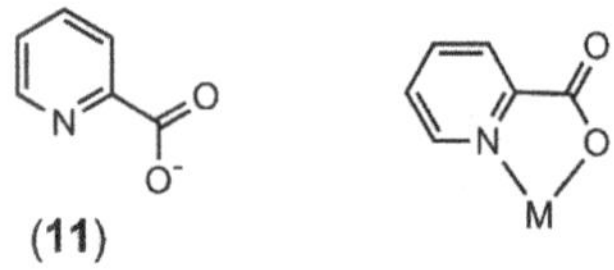

Fig. 15. Chemical structure of picolinato (pic) and its coordination to a metal center.

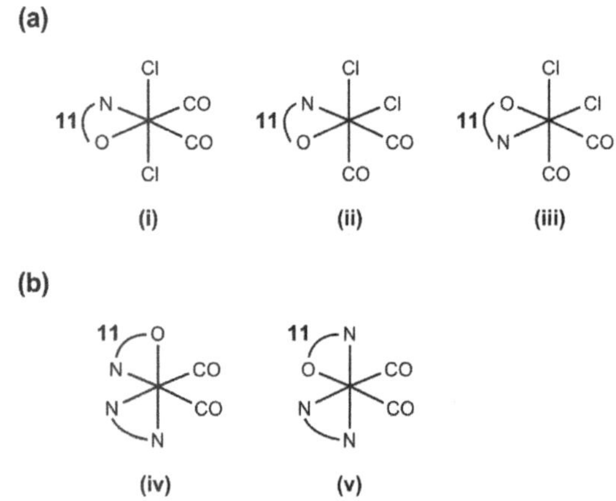

Fig. 16. Possible geometrical isomers containing 11. (a): $[Ru(11)(CO)_2Cl_2]^-$, (b): *cis*-$[Ru(bpy)(11)(CO)_2]^+$.

Two geometrical isomers (Fig. 18) of $[Ru(tpy)(bpyO)(CO)]^+$ (bpyO = 2,2′-bipyridin-6-onato; 12) were selectively synthesized by different synthetic routes (Tomon, Koizumi & Tanaka, 2005). These isomers bearing 12 were determined by X-ray crystal structure analysis. Based on their pK_a's, redox potentials and IR spectra, the electron density at the ruthenium center of the distal isomer in Fig. 18 is higher than that of the proximal isomer. Such a difference is associated with the stronger *trans* influence of CO compared with the central pyridine of tpy, because the electron-withdrawing ability of CO induces the pyridonato structure of the resonance between ruthenium-pyridionate (Ru–bpyO⁻) and -pyridonato (Ru–bpyO) forms. The two-electron reductions of the distal isomer are followed by a partial Ru–CO bond cleavage, whereas the 12-based reduction of the proximal isomer causes cyclometallation by an attack of the pyridonato oxygen at the carbonyl carbon. Thus, cyclometallation caused by 12-based reduction effectively suppresses the reductive cleavage of a Ru–CO bond.

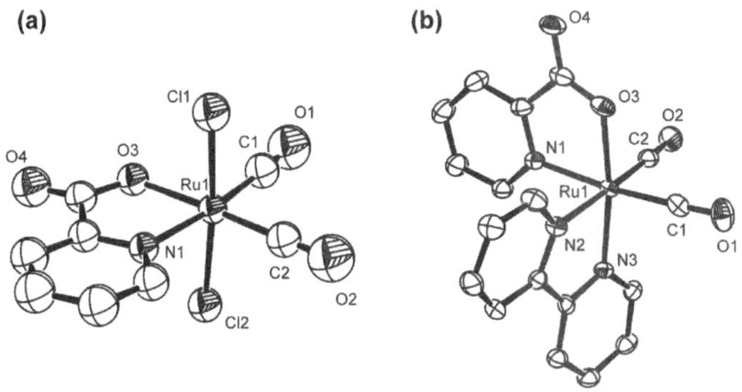

Fig. 17. Molecular structures of (a): $[Ru(11)(CO)_2Cl_2]^-$ (isomer (i) in Fig. 16(a)), (b): $[Ru(bpy)(11)(CO)_2]^+$ (isomer (iv) in Fig. 16(b)).

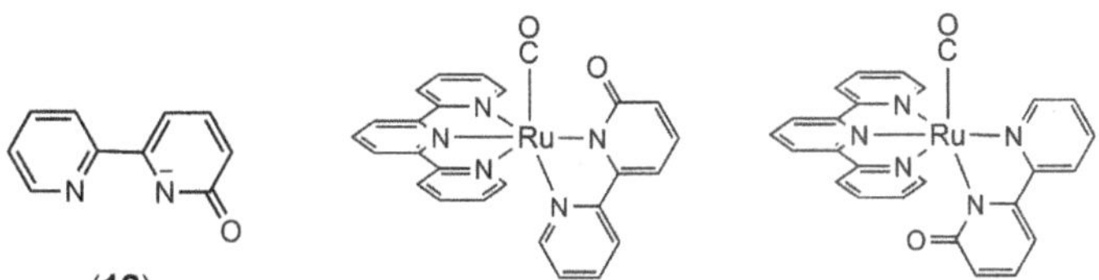

(12) ***(proximal)*** ***(distal)***

Fig. 18. Chemical structure of 2,2′-bipyridin-6-onato (bpyO) and two geometrical isomers in $[Ru(tpy)(12)(CO)]^+$.

2.3 Coordination modes

It is known that not only some acido anions such as NO_3^-, CO_2R^- and CO_3^{2-}, but also organic ligands having multi donor atoms, can coordinate to metals as monodentates, chelating bidentates, or bridging bidentate ligands. In this section, such examples are presented.

In coordination compounds, the NO_3^- ion can be noncoordinating (counter ion), or it can coordinate in a monodentate (NO_3-*O*) or bidentate (NO_3-*O,O′*) fashion. The N-O stretching vibrations, for example, are observed at 1475, 1272 and 991 cm^{-1} in the IR spectrum of *cis*-$[Ru(bpy)_2(CO)(NO_3)]PF_6$ (Oyama et al., 2011). It is rather difficult to differentiate between these structures by vibrational spectroscopy because the symmetry of the NO_3^- ion differs very little between them (C_s vs. C_{2v} symmetry) (Nakamoto, 1986). However, X-ray data clearly indicates the monodentate coordination of the NO_3^- ion. The sums of the three O-N-O bond angles in the nitrato ligand are 360.0°, indicating that the nitrato moiety has a planar structure.

Extensive studies have been made on metal complexes of carboxylic acids. The carboxylate ion may also coordinate to a metal in a monodentate (CO_2R-*O*) or bidentate (CO_2R-*O,O′*) mode. In general, the coordination mode of the carboxylate ion is distinguishable using vibrational spectroscopy (Deacon & Phillips, 1980). Monodentate complexes exhibit

Δν values ($\nu_{as}(CO_2^-)-\nu_s(CO_2^-)$) which are much greater than the ionic compounds, whereas chelating (bidentate) complexes exhibit Δν values which are significantly less than the ionic values. For example, the *cis*-[Ru(bpy)$_2$(CO)(CO$_2$H)]$^+$ complex shows the large formato Δν value of 339 cm^{-1} (Gibson et al., 1999), and this value is clearly greater than the ionic value (Δν = 201 cm^{-1}) (Ito & Bernstein, 1956). X-ray structural data for this complex clearly identified the monodentate formato ligand (CO$_2$H–*O*) bound to an octahedral ruthenium center (Gibson et al., 1999).

As stated above, the azopyridine ligand (3) generally coordinates to a metal ion through both pyridyl and azo nitrogens (bidentate chelate fashion). However, the crystal structure of a ruthenium complex containing 3 as a monodentate ligand ([Ru(bpy)$_2$(3–*N*)(CO)]$^{2+}$) has been reported (Fig. 19(a)) (Oyama, Fujita & Yui, 2008). The ligand 3 coordinates to the ruthenium center through only the pyridyl nitrogen atom. The remarkable feature in the structure of this complex concerns the azo moiety of 3, which is directed toward the adjacent terminal carbonyl ligand. Taking into account the fact that the nitrogen atom of the azo group is located just to the side of the carbonyl carbon, the existence of the CO ligand could influence the orientation of the azo moiety of 3. The redox reactions of [Ru(bpy)$_2$(3–*N*)(CO)]$^{2+}$ using solution IR spectra under electrolysis conditions revealed a nucleophilic attack of the azo nitrogen atom of 3 on the carbonyl carbon (Fig. 20(a)), because of the short interatomic distance between the azo nitrogen and the carbonyl carbon atoms (2.865 Å). In contrast, the dissociation of CO from the two-electron reduced species easily occurs in the analogue with the bidentate 3 ([Ru(tpy)(3–*N,N'*)(CO)]$^{2+}$; Fig. 19(b)), because the ligand is unable to interact with CO (Fig. 20(b)).

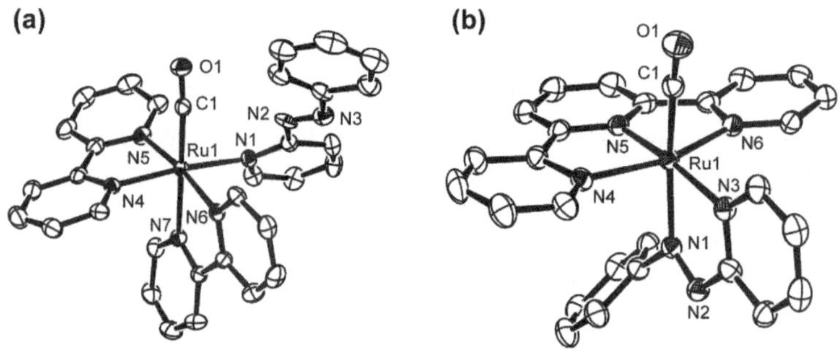

Fig. 19. Molecular structures of (a): [Ru(bpy)$_2$(3–*N*)(CO)]$^{2+}$, (b): [Ru(tpy)(3–*N,N'*)(CO)]$^{2+}$.

The diphenyl-2-phosphinopyridine ligand (dppy; 13) has been used to construct interesting metal complexes because it possesses both the soft (P) and hard (N) donor atoms (Newkome, 1993). Although both the *P*-mono and *P,N*-bidentate coordination modes have been known for mononuclear systems (Fig. 21) (Moldes et al., 1998), 13 coordinates to the ruthenium center through only the P atom in [Ru(bpy)(13)(CO)$_2$Cl]$^+$ (Ooyama & Sato, 2003). A prominent feature of the structure is that the pyridyl nitrogen atom of 13 directs toward the plane that includes two carbonyl ligands (Fig. 22(a)). The existence of carbonyl ligands may influence the orientation of the aromatic rings of 13 because relatively short distances between the noncoordinating nitrogen atom of 13 and the carbonyl carbons are observed (3.068 Å): the analogous complex situating two 13 moieties in mutually *trans* positions also exhibits a similar trend as shown in Fig. 22(b) (Ooyama & Sato, 2004).

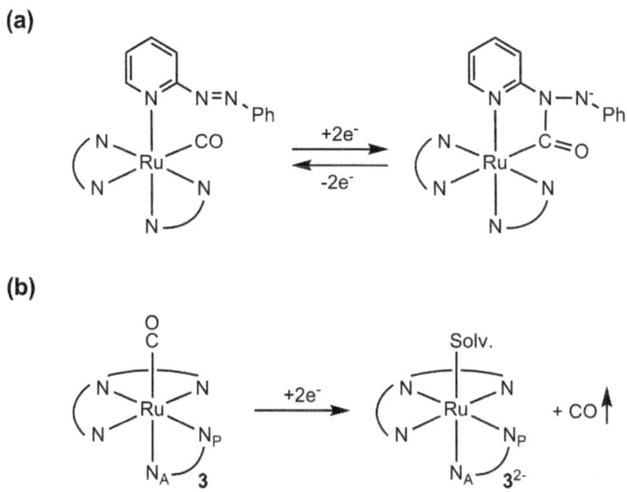

Fig. 20. The different redox behavior of the ruthenium complexes are dependent on the coordination mode of 3. (a): $[Ru(bpy)_2(3\text{-}N)(CO)]^{2+}$ (monodentate 3), (b): $[Ru(tpy)(3\text{-}N,N')(CO)]^{2+}$ (bidentate 3).

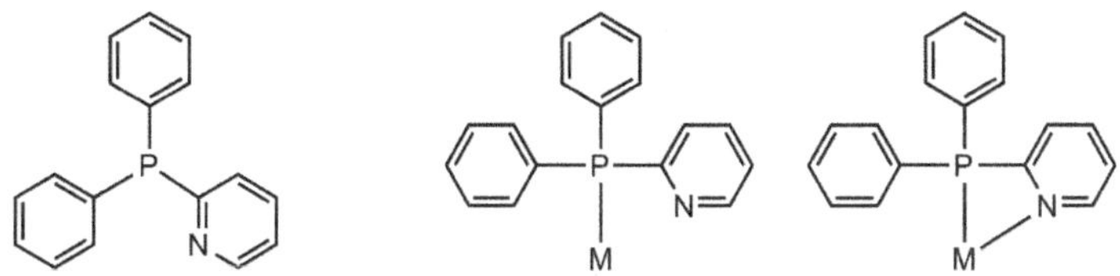

(**13**) dppy-*P* dppy-*P,N*

Fig. 21. Two coordination modes of diphenyl-2-phosphinopyridine (dppy).

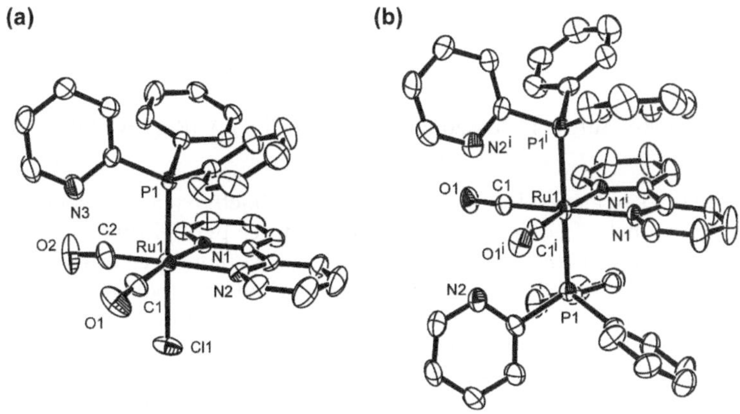

Fig. 22. Molecular structures of (a): $[Ru(bpy)(13\text{-}P)(CO)_2Cl]^+$, (b): $([Ru(bpy)(13\text{-}P)_2(CO)_2]^{2+}$.

3. Comparison of the structures and structural parameters

When chemists simply recognize the molecular structure of a compound, they usually pay particular attention to the points described above. However, it is also very helpful for chemists to understand how intermolecular forces (packing forces) act between molecules in crystals when they investigate the constructions of materials and medicines. In addition, various chemical and physical properties can be predicted by comparing bond parameters (e.g., bond distances, angles or torsion angles) in a compound.

3.1 Comparison of structures

Almost all structurally characterized coordination compounds of azopyridines have the ligand bonded to one metal center via the azoimine (N=N–C=N) chelate arrangement. In the case of 3 and 4, formation of a chelate complex creates the free aromatic ring (phenyl for 3 or 2-pyridyl for 4). The orientation of the pendant aromatic ring is dependent on its coordination environment: the dihedral angle between the two aromatic rings of $[Ru(3)(CO)_2Cl_2]$ is 35.29° (Fig. 23(a)) whereas the dihedral angle between the two pyridyl rings in $[Ru(4)(CO)_2Cl_2]$ is 7.99° (Fig. 23(b)), which shows planarity of the coordinated 4 (Oyama et al., 2009). The distinction observed in solid state can be explained on the basis of a possible weak interaction between the noncoordinating nitrogen atom and the adjacent carbonyl carbon in $[Ru(4)(CO)_2Cl_2]$ rather than a different bulkiness between the CH in phenyl ring of 3 and the N in pyridyl ring of 4, because the interatomic distance between the N atom and the adjacent C atom in $[Ru(4)(CO)_2Cl_2]$ is fairly short (2.682(3) Å). This weak interaction has often been observed in ruthenium carbonyl complexes with noncoordinating pyridyl moieties (Mizukawa et al., 1999; Tomon, Koizumi & Tanaka, 2005), which contain both a pyridyl nitrogen (δ-) and a carbonyl carbon (δ+).

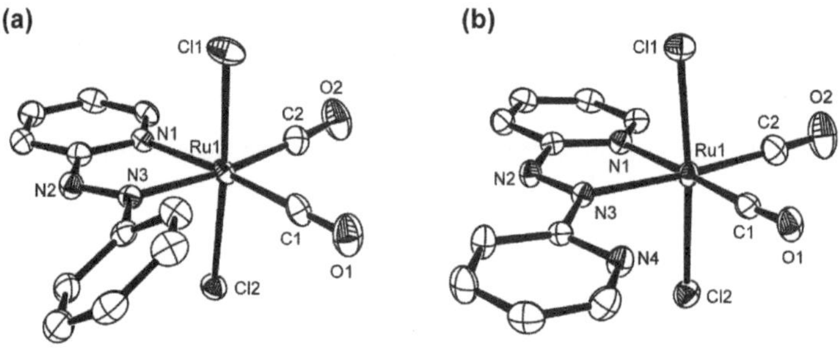

Fig. 23. Molecular structures of $[Ru(azopyridine)(CO)_2Cl_2]$. (a): $[Ru(3)(CO)_2Cl_2]$, (b): $[Ru(4)(CO)_2Cl_2]$.

The X-ray structures of terpyridyl complexes with the dimethoxyphenyl pendant have been determined. In the metal-free ligand 14 (Fig. 24), the three pyridyl rings are approximately coplanar, whereas the dimethoxyphenyl substituent in 14 is not coplanar with the terpyridyl moiety, making an angle of 50.2° with the central pyridyl ring (Storrier, Colbran & Craig, 1998). The dihedral angle between the central pyridyl ring and the dimethoxyphenyl ring in the ruthenium complexes with 14 are in the range of 43 to 56°, which are close to that of metal-free ligand 14 (Oyama et al., 2009; Oyama et al., 2010). Interestingly, the OCH_3 orientation of two methoxy moieties are different in two types of habits (yellow needles and red blocks) of $[Ru(14)(bpy)(CO)](PF_6)_2$. As shown in Fig. 25, two OCH_3 moieties point in the same orientation in the yellow crystal (Fig. 25(a)), whereas in the red crystal they point in opposite directions (Fig. 25(b)) (Oyama et al., 2010). The former structure is identical to the corresponding chlorido and acetonitrile complexes. On the other hand, the latter form is consistent with that observed in the metal-free 14. These differences are most likely due to different packing or intermolecular effects on the orientation of methoxy groups of 14.

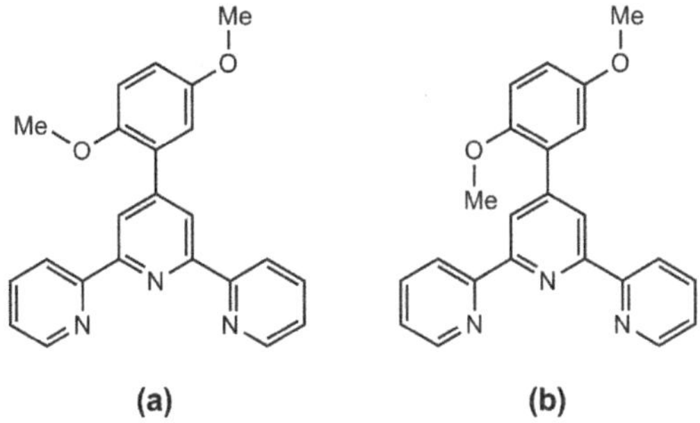

Fig. 24. Chemical structures of 14. (a): same orientation, (b): opposite orientation, regarding two methoxy groups.

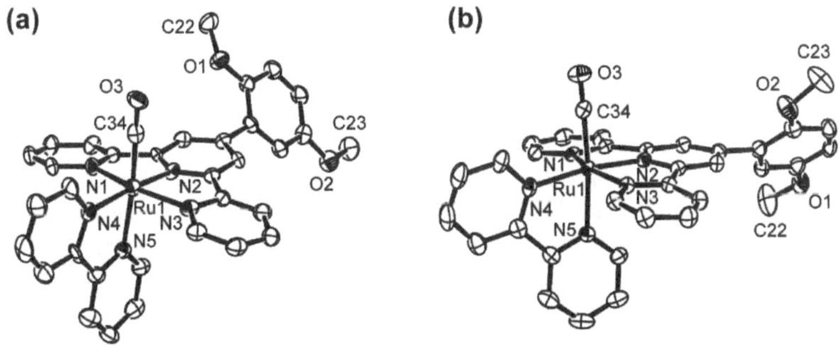

Fig. 25. Molecular structures of [Ru(14)(bpy)(CO)]2+. (a): the yellow crystal, (b): the red crystal.

3.2 Comparison of structural parameters

Physical data of terminal carbonyl groups are generally useful indicators for electronic states around metal centers in metal complexes. In particular, bond distances and vibrational spectra are considered important (Cotton et al., 1999). In order to distinguish the electron states in the ruthenium mono carbonyl complexes with polypyridyl ligands, the stretching force constants of the CO group using IR spectral data were calculated in addition to bond distances and stretching frequencies of the $\{Ru\text{-}CO\}^{2+}$ moiety (Oyama et al., 2009). Although obvious characteristics about the bond distances (Ru-C and C-O) are not shown in the complexes, the force constant (*k*) and the stretching frequencies of the CO groups can clearly be divided into two groups (Table 1): the *k* values of the azo-containing complexes are higher ca. 0.4–1.1 N cm^{-1} than of the values for the complexes without azopyridines. It is presumed that the electron density of the ruthenium center considerably decreases when an azo group (which has greater π–acidity) coordinates to the ruthenium atom, because the decrease of the *k* value caused by ligand-based one-electron reduction corresponds to 0.56 N cm^{-1} in $[Ru(bpy)_2(CO)(quinoline)]^{2+}$ (Wada et al., 2004). Thus, it can be concluded that complex formation with azopyridyl ligands induces relatively high *k* values. This observation can be ascribed to the decrease of the π back-donation from the ruthenium center to the CO group. As a result, this suggests that the azopyridyl ligands serve as prominent electron reservoirs compared with other polypyridyl ligands such as bpy.

polypyridines*	*d*(Ru–C)	*d*(C–O)	νCO	*k*
3/3	1.897(5)	1.136(6)	2031	16.67
4/bpy	1.889(3)	1.116(4)	2002	16.20
3/bpy	1.891(2)	1.122(3)	2001	16.18
4,4′-Me₂bpy/5,6′-Me₂phen	1.867(5)	1.112(7)	1978	15.81
dpk/4,7′-Me₂phen	1.820(11)	1.146(13)	1976	15.80
10/bpy	1.867(5)	1.145(6)	1975	15.76
dpa/5,5′-Me₂bpy	1.917(4)	1.026(6)	1972	15.72
bpy/dpk	1.880(5)	1.119(7)	1970	15.68
4,4′-Me₂bpy/4,7′-Me₂phen	1.95(2)	1.05(2)	1967	15.64
bpy/dpa	1.883(16)	1.08(2)	1967	15.64
bpy/phen	1.857(4)	1.126(6)	1966	15.62
bpy/bpy	1.861(29)	1.122(32)	1966	15.62
biq/biq	1.91(2)	1.01(2)	1964	15.59

* 4,4′-Me_2bpy = 4,4′-dimethyl-2,2′-bipyridine; 5,5′-Me_2bpy = 5,5′- dimethyl-2,2′-bipyridine; 5,6′-Me_2phen = 5,6′-dimethyl-1,10-phenanthroline; 4,7′-Me_2phen = 4,7′- dimethyl-1,10-phenanthroline; dpk = di(2-pyridyl)ketone; dpa = di(2-pyridyl)amine; phen = 1,10-phenanthroline; biq = 2,2′-biquinoline.

Table 1. Bond distances (*d*: Å), stretching frequencies (ν: cm^{-1}) and force constant (*k*: N cm^{-1}) of the CO groups in *cis*-$[Ru(polypyridine)_2(CO)Cl]^+$.

The N-N bond distance in an azo moiety is an excellent indicator of the charge on an azo group (Kaim, 2001). The values for unreduced azo N-N bonds are 1.22–1.31 Å in metal complexes (1.23–1.26 Å for metal-free ligands). The one-electron reduced (anion radical) ligands have a bond distances of 1.31–1.41 Å, whereas the two-electron reduced (hydrazido) forms have single bonds with 1.41–1.50 Å distances (Fig. 26) (Sarkar et al., 2008). As shown in Table 2, the azo N–N bond distance of $[Ru(bpy)_2(CO)(3\text{–}N)]^{2+}$ is dramatically shorter than those of other complexes: the distance (1.188(4) Å) is shorter than that of a typical N=N double bond (1.23 Å) (Oyama, Fujita & Yui, 2008).

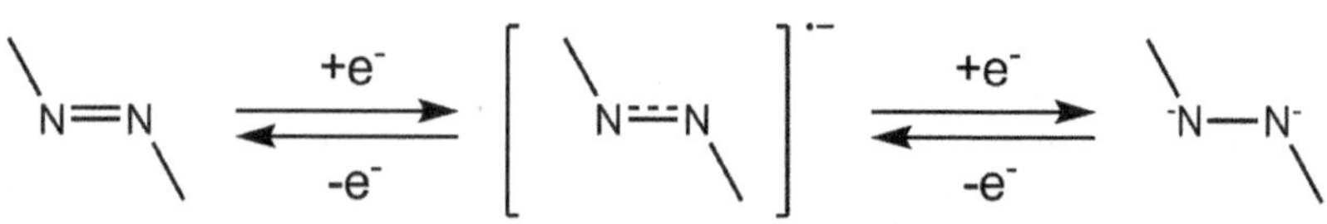

Fig. 26. Three oxidation states of an azo moiety.

compound	d(N–N)	d(M–N_{azo})	d(M–N_{py})
[Ru(3)$(CO)_2Cl_2$]	1.264(3)	2.119(2)	2.099(2)
[Ru(4)$(CO)_2Cl_2$]	1.265(3)	2.107(2)	2.108(2)
[Ru(3)(bpy)(CO)Cl]PF_6	1.287(3)	2.015(2)	2.095(2)
[Ru(4)(bpy)(CO)Cl]PF_6	1.291(4)	1.998(2)	2.114(2)
[Ru(3)(tpy)(CO)]$(ClO_4)_2$	1.265(5)	2.099(3)	2.071(3)
[Ru(4)(tpy)Cl]PF_6	1.299(3)	1.964(2)	2.054(2)
[Ru(3-κ*N*)$(bpy)_2$(CO)]$(ClO_4)_2$	1.188(4)		2.131(2)
$(3H^+)BF_4$	1.248(4)		
$(3H^+)ClO_4$	1.258(5)		
4	1.246(2)		

Table 2. Bond distances (Å) in the complexes of mono-azopyridines and metal-free ligands.

In the case of bis-azopyridyl complexes, bivalent azopyridine properties are observed by the different azo N-N bond distances. For example, the structure of the monoradical [Ru(4)$(4^{\bullet-})$(CO)(PPh_3)]$^+$ has been determined (Shivakumar et al., 2000). The most remarkable feature of the structure is the difference in the two N-N bond distances: one is 1.284(6) Å, the other N-N bond is 1.336(6) Å. Accordingly, both unreduced and one-electron reduced (anion radical) types are present within the same molecule, which represents a model case of ground-state radical localization in one of the two ligands. The stabilization of radical on one ligand might be supported by the *trans* ligand. A similar example is also reported in [Ru$(7^{2-})_2$] (Samanta et al., 2008). The [Ru$(7)_2$] molecule has longer N–N bond distances (1.324(7) and 1.327(6) Å), clearly suggesting the radical dianion form of the ligand (one charge from the deprotonation and another charge from the one-electron reduction centered on the azo group). This diamagnetic molecule may thus be viewed as a singlet diradical species. In this situation, the electrons from a d^4 configuration at the metal couple with the singly occupied ligand molecular orbitals to create a spin-paired entity. This situation has been confirmed by DFT (density functional theory) calculations.

4. Conclusion

Single crystal X-ray diffraction studies provide a valuable probe to visualize molecules. Although some problems such as disorder and twining exist in measurements and analyses, it still represents the most important analytical method for coordination chemists. The author believes that advances in this technology will lead to an increase in the use of single crystal X-ray diffraction, including the X-ray snapshot technique which enables the capture of frame-by-frame movies of chemical reactions as they proceed *in situ*, now ubiquitous in NMR (Inokuma, Kawano & Fujita, 2011).

5. Acknowledgment

The endeavors of past and present co-workers from my research group in obtaining the results presented in this chapter are gratefully acknowledged, their names appear in the reference list. The Institute for Molecular Science, Bruker AXS Japan and Rigaku Corporation are gratefully acknowledged for performing X-ray measurements of some complexes. Finally, I thank Mr. Takashi Yamanaka for his contribution in writing this chapter.

6. References

Abe, T.; Suzuki, T. & Shinozaki, K. (2010). Luminescence Change by the Solvent of Crystallization, Solvent Reorganization, and Vapochromism of Neutral Dicyanoruthenium(II) Complex in the Solid State. *Inorganic Chemistry*, Vol.49, No.4, pp. 4794-4800

Bao, T.; Krause, K. & Krause, R. A. (1988). Hydroxide-assisted Stereospecific Isomerization of a *trans*-Dichlorobischelate of Ruthenium(II). *Inorganic Chemistry*, Vol.27, No.4, pp. 759-761

Cheng, J. Y. K.; Cheung, K.-K. & Che, C.-M. (1997). Synthesis, Characterization and Photophysical Properties of $[OsO_2(mes)_2(NC)Ru(bpy)_2(CN)]$. *Chemical Communications*, No.5, pp. 501-502

Chow, C.-F.; Lam, M. H. W.; Sui, H. & Wong, W.-Y. (2005). Design and Synthesis of Heterobimetallic Donor-Acceptor Chemodosimetric Ensembles for the Detection of Sulfhydryl-containing Amino Acids and Peptides. *Dalton Transactions*, No.3, pp. 475-484

Chowdhury, A. D.; Das, A.; Irshad, K.; Mobin, S. M. & Lahiri, G. K. (2011). Isomeric Complexes of $[Ru^{II}(trpy)(L)Cl]$ (trpy = 2,2':6',2''-Terpyridine and HL = Quinaldic Acid): Preference of Isomeric Structural Form in Catalytic Chemoselective Epoxidation Process. *Inorganic Chemistry*, Vol.50, No.5, pp. 1775-1785

Clear, J. M.; Kelly, J. M.; O'Connell, C. M.; Vos, J. G.; Cardin, C. J.; Costa, S. R. & Edwards, A. J. (1980). Isolation, Structure, Chemistry, and Photochemistry of *cis*-Bis(2,2'-bipyridyl)carbonylchlororuthenium(II) Perchlorate. *Journal of Chemical Society, Chemical Communications*, No.16, pp. 750-751

Corral, E.; Hotze, A. C. G.; Tooke, D. M.; Spek, A. L. & Reedijk, J. (2006). Ruthenium Polypyridyl Complexes Containing the Bischelating Ligand 2,2'-Azobispyridine. Synthesis, Characterization and Crystal Structures. *Inorganica Chimica Acta*, Vol.359, No.3, pp. 830-838

Cotton, F. A.; Wilkinson, G.; Murillo, C. A. & Bochmann, M. (1999). *Advanced Inorganic Chemistry, 6th Ed.*, Wiley-Interscience, ISBN 0-471-19957-5, New York, USA

Dakkach, M.; López, M. I.; Romero, I.; Rodríguez, M.; Atlamsani, A.; Parella, T.; Fontrodona, X. & Llobet, A. (2010). New Ru(II) Complexes with Anionic and Neutral N-Donor Ligands as Epoxidation Catalysts: An Evaluation of Geometrical and Electronic Effects. *Inorganic Chemistry*, Vol. 49, No.15, pp. 7072-7079

Das, C.; Saha, A.; Hung, C.-H.; Lee, G.-H.; Peng, S.-M. & Goswami, S. (2003). Ruthenium Complexes of 2-[(4-(Arylamino)phenyl)azo]pyridine Formed via Regioselective Phenyl Ring Amination of Coordinated 2-(Phenylazo)pyridine: Isolation of Products, X-ray Structure, and Redox and Optical Properties. *Inorganic Chemistry*, Vol.42, No.1, pp. 198-204

Das, D.; Mondal, T. K.; Mobin, S. M. & Lahiri, G. K. (2009). Sensitive Valence Structures of $[(pap)_2Ru(Q)]^n$ (n = +2, +1, 0, -1, -2) with Two Different Redox Noninnocent Ligands, Q = 3,5-Di-*tert*-butyl-*N*-aryl-1,2-benzoquinonemonoimine and pap = 2-Phenylazopyridine. *Inorganic Chemistry*, Vol.48, No.20, pp. 9800-9810

Deacon, G. B. & Phillips, R. J. (1980). Relationships between the Carbon-Oxygen Stretching Frequencies of Carboxylato Complexes and the Type of Carboxylate Coordination. *Coordination Chemistry Reviews*, Vol.33, No.3, pp. 227-250

Gibson, D. H.; Ding, Y.; Miller, R. L.; Sleadd, B. A.; Mashuta, M. S. & Richardson, J. F. (1999). Synthesis and Characterization of Ruthenium, Rhenium and Titanium Formate, Acetate and Trifluoroacetate Complexes. Correlation of IR Spectral Properties and Bonding Types. *Polyhedron*, Vol.18, Nos.8-9, pp.1189-1200

Goswami, S.; Chakravarty, A. R. & Chakravorty, A. (1981). Chemistry of Ruthenium. 2. Synthesis, Structure, and Redox Properties of 2-(Arylazo)pyridine Complexes. *Inorganic Chemistry*, Vol.20, No.7, pp. 2246-2250

Hotze, A. C. G.; Velders, A. H.; Ugozzoli, F.; Biagini-Cingi, M.; Manotti-Lanfredi, A. M.; Haasnoot, J. G. & Reedijk, J. (2000). Synthesis, Characterization, and Crystal Structure of α–[Ru(azpy)$_2$(NO$_3$)$_2$] (azpy = 2-(Phenylazo)pyridine) and the Products of Its Reactions with Guanine Derivatives. *Inorganic Chemistry*, Vol.39, No.17, pp. 3838-3844

Hotze, A. C. G.; Van der Geer, E. P. L.; Kooijman, H.; Spek, A. L.; Haasnoot, J. G. & Reedijk, J. (2005). Characterization by NMR Spectroscopy, X-ray Analysis and Cytotoxic Activity of the Ruthenium(II) Compounds [RuL$_3$](PF$_6$)$_2$ (L = 2-Phenylazopyridine or *o*-Tolylazopyridine) and [RuL′$_2$L″](PF$_6$)$_2$ (L′, L″ = 2-Phenylazopyridine, 2,2′-Bipyridine). *European Journal of Inorganic Chemistry*, No.13, pp. 2648-2657

Inokuma, Y.; Kawano, M. & Fujita, M. (2011). Crystalline Molecular Flasks. *Nature Chemistry*, Vol.3, No.5, pp. 349-358

Ito, K. & Bernsrein,H. J. (1956). The Vibrational Spectra of the Formate, Acetate, and Oxalate Ions. *Canadian Journal of Chemistry*, Vol.34, No.2, pp. 170-178

Kaim, W. (2001). Complexes with 2,2′-Azobispyridine and Related 'S-frame' Bridging Ligands Containing the Azo Function. *Coordination Chemistry Reviews*, Vols.219-221 (October), pp.463-488

Kepert, C. M.; Deacon, G. B.; Sahely, N.; Spiccia, L.; Fallon, G. D.; Skelton, B. W. & White, A. H. (2004). Synthesis of Heteroleptic Bis(diimine)carbonylchlororuthenium(II) Complexes from Photodecarbonylated Precursors. *Inorganic Chemistry*, Vol.43, No.9, pp. 2818-2827

Mondal, B.; Walawalkar, M. G. & Lahili, G. K. (2000). Ruthenium Terpyridine Complexes Incorporating Azo-imine Based Ancillary Ligands. Synthesis, Crystal Structure, Spectroelectrochemical Properties and Solution Reactivities. *Journal of Chemical Society, Dalton Transactions*, No.22, pp. 4209-4217

Mondal, B.; Paul, H.; Puranik, V. G. & Lahiri, G. K. (2001). Ruthenium Mononitro and Mononitroso Terpyridine Complexes Incorporating Azoimine Based Ancillary Ligands. Synthesis, Crystal Structure, Spectroelectrochemical Properties and Kinetic Aspects. *Journal of Chemical Society, Dalton Transactions*, No.4, pp. 481-487

Krause, R. A. & Krause, K. (1980). Chemistry of Bipyridyl-like Ligands. Isomeric Complexes of Ruthenium(II) with 2-(Phenylazo)pyridine. *Inorganic Chemistry*, Vol.19, No.9, pp. 2600-2603

Krause, R. A. & Krause, K. (1982). Chemistry of Bipyridyl-like Ligands. 2. Mixed Complexes of Ruthenium(II) with 2-(Phenylazo)pyridine: A New π–Bonding Probe. *Inorganic Chemistry*, Vol.21, No.5, pp. 1714-1720

Mizukawa, T.; Tsuge, K.; Nakajima, H. & Tanaka, K. (1999). Selective Production of Acetone in the Electrochemical Reduction of CO_2 Catalyzed by a Ru-Naphthyridine Complex. *Angewandte Chemie International Edition*, Vol.38, No.3, pp. 362-363

Moldes, I.; Encarnación, E.; Ros, J.; Alvarez-Larena, Á. & Piniella, J. F. (1998). Rutheium(II) Complexes Containing Both Arene and Functionalized Phosphines. Synthesis and Catalytic Activity for the Hydrogenation of Styrene and Phenylacetylene. *Journal of Organometallic Chemistry*, Vol.566, Nos.1-2, pp. 165-174

Nakamoto, K. (1986). *Infrared and Raman Spectra of Inorganic and Coordination Compounds, 4th Ed.*, Wiley-Interscience, ISBN 0-471-01066-9, New York, USA

Newkome, G. R. (1993). Pyridylphosphines. *Chemical Reviews*, Vol.93, No.6, pp. 2067-2089

Ooyama, D.; Kobayashi, T.; Shiren, K. & Tanaka, K. (2003). Regulation of Electron Donating Ability to Metal Center: Isolation and Characterization of Ruthenium Carbonyl Complexes with *N,N*- and/or *N,O*-donor Polypyridyl Ligands. *Journal of Organometallic Chemistry*, Vol.665, Nos.1-2, pp. 107-113

Ooyama, D. & Sato, M. (2003). Crystal Structure of (2,2′-Bipyridine-*N,N′*)(diphenyl-2-phosphinopyridine-*P*)chloro(dicarbonyl)ruthenium(II) Hexafluorophosphate. *Analytical Sciences*, Vol.19, No.4, pp. x39-x40

Ooyama, D. & Sato, M. (2004). A Synthetic Precursor for Hetero-binuclear Metal Complexes, $[Ru(bpy)(dppy)_2(CO)_2](PF_6)_2$ (bpy = 2,2′-Bipyridine, dppy = 2-(Diphenylphosphino)pyridine). *Applied Organometallic Chemistry*, Vol.18, No.8, pp. 380-381

Ooyama, D. & Asaha, S. (2005). Crystal Structure of a Chlororuthenium(II) Complex with Polypyridyl Ligands, 2-Pyridyl-2-azopyridine and 2,2′:6′,2′′-Terpyridine. *Analytical Sciences*, Vol. 21, No.3, pp. x43-x44

Oyama, D.; Fujita, R. & Yui, S. (2008). Mono- and Bidentate Azopyridyl Ruthenium(II) Complexes: Their Crystal Structures and Different Redox Behavior. *Inorganic Chemistry Communications*, Vol.11, No.3, pp. 310-313

Oyama, D.; Asuma, A. & Takase, T. (2008). Stereospecific Formation of Polypyridylruthenium(II) Complexes Incorporating an Asymmetrical Bidentate Ligand: Influence of Coligands. *Inorganic Chemistry Communications*, Vol. 11, No.9, pp. 1097-1099

Oyama, D.; Asuma, A.; Hamada, T. & Takase, T. (2009). Novel $[Ru(polypyridine)(CO)_2Cl_2]$ and $[Ru(polypyridine)_2(CO)Cl]^+$-type Complexes: Characterizing the Effects of Introducing Azopyridyl Ligands by Electrochemical, Spectroscopic and Crystallographic Measurements. *Inorganica Chimica Acta*, Vol.362, No.8, pp. 2581-2588

Oyama, D.; Kido, M.; Orita, A. & Takase, T. (2009). (2,2′-Bipyridine-κ^2*N,N′*)chlorido[4′-(2,5-dimethoxyphenyl)-2,2′:6′,2′′-terpyridine-κ^3*N,N′,N′′*]ruthenium(II) Hexafluoridophosphate Acetonitrile Monosolvate. *Acta Crystallographica Section E*, Vol.E65, No.9, pp. m1117-m1118

Oyama, D.; Kido, M.; Orita, A & Takase, T. (2010). Synthesis, Structure, Redox Property and Ligand Replacement Reaction of Ruthenium(II) Complexes Containing a Terpyridyl Ligand with a Redox Active Moiety. *Polyhedron*, Vol.29, No.4, pp. 1337-1343

Oyama, D.; Takatsuki, Y. & Fujita, R. (2010). Azopyridylruthenium(II) Complexes Containing Ru-C Bonds: Synthesis, Characterization and Reactivity. *Trends in Inorganic Chemistry*, Vol.12, pp. 31-40

Oyama, D.; Hamada, T. & Takase, T. (2011). Stereospecific Synthesis and Redox Properties of Ruthenium(II) Carbonyl Complexes Bearing a Redox-active 1,8-Naphthyridine Unit. *Journal of Organometallic Chemistry*, Vol.696, No.10, pp.2263-2268

Oyama, D.; Yuzuriya, K. & Takase, T. (2011). Aqua[2-(2-pyridyl)-1,8-naphthyridine-κ^2*N*^{1},*N*^{2}](2,2′:6′,2′′-terpyridine-κ^3*N*,*N*′,*N*′′)ruthenium(II) Bis(hexafluoridophosphate) Acetone Sesquisolvate. *Acta Crystallographica Section E*, Vol.E67, No.6, pp. m737-m738

Oyama, D.; Suzuki, K.; Yamanaka, T. & Takase, T. (2011). One-pot Synthesis of *cis*-Bis(2,2′-bipyridine)carbonylruthenium(II) Complexes from the Carbonato Precursor: X-ray Crystal Structures and Evaluation of Electron-Transfer Processes of the Series Complexes. *Submitted for publication*

Reedijk, J. (1996). Improved Understanding in Platinium Antitumour Chemistry. *Chemical Communications*, No.7, pp. 801-806

Rosenberg, B.; Van Camp, L.; Trosko, J. E. & Mansour, V. H. (1969). Platinum Compounds: a New Class of Potent Antitumour Agents. *Nature*, Vol.222, No.5191, pp. 385-386

Samanta, R.; Mondal, B.; Munshi, P. & Lahiri, G. K. (2001). Mononuclear and Dinuclear Ruthenium(II)/(III) Salicylates Incorporating Azoimine Functionalities as Ancillary Ligands. Synthesis, Spectroscopic and Electron-Transfer Properties. *Journal of Chemical Society, Dalton Transactions*, No.12, pp. 1827-1833

Samanta, S.; Singh, P.; Fiedler, J.; Záliš, S.; Kaim, W. & Goswami, S. (2008). Signal Diradical Complexes of Ruthenium and Osmium: Geometrical and Electronic Structures and Their Unexpected Changes on Oxidation. *Inorganic Chemistry*, Vol.47, No.5, pp. 1625-1633

Sarkar, B.; Patra, S.; Fiedler, J.; Sunoj, R. B.; Janardanan, D.; Lahiri, G. K. & Kaim, W. (2008). Mixed-Valent Metals Bridged by a Radical Ligand: Fact or Fiction Based on Structure-Oxidation State Correlations. *Journal of the American Chemical Society*, Vol.130, No.11, pp. 3532-3542

Seal, A. & Ray, S. (1984). Structures of Two Isomers of Dichlorobis(2-phenylazopyridine)ruthenium(II), $[RuCl_2(C_{11}H_9N_3)_2]$. *Acta Crystallographica Section C*, Vol.C40, No.6, pp. 929-932

Shivakumar, M.; Pramanik, K.; Bhattacharyya, I. & Chakravorty, A. (2000). Chemistry of Metal-Bound Anion Radicals. A Family of Mono- and Bis(azopyridine) Chelates of Bivalent Ruthenium. *Inorganic Chemistry*, Vol.39, No.19, pp. 4332-4338

Storrier, G. D.; Colbran, S. B. & Craig, D. C. (1998). Transition-metal Complexes of Terpyridine Ligands with Hydroquinone or Quinone Substituents. *Journal of Chemical Society, Dalton Transactions*, No.8, pp. 1351-1363

Tomon, T.; Koizumi, T. & Tanaka, K. (2005). Stabilization and Destabilization of the Ru-CO Bond During the 2,2′-Bipyridin-6-onato (bpyO)-Localized Redox Reaction of $[Ru(terpy)(bpyO)(CO)](PF_6)$. *European Journal of Inorganic Chemistry*, No.2, pp. 285-293

Tomon, T.; Koizumi, T. & Tanaka, K. (2005). Electrochemical Hydrogenation of $[Ru(bpy)_2(napy\text{-}\kappa N)(CO)]^{2+}$: Inhibition of Reductive Ru-CO Bond Cleavage by a Ruthenacycle. *Angewandte Chemie International Edition*, Vol.44, No.15, pp. 2229-2232

Tseng, H.-W.; Zong, R.; Muckerman, J. T. & Thummel, R. (2008). Mononuclear Ruthenium(II) Complexes That Catalyze Water Oxidation. *Inorganic Chemistry*, Vol.47, No.24, pp. 11763-11773

Velders, A. H.; Kooijman, H.; Spek, A. L.; Haasnoot, J. G.; de Vos, D. & Reedijk, J. (2000). Strong Differences in the in Vitro Cytotoxicity of Three Isomeric Dichlorobis(2-phenylazopyridine)ruthenium(II) Complexes. *Inorganic Chemistry*, Vol.39, No.14, pp. 2966-2967

Velders, A. H.; Van der Schilden, K.; Hotze, A. C. G.; Reedijk, J.; Kooijman, H. & Spek, A. L. (2004). Dichlorobis(2-phenylazopyridine)ruthenium(II) Complexes: Characterisation, Spectroscopic and Structural Properties of Four Isomers. *Dalton Transactions*, No.3, pp. 448-455

Wada, T.; Fujihara, T.; Tomori, M.; Ooyama, D. & Tanaka, K. (2004). Strong Interaction between Carbonyl and Dioxolene Ligands Caused by Charge Distribution of Ruthenium-Dioxolene Frameworks of Mono- and Dicarbonylruthenium Complexes. *Bulletin of the Chemical Society of Japan*, Vol.77, No.4, pp. 741-749

Zhou, Y.; Xiao, H.-P.; Kang, L.-C.; Zuo, J.-L.; Li, C.-H. & You, X.-Z. (2009). Synthesis and Characterization of Neutral Iron(II) and Ruthenium(II) Complexes with the Isocyanotriphenylborate Ligand. *Dalton Transactions*, No.46, pp. 10256-10262

11

X-Ray Structural Characterization of Cyclometalated Luminescent Pt(II) Complexes

Viorel Cîrcu and Marin Micutz
University of Bucharest, Romania

1. Introduction

The cyclometalated complexes represent one of the most interesting and broadly studied class of organotransition metal complexes. Although there is a strong interest in studying the mechanism of this bond-activation process, cyclometalation is a highly attractive and versatile synthetic approach for generating organometallic systems, with very important application potential (Crabtree, 2005). There are both mononuclear and dinuclear species, but also polynuclear cyclometalated complexes are known (Diez et al., 2011). Many reviews and books have been dedicated to this topic over the past years and one of the most recent can be found here (Albrecht, 2010).

The cyclometalation process consists of a transition metal-mediated activation of a C-R bond to form a metallacycle that contains a metal-carbon σ bond (Hill, 2002). On the other hand, cyclometalation can be regarded as a special case of oxidative addition, in which a C−R (in most cases, C-H) bond in a ligand oxidatively adds to a metal to give rise to a ring.

Although many examples are described, by far most of cyclometalation reactions occur via C-H bond activation. The reaction product is a metallacycle in which the metal is bound by a chelate C^- donor ligand. It is important to note that such chelation leads to organometallic compounds with increased stability. Altogether, the cyclometalation reaction has been widely studied because it represents probably one of the mildest route for activating strong C-H and C-R bonds. The tendency of transition metal salts to undergo cyclometalation reaction, and, in particular, ortho-metalation reaction, with heteroaromatic ligands (mostly including nitrogen donors, but oxygen-, sulfur- and phosphorus-containing ligands have also been cyclometalated) to give five-memberd metallocycles has been demonstrated with various metals, including, for instance, Re(I), Pt(II), Pd(II). This review will take into account only the cyclometalated Pt(II) complexes with nitrogen-containing ligands.

In comparison with Pd(II), which is by far the most used metal in cyclometalation reactions, the cycloplatination reaction is not so intensively studied and not very easy to accomplish (cycloplatination reactions which took about four weeks or required relatively forcing conditions, e.g., refluxing toluene, with poor yields, have been reported). However, it is possible to increase the yields and reduce the time of reaction by using different starting materials such as bis(η^3-allyl)-di-μ-chlorodiplatinum(II) or $PtCl_2(DMSO)_2$, etc, although K_2PtCl_4 or $[Pt_2Me_4(\mu\text{-}SMe_2)_2)]$ are commonly used to yield cyclometalated species. In most of the cases, the reaction products are halo-bridged dimers, that can be used further to form

mononuclear cyclometalates of the type [MX(C^N)L^1] or [M(C^N)L^2] (M = Pt, Pd; C^N = orthometalated ligand; L^1 = neutral monodentate ligand such as pyridine and phosphines; L^2 = bidentate uninegative ligand such as acetylacetonate derivatives; X = halide).
The Pt(II) ions adopt a square-planar geometry, being part of the major group of exceptions to the otherwise very successful model of Kepert. They show coordinative unsaturation which can allow different interactions such as: excimer formation, chemical quenching, interactions with Lewis bases, and oxidative addition. The single-crystal X-ray diffraction is a very powerful technique for characterization of cyclometalated platinum(II) complexes and strong correlations between the structure and luminescence properties can be made as, for example, the solid-state emission is greatly influenced by the crystal packing and the presence of Pt···Pt or π···π interactions is favoured by the coordinative unsaturation. This review is intended to cover the most important structural types of cyclometalated Pt(II) complexes that have been investigated by X-ray diffraction with an emphasis on their luminescence properties.

2. Photophysical properties of cyclometalated platinum(II) complexes

Cyclometalated Pt(II) complexes have been extensively investigated in the past years because of their interesting photophysical properties and several reviews are dealing with this topic (Evans et al., 2006; Williams et al., 2008; Balashev et al., 1997; Ma et al., 2005; Williams, 2007; McGuire Jr. et al., 2010). Although there is a particular fundamental interest in studying their intrinsic emissive states, their luminescence properties find applications in optoelectronic devices (such as OLED, Williams et al., 2008), luminescent probes for biomolecules (cell imaging, biochemistry, Siu et al., 2005) and chemical sensors (for a recent example see Li et al., 2011). Figure 1 shows a simplified schematic representation (Perrin-Jablonski diagram) of the most important processes that take place through the interaction of matter with light: photon absorption, internal conversion, fluorescence, intersystem crossing and phosphorescence (Valeur, 2001). The emission of photons that accompanies the $S_1 \rightarrow S_0$ relaxation is called *fluorescence*. Another process visible on this diagram is the *internal conversion* that is a non-radiative transition between two electronic states of the same spin multiplicity. Also, still a non-radiative process is the *intersystem crossing* that represents a transition between two isoenergetic vibrational levels belonging to electronic states of different multiplicities. Crossing between states of different multiplicity is in principle forbidden, but spin-orbit coupling (i.e. coupling between the orbital magnetic moment and the spin magnetic moment) can be large enough to make it possible. The probability of intersystem crossing depends on the singlet and triplet states involved. If the transition $S_0 \rightarrow S_1$ is of $n \rightarrow \pi^*$ type, for instance, intersystem crossing is often efficient. It should also be noted that the presence of heavy atom (i.e. whose atomic number is large, as it is the case of Pt) increases spin-orbit coupling and thus favors intersystem crossing. It was found that in solution, at room temperature, non-radiative de-excitation from the triplet state, T_1, is preferred rather than the radiative de-excitation called *phosphorescence*. This happens because the transition $T_1 \rightarrow S_0$ is forbidden (but it can be observed because of spin-orbit coupling), and the radiative rate constant is thus very low. On the contrary, at low temperatures and/or in a rigid medium, phosphorescence can be observed. The lifetime of the triplet state may, under these conditions, be long enough to observe phosphorescence on a time-scale up to seconds, even minutes or more. Fluorescence and phosphorescence are particular cases of *luminescence* (emission of light from an electronically excited species).

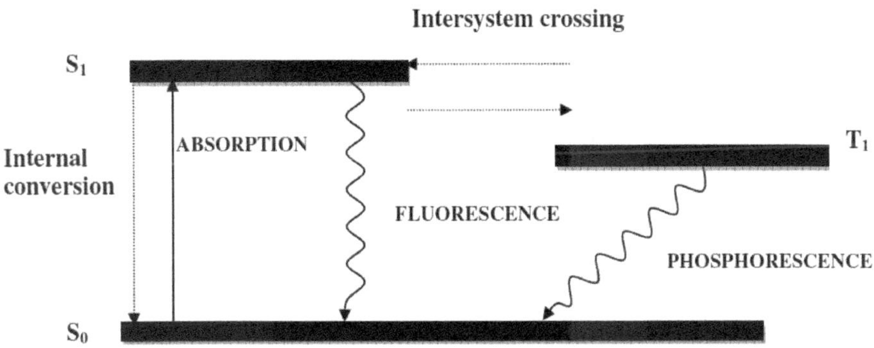

Fig. 1. Schematic representation of the radiative and non-radiative pathways

Thus, in the case of transition metal complexes, there are four types of electronic states or transitions responsible for their luminescent properties, as follows: *1)* dd states (metal-centred, MC, transition). By coordination of the ligands to the metal centre, the d orbitals are split according to the symmetry of the complex. These excited dd states arise from promotion of an electron within d orbitals; *2)* dπ* states (metal-to-ligand-charge-transfer, MLCT). These involve excitation of a metal-centred electron to a π* anti-bonding orbital located on the ligand system; *3)* ππ* or nπ* states (intraligand, IL, transition). Promotion of an electron from a π-bonding or non-bonding orbital to a higher energy anti-bonding orbital gives rise to these states; *4)* πd states (ligand-to-metal-charge-transfer, LMCT). These states arise from the transfer of electronic charge from the ligand π system to a metal-centred orbital.

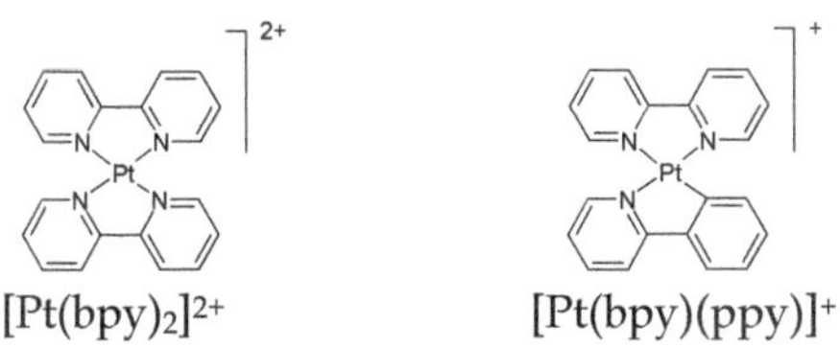

Scheme 1.

The luminescence of Pt(II) complexes is assigned to either ligand-centred (LC) or metal-to-ligand charge transfer (MLCT) states or a mixed of both. The most useful strategy to promote luminescence in platinum(II) complexes is to employ ligands with a very strong ligand field in order to raise the dd states, so that they are not easily accessible for radiationless deactivation process, and this can be achieved by using cyclometalating ligands, mostly with 2-arylpyridine or 2-thienylpyridine derivatives, resulting both homoleptic (metal complexes with identical ligands) or heteroleptic (metal complexes with different ligands) complexes. The influence of cyclometalated ligands on the photophysical properties of square-planar Pt(II) complexes is clearly seen when the Pt(II) complex with 2,2′-bipyridine (bpy) is compared to the cyclometalated Pt(II) complex with 2-phenylpyridine (ppy) (Scheme 1). While the $[Pt(bpy)_2]^{2+}$ is almost non-emissive at room temperature, the Pt(II) cyclometalated complex $[Pt(bpy)(ppy)]^{+}$ shows emission with a quantum yield nearly 0.02 at 298K (Kvam, Jongstad, 1995; Kvam et al., 1995).

3. X-ray structural characterization of cyclometalated Pt(II) complexes

3.1 Pt(II) complexes with 2-phenylpyridine (ppy) and its derivatives as cyclometalating ligand

The homoleptic Pt(II) complex with 2-phenylpyridine was prepared by reacting $[PtCl_2(Et_2S)_2]$ with the lithiated ligand as the *cis* isomer and its crystal structure was reported in 1984 (Chassot et al., 1984).

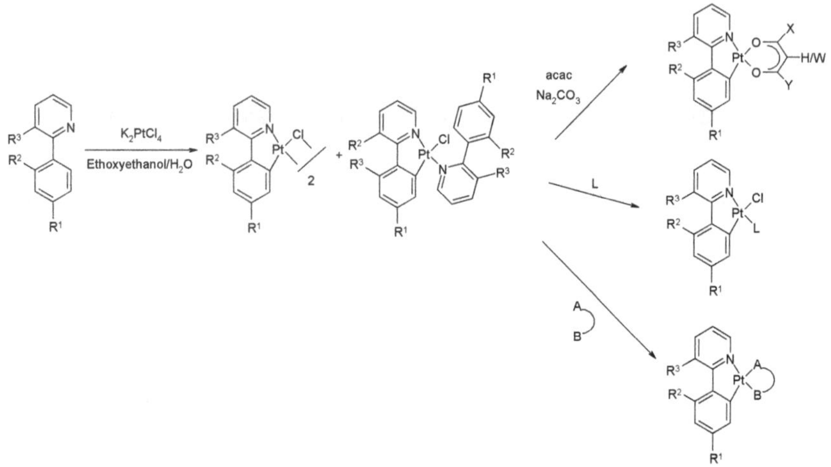

R^1=NPh_2, R^2=R^3=H, X=Y=Me **1a**
R^1=-oxadiazole-4-Ph-p-NPh_2, R^2=R^3=H, X=Y=Me **1b**
R^1=-oxadiazole-4-Ph-p-$NTol_2$, R^2=R^3=H, X=Y=Me **1c**
R^1=R^2=R^3=H, X=Y=Ph **1d**
R^1=R^2=F, R^3=H, X=Y=Me, W=-$(CH_2)_2$-norbornene **1e**
R^1=R^2=R^3=H, L=CO **2a**
R^1=R^2=R^3=H, L=DMSO **2b**
R^1=F, R^2=R^3=H, L=DMSO **2c**
R^1=F, R^2=R^3=H, L=py-2-Ph-4'-F **2d**
R^1=R^2=F, R^3=H, L=py-2-Ph-2',4'-F_2 **2e**
R^1=R^2=F, R^3=H, A^B=3,5-di-*t*Bu-catechol **3**
R^1=R^2=H,R^3=-OHex, A^B=5-NO_2-8-quinoline **4**
R^1=R^2=R^3=H, A^B=$Et_2NC(S)NCOPh$ **5a**
R^1=R^2=R^3=H, A^B=MeO-4-$C_6H_4NHC(S)NCOPh$ **5b**
R^1=R^2=R^3=H, A^B=2-pyridyl hexafluoropropoxide **6**
R^1=R^2=F, R^3=H, A^B=$(pz)_2BEt_2$ **7**
R^1=R^2=R^3=H, A^B=(3,5-$Me_2pz)BH_2$ **8**
R^1=R^2=R^3=H, A^B=dppm, PF_6^- **9**

Scheme 2. The preparation of mononuclear cyclometalated Pt(II) complexes with 2-phenylpyridine derivatives

Generally, the heteroleptic complexes are prepared by the cyclometalation of the coresponding ligand using K_2PtCl_4 in ethoxyethanol to give, in the first step, as a major product, the chloro-bridged dinuclear complex and, in some cases, as a minor product, the mononuclear complex having two molecules of ppy ligand. In the next step, the bridge-splitting reaction, various monodentate (L) or bidentate (A^B) ligands can be used to yield the mononuclear neutral or ionic heteroleptic complexes (see Scheme 2). By far, the most studied Pt(II) cyclometalated complexes bearing ppy ligand for their emission properties are the ones containing acetylacetonate (acac) derivatives as ancillary ligands (Brooks et al, 2002). Various other R^1-R^3 substituents were attached on the ppy core with the aim of conferring different multifunctionalities to the resulting molecules. For instance, He et al.,

2006, reported the synthesis and structural, photophysical, electrochemical, and electroluminescent properties of a novel class of trifunctional Pt(II) cyclometalated complexes incorporating the hole-transporting triarylamine, electron-transporting oxadiazole, and electroluminescent metal components into a single molecule (**1a-c**). Other studies focused on functionalizing the acac derivatives, as it is the example reported by Cho et al., 2007, in which the norbornene-functionalized derivative of acetylacetone has been used to synthesize a series of new polymerizable norbornene-derivatized phosphorescent platinum complexes (**1e**). For these acac Pt(II) complexes, it was found that the two Pt-O bonds are different, with the Pt-O1 trans to C atom bond longer than the Pt-O2 trans to N bond due to the strong trans influence of C-ppy donor atom (Table 1).

Several crystal structures of cyclometalated Pt(II) complexes bearing monodentate neutral ligands such as DMSO or pyridine derivatives were reported (**2a-e**). Another strategy to obtain mononuclear Pt(II) species is to use bidentate mononegative ligands and various such examples were reported so far (**3** – **8**).

By using sodium salts of N-benzoylthiourea derivatives, a series of luminescent Pt(II) ppy complexes were prepared and investigated by single-crystal X-ray diffraction (**5a**, **b**) (Figure 2).

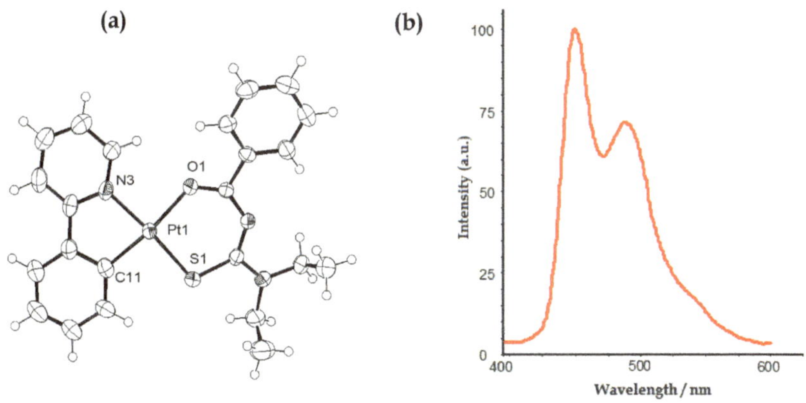

Fig. 2. ORTEP view of complex **5a** (a); the emission spectrum of **5a** recorded in CH_2Cl_2 solution at room temperature (b)

An interesting feature of the structure of complex **5b** is the orientation adopted by the *p*-anisyl ring of the *N*-benzoyl thiourea ligand with a twist of 65.1° with respect to the core plane, leading to the formation of weak intermolecular NH···Pt interactions (H···Pt 2.74 Å; N-H···Pt 156.8°) compared to those found in the salt $(N^nPr_4)_2[PtCl_4][PtCl_2(NH_2Me)_2]$. In this case, the shortest Pt-Pt distance between two neighbouring molecules is 4.18 Å.

Complex	Pt-C	Pt-N	Pt-O1	Pt-O2	Reference
1a	1.957(4)	1.995(4)	2.094(3)	2.001(3)	He et al., 2006
1b	1.921(13)	1.955(11)	2.088(9)	1.991(9)	He et al., 2006
1c	1.969(3)	1.996(3)	2.079(2)	2.002(2)	He et al., 2006
1d	1.954(5)	1.978(5)	2.082(4)	2.006(4)	Liu et al., 2009
1e	2.002(8)	2.020(9)	2.070(7)	2.058(6)	Cho et al., 2007
	Pt-C	Pt-N	Pt-Cl	Pt-L	
2a	1.981(24)	2.114(19)	2.391(11)	1.84(3)	Mdleleni et al., 1995
2b	2.001(4)	2.055(3)	2.4187(1)	2.2181(1)	Godbert et al., 2007
2c	1.998(5)	2.065(4)	2.3962(17)	2.2161(16)	Newman et al., 2007
2d	1.941(10)	2.002(7)	2.412(3)	2.023(8)	Newman et al., 2007
2e	1.987(6)	2.014(7)	2.383(2)	2.045(5)	Rausch et al., 2010
	1.983(6)	2.021(5)	2.3912(16)	2.046(5)	
	Pt-C	Pt-N	Pt-A	Pt-B	
3	1.976(10)	1.969(10)	2.123(8)	1.981(8)	Hirani et al., 2007
4	1.988(2)	1.989(2)	2.0969(18)	2.028(2)	Niedermair et al., 2008
5a	1.973(5)	2.045(5)	2.076(3)	2.2389(15)	Cîrcu et al., 2009a
5b	1.979(6)	2.035(5)	2.075(4)	2.2447(14)	Cîrcu et al., 2009a
6	1.992(4)	1.989(3)	2.082(3)	2.033(3)	Chang et al., 2008
7	1.981(5)	2.006(5)	2.074(4)	2.010(4)	Ma et al., 2005
8	1.997(2)	2.013(2)	2.1092(16)	2.0297(19)	Niedermair et al., 2007
9	2.038(7)	2.073(6)	2.318(2)	2.228(2)	DePriest et al., 2000

Table 1. Selected bond lengths for compounds **1 - 9**

3.2 Cyclometalated Pt(II) complexes with "pincer" ligands

The cyclometalated Pt(II) complexes with terdentate "pincer" type ligands were investigated intensively, in part due to the fact that such ligands may provide additional rigidity to the molecule that could be responsible for favouring the luminescence properties. Many types of pincer ligands have been used in the preparation of both mono or polynuclear organometallic Pt(II) complexes and they are divided according to the donor atom set involved in bond formation with the metal centre, as follows: N^C^N, C^N^N and C^C^N complexes.

3.2.1 N^C^N ligands

Several crystal structures of N^C^N Pt(II) complexes were reported and these are summarised in Scheme 3, while the selected bond lengths are included in Table 2. The chloride Pt(II) complexes (**10a-c**, **12**, **13**) can be easily obtained by orthometalation reaction of the corresponding N^C^N pincer ligand with K_2PtCl_4 in acetic acid with high yields, while the acetylide complexes **11a-c** were prepared in quantitative yields starting from the corresponding chloride precursor, with an excess of aryl acetylene in methanol in the presence of NaOH for one day at room temperature. The crystal structures of complexes **11b** and **11c** show several interesting features such as the intermolecular Pt···Pt, π···π and C-H···π(C≡C) interactions in an orthogonal configuration. On the other hand, the crystal structure of **11a** shows the formation of dimers in a head-to-tail fashion with an interplanar distance of approximately 3.4 Å and a shortest intermetallic separation of 4.886 Å, indicating the presence of π···π interaction, but not a Pt···Pt interaction. The presence of such intermetallic interactions has a marked impact upon the solid state emission properties.

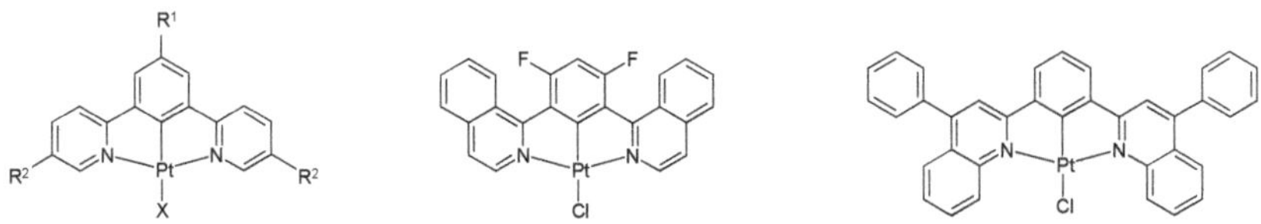

$R^1=R^2=H$, X=Cl **10a** **12** **13**
$R^1=H$, $R^2=Me$, X=Cl **10b**
$R^1=-C(O)OCH_3$, $R^2=H$, X=Cl **10c**
$R^1=R^2=H$, X=-C≡C-Ph **11a**
$R^1=CF_3$, X=-C≡C-Ph **11b**
$R^1=CF_3$, X=-C≡C-Ph-4-NMe_2 **11c**

Scheme 3.

Complex	Pt - C	Pt-N1	Pt-N2	Pt-X	Reference
10a	1.907(8)	2.033(6)	2.041(6)	2.417(2)	Cardenas et al., 1999
10b	1.915(3)	2.041(2)	2.027(2)	2.409(1)	Abe et al., 2007
10c	1.903(4)	2.036(3)	2.035(3)	2.4050(10)	Williams et al., 2003
12	1.918(8)	2.021(7)	2.009(7)	2.412(2)	Wang et al., 2010
13	1.908(3)	2.075(3)	2.060(3)	2.4343(9)	Baik et al., 2006

Table 2. A selection of bond lengths for some N^C^N Pt(II) complexes

The room temperature structureless emissions at λ_{max} 819 (**11b**) and 702 nm (**11c**), respectively, strongly red-shifted compared to solution emission (λ_{max} = 486 and 480, respectively), can be assigned to triplet metal-metal-to-ligand charge-transfer (^{3}MMLCT) excited states. The photophysical properties of chloride complexes (**10a-c**) and analogous with various other R^1 substituents were systematically studied by Farley et al., 2005. It is interesting to note that these N^C^N Pt(II) complexes show relatively shorter Pt-C distances than those found in C^N^N Pt(II) complexes (see comparatively Tables 2 and 3).

3.2.2 C^N^N ligands

The C^N^N type ligands were used in the preparation of both neutral or ionic complexes (see Scheme 4).

An important feature of these complexes is that the central Pt-N1 bond is shorter than the marginal Pt-N2 bond (Table 3) due to the *trans* influence of the strong field cyclometalated ligand. Two crystal structures of **14a** were determined (yellow and red polymorphs). The red colour of the latter polymorph was attributed to the Pt···Pt intermolecular interactions (3.366 Å).

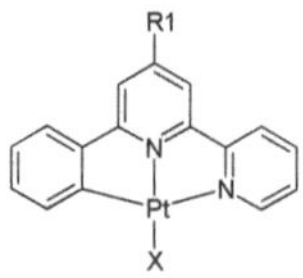

R^1=H, X=Cl **14a**
R^1=Tol, X=Cl 14**b**
R^1=Ph-4-COOMe, X=Cl **14c**
R^1=H, X=-C≡C-Ph **15a**
R^1=Ph, X=-C≡C-*n*Pr **15b**
R^1=H, X=-C≡C-Me_2-fluorene **15c**

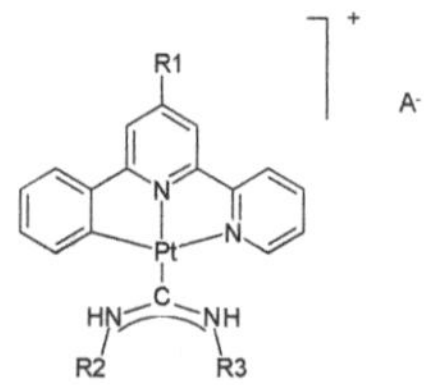

R^1=H, X=-C≡C-Me-carbazole **15d**
R^1=H, X=N≡CXyl, A=PF_6^- **16**
R^1=H, R^2=Me, R^3=*t*Bu, A=ClO_4^- **17a**
R^1=H, R^2=NH_2, R^3=*t*Bu, A=ClO_4^- **17b**
R^1=H, R^2=CH_2Ph, R^3=*t*Bu, A=ClO_4^- **17c**
R^1=H, R^2=Me, R^3=Xyl, A=ClO_4^- **17d**

Scheme 4.

Complex	Pt - C	Pt - N1	Pt-N2	Pt-X	Reference
14a	2.014(9)	1.946(6)	2.124(8)	2.316(2)	Hofmann et al., 2003
	1.995(8)	1.950(6)	2.116(6)	2.307(2)	
14b	1.993(3)	1.954(3)	2.109(3)	2.3657(7)	Schneider et al., 2009
14c	2.10(2)	2.01(2)	1.99(2)	2.323(7)	Schneider et al., 2009
	2.14(2)	2.05(3)	1.96(2)	2.306(7)	
14d	1.987(9)	1.943(8)	2.091(9)	2.315(3)	Qiu et al., 2009
14e	2.021(4)	1.868(3)	2.082(4)	2.2807(15)	Qiu et al., 2010
15a	1.992(5)	1.987(4)	2.123(4)	1.970(5)	Lu et al., 2004
15b	2.060(13)	1.986(10)	2.109(12)	1.982(16)	Shao et al., 2008
15c	2.068(4)	1.984(4)	2.042(4)	1.956(5)	Seneclauze et al., 2007
15d	1.988(9)	1.992(6)	2.119(7)	1.973(9)	Seneclauze et al., 2007
16	2.079(19)	2.004(25)	2.090(13)	1.839(26)	Yuen et al., 2008
17a	2.001(8)	1.995(6)	2.117(6)	1.997(7)	Lai et al., 1999
17b	2.010(5)	1.998(4)	2.109(4)	1.992(4)	Lai et al., 1999
17c	1.989(6)	1.999(5)	2.087(6)	1.989(6)	Lai et al., 1999
17d	2.033(9)	2.006(6)	2.110(8)	1.996(8)	Lai et al., 1999

Table 3. Selected bond lengths for C^N^N Pt(II) complexes **14 - 17**

The chloride ligand can be easily replaced by acetylide ligands to yield various neutral Pt(II) complexes that could be investigated by X-ray crystallography. Thus, such acetylide complexes show high efficient luminescence properties due to the combination of the strong field cyclometalated ligand with the strong field acetylide ligands and the rigidity of the terdentate ligands. Important, their emission properties could be tuned by varying either R^1 substituent on the C^N^N ligand or the acetylide ligand and several studies were reported on this topic (Lu et al., 2004). A series of luminescent cyclometalated platinum(II) diamino-carbene complexes (**17a-d**) were prepared by nucleophilic attack of amines at the coordinated isocyanide ligands of $[(C\text{^}N\text{^}N)PtC{\equiv}NR]^+$ (R = *t*Bu or Ar). Weak π···π stacking interactions between the cyclometalated ligands were observed in the crystal lattice (range 3.5−3.6 Å). These complexes, **17a-d,** show structureless emissions, with λ_{max} ranging from

528 to 558 nm in acetonitrile at 298 K, which were assigned to ^{3}MLCT excited states. Also, dinuclear phosphorescent C^N^N Pt(II) cyclometalated were prepared (also investigated by X-ray crystallography) by using diacetylide - carbazole bridging ligand (Seneclauze et al., 2007).

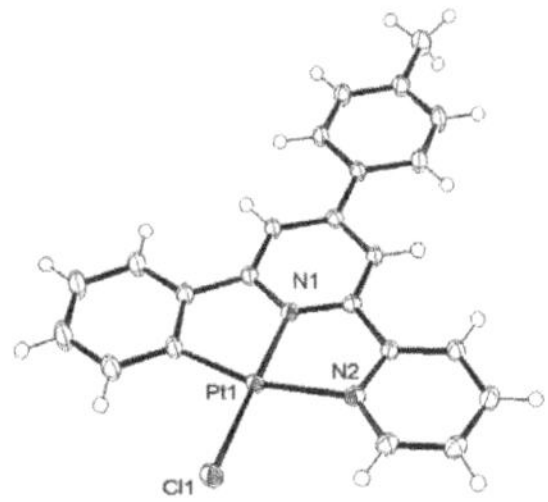

Fig. 3. ORTEP view of complex **14b**

3.2.3 C^N^C ligands

The C^N^C cyclometalated Pt(II) complexes are synthesised by using K_2PtCl_4 and the corresponding 2,6-diphenyl-pyridine derivative, in two consecutive cyclometalation reactions. In the first step, a dinuclear chloride-bridged C^N cyclometalated Pt(II) complex is obtained, followed by reaction with a monodentate neutral ligand to give mononuclear neutral C^N^C species. Several crystallographic studies were carried out on such systems and these are summarised in Scheme 5 and Table 4.

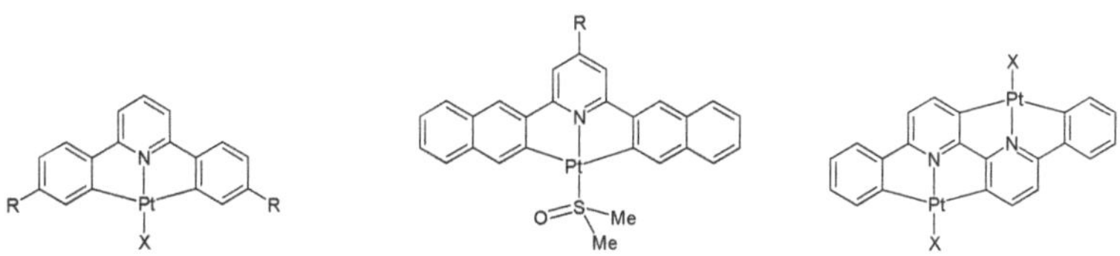

X=pz (pyrazine), R=H **18a**
X=py-4-NMe$_2$, R=H **18b**
R=OC$_6$H$_{13}$, X=CO **18c**

R=H **19a**
R=Ph **19b**
R=Ph-4-Br **19c**

X=3,5-Me$_2$-py **20**

Scheme 5.

Complexes **19a-c** were prepared by reacting K_2PtCl_4 with the terdentate 2,6-di-(2′-naphthyl)-pyridine ligands in glacial acetic acid, followed by heating in DMSO. Their crystal structures reveal that the molecules are paired in a head-to-tail orientation with Pt···Pt separations longer than 6.3 Å, with extensive close C-H···π (d = 2.656–2.891 Å) and π···π (d = 3.322–3.399 Å) interactions. An interesting feature of such C^N^C Pt(II) complexes is that the Pt-C bonds are longer compared to the N^C^N or N^N^C Pt(II) complexes and this is due the *trans* influence of C atoms on each other. These complexes are emissive in both solid state, with the maxima in the range 600 - 650 nm (**19a-c**), and in solution at 77 K (**18a**). Additionally, several dinuclear C^N^C cyclometalated Pt(II) complexes, either by using a dicyclometalated ligand (Zucca et al., 2006) or a diphosphine ligand as a bridge (dppm, diphenylphosphinomethane, Kui et al., 2006), were investigated by X-ray crystallography.

Complex	Pt-C1	Pt-C2	Pt-N	Pt-X	Reference
18a	2.05(1)	2.08(5)	-	1.99(7)	Kulikova et al., 2003
18b	2.06(1)	2.05(1)	1.98(1)	2.04(1)	Lusby et al., 2009
18c	2.070(6)	2.055(6)	2.007(5)	1.827(6)	Newman et al., 2001
19a	2.095(5)	2.074(4)	2.011(4)	2.1915(13)	Kui et al., 2006
19b	2.082(6)	2.058(6)	2.008(4)	2.1927(12)	Kui et al., 2006
19c	2.101(6)	2.070(6)	2.016(5)	2.1959(18)	Kui et al., 2006
20	2.054(6)	2.104(6)	1.983(5)	2.036(5)	Zucca et al., 2006

Table 4. Selected bond lengths for C^N^C Pt(II) compounds **18 - 20**

3.3 Pt(II) complexes with benzo(h)quinoline (bzq) as cyclometalating ligand

The homoleptic Pt(II) complex with benzo(h)quinoline was prepared by reacting $[PtCl_2(Et_2S)_2]$ with the lithiated ligand, its crystal structure being reported in 1996 (Jolliet et al., 1996). It was found that the crystal structure shows an important distortion from the square planar geometry towards a two-bladed helix with average Pt–C and Pt–N distances being 1.988 and 2.151 Å, respectively.

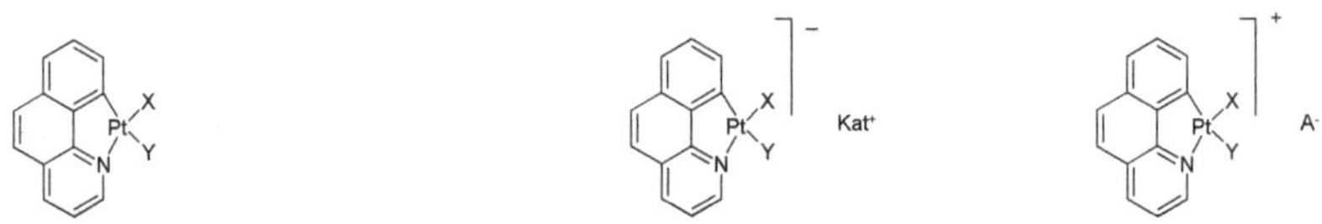

X=Y=-C≡C-Ph, Kat=NBu_4^+ **21**
X=Y=-C≡C-Py-2, Kat=NBu_4^+ **22**
X=Y=-C≡C-Ph, Kat=Pb^{2+} **23**
X=-C≡C-Ph, Y=PPh_2(C≡CPh) **24**
X=Y=-C≡N-Xyl, A=ClO_4^- **25a**
X=Y=-C≡N-Xyl, A=PF_6^- **25b**
X=-C≡N-Xyl, Y=Cl **26**
X=-C≡N-*t*Bu, Y=Cl **27**
X=-C≡N-2-Np, Y=Cl, Np=Naphtyl **28**
X=$PHPh_2$, Y=Cl **29**
X=Y=2-pyridyl hexafluoropropoxide **36**

X=Y=PPh_2 **30**
X=Y=dppm, PF_6^- **31a**
X=Y=dppe, A=PF_6^- **31b**
X=Y=dppp, A=PF_6^- **31c**
X=dppm, Y=4-Tol, Tol=tolyl **32**
X=DMSO, Y=Cl **33a**
X=4-MeO-Py, Y=Cl **33b**
X=SMe_2, Y=4-Tol **34**
X=Y=[9]aneS_3,
1,4,7-trithiacyclononane, A=PF_6^- **35**

Scheme 6. Heteroleptic cyclometalated Pt(II) complexes containing the benzo(h)quinoline ligand

Various heteroleptic cyclometalated Pt(II) complexes either ionic or neutral were synthesized starting from the binuclear μ-chloro bridged complex, and analysed through single-crystal X-ray diffraction. They are summarised in Scheme 6 and their crystallographic data are included in Table 5.

An interesting example showing how the Pt···M interactions affect the photoluminescence properties for such complexes bearing the cyclometalated bzq together with acetylide ligands is represented by complex **23**, whose molecular structure is depicted in Figure 4, which has very short $Pt^{II}\cdots Pb^{II}$ (2.9758(5) and 2.9182(5) Å) and Pt···Pt distances (3.579 Å). For this complex, the emissive state in solid state (77 K) is attributed to a ^{3}MLM'CT [Pt(1)π(C≡CPh)→Pt(2)/Pb(sp)π*(C≡CPh)] state mixed with some ππ* excimeric character.

Another interesting group of cyclometalated Pt(II) complexes bearing benzoquinolinate ligand is represented by those complexes containing isocyanide ligands. They have been found to show interesting photophysical properties, including low-energy emissions in fluid solution, depending on either the counteranion, isocyanide (CNR) ligand or crystal packing. Diez et al. recently reported the X-ray crystal structures of the isocyanide benzoquinolate Pt(II) complexes [Pt(bzq)(CNR)$_2$]X (R = Xyl, X = ClO_4^-, PF_6^-, **25a,b**) as well as the influence of the counteranion and concentration on their luminescent properties (Diez et al., 2009a). The crystal packing of the two complexes containing the cation [Pt(bzq)(CN-Xyl)$_2$]$^+$ and different counteranions (**25a** and **25b**) was found to show significant differences, although in both cases π···π intermolecular interactions are present. The same authors (Diez et al., 2010a) reported the crystal structures and the photophysical properties of the neutral complexes [Pt(bzq)Cl(CNR)] (R= Xyl, 8, ^{t}Bu 9, 2-Np 10), where Xyl=2,6-dimethyl-phenyl or xylyl and Np = naphtyl. It was found that compound **26** forms only yellow crystals in which two monomers are weakly contacting through π···π (bzq) interactions. The complex **27** with tert-butylisocyanide ligand shows solid-state pseudopolymorphic behaviour. Thus, two X-ray structures are reported: a red form, which exhibits an infinite 1-D chain network, including a molecule of chloroform as crystallizing solvent, [27·CHCl$_3$]$_\infty$ (**27a**), and a yellow form which consists of discrete dimers, [27·0.5H$_2$O]$_2$ (**27b**). The crystal packing of these two forms is stabilised by short Pt···Pt distances and π···π (bzq) interplanar bonding interactions. Indeed, the extended 1D-chain of **27a** exhibits equivalent Pt(II)-Pt(II) distances of 3.3547(2) Å and a nearly linear Pt···Pt···Pt angle [169.12(2)°], thus indicating some degree of Pt···Pt interactions along the chain.

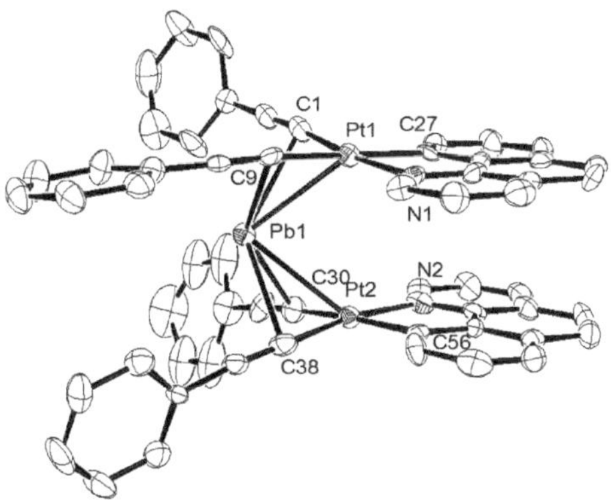

Fig. 4. ORTEP view of complex **23**

When the 2-naphthylisocyanide ligand was employed, again, two isomeric species have been isolated, a neutral yellow derivative, which crystallizes as a Pt···Pt dimer, [Pt(bzq)Cl(CN-2-Np)]$_2$ (**28a**), and a double salt [Pt(bzq)(CN-2-Np)$_2$]$^+$[Pt(bzq)Cl$_2$]$^-$ (**28b**).
Another category of ancillary ligands widely used in the preparation of cyclometalated Pt(II) benzoquinolinate complexes are the phosphine ligands, both in mono and bidentate form. The simplest one is the neutral complex **29** in which the diphenylphosphine (PHPh$_2$) ligand is bound to the platinum center while the coordination sphere is completed with a chloride anion. It is interesting that when the [Pt(bzq)(μ-Cl)]$_2$ was reacted with PHPh$_2$ and excess of NEt$_3$, then a phosphide-bridge platinum dimer [Pt(bzq)(μ-PPh$_2$)]$_2$ (**30**) was obtained.

Complex	Pt - C	Pt - N	Pt - X	Pt - Y	Reference
21	2.055(7)	2.065(8)	2.007(9)	1.985(11)	Fernandez et al, 2003
22	2.050(4)	2.074(4)	1.976(6)	1.992(5)	Fernandez et al., 2003
	2.047(4)	2.063(4)	1.975(6)	1.975(5)	
23	2.060(9)	2.088(7)	1.97(1)	2.02(1)	Berenguer et al., 2008
	2.06(1)	2.058(8)	1.97(1)	2.02(1)	
24	2.061(4)	2.111(4)	1.965(5)	2.316(1)	Diez et al., 2005
25a	2.034(3)	2.059(2)	1.916(2)	2.013(3)	Diez et al, 2009a
25b	2.051(2)	2.049(2)	1.956(2)	1.970(3)	Diez et al, 2009a
26	1.998(5)	2.075(4)	1.895(5)	2.393(1)	Diez et al., 2010a
27a	2.012(10)	2.049(8)	1.899(12)	2.427(3)	Diez et al., 2010a
27b	2.011(7)	2.075(6)	1.896(8)	2.398(2)	Diez et al., 2010a
28	1.972(11)	2.064(9)	1.913(14)	2.404(3)	Diez et al., 2010a
	2.011(11)	2.090(9)	1.915(1)	2.390(3)	
29	2.014(5)	2.090(4)	2.210(1)	2.387(1)	Diez et al., 2005
30	2.092(3)	2.092(4)	2.298(1)	2.298(1)	Diez et al., 2005
31a	2.09(2)	2.08(2)	2.266(5)	2.321(6)	DePriest et al., 1997
31b	2.02(2)	2.09(2)	2.235(6)	2.334(6)	DePriest et al., 1997
31c	2.100(9)	2.084(10)	2.297(3)	2.301(3)	DePriest et al., 1997
32	2.038(2)	2.139(2)	2.3095(6)	2.003(2)	Nabavizadeh et al., 2010
33a	2.006(3)	2.064(3)	2.2074(1)	2.4032(1)	Godbert et al., 2007
33b	1.981(2)	2.024(2)	2.029(2)	2.390(7)	Godbert et al., 2007
34	2.05(1)	2.141(8)	2.358(3)	2.030(9)	Nabavizadeh et al. 2010
35	2.029(9)	2.052(7)	2.334(3)	2.268(2)	Janzen et al., 2008
36	2.005(9)	2.028(6)	1.996(6)	2.041(7)	Chang et al. 2008

Table 5. Crystallographic data of the cyclometalated Pt(II) complexes containing benzoquinoline ligand

A series of cyclometalated Pt(II) complexes with several bidentate bis-diphenylphosphine ligands where the bridge length between the two phosphine units was varied (bis-diphenylphosphinomethane (dppm), bis-diphenylphosphinoethane (dppe) and bis-diphenylphosphinopropane (dppp), respectively, **31a-c**) were prepared and investigated structurally by DePriest et al., 1997. These complexes are a beautiful examples where the C and N atoms can be distinguished based on their different *trans* influence: the anionic C is a better σ-donor and thus, has a greater *trans* influence compared to N atom. This difference can be easily seen in the Pt-P bond lengths, with the Pt-P *trans* to C atom bond being longer than Pt-P trans to N atom bond. All these complexes show emission properties in solution (ethanol/methanol = 4/1, v/v) at low temperatures (77 - 180 K), assigned to ligand-centred (LC) transitions.

Several other cycloplatinated clusters containing bridged phosphine or phosphido ligands have been prepared. If the neutral binuclear phosphido complex $[Pt(bzq)(\mu\text{-}PPh_2)]_2$ (**30**) reacts with cis-$[Pt(C_6F_5)_2(thf)_2]$ in CH_2Cl_2 at 1:2 molar ratio, then a bent neutral triplatinum cluster $[Pt_3(bzq)(\mu\text{-}PPh_2)_2(C_6F_5)_3]$ can be isolated. Its crystal structure was reported by Diez et al, 2006. The crystal structure of complex **35** reveals the two different Pt-S bonds due to the different *trans* influence of the C and N atoms of the cyclometalated ligands, with Pt-S trans to C atom longer than the Pt-S trans to N atom.

Interestingly, complex **36** bearing chelating 2-pyridyl hexafluoropropoxide ancillary ligand shows evidence for the occurrence of π···π stacking between the cyclometalated ligands, but a lack of intermolecular Pt···Pt interaction. The π···π stacking was also confirmed by the observation of additional large Stokes shifted emission attributed to the aggregated counterparts in solid thin film.

3.4 Cyclometalated Pt(II) complexes with imine ligands

Although various cyclometalated Pt(II) species are intensively studied for their luminescent properties, surprisingly, almost no attention was given so far to the luminescent cycloplatinated imino species. It is almost very recent that these types of complexes started to be investigated for their photophysical properties as well (Scaffidi-Domianello et al., 2007).

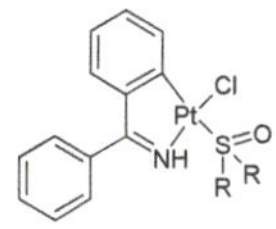

Scheme 7. Cyclometalated Pt(II) complex with benzophenone imine ligands. R = Me, (**37a**), (Pt-C = 2.006(9)Å, Pt-N = 2.019(6)Å, Pt - S = 2.215(2)Å, Pt - Cl = 2.392(2)Å); R = $(CH_2)_4$, (**37b**), (Pt-C = 2.016(4)Å, Pt-N = 2.019(4)Å, Pt - S = 2.2063(13)Å, Pt - Cl = 2.3925(11)Å)(Scaffidi-Domianello et al., 2007)

In **37a**, the asymmetric unit contains two independent Pt complexes, while in **37b**, it includes four Pt complexes linked by the intermolecular hydrogen-bonding network between the NH group and Cl atoms (Figure 5). The Pt - X distances for only one molecule are indicated in Scheme 7. There is no significant Pt···Pt interactions in the solid state. The two Pt(II) complexes show an emission band with the maximum located at 535 nm, along with another less intense emission at 565 nm. The solution quantum yields of the complexes are rather low, with values smaller than 0.0071.

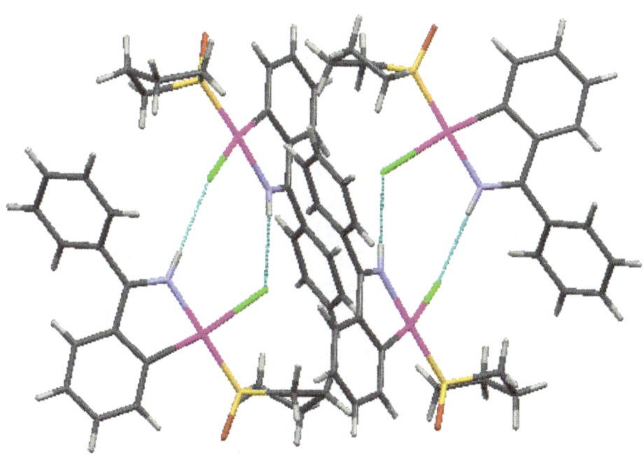

Fig. 5. The hydrogen bonding network connecting four independent molecules in the unit cell of **37b** (this figure was generated by using the Mercury 2.2 software)

The solid-state emission spectra for these complexes show emission maxima at 571 and 559 nm, respectively. These two complexes also exhibit some structure of the emission bands with high and low energy shoulders at 542 and 624 nm and at 525 and 610 nm, respectively.The excited-state emission lifetimes at 575 nm are in the range of 320-615 ns, consistent with the phosphorescence emissive mechanism.

3.4.1 Cyclometalated Pt(II) complexes with N-Benzylidenebenzylamine ligands

This class of cycloplatinated complexes was intensively investigated, but almost for mechanistic studies and not for emission properties. The most representative structural types are presented in Scheme 8.

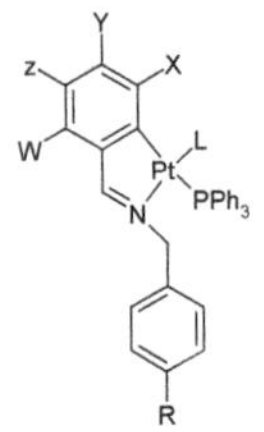

X=R=H, Y=W=Z=F, L=Me **38a**
X=Y=R=H, W=Z=F, L=Me **38b**
X=Y=Z=F, W=R=H, L=Me **38c**
X=W=Z=R=H, Y=Ph, L=Me **39a**
X=Y=Z=R=H, W=Ph, L=Me **39b**

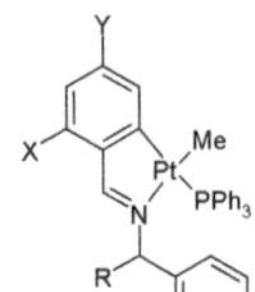

X=Y=R=H, Z=Me, L=Me **40a**
X=Y=R=H, Z=CF_3, L=Me **40b**
X=Y=Z=H, R=Cl, W=-C_6F_5, L=Br **41**
R=Me, X=F, Y=H **42a**
R=-COOMe, X=H, Y=Cl **42b**

Scheme 8. Pt(II) compexes with N-benzylidenebenzylamine ligands

These complexes were prepared by reacting $[Pt_2Me_4(\mu\text{-}SMe_2)_2]$ with the imine ligand in the presence of sodium acetate followed by the exchange reaction with PPh_3 ligand. They were studied intesively from the cycloplatination reaction mechanism point of view and none of them were investigated for their potential luminescent properties.

The stacked structures are well-known for platinum complexes, and they become increasingly favoured with increasing aromatic core. Despite the presence of biphenyl unit in complexes **39a,b**, no significant $\pi \cdots \pi$ interactions have been found for these compounds.

Complex	Pt - C	Pt - N	Pt-L	Pt - P (PPh_3)	Reference
38a	2.026(8)	2.156(8)	2.118(9)	2.303(2)	Crespo et al., 1995
38b	2.051(7)	2.153(5)	2.046(8)	2.299(2)	Crespo et al., 1995
38c	2.064(5)	2.144(4)	2.053(6)	2.2815(12)	Crespo et al., 2002a
39a	2.059(4)	2.128(3)	2.062(4)	2.3022(12)	Crespo et al., 2006a
39b	2.056(6)	2.134(5)	2.054(8)	2.2979(17)	Crespo et al., 2006a
40a	2.018(5)	2.141(4)	2.028(5)	2.2717(14)	Crespo et al., 1996
40b	2.053(4)	2.148(3)	2.049(4)	2.2955(1)	Crespo et al., 1998a
41	2.072(5)	2.098(5)	2.480(2)	2.245(2)	Calvet et al., 2009
42a	2.057(6)	2.119(5)	2.058(7)	2.305(2)	Crespo et al., 1998b
42b	2.072(12)	2.131(9)	2.003(10)	2.308(3)	Rodriguez et al., 2009

Table 6. Selected bond lengths for cyclometalated Pt(II) complexes containing N-benzylidenebenzylamine

3.4.2 Cyclometalated Pt(II) complexes with terdentate imine ligands

The relevant Pt(II) systems bearing terdentate imine ligands are shown in Scheme 9. The cyclometalated Pt(II) complexes with thiosemicarbazone ligands were included here. The crystal structures of such complexes with terdentate imines reveal a strong *trans* influence of the carbon donor ligands. For instance, in complex **47**, the Pt-NMe_2 bond length of 2.162(14) Å is longer than the imine Pt-N bond length of 2.04(2) Å as a consequence of *trans* aryl group which has a high *trans* influence. On the other hand, for such complexes, the longer Pt-N bond is also consistent with the weaker ligating ability of tertiary amines for platinum. The Pt-C bond lengths are found to be in the expected range. It is interesting to note that the complex **52** represents the first luminescent platinum(II) compound with a $[C^\wedge N^\wedge S]^-$ terdentate ligand. The long intermolecular S···S distance precludes the existence of any direct interaction between these atoms. Also, the Pt···Pt distance between neighbouring molecules is 4.216(3) Å which is larger, for instance, than the values reported for related platinacycles having $[C(sp^2)^\wedge N^\wedge N]^-$ and $[C(sp^2)^\wedge N^\wedge S]^-$ terdentate ligands from the "pincer" family. The emission spectrum of this compound recorded at 298K in CH_2Cl_2 solution shows a maximum at 578 nm when excited at 388 nm with no additional information regarding the quantum yield and lifetime provided. For the rest of complexes (**43 - 51**), no data regarding the luminescence properties were reported. Interestingly, the thiosemicarbazone ligands can be used in cycloplatination reaction to give orthometalated complexes in which the ligand acts as terdentate through the C, N and S atoms and two such examples are presented in Scheme 9 (**53a** and **53b**). **53a** was obtained by bridge-cleavage with PPh_3 of the first Pt(II) cyclometalated tetranuclear complexes bearing thiosemicarbazone ligands which were prepared using K_2PtCl_4 as starting material.

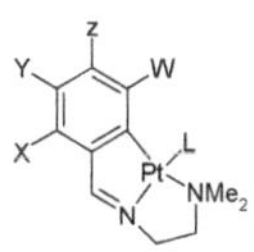

X=Y =W=Z=H, L=Ph **43**
X=Y=W=Z=H, L=Br **44**
X=Z=W=F, Y=H, L=Cl **45**
X=Y=Z=F, W=H, L=Me **46**
X=W=H, Y=Me, Z=OMe, L=Me **47**
X=Ph, Y=W=Z=H, L=Me **48a**
X=Ph, Y=W=Z=H, L=Cl **48b**
X=Y=W=H, Z=Cl, L=4-Tol **49**

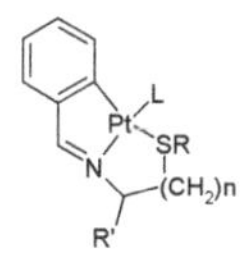

L=Cl, R=Et, R'=H, n = 1 **50** (Riera et al., 2000)
L=Cl, R=Me, R'=COOMe, n =2 **51**

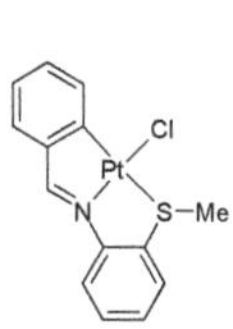

52

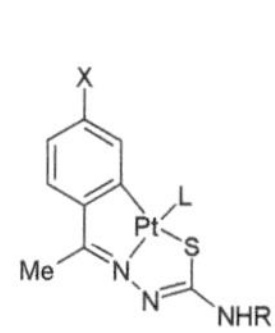

X=Me, R=H, L=PPh_3 **53a**
X=NO_2, R=Me, L=thiosemicarbazone **53b**

Scheme 9. Pt(II) complexes with terdentate imine ligands

Complex	Pt - C	Pt - N	Pt - E (E=N or S)	Pt - L	References
43	1.986(6)	2.016(4)	2.139(6)	2.009(4)	Crespo et al., 2004a
44	2.008(15)	1.970(10)	2.141(12)	2.301(4)	Crespo et al., 2002c
45	1.997(10)	1.970(14)	2.135(14)	2.296(4)	Crespo et al., 2004b
46	1.996(8)	2.009(60	2.160(7)	2.040(8)	Lopez et al., 1997
47	1.97(2)	2.04(2)	2.162(14)	2.154(13)	Vila et al., 1998
48a	1.964(3)	1.990(3)	2.138(3)	2.064(3)	Crespo et al., 2006a
48b	2.009(6)	1.967(6)	2.168(7)	2.302(2)	Crespo et al., 2006b
49	1.976(8)	2.014(5)	2.177(8)	2.026(4)	Martin et al., 2009
51	1.954(14)	1.984(11)	2.349(3)	2.308(4)	Riera et al., 2001
52	2.016(6)	1.928(6)	2.349(2)	2.292(2)	Caubet et al., 2003
53a	2.02(2)	2.03(2)	2.335(5)	2.235(5)	Vasquez-Garcia et al., 2000
53b	2.021(4)	2.006(4)	2.3296(11)	2.2832(12)	Quiroga et al., 2008

Table 7. Selected bond legths for cyclometalated Pt(II) complexes with terdentate imine ligands

On the other hand, complex **53b** was obtained by reacting the thiosemicarbazone ligand with the Pt dinuclear allyl complex $[Pt(\mu\text{-}Cl)(\eta^3\text{-}C_4H_7)]_2$ in refluxing acetone, when, instead of the expected tetranuclear complex, a mononuclear cyclometalated Pt(II) complex containing two thiosemicarbazone molecules was formed. One of the molecule acts as a $[C^\wedge N^\wedge S]^-$ terdentate ligand while the second one is coordinated in a monodentate fashion through the S atom, thus completing the coordination sphere of Pt(II) metal. Bis(thiosemicarbazone) ligands were used for cycloplatination reaction and the first X-ray structure of a cyclometalated Pt(II) complex with such ligands was reported by Lopez-Torres & Mendiola, 2010. No emission properties were reported for such complexes.

3.4.3 Pt(II) complexes bearing cyclometalated oxime ligands

There are several X-ray crystallographic structures reported for a series of cycloplatinated complexes with oxime ligands (see Scheme 10). They were prepared starting from $[PtCl_2(RR'SO)_2]$ and the corresponding oxime ligand. Although they show an interesting structural feature, as it is the case of complex **54b** that reveals an extremely short Pt···Pt contact of 3.337 Å, these complexes were studied mostly for mechanistic and catalytic purposes and no emission properties were investigated (Ryabov et al., 1995).

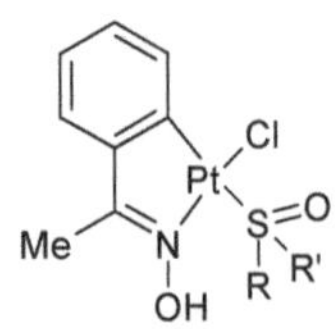

R = R′ = Me (Pt-C = 1.998(4) Å, Pt-N = 2.013(3) Å, Pt – S = 2.2677(11) Å, Pt – Cl = 2.4114(11) Å), (**54a**); R = Me, R′ = Ph (Pt-C = 2.010(4) Å, Pt-N = 2.010(3) Å, Pt – S = 2.2192(12) Å, Pt – Cl = 2.3806(13) Å) , (**54b**) (Ryabov et al., 2002)

Scheme 10.

3.4.4 Bis(imine) ligands – N^C^N pincer ligands

Several crystallographic studies concerning the cycloplatinated complexes with bis(imine) ligands (where the N atom is not part of a ring) were reported (see Scheme 11). Complex **55** was prepared by reacting LiC≡CSiMe$_3$ in THF with the first platinum(II) halide compounds containing the (N^C^N) isophtalaldimine ligands.

R=2,6-(^{i}Pr)$_2$Ph, X=-C≡CSiMe$_3$ **55**
R=2,6-(^{i}Pr)$_2$Ph, X=CH$_3$ **56**
R=-CH(CH$_3$)Ph, X=Cl **57a**
R=^{t}Bu, X=Cl **57b**
R=Ph, X=Cl **57c**

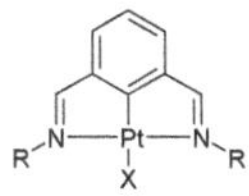

Scheme 11.

On the other hand, complex **56** represents the first X-ray crystal structure of stable trans-arylplatinum methyl complexes [PtMeN^C^N] with imine-type ligands. It is well-known that due to the strong C(sp^2)−C(sp^3) bond, only very few transition-metal compounds having an aryl as well as a methyl group bonded to the same metal atom are reported. The reason is that in such cases reductive elimination occurs. For complex **56** this reaction is prevented due to the trans disposition of the methyl and aryl groups and the rigid coplanarity of the chelate rings. Another representative example is complex **57a** that is the first chiral bis-aldimine (N^C^N)-pincer complexes. Unfortunately, there was no emission data reported for such complexes.

Complex	Pt-C	Pt-N1	Pt-N2	Pt-X	Reference
55	1.945(2)	2.043(2)	2.050(2)	2.065(3)	Hoogervorst et al., 2001
56	1.944(2)	2.036(2)	2.045(2)	2.156(3)	Hoogervorst et al., 2004
57a	1.901(7)	2.027(5)	2.049(6)	2.397(2)	Fossey et al., 2007
57b	1.915(6)	2.093(5)	2.097(5)	2.4192(19)	Fossey et al., 2002
57c	1.908(6)	2.061(5)	2.064(5)	2.3914(18)	Fossey et al., 2002

Table 8. Selected bond lengths for cyclometalated Pt(II) complexes with bis(imine) N^C^N pincer ligands

3.4.5 Cyclometalated Pt(II) complexes containing ferrocene based imine ligands

Several crystal structures of cyclometalated Pt(II) complexes containing imine ligands with ferrocene fragment were reported (see Scheme 12).

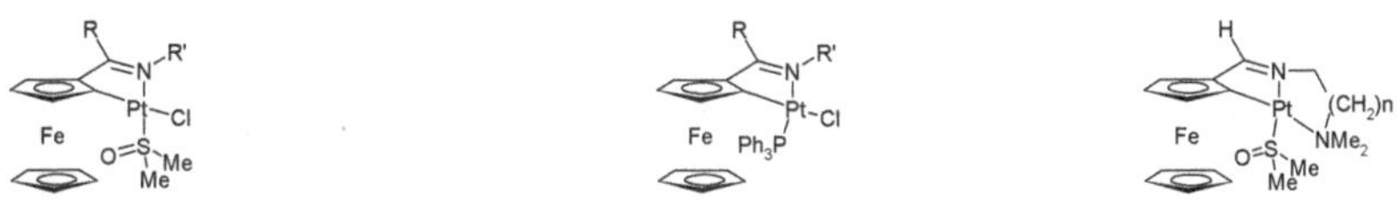

R=Me, R'=Ph **58**
R=H, R'=-CHMeCH$_2$OH **59a**
R=H, R'=-CH(CHMe$_2$)CH$_2$OH **59b**
R=Me, R'=OH **60**
R=H, R'=-CH$_2$CH$_2$OH **61a**

R=H, R'=-CH$_2$CH$_2$OH **61b**

n = 0 **62a**
n = 1 **62b**

Scheme 12. Cycloplatinated complexes bearing ferrocene imine ligands

Complex	Pt - C	Pt - N	Pt-S	Pt-Cl	Reference
58	1.980(5)	2.095(5)	2.195(1)	2.389(1)	Wu et al., 1997
59a	2.01(3)	2.05(2)	2.214(2)	2.427(8)	Lopez et al., 2006
	2.03(3)	2.01(3)	2.197(8)	-	
59b	2.018(1)	2.090(9)	2.2182(3)	2.380(3)	Lopez et al., 2004
60	1.976(7)	2.042(6)	2.186(2)	2.395(2)	Ryabov et al., 1997
61a	2.017(7)	2.078(4)	2.198(2)	2.395(2)	Lopez et al., 2010
	Pt - C	Pt - N	Pt-P	Pt-Cl	
61b	2.03(1)	2.189(7)	2.243(2)	2.376(3)	Lopez et al., 2010
	Pt - C	Pt - N1	Pt-S	Pt-N2	
62a	1.993(8)	1.977(5)	2.190(4)	2.3083(16)	Lopez et al., 2005
62b	1.939(8)	2.020(7)	2.182(7)	2.327(2)	Perez et al., 2003

Table 9. Selected bond lengths for cyclometalated Pt(II) complexes containing imine ligands based on ferrocene fragment

Most of these complexes were prepared by reacting the imine ligand with $[PtCl_2(DMSO)_2]$ in refluxing toluene for a long period of time, with or without sodium acetate as a base agent. They can react further with phosphine ligands to replace the labile DMSO ligand (61b). Several interesting structural features are presented in Table 9. It is important to note that these complexes were studied more from mechanistic, electrochemical and catalytic properties point of view and less, or not at all in most of the cases, for emissive properties.

3.5 Heteropolynuclear cycloplatinated complexes

Due to the presence of metallophilic interactions that have serious consequences on luminescence properties, the study of different heteropolynuclear cycloplatinated complexes has dramatically increased in the last years and this topic has been very recently reviewed (Diez et al., 2011). It has been shown that the cycloplatinated complexes can be useful building blocks in the design of heteropolynuclear and/or multicomponent architectures. Although several interesting heteropolymetallic cycloplatinated systems have been reported, the photoluminescence properties have been studied only in few cases.

3.6 Cyclometalated Pt(II) complexes as liquid crystals

Liquid crystals are, by far, *the most important* molecular electronic materials of the present day. They were discovered more than 100 years ago and are often thought of as the high technology materials found in high content, low power, flat-panel displays known to the whole world as LCDs, their main application (O'Neill & Kelly, 2011). Stable inorganic phosphors showing liquid crystal properties are very promising multifunctional materials because, in contrast with pure organic materials, they are not subjected to photobleaching (loss of luminescence properties upon irradiation over the time), provide anisotropic long range order and thus polarised emission that should improve display performance parameters such as brightness, contrast, energy efficiency and, in some cases, the viewing angle. Potential new materials that fulfill these requirements are the metallomesogens (liquid crystalline materials incorporating metal ions). The first option, lanthanide mesogenic complexes (lanthanidomesogens) have been proposed as potential candidates for bright luminescent emitters (Eliseeva & Bünzli, 2010), but they have low quantum yield efficiency. Thus, by far, due to the possibility of harvesting both triplet and singlet states, as well as the emission in the red - NIR range, the late transition metallomesogens are the most promising candidates and could be employed for the preparation of highly effective phosphorescent OLEDs. In this respect, several studies have been reported recently including luminescent metallomesogens based on Pt(II), Ir(III) or Ru(II), Ag(I) or Au(I) (Binnemans, 2009). There are several studies dealing with light-emitting metallomesogens based on platinum(II) (Liao et al., 2011; Venkatesan et al., 2008; Mocanu et al., 2010) complexes, most of them containing the metal in a cyclometalating environment (Damm et al., 2006; Wang et al., 2011) with 2-arylpyridine or 2-thienylpyridine derivatives. Bruce et. al. reported phosphorescent liquid crystalline complexes of platinum(II) showing a stimulus-dependant emission (Kozhevnikov et al., 2008) as well as highly luminescent (yields higher than 0.5) Pt(II) containing metallomesogens (Santoro et al., 2009). In terms of quantum yields, these later examples were exceeded only by the recently reported Pt(II) metallomesogens bearing pyridyl pyrazole chelates that show quantum yields nearly 1, when recorded in degassed dichloromethane (Liao et al., 2011).

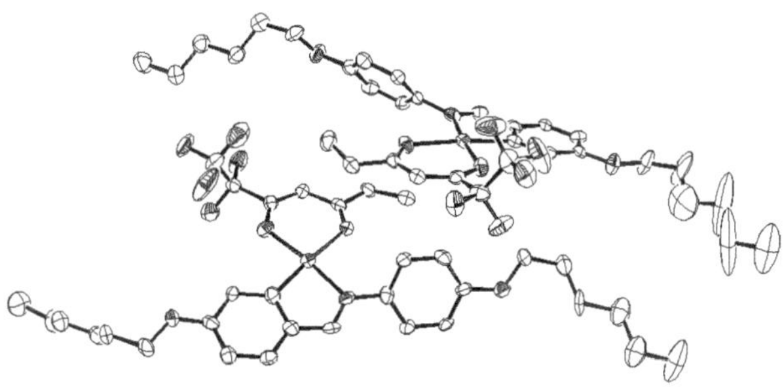

Fig. 6. Molecular structure of a mononuclear cycloplatinated metallomesogen containing an imine and an acetylacetonate ligand (**63**). The two independent molecules contained in the unit cell are shown (H atoms were omitted for clarity).(Pt-C = 1.983(10) and 1.976(11) Å, Pt-N = 1.986(8) and 2.008(7) Å, Pt-O1 = 2.101(6) and 2.080(6) Å, Pt-O2 = 1.992(7) and 2.004(6) Å (Cîrcu et al., 2007)

In most cases, luminescent Pt(II) complexes contain heterocyclic ligands, usually with one or two pyridine rings, while imine ligands were completely ignored from this point of view. Several examples of luminescent cycloplatinated (Scaffidi-Domianello et al., 2007; Pandya et al., 2010) or platinum(II) complexes with N,N-donor diimine ligands are known, though this class of ligands is widely spread in the design of Pt(II) metallomesogens (Cîrcu et al., 2009b; Buey et al., 1996; Diez, L. et al., 2002). Although several important results have been achieved in the design and the synthesis of luminescent liquid crystalline materials based on Pt(II) cyclometalated complexes, only few X-ray structural investigations have been made on such systems comprising either mononuclear (Cîrcu et al., 2007) or dinuclear species (Praefcke et al., 1994; Bilgin Eran et al., 2001). One reason could be that such molecules have long alkyl chains that make single crystals suitable for X-ray crystallography difficult to obtain. The only example of a crystal structure of a mononuclear cycloplatinated metallomesogen (**63**) is depicted in Figure 6. The Pt–C distance is similar to those found in dinuclear chloro- or thiocyanato-bridged, orthometalated platinum compounds with Schiff bases reported by Praefcke et al., 1994, whereas the Pt-N distances are slightly smaller (2.19 and 2.064 Å in Praefcke's work), which could be assigned to the difference in trans-effect of the atoms coordinated to platinum. It is interesting to note that an *anti* configuration (the perfluoroethyl group is trans to nitrogen atom of the imine group) is adopted by the two ligands (Schiff base and acetylacetonate derivative) around platinum centre. This complex **63** shows a monotropic nematic phase on cooling at 37°C.

4. Conclusion

Various cyclometalated Pt(II) complexes were structurally investigated by X-ray crystallography. The use of strong field C donor ligands provides an important route to Pt(II) complexes with interesting photophysical properties, with different applications, mainly in OLED devices. There are several classes of cyclometalated Pt(II) complexes which were not investigated so far for their emission properties, such as the cycloplatinated imino complexes, but this field continuously expands and a growing number of reports dealing with this topic are being published.

5. Acknowledgment

The authors acknowledge the financial support from CNCSIS -UEFISCSU, project number PNII ID_954/2007. The molecular structures were generated by using ORTEP-3 for Windows (L. J. Farrugia, University of Glasgow, UK).

6. References

Abe, T.; Shinozaki, K.; Ikeda, N. & Suzuki, T. (2007). [2,6-Bis(5-methyl-2-pyridyl)phenyl-κ^3*N*,*C*^{1},*N*']chloridoplatinum(II). *Acta Crystallogr., Sect. C: Cryst. Struct. Commun.*, 63, M456 - M458

Albrecht, M. (2010). Cyclometalation using d-block transition metals: fundamental aspects and recent trends. *Chem. Rev.*, 110, 576 - 623

Baik, C.; Han, W.-S.; Kang, Y.; Kang, S. O. & Ko, J. (2006). Synthesis and photophysical properties of luminescent platinum(II) complexes with terdentate polypyridine ligands: [Pt(bpqb)X] and [Pt(tbbpqpy)X](PF_6) (bpqb-H=1,3-bis(4'-phenyl-2'-quinolinyl) benzene; tbbpqpy =4-tert-butyl-1,3-bis(4'-phenyl-2'-quinolinyl) pyridine; X=Cl, C≡CC_6H_5, C≡C$C_6H_4NMe_2$, C≡C$C_6H_4NO_2$). *J. Organomet. Chem.*, 691, 5900 - 5910

Balashev, K. P.; Puzyk, M. V.; Kotlyar, V. S. & Kulikova, M. V. (1997). Photophysics, photochemistry and electrochemistry of mixed-ligand platinum(II) complexes with 2-phenylpyridine and 2-(2'-thienyl)pyridine as cyclometalating ligands. *Coord. Chem. Rev.*, 159, 109 - 120

Berenguer, J. R.; Diez, A.; Fernandez, J.; Fornies, J.; Garcia, A.; Gil, B.; Lalinde, E. & Moreno, M. T. (2008). $[Pt]_2Pb$ Trinuclear systems: impact of the anionic platinum fragment on the lead environment and photoluminescence. *Inorg. Chem.*, 47, 7703 - 7716

Bilgin Eran, B.; Singer, D.; Pickardt, J. & Praefcke, K. (2001). Thiocyanato-bridged platinum heterocycles: structure and properties of disc-like metallomesogens. *J. Organomet. Chem.*, 620, 249 - 255

Binnemans, K. (2009). Luminescence of metallomesogens in the liquid crystal state. *J. Mater. Chem.*, 19, 448 - 453

Brooks, J.; Babayan, Y.; Lamansky, S.; Djurovich, P. I.; Tsyba, I.; Bau, R. & Thompson, M. E. (2002). Synthesis and characterization of phosphorescent cyclometalated platinum complexes. *Inorg. Chem.*, 41, 3055 - 3066

Buey, J.; Diez, L.; Espinet, P.; Kitzerow, H.-S. & Miguel, J. A. (1996). Platinum orthometalated liquid crystals compared with their palladium analogues. First optical storage effect in an organometallic liquid crystal. *Chem. Mater.* , 8, 2375 - 2381

Cardenas, D. J. & Echavarren, A. M.; de Arellano, M. C. R. (1999). Divergent behavior of palladium(II) and platinum(II) in the metalation of 1,3-di(2-pyridyl)benzene. *Organometallics*, 18, 3337 - 3341

Caubet, A.; Lopez, C.; Solans, X. & Font-Bardia, M. (2003). Synthesis, characterisation and study of the first luminescent platinum(II) compound with a $[C,N,S]^-$ terdentate ligand. X-ray crystal structure of $[Pt\{C_6H_4\text{-}CH{=}N\text{-}(C_6H_4\text{-}2\text{-}SMe)\}Cl]$. *J. Organomet. Chem.*, 669, 164-171

Chang, S.-Y.; Cheng, Y.-M.; Chi, Y.; Lin, Y.-C.; Jiang, C.-M.; Lee, G.-H. & Chou, P.-T. (2008). Emissive Pt(II) complexes bearing both cyclometalated ligand and 2-pyridyl hexafluoropropoxide ancillary chelate. *Dalton Trans.*, 6901 - 6911

Chassot, L.; Mueller, R. & von Zelewsky, A. (1984). cis-Bis(2-phenylpyridine)platinum(II) (CBPPP): a simple molecular platinum compound. *Inorg. Chem.*, 23, 4249 - 4253

Cho, J.-Y.; Domercq, B.; Barlow, S.; Suponitsky, K. Yu.; Li, J.; Timofeeva, T. V.; Jones, S. C.; Hayden, L. E.; Kimyonok, A.; South, C. R.; Weck, M.; Kippelen, B. & Marder, S. R. (2007). Synthesis and characterization of polymerizable phosphorescent platinum(II) complexes for solution-processible organic light-emitting diodes. *Organometallics*, 26, 4816 - 4829

Cîrcu, V.; Horton, P. N.; Hursthouse, M. B. & Bruce, D. W. (2007). Mesomorphism of ortho-metallated imine complexes of the β-diketonatoplatinum(II) fragment. Crystal and molecular structure of three platinum imine complexes. *Liq. Cryst.*, 34, 1463 - 1472

Cîrcu, V.; Ilie, M.; Ilis, M; Dumitrascu, F.; Neagoe, I. & Pasculescu, S. (2009). Luminescent cyclometallated platinum(II) complexes with N-benzoyl thiourea derivatives as ancillary ligands. *Polyhedron*, 28, 3739 - 3746

Cîrcu, V.; Manaila-Maximean, D.; Rosu, C.; Iliş, M.; Molard, Y. & Dumitraşcu, F. (2009). Mesomorphic behaviour and TSDC measurements of ortho-metallated palladium(II) and platinum(II) complexes with S,O-donor co-ligands. *Liq. Cryst.*, 36, 123 - 132

Crabtree, R. H. (2005). *The organometallic chemistry of the transition metals* (4th Edition), Wiley Interscience, New Jersey

Crespo, M.; Solans, X. & Font-Bardia, M. (1995). Cyclometalated Platinum(II) Compounds with Fluorinated Iminic Ligands: Synthesis and Reactivity Tuning. Crystal Structures of the Compounds $[PtMe(RCH{=}NCH_2C_6H_5)(PPh_3)]$ (R = 2,3,4-C_6HF_3 and 2,3-$C_6H_2F_2$). *Organometallics*, 14, 355 - 364

Crespo, M.; Solans, X. & Font-Bardia, M. (1996). Steric and electronic effects on the regioselective formation of platinum(II) metallacycles: crystal structure of $[PtMe(3\text{-}MeC_6H_3CH{=}NCH_2C_6H_5)(PPh_3)]$. *J. Organomet. Chem.*, 509, 29 - 36

Crespo, M.; Solans, X. & Font-Bardia, M. (1998). Cyclometallated platinum compounds with N-benzylidenebenzylamines bearing trifluoromethyl groups. Crystal structure of [PtMe{3-(CF_3)C_6H_3CH=NCH_2Ph}(PPh_3)]. *Polyhedron*, 17, 1651 - 1658

Crespo, M.; Solans, X. & Font-Bardia, M. (1998). Cyclometallated platinum compounds with chiral imines. Crystal structure of [PtMe(2-FC_5H_2CH=N-(S)-CHMePh(PPh_3)]. *Polyhedron*, 17, 3927 - 3934

Crespo, M.; Font-Bardia, M. & Solans, X. (2002). Formation and cleavage of platinacycles containing a fluorinated imine. Crystal structure of [PtMe(3,4,5-C_6HF_3CH=N$CH_2C_6H_5$)PPh_3]. *Polyhedron*, 21, 105 - 114

Crespo, M.; Granell, J.; Solans, X. & Font-Bardia, M. (2002). Preparation of metallacycles with anionic terdentate [C,N,N'] ligands by intramolecular oxidative addition of C-X (X = Br, Cl) bonds to [Pt(dba)$_2$]. An unexpected effect of chloro substituents. *Organometallics*, 21, 5140 - 5143

Crespo, M.; Font-Bardia, M. & Solans, X. (2004). Compound [$PtPh_2$(SMe_2)$_2$] as a versatile metalating agent in the preparation of new types of [C,N,N'] cyclometalated platinum compounds. *Organometallics*, 23, 1708 - 1713

Crespo, M.; Granell, J.; Font-Bardia, M. & Solans, X. (2004). Intramolecular oxidative addition of C–F and C–H bonds to [Pt(dba)$_2$]. Crystal structure of [PtCl{Me_2NCH_2CH_2NCH(2,4,5-C_6HF_3)}]. *J. Organomet. Chem.*, 689, 3088 - 3092

Crespo, M.; Font-Bardia, M. & Solans, X. (2006). Synthesis, reactivity and crystal structures of platinum (II) and platinum (IV) cyclometallated compounds derived from 2- and 4-biphenylimines. *J. Organomet. Chem.*, 691, 444 - 454

Crespo, M.; Font-Bardia, M. & Solans, X. (2006). A comparative study of metallating agents in the synthesis of [C,N,N']-cycloplatinated compounds derived from biphenylimines. *J. Organomet. Chem.*, 691, 1897 - 1906

Damm, C.; Israel, G.; Hegmann, T. & Tschierske, C. (2006). Luminescence and photoconductivity in mononuclear ortho-platinated metallomesogens. *J. Mater. Chem.*, 16, 1808 - 1816

DePriest, J.; Zheng, G. Y.; Woods, C.; Rillema, D. P.; Mikirova, N. A. & Zandler, M. E. (1997). Structure, physical and photophysical properties of platinum(II) complexes containing 7,8-benzoquinoline and various bis (diphenylphosphine) ligands. *Inorg. Chim. Acta*, 264, 287 - 296

DePriest, J.; Zheng, G. Y.; Goswami, N.; Eichhorn, D. M.; Woods, C. & Rillema, D. P. (2000). Structure, physical, and photophysical properties of platinum(II) complexes containing bidentate aromatic and bis(diphenylphosphino)methane as ligands. *Inorg. Chem.*, 39, 1955 - 1963

Diez, A.; Fornies, J.; Garcia, A.; Lalinde, E. & Moreno, M. T. (2005). Synthesis, structural characterization, and photophysical properties of palladium and platinum(II) complexes containing 7,8-benzoquinolinate and various phosphine ligands. *Inorg. Chem.*, 44, 2443 - 2453

Diez, A.; Fornies, J.; Garcia, A.; Lalinde, E. & Moreno, M. T. (2006). Formation and characterization of a benzoquinolate triplatinum cluster containing a μ_3-PPh_2 bridging system and two Pt–Pt bonds. *Inorg. Chem. Commun.*, 9, 255 - 258

Diez, A.; Fornies, J.; Fuertes, S.; Larraz, C.; Lalinde, E.; Lopez, J. A.; Martin, A.; Moreno, M. T. & Sicilia, V. (2009). Synthesis and luminescence of cyclometalated compounds with nitrile and isocyanide ligands. *Organometallics*, 28, 1705 -1718

Diez, A.; Fornies, J.; Larraz, C.; Lalinde, E. & Lopez, A. J. (2010). Structural and luminescence studies on π···π and Pt···Pt interactions in mixed chloro-isocyanide cyclometalated platinum(II) complexes. *Inorg. Chem.*, 49, 3239 - 3251

Diez, A.; Lalinde, E. & Moreno, M. T. (2011). Heteropolynuclear cycloplatinated complexes: Structural and photophysical properties. *Coord. Chem. Rev.*, in press, doi:10.1016/j.ccr.2010.12.024

Diez, L.; Espinet, P.; Miguel, J. A. & Ros, M. B. (2002). Ferroelectric liquid crystal behavior in platinum orthometallated complexes. *J. Mater. Chem.*,12, 3694 - 3698

Eliseeva, S. V. & Bünzli, J. C. G. (2010). Lanthanide luminescence for functional materials and bio-sciences. *Chem. Soc. Rev.*, 39, 189-227

Evans, R. C.; Douglas, P. & Winscom, C. J. (2006). Coordination complexes exhibiting room-temperature phosphorescence: Evaluation of their suitability as triplet emitters in organic light emitting diodes. *Coord. Chem. Rev.*, 250, 2093 - 2126

Farley, S.J.; Rochester D. L.; Thompson, A. L.; Howard, J. A. K. & Williams, J. A. G. (2005). Controlling emission energy, self-quenching, and excimer formation in highly luminescent N^C^N-coordinated platinum(II) complexes. *Inorg. Chem.*, 44, 9690 - 9703

Fernandez, S.; Fornies, J.; Belen, G.; Gomez, J. & Lalinde, E. (2003). Synthesis, structural characterisation and photophysics of anionic cyclometalated bis(alkynyl) (benzo[*h*]quinolinate) platinate(II) species. *Dalton Trans*, 822 - 830

Fossey, J. S. & Richards, J. C. (2002). A direct route to platinum NCN-pincer complexes derived from 1,3-bis (imino)benzenes and an investigation into their activity as catalysts for carbon–carbon bond formation. Organometallics, 21, 5259 - 5264

Fossey, J. S.; Russell, M. L.; Malik, K. M. A. & Richards, C. J. (2007). Synthesis and crystal structures of the first C2-symmetric bis-aldimine NCN–pincer complexes of platinum and palladium. *J. Organomet. Chem.*, 692, 4843 - 4848

Godbert, N.; Pugliese, T.; Aiello, I.; Bellusci, A.; Crispini, A. & Ghedini, M. (2007). Efficient, ultrafast, microwave-assisted syntheses of cycloplatinated complexes. *Eur. J. Inorg. Chem.*, 5105 - 5111

He, Z.; Wong, W.-Y.; Yu, X.; Kwok, H.-S. & Lin, Z. (2006). Phosphorescent platinum(II) complexes derived from multifunctional chromophores: synthesis, structures, photophysics, and electroluminescence. *Inorg. Chem.*, 45, 10922 - 10937

Hill, A. F. (2002). *Organotransition metal chemistry*, Royal Society of Chemistry.

Hirani, B.; Li, J.; Djurovich, P. I.; Yousufuddin, M.; Oxgaard, J.; Persson, P.; Wilson, S. R.; Bau, R.; Goddard, W. A. & Thompson, M. E. (2007). Cyclometallated iridium and platinum complexes with noninnocent ligands. *Inorg. Chem.*, 46, 3865 - 3875

Hofmann, A.; Dahlenburg, L. & van Eldik, R. (2003). Cyclometalated analogues of platinum terpyridine complexes: kinetic study of the strong σ-donor cis and trans effects of carbon in the presence of a π-acceptor ligand backbone. *Inorg. Chem.*, 42, 6528 - 6538

Hoogervorst, J. W.; Elsevier, J. C.; Lutz, M. & Spek, A. L. (2001). New cis- and trans-arylplatinum(II) acetylide compounds containing a bis(imino)aryl [NCN] ligand. *Organometallics*, 20, 4437 - 4440

Hoogervorst, J. W.; Koster, A. L.; Lutz, M.; Spek, A. L & Elsevier, J. C. (2004). trans-Arylplatinum(II) methyl compounds containing a bis(imino)aryl [NCN] ligand. *Organometallics*, 23, 1161 - 1164

Janzen, E. D.; Vanderveer, D. G.; Mehne, L. F.; Silva, F.; da Demétrio, A.; Brédas, J. - L. & Grant, G. J. (2008). Cyclometallated Pt(II) and Pd(II) complexes with a trithiacrown ligand. *Dalton Trans.*, 1872 - 1882

Jolliet, P.; Gianini, M.; von Zelewsky, A.; Bernardinelli, G. & Stoeckli-Evans, H. (1996). Cyclometalated Complexes of Palladium(II) and Platinum(II): cis-Configured Homoleptic and Heteroleptic Compounds with Aromatic C^N Ligands. *Inorg. Chem.*, 35, 4883 - 4888

Kozhevnikov, V. N.; Donnio, B. & Bruce, D. W. (2008). Phosphorescent, terdentate, liquid-crystalline complexes of platinum(II): stimulus-dependent emission. *Angew. Chem., Int. Ed.*, 47, 6286-6289

Kui, S. C. F.; Chui, S. S.-Y.; Che, C.-M. & Zhu, N. (2006). Structures, photoluminescence, and reversible vapoluminescence properties of neutral platinum(II) complexes containing extended π-conjugated cyclometalated ligands. *J. Am. Chem. Soc.*, 128, 8297 - 8309

Kulikova, M. V.; Balashev, K. P. & Erzin, Kh. (2003). Synthesis and photophysical properties of a series of biscyclometalated platinum(II) complexes on the basis of a tridentate 2,6-diphenylpyridine. *Russ. J. Gen. Chem.*, 73, 1839 - 1845

Kvam, P. I. & Songstad, J. (1995). Preparation and characterization of some cyclometalated Pt(II) complexes from 2-phenylpyridine and 2-(2′-thienyl)pyridine. *Acta Chem. Scand.*, 49, 313 - 324

Kvam, P. I.; Puzyk, M. V.; Balashev, K. P. & Songstad, J. (1995). Spectroscopic and electrochemical properties of some mixed-ligand cyclometalated platinum(II) complexes derived from 2-phenylpyridine. *Acta Chem. Scand.*, 49, 335 - 343

Lai, S.-W.; Chan, M. C.-W.; Cheung, K.-K. & Che, C.-M. (1999). Carbene and isocyanide ligation at luminescent cyclometalated 6-phenyl-2,2′-bipyridyl platinum(II) complexes: structural and spectroscopic studies. *Organometallics*, 18, 3327 - 3336

Li, K.; Chen, Y.; Lu, W.; Zhu, N. & Che, C. - M. (2011). A cyclometalated platinum(II) complex with a pendent pyridyl motif as solid-state luminescent sensor for acidic vapors. *Chemistry – A Eur. J.*, 17(15), 4109 - 4112

Liao, C.-T.; Chen, H.-H.; Hsu, H.-F.; Poloek, A.; Yeh, H.-H.; Chi, Y.; Wang, K.-W.; Lai, C.-H.; Lee, G.-H.; Shih, C.-W. & Chou, P.-T. (2011). Mesomorphism and luminescence properties of platinum(II) complexes with tris(alkoxy)phenyl-functionalized pyridyl pyrazolate chelates. *Chemistry – Eur. J.*, 17, 546 - 556

Liu, J.; Yang, C.-J.; Cao, Q.-Y.;, Xu, M.; Wang, J.; Peng, H.-N.;, Tan, W.-F.; Lü, X.-X. & Gao, X.-C. (2009). Synthesis, crystallography, phosphorescence of platinum complexes coordinated with 2-phenylpyridine and a series of β-diketones. *Inorg. Chim. Acta*, 362, 575 - 579

Lopez, C.; Caubet, A.; Perez, S.; Solans, X. & Font-Bardia, M. (2004). Easy access to diastereomerically pure platinacycles. *Chem. Commun.*, 540 - 541

Lopez, C.; Solans, X. & Font-Bardia, M. (2005). The importance of the length of the $-(CH_2)_n-$ chain on the cycloplatination of the $[(\eta^5\text{-}C_5H_5)Fe\{(\eta^5\text{-}C_5H_4)\text{-}CH{=}N\text{-}(CH_2)_n\text{-}NMe_2\}]$ (n = 2 or 3) ligands and the properties of the platinacycles. *Inorg. Chem. Commun.*, 8, 631 - 634

Lopez, C.; Caubet, A.; Perez, S.; Solans, X.; Font-Bardia, M. & Molins, E. (2006). Chiral platinum(II) compounds containing ferrocenyl Schiff bases acting as (N), $(N,O)^-$,

[C(sp^2,ferrocene),N]$^-$ or [C(sp^2,ferrocene),N,O]$^{2-}$ ligands. *Eur. J. Inorg. Chem.*, 3974 - 3984

Lopez, C.; Perez, S.; Solans, X.; Font-Bardia, M. & Calvet, T. (2010). Influence of the substituent R_1 on the reactivity of [(η^5- C_5H_5)Fe{(η^5-C_5H_4)-CH=N (R_1)-OH}] {R_1 =-CH_2-CH_2- or 1,2-C_6H_4} with platinum(II) and on the properties of the complexes. *New J. Chem.*, 34, 676 - 685

Lopez, O.; Crespo, M.; Font-Bardia, M. & Solans, X. (1997). Activation of C−F and C−H Bonds by platinum in trifluorinated [C,N,N′] ligands. Crystal structures of [PtFMe_2{Me_2N$CH_2$$CH_2$NHCH($CH_2$COMe)(2,4-$C_6H_2F_2$)}] and [PtMe{$Me_2NCH_2$-$CH_2$N=CH(2,3,4-$C_6HF_3$)}]. *Organometallics*, 16, 1233 - 1240

Lopez-Torres, E. & Mendiola, M. A. (2010). Orthometallated versus coordination compounds for reactions of platinum(II) and palladium(II) with the ligand benzil bis(4-methyl-3-thiosemicarbazone). *Inorg. Chim. Acta*, 363, 1735 - 1740

Lu, W.; Mi, B.-X.; Chan, M. C. W. ;, Hui, Z.; Che, C.-M.; Zhu, N. & Lee, S.-T. (2004). Light-emitting tridentate cyclometalated platinum(II) complexes containing σ-alkynyl auxiliaries: tuning of photo- and electrophosphorescence. *J. Am. Chem. Soc.*, 126, 4958 - 4971

Lusby, P. J.; Mueller, P.; Pike, S. J. & Slawin, A. M. Z. (2009). Stimuli-responsive reversible assembly of 2D and 3D metallosupramolecular architectures. *J. Am. Chem. Soc.*, 131, 16398 - 16400

Ma, B.; Djurovich, P. I. & Thompson, M. E. (2005). Excimer and electron transfer quenching studies of a cyclometalated platinum complex. *Coord. Chem. Rev.*, 249, 1501 - 1510

Ma, B.; Li, J.; Djurovich, P. I.; Yousufuddin, M.; Bau, R. & Thompson, M. E. (2005). Synthetic control of Pt···Pt separation and photophysics of binuclear platinum complexes. *J. Am.Chem. Soc.*, 127, 28-29

Martin, R.; Crespo, M.; Font-Bardia, M. & Calvet, T. (2009). Five- and seven-membered metallacycles in [C,N,N′] and [C,N] cycloplatinated compounds. *Organometallics*, 28, 589 - 597

McGuire Jr., R.; McGuire, M. C. & McMillin, D. R. (2010). Platinum(II) polypyridines: A tale of two axes. *Coord. Chem. Rev.*, 254, 2574 - 2583

Mdleleni, M. M.; Bridgewater, J. S.; Watts, R. J. & Ford, P. C. (1995). Synthesis, structure and spectroscopic properties of ortho-metalated platinum(II) complexes. *Inorg. Chem.*, 34, 2334 -2342

Mocanu, A. S.; Ilis, M; Dumitrascu, F.; Ilie, M. & Cîrcu, V. (2010). Synthesis, mesomorphism and luminescence properties of palladium(II) and platinum(II) complexes with dimeric Schiff base liquid crystals. *Inorg. Chim. Acta*, 363, 729 - 736

Nabavizadeh, S. M.; Haghighi, G. M.; Esmaeilbeig, A. R.; Raoof, F.; Mandegani, Z.; Jamali, S.; Rashidi, M. & Puddephatt, R. J. (2010). Assembly of symmetrical or unsymmetrical cyclometalated organoplatinum complexes through a bridging diphosphine ligand. *Organometallics*, 29, 4893 - 4899

Newman, C. P.; Cave, G. W. V.; Wong, M.; Errington, W.; Alcock, N. W. & Rourke, J. P. (2001). Di-metallated platinum carbonyl complexes: platinum–platinum interactions in the solid state. *Dalton Trans.*, 2678 - 2582

Newman, C. P.; Casey-Green, K.; Clarkson, G. J.; Cave, G. W. V.; Errington, W. & Rourke, J. P. (2007). Cyclometallated platinum(II) complexes: oxidation to, and C–H activation by, platinum(IV). *Dalton Trans.*, 3170 - 3182

Niedermair, F.; Waich, K.; Kappaun, S.; Mayr, T.; Trimmel, G.; Mereiter, K. & Slugovc, C. (2007). Heteroleptic κ2(N,C2)-2-phenylpyridine platinum complexes: the use of bis(pyrazolyl)borates as ancillary ligands. *Inorg. Chim. Acta*, 360, 2767-2777

Niedermair, F.; Kwon, O.; Zojer, K.; Kappaun, S.; Trimmel, G.; Mereiter, Kurt. & Slugovc, C.(2008). Heteroleptic platinum(II) complexes of 8-quinolinolates bearing electron withdrawing groups in 5-position. *Dalton Trans.*, 4006 - 4014

O'Neill, M. & Kelly, S. M. (2011). Ordered Materials for Organic Electronics and Photonics. *Adv. Mater.*, 21, 556 – 584

Pandya, S. U.; Moss, K. C.; Bryce, M. R.; Batsanov, A. S.; Fox, M. A.; Jankus, V.; Al Attar, H. A. & Monkman, A. P. (2010). Luminescent platinum(II) complexes containing cyclometallated diaryl ketimine ligands: synthesis, photophysical and computational properties. *Eur. J. Inorg. Chem.*, 1963 – 1972

Perez, S.; Lopez, C.; Caubet, A.; Solans, X. & Font-Bardia, M. (2003). Synthesis, characterisation and study of the reactivity of the first platinum(II) complex having a [C(sp2, ferrocene),N,N']- terdentate ligand. *New J. Chem.*, 27, 975 – 982

Praefcke, K.; Bilgin, B.; Pickardt, J. & Borowski, M. (1994). Liquid-crystalline compounds, 86. The first disc-shaped dinuclear platinum mesogen. Chem. Ber., 127, 1543 – 1546

Qiu, D.; Wu, J.; Xie, Z.; Cheng, Y. & Wang, L. (2009). Synthesis, photophysical and electrophosphorescent properties of mononuclear Pt(II) complexes with arylamine functionalized cyclometalating ligands. *J. Organomet. Chem.*, 694, 737 – 746

Qiu, D.; Feng, Y.; Wang, H.; Bao, X.; Guo, Y.; Cheng, Y. & Wang, L. (2010). Synthesis, crystal structure and photophysical properties of a novel Pt(II) complex with multi-functionalized cyclometalating ligand. *Inorg. Chem. Commun.*, 13, 613 - 617

Quiroga, A. G.; Cubo, L.; Sanz Miguel, P. J.; Moneo, V.; Carnero, A. & Navarro-Ranninger, C. (2008). Isolation of an intermediate in the platination of p-nitroacetophenone 4-methylthiosemicarbazone: potential application as an antitumor drug. *Eur. J. Inorg. Chem.*, 1183 – 1187

Rausch, A. F.; Monkowius, U. V.; Zabel, M. & Yersin, H. (2010). Bright sky-blue phosphorescence of [*n*-Bu_4N][Pt(4,6-dFppy)$(CN)_2$]: synthesis, crystal structure, and detailed photophysical studies. *Inorg. Chem.*, 49, 7818- 7825

Raybov, A. D.; Ott, S.; Samuleev, P. V.; Polyakov, V. A.; Alexandrova, L.; Kazankov, G. M.; Shova, S.; Revenko, M.; Lipkowski, J. & Johansson, M. H. (2002). Structural and mechanistic look at the orthoplatination of aryl oximes by dichlorobis(sulfoxide or sulfide)platinum(II) complexes. *Inorg. Chem.*, 41, 4286 - 4294

Riera, X.; Caubet, A.; Lopez, C.; Moreno, V.; Solans, X. & Font-Bardia, M. (2000). Activation of σ(C−H) Bonds in $C_6H_5CH{=}NCH_2CH_2SEt$ induced by platinum(II). X-ray crystal structure of [Pt{$C_6H_4CH{=}NCH_2CH_2SEt$}Cl]. *Organometallics*, 19, 1384 – 1390

Riera, X.; Lopez, C.; Caubet, A.; Moreno, V.; Solans, X. & Font–Bardia, M. (2001). Platinum(II) and palladium(II) compounds containing chiral thioimines. *Eur. J. Inorg. Chem.*, 2135 – 2142

Rodriguez, J.; Zafrilla, J.; Albert, J.; Crespo, M.; Granell, J.; Calvet, T. & Font-Bardia, M. (2009). Cyclometallated platinum(II) compounds with imine ligands derived from amino acids: Synthesis and oxidative addition reactions. *J. Organomet. Chem.*, 694, 2467 – 2475

Ryabov, A. D.; Kazankov, G. M.; Panyashkina, I. M.; Grozovsky, O. V.; Dyachenko, O. G.; Polyakov, V. A. & Kuz'mina, L. G. (1997). Cycloplatination of aryl and ferrocenyl oximes by *cis*-[$PtCl_2(OSMe_2)_2$] affording expected platinum(II) and unexpected platinum(IV) products. *J. Chem. Soc., Dalton Trans.*, 4385 - 4392

Ryabov, A. D.; Kuz'mina, L. G.; Polyakov, V. A.; Kazankov, G. M.; Ryabova, E. S.; Pfeffer, M. & van Eldik, R. (1995). Kinetics and mechanism of halogen-bridge cleavage in dimethylaminomethylphenyl-C^1,*N* pallada- and platinacycles by pyridines. pressure effects, and crystal structures of the *N,N-cis* reaction product, its *N,N-trans* orthometallated analogue and a dimer of similar reactivity. *J. Chem. Soc., Dalton Trans.: Inorg. Chem.*, 999 - 1006

Santoro, A.; Whitwood, A. C.; Williams, J. A. G.; Kozhevnikov, V. N. & Bruce, D. W. (2009). Synthesis, mesomorphism, and luminescent properties of calamitic 2-phenylpyridines and their complexes with platinum(II). *Chem. Mater.*, 21, 3871 - 3882

Scaffidi-Domianello, Y. Y.; Nazarov, A. A.; Haukka, M.; Galanski, M.; Keppler, B. K.; Schneider, J.; Du, P.; Eisenberg, R. & Kukushkin, V. Y. (2007). First example of the solid-state thermal cyclometalation of ligated benzophenone imine giving novel luminescent platinum(II) species. *Inorg. Chem.*, 46, 4469 - 4482

Schneider, J.; Du, P.; Wang, X.; Brennessel, W. W. & Eisenberg, R. (2009). Synthesis, electrochemistry, photophysics, and solvatochromism in new cyclometalated 6-phenyl-4-(*p*-R-phenyl)-2,2'-bipyridyl (R = Me, COOMe, $P(O)(OEt)_2$) (C^N^N) platinum(II) thiophenolate chromophores. *Inorg. Chem.*, 48, 1498 - 1506

Seneclauze, J. B.; Retailleau, P. & Ziessel, R. (2007). Design and preparation of neutral substituted fluorene- and carbazole-based platinum(II)–acetylide complexes. *New J. Chem.*, 31, 1412 - 1416

Shao, P.; Li, Y. & Sun, W. (2008). Cyclometalated platinum(II) complex with strong and broadband nonlinear optical response. *J. Phys. Chem. A.*, 112, 1172 - 1179

Sirous, J.; Zahra, M.; Nabavizadeh, S. M.; Milic, D.; Kia, R. & Rashidi, M. (2010). Cyclometalated cluster complex with a butterfly-shaped Pt_2Ag_2 core. *Inorg. Chem.*, 49, 2721 - 2726

Siu, P. K.-M.; Ma, D.-L. & Che, C.-M. (2005). Luminescent cyclometalated platinum(II) complexes with amino acid ligands for protein binding. *Chem. Commun.*, 1025 - 1027.

Valeur, B. (2001). *Molecular fluorescence: principles and applications*, Wiley-VCH, Verlag GmbH, Weinheim

Vasquez-Garcia, D.; Fernandez, A.; Fernandez, J. J.; Lopez-Torres, M.; Suarez, A.; Ortigueira, J. M.; Vila, J. M. & Adams, H. (2000). New cyclometallated platinum(II) compounds with thiosemicarbazones: crystal and molecular structure of [Pt{4-$MeC_6H_3C(Me)$=NN=C(S)NH_2}(PPh_3)]. *J. Organomet. Chem.*, 595, 199 - 207

Venkatesan, K.; Kouwer, P. H. J.; Yagi, S.; Muller, P. & Swager, T. M. (2008). Columnar mesophases from half-discoid platinum cyclometalated metallomesogens. *J. Mater. Chem.*, 18, 400 - 407

Vila, M. J.; Pereira, M. T.; Ortigueira, J. M.; Lata, D.; Torres, M. L.; Fernandez, J. J.; Fernandez, A. & Adams, H. (1998). Synthesis of complexes of platinum (II) with *C,N,N'*-terdentate Schiff base donor ligands. Crystal and molecular structure of [Pt{3-Me-4-$MeOC_6H_2C$(H)=$NCH_2CH_2NMe_2$}(Me)]. *J. Organomet. Chem.*, 566, 93 - 102

Wang, Z.; Turner, E.; Mahoney, V.; Madakuni, S.; Groy, T. & Li, J. (2010). Facile synthesis and characterization of phosphorescent Pt(N C N)X complexes. *Inorg. Chem.*, 49, 11276 - 11286

Wang, Y. F.; Liu, Y.; Luo, J.; Qi, H. R.; Li, X. S.; Nin, M. J.; Liu, M.; Shi, D. Y.; Zhu, W. G. & Cao, Y. (2011). Metallomesogens based on platinum(II) complexes: synthesis, luminescence and polarized emission. Dalton Trans., 40, 5046 – 5051

Williams, J. A. G.; Beeby, A.; Davies, E. S.; Weinstein, J. A. & Wilson, C. (2003). An alternative route to highly luminescent platinum(II) complexes: cyclometalation with N^C^N-coordinating dipyridylbenzene ligands. *Inorg. Chem.*, 42, 8609 - 8611

Williams, J. A. G. (2007). Photochemistry and photophysics of coordination compounds: platinum. *Top. Curr. Chem.*, 281, 205 – 268

Williams, J. A. G.; Develay, S.; Rochester, D. L. & Murphy, L. (2008). Optimising the luminescence of platinum(II) complexes and their application in organic light emitting devices (OLEDs), *Coord. Chem. Rev.*, 252, 2596 - 2611

Wu, Y. J.; Ding, L.; Wang, H. X.; Liu, Y. H.; Yuan, H. Z. & Mao, X. A. (1997). Synthesis, characterization and structure of ferrocenylketimine complexes of platinum(II). *J. Organomet. Chem.*, 535, 49 – 58

Yuen, M.-Y.; Roy, V. A. L.; Lu, W.; Kui, S. C. F.; Tong, G. S. M.; So, M.-H.; Chui, S. S.-Y.; Muccini, M.; Ning, J. Q.; Xu, S. J.; & Che, C.-M. (2008). Semiconducting and electroluminescent nanowires self-assembled from organoplatinum(II) complexes. *Angew. Chem., Int. Ed.*, 47, 9895 – 9899

Zucca, A.; Petretto, G. L.; Stoccoro, S.; Cinellu, M. A.; Minghetti, G.; Manassero, M.; Manassero, C.; Male, L. & Albinati, A. (2006). Dinuclear C,N,C cyclometalated platinum derivatives with bridging delocalized ligands. Fourfold deprotonation of 6,6′-diphenyl- 2,2′-bipyridine, H_4L, promoted by "$Pt(R)_2$" fragments (R = Me, Ph). Crystal structures of $[Pt_2(L)(3,5\text{-}Me_2py)_2]$ and $\{Pt_2(L)(dppe)\}_2$ (dppe =1,2-bis(diphenylphosphino)ethane). X-ray powder diffraction of $[Pt_2(L)(CO)_2]$. *Organometallics*, 25, 2253 - 2265

Part 2

Macromolecules

12

Crystallization, Structure and Functional Robustness of Isocitrate Dehydrogenases

Noriyuki Ishii
National Institute of Advanced Industrial Science and Technology (AIST)
Japan

1. Introduction

Understanding in detail the function of proteins and complexes requires knowledge of their three-dimensional structure. Historically, early development of protein crystallization, only provided a means for the purification of specific proteins from an impure mixture and an index that a protein had been purified (McPherson, 2004). X-ray diffraction analysis in combination with crystallization has become indispensable techniques for establishing the properties and nature of catalytic macromolecules. In spite of remarkable progress over the last two decades in the overexpression of macromolecules, in sophisticated screening of crystallization conditions, and X-ray data collection and analysis, the determination of novel structures by X-ray crystallography is still largely limited by the difficulty of obtaining high-quality crystals of interest and maintaining their quality throughout the data collection stage.

The property of self-assembly exhibited by biological macromolecules plays a central role in biology. Most of the primary stage of the self-associated construction of supramolecular structures such as macromolecular complexes, assemblies, organelles, cell membranes, cytoskeletons, and so on, involves only the establishment of weak and oriented interactions between homologous molecules. The formation of protein crystals can be described as the expression of the self-assembly properties of the constituent molecules placed under favorable conditions. In order to engineer proteins that possess various kinds of physicochemical properties as well as biochemical functions within nano-assemblies, we need to understand features of the intermolecular interactions and the capacities for macromolecules to self-associate that govern the integrating of protein assemblies, such as seen in the primary stage of crystallization and crystal growth processes of protein molecules (Akiba et al., 2005; N. Ishii et al., 2001). It further requires information to specify how the building block molecules are joined into higher order structures. The fundamental and practical importance of these processes motivates the interest of studying self-assembly (self-organization). Crystals of a certain kind of protein belonging to different space groups may provide a better understanding of intermolecular interactions which can guide the development of techniques to manipulate the orientation of each protein molecule and arrangements in the construction of nano-architectures using the desired protein. The conformation of protein molecules as well as configurations of amino-acid residues are often stabilized in the supramolecular complex through the cumulative effects of various intermolecular interactions, such as salt bridges and hydrogen bonds, on the surface. Even in monomeric or dimeric proteins under physiological conditions sometimes seen as highly oligomeric complexes during crystal structure determination.

2. Isocitrate dehydrogenase

Isocitrate dehydrogenase (ICDH, EC 1.1.1.42) is a metal dependent (Mg^{2+} or Mn^{2+}) enzyme that plays an important role in the tricarboxylic acid cycle. It lies at a critical juncture between the cycle and the glyoxylate pathway to the biosynthesis of glutamate. The enzyme catalyzes the subsequent oxidative decarboxylation reaction of 2R,3S-isocitrate to yield 2-oxoglutarate and carbon dioxide with the protonation of NAD or NADP in the cycle. The 2-oxoglutarate is known to be a key substrate in the biosyntheses of cell constituents via reductive amination to glutamate. These pathways are among the first to have evolved in the history of life (Melendez-Hevia et al., 1996). The ICDHs have been distinguished into three subfamilies based on sequence comparisons (Steen et al., 1997, 2001). All of the archaeal and most of the bacterial ICDHs are classified together into subfamily I, eukaryotic homodimeric ICDHs and some bacterial ICDHs are categorized as subfamily II, eukaryotic hetero oligomeric ICDHs constitute subfamily III. In contrast to these homologous proteins, another type of $NADP^+$-dependent monomeric ICDHs with molecular mass of 80-100 kDa have been found (Chen & Gadal, 1990). The active site of these enzymes in this category must be constructed from the side chains of residues within a single polypeptide chain. Although the monomeric ICDH catalyzes a reaction identical to that of the dimeric ICDH, no homology in the primary sequence has been found between the monomeric and dimeric ICDHs (Sahara et al., 2002). In addition, immunological studies suggest that monomeric and dimeric ICDHs are not structurally homologous (Fukunaga et al., 1992; Leyland & Kelly, 1991). A certain bacterium such as *Calwellia maris* possesses both monomeric and dimeric ICDHs (Yoneta et al., 2004). It seems that transcription of both genes is regulated in response to environmental factors.

3. Crystallization and study with the X-ray method

For comprehensive understanding of the dynamics within the cell and the mechanisms responsible for the dynamics, we need to draw how the constituting molecules respond to chemical and physical forces, how the responses are regulated, and how the responses are transmitted through the hierarchy of assemblies and higher order structures. Although there have been techniques which reveal protein structures such as nuclear magnetic resonance (NMR), and cryogenic transmission electron microscopy (cryo-TEM) in combination with computer tomography methods, the dynamical properties discussed above require a level of atomic resolution which can only be addressed by X-ray crystallography.

The search stage for discovering appropriate crystallization condition for the macromolecules of interest is still difficult and time consuming. Figure 1 shows a conceptual diagram that shows such a crystallization condition search stage (screening on the landscape of potential free energy). Various parameters are derived from certain starting conditions, and the condition under which crystals suitable for X-ray crystallography are formed is sought by trial and error so as not to fall into local minima. In the parameters, there are many factors effecting crystallization such as temperature, gravity, magnetic field, pH, precipitant type and concentration, ionic strength, reductive or oxidative environment, concentration of the sample protein, ligands, inhibitors, genetic or chemical modifications, and so on.

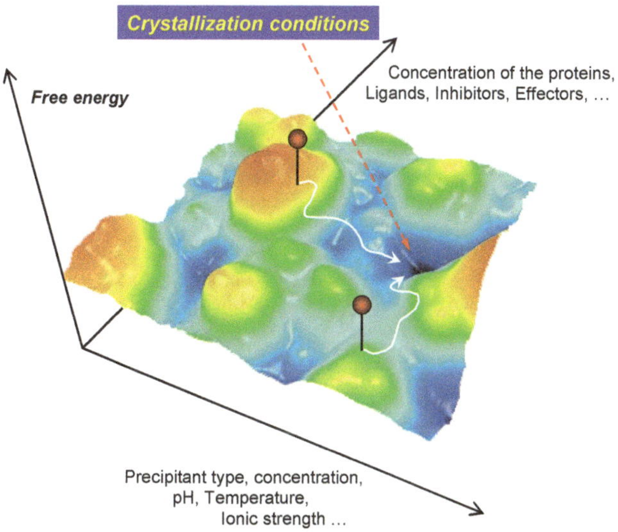

Fig. 1. Conceptual diagram of crystallization condition search screening on the landscape of potential free energy. Various parameters are shaken from certain starting conditions.

In our crystallization study of ICDH from a thermophile, *Thermus thermophilus* HB8 (*Tth* ICDH), *Tth* ICDH was overexpressed in *E. coli* MV1190 which carried plasmids pKID1, and the gene product was purified according to reported methods (Miyazaki et al., 1994). Purity of the yielded protein was checked with the polyacrylamide gel electrophoresis in the presence of sodium dodecyl sulfate (SDS-PAGE) (Laemmli, 1970). After dialysis against pure water, the protein was stored at 277 K until use. The crystallization experiments were carried out by the hanging drop vapor diffusion method in a 24-well tissue-culture linbro plate (Iwaki Glass Co., Ltd., Ciba, Japan) at 298 and/or 277 K. A random-screening protocol with screens developed in-house was used. The initial hits were optimized with further finer grid search. One screen package is similar to the Hampton Crystal Screen and Crystal Screen II (Hampton Research, Aliso Viejo, CA), and the other screen package contains various additives such as cofactors, inhibitors, nucleotides, minerals, salts and buffers with pH range 4 - 9. *Tth* ICDH was dialyzed against 20 mM Tris-HCl, pH 7.2, at 277 K, and incubated at 333 K for 10 min before the crystallization experiments. The initial protein solution contained 10.2 mg/ml of *Tth* ICDH in 20 mM Tris-HCl, pH 7.2. To a droplet of the protein solution, an equal amount of reservoir solution was added and then the droplet was equilibrated over 0.5 ml reservoir solution. The resulting microcrystals were obtained at the conditions with the reservoir commonly containing 100 mM sodium cacodylate and 1.4 M sodium acetate. The crystallization conditions were further optimized with regards to pH, co-existence of DL-isocitric acid, citric acid and/or cations like Mg^{2+} and Mn^{2+}. During the survey for crystallization conditions, information on protein crystallization such as

Species	Mol. mass (kDa)	3D structure	Crystallization condition			Space group	Unit cell dimen-sion (Å)	Reference
			Method	Protein solution	Precipitant			
Subfamily I								
Acidithiobacillus thiooxidans		available at 1.9Å	hdvd	25mM NAD^+	0.95-1.05M Na citrate, pH:4.6 10% Glycerol	$P4_32_12$	a=b=126 c=268	Imada *et al.* (2007)
Aeropyrum pernix		available at 2.20Å			12% PEG6000 60mM $MgCl_2$ 100mM Na-citrate, pH:5.6	$P4_32_12$	a=b= 107.6 c=171.1 $\alpha=\beta=\gamma=$ 90°	Karlström *et al.* (2002, 2005)
	47.9	available at 2.28Å	sdvd	5mM Tris-HCl, pH:8.0	20% PEG3350 200mM Diammonium citrate, pH:5.0	$P4_32_12$	a=b= 107.9 c=172.9 $\alpha=\beta=\gamma=$ 90°	Jeong *et al.* (2004)
Archaeglobus flugidus		available at 2.5Å	hdvd		0.6M $ZnSO_4$ 100mM Na-cacodylate	$P2_1$	a=81.6 b=65.4 c=87.2 β=95.3°	Stokke *et al.* (2007)
Bacillus subtilis	46.4	available at 1.55Å	hdvd	20mM Tris-HCl, pH:7.4 1mM Citrate 5mM $MgCl_2$, 5mM 2-Mercapto-ethanol 0.5mM PMSF 10% Glycerol	23% PEG4000 18% Propylene glycol 100mM Citrate, pH:4.9	$P2_1$	a=73.7 b=73.3 c=80.9 $\alpha=\gamma$=90° β=109°	Singh *et al.* (2001)
Escherichia coli	51	available at 2.5Å		34% Ammonium sulfate 100mM NaCl, 35mM Na_2HPO_4 9mM Citric acid 0.2mM DTT, pH:5.4		$P4_32_12$	a=b= 105.1 c=150.3	Hurley *et al.* (1989)
Subfamily II								
Desulfotalea psychrophila		available at 1.75Å	sdvd	10mM DL-isocitrate 0.25mM $NADP^+$ 20mM Pi-buffer, pH:7.0 10mM $MgCl_2$	100mM Tris-HCl, pH:7.4 1.7-1.9M Ammonium sulfate 2% PEG400 60mM $MgSO_4$	*P1*	a=59.3 b=73.3 c=126.4 α=98.9° β=99.0° γ=113.9°	Fedøy *et al.* (2007)

(*Continued*)

Species	Mol. mass (kDa)	3D structure	Crystallization condition: Method	Protein solution	Precipitant	Space group	Unit cell dimen-sion (Å)	Reference
Human cytosol	46.7	available at 2.7Å	hdvd		100mM MES pH:6.5 12% PEG20000	$P4_32_12$	a=b= 82.7 c=308.0	Xu *et al.* (2004)
		available at 2.41Å	hdvd	10mM DL-isocitrate 10mM $CaCl_2$ 10mM NADP	100mM MES, pH:5.9 20% PEG6000	$P2_1$	a=103.3 b=86.7 c=115.8 β=107.2°	
Thermotoga maritime (homotetrameric)	45.4	available at 2.2Å	sdvd		16% PEG6000 200mM Na- acetate 50mM NaCl 100mM Succinate, pH:6.0	$P2_12_12_1$	a=62.5 b=88.1 c=180.9 α=β=γ= 90°	Karlström *et al.* (2006)
Subfamily III								
porcine heart mitochondria ($NADP^+$-dependent)	45			100mM Triethanol-amine-HCl, pH:7.7 100 or 400mM Na_2SO_4 4mM Isocitrate 2mM $MnSO_4$	16-20% PEG6000 100mM Na_2SO_4 100mM Triethanol-amine acetate, pH:7.7	C2	a=137.0 b=113.4 c=65.0 β=98.5°	Endrizzi *et al.* (1996)
		available at 1.85 Å	hdvd	100mM Triethanol-amine-HCl, pH:7.7 150mM Na_2SO_4 8mM DL-isocitrate 4mM $MnSO_4$	100mM Triethanol-amine-HCl, pH:7.7 150mM Na_2SO_4 20% PEG6000 3% Glycerol	C2	a=138.0 b=113.8 c=66.8 β=97.6°	Ceccarelli *et al.* (2002)
Thermus thermophilus HB8	54.2				10% PEG6000 100mM Tris-HCl, pH:8.5	*I*222 or $I2_12_12_1$	a=100.1 b=150.4 c=87.4	Ohzeki *et al.* (1995)
			hdvd		100mM Na- cacodylate 1.4M Na- acetate pH:7.0	C2	a=495.5 b=189.2 c=336.2 β=126.4°	Ishii *et al.* (2008)
		available at 1.80 Å				$P2_1$	a= c= 73.0 b=95.2 β=92.1°	Lokanath *et al.* (2006)

(*Continued*)

Species	Mol. mass (kDa)	3D structure	Crystallization condition: Method	Protein solution	Precipitant	Space group	Unit cell dimension (Å)	Reference
yeast (in complex with NAD^+, octamer)				10mM Tris-HCl, pH:7.4 40mM NaCl 10mM Na-citrate 4mM $MgCl_2$ 5% Glycerol	100mM Hepes 900mM Na-citrate, pH:7.5	*R*3	a=302.0 c=112.1	Hu *et al.* (2005)
Other family, monomeric								
Azotobacter vinelandii (in complex with DL-isocitrate and Mg^{2+})				2.0-2.2M Ammonium sulfate w/o 1mM Isocitrate 1mM $MgCl_2$ or 1.8M Phosphate	1.6M Ammonium sulfate or 1.4M Na-K phosphate	$P4_22_12$	a=b= 122.1 c=163.9 $\alpha=\beta=\gamma=90°$	Czerwizski *et al.* (1977)
	80-100	available at 1.95 Å		20mM K-phosphate, pH:6.8 2mM $MgCl_2$ 10% Glycerol 10mM 2-Mercapto-ethanol	100mM Hepes-NaOH, pH:7.0 24% PEG6000 20% Glycerol 4mM $MnCl_2$ 4mM DL-isocitrate	$P2_12_12_1$	a=108.4 b=121.7 c=129.7	Yasutake *et al.* (2001)
Corynebacterium glutamicum				2.5M MES, pH:6.8 1.25mM $MnCl_2$ 1.25mM DTT 5% Glycerol	25% PEG2000 monomethyl-ether 180mM $MgCl_2$ 100mM Tris-HCl, pH:7.2	*C*2	a=137.1 b=54.6 c=126.4 β=108°	Audette *et al.* (1999)
	80	available at 1.75 Å			25% PEG2000 monomethyl-ether 200mM $MgCl_2$ 100mM Tris-HCl, pH:7.2	*C*2	a=129.0 b=52.7 c=124.0 β=108.9°	Imabayashi *et al.* (2006)

*hdvd: hanging drop vapor diffusion method, sdvd: sitting drop vapor diffusion method.

Table 1. Summary of crystallization conditions for ICDHs reported so far.

Hofmeister series, the order of effectiveness of the salts (Kunz et al., 2004), the crystallization conditions reported for ICDHs, and so on, were taken into account. Crystallization conditions for ICDHs from several different microorganisms reported so far are summarized in Table 1 with classification to the subfamilies. It may be instructive to show some more details observed on our screening stages. Figure 2 shows the results when 35% saturated ammonium sulfate was used as the precipitant. The reservoir solution contained 35% saturated ammonium sulfate, 0.1 M sodium chroride, 35 mM disodium hydrogenphosphate, 9 mM citric acid, and 0.2 mM dithiothreitol. Although no crystalline substances were recognized at pH 5.0, pillar-shaped thin crystals gathered together forming a dumbbell-like architecture were seen. These dumbbell-like architectures were frequently observed over the wide pH range between pH 5.2 and 7.5. Sometimes those were aggregated in a higher pH region. *Tth* ICDH showed a strong tendency to form pillar-like crystals under the conditions in which ammonium sulfate or 2-methyl-2,4-pentanediol was used as a precipitant. These are shown in Fig. 3. Even under conditions that contained DL-isocitric acid or citric acid, *Tth* ICDH formed thin stick-like or pillar-like crystals (Fig. 4). We have found conditions under which *Tth* ICDH forms crystals that diffract X-ray beyond 4 Å resolution, as such in a rod-like shaped (crystal form I) (Fig. 4) and a monoclinic diamond shaped (crystal form II) (Fig. 5). In the condition for crystal form I, the reservoir solution contained 100 mM sodium cacodylate, 1.4 M sodium acetate, 10 mM citric acid and 10 mM $MnCl_2$ with pH of 6.5 through 7.8. The reservoir solution containing 10 mM DL-isocitric acid instead of 10 mM citric acid is similar to the condition for crystal form I, the remainder of

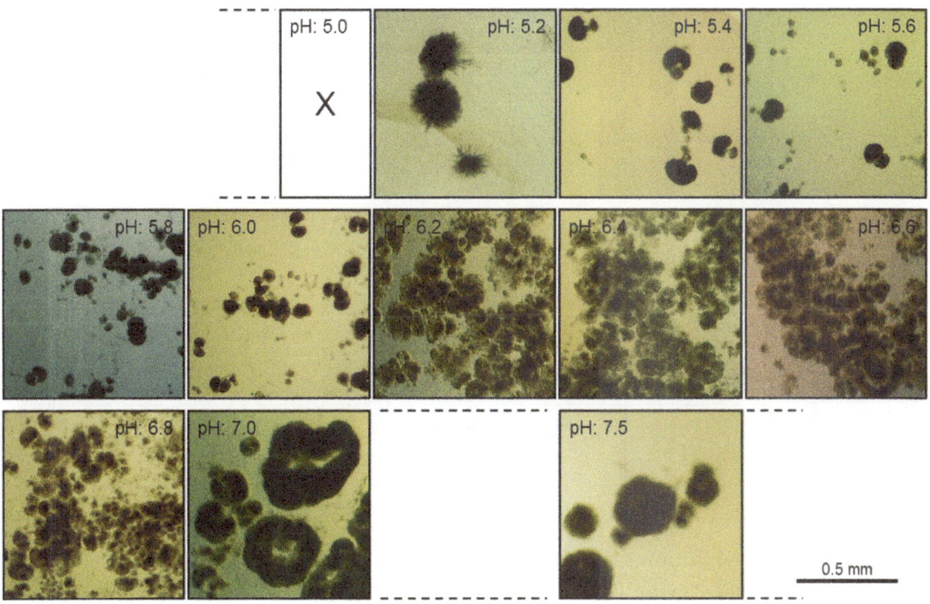

Fig. 2. Crystallization condition screening results of *Tth* ICDH using saturated ammonium sulfate as the precipitant. Pillar-shaped thin crystals forming dumbbell-like architectures were seen over the wide pH range. The 'X' means that no crystals were observed.

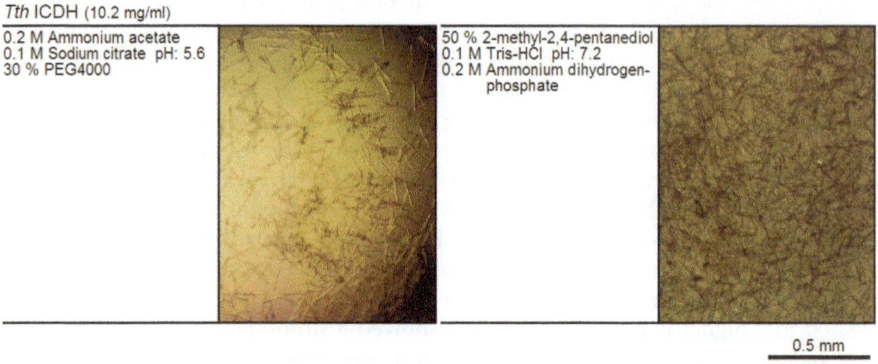

Fig. 3. Pillar-shaped crystals of *Tth* ICDH obtained by using PEG4000 or MPG as the precipitant. Each reservoir condition is indicated. A lot of thin rod-shaped crystals were observed under each condition. However, those were too thin and not suitable for X-ray diffraction experiment.

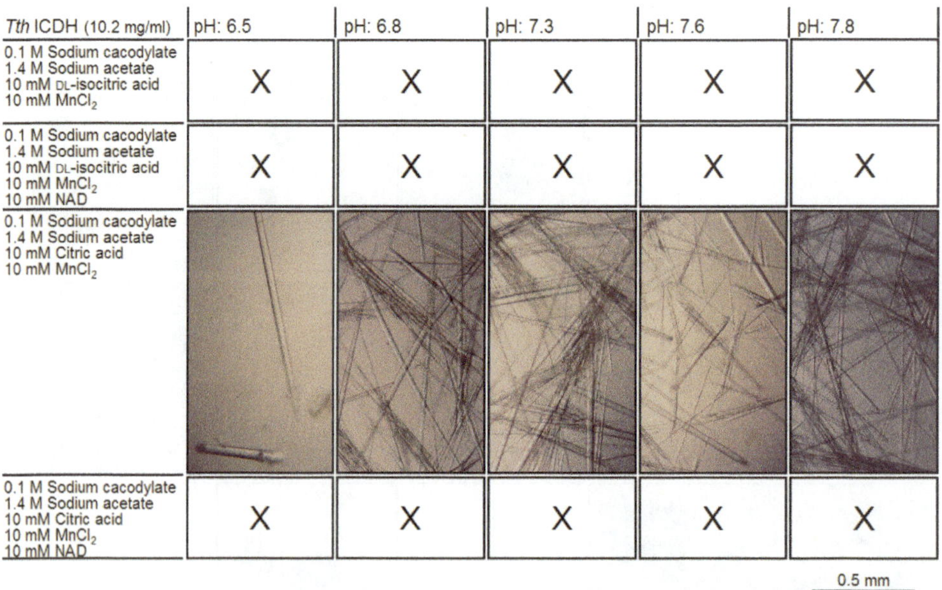

Fig. 4. Crystallization condition screening results of *Tth* ICDH using sodium acetate as the precipitant. Rod-shaped crystals (crystal form I) were obtained in the reservoir solution containing 0.1 M sodium cacodylate, 1.4 M sodium acetate, 10 mM citric acid, and 10 mM $MnCl_2$. The 'X' means that no crystals were observed.

100 mM sodium cacodylate, 1.4 M sodium acetate, and 10 mM $MnCl_2$ was common, and no crystals appeared in the pH range of 6.5 through 7.8. An addition of 10 mM NAD was also

useless. Furthermore, the addition of 10 mM NAD to the exact condition for crystal form I resulted in no crystal formation (Fig. 4). In the condition for crystal form II, the reservoir solution contained 100 mM sodium cacodylate and 1.4 M sodium acetate with wide range of pH condition of 6.1 through 8.4 (Fig. 5). As one can see at pH 8.1 in Fig. 5, multiple crystal forms are sometimes seen coexisting in the same sample of mother liquor. Crystals in the size up to 0.3 × 0.3 × 0.1 mm^3 were frequently observed within one month at 298 K. The crystals seemed to grow faster and larger in a fairly basic pH region. As shown in Fig. 6, the addition of 10 mM DL-isocitric acid or 10 mM citric acid facilitated the aggregate formation of pillar-shaped crystals. No crystals were formed upon the supplement of 10 mM $MgCl_2$ to each of the above conditions (Fig. 6). *Tth* ICDH crystals in crystal form II incubated at the condition for 8 years showed clear edges even though amorphous precipitates were increased (data not shown). It has been said that a protein from a thermophilic bacterium is proportionally thermostable. We are convinced that this is true even in the crystal state.

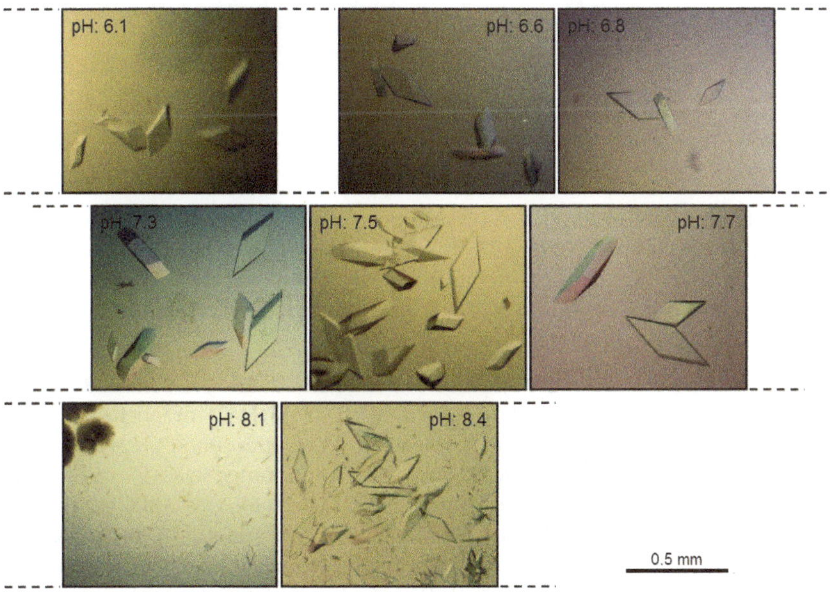

Fig. 5. Crystallization condition screening results of *Tth* ICDH using sodium acetate as the precipitant. The thick diamond-shaped crystals (crystal form II) were obtained in the reservoir solution containing 100 mM sodium cacodylate and 1.4 M sodium acetate with wide range of pH condition of 6.1 through 8.4.

X-ray diffraction experiments were carried out at three different facilities; The diffraction data from both crystal form I and II up to middle range resolution were collected at 286 K on a Rigaku R-AXIS IIc image plate detector (Rigaku Corp., Tokyo, Japan) in our laboratory (AIST, Tsukuba, Japan) equipped with a Rotaflex FR rotating anode generator operated at 45 kV, 50 mA with focal spot size of 0.1 mm. The image data obtained were processed with a program set incorporated in the R-AXIS IIc software package. The cryogenic X-ray diffraction experiments with crystal form I were carried out at the BL-6A beamline station of

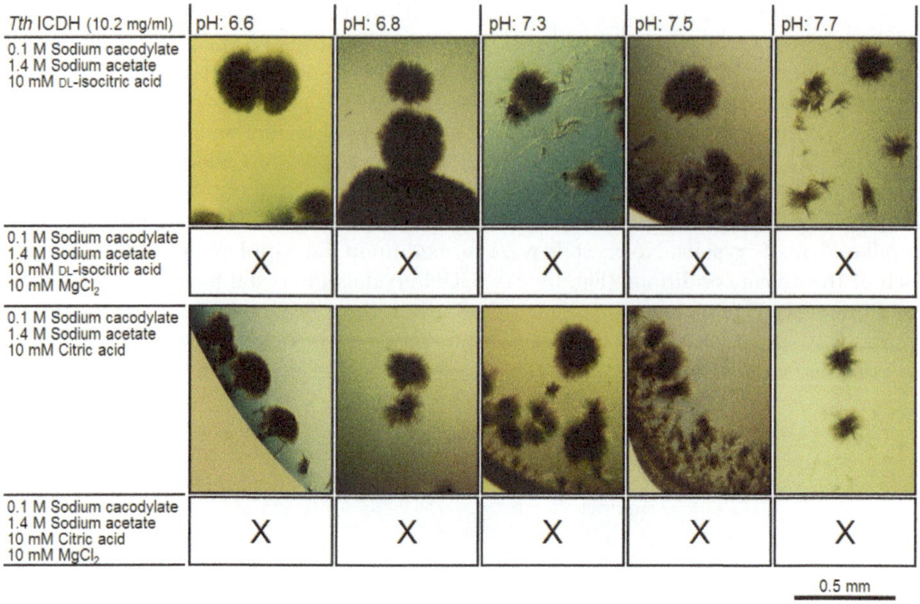

Fig. 6. Crystallization condition screening results of *Tth* ICDH using sodium acetate as the precipitant. In the presence of 10 mM DL-isocitric acid, or 10 mM citric acid, pillar-shaped thin crystals forming dumbbell-like architectures were seen over the wide pH range. However, in a co-addition of 10 mM $MgCl_2$ to each condition, no crystalline substances were observed. The 'X' means no crystals observed.

the Photon Factory (KEK, Tsukuba, Japan). The intensity data were collected at 105 K. The X-ray beam was monochromatized to 1.00 Å with an Si (111) monochromator, and an aperture collimator of 0.10 mm diameter was used. Oscillation photographs were taken by the ADSC Quantum 4R CCD detector. The oscillation range was 5° during a 10-min exposure and the distance from the crystal to the CCD detector was 300 mm. The data were processed using DPS/MOSFLM (Leslie, 1992) and programs from the CCP4 suite (Collaborative Computing Project, 1994). Another measurement with crystal form II was performed at 293 K on a Rigaku R-AXIS IV (Rigaku Corp., Tokyo, Japan) using synchrotron radiation at the BL-24 station in the SPring-8 (JASRI, Hyogo, Japan) with X-rays of wavelength 0.835 Å. The oscillation range during a 3-min exposure was 0.5° and the distance from the crystal to the CCD detector was 300 mm. The exposed images were automatically analyzed with an incorporated R-AXIS IV software package.

The cryogenic X-ray diffraction data were obtained from the crystal form I at BL-6A beamline station in KEK. Preliminary intensity data were collected in which the diffraction extended beyond 3.4 Å resolution. The crystal in form I was assumed to be hexagonal or trigonal, and the unit dimensions were a = b = 163.1 Å, c = 269.1 Å, $\alpha = \beta = 90.0°$, $\gamma = 120.0°$. Further measurement has been hampered by the cryoprotectants selected and used. The crystal form II diffracted X-rays beyond 4 Å at the BL-24 station in SPring-8 using the synchrotron X-ray source. Including diffraction data obtained with Rigaku R-AXIS IIc in the laboratory *Tth* ICDH crystal was determined to be monoclinic and belonged to *C*2 space group. The unit cell dimensions were a = 495.5 Å, b = 189.2 Å, c = 336.2 Å, β = 126.4°. According to the normal V_m range of 1.7 - 3.5 $Å^3$ Da^{-1} (Matthews, 1968), the asymmetric unit was estimated to contain between 33 and 68 *Tth* ICDH molecules with a molecular mass of 54.2 kDa. The calculation indicates that *Tth* ICDH crystal form II contains large number of molecules in the reiterative unit in the crystalline arrays. This fact can be reconciled with the result observed in the non-denaturing PAGE; several bands were stained in the higher molecular mass region in addition to the native band corresponding to *Tth* ICDH dimer in the non-denaturing gel, which were performed on the crystals gathered by centrifugation followed by rapid loading to the stacking gel prior to the application of voltage (data not shown). However, in the PAGE with moderate treatment for the crystals gathered by the centrifugation, the band that corresponded to the molecular mass of the dimer became dominant. These observations can be understood as follows: the crystals are made of large preformed homo-complexes of *Tth* ICDH molecules, which stay stable in the reservoir solution, but soon dissociate into the sub-clusters or singler molecules (*Tth* ICDH dimers) out of the range of the critical crystallization condition.

As to the manner of interaction between the possible supramolecules' packing in the crystals, interesting results were obtained. The thick diamond-shaped crystals grown at around neutral pH region, pH 7.5 for example, was found to maximally diffract X-rays at around 7.0 Å at 95 K after treatment with reservoir solution plus 15 % glycerol as a cryoprotectant. Furthermore, the crystals in form II, having the same appearance, grown in the slightly basic pH region, pH 7.8, 8.1 and 8.4, for example, could never diffract X-rays at 95 K. These observations could be understood in that the formation of the supramolecular units and the interaction between the units were suitable enough to form the form II crystal shape at room temperature, which could be further inferred from the diffraction images in high resolution range (data not shown). When the crystals were treated at cryogenic condition, the intermolecular interactions should have been altered in the direction of increasing entropy (N. Ishii et al., 2008).

Tth ICDH molecules were placed under the crystallization condition for about three months and the protein forming crystal form II were examined as they were at the state with HPLC gel filtration chromatography using a TSKgel G3000SWxl (Tosoh, Tokyo, Japan). The elution profile of the above protein solution co-existing with form II crystals is shown in Fig. 7. There appeared a few peaks, which were labeled 1 (~400 kDa), 2 (~300 kDa), and 3 (~220 kDa) in the molecular mass region larger than the intact *Tth* ICDH dimer (peak 4 (98 kDa)). According to the molecular mass calibration standard, it is comprehensive that the peak 1, 2, and 3 correspond to octamer (presumably 4 dimers), hexamer (3 dimers), and tetramer (2 dimers), respectively. There seems to be hierarchies divisible by integers of the dimer as a unit.

Atomic force microscopy (AFM) gives us useful information on the growth and disorder of macromolecular crystal, and when combined with X-ray diffraction study can bring further insights into the improvement of the macromolecular crystallization protocols (Malkin & Thorne, 2004; Scabert et al., 1995). Separately, we have performed AFM scanning on the crystalline surface of crystal form II. A lot of ellipsoidal bodies were observed (N. Ishii et al., 2008). The average values of the short and long axes of the ellipsoidal bodies detected in AFM imaging are 10.87 ± 1.47, 18.61 ± 2.58 nm, respectively. Therefore, the average volume of the body should be 1151.34 ± 2.92 nm^3. According to the normal V_m range of 1.7 - 3.5 $Å^3$ Da^{-1} (Matthews, 1968), the molecular mass of the ellipsoidal body should fall in between 340 kDa and 700 kDa. These values can be ascribed to hexamer and dodecamer of *Tth* ICDH

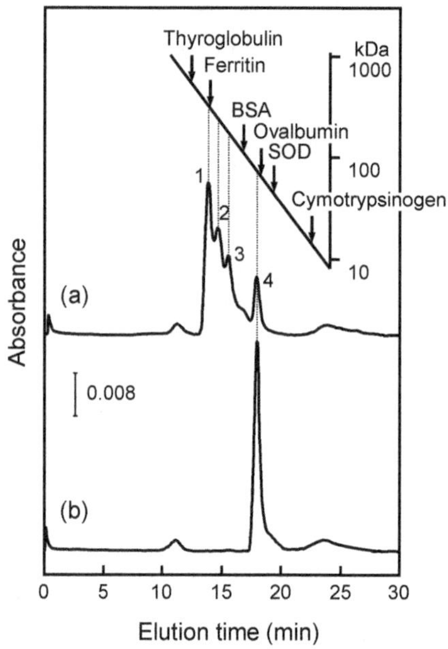

Fig. 7. HPLC gel permeation profiles of *Tth* ICDH solution with and without the thick diamond-shaped crystals (crystal form II). (a) HPLC gel permeation profile of *Tth* ICDH being incubated under the crystallization condition for crystal form II; (b) that of intact *Tth* ICDH. The peaks numbered 1, 2, 3, and 4 correspond to about 400 k, about 300 k, about 220 k, and about 98 kDa respectively.

(monomer) molecules, respectively. Since protein crystals usually include solvent molecules at high content it is difficult to speculate how many *Tth* ICDH molecules constitute the one ellipsoidal body. Taking these results obtained from AFM imaging and HPLC gel filtration into account, one can infer that *Tth* ICDH crystal form II should be comprised of oligomeric building blocks piled one on top of another. The building unit is most likely an octamer (4 dimers), and the next likely to be a hexamer (3 dimers) from the HPLC profile (Fig. 7), where both are made of *Tth* ICDH dimer as a basic unit. In crystal form II of *Tth* ICDH the exact arrangement and manner of the formation are still obscure, although there is possibility that *Tth* ICDH supramolecular complex acts as a block that interacts together in the process of spontaneous building up of form II crystals under the favorable crystallization condition described above. Needless to say, in order to determine a crystal structure of a certain protein species by X-ray method, crystals that well diffract X-rays to high resolution, and at the same time, that contain possibly small number of molecules in crystallographic asymmetric unit are prerequisite (N. Ishii et al., 2000a, 2000b; Shimamura et al., 2004). On the other hand, to construct some architectures in the nano-scale using protein molecules as building blocks we have to understand the nature of interactions between protein molecules, namely, how the ability to self-assembly emerges, including effects mediated through solvent conditions such as pH, temperature and ionic strength, and so on (Biswas et al., 2009; D. Ishii et al., 2003). The result mentioned above have shed light on the usefulness of *Tth* ICDH, a thermostable protein and its form II crystals, in the study of supramolecular complexes and crystal formation by self-assembly. It is apparent that research in this avenue needs to be continued further for elucidation of useful tools useable in protein nanotechnology.

4. Crystal structure of ICDH

ICDH evolved early and is widely distributed among archaea, bacteria, and eukarya. Such an evolutional trace can be found in diverse primary structures, various oligomeric forms taken, and different specificity as to cofactors (Steen et al., 2001). It has been proposed that NAD^+-specific ICDH may be an ancestor enzyme that functions in CO_2 fixation in an early stage of evolution of the Krebs cycle (Shiba et al., 1985).

We have unexpectedly obtained the crystals (form II) of the supramolecular complex of *Tth* ICDH and concentrated on surveying how these building block molecules pile up and self-assemblize into the crystal form II. Finally we have revealed the mechanism of the hierarchical formation that *Tth* ICDH molecules reside, being piled one on top another as a preformed supramolecular nano-architecture in the crystal lattice. In the mean time, Lokanath & Kunishima (2006) successfully determined the structure of *Tth* ICDH at 1.8 Å resolution. It should be instructive to mention an overview of typical ICDH structures, and what are still obscure and open to discussion from the view of structural biology and enzymology of ICDHs picking up some representatives.

4.1 *Escherichia coli* ICDH

The ICDH are usually dimeric proteins with two identical subunits of molecular mass of 40 - 50 kDa per subunit (Chen & Gadal, 1990). In *E. coli*, ICDH is a homodimeric enzyme and its inactivation mechanism by phosphorylation has been reported in detail with regards to the crystal structure (Hurley et al., 1990). The crystal structure of ICDH from *E. coli* (*Ec* ICDH) shows that the substrate binding pockets and catalytic sites of the dimeric enzymes are formed from side chains of residues donated asymmetrically both subunits (Hurley et al., 1989). The tertiary structure of *Ec* ICDH is depicted in Fig. 8. The enzyme is composed of 13 α-helices and 14 β-strands. It contains three domains consisting of a large domain, a small domain, and a clasp domain. This manner is common to ICDH from *Bacillus subtilis* (*Bs* ICDH) and *Tth* ICDH. This enzyme has an active site in a cleft between the large and small domains (Hurley et al., 1994; Stoddard et al., 1993). The reaction mechanism of ICDH has been extensively studied in *E. coli* ICDH. A conformational change occurs from an open to closed form upon the binding of $NADP^+$ and the substrate. In the proposed mechanism, a proton is removed from the α-hydroxyl group of isocitrate, and then, a hydride ion is transferred in a stereospecific way from the α-carbon atom of the substrate to C-4 of the nicotinamide ring of $NADP^+$, oxidizing isocitrate to oxalosuccinate. In the following step, the β-carboxylate group of oxalosuccinate is removed as CO_2, and is replaced by a proton in a stereospecific way to form 2-oxoglutarate. During both transitions the negative charge on the hydroxyl oxygen atom of isocitrate is stabilized by a magnesium ion. There are still controversies as to the mechanisms of the initial proton abstraction and the final proton donation.

The crystal structure of *Ec* ICDH determined with its substrates revealed that the malate moiety of isocitrate, namely 1-carboxyl and 3-carboxyl groups and 2-hydroxyl group, is recognized by three arginine, three aspartate, one lysine, and one tyrosine residues of ICDH, which are the residues conserved between the enzymes (Hurley et al., 1990). As shown in Fig. 9, in the substrate binding site, the α-carboxyl group of isocitrate is bound to three arginine residues, Arg-119, Arg-129, and Arg-153. The β-carboxylate group is bound to Arg-153 and Tyr-160. The α-position of isocitrate forms hydrogen bonds with the Lys-230′, and Asp-283′ residues of the other subunit of the dimer (Doyle et al., 2001; Mesecar et al., 1997).

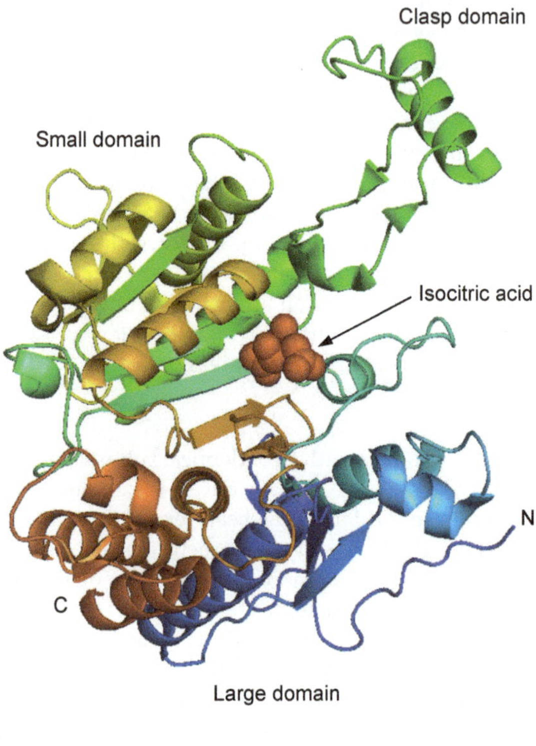

Fig. 8. Diagram of *Ec* ICDH monomer with isocitric acid. (PDB ID: 1P8F). The large domain containing the N and C termini, the small domain, and the clasp domain are indicated. Atomic coordinates were obtained from the RCSB Protein Data Bank (www.rcsb.org/pdb/home/home.do) and imaged using PyMol (The PyMOL Molecular Graphics System, version 0.99, DeLano scientific, LLC).

Mesecar and Koshland Jr. have made precisely explanation as to the mechanism for stereospecificity, that is, to distinguish between the two enantiomers on the basis revealed by electron density maps of the crystal structures of ICDH with L- and D-isomer isocitrate (Mesecar & Koshland Jr., 2000). In the metal-free crystal of ICDH with a racemic mixture of isocitrate, only the L-isomer (2S,3R) is seen bound to the enzyme. When enzyme crystals are presented with a racemic mixture of isocitrate in the presence of Mg^{2+}, only D-isomer is seen in the active site in the crystal structure. They indicate the importance of the -OH group of L-isocitrate association with arginine residue at position 119 in the metal-free enzyme, in contrast to the -OH group of D-isocitrate association with the metal and with two aspartate residues at position 283′ and 307.

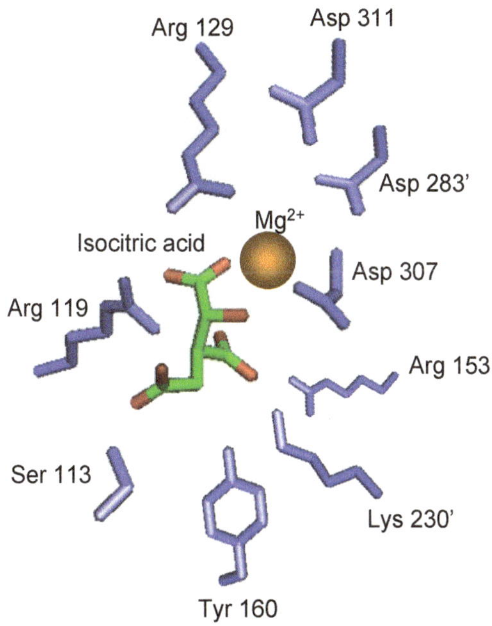

Fig. 9. Overview of the ICDH substrate binding site.

4.2 *Bacillus subtilis* ICDH

Recently, the crystal structure of *Bs* ICDH has been reported and discussed on the difference between *Ec* ICDH. Both have very similar 3-D structures not only because *Bs* ICDH is 67% identical to *Ec* ICDH but also that both are 100% identical in the primary sequence around the phosphorylation site (Singh et al., 2001). The tertiary structure of *Bs* ICDH monomer A is illustrated in Fig. 10, and the *Bs* ICDH dimer is shown in Fig. 11. Each subunit is composed of 15 α-helices and 13 β-strands. Although *Bs* ICDH is a homodimer, interestingly, the crystal structures of the individual monomers are not identical. In the dimerization, so-called clasp domain is formed where constituent of two β-strands and connecting α-helix of each subunit interlock forming a hydrophobic core. The manner of intersubunit interactions in *Bs* ICDH dimer is reported the same for the most part with *Ec* ICDH. There are several reported differences between *Bs* ICDH and *Ec* ICDH, suggesting the robustness of the enzyme in preserving the principle function.

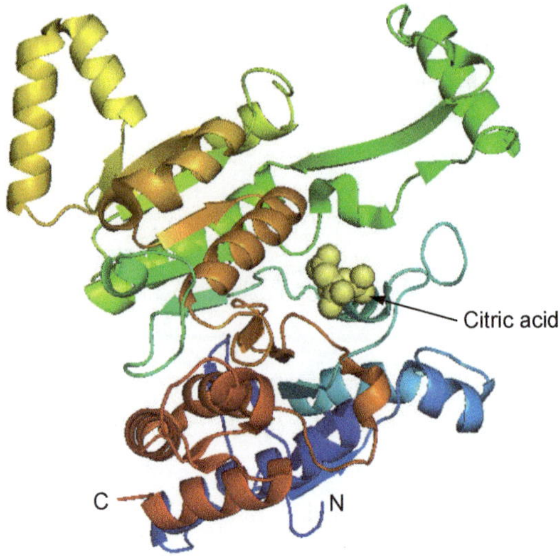

Fig. 10. Cα trace of *Bs* ICDH monomer (monomer A) with citric acid. (PDB ID: 1HQS). Overall folding manner of the monomer resembles that of *Ec* ICDH, but has several differences in the secondary structures.

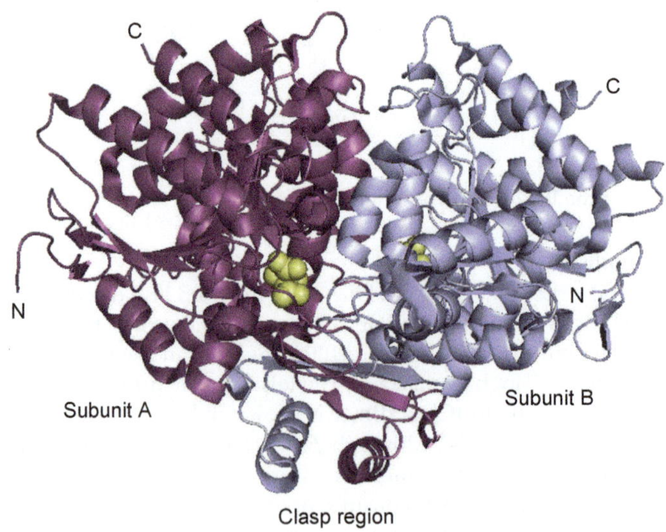

Fig. 11. Diagram of *Bs* ICDH dimer. Each of monomer A, and monomer B having slightly different conformation are related by dimerization mediated through expansive portion of small and clasp domains. (PDB ID: 1HQS).

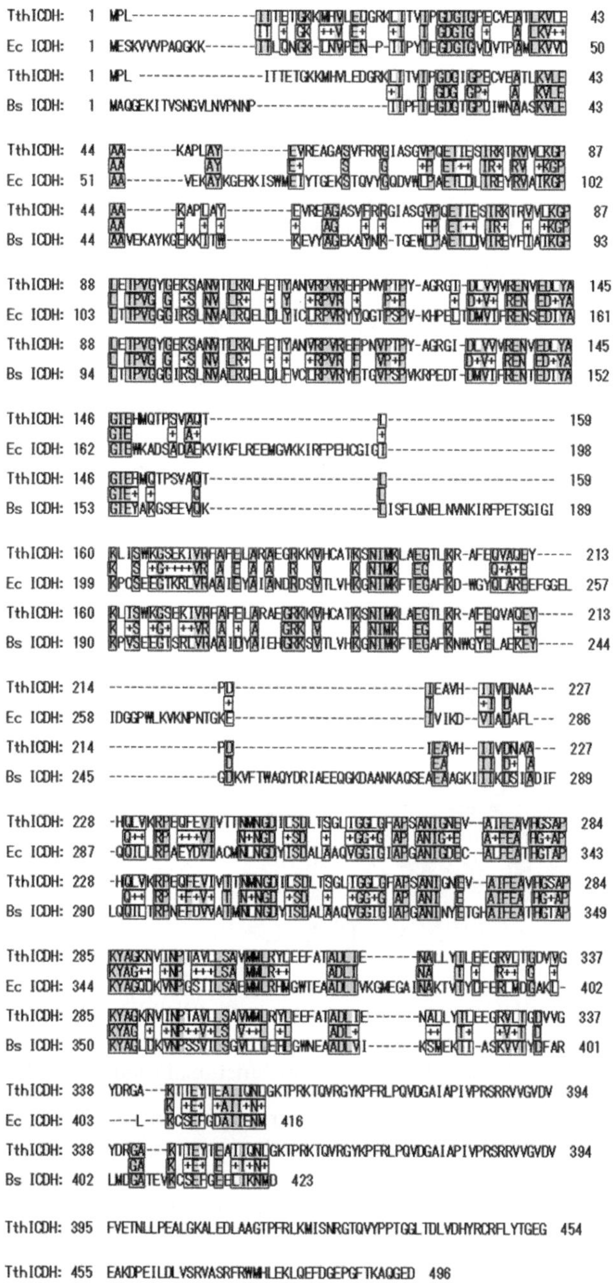

Fig. 12. Amino acid sequence alignment of *Tth* ICDH, *Ec* ICDH, and *Bs* ICDH. The shaded and boxed sequences represent those identical to *Tth* ICDH sequence, and boxed ones represent those in similar to *Tth* ICDH sequence.

4.3 *Thermus thermophilus* ICDH

Between *T. thermophilus* and *E. coli* ICDHs, the residues of 37% are identical and those of 51% show similarity. The typical difference in the primary structure between the two enzymes is the presence of 141 extra residues at the C terminus in *Tth* ICDH. The region may contribute to the folding of the enzyme and the acquired thermostability of the enzyme (Miyazaki et al., 1992). Between *Tth* ICDH and *Bs* ICDH, the residues of 35% are identical and those of 50% have similarity. The primary sequence alignment of *Tth* ICDH, *Ec* ICDH, and *Bs* ICDH is shown in Fig. 12. Enzymes from thermophiles are often highly homologous to the mesophilic counterparts and the catalytic mechanisms are usually identical. The thermophilic enzymes have to be stable enough to withstand denaturation at elevated temperature where the thermophile optimally grows and possess simultaneously the flexibility required for enzymatic activity. Comparison among these enzymes of functions and structural similarities suggest functional robustness. The crystal structure of *Tth* ICDH is shown in Fig. 13. The overall core structure resembles those of *Ec* ICDH and *Bs* ICDH. As shown in Fig. 14, it can be seen that the string of the extra residues constitutes clasp domain, and it appears playing a role in the formation of a stable dimer.

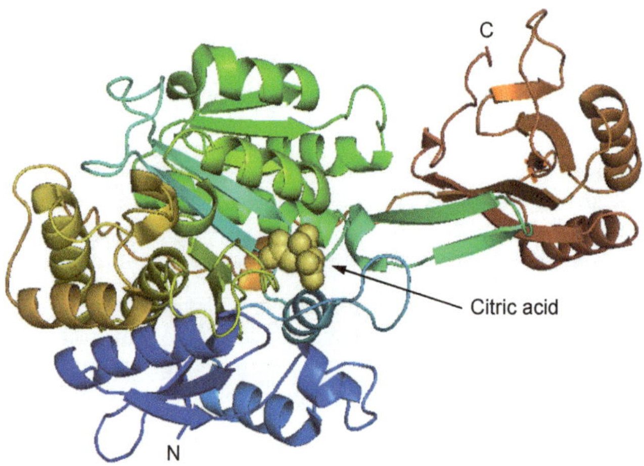

Fig. 13. Cα trace of *Tth* ICDH monomer with citric acid. (PDB ID: 2D1C). Although there is a string of the extra sequence at the C terminus region, overall folding manner is quit similar to *Ec* ICDH and *Bs* ICDH. It may the reflection of functional robustness of the protein.

Although the preliminary X-ray study and crystallization of *Tth* ICDH were already reported by Ohzeki et al. (1995), and quite recently the crystal structure was determined by Lakonath & Kunishima (2006), we have described the other two curious crystal forms (crystal form I, and II) obtained for *Tth* ICDH under different crystallization conditions. Observation on the surface of one of the crystal forms (form II) by AFM motivated extended analysis on the possibility of *Tth* ICDH molecule taking the form of a supramolecular architecture under the condition and crystallization state. Study in this line leads to the important subject which concerns polymorphism of protein crystals, and has focused on the importance of spontaneous hierarchical construction with supramolecular assemblies as a block towards nano-composites.

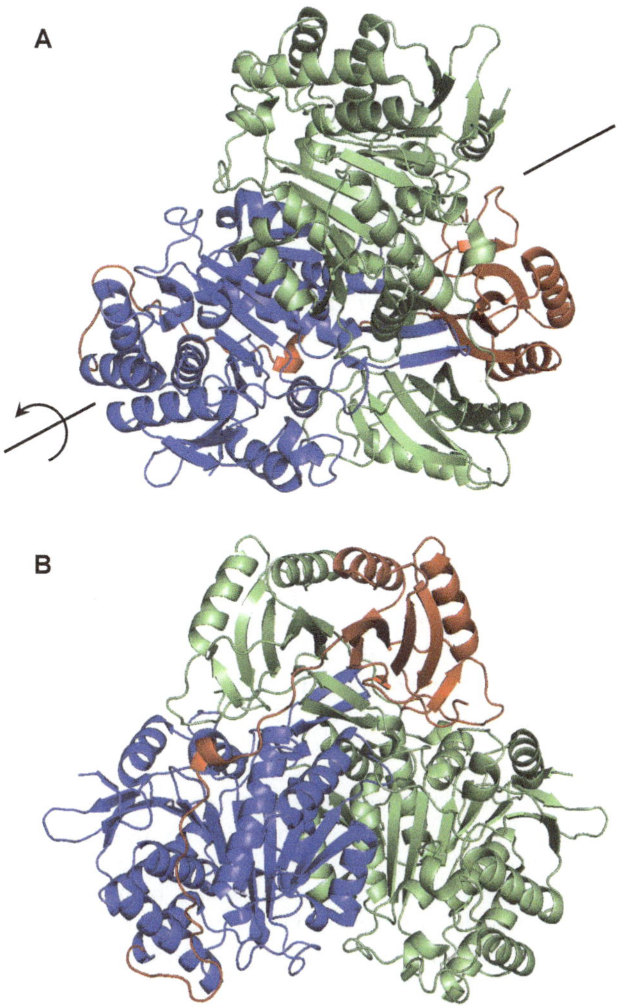

Fig. 14. Diagram of *Tth* ICDH dimer. (A) Monomer A is shown in blue with its 141 extra residues at the C terminus region in red. Monomer B is shown in pale-green. (B) in the view, *Tth* ICDH has been rotated ~180° around the axis indicated in (A). (PDB ID: 2D1C).

5. Conclusion

Although more than 40 structures including many mutants of homo-dimeric ICDH are available from Protein Data Bank (RCSB PDB), there are few structures with as clearly drawn electron density as the nicotinamide ribose moiety for example. It is understood that there lies the difficulty due to disorder, hence, further structural information is still requisite to elucidate the catalytic mechanism. Observation by AFM of the surface of the crystal form II of *Tth* ICDH suggested that the crystals consisted of huge ellipsoidal bodies of a homo-complex of ICDH, of which long axis' diameter was 18.6 nm and short one was 10.9 nm. The

HPLC gel filtration column chromatography of the protein co-existed with its form II crystals which had been incubated under the critical crystallization condition, thus supporting the line of the above assumption. The calculation using the preliminary X-ray diffraction data indicated that an asymmetric unit contains lots of oligomeric *Tth* ICDH in crystal form II. We have found huge supramolecular complex formation under the appropriate crystallization condition among the ICDH subfamilies, which concerns polymorphism of protein crystals.

6. Acknowledgment

The author thanks Dr. Kenneth S. Kim for discussion and critical reading the manuscript.

7. References

Akiba, T.; Ishii, N.; Rashid, N.; Morikawa, M.; Imanaka, T. & Harata, K. (2005). Structure of RadB recombinase from a hyperthermophilic archaeon, *Thermococcus kodakaraensis* KOD1: an implication for the formation of a near-7-fold helical assembly. *Nucleic Acids Res.*, Vol.33, No.10, (June 2005), pp. 3412-3423, ISSN 0305-1048

Audette, G.F.; Quail, J.W.; Hayakawa, K.; Bai, C.; Chen, R. & Delbaere, L.T. (1999). Crystallization and preliminary X-ray diffraction studies of monomeric isocitrate dehydrogenase from *Corynebacterium glutamicum*. *Acta Crystallogr. D Biol. Crystallogr.*, Vol.55, No.9, (September 1999), pp. 1584-1585, ISSN 0907-4449

Biswas, S.; Kinbara, K.; Oya, N.; Ishii, N.; Taguchi, H. & Aida, T. (2009). A tubular biocontainer: metal ion-induced 1D assembly of a molecularly engineered chaperonin. *J. Am. Chem. Soc.*, Vol.131, No.22, (June 2009), pp. 7556-7557, ISSN 0002-7863

Ceccarelli, C.; Grodsky, N.B.; Ariyaratne, N.; Colman, R.F. & Bahnson, B.J. (2002). Crystal structure of porcine mitochondrial $NADP^+$-dependent isocitrate dehydrogenase complexed with Mn^{2+} and isocitrate. Insights into the enzyme mechanism. *J. Biol. Chem.*, Vol.277, No.45, (November 2002), pp. 43454-43462, ISSN 0021-9258

Chen, R.D. & Gadal, P. (1990). Structure, functions and regulation of NAD and NADP dependent isocitrate dehydrogenases in higher plants and in other organisms. *Plant Physiol. Biochem.*, Vol.28, pp. 411-427, ISSN 0981-9428.

Collaborative Computing Project Number 4 (1994). The CCP4 suite: programs for protein crystallography. *Acta Crystallogr. D*, Vol.50, No.5, (September 1994), pp. 760-763, ISSN 0907-4449

Czerwinski, E.W.; Bethge, P.H.; Mathews, F.S. & Chung, A.E. (1977). A preliminary crystallographic study of isocitrate dehydrogenase from *Azotobacter vinelandii*. *J. Mol. Biol.*, Vol.116, No.1, (October 1977), pp. 181-187, ISSN 0022-2836

Doyle, S.A.; Beernink, P.T. & Koshland Jr., D.E. (2001). Structural basis for a change in substrate specificity: Crystal structure of S113E isocitrate dehydrogenase in a complex with isopropylmalate, Mg^{2+}, and NADP. *Biochem.*, Vol.40, No.14, (March 2001), pp. 4234–4241, ISSN 0006-2960

Endrizzi, J.; Zhang, G.; Xiong, W.C.; Hurley, J.H.; Remington, S.J. & Colman, R.F. (1996). Crystallization and preliminary diffraction analysis of porcine heart mitochondrial $NADP^+$-dependent isocitrate dehydrogenase. *Acta Crystallogr. D Biol. Crystallogr.*, Vol.52, No.5, (September 1996), pp. 1024-1026, ISSN 0907-4449

Fedøy, A.E.; Yang, N.; Martinez, A.; Leiros, H.K. & Steen, I.H. (2007). Structural and functional properties of isocitrate dehydrogenase from the psychrophilic bacterium *Desulfotalea psychrophila* reveal a cold-active enzyme with an unusual high thermal stability. *J. Mol. Biol.*, Vol.372, No.1, (September 2007), pp. 130-149, ISSN 0022-2836

Fukunaga, N.S.; Sahara, I.T.; Ishii, S.A. & Suzuki, M. (1992). Purification and characterization of monomeric isocitrate dehydrogenase with $NADP^+$-specificity from *Vibrio parahaemolyticus* Y-4. *J. Biochem.*, Vol.112, No.6, (December 1992), pp. 849-855 ISSN 0021-924X

Hu, G.; Taylor, A.B.; McAlister-Henn, L. & Hart, P.J. (2005). Crystallization and preliminary X-ray crystallographic analysis of yeast NAD^+-specific isocitrate dehydrogenase. *Acta Crystallogr. F Struct. Biol. Cryst. Commun.*, Vol.61, No.5, (May 2005), pp. 486-488, ISSN 1744-3091

Hurley, J.H. & Dean, A.M. (1994). Structure of 3-isopropylmalate dehydrogenase in complex with NAD^+: ligand-induced loop closing and mechanism for cofactor specificity. *Structure*, Vol.2, No.11, (November 1994), pp. 1007-1016, ISSN 0969-2126

Hurley, J.H.; Dean, A.M.; Sohl, J.L. & Koshland Jr, D.E. (1990). Regulation of an enzyme by phosphorylation at the active site. *Science*, Vol.249, No. 4972, (August 1990), pp. 1012-1016, ISSN 0036-8075

Hurley, J.H.; Thorsness, P.E.; Ramalingam, V.; Helmers, N.H. Koshland Jr, D.E. & Stroud, R.M. (1989). Structure of a bacterial enzyme regulated by phosphorylation, isocitrate dehydrogenase. *Proc. Natl. Acad. Sci. USA*, Vol.86, No.22, (November 1989), pp.8635-8639, ISSN 0027-8424

Imabayashi, F.; Aich, S.; Prasad, L. & Delbaere, L.T. (2006). Substrate-free structure of a monomeric NADP isocitrate dehydrogenase: an open conformation phylogenetic relationship of isocitrate dehydrogenase. *Proteins*, Vol.63, No.1, (April 2006), pp. 100-112, ISSN 0887-3585

Imada, K.; Tamura, T.; Takenaka, R.; Kobayashi, I.; Namba, K. & Inagaki, K. (2008). Structure and quantum chemical analysis of NAD^+-dependent isocitrate dehydrogenase: hydride transfer and co-factor specificity. *Proteins*, Vol.70, No.1, (January 2008), pp. 63-71, ISSN 0887-3585

Ishii, D.; Kinbara, K.; Ishida, Y.; Ishii, N.; Okochi, M.; Yohda, M. & Aida, T. (2003). Chaperonin-mediated stabilization and ATP-triggered release of semiconductor nanoparticles. *Nature*, Vol.423, No.6940, (June 2003), pp. 628-632, ISSN 0028-0836

Ishii, N.; Haga, K.; Yamane, K. & Harata, K. (2000a). Crystal structure of asparagine 233-replaced cyclodextrin glucanotransferase from alkalophilic *Bacillus* sp. 1011 determined at 1.9 Å resolution. *J. Mol. Recognit.*, Vol.13, No.1, (January-February 2000), pp. 35-43 ISSN 0952-3499

Ishii, N.; Haga, K.; Yamane, K. & Harata, K. (2000b). Crystal structure of alkalophilic asparagine 233-replaced cyclodextrin glucanotransferase complexed with an inhibitor, acarbose, at 2.0 Å resolution. *J. Biochem.*, Vol.127, No.3, (March 2000), pp. 383-391, ISSN 0021-924X

Ishii, N.; Saijo, S.; Sato, T; Tanaka, N. & Harata K. (2001). Crystallization and preliminary X-ray studies of V_1-ATPase of *Thermus thermophilus* HB8 complexed with Mg-ADP. *J. Struct. Biol.*, Vol.134, No.1, (April 2001), pp. 88-92, ISSN 1047-8477

Ishii, N.; Umemura, K. & Miyazaki, K. (2008). Supramolecular complex formation and crystallization of isocitrate dehydrogenase from *Thermus thermophilus* HB8:

Preliminary studies with X-ray crystallography and atomic force microscopy. *Biosci. Biotechnol. Biochem.*, Vol.72, No.9, (September 2008), pp. 2369-2376, ISSN 0916-8451

Jeong, J.J.; Sonoda, T.; Fushinobu, S.; Shoun, H. & Wakagi, T. (2004). Crystal structure of isocitrate dehydrogenase from *Aeropyrum pernix*. *Proteins*, Vol.55, No.4, (June 2004), pp. 1087-1089, ISSN 0887-3585

Karlström, M.; Steen, I.H.; Madern, D.; Fedöy, A.E.; Birkeland, N.K. & Ladenstein, R. (2006). The crystal structure of a hyperthermostable subfamily II isocitrate dehydrogenase from *Thermotoga maritima*. *FEBS J.*, Vol.273, No.13, (July 2006), pp. 2851-2868, ISSN 1742-464X

Karlström, M.; Steen, I.H.; Tibbelin, G.; Lien, T.; Birkeland, N.K. & Ladenstein, R. (2002). Crystallization and preliminary X-ray structure analysis of isocitrate dehydrogenase from two hyperthermophiles, *Aeropyrum pernix* and *Thermotoga maritima*. *Acta Crystallogr. D Biol. Crystallogr.*, Vol.58, No.12, (November 2002), pp. 2162-2164, ISSN 0907-4449

Karlström, M.; Stokke, R.; Steen, I.H.; Birkeland, N.K. & Ladenstein, R. (2005). Isocitrate dehydrogenase from the hyperthermophile *Aeropyrum pernix*: X-ray structure analysis of a ternary enzyme-substrate complex and thermal stability. *J. Mol. Biol.*, Vol. 345, No.3, (January 2005), pp. 559-577, ISSN 0022-2836

Kunz, W.; Henle, J. & Ninham, B.W. (2005). 'Zur Lehre von der Wirkung der Salze' (about the science of the effect of salts): Franz Hofmeister's historical papers. *Curr. Opin. Coll. Interface Sci.*, Vol.9, No.1-2, (August 2004), pp. 19-37, ISSN 1359-0294

Laemmli, U.K. (1970). Cleavage of structural proteins during the assembly of the head of bacteriophage T4. *Nature*, Vol.227, No.5259, (August 1970), pp. 680-685, ISSN 0028-0836

Leslie, A.G.W. (1992). Joint CCP4 and ESF-EACMB Newsletter on Protein Crystallography No.26, Daresbury Laboratory, Warrington, UK., 1992

Leyland, M.L. & Kelly, R. (1991). Purification and characterization of a monomeric isocitrate dehydrogenase with dual coenzyme specificity from the photosynthetic bacterium *Rhodomicrobium vannielii*. *Eur. J. Biochem.*, Vol.202, No.1, (November 1991), pp. 85-93, ISSN 1432-1033

Lokanath, N.K. & Kunishima, N. (2006). Crystal structure of TT0538 protein from *Thermus thermophilus* HB8, In: *RCSB PDB (Protain Data Bank)*, July 2011, Available from: < http://www.rcsb.org/pdb/explore/explore.do?structureId=2D1C >

Malkin, A.J. & Thorne, R.E. (2004). Growth and disorder of macromolecular crystals: insights from atomic force microscopy and X-ray diffraction studies. *Methods*, Vol.34, No.3, (November 2004), pp. 273-299, ISSN 1046-2023

Matthews, B.W. (1968). Solvent content of protein crystals. *J. Mol. Biol.*, Vol.33, No.2, (April 1968), pp. 491-497, ISSN 0022-2836

McPherson, A. (2004). Introduction to protein crystallization. *Methods*, Vol.34, No.3, (November 2004), pp. 254-265, ISSN 1046-2023

Melendez-Hevia, E.; Waddell, T.G. & Cascante, M. (1996). The puzzle of the Krebs citric acid cycle: assembling the pieces of chemically feasible reactions, and opportunism in the design of metabolic pathways during evolution. *J. Mol. Evol.*, Vol.43, No.3, pp. 293-303, ISSN 0022-2844

Mesecar, A.D. & Koshland Jr., D.E. (2000). A new model for protein stereospecificity. *Nature*, Vol.403, No.6770, (February 2000), pp. 614-615, ISSN 0028-0836

Mesecar, A.D.; Stoddard, B.L. & Koshland Jr., D.E. (1997). Orbital steering in the catalytic power of enzymes: Small structural changes with large catalytic consequences. *Science*, Vol.277, No.202, (July 1997), pp. 202-206, ISSN 0036-8075

Miyazaki, K.; Eguchi, H.; Yamagishi, A.; Wakagi, T. & Oshima, T. (1992). Molecular cloning of the isocitrate dehydrogenase gene of an extreme thermophile, *Thermus thermophilus* HB8. *Appl. Environ. Microbiol.*, Vol.58, No.1, (January 1992), pp. 93-98, ISSN 0099-2240

Miyazaki, K.; Yaoi, T. & Oshima, T. (1994). Expression, purification, and substrate specificity of isocitrate dehydrogenase from *Thermus thermophilus* HB8. *Eur. J. Biochem.*, Vol.221, No.3, (May 1994), pp. 899-903, ISSN 1432-1033

Ohzeki, M.; Yaoi, T.; Moriyama, H.; Oshima, T. & Tanaka, N. (1995). Crystallization and preliminary X-ray studies of isocitrate dehydrogenase from *Thermus thermophilus* HB8. *J. Biochem.*, Vol.118, No.4, (October 1995), pp. 679-680, ISSN 0021-924X

Sahara, T.; Takada, Y.; Takeuchi, Y.; Yamaoka, N. & Fukunaga, N. (2002). Cloning, sequencing, and expression of a gene encoding the monomeric isocitrate dehydrogenase of the nitrogen-fixing bacterium, *Azotobacter vinelandii*. *Biosci. Biotechnol. Biochem.*, Vol.66, No.3, (March 2002), pp. 489-500, ISSN 0916-8451

Scabert, F.A.; Henn, C. & Engel, A. (1995). Native *Escherichia coli* OmpF porin surfaces probed by atomic force microscopy. *Science*, Vol.268, No.5207, (April 1995), pp.92-94 ISSN 0036-8075

Shiba, H.; Kawasumi, T.; Igarashi, Y.; Kodama, T. & Minoda, Y. (1985). The CO_2 assimilation via the reductive tricarboxylic acid cycle in an obligately autotrophic, aerobic hydrogen-oxidizing bacterium, *Hydrogenobacter thermophilus*. *Arch. Microbiol.*, Vol.141, No.3, (April 1985), pp. 198-203, ISSN 0302-8933

Shimamura, T.; Koike-Takeshita, A.; Yokoyama, K.; Masui, R.; Murai, N.; Yoshida, M.; Taguchi, H. & Iwata, S. (2004). Crystal structure of the native chaperonin complex from *Thermus thermophilus* revealed unexpected asymmetry at the cis-cavity. *Structure*, Vol.12, No.8, (August 2004), pp. 1471–1480, ISSN 0969-2126

Singh, S.K.; Matsuno, K.; LaPorte, D.C. & Banaszak, L.J. (2001). Crystal structure of *Bacillus subtilis* isocitrate dehydrogenase at 1.55 Å. *J. Biol. Chem.*, Vol.276, No.28, (July 2001), pp. 26154-26163, ISSN 0021-9258

Steen, I.H.; Lien, T. & Birkeland, N.K. (1997). Biochemical and phylogenetic characterization of isocitrate dehydrogenase from a hyperthermophilic archaeon, *Archaeoglobus fulgidus*. *Arch. Microbiol.*, Vol.168, No.5, (November 1997), pp. 412-420, ISSN 0302-8933

Steen, I.H.; Madern, D.; Karlström, M.; Lien, T.; Ladenstein, R. & Birkeland, N.K. (2001). Comparison of isocitrate dehydrogenase from three hyperthermophiles reveals differences in thermostability, cofactor specificity, oligomeric state, and phylogenetic affiliation. *J. Biol. Chem.*, Vol.276, No.47, (August 2001), pp. 43924-43931, ISSN 0021-9258

Stoddard, B.L.; Dean, A. & Koshland Jr, D.E. (1993). Structure of isocitrate dehydrogenase with isocitrate, nicotinamide adenine dinucleotide phosphate, and calcium at 2.5-Å resolution: a pseudo-Michaelis ternary complex. *Biochem.*, Vol.32, No.36, (September 1993), pp. 9310-9316 ISSN 0006-2960

Stokke, R.; Karlström, M.; Yang, N.; Leiros, I.; Ladenstein, R.; Birkeland, N.K. & Steen, I.H. (2007). Thermal stability of isocitrate dehydrogenase from *Archaeoglobus fulgidus* studied by crystal structure analysis and engineering of chimers. *Extremophiles*, Vol.11, No.3, (May 2007), pp. 481-493, ISSN 1431-0651

Xu, X.; Zhao, J.; Xu, Z.; Peng, B.; Huang, Q.; Arnold, E. & Ding, J. (2004). Structures of human cytosolic NADP-dependent isocitrate dehydrogenase reveal a novel self-regulatory mechanism of activity. *J. Biol. Chem.*, Vo.279, No.32, (August 2004), pp. 33946-33957, ISSN 0021-9258

Yasutake, Y.; Watanabe, S.; Yao, M.; Takada, Y.; Fukunaga, N. & Tanaka, I. (2001). Crystallization and preliminary X-ray diffraction studies of monomeric isocitrate dehydrogenase by the MAD method using Mn atoms. *Acta Crystallogr. D Biol. Crystallogr.*, Vol.57, No.11, (November 2001), pp. 1682-1685 ISSN 0907-4449

Yoneta, M.; Sahara, T.; Nitta, K. & Takada, Y. (2004). Characterization of chimeric isocitrate dehydrogenases of a mesophilic nitrogen-fixing bacterium, *Azotobacter vinelandii*, and a psychrophilic bacterium, *Colwellia maris*. *Curr. Microbiol.*, Vol.48, No.5, (May 2004), pp. 383-388, ISSN 0343-8651

13

Crystallographic Studies on Autophagy-Related Proteins

Nobuo N. Noda[1], Yoshinori Ohsumi[2] and Fuyuhiko Inagaki[3]
[1]*Institute of Microbial Chemistry, Tokyo*
[2]*Tokyo Institute of Technology*
[3]*Hokkaido University*
Japan

1. Introduction

Autophagy is an intracellular bulk degradation system conserved in most eukaryotes. Autophagy plays various physiological roles such as intracellular clearance, development, differentiation and intracellular immunity, and its dysfunction is related to severe diseases such as neurodegeneration and cancer (Mizushima, 2007; Mizushima et al., 2008). Upon induction of autophagy, isolation membranes appear in the cytoplasm and expand to enclose a portion of the cytoplasm to be double membrane structures, autophagosomes (Mizushima et al., 2011; Nakatogawa et al., 2009). In some cases, organelles such as mitochondria and peroxisomes, and even invading microbes are also enclosed. All the matters enclosed by an autophagosome are named cargoes. The autophagosome fuses with the lysosome in mammals and with the vacuole in yeast and plants and exposes its inner membrane containing cargoes to vacuolar/lysosomal hydrolases, which degrade cargoes. Using the budding yeast, *Saccharomyces cerevisiae,* as a model organism, many *ATG* genes required for autophagy, especially in the step of autophagosome formation, have been identified (Klionsky et al., 2003; Tsukada and Ohsumi, 1993). Most *ATG* genes are evolutionarily conserved among higher eukaryotes including mammals, suggesting that the basic molecular mechanisms of autophagy are also evolutionarily conserved. Of the 35 Atg proteins identified thus far, 18 have been shown as crucial for autophagosome formation (Nakatogawa et al., 2009). However, it remains to be elucidated how these Atg proteins mediate autophagosome formation. In order to establish the molecular functions of each Atg protein and unveil the mechanism of autophagy, we started comprehensive structural studies of Atg proteins using X-ray crystallography. In this chapter, we review X-ray crystallographic studies of Atg proteins focusing on experimental details. For a more detailed description of the structure and function of Atg proteins, please refer to other recent reviews (Mizushima et al., 2011; Nakatogawa et al., 2009; Noda et al., 2009, 2010).

2. X-ray crystallographic studies of Atg proteins involved in ubiquitin-like conjugation reactions

Of the 18 Atg proteins essential for autophagosome formation, eight, namely, Atg3, Atg4, Atg5, Atg7, Atg8, Atg10, Atg12 and Atg16, constitute two ubiquitin-like conjugation systems: the Atg8 system and the Atg12 system (Nakatogawa et al., 2009) (Figure 1). In the Atg8 system, the ubiquitin-like protein Atg8 is first processed by the cysteine protease Atg4 (Kirisako et al., 2000). The exposed C-terminal Gly116 of Atg8 is activated by Atg7, an E1-like enzyme, to form an Atg7~Atg8 thioester intermediate (Tanida et al., 1999). Then, Atg8 is transferred to Atg3, an E2-like enzyme, to form an Atg3~Atg8 thioester intermediate (Ichimura et al., 2000). Finally, Gly116 of Atg8 is conjugated to a phosphatidylethanolamine (PE) to form Atg8—PE conjugates (Ichimura et al., 2000), in which Atg12—Atg5 conjugates (mentioned below) play crucial roles (Fujioka et al., 2008a; Hanada et al., 2007). Atg8—PE conjugates localize to isolation membranes and autophagosomes (Kirisako et al., 1999) and play crucial roles in both autophagosome formation and cargo recognition during autophagy. In the Atg12 system, the C-terminal Gly186 of another ubiquitin-like protein, Atg12, is activated by Atg7 to form an Atg7~Atg12 thioester intermediate. Then, Atg12 is transferred to Atg10, an E2-like enzyme, to form an Atg10~Atg12 thioester intermediate (Shintani et al., 1999). Finally, Gly186 of Atg12 is conjugated to the lysine side chain of Atg5 to form Atg12—Atg5 conjugates (Mizushima et al., 1998). Atg12—Atg5 conjugates form a complex with Atg16 through the direct interaction between Atg5 and Atg16 and localize to isolation membranes in the form of the Atg12—Atg5-Atg16 ternary complex (Mizushima et al., 2003; Mizushima et al., 1999). The Atg12—Atg5-Atg16 ternary complex also plays crucial roles in autophagy, one of which is an E3-like function in the Atg8 system.

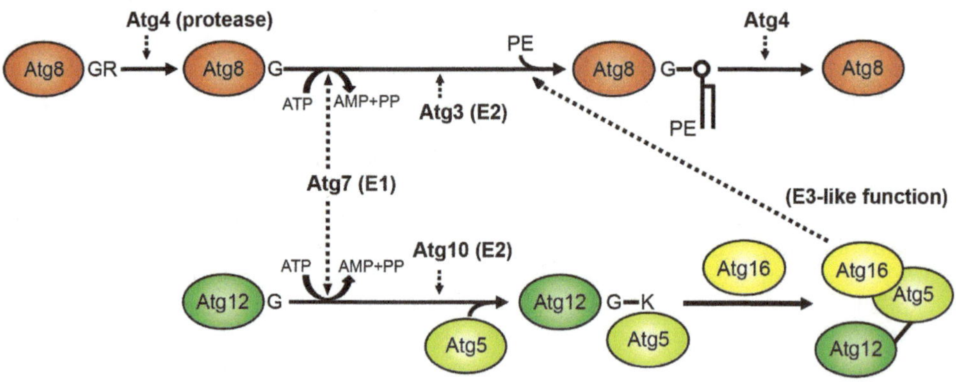

Fig. 1. Atg conjugation systems. Atg12 is irreversibly conjugated to Atg5 by Atg7 and Atg10, and the Atg12—Atg5 conjugate further forms a complex with Atg16. Atg8 is reversibly conjugated to PE by Atg3, Atg4, Atg7 and Atg12—Atg5-Atg16 complex.

The molecular mechanism of conjugation reactions in the Atg8 and Atg12 systems has been considered similar to that of ubiquitylation. However, all Atg proteins involved in such reactions have a low sequence homology with proteins involved in ubiquitylation and other ubiquitin-like modification systems. Moreover, the Atg8 system mediates a unique protein-lipid conjugation reaction, whereas the Atg12 system mediates the modification of only a single target protein. Owing to these unique features of the Atg conjugation systems, it is difficult to speculate the precise molecular mechanisms of the above conjugation reactions on the basis of our knowledge of other modification systems. In addition to the conjugation reaction mechanism, the molecular roles of Atg8—PE and Atg12—Atg5-Atg16 complex in autophagy also present critical problems that require elucidation. In order to answer these problems, we started comprehensive structural studies of Atg proteins involved in the Atg8 and Atg12 systems using mainly X-ray crystallography.

2.1 Atg8-family proteins

Atg8 is a ubiquitin-like modifier involved in autophagy, and is conjugated to the amino group of PE through ubiquitin-like reactions as mentioned above. Atg8—PE conjugates localize to autophagic membranes, and play at least two critical roles in autophagy: one is autophagosome formation (Nakatogawa et al., 2007), and the other is selective cargo recognition (Noda et al., 2008b). Atg8 was the first Atg protein we attempted to crystallize; however, we could not obtain any crystals for Atg8. At that time, LC3 was identified as a mammalian Atg8 ortholog and was shown to play crucial roles in mammalian autophagy. Therefore, we switched the crystallization target from Atg8 to LC3 and performed an X-ray crystallographic study of LC3 at first. Much later, we succeeded in obtaining crystals of Atg8 and determined the crystal structure of Atg8.

2.1.1 LC3

Microtubule-associated protein light chain 3 (LC3) is a mammalian ortholog of Atg8 and is conjugated to PE through ubiquitin-like reactions similarly to Atg8 (Kabeya et al., 2000). Since LC3—PE localizes to isolation membranes and autophagosomes, it is widely used as a marker protein for autophagic membranes (Kabeya et al., 2000; Mizushima et al., 2010). Rat LC3 is composed of 142 amino acids. The C-terminal 22 residues (121-142), which are processed by a mammalian Atg4 homolog (Atg4B) (Kabeya et al., 2004), are not important for the function of LC3. Thus, we performed a structural study of the processed form of LC3 (residues 1-120). LC3 was well expressed in *E. coli* as a soluble protein, and was easily crystallized into forms I and II crystals (Sugawara et al., 2003). Although the form I crystal diffracted poorly (~3.5 Å), the form II crystal diffracted up to 2.05 Å resolution. Thus, using the form II crystal, we determined the crystal structure of LC3 by the molecular replacement method (Sugawara et al., 2004). As a search model, the crystal structure of GATE-16, another mammalian homolog of Atg8, was used. As shown in Figure 2A, the presence of two α-helices (α1 and α2) at the N-terminal side of the ubiquitin fold is a structural feature of LC3 and all Atg8-family members (Noda et al., 2009, 2010).

2.1.2 Atg8

Atg8 is composed of 117 amino acids, and immediately after translation, it is processed by Atg4 to expose Gly116 at the C-terminus. The processed form of Atg8 (1-116) was well expressed in *E. coli* as a soluble protein and highly purified; however, it was never crystallized. An NMR study of Atg8 suggested that the N-terminal region of Atg8, which corresponds to α1 and α2 of LC3, does not have a rigid uniform conformation (Kumeta et al., 2010), which was suggested to hamper the crystallization of Atg8. Although we could not crystallize Atg8 alone, we succeeded in obtaining good crystals of Atg8 as a complex with its binding peptide. The used 4-residue peptide, Trp-Glu-Glu-Leu, corresponds to the C-terminal four residues of Atg19, an autophagic receptor protein involved in the selective transport of vacuolar enzymes into the vacuole through autophagic processes (Scott et al., 2001). Thus we determined the crystal structure of Atg8 as a complex with a peptide by the molecular replacement method (Noda et al., 2008b). As shown in Figure 2B, the peptide has an extended conformation and forms an intermolecular β-sheet with β2 of Atg8.

Furthermore, the hydrophobic side-chains of Trp and Leu of the peptide bind to the two hydrophobic pockets, namely, theW-site and L-site, respectively, and the acidic side-chains of two Glu residues of the peptide form ionic interaction with basic Arg28 and Arg67 of Atg8. These interactions significantly stabilize the conformations of α1 and α2 of Atg8, and thus promote the crystallization of Atg8.

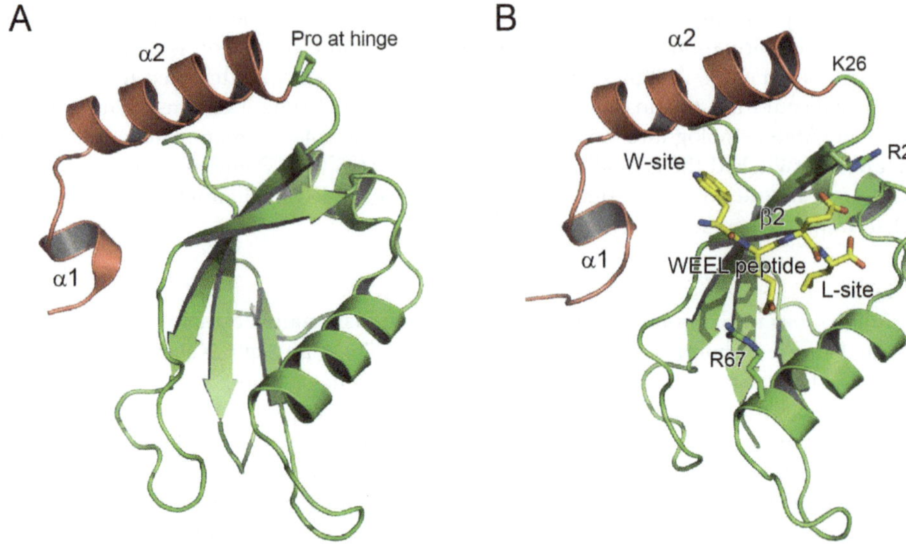

Fig. 2. Structure of LC3 (left) and Atg8 in complex with a WEEL peptide (right). The N-terminal region unique to Atg8-family proteins are colored salmon pink.

In contrast to ligand-free Atg8, which is difficult to crystallize, three mammalian Atg8 homologs (i.e., LC3, GATE-16 and GABARAP) are easily crystallized in free form. We noticed that in the hinge region between α2 and the ubiquitin fold moiety, a Pro residue is conserved among mammalian Atg8 homologs but not in Atg8. We thought that the introduction of a Pro residue into the hinge will stabilize the N-terminal region of Atg8. As expected, the substitution of Lys26 with Pro increased the stability of Atg8 and significantly improved the NMR spectra obtained from the mutant protein. Using this mutant, we determined the solution structure of Atg8 by NMR, showing that the N-terminal region of Atg8 has a convergent conformation similar to that of mammalian Atg8 homologs (Kumeta et al., 2010). This mutation affected neither the conjugation of Atg8 with PE nor the autophagic activity in yeast cells (Kumeta et al., 2010), suggesting that the flexible nature of the N-terminal region of Atg8 is not important for the function of Atg8 and that Atg8 K26P can be used for structural study as a fully active protein. We actually used this mutant for analyzing the interaction between Atg8 and Atg3 with NMR spectroscopy (Yamaguchi et al., 2010).

2.2 Atg12

Atg12 is another ubiquitin-like modifier involved in autophagy and is conjugated to the lysine side chain of Atg5 through ubiquitin-like reactions. Atg12 has several unique features compared with other ubiquitin-like modifiers. First, *S. cerevisiae* Atg12 is composed of 186 amino acids and is much larger than ubiquitin and other ubiquitin-like proteins including Atg8. Second, no processing/deconjugating enzyme for Atg12 has been reported, and thus its conjugation reaction appears to be irreversible. Third, the modification target for Atg12 is restricted to a single protein, Atg5 (although Atg3 was reported to be another modification target for Atg12 in mammals (Radoshevich et al., 2010)). Finally, Atg12 has little sequence homology with other ubiquitin-like proteins. To establish the structure and function of Atg12, we started its structural study.

Unlike ubiquitin and other ubiquitin-like modifiers, soluble recombinant Atg12 could not be obtained from *E. coli*. We also tried to prepare recombinant proteins for a mammalian Atg12 homolog but failed for the same reason as above. We then performed multiple sequence alignment among Atg12 homologs and noticed that the *Arabidopsis thaliana* Atg12 homologs AtAtg12a and AtAtg12b are composed of only 96 and 94 amino acids, respectively, and thus are much smaller than other Atg12 homologs. The smaller size of AtAtg12s seemed advantageous for the preparation of recombinant proteins and crystallization, so we switched the crystallization target from Atg12 to AtAtg12s.

Although every component of the Atg12 system has already been found in the Arabidopsis genome at that time, it remains to be shown that they are functional orthologs of yeast *ATG* genes and crucial for plant autophagy. Thus, before our structural study, we performed functional analyses of AtAtg12a and AtAtg12b (Suzuki et al., 2005). We first attempted to detect AtAtg12-AtAtg5 conjugates in *Arabidopsis* using antibodies against AtAtg12s. We detected AtAtg12-AtAtg5 conjugates in wild-type plants. In contrast, we could not detect them in *Atatg5-1/Atatg5-1* and *Atatg10-1/Atatg10-1* plants. Consistently, *Atatg5-1/Atatg5-1* or *Atatg10-1/Atatg10-1* plants showed no autophagic activity. Taken together, these data suggest that the Atg12 system is also conserved in *Arabidopsis* and plays an essential role in plant autophagy.

We then started structural studies of AtAtg12s. Although AtAtg12a and AtAtg12b have a similar size (96 versus 94 amino acids) and high sequence identity (93%), only AtAtg12b was obtained from *E. coli* as a soluble protein. During purification, AtAtg12b behaved as both a monomer and a dimer, and an irreversible change from a monomer into a dimer was observed during purification. The dimeric form of AtAtg12b showed higher solubility and stability than the monomeric form, and only the dimeric form crystallized. The obtained crystals diffracted up to 1.8 Å resolution, and the crystal structure of AtAtg12b was determined by the multiple isomorphous replacement method (MIR) (Suzuki et al., 2005).

Surprisingly, AtAtg12b existed as an intertwined dimer in the crystal (Figure 3). The loop connecting residues 58 and 61 acted as a hinge for domain swapping, and the 61-94 region was exchanged between the two molecules. In spite of domain swapping, the split half of the hinge has a canonical ubiquitin fold. The intertwined dimer structure in the crystal presumably represents the dimer structure in solution; however, dimer formation itself appears to be an *in vitro* artifact. *In vivo*, most Atg12 exists as a conjugate with Atg5, and in the absence of Atg16, Atg12—Atg5 conjugates exist as a monomer (Kuma et al., 2002). We also prepared recombinant proteins for Atg12—Atg5 conjugates (Noda et al., 2008a), which existed as a monomer in solution. These observations suggest that free Atg12 is unstable and significantly stabilized by conjugation with Atg5 both *in vitro* and *in vivo*. On the other hand, AtAtg12b appears to be artificially stabilized by domain swapping *in vitro*. The unstable nature of Atg12, which is unique compared with ubiquitin and other ubiquitin-like proteins, might hold the key to Atg12-specific functions.

The monomeric structure of AtAtg12b (the split half of the intertwined dimer) consists of an ubiquitin fold and has no unique insertions. However, the ubiquitin fold has a hydrophobic patch that is conserved among Atg12 homologs but not among other ubiquitin-like modifiers (Suzuki et al., 2005). Mutation at this patch of yeast Atg12 causes a severe defect in autophagy (Hanada and Ohsumi, 2005), suggesting that the conserved patch is crucial for the Atg12 function in autophagy.

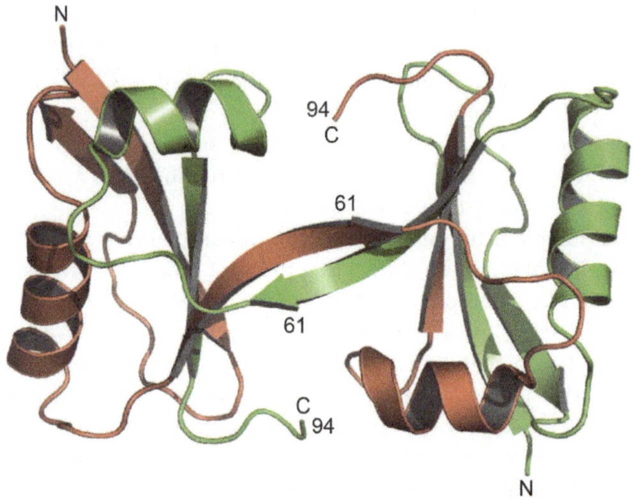

Fig. 3. Crystal structure of the intertwined dimer of AtAtg12b. Two polypeptides are colored green and salmon pink, respectively. N- and C-termini are labeled N and C, respectively.

2.3 Atg3

Atg3 is an E2-like enzyme in the Atg8 system. Atg3 receives Atg8 from an Atg7~Atg8 thioester intermediate to form an Atg3~Atg8 thioester intermediate. Then, with the help of the Atg12—Atg5-Atg16 complex (see below), Atg3 transfers Atg8 to PE. Atg3 is unique among E2-family proteins in that it mediates the conjugation reaction between a protein and a lipid. Atg3 has a low sequence homology with canonical E2 enzymes, and at 36 kDa is larger than them (typically around 20 kDa). Owing to these unique features of Atg3, it was difficult to speculate about the structure of Atg3 without experimental information.

Atg3 is composed of 310 amino acids. Full-length Atg3 was expressed as a GST-fusion protein in *E. coli*, in which two plasmids, pGEX-4T and pGEX-6P (GE Healthcare), were used. From the pGEX-4T and pGEX-6P vectors, GST-fused Atg3 with a thrombin and a PreScission protease-cleavage sequence, respectively, between GST and Atg3 were expressed. After affinity chromatography, GST was cleaved from GST-Atg3 using either thrombin or PreScission protease, resulting in Atg3 with a Gly-Ser (thrombin digestion) or Gly-Pro-Leu-Gly-Ser (PreScission digestion) artificial sequence at the N-terminus. These two Atg3s were subjected to crystallization screening. Intriguingly, crystals were obtained from Atg3 with a Gly-Ser artificial sequence, but not from Atg3 with a Gly-Pro-Leu-Gly-Ser artificial sequence (Yamada et al., 2006). Thus, we used pGEX-4T-Atg3 for further studies, despite the fact that we usually use the pGEX-6P vector rather than the pGEX-4T vector since a PreScission protease has a higher activity than a thrombin protease at low temperature.

Using the obtained crystals, the crystal structure of Atg3 was determined by MIR (Yamada et al., 2007). Figure 4 shows the overall structure of Atg3. Atg3 has a unique hammer like shape consisting of a head and a handle. The head moiety has a topology similar to that of canonical E2 enzymes, whereas the handle region (HR) is unique to Atg3. In addition to HR, Atg3 has another large, unique insertion (~80 amino acids) in its head moiety. Most of the residues constituting this region were not modelled owing to a low electron density; thus, this region was named the flexible region (FR). NMR studies showed that Atg3 FR actually has a flexible conformation in solution (Yamada et al., 2007). Truncational studies showed that HR and FR are responsible for the interaction with Atg8 and Atg7, respectively, and play crucial roles in Atg8–PE formation (Yamada et al., 2007).

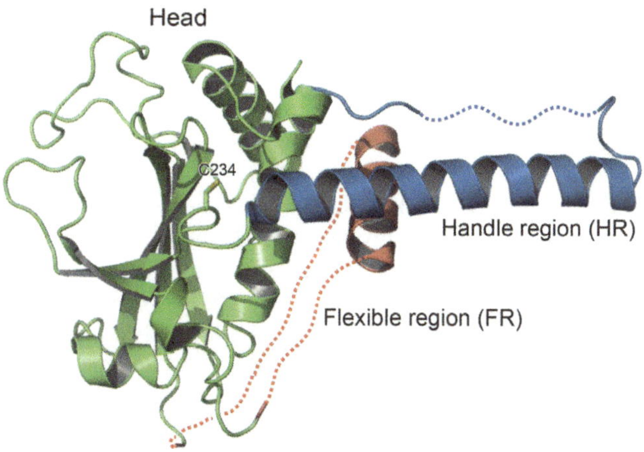

Fig. 4. Structure of Atg3. Head, FR and HR of Atg3 are colored green, salmon pink and blue, respectively. Disordered region in FR and HR is indicated as a broken line.

2.4 Atg4

Atg4 is the sole protease among all Atg proteins and mediates both the processing of the precursor Atg8 and the deconjugation of Atg8–PE. The reversible modification of Atg8 with PE is crucial for autophagy; therefore, Atg4 can be a target for developing autophagy-specific inhibitors. Although Atg4 was predicted to be a cysteine protease, it has little sequence homology with any known cysteine proteases. Thus, structural information on Atg4 was strongly desired for designing the specific inhibitors of this protein.

2.4.1 HsAtg4B as a free form

We first attempted to crystallize *S. cerevisiae* Atg4. However, we never obtained its crystals. Humans have four Atg4 homologs (4A-4D), among which Atg4B was reported to be responsible for LC3 processing and delipidation (Kabeya et al., 2004). Therefore, we changed the crystallization target from yeast Atg4 to human Atg4B (HsAtg4B). Full-length

HsAtg4B, which consists of 393 amino acids, crystallized easily, and the obtained crystals diffracted beyond 2 Å. The crystal structure of HsAtg4B was determined by the multi-wavelength anomalous dispersion method (MAD) using selenomethionine-substituted crystals (Figre 5A) (Sugawara et al., 2005).
The overall structure of HsAtg4B is composed of a papain-like fold conserved among cysteine proteases and a unique region that we named "the short Fingers domain" since its insertion site is similar to the "Fingers domain" of the herpesvirus-associated ubiquitin-specific protease, a cysteine protease structurally similar to HsAtg4B. Cys74 (catalytic cysteine), Asp278 and His280 form a catalytic triad in the crystal. The primary sequence order (Cys-Asp-His) of the catalytic triad is distinct from that of most cysteine proteases (Cys-His-Asp). Cys74 is buried under a loop named the regulatory loop and Trp142, suggesting that free HsAtg4B is autoinhibited. Consistent with the autoinhibited structure, a peptide corresponding to residues 116-124 of LC3, which include a scissile site of HsAtg4B (Gly120), was not processed by HsAtg4B (Sugawara et al., 2005). It was therefore proposed that in order to access the catalytic site of HsAtg4B and to be processed, the substrate (LC3) should induce a conformational change of HsAtg4B to release the autoinhibited structure.

2.4.2 HsAtg4B-LC3 complex

We next attempted to determine the structure of HsAtg4B in complex with LC3. First, we crystallized full-length HsAtg4B in complex with LC3, but obtained no crystal. In the free HsAtg4B structure, the C-terminal 39 residues (residues 355-393) showed a low electron density, suggesting that they are highly flexible. Therefore, we used a truncated form of HsAtg4B (1-354) for co-crystallization with LC3, and obtained good crystals of the HsAtg4B(1-354)-LC3(1-120) complex (Satoo et al., 2007). Furthermore, we obtained two types of the HsAtg4B-LC3 precursor complex: HsAtg4B(1-354, H280A)-LC3(1-124) and HsAtg4B(1-354, C74S)-LC3(1-124) complexes. H280A and C74S mutations were introduced to prevent the cleavage of the LC3 precursor. Using these crystals, we determined three complex structures by the molecular replacement method (Satoo et al., 2009).
Figure 5B shows the structure of the HsAtg4B(1-354, C74S)-LC3(1-124) complex. A structural comparison between the free and complex structures suggests that LC3 induces a large conformational change in two regions in HsAtg4B: one is the regulatory loop and the other is the N-terminal tail. The regulatory loop and Trp142, which block the catalytic site of the free HsAtg4B structure, are lifted up by LC3 Phe119, forming a narrow groove under the loop, through which LC3 Gly120 has access to Cys74 of HsAtg4B. This structure explains why the Phe-Gly sequence of LC3 is essential for the hydrolysis by HsAtg4B. In addition to the C-terminal tail, the ubiquitin-fold core of LC3 also interacts with HsAtg4B, which may also be required for releasing the autoinhibited conformation of HsAtg4B. On the other hand, the N-terminal tail, which interacts with the papain-fold moiety of free HsAtg4B, is bound to the crystallographically adjacent LC3 molecule in the complex structure using a Tyr-X-X-Leu sequence. This conformational change exposes a large flat surface at the exit of the catalytic site. It is unclear whether this interaction is a crystallization artifact or whether it actually occurs in solution. Therefore, we performed NMR analysis, which clearly showed that the interaction observed in the crystal actually occurs in solution (Satoo et al., 2009). Although the biological significance of this interaction has not been proved yet, we propose

that the conformational change of the N-terminal tail of HsAtg4B regulates the accessibility of HsAtg4B to the membrane where LC3–PE is embedded.

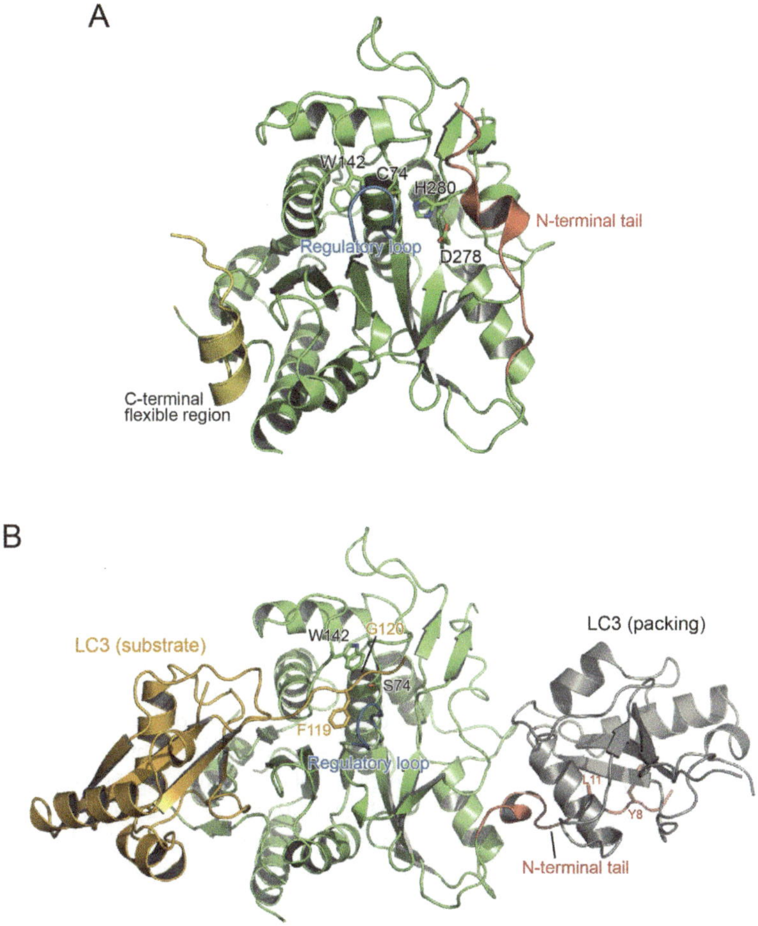

Fig. 5. Structure of HsAtg4B. (A) Free HsAtg4B structure. N-terminal region, regulatory loop as well as C-terminal flexible region are colored salmon pink, blue and yellow, respectively. (B). HsAtg4B-LC3 precursor complex structure. LC3 bound to the catalytic site of HsAtg4B is colored light orange, while LC3 bound to the N-terminal tail of HsAtg4B is colored gray. The other coloring is as in (A).

We obtained two HsAtg4B-LC3 precursor complex structures: one has a C74S mutation and the other has an H280A mutation in HsAtg4B. Each mutation completely abolishes the activity of HsAtg4B, and the LC3 precursor in both crystals actually remains intact. However, the two mutations have different effects on the conformation of the catalytic site: the H280A mutation affects the conformation of Cys74 and the bound C-terminal tail of LC3, whereas the C74S mutation does not. Thus, it is important for us to select an appropriate mutant for the structural study of the enzyme-substrate complex.

2.5 Atg5-Atg16 complex

Atg5 is the sole target for Atg12 modification. The side-chain of Atg5 Lys149 forms an isopeptide bond with Atg12 Gly186. Atg5 also interacts with Atg16 non-covalently, on which Atg12 conjugation has no effect. In yeast, Atg5 is presumed to function only as the Atg12—Atg5-Atg16 complex. The Atg12—Atg5-Atg16 complex localizes to isolation membranes and play crucial roles in autophagosome formation. Using an *in vitro* reconstitution system, we showed that the Atg12—Atg5 conjugate functions as an E3-like enzyme and accelerates the conjugation reaction between Atg8 and PE (Fujioka et al., 2008a; Hanada et al., 2007). We showed that Atg12 has a ubiquitin-like structure (see above); however, Atg5 and Atg16 have little sequential homology not only with known E3 enzymes but also with protein whose structure has already been reported. Therefore, in order to elucidate the overall architecture of the Atg12—Atg5-Atg16 complex, we started structural studies on Atg5 and Atg16.

2.5.1 Atg5 complexed with the N-terminal region of Atg16

S. cerevisiae Atg5 and Atg16 consist of 294 and 150 amino acids, respectively. We first attempted to crystallize full-length Atg5 alone, but failed owing to the low expression level of Atg5 in *E. coli* and its aggregate-prone nature. It was reported that Atg5 forms a stable complex with Atg16 *in vivo*. Therefore, we considered that Atg5 can be stabilized in *E. coli* by coexpression with Atg16. In fact, coexpression of Atg16 markedly increased the expression level of Atg5, thereby improving the solubility of Atg5. We then succeeded in obtaining a stable Atg5-Atg16 complex. It behaved as a complex throughout the entire purification process, suggesting that the affinity between them is significantly high. Using the purified complex, we started crystallization trials. However, the full-length Atg5-Atg16 complex did not crystallize at all. Atg16 alone was easily cleaved at the C-terminal side of Phe46 by contaminated proteases, and the N-terminal portion of the cleaved sample (residues 1-46) also formed a stable complex with Atg5. Therefore, we next used the Atg5-Atg16(1-46) complex for crystallization, and succeeded in obtaining good crystals (Matsushita et al., 2006). The crystal structure of the Atg5-Atg16(1-46) complex was determined by MIR in combination with MAD (Matsushita et al., 2007). We also attempted to crystallize the Atg5-Atg16 complex with Atg16 of various lengths. We obtained the Atg5-Atg16(1-57) complex crystal, but failed to obtain complex crystals containing Atg16 longer than 57 amino acids. The crystal structure of the Atg5-Atg16(1-57) complex was determined by the molecular replacement method.

The two complex structures are essentially identical with each other except for the non-overlapping region (residues 47-57 of Atg16). Atg5 has a unique structure consisting of two ubiquitin-like domains and a helix-rich domain between them (Figure 6A). These three domains interact with each other to form a globular fold. Lys149, the conjugation site for Atg12, is located at the helix-rich domain and has an exposed side chain. The N-terminal region of Atg16 is composed of an α-helix (resides 22-40) and its downstream loop (residues 41-57). The α-helix moiety binds to the groove formed by the three domains of Atg5, while the loop region binds to the N-terminal ubiquitin-like domain. These two structures and results of a mutational analysis suggested that residues 22-46 of Atg16 are the minimum region required for Atg5 binding. Lys149 and the residues involved in Atg16 binding are located distally to each other, which is consistent with the observation that the conjugation of Atg12 to Atg5 does not affect the Atg5-Atg16 interaction.

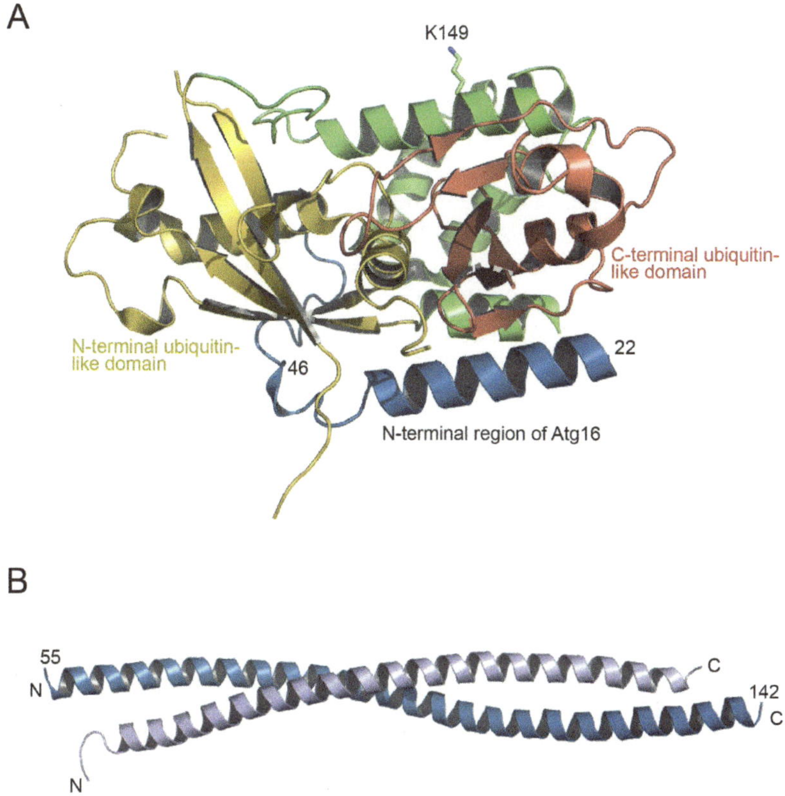

Fig. 6. Structure of Atg5 and Atg16. (A) Structure of Atg5 complexed with the N-terminal region (residues 1-57) of Atg16. N- and C-terminal ubiquitin-like domains and the other region are colored yellow, salmon pink and green, respectively, while N-terminal region of Atg16 is colored blue. (B) Structure of the coiled-coil dimer of full-length Atg16. One molecule is colred blue, while the other is colored light blue. N- and C-termini are labeled N and C, respectively.

2.5.2 Atg16

Although the structure of the N-terminal Atg5-binding region of Atg16 was determined as a complex with Atg5, that of the C-terminal region of Atg16 lacked structural information. From the primary sequence, the C-terminal region of Atg16 was predicted to contain a coiled-coil motif. Furthermore, on the basis of the results of gel-filtration chromatography experiments, Atg16 was suggested to form a multimer, possibly a tetramer, through the coiled-coil motif (Kuma et al., 2002). In order to determine the overall architecture of Atg16, we performed crystallization trials on Atg16 constructs of various lengths containing the predicted coiled-coil motif. Most of the tested constructs crystallized; however, these crystals diffracted poorly (6~8 Å resolution), except for one corresponding to residues 50-123 of Atg16 (Fujioka et al., 2008b). Since Atg16(50-123) does not have a Met residue, Leu103 was replaced with Met and using this construct, selenomethionine-labeled crystals were prepared and used for single anomalous dispersion (SAD) phasing. The crystal structure of Atg16(50-123) is composed of one α-helix of 90 Å length (Fujioka et al., 2010). The asymmetric unit contains six Atg16(50-123) molecules, which form three similar parallel coiled-coil dimers. These coiled-coil dimers expose two acidic residues (Asp101 and Glu102). By replacing these acidic residues with alanine, we succeeded in obtaining good crystals of full-length Atg16 and determined the structure by the molecular replacement method (Fujioka et al., 2010). The crystal structure of full-length Atg16 (D101A+E102A) is composed of one α-helix of 130 Å length. The structure lacked the electron density of the N-terminal region (residues 1-54), suggesting that the N-terminal region has various orientations relative to the C-terminal region. The asymmetric unit contained two Atg16 molecules, which formed a parallel coiled-coil dimer similar to the Atg16(50-123) dimer (Figure 6B). In both crystals, various interactions were observed between the coiled-coil dimers. In order to establish the oligomerization state of Atg16 in solution, we performed analytical ultracentrifugation experiments, which showed that Atg16 exists as a homodimer in solution (Fujioka et al., 2010) and that the Atg5-Atg16 complex exists as a 2:2 heterotetramer in solution. These data are consistent with the coiled-coil dimer structure of Atg16 observed in both crystals, suggesting that Atg16 forms a dimer but not a tetramer in solution.

The obtained structural and biochemical information suggested that the overall architecture of the Atg5-Atg16 complex is composed of two sets of the N-terminal short α-helix of Atg16 bound to Atg5, the C-terminal parallel coiled-coil homodimer of Atg16, and flexible linkers connecting them, resulting in a 2:2 complex. Since Atg12 is conjugated to Lys149 of each Atg5, the Atg12—Atg5-Atg16 complex forms a 2:2:2 complex. The overall architecture of the Atg12—Atg5-Atg16 complex, which is quite unique compared with any structure-reported proteins including other E3 enzymes, will be one basis for studying the molecular functions of this unique protein complex in autophagy.

3. Conclusion

Obtaining good crystals is the rate-limiting step of X-ray crystallography. We have no versatile method for obtaining good crystals and thus have attempted the crystallization of each protein through trial and error. However, accumulated experience will increase the probability of obtaining good crystals and accelerate structural studies. We started a comprehensive structural study of Atg proteins about ten years ago, and have already obtained good crystals for more than ten Atg proteins that include not only those involved in conjugation reactions but also those involved in the construction of the pre-autophagosomal structure (Suzuki and Ohsumi, 2010). Continuous trials will surely provide us with good crystals for all the Atg proteins, and through the comprehensive structural study of these protein crystals, the molecular mechanism of autophagy will be established in the near future.

4. Acknowledgment

This work was supported by Grants-in-Aid for Scientific Research on Priority Areas and Targeted Proteins Research Program from the Ministry of Education, Culture, Sports, Science and Technology of Japan.

5. References

Fujioka, Y., Noda, N.N., Fujii, K., Yoshimoto, K., Ohsumi, Y. & Inagaki, F. (2008a). In vitro reconstitution of plant Atg8 and Atg12 conjugation systems essential for autophagy. *J Biol Chem*, Vol.283, No.4, Jan 25, pp. 1921-1928, 0021-9258

Fujioka, Y., Noda, N.N., Matsushita, M., Ohsumi, Y. & Inagaki, F. (2008b). Crystallization of the coiled-coil domain of Atg16 essential for autophagy. *Acta Crystallogr Sect F Struct Biol Cryst Commun*, Vol.64, No.Pt 11, Nov 1, pp. 1046-1048, 1744-3091

Fujioka, Y., Noda, N.N., Nakatogawa, H., Ohsumi, Y. & Inagaki, F. (2010). Dimeric coiled-coil structure of Saccharomyces cerevisiae Atg16 and its functional significance in autophagy. *J Biol Chem*, Vol.285, No.2, Jan 8, pp. 1508-1515, 0021-9258

Hanada, T., Noda, N.N., Satomi, Y., Ichimura, Y., Fujioka, Y., Takao, T., Inagaki, F. & Ohsumi, Y. (2007). The Atg12-Atg5 conjugate has a novel E3-like activity for protein lipidation in autophagy. *J Biol Chem*, Vol.282, No.52, Dec 28, pp. 37298-37302, 0021-9258

Hanada, T. & Ohsumi, Y. (2005). Structure-function relationship of Atg12, a ubiquitin-like modifier essential for autophagy. *Autophagy*, Vol.1, No.2, Jul, pp. 110-118, 1554-8627

Ichimura, Y., Kirisako, T., Takao, T., Satomi, Y., Shimonishi, Y., Ishihara, N., Mizushima, N., Tanida, I., Kominami, E., Ohsumi, M., *et al.* (2000). A ubiquitin-like system mediates protein lipidation. *Nature*, Vol.408, No.6811, Nov 23, pp. 488-492, 0028-0836

Kabeya, Y., Mizushima, N., Ueno, T., Yamamoto, A., Kirisako, T., Noda, T., Kominami, E., Ohsumi, Y. & Yoshimori, T. (2000). LC3, a mammalian homologue of yeast Apg8p, is localized in autophagosome membranes after processing. *EMBO J*, Vol.19, No.21, Nov 1, pp. 5720-5728, 0261-4189

Kabeya, Y., Mizushima, N., Yamamoto, A., Oshitani-Okamoto, S., Ohsumi, Y. & Yoshimori, T. (2004). LC3, GABARAP and GATE16 localize to autophagosomal membrane depending on form-II formation. *J Cell Sci*, Vol.117, No.Pt 13, Jun 1, pp. 2805-2812, 0021-9533

Kirisako, T., Baba, M., Ishihara, N., Miyazawa, K., Ohsumi, M., Yoshimori, T., Noda, T. & Ohsumi, Y. (1999). Formation process of autophagosome is traced with Apg8/Aut7p in yeast. *J Cell Biol*, Vol.147, No.2, Oct 18, pp. 435-446, 0021-9525

Kirisako, T., Ichimura, Y., Okada, H., Kabeya, Y., Mizushima, N., Yoshimori, T., Ohsumi, M., Takao, T., Noda, T. & Ohsumi, Y. (2000). The reversible modification regulates the membrane-binding state of Apg8/Aut7 essential for autophagy and the cytoplasm to vacuole targeting pathway. *J Cell Biol*, Vol.151, No.2, Oct 16, pp. 263-276, 0021-9525

Klionsky, D.J., Cregg, J.M., Dunn, W.A., Jr., Emr, S.D., Sakai, Y., Sandoval, I.V., Sibirny, A., Subramani, S., Thumm, M., Veenhuis, M., *et al.* (2003). A unified nomenclature for yeast autophagy-related genes. *Dev Cell*, Vol.5, No.4, Oct, pp. 539-545, 1534-5807

Kuma, A., Mizushima, N., Ishihara, N. & Ohsumi, Y. (2002). Formation of the approximately 350-kDa Apg12-Apg5.Apg16 multimeric complex, mediated by Apg16 oligomerization, is essential for autophagy in yeast. *J Biol Chem*, Vol.277, No.21, May 24, pp. 18619-18625, 0021-9258

Kumeta, H., Watanabe, M., Nakatogawa, H., Yamaguchi, M., Ogura, K., Adachi, W., Fujioka, Y., Noda, N.N., Ohsumi, Y. & Inagaki, F. (2010). The NMR structure of the autophagy-related protein Atg8. *J Biomol NMR*, Vol.47, No.3, Jul, pp. 237-241, 1573-5001

Matsushita, M., Suzuki, N.N., Fujioka, Y., Ohsumi, Y. & Inagaki, F. (2006). Expression, purification and crystallization of the Atg5-Atg16 complex essential for autophagy. *Acta Crystallogr Sect F Struct Biol Cryst Commun*, Vol.62, Pt 10, Oct 1, pp. 1021-1023, 1744-3091

Matsushita, M., Suzuki, N.N., Obara, K., Fujioka, Y., Ohsumi, Y. & Inagaki, F. (2007). Structure of Atg5.Atg16, a complex essential for autophagy. *J Biol Chem*, Vol.282, No.9, Mar 2, pp. 6763-6772, 0021-9258

Mizushima, N. (2007). Autophagy: process and function. *Genes Dev*, Vol.21, No.22, Nov 15, pp. 2861-2873, 0890-9369

Mizushima, N., Kuma, A., Kobayashi, Y., Yamamoto, A., Matsubae, M., Takao, T., Natsume, T., Ohsumi, Y. & Yoshimori, T. (2003). Mouse Apg16L, a novel WD-repeat protein, targets to the autophagic isolation membrane with the Apg12-Apg5 conjugate. *J Cell Sci*, Vol.116, No.Pt 9, May 1, pp. 1679-1688, 0021-9533

Mizushima, N., Levine, B., Cuervo, A.M. & Klionsky, D.J. (2008). Autophagy fights disease through cellular self-digestion. *Nature*, Vol.451, No.7182, Feb 28, pp. 1069-1075, 1476-4687

Mizushima, N., Noda, T. & Ohsumi, Y. (1999). Apg16p is required for the function of the Apg12p-Apg5p conjugate in the yeast autophagy pathway. *EMBO J*, Vol.18, No.14, Jul 15, pp. 3888-3896, 0261-4189

Mizushima, N., Noda, T., Yoshimori, T., Tanaka, Y., Ishii, T., George, M.D., Klionsky, D.J., Ohsumi, M. & Ohsumi, Y. (1998). A protein conjugation system essential for autophagy. *Nature*, Vol.395, No.6700, Sep 24, pp. 395-398, 0028-0836

Mizushima, N., Yoshimori, T. & Levine, B. (2010). Methods in mammalian autophagy research. *Cell*, Vol.140, No.3, Feb 5, pp. 313-326, 1097-4172

Mizushima, N., Yoshimori, T. & Ohsumi, Y. (2011). The Role of Atg Proteins in Autophagosome Formation. *Annu Rev Cell Dev Biol*, Vol.27, Nov 10, pp. 107-132, 1530-8995

Nakatogawa, H., Ichimura, Y. & Ohsumi, Y. (2007). Atg8, a ubiquitin-like protein required for autophagosome formation, mediates membrane tethering and hemifusion. *Cell*, Vol.130, No.1, Jul 13, pp. 165-178, 0092-8674

Nakatogawa, H., Suzuki, K., Kamada, Y. & Ohsumi, Y. (2009). Dynamics and diversity in autophagy mechanisms: lessons from yeast. *Nat Rev Mol Cell Biol*, Vol.10, No.7, Jul, pp. 458-467, 1471-0080

Noda, N.N., Fujioka, Y., Ohsumi, Y. & Inagaki, F. (2008a). Crystallization of the Atg12-Atg5 conjugate bound to Atg16 by the free-interface diffusion method. *Journal of synchrotron radiation*, Vol.15, No.Pt 3, May, pp. 266-268, 0909-0495

Noda, N.N., Kumeta, H., Nakatogawa, H., Satoo, K., Adachi, W., Ishii, J., Fujioka, Y., Ohsumi, Y. & Inagaki, F. (2008b). Structural basis of target recognition by Atg8/LC3 during selective autophagy. *Genes Cells*, Vol.13, No.12, Dec, pp. 1211-1218, 1365-2443

Noda, N.N., Ohsumi, Y. & Inagaki, F. (2009). ATG systems from the protein structural point of view. *Chem Rev*, Vol.109, No.4, Apr, pp. 1587-1598, 1520-6890

Noda, N.N., Ohsumi, Y. & Inagaki, F. (2010). Atg8-family interacting motif crucial for selective autophagy. *FEBS Lett*, Vol.584, No.7, Apr 2, pp. 1379-1385, 1873-3468

Radoshevich, L., Murrow, L., Chen, N., Fernandez, E., Roy, S., Fung, C. & Debnath, J. (2010). ATG12 conjugation to ATG3 regulates mitochondrial homeostasis and cell death. *Cell*, Vol.142, No.4, Aug 20, pp. 590-600, 1097-4172

Satoo, K., Noda, N.N., Kumeta, H., Fujioka, Y., Mizushima, N., Ohsumi, Y. & Inagaki, F. (2009). The structure of Atg4B-LC3 complex reveals the mechanism of LC3 processing and delipidation during autophagy. *EMBO J*, Vol.28, No.9, May 6, pp. 1341-1350, 1460-2075

Satoo, K., Suzuki, N.N., Fujioka, Y., Mizushima, N., Ohsumi, Y. & Inagaki, F. (2007). Crystallization and preliminary crystallographic analysis of human Atg4B-LC3 complex. *Acta Crystallogr Sect F Struct Biol Cryst Commun*, Vol.63, No.Pt 2, Feb 1, pp. 99-102, 1744-3091

Scott, S.V., Guan, J., Hutchins, M.U., Kim, J. & Klionsky, D.J. (2001). Cvt19 is a receptor for the cytoplasm-to-vacuole targeting pathway. *Mol Cell*, Vol.7, No.6, Jun, pp. 1131-1141, 1097-2765

Shintani, T., Mizushima, N., Ogawa, Y., Matsuura, A., Noda, T. & Ohsumi, Y. (1999). Apg10p, a novel protein-conjugating enzyme essential for autophagy in yeast. *EMBO J*, Vol.18, No.19, Oct 1, pp. 5234-5241, 0261-4189

Sugawara, K., Suzuki, N.N., Fujioka, Y., Mizushima, N., Ohsumi, Y. & Inagaki, F. (2003). Crystallization and preliminary X-ray analysis of LC3-I. *Acta Crystallogr D Biol Crystallogr*, Vol.59, No.Pt 8, Aug, pp. 1464-1465, 0907-4449

Sugawara, K., Suzuki, N.N., Fujioka, Y., Mizushima, N., Ohsumi, Y. & Inagaki, F. (2004). The crystal structure of microtubule-associated protein light chain 3, a mammalian homologue of Saccharomyces cerevisiae Atg8. *Genes Cells*, Vol.9, No.7, Jul, pp. 611-618, 1356-9597

Sugawara, K., Suzuki, N.N., Fujioka, Y., Mizushima, N., Ohsumi, Y. & Inagaki, F. (2005). Structural basis for the specificity and catalysis of human Atg4B responsible for mammalian autophagy. *J Biol Chem*, Vol.280, No.48, Dec 2, pp. 40058-40065, 0021-9258

Suzuki, N.N., Yoshimoto, K., Fujioka, Y., Ohsumi, Y. & Inagaki, F. (2005). The crystal structure of plant ATG12 and its biological implication in autophagy. *Autophagy*, Vol.1, No.2, Jul, pp. 119-126, 1554-8627

Suzuki, K. & Ohsumi, Y. (2010). Current knowledge of the pre-autophagosomal structure (PAS). *FEBS Lett*, Vol.584, No.7, Apr 2, pp. 1280-1286, 1873-3468

Tanida, I., Mizushima, N., Kiyooka, M., Ohsumi, M., Ueno, T., Ohsumi, Y. & Kominami, E. (1999). Apg7p/Cvt2p: A novel protein-activating enzyme essential for autophagy. *Mol Biol Cell*, Vol.10, No.5, May, pp. 1367-1379, 1059-1524

Tsukada, M. & Ohsumi, Y. (1993). Isolation and characterization of autophagy-defective mutants of Saccharomyces cerevisiae. *FEBS Lett*, Vol.333, No.1-2, Oct 25, pp. 169-174, 0014-5793

Yamada, Y., Suzuki, N.N., Fujioka, Y., Ichimura, Y., Ohsumi, Y. & Inagaki, F. (2006). Crystallization and preliminary X-ray analysis of Atg3. *Acta Crystallographica Section F: Structural Biology and Crystallization Communication*, Vol.62, Pt 10, Oct 1, pp. 1016-1017, 1744-3091

Yamada, Y., Suzuki, N.N., Hanada, T., Ichimura, Y., Kumeta, H., Fujioka, Y., Ohsumi, Y. & Inagaki, F. (2007). The crystal structure of Atg3, an autophagy-related ubiquitin carrier protein (E2) enzyme that mediates Atg8 lipidation. *J Biol Chem*, Vol.282, No.11, Mar 16, pp. 8036-8043, 0021-9258

Yamaguchi, M., Noda, N.N., Nakatogawa, H., Kumeta, H., Ohsumi, Y. & Inagaki, F. (2010). Autophagy-related protein 8 (Atg8) family interacting motif in Atg3 mediates the Atg3-Atg8 interaction and is crucial for the cytoplasm-to-vacuole targeting pathway. *J Biol Chem*, Vol.285, No.38, Sep 17, pp. 29599-29607, 0021-9258

14

Protein-Noble Gas Interactions Investigated by Crystallography on Three Enzymes - Implication on Anesthesia and Neuroprotection Mechanisms

Nathalie Colloc'h[1], Guillaume Marassio[1] and Thierry Prangé[2]
[1]*CI-NAPS UMR 6232 Université de Caen Basse-Normandie – CNRS, Centre Cyceron, Caen,*
[2]*LCRB, UMR 8015 – CNRS – Université Paris Descartes, Faculté de Pharmacie, Paris, France*

1. Introduction

General anesthetics have been used in clinical practice since the middle of the XIXth century, but their molecular mechanisms of action on the nervous system remains poorly understood. The main targets of most inhalational anesthetic are pentameric ligand-gated ion channels such as inhibitory $GABA_A$ (γ-amino-butyric acid) receptors whose activity is potentiated by general anesthetics. However, the main targets of gaseous anesthetics like xenon or nitrous oxide are excitatory NMDA (N-methyl-D-aspartate) receptors and nicotinic acetylcholine receptors whose activity is inhibited by gaseous anesthetics (Campagna et al., 2003; Franks, 2008). The use of gaseous anesthetics became widely applicable in the nineteen-fifties, nitrous oxide being often administrated in complement of halogenated anesthetics. Xenon anesthetic properties have been described by Cullen in 1951 (Cullen et al., 1951), and is used in anesthesia since mid-2000 (Sanders et al., 2004; Sanders et al., 2005) in spite of its excessive cost, a major obstacle to its widespread use (Kennedy et al., 1992).
Anesthesia is a complex process that refers to several physiologically altered functions. Early stages of anesthesia such as amnesia and hypnosis required anesthetic concentrations lower than those required to produce deep sedation and reduction of motor and automomic responses to noxious stimuli (Campagna et al., 2003). Scales that assess the *in-vivo* potency of inhaled anesthetics in humans are based on the minimum alveolar anesthetic concentrations (MAC) that are associated with well-defined behavioural endpoints. Following this, MAC-awake defines the MAC that induces the first stages of anesthesia such as amnesia and hypnosis, and MAC-immobility defines the MAC that produces deep sedation and suppresses movement in response to a noxious stimuli.
Anesthesia mechanisms were for a long time though to be mediated by non-specific membranous perturbation (Trudell, 1977). This membranous theory was based on the Meyer-Overton rule that showed an almost perfect relationship between the anesthetic property of a chemical compound and its solubility in olive oil or benzene. However, more and more exceptions were found to the Meyer-Overton rule such as the non-immobilizers

predicted to be potent anesthetics which indeed produced amnesia but not anesthesia or the eniantomers which have the same solubility in lipids but different anesthetic potencies (Campagna et al., 2003). Since the mid-1980, general anesthetics are though to act by disrupting protein functions (Franks et al., 1984; Franks et al., 1994). Chemical compounds that cross the blood brain barrier are generally soluble in lipids and could explain the Meyer-Overton rule (Franks, 2008). Anesthetics act mainly at ionotropic receptors which play a key role in regulation of ions' concentration on each side on the cytoplasmic membrane (Campagna et al., 2003). For a review of the major targets of general anesthetics and their effects, see (Yamakura et al., 2000).

Xenon and nitrous oxide have been shown in 1998 to be effective inhibitors of the NMDA receptor (Franks et al., 1998; Jevtovic-Todorovic et al., 1998) which gave them potentially interesting neuroprotective properties for treating major brain insults such as cerebral ischemia. Cerebral ischemia is indeed a major health problem, constituting the third cause of mortality and the first cause of morbidity in industrialized countries. Cerebral stroke is provoked by an acute interruption of the cerebral blood flow, leading to an oxygen and glucose deprivation for the brain, inducing dramatic dysfunctions in the excitatory glutamatergic neurotransmission. The resulting toxic accumulation of glutamate leads to an over-stimulation of glutamatergic receptors like the NMDA receptors, this process is called glutamate excitotoxicity (Choi et al., 1988; Dirnagl et al., 1999; Lo et al., 2005). The use of NMDA receptor antagonists to block the neurotoxic cascade initiated by the glutamate has not yet been proven to be clinically efficient in humans, because of the intrinsic neurotoxicity of these chemical compounds (Olney et al., 1991). The only treatment today is the use of a thrombolytic agent, the tissue-type plasminogen activator (tPA), which degrades the insoluble fibrin clots (NINDS, 1995). However, this treatment is only applicable in a few number of cases, due to the hemorrhagic risk which shorten the therapeutic window (Lee et al., 1999; Ahmed et al., 2010). The research of new neuroprotective drugs thus constitutes a major therapeutic goal. Inert gases are a new class of therapeutic agents which have a remarkably safe clinical profile and readily cross the blood brain barrier. Moreover, they have low solubility in blood, which is advantageous in terms of rapid inflow and washout. Gases have thus the great advantages to present a reducing risk of neurotoxic side effects, compared to chemical neuroprotective drugs, especially at the low concentrations used for neuroprotection.

Nitrous oxide and xenon reduce ischemic neuronal death in an *in-vivo* model of transient cerebral ischemia in rats and decrease the NMDA-induced Ca^{2+} influx on neuronal cell cultures studied by *in-vitro* calcium video microscopy (David et al., 2003). These two gases produce the same effect than memantine, a low-affinity antagonist of NMDA receptor which is already used in clinics for neurodegenerative disease treatments (David et al., 2006). Investigations in rodents have confirmed that xenon at subanesthetic concentrations of about 50 vol% provides maximal neuroprotection, even when given 2 to 4 h after the insult onset (Ma et al., 2005; Dingley et al., 2006; David et al., 2008). Nitrous oxide (Haelewyn et al., 2008) and argon (David et al., submitted) which are less expensive gases than xenon, possess mild-to-moderate neuroprotective properties against excitotoxic insults and hypoxic-ischemic injuries.

However, xenon and nitrous oxide inhibit tPA-induced thrombolysis, preventing their use during the intra-ischemic period. When administrated after the reperfusion, xenon has beneficial effect by suppressing ischemic brain damage and tPA-induced brain hemorrhages

(David et al., 2010) while nitrous oxide reduces ischemic brain damage but increases tPA-induced brain hemorrages (Haelewyn et al., 2011).
Xenon is thus a very promising neuroprotective drug with few or no adverse side effects in models of acute ischemic stroke or perinatal hypoxia-ischemia (Homi et al., 2003; Ma et al., 2003; Abraini et al., 2005; David et al., 2008; Luo et al., 2008). Despite this, the widespread clinical use of xenon is limited by its scarceness and excessive cost of production, even if close xenon delivery systems are now being developed.
Using a mixture of xenon and another anesthetic gas like nitrous oxide (Marassio et al., 2011), argon (David et al., submitted), or helium (David et al., 2009) could combine the efficiency of xenon and the low cost and availability of the second gas and is thought to be a cost-efficient strategy.
Argon is an inert gas which is easily available and has no narcotic nor anesthetic action at ambient pressure. It presents some mild to moderate neuroprotective properties (David et al., submitted). Argon, contrary to xenon and nitrous oxide, may act directly by potentiating GABA neurotransmission at the $GABA_A$ receptor (Abraini et al., 2003).
Krypton is significantly less potent as an anesthetic agent than xenon, consistently with the Meyer-Overton rule which shows that krypton anesthetic potency is four fold less than xenon potency (Cullen et al., 1951; Kennedy et al., 1992).
Xenon, which has the highest solubility in lipids, also has the highest anesthetic potency (i.e. the lowest MAC-immobility) compared to krypton and argon (Table 1). Xenon also has the highest polarizability due to its high number of electron, compared to krypton and argon, so is predicted to be the gas which interact the most with proteins (Quillin et al., 2000).

Gas	Number of electrons	Polarizability ($Å^3$)	van der Waals radius (Å)	Solubility in lipids	MAC-immobility (bar)
Ar	18	1.64	1.91	0.14	27
Kr	36	2.48	2.03	0.43	7.31
Xe	54	4.04	2.21	1.17	1.61

Table 1. Physical and anesthetic properties of argon, krypton and xenon (from (Koblin et al., 1998; Quillin et al., 2000; Ruzicka et al., 2007)).

2. Determination of crystallographic structures of proteins under inert gases pressure

To investigate the mechanism of interaction of gases with proteins, a structural approach using protein crystallography under gas pressure was developed. Xenon binds reversibly to proteins through non-covalent, weak energy van der Waals forces (Ewing et al., 1970). The first structures of protein - xenon complexes were solved in 1965 with myoglobin and haemoglobin under a xenon pressure of 2.5 bar, evidencing a xenon binding site in these two globins (Schoenborn, 1965; Schoenborn et al., 1965). At a pressure of 7 bar, four xenon binding sites were found in myoglobin indicating that the number of xenon binding sites rises with pressure (Tilton et al., 1984).
Since then, many structures of protein-xenon complexes were solved, with xenon used as a heavy atom in isomorphous replacement phasing method (MIR), because xenon has a high

number of electrons (54 e-) and binds with very little perturbation of the protein structure (Vitali et al., 1991; Schiltz et al., 1994; Bourguet et al., 1995; Colloc'h et al., 1997). On the other hand, krypton, though lighter than xenon, was popularized as an internal reference in anomalous phasing techniques MAD or SAD (Schiltz et al., 1997; Cohen et al., 2001) thanks to its absoption K edge at a convenient and useful wavelength easy to tune at all synchrotron places; for a review, see (Schiltz et al., 2003).
Xenon and other noble gas binds primarily in pre-existing hydrophobic cavities or pockets, very often empty in the native gas-less structures (Prangé et al., 1998). Xenon diffusing through protein atoms reaches easily its completely buried binding sites. Xenon was also used as an oxygen probe, based on the hypothesis that xenon and dioxygen would have equivalent binding sites (Duff et al., 2004). The comparison of the binding mode of xenon, krypton and argon was done on the phage T4 lysozyme, showing that gas occupancy rises with gas size and polarizability (Xe > Kr > Ar) (Quillin et al., 2000).
X-ray diffraction data of a protein under xenon pressure are collected either at liquid nitrogen temperature (100 K) or at room temperature. In the first case, the crystal inserted in a cryo-loop is placed in a xenon pressure chamber for a given time, then immediately after frozen in liquid nitrogen, to minimize the amount of xenon which could escape the protein crystal. The determination of the gas pressure within the crystal is thus quite imprecise. For the present study which needs the determination of protein structures under a large range of gas pressures, we have used a pressurisation cell in capillary, designed and developed for the preparation of isomorphous xenon derivatives (Schiltz et al., 1994; Schiltz et al., 2003).

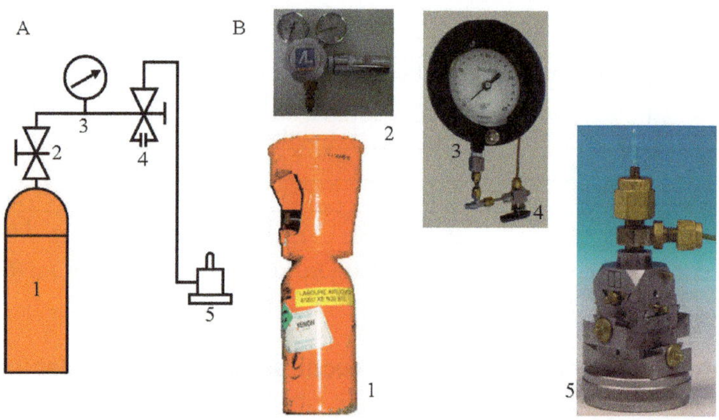

Fig. 1. The pressurisation cell setting. A. Connection between the five elements shown in B. 1- Xenon bottle, 2- Gas regulator, 3- High precision gauge, 4- Bleeding valve, 5- Pressurisation cell.

Typically, a crystal of protein is placed inside a quartz capillary mounted on the pressurisation cell. The pressurisation cell is fixed on a standard goniometer head, and connected to a gas bottle. The pressure within the cell is determined precisely with a calibrated Ashcroft precision gauge (Figure 1). The pressure is maintained constant during all the data collection.
For the present study, we have investigated three different enzymes, urate oxidase, elastase and lysozyme in complex with three gases, xenon, krypton and argon. In urate oxidase,

xenon binds primarily in a large buried hydrophobic cavity close to the active site (Colloc'h et al., 2007; Marassio et al., 2011). Xenon was used as an isomorphous derivative during the determination of urate oxidase structure (Colloc'h et al., 1997). In elastase, like in most of the serine proteases, xenon binds within the specificity pocket S1 of the active site (Schiltz et al., 1995). In lysozyme, xenon binds weakly in an internal cavity and mainly in a pocket located at a crystallographic interface (Schiltz et al., 1997; Prangé et al., 1998).

One of the drawbacks of using X-ray crystallography is the requirement to have a high gas pressure to be able to observe it in the electron density map. A gas pressure about 5 to 10 fold the physiological concentration is estimated to correspond to physiological conditions (Miller, 2002). In the present study, gas pressure ranges from 1 to 40 bar in order to reach a maximum occupancy at saturation, however, only the data between 5 and 10 bar can be compared to physiological conditions.

In the present study, diffraction data were collected at room temperature at the BM14, BM16 and BM30A beamlines at the European Synchrotron Radiation Facility (Grenoble, France). Detectors used were a MAR CCD detector for BM14, an ADSC Q210r CCD detector for BM16 and an ADSC Q315r CCD detector for BM30A. Data were indexed and integrated by *DENZO* and scaled independently and reduced using *SCALEPACK*, both programs from the *HKL* package (Otwinowski et al., 1997) or indexed and integrated by *MOSFLM* (Leslie, 2006) or *XDS* (Kabsch, 2010) and scaled by *SCALA*; intensities were converted in structure factor amplitudes and put on absolute scale using *TRUNCATE* and structure refinements were carried out by *REFMAC* (Murshudov et al., 1997), all programs from the *CCP4* package (Collaborative Computational Project, 1994). The graphics program *COOT* (Emsley et al., 2004) was used to visualize $|2Fobs - Fcalc|$ and $|Fobs - Fcalc|$ electron density maps and for manual rebuilding. Cavity volume were calculated with the program *CastP* (Dundas et al., 2006) with a probe radius of 1.3 Å. Structural figures were prepared using *PyMol* (deLano W.L., DeLano Scientific, Palo Alto, CA, USA).

3. Structure of urate oxidase under inert gas pressure

3.1 Structure of urate oxidase under pressure of xenon and nitrous oxide and comparison with *in-vivo* pharmacology effects

Aspergillus flavus urate oxidase (EC 1.7.3.3) is a homotetrameric enzyme of 301 residues per subunit which is involved in the oxidation of uric acid in presence of molecular oxygen. It crystallizes in the orthorhombic space group I222 with one monomer per asymmetric unit (cell: a = 79.8 Å, b = 96.2 Å, c = 105.4 Å, $\alpha = \beta = \gamma = 90°$). X-ray structures of urate oxidase under various pressures of xenon and nitrous oxide have been determined. Both gases were bound mainly in an internal cavity close to the active site of the enzyme, this cavity being empty in the native gas-less structure (Figure 2). This cavity, completely buried within the monomer, is highly hydrophobic, with 86 % of the atoms lining the cavity being carbons. Both gases were bound also very weakly to a second location, a small extension of a solvent-accessible pocket quite hydrophobic (lined by 75 % carbons). The gas occupancy in this second binding site remained very low (less than 30 % at 30 bar of pressure). Gas occupancies in the main binding site were high, reaching saturation at 100 % for xenon and 60 % for nitrous oxide (Table 2). The main effect of the gas was to expand the volume of the cavity where it binds. This expansion increased with gas occupancy and hence with gas pressure.

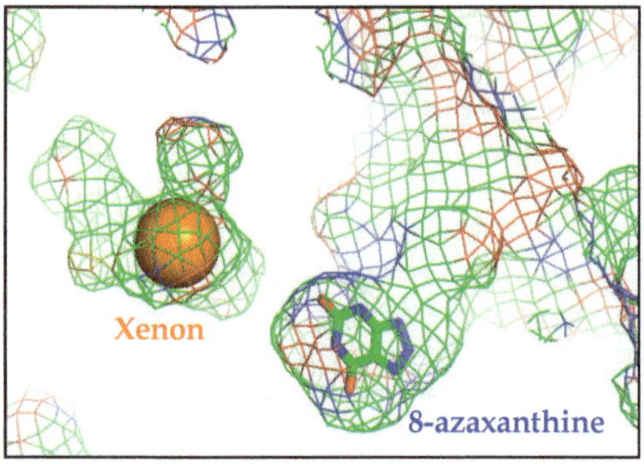

Fig. 2. Hydrophobic cavity in urate oxidase where xenon is bound (shown as an orange sphere). This cavity is close to the active site, where the competitive inhibitor 8-azaxanthine is located (colored in stick by atomic type). The solvent-accessible surface is shown in mesh representation colored by atomic type.

The ratio of gas-induced expansion of the hydrophobic cavity volume for xenon and nitrous oxide in urate oxidase, which could be considered as a model of globular proteins where inert gases bind and whose activity is disrupted by their presence, ranged between 1.1 and 1.5, depending on the applied pressure (Table 2). For the pressures estimated to correspond to physiological conditions (i.e. 5-10 bar), this ratio ranged between 1.3 and 1.5.

Xenon pressure (bar)	Xe occ. (%) Main	Xe-induced expansion	N_2O occ. (%) main	N_2O-induced expansion (%)	Ratio Xe/N_2O-induced expansion	Xe occ. (%) 2nd	N_2O occ. (%) 2nd
5	18	10.8	0	8.5	1.3	0	0
10	60	18.8	40	12.4	1.5	10	0
15	100	19.5	50	17.8	1.1	20	0
20	100	23.1	60	18.4	1.3	22	0
30	100	23.2	60	20.1	1.2	27	25

Table 2. Gas pressure, xenon and nitrous oxide occupancies in the main binding site, gas-induced expansion of the main gas binding site, ratio of expansion, and xenon and nitrous oxide occupancies in the secondary binding site.

If we compared these data with *in-vivo* pharmacology studies, we noticed that this ratio corresponded to the ratio of the narcotic potency of xenon compared to nitrous oxide (about 1.38) as estimated by the concentration of gas necessary to induce loss of righting reflex in rodents (Koblin et al., 1998; David et al., 2003), considered to be a behavioural endpoint closely related to MAC-awake (Campagna et al., 2003).

In comparison, the ratio of gas-induced volume expansion for xenon and nitrous oxide in annexin V, a protein which could be considered as a prototype of NMDA receptor for its properties of ion selectivity and voltage gating (Demange et al., 1994), did not correspond to the ratio of anesthetic potency of xenon and nitrous oxide. However, when considering urate oxidase and annexin V together as a model of simultaneous occupancy of globular proteins and ion-channel receptors, the ratio of gas-induced expansion for xenon and nitrous oxide was close to 1, a value that corresponded to the ratio of anesthetic potency of xenon compared to nitrous oxide, as assessed by their MAC-immobility (Russell et al., 1992; Koblin et al., 1998).

These relationships between gas-induced structural effect and gas-induced narcotic effect allowed proposing a step-by-step mechanism of anesthesia. Gas would first bind to globular cytosolic or extracellular proteins which possess suitable gas binding site easily accessible. Gas-induced disruption of their function would lead to the early stages of anesthesia, i.e. amnesia and hypnosis. When all easily accessible gas binding sites are occupied, gas would then bind to neuronal channel which possess smaller gas binding sites, the disruption of their function would lead to surgical anesthesia, i.e. deep sedation and lack of responses to noxious stimuli (Colloc'h et al., 2007).

3.2 Structure of urate oxidase under pressure of xenon and nitrous oxide and comparison with *in-vitro* activity assays

As mentioned above, the main gas binding site in urate oxidase is very close to the active site (Figure 2). To investigate if gas occupancy and gas-induced volume expansion may have some functional relevance, we performed activity assays on urate oxidase in presence either of air, and either of a mixture of 75 vol % xenon or nitrous oxide and 25 vol % oxygen. To evaluate the alternative therapeutic strategy of using a mixture of xenon and nitrous oxide to combine the efficiency of xenon and the low cost and availability of nitrous oxide, we have also determined the structure of urate oxidase with various pressure of an equimolar mixture of xenon and nitrous oxide, and we have performed an activity assay in presence of a mixture of 37.5 vol% xenon, 37.5 vol% N_2O and 25 vol% oxygen. We found that Xe:N_2O induced a higher expansion of the cavity volume than pure xenon, which in turn induced a higher expansion than N_2O as seen above. *In-vitro* activity assays revealed that Xe:N_2O-induced inhibition was higher than Xe-induced inhibition, itself higher than N_2O-induced inhibition. The relationship between structural effect of the gas, i.e. gas-induced volume expansion, and the functional effect of the gas, i.e. gas-induced inhibition of the enzymatic reaction, highlighted the way by which gases might disrupt protein function through an indirect mechanism (Marassio et al., 2011).

The role of the void hydrophobic cavity in the catalytic mechanism was thus demonstrated by the activity assays in presence of gas. This functional role was also suggested by the high-pressure structural and functional study of urate oxidase. Under high hydrostatic pressure (150 Ma; 1500 bar) the volume of the cavity was reduced as expected, while the volume of the active site was expanded. High pressure also inhibited the catalytic mechanism of urate oxidase, this loss of activity being a loss of substrate affinity (Girard et al., 2010).

In both cases (gas or pressure), there was a loss of flexibility of the cavity, either by the gas presence which induced an expansion and inhibited its contraction, either by high pressure which induced a cavity contraction but inhibited its expansion. The role of the cavity in the functional mechanism of urate oxidase seems then to give some flexibility to the active site to allow a structural fit for the ligand in the active site.

3.3 Structure of urate oxidase under pressure of krypton and comparison with gas solubility in lipids

Structures of urate oxidase under krypton pressure of 2 to 30 bar were determined in the present study. Krypton was bound to the exact same location than xenon, in the

hydrophobic cavity close to the active site. Krypton occupancy increased with the applied pressure up to 45 % at 30 bar (Table 3). Like xenon, krypton was also weakly bound to a secondary binding site at the bottom of a solvent-accessible pocket, but only at pressure above 20 bar.
For an identical pressure, krypton occupancy was always lower than xenon occupancy (Figure 3A). Xenon which has a higher number of electrons than krypton has a higher polarizability (Table 1) and binds thus with a higher occupancy, as already observed in the case of phage T4 lysozyme (Quillin et al., 2000).

Krypton pressure (bar)	Resolution (Å)	Occ. in main binding site (%)	Kr-induced expansion (%)	Xe-induced expansion (%)	Ratio Xe / Kr induced expansion	Occ. in 2nd binding site (%)
2	1.60	10	3.2	3.3	1.0	0
5	1.60	15	4.1	10.8	2.6	0
10	1.55	20	11.4	18.8	1.6	0
20	1.65	40	12.8	23.1	1.8	10
30	1.65	45	15.2	23.2	1.5	15

Table 3. Krypton pressure, resolution of the crystallographic structure, krypton occupancy in the main binding site, krypton-induced and xenon-induced volume expansion of the main binding site, ratio of the Xe and Kr-induced volume expansion and krypton occupancy in the secondary binding site.

If one refers to the Meyer-Overton rule, the narcotic potency of a gas would be related to its solubility in lipids. The ratio of solubility in lipids of xenon compared to krypton is 1.17 / 0.14 = 2.7 (Table 1), a value which correspond to the ratio of Xe and Kr-induced volume expansion at the pressure of 5 bar, well within the range of pressure estimated to correspond to physiological condition. This result confirmed what was shown previously when comparing the structural-induced effect of xenon and nitrous oxide on urate oxidase to their in-vivo effect as evaluated by their MAC-awake (Colloc'h et al., 2007). However, the MAC-immobility which prevents response to noxious stimuli for xenon in man is about 4.5 higher than the MAC-immobility of krypton (Table 1), which does not correspond to the structural Xe- and Kr-induced structural effect in urate oxidase, considered as a model for globular protein whose function is disrupted by the presence of gas.

3.4 Structure of urate oxidase under pressure of argon and comparison with *in-vivo* pharmacology study

Structures of urate oxidase under argon pressure of 10 to 65 bar were determined in the present study. Argon was bound to the exact same location than xenon and krypton, in the large hydrophobic cavity close to the active site. Argon became visible in the electron density map at a pressure of 30 bar and above, with an occupancy factor of 40 % at a pressure of 65 bar (Table 4). 65 bar is the maximum pressure which could be reach in the quartz capillary; above that pressure, the risk of breakage of the capillary became very high. At the same pressure, argon occupancy was always lower than krypton and xenon occupancies (Figure 3A). In the secondary binding site where xenon and krypton bind very weakly, no argon is detectable in the electron density map, even at a pressure of 65 bar.

Argon pressure (bar)	Resolution (Å)	Occ. in main binding site (%)	Ar-induced expansion (%)	Xe-induced expansion (%)	Ratio Xe / Ar Indiced expansion
10	1.65	0	7.1	20.7	2.9
20	1.90	0	11.0	24.6	2.2
25	1.60	20	13.4		
30	1.60	20	12.3	24.6	2
35	1.60	20	14.1		
40	1.75	20	10.3		
45	1.60	25	11.1		
55	1.60	30	10.7		
65	1.60	40	16.4		

Table 4. Argon pressure, resolution of the crystallographic structure, argon occupancy, argon-induced and xenon-induced cavity volume expansion, and ratio of the Xe and Ar-induced volume expansion.

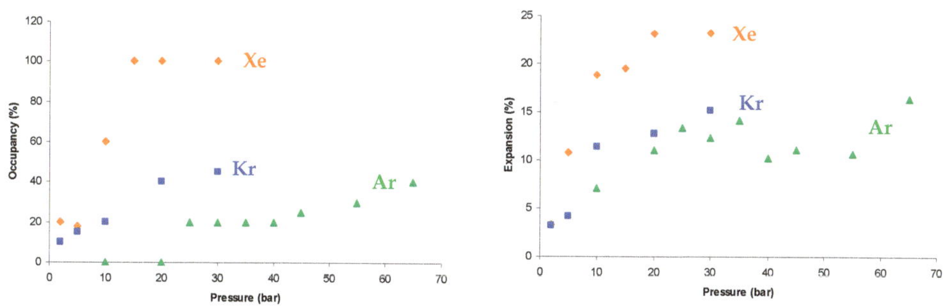

Fig. 3. A. Gas occupancy as a function of pressure. B. Gas-induced expansion of the main binding site as a function of pressure.

As for the two other nobles gas, the main effect of argon was to expand the volume of the cavity where it was bound. However, due to its quite low occupancy factor and its small size, the argon-induced expansion remained low, around 10 % of expansion except for the pressure of 65 bar where expansion reached 16 % (Table 4). For pressure of 10 and 20 bar where argon was not detectable in the electron density map, there was already a volume expansion indicating with little doubt the presence of argon within the cavity (Table 4, Figure 3B).

The ratio of Xe- and Ar-induced expansions of the cavity volume was 2.9 for a pressure of 10 bar, which did not correspond to their inverse ratio of MAC-immobility (27/1.61 = 16.8) nor their ratio of solubility in lipids (1.17 / 0.14 = 8.4) (Table 1). However, argon is not narcotic at ambient pressure and needs to be pressurised to have some narcotic action.

In order to allow comparison of the effects of argon and xenon in urate oxidase, we further calculated the theoretical expansion of the gas binding cavity produced by argon at 100 % occupancy according to a linear regression model. For occupancy of 100% of argon, the corresponding volume expansion would be of 23.3 % (Figure 4A). According to a linear regression model, this expansion would be reach for a pressure of 164 bar (Figure 4B). This pressure corresponds to about ten fold the pressure of 14 to 17 bar at which argon is known to produces narcosis in rodents (Abraini et al., 1998; Koblin et al., 1998).

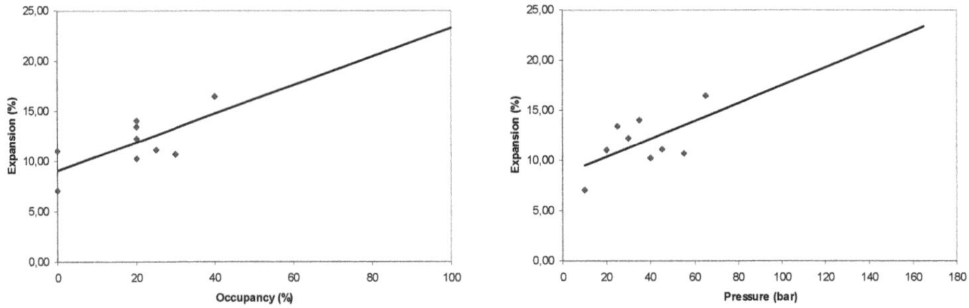

Fig. 4. Linear regression model for cavity volume expansion as a function of occupancy (A) and as a function of pressure (B).

In addition, it should be mentioned that these estimated values for argon were also consistent with crystallographic data that have demonstrated that xenon at full occupancy (pressure of about 20 bar) produces a similar maximal expansion around 23-25 % of the gas binding site (Table 2). This is consistent with the fact that the ratio between the efficient estimated pressure of argon and the efficient experimental pressure of xenon at producing full occupancy and maximal expansion of the gas binding site (164 / 20 = 8.2) is similar to the ratio of their solubility in lipids (1.17 / 0.14 = 8.4) as predicted by the Meyer-Overton rule (Abraini et al., 2003; Campagna et al., 2003).

Argon is narcotic only in hyperbaric condition. At ambient pressure, argon may thus have a very limited influence on its target function. Since one of the major effect of hydrostatic pressure is to contract the volume of internal cavities (Girard et al., 2010), it may explain why argon needs hyperbaric condition to exert its influence.

3.5 Conclusion on urate oxidase structures under inert gas pressure

The three noble gases were bound to an identical location in urate oxidase, within an internal hydrophobic cavity. The gas occupancies increased in the sequence argon < krypton < xenon, as it was the case for T4 lysozyme (Quillin et al., 2000), who noticed that smaller gases do not bind as well as larger ones as a result of their attenuated polarizability. Xenon and krypton were bound also weakly in a secondary binding site, while argon was not observed even at high pressure.

The main effect of the gas was to expand the cavity volume where it binds. The ratio of expansion was related to the narcotic potency of the gas, as evaluated by their MAC-awake or their solubility in lipids. The presence of xenon within the cavity induced an inhibition of the catalytic mechanism, with a relationship between gas-induced expansion and gas-induced inhibition, as shown by the comparison between xenon and nitrous oxide structural and functional effects. No activity assays were performed in presence of krypton or argon, but we can predict, based on the present structural results, that krypton should induce an inhibition of the catalytic mechanism of urate oxidase. However, krypton-induced inhibition should be lower than xenon-induced inhibition, according to their relative induced expansion in the range 5-10 bar (Table 3). Argon-induced inhibition should be very low or inexistent, according to the very low gas occupancy and argon-induced expansion at a pressure of 10 bar (Table 4).

4. Structure of elastase under inert gas pressure

4.1 Introduction on elastase structures under inert gas pressure

Pancreatic porcine elastase (EC 3.4.21.36) is a serine protease of 266 residues which hydrolyzes peptide bonds in proteins, its main substrate being elastine. The catalytic triad of elastase is composed, as for all serine proteases, of an activated serine (Ser 195) assisted by a proton relay (His 57), which acts as a general base, and stabilized through an hydrogen bond by an aspartic acid (Asp 102). Pancreatic porcine elastase crystallizes in the orthorhombic space group $P2_12_12_1$ with one monomeric enzyme per asymmetric unit (cell : a = 51.4 Å, b = 58.0 Å, c = 75.3 Å, $\alpha = \beta = \gamma = 90°$). In the crystallographic structure, a sulphate or an acetate ion is bound in the oxyanion hole, depending on the concentration of the precipitating agents. The primary specificity pocket S1 is a hydrophobic pocket located below the oxyanion hole and the Ser 195 which is specific for recognition of the peptidic substrate.

In the crystallographic structures of apo elastase, the S1 pocket was either empty, either filled by a water molecule hydrogen-bonded to a water molecule outside of the S1 pocket and to an oxygen atom of the sulphate ion (Figure 5A). When this water molecule (termed W-S1) was present, its B-factor was quite elevated. In the different crystallographic structures of elestase in complex with xenon deposited in the Protein Data Bank, 1C1M (Schiltz et al., 1995), 1L1G and 1L0Z (Panjikar et al., 2002), 1UO6 and 2A7C (Mueller-Dieckmann et al., 2004) and 2OQU (Kim et al., 2007), xenon was bound within the specificity pocket S1 in the active site of elastase (Figure 5B). This xenon binding site is moderately hydrophobic, lined with 60% carbons.

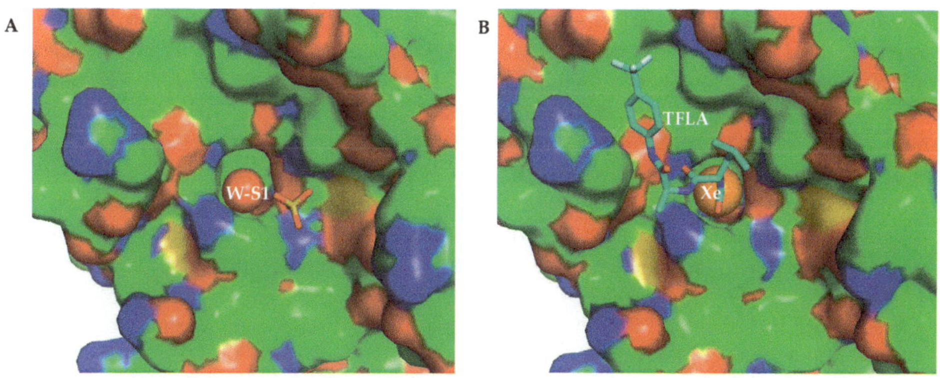

Fig. 5. Elastase shown with its solvent-accessible surface colored by atomic type. A. Native gas-less elastase with the water molecule W-S1 in the S1 pocket. B. Elastase in complex with xenon (shown as an orange sphere) or in complex with a peptidic inhibitor TFLA (shown in cyan in stick representation).

In the present study, structures of native elastase (gas-less) were determined in the same conditions than structures under inert gas pressure, i.e. at room temperature in a quartz capillary. Three structures have been solved, at 1.38 Å, 1.7 Å and 1.45 Å resolution. In two native structures, the S1 pocket was empty while in one of them, there was the water molecule W-S1 with a B-factor of 36.3 Å^2.

4.2 Structure of elastase under pressure of xenon and comparison with *in-vitro* activity assays

Structures of elastase under xenon pressure from 1 bar to 30 bar were determined in the present study. Whatever the applied pressure, xenon was bound to a unique site, within the specificity pocket S1, with an occupancy which increased with the applied pressure. Occupancy reached 100 % for a pressure of 30 bar (Table 5). The S1 pocket where xenon binds is moderately hydrophobic, with 60 % of atoms lining it being carbons. This gas binding site is less hydrophobic than the main gas binding site in urate oxidase, which was lined by 86 % carbons. The atom the closest to xenon is the side chain atom Oγ of the catalytic Ser 195.

Xenon pressure (bar)	Resolution (Å)	Occ. (%)	Xe-induced expansion (%)
1	1.40	15	3.0
2	1.45	25	3.3
5	1.50	30	7.8
10	1.50	70	9.4
20	1.60	90	12.7
30	1.65	100	13.6

Table 5. Xenon pressure, resolution of the crystallographic structure, occupancy and xenon-induced expansion of the S1 pocket in elastase.

Whatever the pressure, there was no water molecules in the S1 pocket, so if the water molecule W-S1 was present in the native gas-less structure, it was not displaced but replaced by xenon. However, xenon did not take the exact location of the W-S1 and was closer to the Oγ of the catalytic Ser 195 (3.4 Å instead of 3.8 Å).

The presence of xenon within the S1 pocket expanded its volume, its expansion rising with the applied pressure. However, since the gas was bound directly within the active site, the gas-induced inhibition is likely to be a direct inhibition and the expansion by itself has probably no functional relevance. Xenon took indeed the place of peptidic inhibitors, like the trifluroacetyl-leu-ala (TFLA) known to be an excellent inhibitor of elastase (Li de la Sierra et al., 1990) (Figure 5B).

To investigate the direct inhibition by xenon, we performed activity assays on elastase in presence either of air, either of 100 vol % xenon. Initial velocity in presence of xenon when compared to air (taken as 100 %) was 81.5 + 2.1 % revealing an inhibition of the catalytic activity of elastase of around 20 % by xenon. However, this inhibition was lower than xenon occupation in the range 5-10 bar (30 – 70 % occupation).

Tissue-type plasminogen activator (tPA), the only approved treatment for thrombolysis after an ischemic stroke, is also a serine protease. As in the case of elastase, xenon inhibited tPA enzymatic activity (David et al., 2010). This inhibition is likely to be a direct inhibition with xenon binding directly in the S1 pocket in the active site of tPA. Serine proteases have indeed structurally similar active site with the same catalytic triad which can be easily superimposed (Schiltz et al., 1995). The superimposition of the catalytic triad of elastase and of tPA suggested that xenon fitted perfectly in the S1 pocket of tPA, thus inhibiting tPA enzymatic activity.

4.3 Structure of elastase under pressure of krypton

Structures of elastase under pressure of krypton from 2 to 30 bar were determined in the present study. Krypton was bound within the S1 pocket, at the exact same location than xenon. The water molecule W-S1 if it was present was replaced by the krypton atom. Occupancy of krypton reached 70 % at the pressure of 30 bar (Table 6).

Krypton pressure (bar)	Resolution (Å)	Occ. (%)	Kr-induced expansion (%)
2	1.35	20	4.9
5	1.45	25	4.9
10	1.50	35	5.8
20	1.75	50	7.9
30	1.45	70	8.5

Table 6. Krypton pressure, resolution of the crystallographic structure, occupancy and krypton-induced expansion of the S1 pocket in elastase.

Krypton occupancy was always lower than xenon occupancy for a same pressure (Figure 6). Xenon which is more polarizable was bound with a higher occupancy than krypton. Since the gas binding site is only moderately hydrophobic, with a lot of polar atoms lining the gas, it is likely that dipole-induced interactions play an important role in gas binding. Like xenon, krypton induced an expansion of the S1 pocket where it binds. However, this expansion remained quite low, less than 10 %.

4.4 Structure of elastase under pressure of argon

Structures of elastase under pressure of argon of 10 to 60 bar were determined in the present study. Argon was bound within the S1 pocket, at the exact same location than xenon and krypton. Argon occupancy reached 50 % for a pressure of 60 bar (Table 7). For technical conditions, it was not possible to reach higher pressure due to the risk of failure of the quartz capillary. Argon occupancy was always lower than krypton and xenon occupancy for a same pressure, according to a lower polarizablity (Figure 6).

Argon pressure (bar)	Resolution (Å)	Occ. (%)	B factor W-S1
10	1.35	35	42.5
20	1.45	35	missing
30	1.50	40	40.1
40	1.35	40	missing
60	1.50	50	missing

Table 7. Argon pressure, resolution of the crystallographic structure, occupancy and B-factor of the water molecule W-S1 if it is present.

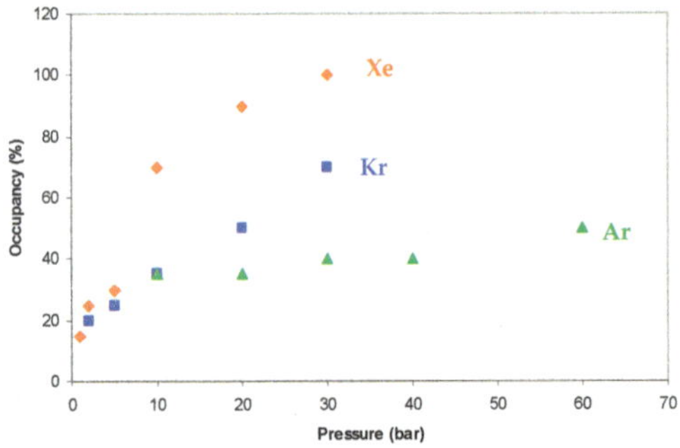

Fig. 6. Xenon, krypton and argon occupancies as a function of pressure.

Argon occupancy was higher in elastase than in urate oxidase (Table 4), probably because of the smaller size of the gas binding site in elastase which would allow a better binding of argon.

Argon-induced expansion of the S1 pocket was very low and not very significant (less than 5 %), probably due to the small size of the argon atom.

Contrary to xenon and krypton cases, when the water molecule W-S1 was present in the native gas-less structure, this water molecule remained visible close to the argon electronic density (Figure 7), with a lower occupation factor. The distance between the argon atom and the W-S1 molecule was about 2.8 Å.

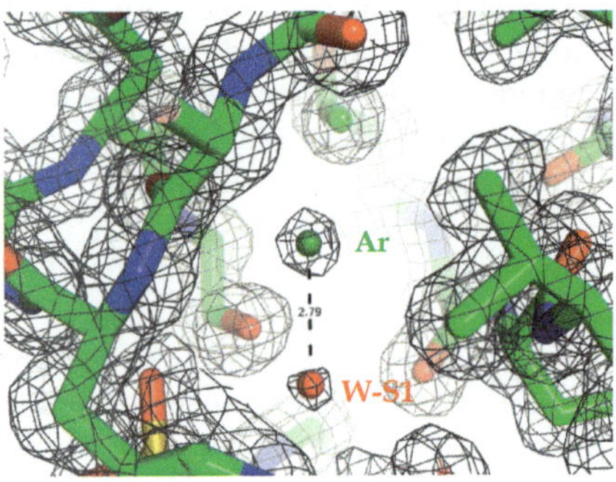

Fig. 7. *2Fo-Fc* electron density map of elastase under an argon pressure of 10 bar, contoured at 1 σ.

4.5 Conclusion on elastase structures under inert gas pressure

In elastase, there is one unique gas binding site, located within the S1 pocket in the active site. Contrary to urate oxidase where the gas binding site was an empty cavity, there might be a water molecule in the moderately hydrophobic S1 pocket. When present, this water molecule was replaced by xenon and krypton, but remained visible close to argon.

Xenon inhibited directly elastase catalytic activity by taking the place of the substrate, even if its inhibition stayed lower than its occupancy. It is also likely that krypton inhibited elastase catalytic activity. Since krypton occupancy was lower than xenon occupancy, krypton-induced inhibition is expected to be lower than xenon-induced inhibition. In the range 5-10 bar, krypton occupancy ranged between 25 and 35%, inducing probably a rather small inhibition of the enzymatic reaction. Argon-induced inhibition of the catalytic reaction is likely to be quite low and similar to krypton-induced inhibition, with an occupation of argon of 35 % at 10 bar.

5. Structure of lysozyme under inert gas pressure

5.1 Introduction on lysozyme structures under inert gas pressure

Egg white lysozyme C (EC 3.2.1.17) is an enzyme of 129 residues which hydrolyse peptidoglycans. It crystallizes in tetragonal space group $P4_32_12$ with one monomeric enzyme per asymmetric unit (cell: a = 79.2 Å, b = 79.2 Å, c = 37.9 Å, $\alpha = \beta = \gamma = 90°$). In the different crystallographic structures of lysozyme in presence of xenon deposited in the Protein Data Bank : 1C10 (Prangé et al., 1998), 2A7D (Mueller-Dieckmann et al., 2004) and 1VAU (Takeda et al., 2004), xenon was bound mainly in a pocket at a crystallographic interface (gas binding site I or GBS I in Figure 8) and weakly in a small internal cavity (GBS II in Figure 8). In the native gas-less structure, a water molecule was present in GBS I, and GBS II was empty. GBS I is moderately hydrophobic, lined with 55 % carbons, and GBS II is very hydrophobic, lined with 83 % carbons. The occupancy of xenon ranged between 30 and 60% in GBS I and stayed beyond 25 % in GBS II.

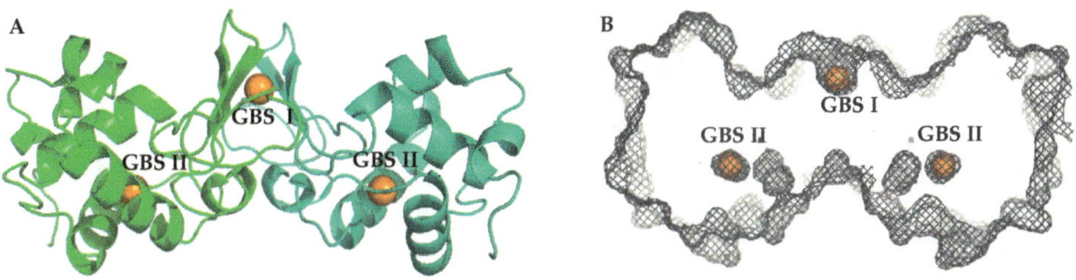

Fig. 8. Xenon binding sites in lysozyme, shown with two symmetrical subunits. A. Xenon was bound at the crystallographic interface between two symmetrical monomers (GBS I) and a second xenon was bound within a small internal cavity buried within each monomer (GBS II) (xenon shown as orange spheres, and lysozyme with its Cα chain as ribbon, one symmetric is colored in green, the second in blue). B. Solvent-accessible surface of two symmetrical lysozymes.

In the structure 1C10, a third xenon was located in the active site, where either two water molecules or one water molecule and one chloride ion were found in the other deposited structures. In the structure 1VAU, a third xenon was located in another crystallographic pocket at the interface between two monomers, where a water molecule was found in the other deposited structures.

In the crystallographic structure of lysozyme under a 55 bar pressure of krypton 1QTK, (Schiltz et al., 1997; Prangé et al., 1998), krypton was bound in the internal cavity (GBS II) but not in the crystallographic pocket (GBS I).

In the present study, the structure of a native gas-less lysozyme was determined in the same condition than structures under inert gas pressure, i.e. at room temperature in quartz capillary. Two native structures have been solved, at 1.6 Å and 1.55 Å resolution. In both structures, the internal cavity GBS II was empty and there was a water molecule in GBS I which had a B-factor between 25 and 29 Å^2. The volume of the internal cavity GBS II is very small, about 40 Å^3 while the volume of GBS I at a crystallographic interface is larger, around 100 Å^3.

5.2 Structure of lysozyme under pressure of xenon

The structures of lysozyme under xenon pressure of 10 to 30 bar were determined in the present study. Xenon was bound in GBS I with an occupancy which did not increase with pressure and remained around 20 and 30 % (which is the double of the occupancy for the monomer since the xenon is located at a crystallographic interface). Xenon was also bound in GBS II with an occupancy which increased with the applied pressure but which remained quite low (20 % occupancy at 30 bar).

Xenon occupancy in GBS I remained stable whatever the applied pressure and induced almost no expansion of its volume (less than 10 %). Xenon location was not exactly superposed to the water molecule location present in the native gas-less structure, with a distance of 0.6 to 0.7 Å between them. This gas binding site being located at a crystallographic interface is likely to be not physiological.

Xenon pressure (bar)	Resolution (Å)	Xe occ. in GBS I (%)	Xe occ. in GBS II (%)	Xe-induced expansion of GBS II (%)
10	1.55	30	10	9.1
15	1.60	20	10	4.6
20	1.55	30	12	6.9
25	1.90	24	15	13.6
30	1.65	30	20	23.4

Table 8. Xenon pressure, resolution of the crystallographic structures, xenon occupancy in GBS I , xenon occupancy in GBS II and xenon-induced expansion of GBS II.

Xenon occupancy in GBS II remained very low, due to the small size of the cavity (volume around 40 Å^3). Xenon van der Waals radius is of 2.21 Å, its theoretical volume is thus of 45 Å^3, close to the volume of GBS II. Xenon induced an expansion of the volume of this small internal cavity which reached a volume around 50 Å^3 for xenon occupancy of 20 %. In the range of pressure corresponding to physiological conditions (5 - 10 bar), the occupation of xenon in GBS II is likely to be very small (less than 10 %).

5.3 Structure of lysozyme under pressure of krypton

The structures of lysozyme under krypton pressure of 5 to 20 bar were determined in the present study. No krypton was visible in GBS I whatever the applied pressure, as it was the case in the structure determined under a krypton pressure of 55 bar. In all cases, there was a water molecule at the exact same location than in the native gas-less structure. Krypton was bound weakly in GBS-II and became visible in the electron density map at a pressure of 20 bar. Krypton induced an expansion of the volume of GBS II which seemed not related to krypton occupancy. However, this krypton-induced expansion of GBS II indicated that krypton was present event when it was not visible in the electron density map (Table 9).

Krypton is smaller than xenon (van der Waals radius of 2.03 Å instead of 2.21 Å, and volume of 35 Å^3 instead of 45 Å^3), it is thus likely than krypton occupancy can reach higher value than xenon occupancy in GBS II (Figure 9).

Krypton pressure (bar)	Resolution (Å)	Kr occ. in GBS I (%)	Kr occ. in GBS II (%)	Kr-induced expansion of GBS II (%)
5	1.55	0	0	12.1
10	1.60	0	0	4.6
20	1.55	0	10	14.9
55 (1QTK)	2.03	0	49	7.1

Table 9. Krypton pressure, resolution of the crystallographic structures, krypton occupancy in GBS I , krypton occupancy in GBS II and krypton-induced expansion of GBS II.

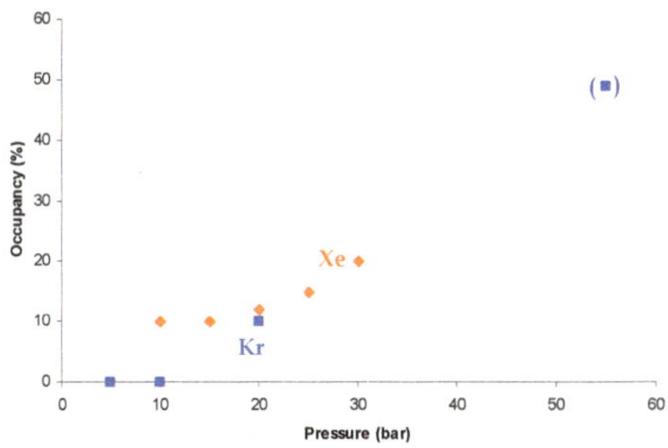

Fig. 9. Xenon and krypton occupancies as a function of pressure.

A theoretical study based on a three-dimensional distribution function theory of molecular liquids was applied to lysozyme in presence of water and noble gases (Imai et al., 2007). This study predicted that krypton had a slightly better affinity for GBS II than xenon. Structures of lysozyme under higher pressure of krypton (30 and 40 bar) would be necessary to confirm this prediction.

5.4 Structure of lysozyme under pressure of argon

The structures of lysozyme under argon pressure of 10 to 50 bar were determined in the present study. Whatever the applied pressure, no argon was visible in the electron density map even at a pressure of 50 bar neither in GBS I or in GBS II. In GBS I, the water molecule was present at the exact place of the water molecule in the native gas-less structure. The smaller size of argon should not prevent argon to bind within GBS II, however, a higher pressure would probably be necessary to be able to visualize argon in the electron density map in GBS II.

In the theoretical study already mentioned (Imai et al., 2007), argon affinity for GBS II was predicted to be very high. However, the present study showed that even at high pressure, argon did not bind within GBS II, even if it smaller size would allow its binding. Argon polarizability is lower than krypton and xenon polarizabilities (Table 1), due to its small number of electron, and could explain its very low affinity for lysozyme.

5.5 Conclusion on lysozyme structures under inert gas pressure

Lysozyme possesses two gas binding sites, one located in a pocket at a crystallographic interface (GBS I) and one located within a quite small internal cavity (GBS II). Both xenon and krypton were bound within GBS II with a quite low occupation, and only xenon was bound within GBS I. Argon did not bind to lysozyme even at a pressure of 50 bar. The rule that showed that gas occupancy rose with gas size and polarizability is almost respected, since the small size of GBS II could prevent xenon binding and thus favour krypton binding. In the pressure range which would correspond to physiological conditions (5 - 10 bar), xenon occupancy is likely to be very low (less than 10 %), and krypton and argon occupancies are null. It is then likely than enzymatic activity of lysozyme is not modified by the presence of an inert gas.

6. Conclusion

Noble gases bind to proteins essentially through non-covalent van der Waals interactions, their binding constant depending on their electronic polarizability. In the three studied enzymes, gas occupancies were in the order of their polarizability, Xe > Kr > Ar, as it was already found for T4 lysozyme (Quillin et al., 2000). The only slight exception was the internal cavity of egg white lysozyme, where its smaller size prevented xenon higher occupancy.

The major physiological targets of gaseous anesthetics are postulated to be neuronal channels receptors, like the NMDA receptor which is inhibited by xenon (Campagna et al., 2003; Franks, 2008), or the $GABA_A$ receptor which could be modulated by argon (Abraini et al., 2003). However, at lower concentration, gas would bind mainly to globular targets, amongst them enzymes, whose functions would be modulated by the presence of inert gas.

From the present study, we can infer different mode of inhibition by gas. In urate oxidase, gas inhibited the catalytic reaction through an indirect mechanism; the presence of the gas within the cavity would prevent the cavity contraction, thus modifying the active site flexibility. In elastase, gas inhibited the catalytic reaction through a direct mechanism; the presence of the gas in the active site would prevent the substrate binding. In lysozyme, gas would not inhibit the catalytic reaction, their occupation being too weak.

Protein activity requires some conformational flexibility (Frauenfelder et al., 1991). In enzymes, the balance between conformational flexibility and rigidity is adjusted to optimize the catalytic efficiency for a given condition (Chiuri et al., 2009). Cavities would facilitate conformational changes and are though to play a key role in protein function (Hubbard et al., 1994). Anesthetics have been postulated to act by stabilizing high-energy conformers inducing altered functions (Eckenhoff, 2001; Johansson et al., 2005). High pressure was also postulated to stabilize high-energy conformers with altered functions (Frauenfelder et al., 1990; Akasaka, 2006; Fourme et al., 2006). Urate oxidase is thus a key example which highlights the effect of anesthetic, since this enzyme is both inhibited by gas presence in a hydrophobic cavity (Marassio et al., 2011) and by high pressure (Girard et al., 2010).

Inert gas binds to proteins through very weak non-covalent van der waals interaction. How such weak interactions could generate such high biological effect such as anesthesia ? It was suggested that anesthesia would arise from small effects at many biological targets (Eckenhoff, 2001). The present study would confirm this hypothesis, showing that some enzymes could be stabilized by the presence of gas in hydrophobic cavity (as in urate oxidase), some enzymes could be directly inhibited by gas (as in elastase case), and some enzymes are not affected by gas (as in the case of lysozyme). The mechanisms of neuroprotection and anesthesia by inert gases are thus very complex processes with many biological targets whose function are modulated (inhibited or potentiated) by the presence of gas.

7. Acknowledgement

The authors thank Jacques H. Abraini (CI-NAPS, UMR 6232, Caen) for advises during this work.

8. References

Abraini, J.H., David, H.N. & Lemaire, M. (2005). Potentially neuroprotective and therapeutic properties of nitrous oxide and xenon. *Ann. N. Y. Acad. Sci.* Vol. 1053: pp. 289-300.

Abraini, J.H., Kriem, B., Balon, N., Rostain, J.C. & Risso, J.J. (2003). Gamma-aminobutyric acid neuropharmacological investigations on narcosis produced by nitrogen, argon, or nitrous oxide. *Anesth. Analg.* Vol. 96 No. 3: pp. 746-749.

Abraini, J.H., Rostain, J.C. & Kriem, B. (1998). Sigmoidal compression rate-dependence of inert gas narcotic potency in rats: implication for lipid vs. protein theories of inert gas action in the central nervous system. *Brain Res.* Vol. 808 No. 2: pp. 300-304.

Ahmed, N., Wahlgren, N., Grond, M., Hennerici, M., Lees, K.R., Mikulik, R., Parsons, M., Roine, R.O., Toni, D. & Ringleb, P. (2010). Implementation and outcome of thrombolysis with alteplase 3-4.5 h after an acute stroke: an updated analysis from SITS-ISTR. *Lancet Neurol.* Vol. 9 No. 9: pp. 866-874.

Akasaka, K. (2006). Probing conformational fluctuation of proteins by pressure perturbation. *Chem. Rev.* Vol. 106 No. 5: pp. 1814-1835.

Bourguet, W., Ruff, M., Chambon, P., Gronemeyer, H. & Moras, D. (1995). Crystal structure of the ligand-binding domain of the human nuclear receptor RXR-alpha. *Nature* Vol. 375 No. 6530: pp. 377-382.

Campagna, J.A., Miller, K.W. & Forman, S.A. (2003). Mechanisms of actions of inhaled anesthetics. *New Engl. J. Med.* Vol. 348 No. 21: pp. 2110-2124.

Chiuri, R., Maiorano, G., Rizzello, A., del Mercato, L.L., Cingolani, R., Rinaldi, R., Maffia, M. & Pompa, P.P. (2009). Exploring local flexibility/rigidity in psychrophilic and mesophilic carbonic anhydrases. *Biophys. J.* Vol. 96 No. 4: pp. 1586-1596.

Choi, D.W., Koh, J.Y. & Peters, S. (1988). Pharmacology of glutamate neurotoxicity in cortical cell culture: attenuation by NMDA antagonists. *J. Neurosci.* Vol. 8 No. 1: pp. 185-196.

Cohen, A., Ellis, P., Kresge, N. & Soltis, S.M. (2001). MAD phasing with krypton. *Acta Crystallogr.* Vol. D57 No. 2: pp. 233-238.

Collaborative Computational Project, Number 4. (1994). The *CCP4* suite : programs for protein crystallography. *Acta Crystallogr.* Vol. D50 No. 2: pp. 760-763.

Colloc'h, N., El Hajji, M., Bachet, B., L'Hermite, G., Schiltz, M., Prangé, T., Castro, B. & Mornon, J.P. (1997). Crystal structure of the protein drug urate oxidase-inhibitor complex at 2.05 Å resolution. *Nat. Struc. Biol.* Vol. 4 No. 11: pp. 947-952.

Colloc'h, N., Sopkova-de Oliveira Santos, J., Retailleau, P., Vivarès, D., Bonneté, F., Langlois d'Estaintot, B., Gallois, B., Brisson, A., Risso, J.J., Lemaire, M., Prangé, T. & Abraini, J.H. (2007). Protein crystallography under xenon and nitrous oxide pressure: comparison with in vivo pharmacology studies and implications for the mechanism of inhaled anesthetic action. *Biophys. J.* Vol. 92 No. 1: pp. 217-224.

Cullen, S.C. & Gross, E.G. (1951). The anesthetic properties of xenon in human and animal beings, with additional observations on krypton. *Science* Vol. 113 No. 2942: pp. 580-582.

David, H.N., Ansseau, M., Lemaire, M. & Abraini, J.H. (2006). Nitrous oxide and xenon prevent amphetamine-induced carrier-mediated dopamine release in a memantine-like fashion and protect against behavioral sensitization. *Biol. Psychiatry* Vol. 60 No. 1: pp. 49-57.

David, H.N., Haelewyn, B., Chazalviel, L., Lecocq, M., Degoulet, M., Risso, J.J. & Abraini, J.H. (2009). Post-ischemic helium provides neuroprotection in rats subjected to middle cerebral artery occlusion-induced ischemia by producing hypothermia. *J. Cereb. Blood Flow Metab.* Vol. 29 No. 6: pp. 1159-1165.

David, H.N., Haelewyn, B., Degoulet, M., Lecocq, M., Colomb, D.G. & Abraini, J.H. (submitted). Argon-induced neuroprotection in in-vivo models of ecitotoxic insult and ischemic brain damage in rats. *Br. J. Anaesth.*

David, H.N., Haelewyn, B., Risso, J.J., Colloc'h, N. & Abraini, J.H. (2010). Xenon is an inhibitor of tissue-plasminogen activator: adverse and beneficial effects in a rat model of thromboembolic stroke. *J. Cereb. Blood Flow Metab.* Vol. 30 No. 4: pp. 718-728.

David, H.N., Haelewyn, B., Rouillon, C., Lecoq, M., Chazalviel, L., Apiou, G., Risso, J.J., Lemaire, M. & Abraini, J.H. (2008). Neuroprotective effects of xenon: a therapeutic window of opportunity in rats subjected to transient cerebral ischemia. *Faseb J.* Vol. 22 No. 4: pp. 1275-1286.

David, H.N., Leveillé, F., Chazalviel, L., MacKenzie, E.T., Buisson, A., Lemaire, M. & Abraini, J.H. (2003). Reduction of ischemic brain damage by nitrous oxide and xenon. *J. Cereb. Blood Flow Metab.* Vol. 23 No. 10: pp. 1168-1173.

Demange, P., Voges, D., Benz, J., Liemann, S., Göttig, P., Berendes, R., Burger, A. & Huber, R. (1994). Annexin V: the key to understanding ion selectivity and voltage regulation? *Trends Biochem. Sci.* Vol. 19 No. 7: pp. 272-276.

Dingley, J., Tooley, J., Porter, H. & Thoresen, M. (2006). Xenon provides short-term neuroprotection in neonatal rats when administered after hypoxia-ischemia. *Stroke* Vol. 37 No. 2: pp. 501-506.

Dirnagl, U., Iadecola, C. & Moskowitz, M.A. (1999). Pathobiology of ischaemic stroke: an integrated view. *Trends Neurosci.* Vol. 22 No. 9: pp. 391-397.

Duff, A.P., Trambaiolo, D.M., Cohen, A.E., Ellis, P.J., Juda, G.A., Shepard, E.M., Langley, D.B., Dooley, D.M., Freeman, H.C. & Guss, J.M. (2004). Using xenon as a probe for dioxygen-binding sites in copper amine oxidases. *J. Mol. Biol.* Vol. 344 No. 3: pp. 599-607.

Dundas, J., Ouyang, Z., Tseng, J., Binkowski, A., Turpaz, Y. & Liang, J. (2006). CASTp: computed atlas of surface topography of proteins with structural and topographical mapping of functionally annotated residues. *Nucleic Acids Res.* Vol. 34 No. Web Server issue: pp. W116-118.

Eckenhoff, R.G. (2001). Promiscuous ligands and attractive cavities: how do the inhaled anesthetics work? *Mol. Interv.* Vol. 1 No. 5: pp. 258-268.

Emsley, P. & Cowtan, K. (2004). Coot: model-building tools for molecular graphics. *Acta Crystallogr.* Vol. D60 No. 12: pp. 2126-2132.

Ewing, J.G. & Maestas, S. (1970). Thermodynamics of absorption of xenon by myoglobin. *J. Phys. Chem.* Vol. 74 No. 11: pp. 2341-2344

Fourme, R., Girard, E., Kahn, R., Dhaussy, A.C., Mezouar, M., Colloc'h, N. & Ascone, I. (2006). High-pressure macromolecular crystallography (HPMX): status and prospects. *Biochim. Biophys. Acta* Vol. 1764 No. 3: pp. 384-390.

Franks, N.P. (2008). General anaesthesia: from molecular targets to neuronal pathways of sleep and arousal. *Nat. Rev. Neurosci.* Vol. 9 No. 5: pp. 370-386.

Franks, N.P., Dickinson, R., de Sousa, S.L., Hall, A.C. & Lieb, W.R. (1998). How does xenon produce anaesthesia? *Nature* Vol. 396 No. 6709: pp. 324.

Franks, N.P. & Lieb, W.R. (1984). Do general anaesthetics act by competitive binding to specific receptors ? *Nature* Vol. 310 No. 5978: pp. 599-601.

Franks, N.P. & Lieb, W.R. (1994). Molecular and cellular mechanisms of general anaesthesia. *Nature* Vol. 367 No. 6464: pp. 607-614.

Frauenfelder, H., Alberding, N.A., Ansari, A., Braunstein, D., Cowen, B.R., Hong, M.K., Iben, I.E.T., Johnson, J.B., Luck, S., Marden, M.C., Mourant, J.R., Ormos, P., reinisch, L., Scholl, R., Schiulte, A., Shyamsunder, E., Sorensen, L.B., Steinbach, P.J., Xie, A., Young, R.D. & Yue, K.T. (1990). Proteins and pressure. *J. Phys. Chem.* Vol. 94 No. 3: pp. 1024-1037.

Frauenfelder, H., Sligar, S.G. & Wolynes, P.G. (1991). The energy landscapes and motions of proteins. *Science* Vol. 254 No. 5038: pp. 1598-1603.

Girard, E., Marchal, S., Perez, J., Finet, S., Kahn, R., Fourme, R., Marassio, G., Dhaussy, A.C., Prangé, T., Giffard, M., Dulin, F., Bonneté, F., Lange, R., Abraini, J.H., Mezouar, M. & Colloc'h, N. (2010). Structure-function perturbation and dissociation of tetrameric urate oxidase by high hydrostatic pressure. *Biophys. J.* Vol. 98 No. 10: pp. 2365-2373.

Haelewyn, B., David, H.N., Colloc'h, N., Colomb, D.G., Risso, J.J. & Abraini, J.H. (2011). Interactions between nitrous oxide and tissue-plasminogen activator in a rat model of thromboembolic stroke. *Anesthesiol.* Vol. 115 No 5: pp. 1044-1053

Haelewyn, B., David, H.N., Rouillon, C., Chazalviel, L., Lecocq, M., Risso, J.J., Lemaire, M. & Abraini, J.H. (2008). Neuroprotection by nitrous oxide: facts and evidence. *Crit. Care Med.* Vol. 36 No. 9: pp. 2651-2659.

Homi, H.M., Yokoo, N., Ma, D., Warner, D.S., Franks, N.P., Maze, M. & Grocott, H.P. (2003). The neuroprotective effect of xenon administration during transient middle cerebral artery occlusion in mice. *Anesthesiol.* Vol. 99 No. 4: pp. 876-881.

Hubbard, S.J., Gross, K.H. & Argos, P. (1994). Intramolecular cavities in globular proteins. *Protein Eng.* Vol. 7 No. 5: pp. 613-626.

Imai, T., Hiraoka, R., Seto, T., Kovalenko, A. & Hirata, F. (2007). Three-dimensional distribution function theory for the prediction of protein-ligand binding sites and affinities: application to the binding of noble gases to hen egg-white lysozyme in aqueous solution. *J. Phys. Chem. B* Vol. 111 No. 39: pp. 11585-11591.

Jevtovic-Todorovic, V., Todorovic, S.M., Mennerick, S., Powell, S., Dikranian, K., Benshoff, N., Zorumski, C.F. & Olney, J.W. (1998). Nitrous oxide (laughing gas) is an NMDA antagonist, neuroprotectant and neurotoxin. *Nat. Med.* Vol. 4 No. 4: pp. 460-463.

Johansson, J.S., Manderson, G.A., Ramoni, R., Grolli, S. & Eckenhoff, R.G. (2005). Binding of the volatile general anesthetics halothane and isoflurane to a mammalian beta-barrel protein. *Febs J.* Vol. 272 No. 2: pp. 573-581.

Kabsch, W. (2010). XDS. *Acta Crystallogr.* Vol. D66 No. 2: pp. 125-132.

Kennedy, R.R., Stokes, J.W. & Downing, P. (1992). Anaesthesia and the 'inert' gases with special reference to xenon. *Anaesth. Intensive Care* Vol. 20 No. 1: pp. 66-70.

Kim, C.U., Hao, Q. & Gruner, S.M. (2007). High-pressure cryocooling for capillary sample cryoprotection and diffraction phasing at long wavelengths. *Acta Crystallogr.* Vol. D63 No. 5: pp. 653-659.

Koblin, D.D., Fang, Z., Eger, E.I., II, Laster, M.J., Gong, D., Ionescu, P., Halsey, M.J. & Trudell, J.R. (1998). Minimum alveolar concentrations of noble gases, nitrogen, and sulfur hexafluoride in rats: helium and neon as nonimmobilizers (nonanesthetics). *Anesth. Analg.* Vol. 87 No. 2: pp. 419-424.

Lee, J.M., Zipfel, G.J. & Choi, D.W. (1999). The changing landscape of ischaemic brain injury mechanisms. *Nature* Vol. 399 No. 6738 Suppl: pp. A7-14.

Leslie, A.G. (2006). The integration of macromolecular diffraction data. *Acta Crystallogr.* Vol. D62 No 1: pp. 48-57.

Li de la Sierra, I., Papamichael, E., Sakarellos, C., Dimicoli, J.L. & Prangé, T. (1990). Interaction of the peptide CF3-Leu-Ala-NH-C6H4-CF3 (TFLA) with porcine pancreatic elastase. X-ray studies at 1.8 A. *J. Mol. Recognit.* Vol. 3 No. 1: pp. 36-44.

Lo, E.H., Moskowitz, M.A. & Jacobs, T.P. (2005). Exciting, radical, suicidal: how brain cells die after stroke. *Stroke* Vol. 36 No. 2: pp. 189-192.

Luo, Y., Ma, D., Ieong, E., Sanders, R.D., Yu, B., Hossain, M. & Maze, M. (2008). Xenon and sevoflurane protect against brain injury in a neonatal asphyxia model. *Anesthesiol.* Vol. 109 No. 5: pp. 782-789.

Ma, D., Hossain, M., Chow, A., Arshad, M., Battson, R.M., Sanders, R.D., Mehmet, H., Edwards, A.D., Franks, N.P. & Maze, M. (2005). Xenon and hypothermia combine to provide neuroprotection from neonatal asphyxia. *Ann. Neurol.* Vol. 58 No. 2: pp. 182-193.

Ma, D., Yang, H., Lynch, J., Franks, N.P., Maze, M. & Grocott, H.P. (2003). Xenon attenuates cardiopulmonary bypass-induced neurologic and neurocognitive dysfunction in the rat. *Anesthesiol.* Vol. 98 No. 3: pp. 690-698.

Marassio, G., Prangé, T., David, H.N., Sopkova-de Oliveira Santos, J., Gabison, L., Delcroix, N., Abraini, J.H. & Colloc'h, N. (2011). Pressure-response analysis of anesthetic gases xenon and nitrous oxide on urate oxidase: a crystallographic study. *Faseb J.* Vol. 25 No. 7: pp. 2266-2275.

Miller, K.W. (2002). The nature of sites of general anaesthetic action. *Brit. J. Anaesthesia* Vol. 89 No. 1: pp. 17-31.

Mueller-Dieckmann, C., Polentarutti, M., Djinovic Carugo, K., Panjikar, S., Tucker, P.A. & Weiss, M.S. (2004). On the routine use of soft X-rays in macromolecular crystallography. Part II. Data-collection wavelength and scaling models. *Acta Crystallogr.* Vol. D60 No. 1: pp. 28-38.

Murshudov, G.N., Vagin, A.A. & Dodson, E.J. (1997). Refinement of macromolecular structures by the Maximum-Likelihood method. *Acta Crystallogr.* Vol. D53: pp. 240-255.

NINDS (1995). Tissue plasminogen activator for acute ischemic stroke. The National Institute of Neurological Disorders and Stroke rt-PA Stroke Study Group. *N. Engl. J. Med.* Vol. 333 No. 24: pp. 1581-1587.

Olney, J.W., Labruyere, J., Wang, G., Wozniak, D.F., Price, M.T. & Sesma, M.A. (1991). NMDA antagonist neurotoxicity: mechanism and prevention. *Science* Vol. 254 No. 5037: pp. 1515-1518.

Otwinowski, Z. & Minor, W. (1997). Processing of X-ray diffraction data collected in the oscillation mode. *Methods Enzymol.* Vol. 276: pp. 307-326.

Panjikar, S. & Tucker, P.A. (2002). Xenon derivatization of halide-soaked protein crystals. *Acta Crystallogr.* Vol. D58 No. 9: pp. 1413-1420.

Prangé, T., Schiltz, M., Pernot, L., Colloc'h, N., Longhi, S., Bourguet, W. & Fourme, R. (1998). Exploring hydrophobic sites in proteins with xenon or krypton. *Proteins* Vol. 30 No. 1: pp. 61-73.

Quillin, M.L., Breyer, W.A., Griswold, I.J. & Matthews, B.W. (2000). Size *versus* polarizability in protein-ligand interactions: binding of noble gases within engineered cavities in phage T4 lysozyme. *J. Mol. Biol.* Vol. 302 No. 4: pp. 955-977.

Russell, G.B. & Graybeal, J.M. (1992). Direct measurement of nitrous oxide MAC and neurologic monitoring in rats during anesthesia under hyperbaric conditions. *Anesth. Analg.* Vol. 75 No. 6: pp. 995-999.

Ruzicka, J., Benes, J., Bolek, L. & Markvartova, V. (2007). Biological effects of noble gases. *Physiol. Res.* Vol. 56 No. Suppl 1: pp. S39-44.

Sanders, R.D., Ma, D. & Maze, M. (2004). Xenon: elemental anaesthesia in clinical practice. *Br. Med. Bull.* Vol. 71: pp. 115-135.

Sanders, R.D. & Maze, M. (2005). Xenon: from stranger to guardian. *Curr. Opin. Anaesthesiol.* Vol. 18 No. 4: pp. 405-411.

Schiltz, M., Fourme, R., Broutin, I. & Prangé, T. (1995). The catalytic site of serine proteinases as a specific binding cavity for xenon. *Structure* Vol. 3 No. 3: pp. 309-316.

Schiltz, M., Fourme, R. & Prangé, T. (2003). Use of noble gases xenon and krypton as heavy atoms in protein structure determination. *Methods Enzymol.* Vol. 374: pp. 83-119.

Schiltz, M., Prangé, T. & Fourme, R. (1994). On the preparation and X-ray data collection of isomorphous xenon derivatives. *J. Appl. Crystallogr.* Vol. 27: pp. 950-960.

Schiltz, M., Shepard, W., Fourme, R., Prangé, T., de la Fortelle, E. & Bricogne, G. (1997). High-pressure krypton gas and statistical heavy-atom refinement: a successful combination of tools for macromolecular structure determination. *Acta Crystallogr.* Vol. D53 No. 1: pp. 78-92.

Schoenborn, B.P. (1965). Binding of xenon to horse haemoglobin. *Nature* Vol. 208 No. 5012: pp. 760-762.

Schoenborn, B.P., Watson, H.C. & Kendrew, J.C. (1965). Binding of xenon to sperm whale myoglobin. *Nature* Vol. 207 No. 992: pp. 28-30.

Takeda, K., Miyatake, H., Park, S.Y., Kawamoto, M., Kamiya, N. & Miki, K. (2004). Multi-wavelength anomalous diffraction method for I and Xe atoms using ultra-high-energy X-rays from SPring-8. *J. Appl. Crystallogr.* Vol. 37: pp. 925-933.

Tilton, R.F., Jr., Kuntz, I.D., Jr. & Petsko, G.A. (1984). Cavities in proteins: structure of a metmyoglobin-xenon complex solved to 1.9 Å. *Biochemistry* Vol. 23 No. 13: pp. 2849-2857.

Trudell, J.R. (1977). A unitary theory of anesthesia based on lateral phase separations in nerve membranes [proceedings]. *Biophys. J.* Vol. 18 No. 3: pp. 358-359.

Vitali, J., Robbins, A.H., Almo, S.C. & Tilton, R.F. (1991). Using xenon as a heavy atom for determining phases in sperm whale met-myoglobin. *J. Appl. Crystallogr.* Vol. 24 No. 5: pp. 931-935.

Yamakura, T. & Harris, R.A. (2000). Effects of gaseous anesthetics nitrous oxide and xenon on ligand-gated ion channels. Comparison with isoflurane and ethan ol. *Anesthesiol.* Vol. 93 No. 4: pp. 1095-1101.

15

Knowledge Based Membrane Protein Structure Prediction: From X-Ray Crystallography to Bioinformatics and Back to Molecular Biology

Alejandro Giorgetti[1,2] and Stefano Piccoli[1]
[1]*Applied Bioinformatics Group, Dept. of Biotechnology, University of Verona,*
[2]*German Research School for Simulation Sciences, Jülich Research Center and RWTH-Aachen University, Jülich*
[1]*Italy*
[2]*Germany*

1. Introduction

Life, or at least as we know it, would not exist without the ability of living organisms to communicate with their surroundings and respond to changes within them. Cells are able to capture and decode environmental stimuli into biologically signals. Indeed, communicating mechanisms, able to detect stimuli coming from the outside world are of fundamental importance for the survival of living beings. A deep understanding of the molecular mechanisms underlying signal transduction is thus needed for a complete characterization of the way our cells communicate with the rest of the world.

Living cells are surrounded by a plasma membrane that forms a boundary between the cell interior and the external physical world. As a consequence, the cellular plasma membrane presents a major target for environmental stimuli acting upon a living cell. The membrane contains protein molecules that confer various functions on it.

Integral membrane proteins play a key role in detecting and conveying outside signals into cells, allowing them to interact and respond to their environment in a specific manner. They are involved as main players in several signaling pathways and therefore, the majority of drug targets are associated to the cell's membrane. The original human genome sequence project estimated 20% of the total gene count of 31,778 genes to code for membrane proteins (Almen et al., 2009). Thus membrane proteins constitute a very large set of yet-to be-characterized proteins mediating all the relevant life-related functions both in prokaryotes and eukaryotes. Moreover, the total amount of membrane proteins for which the three-dimensional structure is known, is just about 842, corresponding to 298 unique proteins, as included in the Membrane Proteins with Known Structure database (http://blanco.biomol.uci.edu/mpstruc/listAll/list).

This extremely low number of membrane proteins with known structure is due to the fact that membrane proteins are very difficult to study because they are inserted into lipid bilayers surrounding the cell and its sub-compartments. In these conditions they expose to the polar outer and inner environments portions of different sizes, completely changing the biophysics with respect to soluble proteins. Thus, when isolated from membranes, membrane proteins are generally less stable than globular ones. It is therefore difficult to purify them in the native, functional form, and more difficult to crystallize them. Thus, crystallization of this type of proteins is yet a very difficult process, given the fact that they expose two different chemical-physical surfaces to the environment: water- and lipid-like. On the other hand, the lipid environment constraints the membrane protein stable folding: it is indeed evident from the structures that have been deposited so far that only all-alpha and beta-barrel structural organizations are present in nature. Indeed most of membrane proteins in the Protein Data Bank (PDB: http://www.pdb.org), i.e. 67%, consist of bundles of transmembrane helices with different tilting with respect to the membrane plane and to each other.

In the last few years, several efforts have been carried out for the determination of new crystal structures of membrane proteins. Although the improvements in the technologies allowed the determination of several structures, the gap between the known sequences and the solved structures is still enormous. Furthermore, albeit their extreme importance, crystal structure only provide a static image of a protein under conditions that sometimes are far from being physiological.

Thus, the combination of existing crystal structures, computational biology techniques and molecular biology validating experiments, may be the key to face the challenges of bridging the gap between the characterized membrane proteins with and without solved structure.

This and other issues may be resolved in the post-genomic era. To do this we should take advantage of all the theoretical efforts aiming at developing tools based on our present knowledge that are capable of extracting selected structural/functional features from known sequences/structures and of computing the likelihood of their presence in never-seen before sequences/structures. Moreover, once generated, the models can be considered as hypothesis to be tested. Thus, it is of fundamental importance to accurately validate the models and the conclusions outlined from their analysis.

Here we will review some of the efforts of the last years aimed at the characterization at the structural level of different membrane proteins for which the crystal structure is not known.

We will introduce state-of-art modelling techniques that use the most recent membrane proteins crystal structures as input for the modelling of membrane proteins in different activation states. For each of the studied cases we will review also the experiments carried out in order to validate the proposed hypothesis.

2. Protein structural bioinformatics

Proteins, as defined in the glossary of Molecular Biology of the cell, are just 'the major macromolecular constituent of cells. A linear polymer of amino acids linked together by peptide bonds in a specific sequence' (Alberts et al., 2002). The latter definition of a protein, albeit technically correct, lacks of the most important part of the definition, that is, proteins are the product of evolution. In fact, proteins are indeed polypeptides, but formed following a very precise relation of sequence-structure-function, modelled by events of random variation and natural selection along millions of years of evolution, that introduced small changes in the protein sequences passing from one generation to the other. The structure/fold of a protein is then determined by its amino acid sequence and, therefore, upon mutation of one amino acid with another may affect the structure in different ways. Taking into account the folded-protein energy landscape, it was observed that, although summing up big numbers coming from a multitude of weak interactions between their atoms, functional proteins have a limited stability, equivalent to just a few hydrogen bonds

(H-bonds). This suggests that this extremely weak folding energy landscape might be easily altered by mutations, giving place to unfolded proteins. However, as all the living beings that we know today are indeed 'alive', it means that during evolution, the structure-function relationship has been preserved, so all the proteins that we observe today can only contain mutations that did not alter too much the global folding energy landscape with respect to their ancestor sequence. From those considerations it was concluded that evolutionarily related proteins, that diverged from a common ancestor via the accumulation of small changes, cannot but have similar structure, where mutations have been accommodated only causing small local rearrangements. If the number of changes, i.e. the evolutionary distance, is high, these local rearrangements can cumulatively affect the protein structure and produce relevant distortions, but the general architecture, i.e. the fold, of the protein has to be conserved. On the other hand, if two proteins have evolved from a common ancestor, it is likely that a sufficient proportion of their sequences has remained unchanged so that an evolutionary relationship can be deduced by their comparative analysis. Therefore if we can ensure that two proteins are homologous, that is evolutionary related, the structure of one can be used as a template for the building up of the structure of the other. This forms the basis of the technique known as comparative or homology modeling (Tramontano, 2006).

In this regard, 25 years ago, Chothia and Lesk, in a seminal article (Chothia & Lesk, 1986) have aligned the sequences and structures of all the proteins with known structure, finding a correlation between the evolutionary distance and structural divergence between evolutionary related proteins. This work settled up the basis of the comparative (or homology) modeling technique, a method that allows the prediction of protein structures using as a template a member of the family for which the three-dimensional structure (3D-structure) is known. So, if we assume that the sequence alignment between two protein sequences, one of unknown (the target) and one of known (the template) structure, reflects the evolutionary relationship between their amino acids, we can assume that most of them have conserved the same relative position in the structure and use the coordinates of the backbone of the template as first approximations of the coordinates of the backbone of the target (Tramontano, 2006). We must then model the conformation of the side chains and the local rearrangements of the structure brought about by the amino acid substitutions (Tramontano, 2006). Templates for a comparative model would be found among the structures present in databases, i.e. PDB and can be extracted by searching the database for proteins putatively homologous to the target protein.

All these ideas were shown to be valid through the years for soluble proteins. Indeed, the comparative modeling technique has not always been considered valid when applied to membrane proteins (Floriano et al., 2006). The main criticism regarded the low amount of membrane proteins with known three-dimensional structure and the enormous evolutionary gap that must be filled in order to produce models of the most studied membrane proteins. However, in the last few years X-ray crystallography reached very high levels of applicability in the field of membrane proteins and a lot of new and more refined structures were solved by several groups. In the applications section we will review some of the most relevant structures and how they allowed a better characterization of the mechanisms underlying the function. Indeed, this great advancement in the crystallographic techniques produced as of today 298 structures of unique membrane proteins for a total of 842 structures. At this point, we have checked for the existence of correlation between evolutionary relationship and structure similarity. We have recently (not published) followed the same protocol as Chothia and Lesk (Chothia & Lesk, 1986) but considering just membrane proteins. Indeed, by the use of the LGA server (http://proteinmodel.org/) we have aligned (the structures and sequences) present in the core of all the membrane proteins with known three-dimensional structure and produced the graph of Figure 1. The structural divergence between two evolutionary correlated proteins is measured as their Root Mean Square Deviation (RMSD).

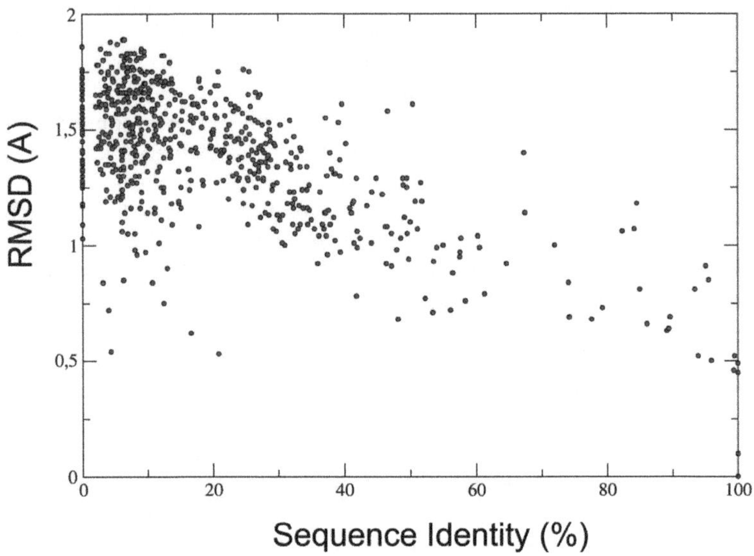

Fig. 1. RMSD versus Percent Sequence Identity of membrane proteins. The core of all the membrane proteins found in the Membrane Proteins with Known Structure Database (http://blanco.biomol.uci.edu/mpstruc/listAll/list) was aligned at the structural level and at the sequence level.

Figure 1 shows the RMSD of the backbone of the core of pairs of evolutionarily related proteins as a function of the percent of identity between their amino acid sequences. The definition of the "core" of the structure differs in different methods. It can be intuitively seen as the internal, closely packed, evolutionary conserved part of the structure that contains most of the repetitive secondary structure elements (Tramontano, 2006). For practical purposes we considered as the core of the proteins, all the amino acids present in secondary structure elements and those regions not diverging for more than 3 Å, as Chothia and Lesk did (Chothia & Lesk, 1986).

As stated before, Comparative Modeling is based on the idea that evolutionary correlated proteins share similar three-dimensional structures. That is, if we want to predict the structure of a protein we can look in database for an evolutionary correlated protein with known structure, and use the latter as template for building up the structural model of our preferred protein (Tramontano, 2006). The important thing is that, based on Figure 1 the procedure is valid also for membrane proteins. In this regard, the astonishing improvements in membrane proteins crystallography together with comparative modeling techniques will allow the characterization of an enormous amount of membrane protein in the near future.

3. Applications

In the following sections we will present two cases in which homology modelling has been applied on membrane proteins, i.e. ion channels and G-protein-coupled receptors. Both cases are representative icons of the difficulties found in the structure solving and modelling of membrane proteins for many years. Fortunately, in the last few years there was an explosion of newly solved crystal structures that completely revolutionised the field. Indeed, several mechanisms were understood and functional features could be extended to several members of the families.

We will review in both cases the advancements in X-ray crystallography and how we have used the recently solved crystal structures combined with homology modelling and molecular biology experiments to characterize functional mechanisms.

3.1 Ion channels

Ion channels are integral membrane proteins that function as molecular sensors of physical and chemical stimuli and convert these stimuli into biological signals vital for the existence of every living organism. In other words, ion channels represent the doors and windows of the cell, that open and close following precise stimuli and leave the entrance/exit of very accurately selected 'visitors'. As molecular transducers of mechanical, electrical, chemical, thermal or electromagnetic (light) stimuli, ion channels contribute to changes in electrical, chemical or osmotic activity within cells by gating between the two basic conformations in which they exist - open and closed. Through the gating mechanisms, i.e. opening and closing, ion channels regulate the permeation of ions (in some cases also other solutes), allowing ions to cross the hydrophobic core of the cell membrane, affecting its activity. Because of the well-known difficulties in obtaining high resolution 3D structures by X-ray crystallography of ion channels, alternative strategies based on computational biology tools are currently used to investigate their biophysical properties (for a review about ion channel modelling see: Giorgetti & Carloni, 2003).

The last two decades have been exceptionally exciting for research in the field of ion channels. Astonishing progress has resulted from the use of multidisciplinary approaches to gain insight into the structure and function of ion channels and their role in various aspects of cell physiology and signal transduction. Molecular biology and genetics have provided the sequences of a very large number of ion channel proteins and have helped identify their contribution to various cellular functions. The patch clamp technique has provided the means to study the functional properties of single ion channels with unprecedented precision. X-ray and electron crystallography have provided structural snapshots of a number of ion channel molecules at near atomic resolution, whereas magnetic resonance spectroscopy and fluorescence spectroscopy have provided means to access the dynamics of these molecules. In detail, as of today we count with 14 unique ion channel structures (http://blanco.biomol.uci.edu/mpstruc/listAll/list) for a total of about 45 crystal structures of ion channels solved in different activation states and co-crystallized with different ligands and ions. Moreover, more than 2/3 of the solved structures were obtained in the last two years, showing an exponential development of the field.

Using the structural and functional information obtained by these experimental techniques, computer-assisted molecular modeling has brought ion channels to life by allowing the features underlying the molecular events that shape their function. Indeed, the multidisciplinary approach to the study of ion channels has yielded an unprecedented wealth of new data.

In the next sections examples of applications on a specific ion channel will be illustrated.

3.1.1 CNG channels

Cyclic nucleotide-gated (CNG) channels are ion channels that generally express in several sensory and non-sensory cells (Kaupp & Seifert, 2002; Matulef & Zagotta, 2003). The most characterized members of the family are those involved in sensory transduction in vertebrate photoreceptors and in olfactory sensory neurons (Kaupp & Seifert, 2002; Matulef & Zagotta, 2003). In their native forms, CNG channels are heterotetramers (Kaupp & Seifert, 2002; Matulef & Zagotta, 2003). CNG channels are differentially sensitive to cyclic nucleotides (CNs): CNG channels from vertebrate rod photoreceptors are selectively activated by cGMP binding, whereas CNG from olfactory neurons are activated by either cAMP or cGMP binding.

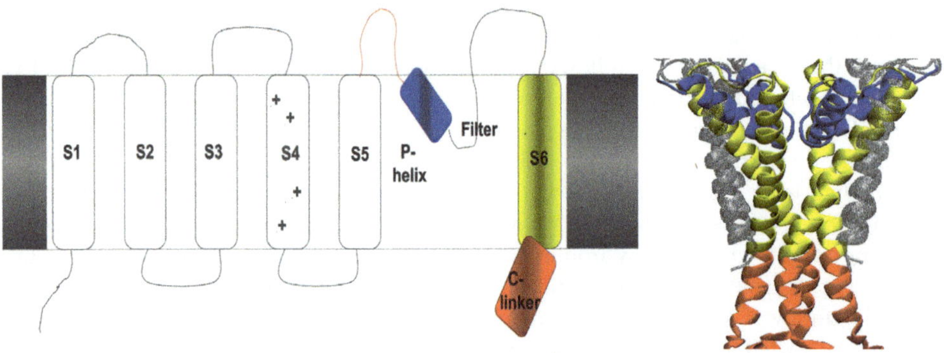

Fig. 2. Left panel: Topology of CNG channels. The plot shows the topological relevant elements of CNG channels analyzed in the main text. In particular P-helix was depicted in blue while the S6 helix and the C-linker, are shown in yellow and red, respectively. The C-linker domain, connects the transmembrane domain with the cyclic nucleotide binding domain. S4 is the positively-charged voltage sensor of the channel. The conformational changes proposed regards C-linker, S6 transmembrane helix and the filter P-helix. Right panel: the figure shows the three dimensional configuration of the tetrameric filter region of CNG channels, as obtained through the modeling procedure.

Several very important groups worked for years, accumulating electrophysiological data for mutants and wild-type CNG channels in different activation states, in order to characterize the biophysical properties underlying the functioning of these channels. In particular, most of the experimental available information regard electrophysiological investigations with mutant channels, performed on the CNGA1 channels from bovine rods (Craven & Zagotta, 2006; Kaupp & Seifert, 2002; Matulef & Zagotta, 2003), a channel with a primary structure composed of 690 residues (Kaupp et al. 1989). Although the analysis of accessibility, based on Cysteine Scanning Mutagenesis (CSM) (Akabas et al., 1992; Karlin & Akabas, 1998) has shown that CNG and potassium channels share the same gross topology (Bechetti et al. 1999; Flynn & Zagotta, 2001; Giorgetti et al., 2005; Liu & Siegelbaum, 2000; Matulef & Zagotta, 2003; Mazzolini et al., 2009; Nair et al., 2009) , these two families of channels have different functional properties. Indeed, while voltage-gated potassium channels are extremely selective (MacKinnon, 2003) and their gating strongly depends on membrane voltage (Bezanilla, 2005; Bezanilla, 2008; Swartz, 2004; Swartz, 2008) , CNG channels have a low ion selectivity and their gating is only poorly voltage dependent (Kaupp & Seifert, 2002).

Here, we review some of the recent years results of an extensive combined experimental/computational structural study on the widely characterized homotetrameric cyclic nucleotide-gated channel (CNG), from *bovine rod*. As stated before, CNG channels are tetrameric and each subunit consists of two domains arranged as shown in Figure 2: (i) a transmembrane domain formed by six transmembrane helices (S1-S6) and a pore helix (P-helix) with the same topology of voltage-gated potassium channels (Beccheti et al., 1999; Sesti et al., 1995); (ii) A cytoplasmic domain formed by the cyclic nucleotide binding domain (CNBD) which is linked to the transmembrane domain through the so called C-linker region. The pore, un-selective for sodium and potassium, is believed to gate via a conformational change of S6 transmembrane helix (TMH) initiated by the binding of cyclic nucleotides to the binding domains. This conformational change is then transmitted to the pore via coupling with the four P-helices (Johnson & Zagotta, 2001; Matulef et al., 1999). During the last 5 years, we and our collaborators have provided the molecular basis for the characterization of the mechanisms underlying the functioning of these channels by constructing homology models of the transmembrane region of the CNGA1 channel. The models include the S6, P-helix-loop (P-helix+pore wall or filter), along with the C-linker N-terminal sections. Indeed, all the modeled regions were extensively characterized by a great amount of experimental data.

Models of P-helix-loop and S6 are based on the KcsA X-ray structure, whose topology has been suggested to be similar to that of CNG channels (Beccheti et al., 1999). On the other hand, the C-linker domain was modeled using the C-linker of the *mouse* Hyperpolarization-activated and Cyclic-nucleotide-modulated (mHCN) channel in its ligand bound state, for which the X-ray structure has been recently solved (Zagotta et al., 2003).

The homology models, were then refined by the inclusion of an extensive dataset of spatial constraints inferred by electrophysiological measurements on cysteine mutants. A large set (about 50) of structural constraints among Cα atoms were inferred from measurements of the electrophysiological properties of the channel in the presence of metal ions (Beccheti et al., 1999; Becchetti & Roncaglia, 2000; Flynn & Zagotta, 2001; Johnson & Zagotta, 2001; Liu & Siegelbaum, 2000; Mazzolini et al., 2010; Nair et al., 2009). For example (i) Cadmium, which can block the channel when it binds to, at least, two cysteine residues. (ii) The mild-oxidizing agent copper phenanthroline (CuP) favors disulfide bridge formation between two cysteines separated by a distance going from 6 to 11 Å. The electrophysiological results were then converted into distance constraints by a statistical analysis of the PDB. Indeed, we have looked for all the proteins co-crystallized with cadmium atoms and we have extracted all the mean distances from cysteines bound to the cadmium atoms. Then, taking into account that the channels are homotetramers, the reversible/irreversible condition of the cadmium blockage was converted into distance restraints by just geometrical considerations. These and other agents were included in the solutions to characterize different features of the channel mechanisms (Mazzolini et al., 2010; Nair et al., 2009). The procedure followed by us and by our experimental collaborators, consisted in a series of iterative steps that extended for more than three years of successive cycles of modeling followed by experiments and vice versa.

This experiment-guided computational model allowed us to gain insights into the structural basis of CNG channel gating mechanism. We have suggested several mechanical features underlying channel functioning. For example we have hypothesized the bending and the counterclockwise rotation of the C-linker N-terminal section. Indeed, this motion is suggested to be transmitted upwards to cause the upper part of S6 to rotate counterclockwise producing the conformational changes needed for the opening and closing of the filter region (Giorgetti et al., 2005). The procedure that allowed to unravel the functional gating mechanisms was characterized by an iterative theoretical/experimental work that permitted not only the hypotheses generation but also their experimental validation. Indeed, on the basis of our models, and using cysteine-scanning mutagenesis, our experimental collaborators were also able to show that in the presence of a mild oxidizing agent, copper phenanthroline (CuP), certain cysteine mutations were able to lock the channel in either the closed or open state, depending on whatever state they happened to be in at the time of CuP application (Nair et al., 2006).
This kind of work has very few precedents due to the general difficulties found in the expression of different mutants. Indeed, some of the suggested mutants included double and triple mutations for a single protein.

3.2 GPCRs

G protein-coupled receptors (GPCRs), or 7 transmembrane helix receptors (Figure 3), are membrane embedded proteins, responsible for the communication between the cell and the environment (Sakmar et al. 2002). Malfunction of these receptors are generally involved in many major diseases, thus making GPCR receptors one of the most exploited targets for the pharmaceutical industry (Schertler, 1998). About 5500 GPCR sequences are publicly available. The total number of GPCRs with and without introns in the human genome has been estimated to be approximately 900, of which 500 are odorant or taste receptors and 450 are receptors for endogenous ligands (Takeda et al., 2002). Binding constants are available for approximately 30000 ligand–receptor combinations (Horn et al., 1998).
This wealth of sequences, ligands, and mutations are in net contrast with the small amount of structural information available. Indeed, till a few years ago, rhodopsin was the only structurally characterized GPCR and is still considered a prototypical member of the superfamily. The first X-ray structure of rhodopsin reflected the dark adapted ground-state of the bovine receptor, captured in four different crystals of 2.8 Å (Palezewski et al., 2000), 2.65 Å (Li et al., 2004) and 2.2 Å (Okada et al., 2004), 3.4 Å (Standfus et al., 2007), 4.15 Å (Salom et al., 2006). Very recently the structure of a GPCR in its empty state, Opsin, has been crystallized (Park et al., 2008), opening in this way an exciting new era in the studies of GPCRs. This new structure of rhodopsin followed another very recent key event, the crystallization of a second GPCR, the beta2-adrenergic receptor (Bokoch et al., 2010; Cherezov et al., 2007; Hanson et al., 2008; Rasmussen 2007).
Nearly all medicines are discovered by trial and error. Nevertheless, most pharmaceutical companies have large research departments that use every imaginable technique to design drugs. Homology modelling, as a tool to obtain structural information, is one of those techniques. In the past, bacteriorhodopsin (Henderson & Schertler, 1990; Luecke et al., 1998; Pebay-Peyroula et al., 1997; Takeda et al., 1998) was often used as a modelling template, but from 2000 the three-dimensional coordinates (Palezewski et al., 2000) of bovine rhodopsin have become available. Along several years, it was shown to be a much better template for GPCR homology modelling than bacteriorhodopsin.

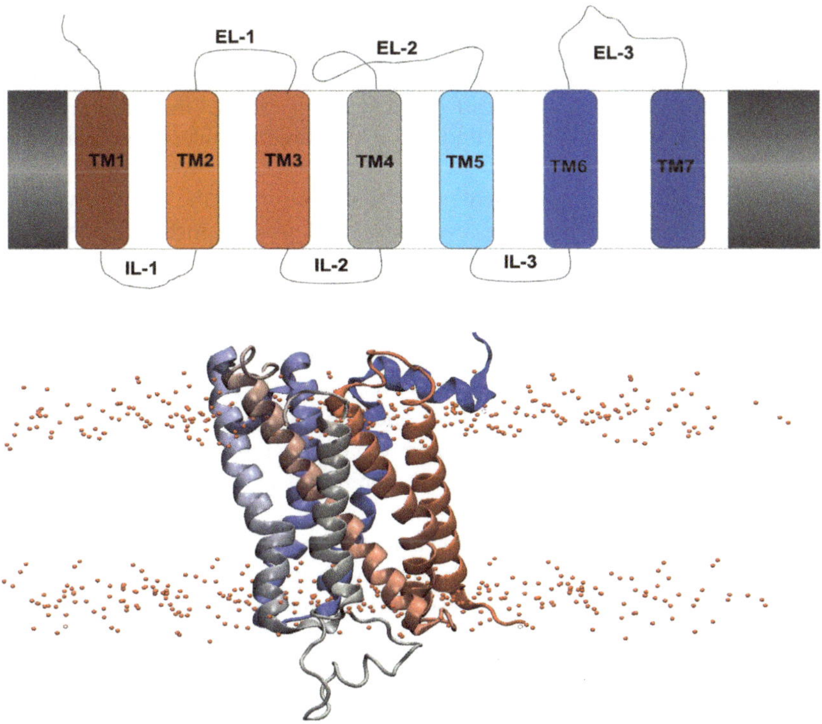

Fig. 3. Upper panel. Topology of GPCR receptors. TM, EL and IL are transmembrane helix, extracellular loops and intracellular loops, respectively. Lower panel: The three dimensional distribution of a GPCR: model of the TAS2R38 receptor.

As can be appreciated from Figure 3, structurally GPCRs are characterized by an extracellular N-terminus, followed by seven transmembrane (7-TM) α-helices (TM-1 to TM-7) connected by three intracellular (IL-1 to IL-3) and three extracellular loops (EL-1 to EL-3), and finally an intracellular C-terminus. The GPCR arranges itself into a tertiary structure resembling a barrel, with the seven transmembrane helices forming a cavity within the plasma membrane that serves a ligand-binding domain that is often covered by EL-2.

Rhodopsin has been for several years an extraordinarily valuable system for understanding the structure and mechanism of activation of G-protein-coupled receptors (GPCRs). Rhodopsin is highly specialized for the detection of light, exhibiting functional and biochemical characteristics that differentiate it from GPCRs expressed in other tissues such as those specialized in detecting diffusible hormones and neurotransmitters. Crystal structures have recently been determined also for the human β2 adrenoreceptor (β2AR) (Bokoch et al., 2010; Cherezov et al., 2007; Hanson et al., 2008; Rasmussen 2007), a receptor for adrenalin and noradrenalin that is involved in the regulation of cardiovascular and pulmonary function by the sympathetic nervous system. β2AR was the first non-rhodopsin GPCR to be cloned and is one of the most extensively studied members of this family (Lefcowitz, 2000). These structures provide the keys for a highly expected way for compare and contrast with rhodopsin.

The structures of the β2AR provide new exciting insights into the mechanisms of activation in several ways. On the other hand, there exist a huge amount of mutagenesis, biophysical and computational data for the β2AR and closely related receptors. The new structures provided us with a structural scaffold that may allow the interpretation and validation/rejection of these studies and for generating testable hypothesis for future studies.

One of the most important achievements of crystallography in the last few years, was to provide the community with a way of comparing rhodopsin against other members of the superfamily with different functions. Indeed, rhodopsin evolved for the efficient detection of light: it is present in only one organ and serves only one purpose. In the dark it has almost no activity toward its corresponding G protein, but just one photon can photoisomerize its covalently bound ligand, retinal, changing it from an inverse agonist to a full agonist. By contrast, the β2AR, like many other GPCRs, has a broader range of signalling behaviour: coupling to more than one G protein and to G protein independent pathways, and responding to a large spectrum of cognate molecules (Lefcowitz & Shenoy, 2005). Comparing the structures can help in gaining insights into the structural basis for these functional differences. In rhodopsin, the binding pocket, specialized in covalently binding of retinal, is hindered by a β sheet lid formed by the second extracellular loop (ECL2), and a small domain formed by the N terminus, protecting *cis*-retinal from hydrolysis. This structure would limit access for diffusible agonists and is not present in the β2AR. By contrast, in the β2AR ECL2 forms a helix that is constrained by two disulfide bonds such that there is open access to the ligand-binding pocket.

Very recently the structures of two new GPCR were solved, i.e. the human adenosine A2 receptor (Jakola et al., 2008; Lebon et al., 2011; Xu et al., 2011;) and the human beta1-adrenergic receptor, β1AR, (Moukhametzianov et al. 2011; Warne et al., 2008; Warne et al., 2011), giving also valuable information that allowed the generalization of several structural/functional features conserved along the families. Indeed, all the available structural information, along with experiments coming from the molecular biology and functional assays combined with extensive computational calculations were applied by us and collaborators in order to unravel the binding site and the gating mechanisms of a very particular family of GPCRs, i.e. the bitter taste receptors (Biarnés et al., 2010).

3.2.1 Bitter taste receptors

Mammals, have been prevented, during evolution, from ingesting toxic compounds because of their strong bitter taste (Behrens & Meyerhof, 2009; Meyerhof, 2005; Mueller et al. 2005; Soranzo et al., 2005). This protection mechanism has been carried out for millions of years by a family of about 30 bitter taste receptors (TAS2Rs) expressed in taste receptor cells (Adler et al., 2000; Behrens et al., 2007; Chandrashekar et al., 2000; Matsunami et al. 2000; Shi & Zhang, 2006). TAS2Rs was shown to belong to the super family of GPCR receptors, albeit their low sequence identity with rhodopsin, for example (Adler et al., 2000; Chandrashekar et al., 2000; Matsunami et al., 2000). The binding of a bitter compound to its cognate target TAS2R, is able to fire a downstream cascade of events inside the cell, typical of GPCRs signaling pathways (Chandrashekar et al., 2006), leading to the production of an electrical signal, i.e. bitter taste perception (Behrens & Meyerhof, 2009). Thus, albeit natural selection decreased its constraints in this sense, i.e. we are not walking around tasting plants that we do not know, the bitter taste is becoming more and more important regarding food taste, cuisine and pharmaceutics matters. Thus, a complete characterization of the events giving rise to taste perception is needed. One of the most interesting bitter taste receptors, also from the evolutionary point of view, is the human receptor for phenylthiocarbamide (PTC) and propylthiouracil (PROP) molecules, i.e. TAS2R38 receptor. Indeed, within the polymorphisms present in this receptor, the most pronounced ones affect its perception of the PROP/PTC. In fact, differences in the perception of PROP and PTC has divided the human population into tasters and non-tasters. Albeit these astonishing results were reported in the early nineties, polymorphisms in the hTAS2R38 gene underlying the observed phenotype were identified only recently by Kim and coworkers (Kim et al., *2003*). Indeed, the taster/non-taster quality originate in three hTAS2R38 non-synonymous polymorphisms, i.e. the haplotypes code for either the amino acids PAV (P49, A262, I296) constituting the taster variant of hTAS2R38 or AVI in the corresponding positions for the non-taster variant (Bufe et al., 2005; Kim et al., 2003). Up to now, the molecular/structural basis of bitter taste sensing were analyzed by very few studies, i.e. on hTAS2R16 and hTAS2R38 relied on computations only (Floriano et al., 2006; Miguet et al., 2006).

In addition, three experimentally guided structure-activity studies are available now, which all addressed hTAS2Rs distantly related to hTAS2R38 (Brockhoff et al., 2010; Pronin et al., 2004; Sakurai et al., 2010). First principle (Floriano et al., 2006) and homology modeling approaches based on bovine rhodopsin (Miguet et al., 2006) have been used to predict the structure of the widely studied bitter taste receptor hTAS2R38 (Bufe et al., 2005; Khafizov et al., 2007; Kim et al., 2003; Kleinau et al., 2007). Both works coincide in the fact that more computational refinement and/or experimental validations are needed.

Very recently (Biarnés et al., 2010) we have used a combined experimental/computational iterative approach with the aim at identifying hTAS2R38 residues involved in binding to one of its main agonists, i.e. PTC, as well as in receptor activation. We used state-of-the-art bioinformatics approaches based on multiple sequence alignment across the whole family of GPCRs combined with structural bionformatics tools; we also used the homology modeling techniques because sequences at their own were not likely to be sufficient to identify residues in the binding site, as ligands pockets vary largely in position and orientation across this family (Jaakola et al., 2008). Furthermore extensive virtual docking experiments were carried out to predict the putative binding cavities for PTC. In fact, homology modeling and molecular docking has been shown to guide satisfactorily the design of site-directed mutagenesis experiments, in spite of the little power of the structural predictions (Ballesteros & Weinstein, 1992). Indeed, the proposed receptor positions were then studied and validated/rejected by site-directed mutagenesis experiments and measurements of receptor activation by recording intracellular calcium levels following agonist administration.

We thus have proposed that hTASR38 activation upon PTC binding is reminiscent of the transition of the G-protein/opsin complex to free rhodopsin (Scheerer et al., 2008). Indeed, we were able to identify some of the residues directly involved in the interaction with the ligand, those that define the shape of the binding cavity and, more important, we were able to identify the residues participating in receptor activation. In our model, TMs 5, 6 and 7 change conformation upon ligand binding, in particular TM6 tilts around the helical bundle upon G-protein binding. Similar sequences of events also have been suggested to play a role for activation of all GPCRs (Altenbach et al., 2008).

4. Conclusion

Membrane proteins, as we saw, are of fundamental importance for the survival of any living being. A deep insight into the molecular mechanisms underlying their function is thus needed for a complete characterization of the way our cells communicate with the rest of the world. Obviously, a complete characterization implies the passage through the structural features of the membrane proteins. Unfortunately, as of today, the structural biology scientific community still finds several inconvenients to systematically solve the structure of membrane proteins, albeit giant steps forward were carried out in the last few years. Moreover, we have to consider that the main challenges for the near future will include the development and application of methods that permit the full description at the molecular/structural level of large protein complexes, most of all including membrane proteins with unknown structure.

In this chapter we have described two examples for which an experimental/computational multidisciplinary approach was shown to be the key for the gaining of insights into complex systems. In both cases not only 'static' structural elements have been identified, but also putative dynamical mechanisms, comprising large conformational changes. Indeed, we have demonstrated that advancements in experimental structural biology, extensively combined with state-of-art computational biology tools and model-guided molecular biology experiments may become extremely effective for the characterization of complex molecular mechanisms including membrane proteins. These observations make us confident that more difficult challenges of structural/functional characterization can be undertaken in short time and that these approaches may provide a great improvement to our understanding of cell and molecular biology events.

5. Acknowledgements

We deeply acknowledge M. Garonzi, M. Zucchelli, A. Pizzolato, C. Grigoli, A. Turati, M. Denitto, A. Atzeni, A. Bazaj, V. Marino, S. Compri, M. I. Muddei and G. Tosadori for the excellent work done for obtaining Figure 1.

6. References

Adler,E. et al. (2000) A novel family of mammalian taste receptors. *Cell*, 100, 693-702.

Akabas,M.H. et al. (1992) Acetylcholine receptor channel structure probed in cysteine-substitution mutants. *Science*, 258, 307-310.

Alberts B., Johnson A., Lewis J., Raff M., Roberts K., and Walter P. (2002) *Molecular Biology of the Cell (4th edition),* Garland Science, ISBN-10: 0-8153-3218-1, New York.

Almén,M.S. et al. (2009) Mapping the human membrane proteome: a majority of the human membrane proteins can be classified according to function and evolutionary origin. *BMC Biol*, 7, 50.

Altenbach,C. et al. (2008) High-resolution distance mapping in rhodopsin reveals the pattern of helix movement due to activation. *Proc. Natl. Acad. Sci. U.S.A*, 105, 7439-7444.

Ballesteros,J A and Weinstein,H. (1992) Analysis and refinement of criteria for predicting the structure and relative orientations of transmembranal helical domains. *Biophys. J*, 62, 107-109.

Becchetti,A. et al. (1999) Cyclic nucleotide-gated channels. Pore topology studied through the accessibility of reporter cysteines. *J. Gen. Physiol*, 114, 377-392.

Becchetti,A. and Roncaglia,P. (2000) Cyclic nucleotide-gated channels: Intra- and extracellular accessibility to Cd2+ of substituted cysteine residues within the P-loop. *Pflugers Archiv European Journal of Physiology*, 440, 556-565.

Behrens, M. and Meyerhof,W. (2009) Mammalian bitter taste perception. *Results Probl Cell Differ*, 47, 203-220.

Behrens, M. et al. (2007) Gustatory expression pattern of the human TAS2R bitter receptor gene family reveals a heterogenous population of bitter responsive taste receptor cells. *J. Neurosci*, 27, 12630-12640.

Bezanilla,F. (2008) How membrane proteins sense voltage. *Nat. Rev. Mol. Cell Biol*, 9, 323-332.

Bezanilla,F. (2005) Voltage-gated ion channels. *IEEE Trans Nanobioscience*, 4, 34-48.

Biarnés,X. et al. (2010) Insights into the binding of Phenyltiocarbamide (PTC) agonist to its target human TAS2R38 bitter receptor. *PLoS ONE*, 5, e12394.

Bokoch,M.P. et al. (2010) Ligand-specific regulation of the extracellular surface of a G-protein-coupled receptor. *Nature*, 463, 108-112.

Brockhoff,A. et al. (2010) Structural requirements of bitter taste receptor activation. *Proc. Natl. Acad. Sci. U.S.A*, 107, 11110-11115.

Bufe,B. et al. (2005) The molecular basis of individual differences in phenylthiocarbamide and propylthiouracil bitterness perception. *Curr. Biol*, 15, 322-327.

Chandrashekar,J et al. (2000) T2Rs function as bitter taste receptors. *Cell*, 100, 703-711.

Chandrashekar,J et al. (2006) The receptors and cells for mammalian taste. *Nature*, 444, 288-294.

Cherezov,V. et al. (2007) High-resolution crystal structure of an engineered human beta2-adrenergic G protein-coupled receptor. *Science*, 318, 1258-1265.

Chothia,C. and Lesk,A.M. (1986) The relation between the divergence of sequence and structure in proteins. *EMBO J*, 5, 823-826.

Craven,K.B. and Zagotta,W. N. (2006) CNG and HCN channels: two peas, one pod. *Annu. Rev. Physiol*, 68, 375-401.

Floriano,W.B. et al. (2006) Modeling the human PTC bitter-taste receptor interactions with bitter tastants. *J Mol Model*, 12, 931-941.

Flynn,G.E. and Zagotta,W. N. (2001) Conformational changes in S6 coupled to the opening of cyclic nucleotide-gated channels. *Neuron*, 30, 689-698.

Giorgetti,A. and Carloni,P. (2003) Molecular modeling of ion channels: structural predictions. *Curr Opin Chem Biol*, 7, 150-156.

Giorgetti,A., Nair,A.V., et al. (2005) Structural basis of gating of CNG channels. *FEBS Lett*, 579, 1968-1972.

Giorgetti,A., Raimondo,D., et al. (2005) Evaluating the usefulness of protein structure models for molecular replacement. *Bioinformatics*, 21 Suppl 2, ii72-76.

Hanson,M.A. et al. (2008) A specific cholesterol binding site is established by the 2.8 A structure of the human beta2-adrenergic receptor. *Structure*, 16, 897-905.

Henderson,R. and Schertler,G.F. (1990) The structure of bacteriorhodopsin and its relevance to the visual opsins and other seven-helix G-protein coupled receptors. *Philosophical transactions of the Royal Society of London. Series B: Biological sciences*, 326, 379-389.

Horn,F. et al. (1998) GPCRDB: An information system for G protein-coupled receptors. *Nucleic Acids Research*, 26, 275-279.

Jaakola,V. P. et al. (2008) The 2.6 angstrom crystal structure of a human A2A adenosine receptor bound to an antagonist. *Science*, 322, 1211-1217.

Johnson,J.P. and Zagotta,William N. (2001) Rotational movement during cyclic nucleotide-gated channel opening. *Nature*, 412, 917-921.

Karlin, A. and Akabas,M.H. (1998) Substituted-cysteine accessibility method. *Meth. Enzymol*, 293, 123-145.

Kaupp,U.B. et al. (1989) Primary structure and functional expression from complementary DNA of the rod photoreceptor cyclic GMP-gated channel. *Nature*, 342, 762-766.

Kaupp,U.B. and Seifert,R. (2002) Cyclic nucleotide-gated ion channels. *Physiol. Rev*, 82, 769-824.

Khafizov,K. et al. (2007) Ligand specificity of odorant receptors. *J Mol Model*, 13, 401-409.

Kim,U. K. et al. (2003) Positional cloning of the human quantitative trait locus underlying taste sensitivity to phenylthiocarbamide. *Science*, 299, 1221-1225.

Kleinau,G. et al. (2007) Implications for molecular mechanisms of glycoprotein hormone receptors using a new sequence-structure-function analysis resource. *Mol. Endocrinol*, 21, 574-580.

Lebon,G. et al. (2011) Agonist-bound adenosine A2A receptor structures reveal common features of GPCR activation. *Nature*, 474, 521-525.

Lefkowitz,R.J. (2000) The superfamily of heptahelical receptors. *Nature Cell Biology*, 2, E133-E136.

Lefkowitz,R.J. and Shenoy,S.K. (2005) Transduction of receptor signals by β-arrestins. *Science*, 308, 512-517.

Li,J. et al. (2004) Structure of bovine rhodopsin in a trigonal crystal form. *J. Mol. Biol*, 343, 1409-1438.

Liu,J. and Siegelbaum,S.A. (2000) Change of pore helix conformational state upon opening of cyclic nucleotide-gated channels. *Neuron*, 28, 899-909.

Luecke,H. et al. (1998) Proton transfer pathways in bacteriorhodopsin at 2.3 angstrom resolution. *Science*, 280, 1934-1937.

MacKinnon,R.(2003) Potassium channels. *FEBS Lett*, 555, 62-65.

Matsunami,H. et al. (2000) A family of candidate taste receptors in human and mouse. *Nature*, 404, 601-604.

Matulef,K. et al. (1999) Molecular rearrangements in the ligand-binding domain of cyclic nucleotide-gated channels. *Neuron*, 24, 443-452.

Matulef,K. and Zagotta,W. N. (2003) Cyclic nucleotide-gated ion channels. *Annu. Rev. Cell Dev. Biol*, 19, 23-44.

Mazzolini,M., Anselmi,C., and Torre,V. (2009a) The analysis of desensitizing CNGA1 channels reveals molecular interactions essential for normal gating. *J. Gen. Physiol*, 133, 375-386.

Mazzolini,M., Marchesi,A., Giorgetti,A., and Torre,Vincent (2010a) Gating in CNGA1 channels. *Pflugers Arch*, 459, 547-555.

Meyerhof,W. (2005) Elucidation of mammalian bitter taste. *Rev. Physiol. Biochem. Pharmacol*, 154, 37-72.

Miguet,L. et al. (2006) Computational studies of ligand-receptor interactions in bitter taste receptors. *J. Recept. Signal Transduct. Res*, 26, 611-630.

Moukhametzianov,R. et al. (2011) Two distinct conformations of helix 6 observed in antagonist-bound structures of a beta1-adrenergic receptor. *Proc. Natl. Acad. Sci. U.S.A*, 108, 8228-8232.

Mueller,K. L. et al. (2005) The receptors and coding logic for bitter taste. *Nature*, 434, 225-229.

Nair,A.V. et al. (2006) Locking CNGA1 channels in the open and closed state. *Biophys. J*, 90, 3599-3607.

Nair,A.V. et al. (2009) Conformational rearrangements in the S6 domain and C-linker during gating in CNGA1 channels. *Eur. Biophys. J*, 38, 993-1002.

Okada,T. et al. (2004) The retinal conformation and its environment in rhodopsin in light of a new 2.2 A crystal structure. *J. Mol. Biol*, 342, 571-583.

Palczewski,K et al. (2000) Crystal structure of rhodopsin: A G protein-coupled receptor. *Science*, 289, 739-745.

Park,J.H. et al. (2008) Crystal structure of the ligand-free G-protein-coupled receptor opsin. *Nature*, 454, 183-187.

Pebay-Peyroula,E. et al. (1997) X-ray structure of bacteriorhodopsin at 2.5 angstroms from microcrystals grown in lipidic cubic phases. *Science*, 277, 1676-1681.

Pronin,A.N. et al. (2004) Identification of ligands for two human bitter T2R receptors. *Chem. Senses*, 29, 583-593.

Rasmussen,S.G.F. et al. (2007) Crystal structure of the human beta2 adrenergic G-protein-coupled receptor. *Nature*, 450, 383-387.

Sakmar,T.P. (2002) Structure of rhodopsin and the superfamily of seven-helical receptors: the same and not the same. *Curr. Opin. Cell Biol*, 14, 189-195.

Sakurai,T. et al. (2010) Characterization of the beta-D-glucopyranoside binding site of the human bitter taste receptor hTAS2R16. *J. Biol. Chem*, 285, 28373-28378.

Salom,D. et al. (2006) Crystal structure of a photoactivated deprotonated intermediate of rhodopsin. *Proc. Natl. Acad. Sci. U.S.A*, 103, 16123-16128.

Scheerer,P. et al. (2008) Crystal structure of opsin in its G-protein-interacting conformation. *Nature*, 455, 497-502.

Schertler,G.F. (1998) Structure of rhodopsin. *Eye (Lond)*, 12 (Pt 3b), 504-510.

Sesti,F. et al. (1995) The multi-ion nature of the cGMP-gated channel from vertebrate rods. *Journal of Physiology*, 487, 17-36.

Shi,P. and Zhang,J. (2006) Contrasting modes of evolution between vertebrate sweet/umami receptor genes and bitter receptor genes. *Mol. Biol. Evol*, 23, 292-300.

Soranzo,N. et al. (2005) Positive selection on a high-sensitivity allele of the human bitter-taste receptor TAS2R16. *Curr. Biol*, 15, 1257-1265.

Standfuss,J. et al. (2007) Crystal structure of a thermally stable rhodopsin mutant. *J. Mol. Biol*, 372, 1179-1188.

Swartz,K.J. (2008) Sensing voltage across lipid membranes. *Nature*, 456, 891-897.

Swartz,K.J. (2004) Towards a structural view of gating in potassium channels. *Nat. Rev. Neurosci*, 5, 905-916.

Takeda,K. et al. (1998) A novel three-dimensional crystal of bacteriorhodopsin obtained by successive fusion of the vesicular assemblies. *Journal of Molecular Biology*, 283, 463-474.

Takeda,S. et al. (2002) Identification of G protein-coupled receptor genes from the human genome sequence. *FEBS Lett*, 520, 97-101.

Tramontano,A. (2006) The role of molecular modelling in biomedical research. *FEBS Letters*, 580, 2928-2934.

Warne,T. et al. (2011) The structural basis for agonist and partial agonist action on a β(1)-adrenergic receptor. *Nature*, 469, 241-244.

Warne,T. et al. (2008) Structure of a beta1-adrenergic G-protein-coupled receptor. *Nature*, 454, 486-491.

Xu,F. et al. (2011) Structure of an agonist-bound human A2A adenosine receptor. *Science*, 332, 322-327.

Zagotta,W.N. et al. (2003) Structural basis for modulation and agonist specificity of HCN pacemaker channels. *Nature*, 425, 200-205.

Part 3

Complimentary Methods

16

Monitoring Preparation of Derivative Protein Crystals *via* Raman Microscopy

Antonello Merlino, Filomena Sica and Alessandro Vergara
Department of Chemistry, University of Naples "Federico II", Naples, Italy Istituto di Biostrutture e Bioimmagini, CNR, Naples
Italy

1. Introduction

Below to crystallographic applications, protein crystals are of great interest in other numerous fields of biology and biotechnology. Cataract, or the loss of transparency of the eye lens, is related to the alteration of physical, chemical, and structural properties of proteins of the crystallin family, that may lead to crystallization under some physiological conditions (Tardieu, 1998). Other pathological states are known to be a consequence of the *in vivo* formation of crystals, made of either proteins or other macromolecular assemblies. Examples are viral proteins stored in plant cells, viral particles in animal cells, hemoglobin C and S causing anemia, or ribosomal particles accumulating in the brain of patients suffering from *presenile dementia* (McPherson, 1999).

Of course, crystals of biological macromolecules that are prepared *in vitro* have important applications: they are tools to obtain atomic models of the molecules and to design specific ligands and new drug formulations. Medicinal formulations composed of either insulin (Richards et al., 1999) or α-interferon crystals (Reichert et al., 1996) are already applied in treatments to ensure the continuous release of protein in blood. Crystallographic analysis of highly ordered crystals with intense X-ray sources provides accurate three dimensional structures (Ducruix & Giegé, 1999). The success of this technique strictly depends on obtaining diffraction-quality crystals. The process of crystallization remains a hit-and-miss affair, typically involving screening hundreds of conditions. The crystallization of biological macromolecules shares many common properties with those of small solute molecules (e.g. growth by 2D nucleation or by screw dislocation mechanisms), but their crystals exhibit several peculiarities: most of them have a high solvent content (e.g. 30-80 vol%), few intermolecular contacts, and a high density of defects (Malkin et al., 1996).

Briefly, protein crystallization requires the formation of a supersaturated protein-precipitant solution. The transition from a stable solution to a supersaturated one can be achieved by increasing the concentration of precipitant and/or that of protein (Vergara et al., 2003). The most frequently used crystallization method is the *vapor diffusion technique*. A drop containing protein, buffer, salt and precipitant is equilibrated against a reservoir (buffer, salt and precipitant). The difference in concentration between the drop (lower) and the reservoir (higher) drives the system toward equilibrium by diffusion through the vapor phase. The drop can either be placed on the underside of the cover slide (*hanging drop*) or placed on a

plastic support above the surface of the reservoir (*sitting drop*). *Batch crystallization* is a method where the sample is mixed with precipitant and additives creating a homogenous crystallization medium. In the *free interface diffusion* the protein sample is stratified over the precipitant solution; over time the sample and precipitant diffuse into one another and crystallization may occur at the interface. In a *microdialysis experiment* the protein solution is equilibrated through a membrane against the precipitant solution over time in a stepwise manner. In some cases, to improve the X-ray diffraction properties the crystallization is performed in a gel medium. Agarose, agar and silica have been successfully used as gel materials to obtain protein crystals (Chayen, 1998; Vergara et al., 2003; Vergara et al., 2009).
It is a frequent task to produce modified protein crystals in order to study the structural modifications undergoing a chemical treatment. These derivative crystals can be prepared *via* co-crystallization, or *via* soaking. Co-crystallization means that crystallization of the chemically modified biopolymers (protein and additive) is conducted from solution phase. Alternatively, protein crystals are first grown and then they are chemically modified *via* diffusion of the additives that are soaked into the solvent channels. Both these preparative procedures can be supported by the application of Raman monitoring, particularly by difference Raman spectra (in co-crystallization) or by a time-resolved Raman microscopy (for soaking procedure).

2. Principles of Raman spectroscopy

2.1 Raman effect

Raman spectroscopy is a vibrational spectroscopy based on the anelastic scattering of a monochromatic light (frequency ν) due to the interaction with the sample. The polarizability tensor of the sample oscillating with a normal frequency ν_0, that is associated to the *omni* present molecular vibrations, interacts with the electric field of the laser light. This produces an induced electric dipole oscillating (then irradiating) with three distinct frequency (ν, $\nu+\nu_0$ and $\nu-\nu_0$). The strong elastic scattering at frequency ν_0 is the Rayleigh scattering that usually is cut in a Raman experiment. The two minor inelastic components are Raman Stokes bands ($\nu-\nu_0$) and Raman antiStokes ($\nu+\nu_0$). At room temperature only Raman stokes bands are intense enough to be recorded. The difference between the wavenumber of the incident light and the scattered Raman band is the *Raman shift* (the x-axis of any Raman spectrum). The y-axis is usually reported in arbitrary units, so that accurate quantitative analysis can be based only on ratio of two distinct bands or by using some internal standard. In the Stokes Raman region (at wavelength lower than the excitation line) also fluorescence is collected. Therefore the counts read in the y-axis is always the sum of Raman and fluorescence, that should be kept as low as possible.

2.2 Off-resonance and resonance Raman

We should point out that two major kinds of Raman spectra can be collected, depending on the laser line (off-resonant Raman and resonant Raman spectra). In off-resonance Raman spectra no relation exists between excitation line and electronic absorption condition. On the contrary, resonance Raman spectra are collected when a particular excitation wavelength is used, namely within one electronic absorption band of the sample. Resonance Raman spectra are much more intense (depending on the extinction coefficients of the electronic transition) and selective (only normal mode that couples with the vibronic transition can be enhanced) than off-resonance Raman spectra. A more extensive presentation of the quantum mechanics of the Raman effect is elsewhere reported (Long, 2002).

3. Handling of protein crystals for Raman microscopy experiments

Raman microscopy is ideal for the monitoring of bioprocesses as it is non-destructive, inexpensive, rapid and quantitative. Its confocal nature makes it possible to focus through transparent capillary or directly on crystals kept in their crystallization reactor allowing straight analysis on the sample.
In co-crystallization experiments two independent Raman spectra are collected on native and derivative crystals, using the same mother liquor with the exception of the reactant. In this case, the comparative analysis of the two spectra can provide differences in a) number, b) position and c) intensity of protein Raman bands. Manipulation of crystals for Raman measurements requires standard handling.

3.1 Crystal sampling

Confocal apparatus allows to get rid of any influence of the cover slip (as in vapour diffusion) or container wells (as in FID or batch crystallization supports). Measurement into drops (both hanging and sitting drop) is particularly feasible. In order to avoid significant scattering from mother liquor, thus reduction in spectral quality, it is better to keep the drop as small as possible, and to use a minimal depth into the crystal (particularly for resonance Raman spectra). Water is a weak Raman scatterer, with a small contribution at 1640 cm^{-1}. Precipitating agents, especially when at high content (PEG, MPD, alcohol) may interfere or not, depending on the spectral region of interest. Agarose and silica gel matrix do not interfere significantly, so crystals grown in gel medium can be used as well.
Despite we deal with solid state we will not consider frequencies of the lattice vibrational modes, that for protein crystals are very low (below the Rayleigh cut). For lysozyme crystals it is 25 cm^{-1}.
We will focus on two distinct experimental setups, namely for *in-situ* and *ex-situ* Raman spectroscopy, referred to simultaneous or not to X-ray diffraction experiments (also reported as on-line and off-line (McGeehan et al, 2011)). Below we present separately these two setups and applications.

3.2 Ex situ Raman experiments

Most of the analysis to monitor chemical modification of protein crystals can be performed by *ex situ* Raman experiments, which involve distinct acquisition of Raman and X-ray diffraction data. The experiments are, indeed, carried out on a Raman microscope that is physically separated from the X-ray diffractometer. This kind of experiments can be easily performed on a commercial or home-built Raman microscope, and it aims to the definition of the experimental conditions (eg soaking time and reactant/protein molar ratio) for the preparation of derivative protein crystals. Raman spectra on very small drops can be recorded also at low temperatures by using a dedicated autostage (Linkham Co), though care must be taken during cooling to ensure high transparency of the drop and to avoid crystal movement within the drop. When flash freezing is adopted, these impediments are overcome. Crystals monitored at room temperature *via* ex situ apparatus can even be frozen and taken to the X-ray diffractometer for the data collection.

3.3 Raman microscopy apparatus

The Raman confocal microscope (Jasco, NRS-3100) currently available for *ex-situ* experiments at the Department of Chemistry of the University of Naples 'Federico II' is depicted in Figure 1. One of the 458, 488 and 514-nm lines of an air-cooled Ar^+ laser (Melles Griot, 35 LAP 431-220), or 406, 413 and 647 nm lines of a water-cooled Kr^+ laser (Coherent, Innova 320) can be injected into an integrated Olympus confocal microscope and focused to a spot size of approximately 2 µm by a 100x or 20x objective. The laser power at the sample depends on the wavelength ranging from 5 to 100 mW. A holographic notch filter is used to reject the excitation laser line. Raman back-scattering is dispersed through a monochromator (2400 or 1200 grooves/mm grating) and collected by a Peltier-cooled 1024 x 128 pixel CCD photon detector (Andor DU401BVI). Frequency shifts are calibrated by using indene, cyclohexane or CCl_4 as standard, depending on the spectral range of interest. An acceptable spectral resolution is usually considered 4 cm^{-1}.

Actually, since Raman shift for off-resonant spectra is independent of the excitation wavelength, in principle also IR or UV excitation could be used, though at least different optics and detector (a different apparatus) is required.

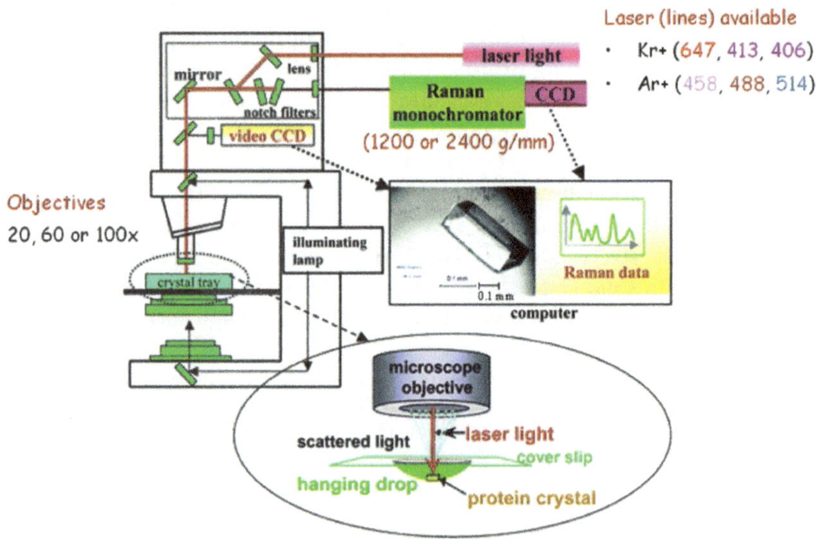

Fig. 1. Scheme of Raman apparatus available at the University of Naples "Ferderico II" (Dept of Chemistry) to collect spectra from protein single crystals. Adapted after (Carey & Dong, 2004).

3.4 In situ Raman experiments

Whether the crystal is on the diffractometer or away from it (*in situ* or *ex situ*), Raman spectra are not affected by it, but of course only *in-situ* analysis reveals radiation damage effects (that could be even reversible (Adam et al, 2009)).

That is the reason for many efforts in the last decades to make Raman microscopes available at synchrotron beamlines: the European Synchrotron Radiation Facility (ESRF) (Carpentier et al. 2007), National Synchrotron Light Source (NSLS), Brookhaven National Laboratory (BNL) (Stoner-Ma et al., 2011) and Swiss Light source (SLS) (Owen et al. 2010) part-time

dedicates one of their beamlines to *in situ* Raman-assisted X-ray biocrystallography (see below).
A recent review on the microRaman apparatus available at synchrotron facility is reported in (McGeehan et al., 2011). For *in situ* setup we intend that the X-ray diffractometer is integrated with a laser (guided by an optical fiber) that is backscattered to a spectrograph. To this end, crystals are mounted on an usual crystallographic head kept into a cryo-loop under a nitrogen flux (typically at 100 K). After the crystal is centered for X-ray diffraction data collection experiments, the optimal orientation for the collection of the best Raman spectra has to be determined, typically by manually re-orienting the crystal. Once the crystal orientation that results in the best signal-to-noise ratio is found, Raman spectra are acquired prior and after the X-ray diffraction data collection, at the same crystal orientation. This allow to check if any photo-chemistry is induced by the X-ray beam exposure. For off-resonance Raman spectra it is particularly important to keep the drop volume low and to avoid interference from the loop (nylon loops should not intersect the Raman laser beam path (Carpentier et al., 2011)). When Raman spectra are collected as a function of the crystal orientation (Raman crystallography), information can be obtained on the orientation order and the variability of the relative intensity of Raman bands can be estimated (Tsuboi et al., 2007). Indeed, both off-resonance (Kudryavtsev et al., 1998) and resonance spectra (Smulevich & Spiro 1990) are mildly affected by crystal orientation. In this case general physical chemistry rather than structural information can be extracted by Raman crystallography.

4. Interpretation of Raman spectra of derivative protein crystals

4.1 Raman spectra analysis

A Raman spectrum provides a lot of information: energy shift, intensity and polarization of the scattered light and width of the peak. We will mostly focus on the Raman shift and intensity features. Each Raman shift corresponds to a permitted vibrational transition that relates to the strength constant k of the harmonic oscillator *via*

$$\nu = \frac{1}{2\pi}\sqrt{\frac{k}{\mu}} \tag{1}$$

where μ is the reduced mass. Therefore high frequency corresponds to either stretching or bending with a high strength constant or involving light atoms, such as hydrogen. The presence of the reduced mass at the denominator in equation 1 justifies the wide use of isotopic replacement (eg H/D) for band assignment. Raman transition corresponds, most likely, to a fundamental transition (from the vibrational ground state to the first excited state). Indeed, overtone and combination bands in Raman spectroscopy are much more unlikely than in Infrared spectroscopy (FT-IR). Furthermore, compared to FT-IR, Raman spectroscopy has very different selection rules, and usually bands that are strong in the IR absorption, are not allowed in Raman scattering. Indeed, the great advantage of Raman when studying biological samples is the very weak contribution from water to Raman spectra, compared to IR spectra.
Once Raman spectrum has been recorded, reduction for the temperature correction can be performed (Pernice et al., in press). Unless strictly necessary, as for a variable fluorescence coming from background, no baseline correction should be performed. The position of the

Raman band can be assigned according to literature or basing on theoretical calculations, isotopic substitution or symmetry considerations (Long, 2002). But usually, the last operations do require experienced spectroscopists (see par 5.1.1) . The comparison of native protein crystal and chemically modified crystal can be performed *via* difference spectra in case of off-resonance spectrum or just *via* a simple comparison of the spectra for resonance Raman (Carey, 2006).

5. Examples of Raman assistance in chemical modifications of protein crystals

Application of complementary techniques that probe different features is an added value to the understanding of the relationship between structure, function and dynamics.
Raman microscopy proved to be a valuable support to protein crystallography in all the steps of the 3D structure determination, from the preparation of the derivative crystals up to the interpretation of the electron density maps. We will first present examples of chemical composition changes (cfr 5.1), and then we will move to possible secondary (cfr. 5.2) and tertiary structure (cfr 5.3) modifications occurring upon a physico-chemical treatment.
In some cases crystallography can assist Raman spectroscopy in defining frequency assignments. In this case we speak about crystallography-assisted Raman spectroscopy.

5.1 Raman helping phase problem solution

Phase problem in crystallography can be solved for new proteins *via* multiple wavelength anomalous diffraction (MAD) or by multiple isomorphous replacement (MIR). Both these approaches usually-require a chemical modification of the protein crystals.

5.1.1 Raman detection of Se-Met incorporation into protein crystals

Se-Met derivatives are expressed for phase determination using multiwavelength anomalous dispersion (MAD) experiments (Cassetta et al., 1999). Se-Met inclusion is a widespread approach for MAD experiments since Met residues are present with an average occurrence of 1 per 59 residues (Hendrikson et al., 1990). Once crystals of a Se-Met derivative protein have been obtained, only then the absorption spectrum can be used to reveal the presence of selenium for determination of the crystal structure. Se-Met derivative crystals should be stored in a reducing environment (usually adding DTT) and diffraction experiments should be carried out using fresh crystals. It has been reported that two-week storage of Se-Met protein crystals can produce significant deterioration of the crystal diffraction power (Doublié, 1997). Since synchrotron beam-time is not always available immediately after growth of the crystals, a tool to check the Se-Met status in stored crystals is beneficial.
A Raman microscopy study was conducted on isomorphous wild-type and Se-Met crystals of the βγ-crystallin from *Geodia cydonium* (geodin) (Vergara et al., 2008). Geodin is a protein of unknown function, whose structural characterization could provide information on how homologous βγ-crystallin monomeric proteins evolved.
Table 1 summarizes the most relevant Raman bands observed in off-resonant Raman spectra for the SeMet-derivative crystals of geodin. Spectra at low frequency region (400-1200 cm^{-1}) are reported in Figures 2. Usually, the Raman spectrum of a polypeptide is subdivided into three main regions of interest: 1) the range between 870–1150 cm^{-1}, associated with the

vibrations of the backbone Cα-C and Cα-N bonds; 2) the range between 1230–1350 cm^{-1}, containing the amide III region vibrations, associated with normal modes of various combinations of the Cα-H and N-H deformations together with the Cα-C and Cα-N stretchings (Asher et al., 2001); 3) the range between 1630–1700 cm^{-1}, associated with C=O stretching modes, defined as the amide I region (Ngarize et al., 2004). Furthermore, the lower and higher regions can also be informative: i) the conformation and detection of disulphide bridges can be investigated in the low frequency (500-540 cm^{-1}) region (Kudryatsev et al., 1998); ii) hydrophobic interactions can be investigated by analysing the C-H stretching region (2800-3200 cm^{-1}) (Chourpa et al., 2006), and the S-H stretching region 2550-2600 cm^{-1} can serve as a valuable probe of local dynamics (Thomas, 1999).

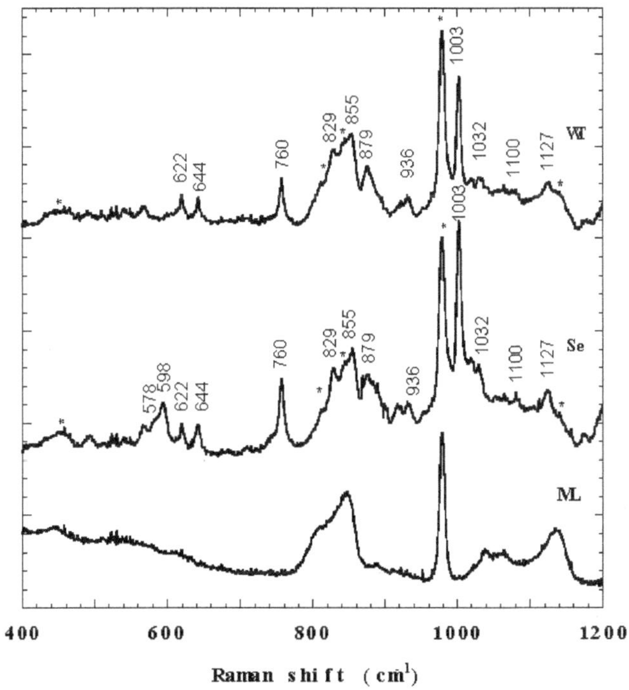

Fig. 2. Low frequency Raman spectra of the wild-type geodin crystals (WT), the Se-Met labelled crystals (Se), the mother liquor from which both kind of crystals grew up (ML). Signals attributed to the mother liquor are tagged by a star. The spectral resolution is 4 cm^{-1}. (after Vergara et al., 2008).

The Raman spectra collected on the isomorphous crystals of wild-type and Se-Met geodin crystals reveal the same secondary structure features (Amide I and III). These findings suggest that the presence of the Se-Met does not alter the structure of geodin, as observed for most of the proteins. The main difference between the two spectra (excluding the slightly different intensity of mother liquor signals) is in one narrow region at low frequency. Indeed, the bands in the 570-600 cm^{-1} region are present only in the Se-Met geodin crystal, and not in the spectra of the mother liquor or in the wild-type geodin crystal (Table 1). This spectral feature can be confidently assigned to the C-Se stretching, in agreement with previous Raman studies on selenium-containing organic compounds (Hamada & Morishima, 1977; Paetzold et al., 1967), and on the selenomethionine aminoacid (Zainal & Wolf, 1995; Lopez et al., 1981).

Band frequency (cm^{-1})	**assignment of vibration mode**	**Primary Structure**	**Secondary structure**
	C=O stretching		**Amide I**
1668			β sheet and random coil
1650 shoulder			α-helix
1614	Ring stretching, side chain	Tyr-Phe	
1605			
1577	Ring stretching	Trp	
1548	Ring stretching	Trp	
1449	$C\text{-}H_2$ scissoring		
1400	COO^- stretching		
	N-H, C-H and CH_2 bending		**Amide III**
1318			α-helix
1245			random coil
1237			β sheet
1205	C-C stretching	Tyr-Phe	
1159			
1127	C-C stretching		
1100	C-C stretching		
1032	C-C stretching	Phe	
1003	Ring breathing	Phe	
936	C-C, skeletal stretching		
889	C-C, C-O deformations	Trp	
855	Fermi resonance doublet	Tyr	
829			
760	C-C, C-O deformations	Trp	
644	ring bending	Tyr	
622	ring bending	Phe	
598	symmetric C-Se stretching	**Se-Met**	
578 shoulder	asymmetric C-Se stretching		

Table 1. Assignment of characteristic Raman bands measured for the Se-Met derivative crystals of the crystalline-like protein from *Geodia cydonium* (geodin).

Using these assignments, SeMet Raman peaks have been observed also in the SeMet derivative of protein SOUL (Rossi et al., 2009). In these crystals, a quantitative evaluation of the relative amount of SeMet replacement was also achieved by comparative analysis.

In principle, Raman microspectra could be also used to reveal post translational modifications, such as phosphorylation, acetylation, trimethylation, ubiquitination and glycosilation, as already done by comparing spectra of native and derivative proteins in solution (Sundararajan, et al. 2006; Brewster et al., 2011).

5.1.2 Raman detection of heavy atom derivatives

As previously stated, heavy atoms can be incorporated into protein crystals for the purposes of phasing. The simplest way to soak a heavy atom into protein crystals is to immerse the crystal straight into a drop containing the heavy atom at the final concentration for a soaking time ranging from 10 min up to several days. Different salts can be used for this purpose: successful experiences have been reported using K_2PtCl_4, $KAu(CN)_2$, K_2HgI_4, $UO_2 (C_2H_3O_2)_2$, $HgCl_2$, $K_3UO_2F_5$ and para-chloromercurybenzoic sulfate (PCMBS) (Boggon & Shapiro, 2000).

During this procedure, typical question are: did actually the ion soak into the crystal? Is the time of soaking long enough for allowing the heavy atom diffusion into the crystal?

Raman microscopy can help answering these questions without waiting for a Patterson map. Reports of Raman-based direct Pt incorporation and indirect Eu^{3+} soaking are available in literature (Carpenter et al., 2007). Indeed, the Pt-Cl stretching appears at 330 cm^{-1} that can be compared to the Trp band at 760 cm^{-1}. From the time-evolution of the relative intensity I330/I760 we can follow Pt^{2+} soaking into the crystal.

Authors have recently defined a protocol to follow Hg^{2+} binding to free Cys residues (unpublished results), that is dependent on the number of Cys residues involved in the binding (1 or 2).

5.2 Raman detection of chemical modifications perturbing secondary structure

In some rare cases chemical modifications, such as pH change or Cys-Cys reduction, can significantly change even secondary structure. By taking advantage from the sensitivity of Amide I band to conformational variations, Thomas revealed a β-sheet to α-helix transition upon TCEP [tris(2-carboxyethyl)phosphine] addition to bovine insulin crystals (Zheng et al., 2004). Unfortunately, upon chemical reduction crystals do not diffract anymore, and no crystallographic counterpart is available. An analogous conformational transition occurred upon pH decrease of 5S subunit of transcarboxylate (Zheng et al., 2004), where modifications in the Amide III region (Mikhonin et al., 2006) suggested a significant modification into the crystal matrixes. Also hydration can affect deeply Amide I, and eventually secondary structure, as observed for a collagen-like polypeptide, for which Raman spectrum of the lyophilized powder showed very different Amide I frequencies when compared to single crystals (Figure 3, from Merlino at al., 2008a).

In particular, in this case, spectra of $(PPG)_{10}$ powders are characterized by three distinct well-resolved amide I bands (1638, 1655, 1690 cm^{-1}), that do not match the frequencies observed in the spectra of $(PPG)_{10}$ crystals (1629, 1645 and 1669 cm^{-1}).

The three amide I bands, and especially the well-resolved band at 1690 cm^{-1}, do not match the frequencies reported by FT-IR solution studies for the unfolded $(PPG)_{10}$ (maximum at 1633 and a shoulder at 1665 cm^{-1}) or for the unfolded polyproline (one band at 1621 cm^{-1}) (Bryan et al., 2007).

The amide I bands of $(PPG)_{10}$ crystals and powders are significantly different, not only for the intensity distribution of the crystal spectrum (depending on the orientation of the polarization of the exciting laser beam with respect to the main axes of the Raman tensor in the respective unit cell), but also for the frequency of the Raman peaks (After Merlino et al., 2008a).

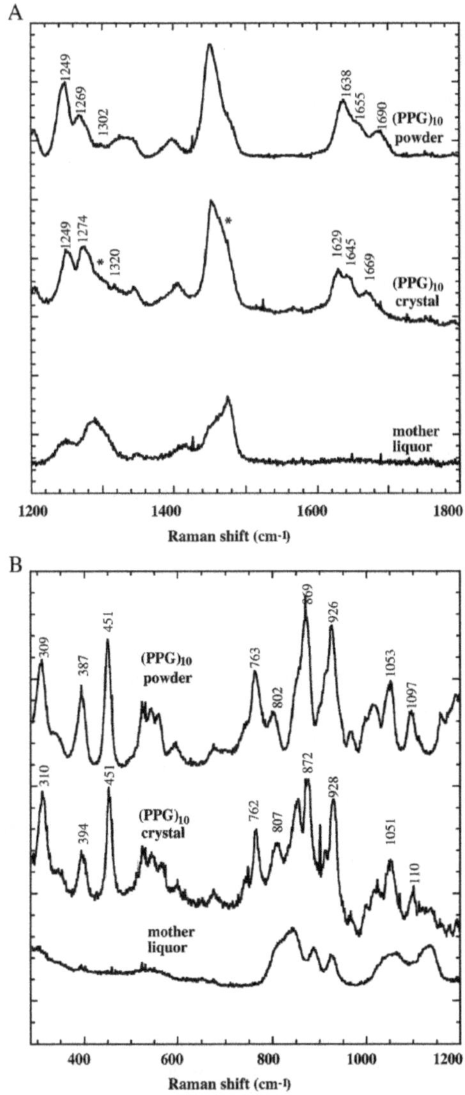

Fig. 3. Medium frequency (A) and low frequency (B) Raman spectra of $(PPG)_{10}$ powders, crystals, and mother liquor from which crystals grew up. In spectra (A) and (B) the signals attributed to the mother liquor are tagged by a star. Spectral resolution is 4 cm^{-1}. After (Merlino et al. 2008a).

5.3 Raman detection of chemical modification perturbing tertiary structure

Certainly, the most common structural modifications are related to tertiary structure. The amino-acidic modification can be followed even by off-resonance Raman spectra, whereas modification at metal center can only be detected *via* resonance Raman spectra.

5.3.1 Off-resonance Raman spectra showing tertiary structure modifications

Many Raman markers can provide information on the modified microenvironment of side chains (particularly free Cys and Tyr residues), disulfide bridges conformation or ligands entry, or metal center.

The relative intensity of Fermi resonance doublet from Tyr (I856/I830) is strictly dependent on the strength of the H-bond formed by tyrosines with the adjacent groups (1.25 exposed and 0.5 for strong H-bond donation of the OH).

Sulfidryl groups exposure of Cys residues can be investigate following the kinetics of H/D exchange *vs* temperature. Also the pKa of Cys can be extracted by titrating S-H stretching at 2550-70 cm^{-1} (mercaptoethanol 2580 cm^{-1}) compared to the envelope at 2800-3000 cm^{-1} (Thomas, 1999).

When ligand soaking is followed, difference Raman spectra can provide indication on both bound ligands and related conformation changes of the proteins. Carey reported a review of Raman-microscopy applications to follow ligand binding (Carey, 2004), and more recently he has applied this technique in kinetic crystallography to monitor time evolution of β-lactamases binding with clinical inhibitors (Carey, 2011). These spectroscopic evidences can be even used as restrains during crystallographic model refinement.

5.3.2 Resonance Raman spectra showing tertiary structure modifications

Metal centers have been well studied *via* Resonance Raman (RR) spectroscopy for hemoprotein and not-containing heme proteins. A recent example was reported for the major haemoglobin from the sub-Antarctic fish *Eleginops maclovinus*, for which a variety of coordination, spin state were observed keeping the same oxidation states (Merlino et al., 2010). In this example, Raman microscopy experiments were conducted on two different carbomonoxy crystals (called Ortho and Hexa) as well as on their ferric and deoxy forms (Figure 4).

The high-frequency region (1300–1700 cm^{-1}) of the RR spectrum includes the porphyrin in-plane vibrational modes (which are sensitive to the electron density of the macrocycle and the oxidation, coordination and spin state of the iron ion).

The deoxy forms of both Ortho and Hexa are pentacoordinated high-spin states (bands at 1355, 1548–1549, 1582 and 1602–1607 cm^{-1}). The ferric form contains a hexacoordinated low-spin hemichrome (bands at 1505, 1559, 1588 and 1640 cm^{-1}) (Merlino et al., 2011). After long laser exposure times (about 10 min), the Ortho but not the Hexa form appears to be unstable under laser irradiation, and it irreversibly converts to a hexacoordinated low-spin haemochrome state (bands at 1361, 1496 and 1587 cm^{-1}, respectively).

RR microscopy can be also used to identify anomaly in the coordination state of hemoprotein (Merlino et al., 2008b; Vergara et al., 2010). A recent and fine case is the unusual deoxy coordination found in the Antarctic fish hemoglobin from *Trematomus newnesi* (Hb1Tn) (Vergara et al., 2010). The crystal structure of deoxy form of this protein reveals distinct coordinations at the α and β hemes, and a high disorder at the EF helices of α heme, hosting the active site. Particularly, the distances His-Fe-His were unusual at the α subunits, raising doubt of hexa-coordination. The medium frequency region ruled out any contribution of bis-histidyl deoxy coordination, confirming a penta coordination. The low frequency regions clearly showed an heterogeneity in the Fe-His stretching, with a broad band attributed to the α/β structural differences in coordination (Figure 5, after Vergara et al., 2010).

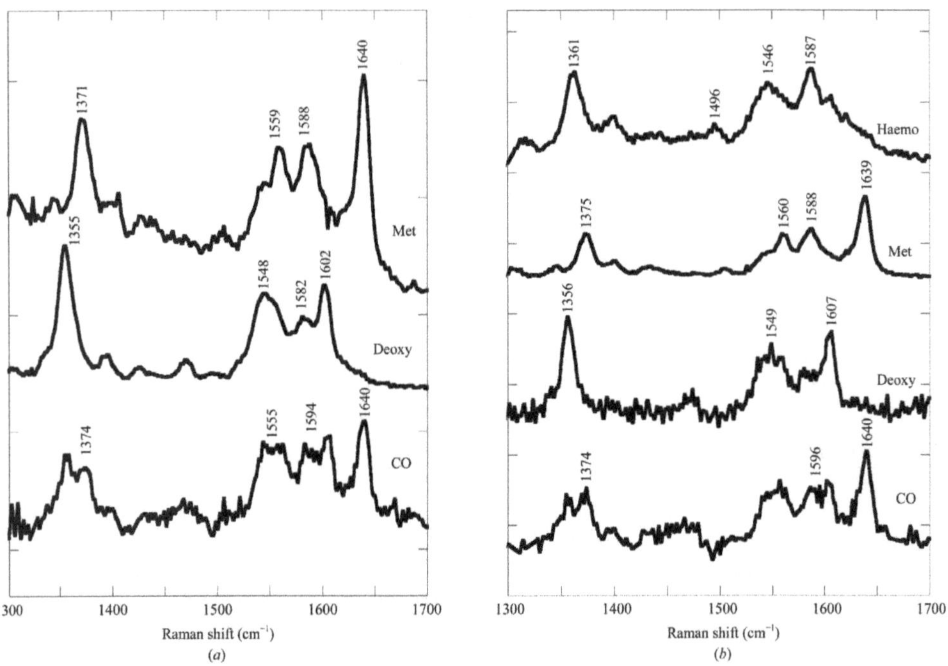

Fig. 4. Resonance Raman spectra of crystals of Hb1Em in 100 mM Tris-HCl buffer pH 8.0 at room temperature in the carbomonoxy (CO), deoxygenated (Deoxy) and ferric (Met) forms for Hexa (a) and Ortho (b) crystals. The Ortho deoxygenated form (b) converts quickly into haemochrome (Haemo) after 10 min laser exposure. Excitation wavelength, 514.5 nm; laser power at the sample 2 mW for the ferric and deoxy forms and 0.1 mW for the carbomonoxy form. All spectra were an average of at least six spectra with 2 min integration time (After Merlino et al., 2010).

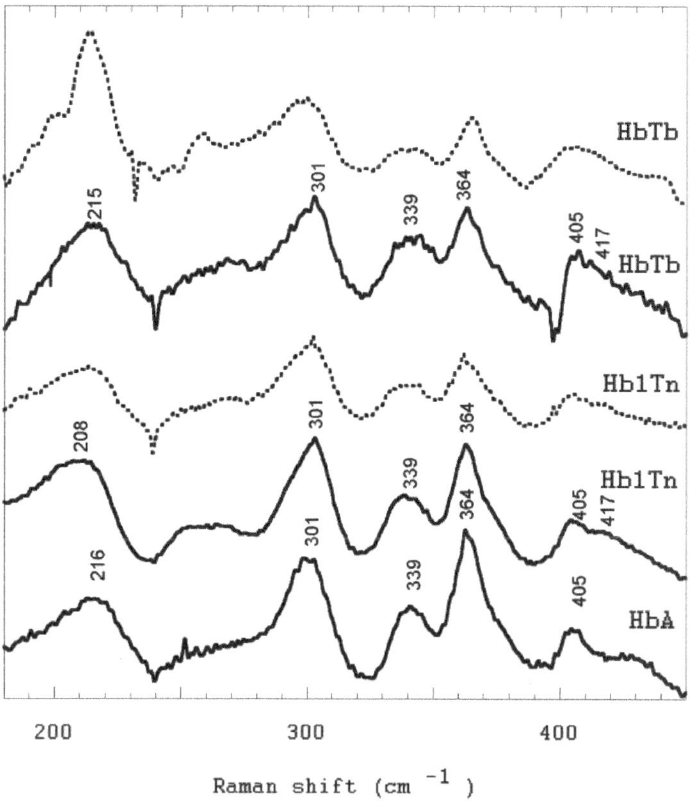

Fig. 5. Resonance Raman spectra in the low-wavenumber region of Hb1Tn (major hemoglobin of *T. newnesi*) both in solution (solid lines) and in the crystal state (dashed lines) in the deoxy state. As reference of usual coordination also HbTb (major hemoglobin of *T. bernacchii*) and human HbA are reported (After Vergara et al., 2010).

6. Raman-assisted biocrystallography to study chemical mechanisms and X-ray radiation damage

Raman assistance can go much beyond the crystal preparation and can provide additional structural information to solve some ambiguities in the electron density map interpretation. Examples include the 1) effect of X-ray damage as a function of the X-ray dose (Garman, 2010). Some examples are breakage of disulphide bonds (Carpentier et al., 2010) and Brominated-DNA photo-dissociation (McGeehan et al., 2007), and even X-ray-induced transient photobleaching in a photoactivatable green fluorescent Protein (Adam et al., 2009). Indeed, third generation synchrotrons brought back the problem of X-ray damage, even at 100 K (Garman, 2010). A complete microspectroscopic picture of the radiation damage is not yet available. The possibility to investigate the different rates of decay for different chemical groups would help crystallographers in making more stable derivative crystals. Authors are currently working on this subject in collaboration with SLS scientists. 2) X-ray induced photo-reduction that may affect metal spin and oxidation state. 3). Identification and freeze-trapping of reaction intermediates (Carpenter et al., 2011). Bourgeois' group was pioneer in this field reporting non-resonance Raman of a trapped iron(III)-(hydro)peroxo species in crystals of superoxide reductase, a nonheme mononuclear iron enzyme that scavenges superoxide radicals (Katona et al., 2007), and further investigations came on RNA polymerase reactions as well (Carey et al., 2011). However this is out of the scope of this chapter, although it represents a formidable challenge for Raman-assisted X-ray biocrystallography.

7. Aknowledgements

This work was financially supported by PNRA.

8. References

Adam, V., Carpentier, P., Violot, S., Lelimousin, M., Darnault, C., Nienhaus, G.U. & Bourgeois, D. (2009) "Structural Basis of X-ray-Induced Transient Photobleaching in a Photoactivatable Green Fluorescent Protein" J. Am. Chem. Soc., 131, 18063

Brewster V. L., Ashton, L. & Goodacre, R. (2011) "Monitoring the glycosylation status of proteins using Raman spectroscopy" Anal. Chem., 83, 6074

Bryan, M., Brauner, J. , Anderle, G. , Flach, C. , Brodsky, B. & Mendelsohn, R. (2007) "FTIR studies of collagen model peptides: complementary experimental and simulation approaches to conformation and unfolding" J. Am. Chem. Soc., 129, 7877

Carey, P.R. (2006) "Raman crystallography and other biochemical applications of Raman microscopy, Ann. Rev. Phys. Chem., 57, 527

Carey, P R, Yuanyuan Chen, Bo Gong & Matthew Kalp, (2011) "Kinetic crystallography by Raman microscopy" Biochim. Biophys. Acta, 1814, 742

Carey, P. R. & J. Dong, (2004) "Following ligand binding and ligand reactions in proteins *via* Raman crystallography" Biochemistry, 43, 8885

Carpentier, P., Royant, A, Ohana & J. Bourgeois, D (2007) "Advances in spectroscopic methods for biological crystals. 2. Raman spectroscopy" J. Appl. Cryst., 40, 1113

Carpentier, P., Royant, A., Weik, M. & Bourgeois, D. (2010) "Raman-Assisted Crystallography Suggests a Mechanism of X-Ray-Induced Disulfide Radical Formation and Reparation" Structure, 18, 1410

Cassetta, A., Deacon, A. M., Ealick, S. E., Helliwell, J. R., & Thompson, A. W. (1999) "Development of instrumentation and methods for MAD and structural genomics at the SRS, ESRF, CHESS and Elettra facilities" J. Synchr. Rad., 6, 822

Chayen, N. (1998) "Comparative studies of protein crystallization by vapour-diffusion and microbatch techniques" Acta Cryst. D: Biol. Cryst., 54, 8

Ducruix, A. & Giegé, R. (1999) "Crystallisation of nucleic acids and proteins. A practical approach" second edition, IRL Press, Oxford.

Garman, E F. (2010) "Radiation damage in macromolecular crystallography: what is it and why should we care?"Acta Cryst. D: Biol. Cryst., 66, 339

Katona, G., Carpentier, P., Nivière, V., Amara, P., Adam, V., Ohana, J., Tsanov, N. Bourgeois, D. (2007) "Raman-assisted crystallography reveals end-on peroxide intermediates in a non-heme iron enzyme" Science, 316, 449

Kudryavtsev, A. B., Mirov, S. B., DeLucas, L. J., Nicolete, C., van der Woerd, M., Brayb, T. L. & Basiev, T. T. (1998) "Polarized Raman spectroscopic studies of tetragonal lysozyme single crystals " Acta Cryst. D: Biol. Cryst., 54, 1216

Long, D.A., (2002), "The Raman effect" John Wiley & Sons, Chichester, England

McGeehan, J. E., Carpentier, P., Royant, A., Bourgeois, D. & Ravellia, R.G.B. (2007) "X-ray radiation-induced damage in DNA monitored by online Raman" J. Synchr. Rad. 14, 99

McGeehan, J.E., Bourgeois, D., Royant, A. & Carpentier, P. (2011) "Raman-assisted crystallography of biomolecules at the synchrotron: Instrumentation, methods and applications" Biochim. Biophys. Acta, 1814, 750

McPherson, (1999) "A. Crystallization of biological macromolecules" Cold Spring Harbor Laboratory Press, New York

Merlino, A., Sica, F., Zagari, A., Mazzarella, L., Vergara, A. (2008a) "Correlation between Raman and x-ray crystallography for the collagen-model peptide $(Pro-Pro-Gly)_{10}$" Biophys. Chem. 137, 24-27

Merlino, A., Verde, C., di Prisco, G., Mazzarella, L. & Vergara, A. (2008b)"Reduction of ferric hemoglobin from Trematomus bernacchii in a partial bis-histidyl state produces a deoxy coordination even when encapsulated into the crystal phase". (2008b). Spectroscopy, Int. J. 2, 143

Merlino, A., Vitagliano, L., Howes, B., Verde, C., di Prisco, G., Smulevich, G., Sica, F. & Vergara, A. (2009) "Combined crystallographic and spectroscopic analysis of Trematomus bernacchii hemoglobin highlights analogies and differences in the peculiar oxidation pathway of Antarctic fish hemoglobins" Biopolymers, 91, 1117

Merlino, A., Vitagliano, L., Howes, B., Nicoletti, F., Balsamo, A., Giordano, D., Coppola, D., Verde, C., di Prisco, G., Smulevich, G., Mazzarella, L. & Vergara, A. (2010) "Crystallization, preliminary X-ray diffraction studies and Raman microscopy of the major haemoglobin from the sub-Antarctic fish Eleginops maclovinus in the carbomonoxy form" Acta Cryst. F: Str. Biol. Cryst., F66, 1536

Merlino, A. , Howes, B.D., di Prisco, G., Verde, C. , Smulevich, G. , Mazzarella, L. & Vergara, A. (2011) "Occurrence and formation of endogenous histidine hexacoordination in cold-adapted hemoglobins".IUBMB Life, 63, 295

Mikhonin, A. V. , Bykov, S. V. Myshakina, N. S. & Asher, S. A. (2006) "Peptide secondary structure folding reaction coordinate: correlation between UV Raman amide III frequency, ψ Ramachandran angle, and hydrogen bonding" J. Phys. Chem. B, 110, 1928

Pernice, P., Vergara, A., Sirleto, L., Aronne, A., Gagliardi, M., Fanelli, E. & Righini, G. (2011) "Large Raman gain in a stable nanocomposite based on niobiosilicate glass", J. Phys. Chem. C., 115, 17314

Reichert, T.L. Nagabhusan, M.M. Long & C.E. Bugg, (1996) "Macroscale production of crystalline interferon alpha-2B in microgravity on STS-52, In Space Technology and Application International Forum (STAIF-96)", (El-Genk, M.S., ed.), American Institute of Physics, Woodbury, N.Y., 361, 139

Richards, J.P., M.P. Stickelmmeyer, B.H. Frank, S. Pye, M. Barbeau, J. Radziuk, G.D. Smith & M.R. Defelippis, (1999) "Preparation of a microcrystalline suspension formulation of Lys(B28)Pro(B29)-human insulin with ultralente properties" Pharm. Sci., 88, 861

Rossi, B., G. Mariotto, E. Ambrosi & H. L. Monaco (2009) "Raman scattering investigation of selenomethionine replacement in protein SOUL crystals" J. Raman Spectroscopy, 40, 1844.

Smulevich, G. & Spiro, T. G. (1993) "Single-crystal micro-Raman spectroscopy" Meth. Enzymol., 226, 397

Stoner-Ma, D., Skinner, J.M., Schneider, D.K., Cowan,M., Sweet, R.M. & Oliver, A.M., (2011) "Single-crystal Raman spectroscopy and X-ray crystallography at beamline X26-C of the NSLS" J. Synchr. Rad., 18, 37

Sundararajan, N. Mao, D. Q., Chan, S., Koo, T. W., Su, X., Sun, L., Zhang, J.W., Sung, K. B., Yamakawa, M., Gafken, P. R., Randolph, T., McLerran, D., Feng, Z. D., Berlin, A. A. & Roth, M. B. (2006) "Ultrasensitive detection and characterization of posttranslational modifications using Surface-Enhanced Raman Spectroscopy" Anal. Chem., 78, 3543

Tardieu, A. (1998) "Alpha-crystallin quaternary structure and interactive properties control eye lens transparency" Int. J. Biol. Macromol., 22, 211

Thomas, G. J. J. (1999) "Raman spectroscopy of protein and nucleic acid assemblies" Ann. Rev. Biophys. Biomol. Struct., 28, 1

Tsuboi, M., Benevides, J.M. &. Thomas, G.J. Jr. (2007) "The Complex of Ethidium Bromide with Genomic DNA: Structure Analysis by Polarized Raman Spectroscopy" Biophys. J., 92 928

Vergara, A., Lorber, B. Zagari, A., Giegé, R. (2003) "Physical aspects of protein crystal growth investigated with the Advanced Protein Crystallization Facility in reduced gravity environments" Acta Cryst. D: Biol. Cryst., 59, 2

Vergara, A., Merlino, A., Pizzo, E., D'Alessio, G. & Mazzarella, L. (2008) "A novel method for detection of seleno-methionine incorporation in protein crystals *via* Raman microscopy" Acta Cryst. D., D64, 167

Vergara, A., Castagnolo, D., Carotenuto, L., Vitagliano, L., Berisio, R., Sorrentino, G., García-Ruiz, J.-M. , Zagari, A. (2009)"Phase behavior and crystallogenesis under counter-diffusion conditions of the collagen-model peptide $(Pro\text{-}Pro\text{-}Gly)_{10}$" J. Cryst. Growth, 311, 304

Vergara, A., Vitagliano, L., Marino, K., Merlino, A., Sica, F., Verde, C., di Prisco G. Mazzarella, L. (2010) "An order-disorder transition plays a role in switching off the Root effect in fish hemoglobins" J. Biol. Chem., 285, 32568

Vitagliano, L., Vergara A., Bonomi, G., Merlino, A., Smulevich, G., Howes, B., di Prisco, G., Verde, C. & L. Mazzarella, (2008) "Spectroscopic and crystallographic analysis of a tetrameric hemoglobin oxidation pathway reveals features of an intermediate R/T state" J. Am. Chem. Soc., 130, 10527

Zheng, R., Zheng, X., Dong, J. & Carey, P.R. (2004) "Proteins can convert to β-sheet in single crystals" Prot. Sci., 13, 1288

17

Investigating Macromolecular Complexes in Solution by Small Angle X-Ray Scattering

Cristiano Luis Pinto Oliveira
Department of Experimental Physics, Institute of Physics, University of São Paulo Brazil

1. Introduction

Macromolecular complexes have a huge interest in molecular biology. The comprehension of the biological processes in living systems is directly related to the knowledge of the shape and structure of the formed complex and the process of formation. Although X-ray diffraction, Nuclear Magnetic Resonance and cryoEM can provide information on the formed structures, there are several cases where none of those techniques can be applicable. Limitations on molecular weight, the necessity of a well ordered crystal, difficulties on sample preparation etc, are some of the bottle necks of those techniques (Svergun, 2007; Oliveira et al, 2010). Most importantly, in several cases the studies have to be performed directly in solution, with minimum interaction with the studied sample in order to avoid biased results. In this respect, scattering techniques are highly recommended since they allow a study directly in solution in a very non-invasive way. Small angle X-Ray scattering (SAXS) is a standard technique that can be applied to the study of particles in solution, providing information on size, shape, polydispersity, flexibility, oligomerization and aggregation state. Also, it allows real time measurements where the system can be monitored directly in solution, enabling the study of the in situ particle formation (Oliveira et al, 2009). The combination of SAXS and microscopy techniques has been used in several applications due to their complementarity (Oliveira et al, 2010; Andersen et al, 2009). In this chapter some general aspects of Small Angle X-Ray scattering and the state of the art modeling methods will be presented, with several applications.

2. Small angle X-Ray scattering

When a collimated beam (assumed as parallel waves) of X-ray photons strikes a sample, a fraction of the incident beam interacts with the electrons clouds of each molecule, and a possible process is the absorption of this photon by the atoms which excites the electrons of the atom to higher energy levels. When the excited electrons decay to ground state another X-Ray photon is reemitted as a spherical wave. In this way, this process can be viewed as the scattering of the incident photon over the electronic cloud. Depending on the energy of the incident photon several processes can happen: Rayleigh scattering, Resonance scattering, Compton effect, Thomson scattering, pair production, etc. It is beyond the scope of this chapter investigate all these possible processes. However, under certain energy limits (~7-12KeV), this scattering is well described by the so called first Born approximation, where the photon interacts only with one atom and the resulting scattered photon has the same energy of the incident photon (elastic scattering). This effect is mostly related the so called Thomson scattering. The complete solution of the scattered beam is a sum of a plane wave plus a spherical wave (Jackson, 1988). Since the information of the scattering process is related with the spherical wave only this part is considered to investigate the structural information of the particle (Guinier and Fournet, 1955; Glatter and Kratky, 1982; Feigin and Svergun, 1987). One possible way to understand the scattering process is to start from the concept of the scattering from a single particle, fixed in space. This is sketched on Fig.1, where an incident beam of wave vector $\vec{k}_0$ strikes the particle at the points O and P, separated by the vector $\vec{r}$.

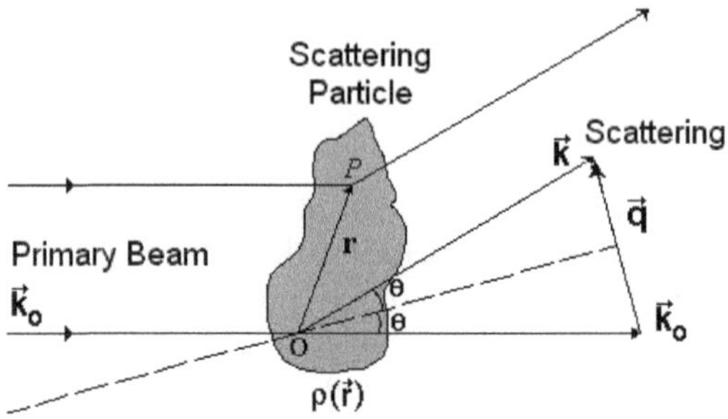

Fig. 1. Representation of the scattering process for a fixed particle.

Since the scattering is assumed to be elastic, the scattered wave with wave vector $\vec{k}$ has the same modulus of the incident wave so the difference between the incident and the scattered beam is given by:

$$\vec{q} = \vec{k} - \vec{k}_0, \qquad |\vec{k}| = |\vec{k}_0|$$
$$\left.\begin{array}{l} q = 2k\sin\theta \\ k = \dfrac{2\pi}{\lambda} \end{array}\right\} \Rightarrow q = \frac{4\pi}{\lambda}\sin\theta \tag{1}$$

which leads to the definition of the reciprocal space momentum transfer vector q. The scattering amplitude $f(\vec{q})$ is given by the Fourier transformation of the particle electron density $\rho(\vec{r})$:

$$f(\vec{q}) = \frac{1}{4\pi}\int_V \rho(\vec{r})\exp(-i\vec{q}\cdot\vec{r})d\vec{r} \tag{2}$$

The measurable quantity is the scattering intensity $I_1(\vec{q})$, which is the square modulus of the scattering amplitude:

$$I_1(\vec{q}) = |f(\vec{q})|^2 = f(\vec{q}) \bullet f(\vec{q})^*$$
$$I_1(\vec{q}) = \int_V\int_V \rho(\vec{r}')\rho(\vec{r}' - \vec{r})\exp(-i\vec{q}\cdot\vec{r})d\vec{r}d\vec{r}' \tag{3}$$

The index "1" indicates that until now this intensity is related to a single particle with fixed orientation. One usual mathematical procedure is to take the convolution integral in r' and define as the so called self-correlation function $\gamma(\vec{r})$:

$$\gamma(\vec{r})=\int_V \rho(\vec{r}')\rho(\vec{r}'-\vec{r})d\vec{r}' \tag{4}$$

Now the scattering from a single fixed particle can be rewritten as,

$$I_1(\vec{q})=\int_V \gamma(\vec{r})\exp(-i\vec{q}\cdot\vec{r})d\vec{r} \tag{5}$$

The self-correlation function $\gamma(\vec{r})$ has several properties and asymptotic limits enabling the retrieving of several general parameters. The interested reader is invited to read the seminal book from Guinier and Fournet (Guinier and Fournet, 1955) and the articles from Ciccariello (Ciccariello, 1985) among others. A theoretical calculation of a 2D scattering profile for a fixed particle in space is shown in Fig2.

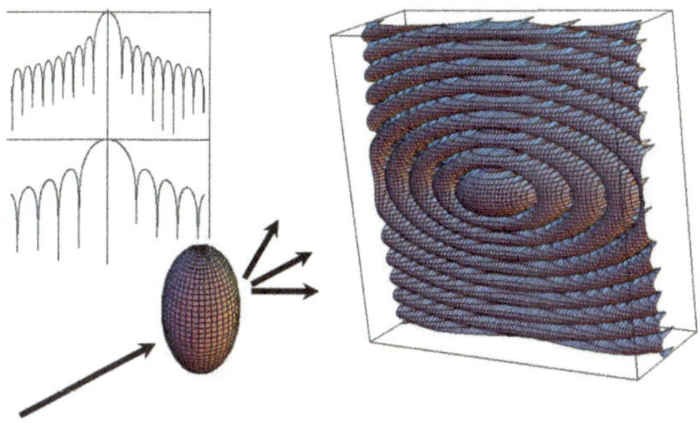

Fig. 2. Theoretical calculation for a two dimensional scattering profile for a fixed ellipsoidal particle. The intensity is given in logarithm scale. Inserts: vertical and horizontal 1D-profiles of the intensity.

Equation (5) is still too general to be used in practice. In real systems, the particles investigated are not fixed in space but instead they might be randomly oriented. The averaging procedure can be made either in real space or in reciprocal space. From the mathematical point of view is easier to perform the average in reciprocal space, by an extra integration over the solid angle Ω:

$$I_1(q)=\left\langle f(q)^2\right\rangle_\Omega=(4\pi)^{-1}\int_\Omega I_1(\vec{q})d\Omega \tag{6}$$

Substituting (5) on (6) we have,

$$\begin{aligned} I_1(q)&=\frac{1}{4\pi}\int_0^\infty\int_0^{4\pi} r^2drd\omega\int_0^{4\pi} d\Omega\gamma(\vec{r})\exp(-i\vec{q}\cdot\vec{r}) \\ &=\int_0^\infty r^2dr\int_0^{4\pi}\gamma(\vec{r})d\omega\frac{1}{4\pi}\int_0^{4\pi}\exp(-i\vec{q}\cdot\vec{r})d\Omega \end{aligned} \tag{7}$$

Where the volume integral element for $d\vec{r}$ was written as $r^2 dr d\omega$, the spherical coordinates in real space. In the last operation the terms were rearranged. The angular integrals can be performed directly:

$$\left\langle \exp(-i\vec{q}\cdot\vec{r}) \right\rangle_\Omega = \frac{\sin(qr)}{qr}$$
$$\gamma(r) = \frac{1}{4\pi}\int_0^{4\pi} \gamma(\vec{r})\, d\omega = \left\langle \rho(\vec{r}) * \rho(-\vec{r}) \right\rangle_\omega \tag{8}$$

where the function $\gamma(r)$ is known as average self-correlation function. With these substitutions the intensity for a single particle randomly oriented is given by

$$I_1(q) = 4\pi \int_0^\infty \gamma(r) \frac{\sin(qr)}{qr} r^2 dr \tag{9}$$

The theoretical intensity of an ellipsoidal particle randomly oriented in space if shown in Fig3. As can be directly seen, now the 2D spectra is angular independent and any cut starting from the center towards a radial direction will have the same profile.

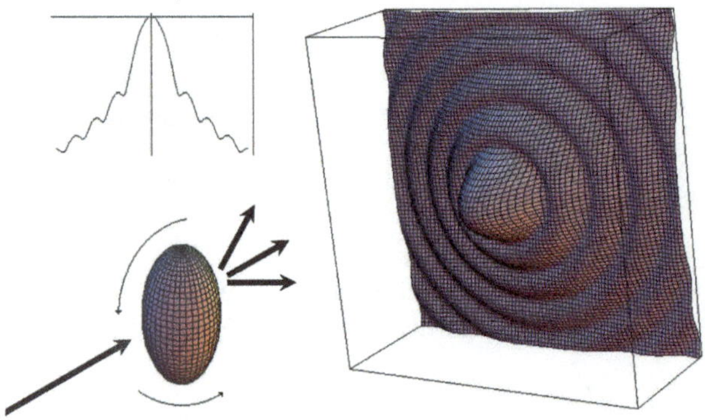

Fig. 3. Theoretical calculation for a two dimensional scattering profile for an ellipsoidal particle randomly oriented. The intensity is given in logarithm scale. Insert: 1D-profiles of the intensity.

One usual procedure is to define the so called pair distance distribution function *p(r)*, $p(r) = r^2\gamma(r)$ which is a histogram of pair distances inside of the particle, weighted by the distance length and by the product of the electron densities of the infinitesimal elements of the pair (Glatter, 2002). The *p(r)* function permits the definition of the Fourier pair:

$$I_1(q) = 4\pi \int_0^\infty p(r) \frac{\sin(qr)}{qr} dr$$
$$p(r) = \frac{r}{2\pi^2} \int_0^\infty q^2 I_1(q) \frac{\sin(qr)}{qr} dq \tag{10}$$

Some interpretations for the $p(r)$ function will be explained in the next sections.
The result given in equation (10) was derived for a single particle randomly oriented in space. In real systems the particles are dispersed in a matrix with electron density ρ_0 and therefore it is necessary to extrapolate this result for a system of particles. One expression for the intensity of a system of particles is given by

$$I(q) = N\left\langle f^2(\vec{q})\right\rangle\left\langle S(\vec{q})\right\rangle \tag{11}$$

Where $< f(\vec{q})^2 > = P(q)$ is the so called particle form factor and the function $S(\vec{q})$ is the system structure factor. For systems composed of identical particles (monodisperse sytems) the form factor is identical to the average scattering intensity of a single particle, $P(q) = I_1(q)$. For polydisperse systems the form factor will be an average over the different sizes, electron densities and particle shapes. In this case, one usual procedure is to assume a known shape and electron density and performs the average over a distribution of sizes (Glatter and Kratky, 1982). On the other hand, the structure factor is related with particle interactions and there are several approaches to describe its behavior (Pedersen, 2002). For very diluted systems the particle interactions can be neglected and the structure factor is equal to 1. Therefore for a system of identical particles in a dilute regime we have

$$I(q) = NI_1(q) \tag{12}$$

indicating that, although the measured intensity is a contribution of a large number of particles, it contains the information of the scattering from a single particle randomly oriented. This shows that in a real case where the intensity $I(q)$ is measured, it might be possible to obtain information about the single particle shape and conformation.
As mentioned above, the particles can be considered to be immersed in a matrix with constant electron density. It can be shown that in this case the scattering event will only happen if there are differences between the electron density of the particles and the matrix. In this way the electron density $\rho(r)$ shown in equations 2-5 should be replaced by the electron density contrast $\Delta\rho(r) = \rho(r)-\rho_0$. In order to make this point clearer, one usual approach is to consider the particle form factor to be normalized and the electronic mass is explicitly shown:

$$I(q) = N\Delta\rho^2 V^2 P_1(q) \tag{13}$$

where $\Delta\rho = \rho - \rho_0$, is the scattering contrast between the particles and the matrix, V is the particle volume and $P_1(q)$ is the normalized form factor ($P_1(q=0)=1$).

2.1 Experimental aspects and absolute calibration

A schematic setup for a typical SAXS experiment is shown in figure 4. Specific technical details about geometries and configurations can be found in several sources (Lindner and Zemb, 2002) and it will not be presented in this chapter. However, some general characteristics have to be addressed. Since only a small fraction of the incident beam is scattered, the detectors should be set to detect reasonably low intensities. Therefore the incoming beam that passes without interaction with the sample has to be blocked by a beam stopper, to avoid possible damaging of the detector. The size of the beam stopper depends on the equipment geometry.

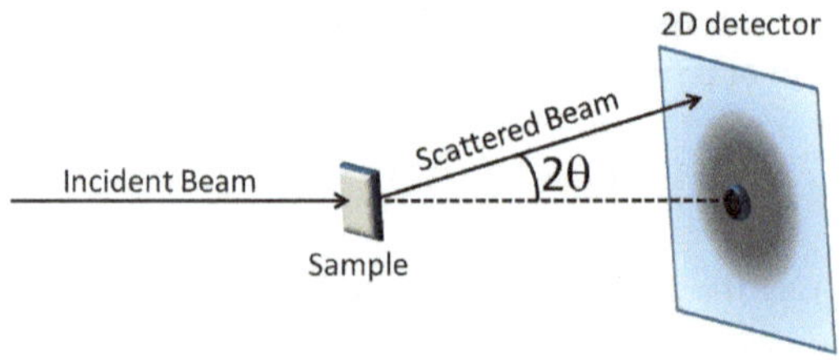

Fig. 4. Schematic setup of a SAXS experiment.

In a typical experiment it is necessary to measure the intensity from the system (sample → matrix+particles) and subtract the intensity from the matrix where the particles are immersed (blank). To normalize the data to absolute scale, scattering standards have to be used. In the applications described in the present chapter two procedures were applied. In one procedure a known protein is measured on the same sample conditions (buffer, temperature, etc) and the forward scattering obtained for this sample is used to normalize the other, unknown data. In other procedure water at 20°C is used as primary standard. This is a convenient standard since the value of the scattering cross section can be calculated with very high accuracy from the fundamental macroscopic properties of water. In both cases, the data has to be normalized by the value obtained from the standard on the same experimental conditions and multiplied by the theoretical intensity value. Assuming that the sample and the blank are measured in the same cell, the treated intensity, normalized to absolute scale can be given by:

$$I_{Treated}(q) = \left(\frac{I_{sample}(q)}{\Phi_s T_s t_s} - \frac{I_{blank}(q)}{\Phi_b T_b t_b} \right) \frac{d\Sigma(0) / d\Omega_{std}}{I(0)_{std}} \tag{14}$$

Where q is modulus of the scattering vector, defined as $q = (4\pi/\lambda)\ sin\ \theta$, where 2θ is the scattering angle as shown in Fig1 and Fig4 and λ is the wavelength of the monochromatic beam; $I_{Treated}(q)$ is the treated scattering intensities for the sample on absolute scale, i.e. the scattering cross section of the sample; $I_{sample}(q)$ is the raw data measured for the sample, $I_{blank}(q)$ is the raw data from the matrix scattering; $I(0)_{std}$ is the value at $q = 0$ of treated standard data (background subtracted and normalized by flux, transmission and acquisition time); Φ_i is the flux of the incident beam; T_i is the sample transmission and t_i is the exposure time, where the index i is s (sample), b (blank); $(d\Sigma/d\Omega)_{std}$ is the theoretical scattering cross section for the standard. For water at 20°C this cross section has the value 0.01632 [cm^{-1}]. For proteins in typical buffers without high amounts of salt, glycerol or other additives, the theoretical cross section for a system of proteins in solution with mass concentration c (in mg/mL) and molecular weight M_W (in kDa) is given by $(d\Sigma/d\Omega)_{std}=6.645x10^{-4}\ c\ M_W$ [cm^{-1}] (see equation 15 below).

Having the data on absolute scale, information about its contrast, particle volume or particle concentration can be obtained, depending on the knowledge about the system. One very important parameter when studying proteins in solution is the determination of the molecular weight, which is a direct indication of the oligomerization state of the protein. Starting from equation (13), multiplying and dividing by the particle specific volume $\overline{v}$ and some simple algebraic manipulations is possible to rewrite the intensity $I(q)$ as:

$$I(q) = c(\Delta\rho_M)^2 (M_W / N_A) P(q) \tag{15}$$

were c is the concentration in mg/mL, $\Delta\rho_M$ is the excess scattering length density per unit mass (cm/g), M_W is the molecular weight in kDa, N_A is the Avogadro's number and $P(q)$ is the normalized form factor ($P(0)=1$). For proteins, a good approximation of $\Delta\rho_M$ is $2x10^{10}$cm/g (Oliveira and Pedersen, unpublished). The above equation directly shows that the molecular weight of the proteins can be directly estimated from the forward intensity $I(0)$:

$$M_W = \frac{I(0)}{c(\Delta\rho_M)^2} N_A \tag{16}$$

In general, the precision on the molecular weight determination has an uncertainty of 10% - 20%, which enables to check the monodispersity of the sample or to indicate the oligomeric state. However, this approach is very dependent on the knowledge of the scattering contrast and sample concentration.

2.2 Modeling methods

From the above considerations it is possible to see that from the analysis of SAXS data it might be possible to obtain structural information about the studied system. There are several methods that can be used, depending on the knowledge about the system. Usually the information that is desired is the scattering length density distribution $\Delta\rho(r)$, which might provide the particle shape, size, etc. This approach is the so called "inverse scattering problem", ie, retrieve real space information from the data in reciprocal space. The modeling is based on the comparison of a given model and experimental SAXS data. From the characteristics of scattering experiments the χ^2 (chi-square) test is a good minimization function for the optimization procedure. Given a set of N experimental points $I_{exp}(q_i)$ with standard deviations $\sigma(q_i)$ an the theoretical intensity $I_{teo}(q_i)$ calculated for the same angular values, q_i, the χ^2 function is defined as:

$$\chi^2 = \sum_{i=1}^{N} \frac{\left(I_{\exp}(q_i) - I_{teo}(q_i)\right)^2}{\sigma(q_i)^2} \tag{17}$$

A common practice is to use the reduced χ^2, $\chi_R^2 = \chi^2 / (N - M)$, where N is the number of experimental points and M is the number of independent parameters used in the theoretical model. If a good fitting is achieved, the differences between the model and the experimental data will have to be lower or equal to the standard deviations $\sigma(q_i)$. Therefore, since χ_R^2 is normalized by $(N-M)$ if N is reasonably larger than M, the χ_R^2 for a good fit should be close to 1. Values considerably larger than 1 might indicate important discrepancies between the model and experimental data. However, it can also indicate underestimated uncertainties.

On the other hand, values considerably lower than 1 can indicate overestimated uncertainty values.

2.2.1 Indirect fourier transform – model independent approach

In the theoretical description shown above, the pair distance distribution function *p(r)* was introduced as a natural step on the equation manipulation and, as indicated in equation (10), it forms a Fourier pair with the scattering intensity of a single particle $I_1(q)$. Since the total intensity from a system is proportional to the scattering of a single particle (equation (12)), this procedure might be used to calculate the real space function *p(r)* from measured scattering data. This procedure has intrinsic limitations since the Fourier transformation involve integrals from 0 to infinity and the measured scattering data is only obtained for a very small region of reciprocal space. As a consequence, direct calculations of the *p(r)* function from the integral of *I(q)* are usually not successful since the truncation of the integral leads to strong oscillations of the *p(r)* function. Another method was introduced by Glatter (Glatter, 1977) and it is known as Indirect Fourier Transformation method (Program ITP and GIFT; Glatter, 1977; Bergmann et al, 2000; Fritz and Glatter, 2006). In this approach one starts from the *p(r)* function, describing it using a set of base functions (in the Glatter method, spline functions) and perform the Fourier transformations on those functions in order to have a similar set of base functions in reciprocal space. Since all operations are linear, the coefficients of the *p(r)* base functions are the same as the ones for the *I(q)* base functions and therefore by the fitting of the experimental data one can direct obtain the best set of coefficients and consequently the best *p(r)* functions. Since the interval of *I(q)* is still limited, this operation also leads to oscillating *p(r)* functions. In order to avoid this problem, Glatter introduced a damping parameter that is selected in the fitting procedure in order to provide a smooth *p(r)* function. A similar approach was used by Svergun and co-workers (Semenyuk and Svergun, 1991) in the program package GNOM. In both cases the fitting process is iterative and the user has to obtain the maximum particle size D_{MAX} that gives the best fit and *p(r)* function. In an interesting development Hansen (Hansen, 2000) proposed a method where the maximum dimension is obtained using Baesyan probabilities. Recently, performing a procedure based on the Glatter method (Pedersen et al, 1994), Oliveira and Pedersen developed a procedure that enabled the calculation of the *p(r)* function from both diluted (program WIFT) and concentrated systems (program WGIFT), where structure factors are taken into account in the optimization (Oliveira et al, 2009). The calculation of the *p(r)* function for concentrated systems was also implemented by Glatter in a new implementation of his approach (Program GIFT) by optimization using simulated annealing.

A common result of all the above program packages is the pair distance distribution function *p(r)*. As mentioned above, this function is a histogram of pair distances inside of the particle, weighted by the distance length and by the product of the electron densities of the infinitesimal elements of the pair. For particles with finite size, it will exists a maximum distance from which the *p(r)* function is zero. This corresponds to the maximum size of the particle. Since the histogram is weighted by the distance length, the *p(r)* function also might starts from zero. In this way, it is easy to see that the *p(r)* should start from zero and ends at zero when reach the maximum particle size. The shape of the function will be a consequence of the particle shape and electron density distribution. A set of theoretical calculations for the *p(r)* function is shown in Fig5, Fig6 and Fig7. In Fig5 one can see that globular particles

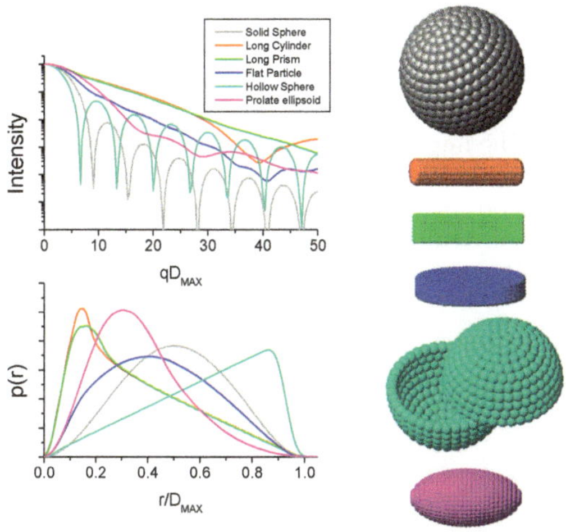

Fig. 5. Theoretical calculations for scattering intensities and corresponding *p(r)* functions for bodies with simple shapes. The form factors were normalized to one.

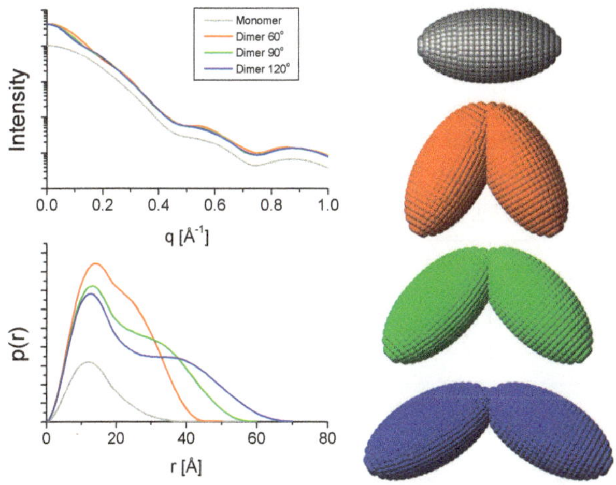

Fig. 6. Theoretical calculations for scattering intensities and corresponding *p(r)* functions for bodies with simple shapes. The form factors were normalized by the forward scattering of the monomer.

will have a *p(r)* function with a bell shape, with the maximum close to $(r/D_{MAX})/2$. Any anisotropy will move the maximum to the left, towards lower r/D_{MAX} values. Elongated (prolate) particles with constant cross-section like cylinders or prisms will have *p(r)* functions with linearly descent regions. Flat (oblate) particles will have *p(r)* functions with shapes different from the two previous cases. Hollow particles will have *p(r)* functions with the maximum moved to the right, towards higher r/D_{MAX} values. Dimeric particles will have *p(r)* with shoulders, as viewed in Fig6. Interestingly, the differences in the opening angle of a dimeric particle are easier to detect in the *p(r)* function than in the intensity *I(q)*. Finally, particles with differences in the scattering length contrast might have *p(r)* functions with negative portions as indicated in Fig7. For a broader and deeper review on the *p(r)* interpretation the reader is invited to read several works in the literature (Glatter, 1979; Glatter and Kratky, 1982). The important point of this modeling approach is that, apart from the assumption that the system is composed of identical particles, no other hypothesis are made and the *p(r)* function provides a direct insight about the particle shape and dimensions. This approach is widely used in analysis of SAXS data because it provides a first guess about the particle shape.

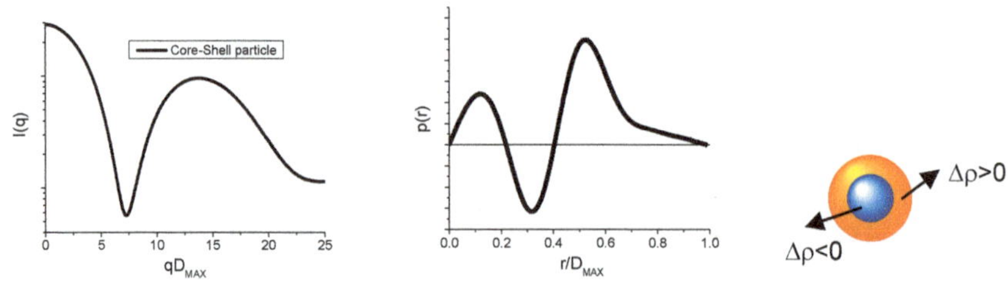

Fig. 7. Theoretical calculations for scattering intensities and corresponding *p(r)* functions for a core-shell particle with different scattering length contrasts.

2.2.2 Model dependent approach – assuming a known form factor

For simple particle shapes it is possible to integrate equation (2) and obtain the amplitude form factor $f(\vec{q})$. Then, performing the angular integral given in equation (6) it is possible to obtain a analytical or semi-analytical expression for $I_1(q)$. Some examples are shown in Table 1. A more complete list of analytical expressions for form factors can be found in

Shape	Normalized Form Factor
Homogeneus sphere with radius R	$I_1(q) = f_1(q)^2 = \left(\frac{3[\sin(qR) - qR\cos(qR)]}{(qR)^3} \right)^2$
Spherical shell with inner radius R1 and outer radius R2	$I_1(q) = \left[\frac{V(R_1)f_1(q,R_1) - V(R_2)f_2(q,R_2)}{V(R_1) - V(R_2)} \right]^2$ $V(R) = 4\pi R^3 / 3$
Homogeneous cylinder with radius R and height L	$I_1(q) = \int_0^{\pi/2} \left[\frac{2J_1(qR\sin\alpha)}{qR\sin\alpha} \frac{\sin(qL\cos\alpha / 2)}{qL\cos\alpha / 2} \right]^2 \sin\alpha d\alpha$
Ellipsoid of revolution with semi axes R, R and εR	$I_1(q) = \int_0^{\pi/2} f_1(q, r(R,\varepsilon,\alpha))\sin\alpha d\alpha$ $r(R,\varepsilon,\alpha) = R\left(\sin^2\alpha + \varepsilon\cos^2\alpha\right)^{1/2}$

Table 1. Few examples of semi-analytical expressions of the scattering intensity calculated for particles with simple shapes.

several reviews (Pedersen, 1997). Also, advanced modeling approaches can be found in specialized articles in the literature (Székely et al, 2010).

The main advantage of the use of analytical or semi analytical expressions describing the form factor is that, usually, there is a low number of parameters to adjust against experimental data, permitting the determination of structural information with reasonable reliability. Also, if the model does not fit the data correctly, this directly indicates that the particle shape is different from the one that is been assumed. One example of application si presented in Fig8 where the model of an elongated cylinder was used to described the SAXS data of mature glucagon fibers. In several cases, the particle possible shape is known but the calculations of the integrals is impractical. In these cases it is possible to use the finite element method which consists of build up the particle shape using known subunits. One approach is to use spherical subunits and apply the Debye formula to calculate the intensity (Debye, 1915; Glatter, 1980):

$$P_{Model}(q) = \frac{f_1(q, r_{bead})^2}{N^2} \sum_{i,j=1}^{N} \frac{\sin(qr_{ij})}{qr_{ij}} \tag{18}$$

This procedure enables the calculation of very complicated models. From this calculation the model parameters can be optimized against experimental data (Oliveira et all, 2009, 2010).

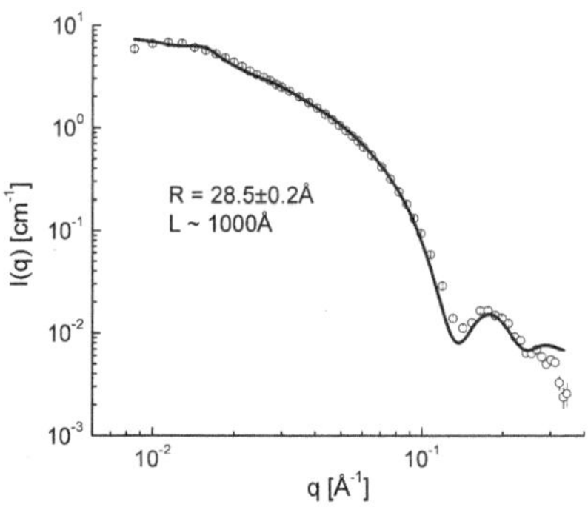

Fig. 8. Example of application of the use of a known form factor . Experimental data (open circles) of a mature fiber of Glucagon (Oliveira et al, 2009) and the theoretical fit (solid line), assuming a form factor of cylinders, with radius R and length L. The SAXS data was measured at the laboratory SAXS instrument Nanostar™, from Professor Jan Skov Pedersen at University of Aarhus, Denmark.

2.2.3 Deconvolution square root – obtaining the electron density profile

In some applications, the particle shape is known but the electron density profile and overall dimensions have to be determined. Amphifilic molecules like surfactants and several types of diblock copolymers self assemble into structures that can be analyzed in this way. Several propositions for the deconvolution square root procedure can be found in the literature (Pape and Kreutz, 1978; Nagle and Wierner, 1989). An initial approach was to take the square root of the scattering intensity, which gives an absolute value for the amplitude function *f(q)*. Then, by Fourier transforming this function is possible to retrieve the electron density distribution *ρ(r)*. This procedure has serious problems since the signals of the *f(q)* function has to be guessed an also the very short interval of data on reciprocal space precludes a trustful calculation of the inverse Fourier transformation. A more stable process was proposed by Glatter (Glatter 1981; Glatter 1984, Bergmann et al, 2000) where the deconvolution is made by the use of the *p(r)* function. Apart of the overall sign of the electron density profile (±1 factor), this procedure enables a correct estimation of the electron density profile, and has been used in several applications (Rathgeber et al, 2002). One example of application of this method is presented in Fig9, where the radial electron density of SDS micelles could be obtained.

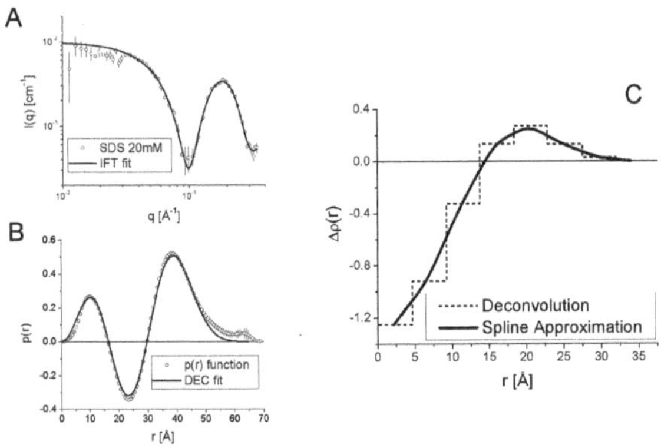

Fig. 9. Example of application of the deconvolution method. A) Experimental SDS data (open circles) and the IFT fit (solid line). B) IFT Calculated *p(r)* function (open circles) and the theoretical *p(r)* obtained from the deconvolution method (solid line). C) Restored radial electron density profile presented as step functions (dashed line) and by the use of a smooth approximation (solid line). The SAXS data was measured at the laboratory SAXS instrument Nanostar™, from Professor Jan Skov Pedersen at University of Aarhus, Denmark.

2.2.4 Ab initio modeling – an overview

The shape of the scattering function is directly related to the three-dimensional shape of the particle. However, since the particles are randomly oriented (equation 6) and there is only a limited measurable region in reciprocal space, the information content in a SAXS curve is very low (Patel and Schimidt, 1971). Nevertheless, even with these limitations important developments occurred in the last decades have proof that it is possible to obtain a 3D model from the 1D SAXS curves. Starting from the seminal work from Sturhmann in late 70's (Sturhmann, 1973; Sturhmann and Miller, 1978), Svergun and co-workers had used a set of spherical harmonics (multipolar expansion) to describe the particle electron density and by the use of a nonlinear minimization procedure it is possible to obtain the set of spherical harmonics coefficients that gives the best fit of the scattering data. Details on the calculations and the representation of the scattering intensity using spherical harmonics can be found in the original articles (Svergun and Sturhmann, 1991). The success of this method has shown that, even though it is not possible to obtain a unique solution for the particle shape, the fitting of the experimental scattering data enabled a direct *ab initio* determination of the three-dimensional shape. Since the representation using spherical harmonics only enables the construction of smooth shapes without sharp edges or corners, this approach provides a rough representation of the particle shape. In this way it can be said that this method provides a very low resolution approximation of the scattering data and usually enables the fitting of only the initial part of the scattering intensity. Program packages that enables the fitting of experimental data are available in the literature (programs ASSA and SASHA, Kozin et al, 1997). An example of this procedure is shown in fig10 where the experimental data for lysozime in solution was adjusted using multipolar expansion by using the program SASHA. It can be directly seen that the correct anisotropy and overall shape can be obtained from this approach. In all the examples shown in figures 10-14 the measurements were performed in the SAXS beamline of Brazilian Synchrotron Light Laboratory.

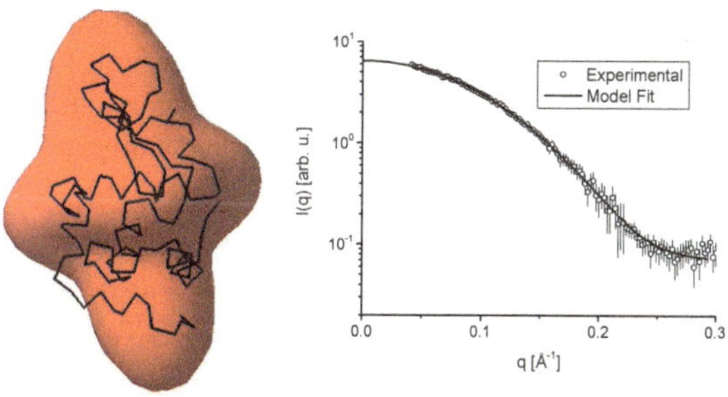

Fig. 10. *Ab initio* modeling of experimental SAXS data using multipolar expansion. Left: on red it is shown the model obtained superposed with the backbone of the protein obtained from its known crystallographic structure (pdb entry *6lyz.pdb*). Right: Fit of experimental data. Open circles - experimental data. Solid line - model fit.

A further improvement on the *ab initio* procedure for modeling SAXS data was proposed initially by Chacón (program DALAI, Chacón et al, 1998), and later by Svergun (Program DAMMIN, Svergun, 1999), Doniach (program SAXS3D, Walter et al, 2000) among others. In this method the particle is build using the finite element approach, by the use of a closed packing arrangement of spherical subunits. Since the number of possible solutions is very large, Monte Carlo based optimization methods are used to obtain the set of spherical subunits that gives the best fitting of the scattering data. The program DAMMIN is widely cited in the literature and starts by creating a spherical search space with diameter equal or slightly larger than the particle maximum diameter D_{MAX} (obtained from the $p(r)$ curve). By the application of a simulated annealing procedure, constrained by penalty functions that ensure particle compactness and smoothness (Volkov and Svergun, 2002), a subset of the initial search space can be obtained providing a three-dimensional model that represents the particle shape. Due to the intrinsic randomness of Monte Carlo approaches, several independent runs of these model procedures will lead to different models. However, it is possible to show that all models might share similar features like overall anisotropy, size, etc. This model approach permits a better representation of the particle shape than the multipolar expansion since it does not have the above mentioned limitations for the shape description. However, since the internal structure is not represented, this method cannot describe data up to high q values (Volkov and Svergun, 2002). One example of this so called "dummy atom modeling" is shown in Fig11 where the *ab initio* model was obtained from SAXS data of lysozyme in solution.

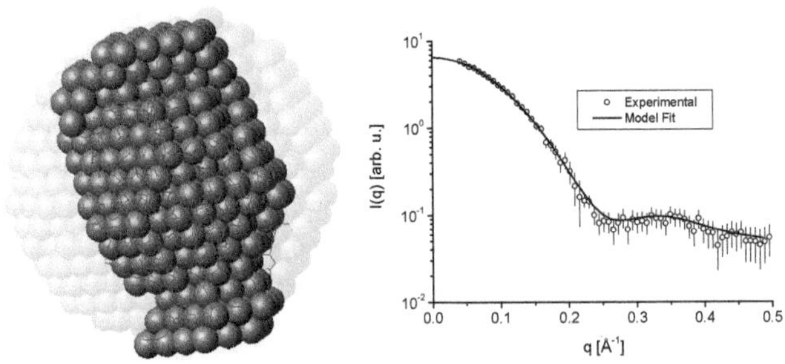

Fig. 11. *Ab initio* modeling of experimental SAXS data using dummy atom modeling. Left: model results. Semitransparent spheres - initial search space. Solid spheres - selected subset that gives the best fit. Blue backbone - protein backbone obtained from its known crystallographic structure (pdb entry *6lyz.pdb*). Right: Fit of experimental data. Open circles - experimental data. Solid line - model fit.

When dealing with SAXS data from proteins, one additional very useful constraint can be used for the model building. Proteins are composed of a sequence of aminoacids, which forms its backbone, known as primary structure. This primary sequence folds into specific patterns like α-helices, β-sheets, turns, etc, composing the secondary structure. Finally, the secondary structure folds into a specific three-dimensional arrangement, known as tertiary structure. In some cases this protein can even be part of a supramolecular complex which comprises the quaternary structure (Voet et al, 2008). Due to the intrinsic low resolution and information of a SAXS data, the information about the atomic resolution structure or secondary structure cannot be accessed but instead, the overall shape and size. However, the information of sequence continuity can be used as a constraint to enable a better modeling of proteins in solution. This procedure was implemented by Svergun, Pethoukov and co-workerks in the "dummy chain model" approach (Program GASBOR, Svergun et al, 2001). In this method a sequence of interconnected chains is used to represent the protein backbone. Each sphere corresponds to one amino acid and therefore the total number of spheres is identical to the number of protein residues. Starting from a spherical arrangement of the backbone the optimization program performs a simulated annealing optimization in which the backbone three-dimensional arrangement is changed in order to provide the best fitting of the scattering data. Similarly to the dummy atom approach, the theoretical intensity is calculated using a variation of the Debye formula (equation 18). The natural constraint imposed by the continuity of the backbone makes leads to better representations of protein structures. Also, this approach can fit experimental data up to higher q values than the previous ones since the internal structure of the protein is somehow represented by the backbone. One example of this so called "dummy chain modeling" is shown in Fig12 where the *ab initio* model was obtained from SAXS data of lysozyme in solution.

The previous examples showed the possibility of apply *ab initio* methods to retrieve the three-dimensional structure. Although the model results for the dummy atom and dummy chain methods are not unique due to the heuristic nature of the optimization methods, the

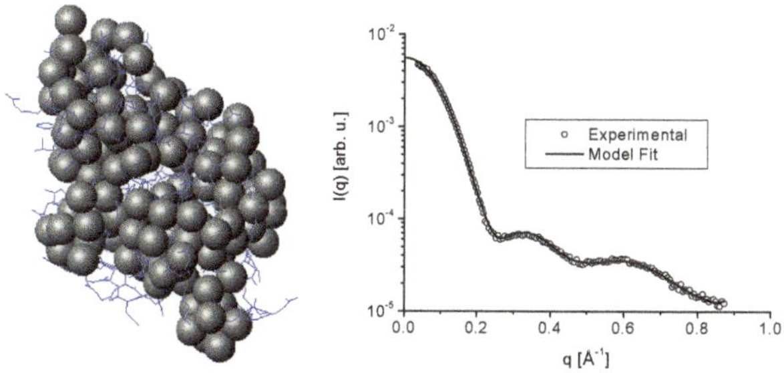

Fig. 12. *Ab initio* modeling of experimental SAXS data using dummy chain modeling. Left: optimized backbone structure (solid spheres) superimposed by the backbone obtained from the protein known crystallographic structure (pdb entry *6lyz.pdb*). Right: Fit of experimental data. Open circles - experimental data. Solid line - model fit.

overall size, shape and anisotropy can be obtained from these approaches. Another very useful application of the study of proteins in solution is the use of known atomic resolution data in connection with SAXS data. If the full atomic model for the protein is known, the comparison of the theoretical scattering intensity against experimental data provides direct information about the conformation of the protein in solution in comparison with the atomic resolution structure. A good fit indicates that the structure of the protein in solution is similar to the given by the atomic resolution model. Discrepancies in the fit indicate differences between the atomic resolution model and the protein structure in solution. A widely cited procedure that enabled a successful comparison between experimental data and atomic resolution structures was developed by Svergun and co-workers (program CRYSOL, Svergun et al, 1995) where it was demonstrated that a hydration shell around the protein with slightly higher electron density than the one from the bulk was necessary to be considered. One example of this comparison between experimental data and theoretical SAXS intensity calculated from atomic coordinates is shown in Fig13 where the crystallographic structure for the protein lysozyme was used for the comparison with experimental data of lysozyme in solution.

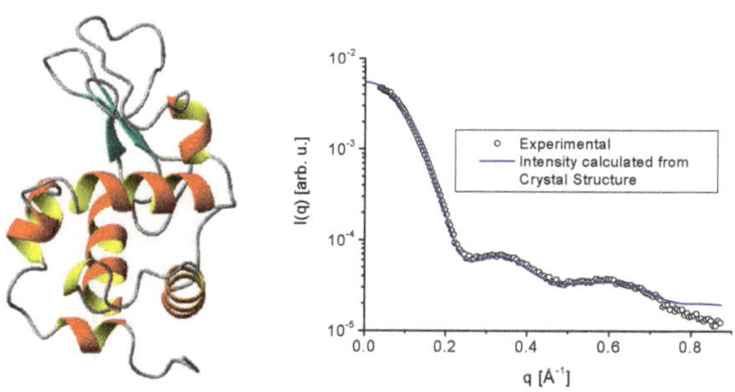

Fig. 13. *Ab initio* modeling of experimental SAXS data using dummy chain modeling. Left: representation of the crystallographic structure of lysozyme (pdb entry *6lyz.pdb*). Right: Fit of experimental data. Open circles - experimental data. Solid line - theoretical fit.

One of the major applicability of the use of SAXS data and the knowledge about atomic resolution models for proteins is for the cases where just part of the structure is known. In these situations the SAXS data can be used to generate (using the dummy chain approach) the missing aminoacid loops in the known structure (program BUNCH, Petoukhov and Svergun, 2005) or/and to obtain the spatial arrangement of known domains in order to form the full structure (program SASREF, Petoukhov and Svergun, 2005). Both the generation of the missing loops and the optimization of domains are performed by the use of Monte Carlo methods which, similarly to the previous cases, do not lead to a unique solution. However, even though the solution is not unique, the obtained model is a very good representation of the overall structure. Test examples are shown in Fig14 and Fig15. In Fig14 part of the lysozyme structure was clipped and as it can be seen in the curve, without the loop the atomic model cannot fit the experimental data correctly. With the addition of a dummy chain loop and its optimization it is possible to obtain a very good fit of the experimental data. The generated loop (blue loop in the model) is a reasonable approximation of the real loop superposed to it.

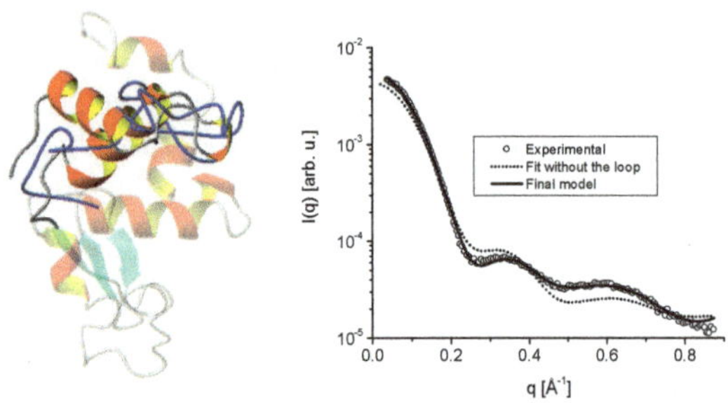

Fig. 14. *Ab initio* modeling of missing loop of a hypothetical structure using experimental SAXS data. Left: crystallographic structure of lysozyme (pdb entry *6lyz.pdb*) and the restored loop. Semitransparent structure - lysozyme structure with a missing part. Blue blackbone - restored loop superposed to the real, clipped loop. Right: Fit of the experimental data. Open circles - experimental SAXS data for lysozyme in solution. Dotted line - fitting of the scattering data for the structure without the loop. Solid line - fitting of the scattering data for the structure with the optimized loop.

In Fig15 a hypothetical situation of a heterodimer is shown. The optimization of the structure components does not give a perfect agreement with the initial structural but there is a remarkable similarity, indicating that SAXS data can also be applied in these cases.
The situations presented here are just a small representation of possibilities for the applications of these modeling tools. Advanced modeling examples based on these procedures can be found in several articles in the literature (Svergun, 2007). An intrinsic problem of any SAXS modeling is the ambiguity that might arises in the results. In general, it is not possible to obtain a unique solution from the modeling procedure. Therefore it is necessary to complement any scattering modeling with additional information in order to reduce the number of possible solutions. There are several ways on doing this. When available, information about binding sites or specific arrangement of domains can be used as constraints in the modeling. Results from biochemical/biophysical techniques can provide useful information about structure change or binding. For example, fluorescence spectroscopy and isothermal titration calorimetry can provide important information on binding and stoichiometry. In recent applications the simultaneous modeling of scattering

data and other experimental data is been tried. A simultaneous modeling of SAXS and NRM data was proposed by Mareuil and co-workers in the program DADIMODO (Mareiul et al, 2007). Also, automatic tools for the use of complementarity between SAXS and NMR is been currently developed in connection with the SAXIER project (Svergun, 2007; Svergun, 2009).

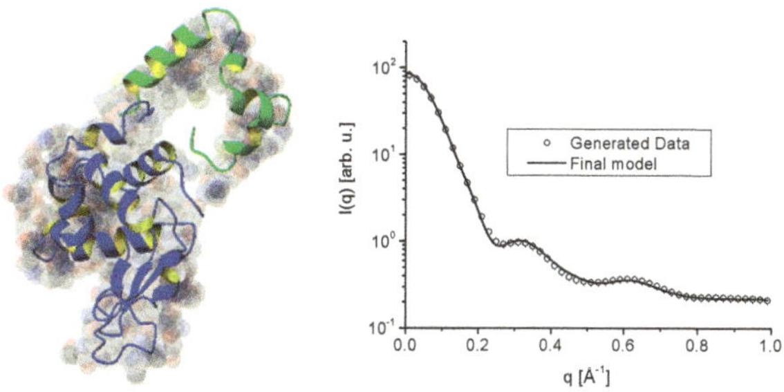

Fig. 15. Rigid body modeling of a hypothetical structure using calculated SAXS data. Left: hypothetical heterodimer built using two atomic resolution structures. Semitransparent spheres: original structure. Blue and green strands: optimized heterodimer. Right: Fit of the generated data. Open circles - generated SAXS data for the heterodimer. The data was created using program CRYSOL from the built model. Standard deviations were added in order to mimic experimental uncertainties. Solid line - fitting of the scattering data for the optimized structure.

3. Applications

Two applications of SAXS analysis will be presented. In the first case, an *in-situ* aggregation study of lysozyme is presented. As a second example a structural characterization of a giant protein complex is described. These two cases are good examples of the application of the SAXS technique to investigate biological systems.

3.1 Lysozyme denaturation and aggregation induced by heat

The structure of proteins is intrinsically related to its shape. The protein shape, on the other hand, is a result of the protein folding. In the native state, proteins are known to adopt hierarchical structures, which might be a result of a multistep folding process. One possible way to investigate this characteristic is to induce protein denaturation. The denaturation or unfolding can be induced by changes in temperature, pH, or even by the addition of denaturant agents like sodium dodecyl sulfate (SDS). A study of denaturation induced by heat will be presented here.

The experiments were performed at the SAXS beamline of the Brazilian Synchrotron Light Laboratory, Campinas, Brazil. The wavelength selected for the experiments was λ =1.49 Å and the distance between the sample and detector was 745 mm. The measurements were performed using a 1D Gabriel-type detector. The samples were exposed in a 1.5mm capillary tube in a thermally controlled sample holder directly connected to the evacuated beam path. These experiments were performed with lysozyme samples at 10 mg/mL and pH 7.0 in a 10mM phosphate buffer with 50mM of NaCl. Indirect Fourier transformations were performed using program package GNOM which enabled the correction of smearing effects. *Ab initio* models were built using program DAMMIN.

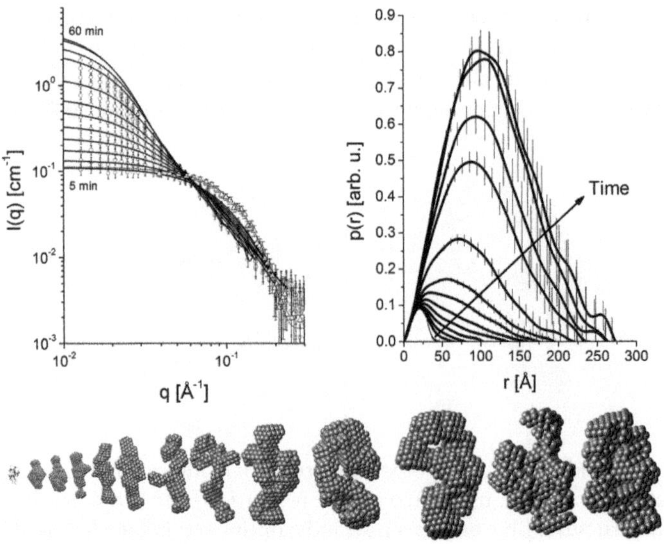

Fig. 16. Aggregation of Lysozyme induced by heat. Top left: scattering data (open circles) and desmeared IFT fits (solid lines). The frames were collected at 80ºC in intervals of 5 minutes (first-5min, last-60min). A frame of lysozyme at room temperature (open triangles and dotted line) was added for comparison. Top right: pair distance distribution functions *p(r)* for each dataset. A frame obtained from the SAXS data for lysozyme at room temperature (dotted line) was added for comparison. Bottom: *Ab initio* models restored for each frame. It is possible to see the increase in size for the average model. For comparison, the crystal structure of lysozyme is shown on the left as ribbons.

The results are shown in Fig16. As can be seen, when the protein solution is subjected to 80ºC an evolution of the SAXS profiles as a function of time is observed. As shown in equation 13 and 14, the forward scattering I(0) can provide an estimation for the molecular weight. For a system that presents the formation of aggregates over time, the obtained molecular weight will be an averaged value since a distribution of sizes can be present in the system. However, because the forward scattering is proportional to the square of the particle volume, large particles will have a higher contribution to the final intensity. If one assumes that in each stage the aggregates have a similar size, since the total mass of proteins is constant, it is possible to write,

$$I(0)_{agg} = \frac{c(\Delta\rho_M)^2}{N_A}\overline{M_{W\,agg}} \tag{19}$$

If we normalize $I(0)_{agg}$ by the forward scattering of the lysozyme measured at room temperature (native state) at the same concentration, this fraction will be a good estimation for the average number of monomers per aggregate:

$$\frac{I(0)_{agg}}{I(0)_{lyz,20^oC}} = N_{mon} \tag{20}$$

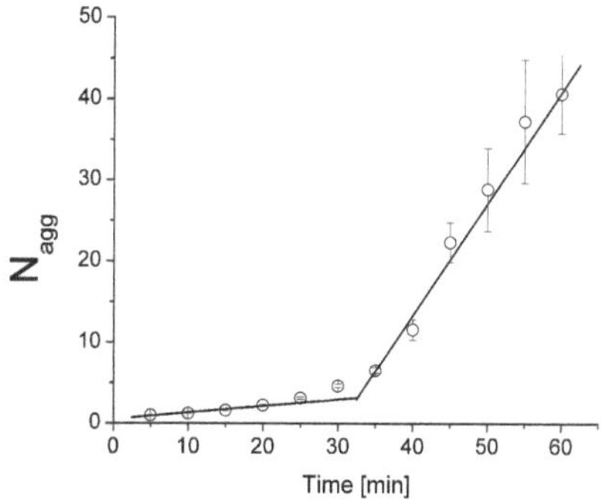

Fig. 17. Average number of monomers per aggregate as a function of time. At least two aggregation rates can be found. The lines are just for eye guide.

The forward scattering for the native protein and the first 5 min frame at 80oC are almost identical indicating that at this stage the protein is still in monomeric state. However the differences in the scattering curve and in the *p(r)* functions when compared to the native state (Fig15) indicates that the protein has adopted a different conformation. This state is known as molten-globule which is a state where the protein is partially denaturated. Interestingly, the protein starts to be denaturated at 80°C, being stable over lower temperatures (data not shown). Using equation 18, it was possible to calculate the average number of monomers per aggregate, which is shown as a function of time in Fig 17. From this graph, at least two aggregation rates can be identified in the graph, which might indicates that initially the aggregation process is slow but after 30 minutes at 80°C and with around 5 monomers per aggregate the aggregation is accelerated reaching a number of around 45 monomers per aggregate for 1h at 80oC. A visualization of the obtained aggregates is shown in Fig 14 which agrees with above conclusions. The obtained results confirms other results from the literature which indicates that the denaturation of the lysozyme can be understood as one stage process (Hirai et al, 1998)

3.2 Shape and low resolution structure of extracellular hemoglobins calculated from SAXS data

Given the inherent difficulties to obtain the crystallization of proteins with high molecular weights, low resolution studies of extracellular hemoglobins in solution have been the main tool of its structural studies. The physicochemical properties of extracellular hemoglobins (erythrocruorins) have been under study since the 1930s. In particular, different oxygen affinities and cooperativities were reported for molecules with very similar heme content, dimensions and molecular weight. This fact has led the investigators to focus attention on the possible structural differences that could explain this diverse functional behavior. Two very comprehensive reviews on the structure of extracellular hemoglobins have been published by Chung (1979) and Weber (2001). The challenge has always been the elucidation of the interaction among the more than 200 subunits of these respiratory proteins, which lead to the spontaneous, self-limited assembly and cooperative oxygen binding, which are not yet completely understood. In this section the results of the study of extracellular hemoglobins from *Glossoscolex paulistus* with molecular weight of ~3,100 kD will be presented. Advanced methods of shape restoration from the X-ray scattering data allowed a description of the subunit arrangement of these molecules as well as the determination of dimensional parameters which could also be confirmed by the results of hydrodynamic measurements and calculations for the models proposed. There are only minor differences in the properties already reported on the subunit structure of *Lumbricus terrestris* hemoglobin (Fushitani et al., 1991) and the previous works on the structural subunits of *G. paulistus* studied by pH induced and high pressure dissociation by Bonafe et al.,1991 and Silva et al., 1989, indicating the similarity of these proteins, spite of the differences in molecular weight.

Samples of *G. paulistus* were purified according to a standard procedure (Silva et al., 1989, Bonafe et al., 1991) in several concentrations. SAXS measurements were made using synchrotron radiation at Brazilian Synchrotron Light Laboratory, with hemoglobin in 0.05 M TRIS-HCl buffer pH 7.5. The hemoglobin concentrations used in the experiments varied from 0.5 to 40 mg/mL and the final combination of the frames enabled the extrapolation to zero concentration. The scattered intensities were recorded with a linear position sensitive detector and the primary data correction was done using standard procedures. The *q* range was from q = 0.005 to 0.1882 $Å^{-1}$, with radiation wavelength of λ = 1.74Å. To collect the low and high angles scattering data, two sample-detector distances were used (1.74m and 0.84 m). The samples were kept in a 1.5 mm diameter capillary tube sample holder, kept at a constant temperature (20°C). Indirect Fourier Transformation was performed using the GNOM program package. *Ab initio* calculations were performed using program DAMMIN. The experimental scattered intensity was normalized to absolute scale using water as a primary standard, which enabled the calculation of the protein molecular weight and volume. Finally, hydrodynamic properties of molecular models can be calculated using an approach initially developed for crystallographic structures (program HYDROPRO, de La Torre et al., 2000), which can be easily extended to dummy atom models when the molecular mass and partial specific volume of the protein are known (Arndt et al., 2002). As a result, several hydrodynamic parameters can be calculated and compared with the values obtained by other experimental methods. This comparison can be very useful in order to check the validity of the molecular conformation represented by the 3D models proposed.

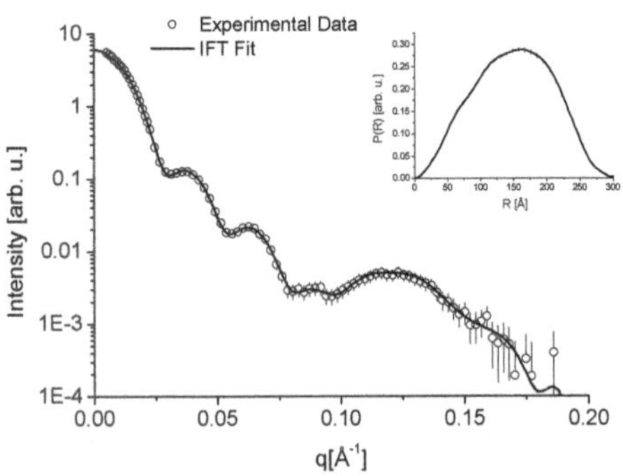

Fig. 18. Scattering curve (open circles) and IFT fitting (solid line) of hemoglobin from *G. paulistus*. Inset - pair distance distribution function (*p(r)*).

From the *p(r)* function we obtain a radius of gyration of 113.6 +/- 0.7 Å and maximum dimension of 300 +/- 10 Å . In Fig. 18 we see the excellent fitting of the intensity curve and the *p(r)* function from which the values of Rg and Dmax were calculated. The molecular mass and particle volume where calculated for *G. paulistus* using the *I(0)* value giving 3.1 ± 0.2 MDa and 3.8 ± 0.1 x 10^6Å^3, respectively. These values are compatible with the dimensions obtained with SAXS and from electron micrografs (EM) from *G. paulistus* (Souza, 1990). The overall shape of the particle as present in the EM analysis shows a P62 symmetry, which can be used as a constraint in the model calculation. The introduction of symmetry constraints decreases the number of degrees of freedom, and consequently leads to the restoration of a better three-dimensional model (Svergun, 2000; Oliveira, 2001). In this way it was used a P62 symmetry in the model optimization. In fig. 16 we present one of the best results of the three-dimensional molecular models. Several runs of the optimization program were performed. For each obtained model hydrodynamic parameters were calculated and the ones that provided values not in agreement with the experimental results were excluded. As a result, a model based in the SAXS results and also in agreement with other experimental data could be selected (Table 2). For comparison it is shown in Fig16 the result obtained by Royer et al. (Royer et al,2000) for the hemoglobin of *Lumbricus terrestris* using protein crystallography and electron microscopy. These results showed that the proteins are quite similar in quaternary structure.

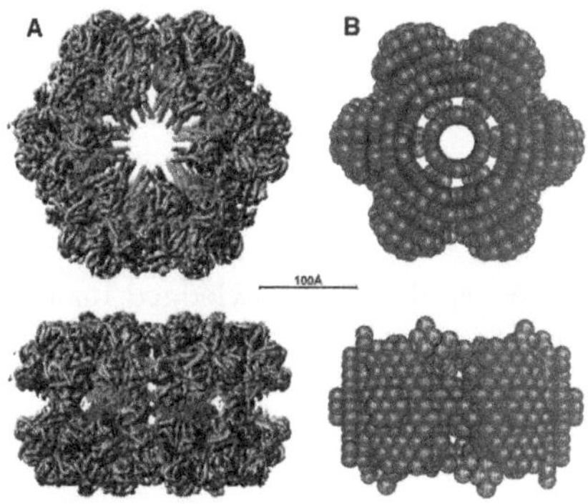

Fig. 19. A) Crystallographic structure of *Lumbricus terrestris* - from Royer et al.(2000). B) Calculated dummy atom models for the hemoglobin from *G. paulistus* with the computer program DAMMIN using a P62 symmetry. The models are in the same scale.

Parameter	Experimental Data	Model
Molecular Weight [MDa]	3.1±0.2 *	---
Sedimentation Coefficient [S]	58 **	57
Stokes Radius [Å]	139***	138
Diffusion Coefficient [10-7 cm2/s]	1.56***	1.55
f/fmin	1.42***	1.41

Table 2. Hydrodynamic Parameters and molecular weight of the hemoglobin from G. *Paulistus* obtained from SAXS models and experimental techniques. * from experiment, ** from Costa MCP 1988, *** from S value.

Due to the inherent difficult of make crystals of proteins, particularly for large proteins like the *G. paulistus*, the presented results demonstrated the capability of SAXS technique and the new optimization methods to provide a fast and reliable procedure to investigate the shape and quaternary structures of large protein complexes. Also, the correlation of SAXS results with the hydrodynamic properties increases the reliability of the results and makes possible to perform a model search integrated with hydrodynamic calculations.

4. Conclusions

General aspects of Small Angle X-Ray applied to the study of colloidal particles in solution were presented. Two examples of application were shown, demonstrating the versatility of the SAXS technique. One of the main strengths of this technique is the possibility of investigate systems directly in solution, close to the native state, in a broad range of sizes and molecular weights. On the other hand, due to the low information content of a typical SAXS data, the scattering data has to be correlated and supported by additional information, obtained from other experimental techniques. In this way, even thought SAXS data can provide a valuable and important structural information, the technique and the modeling methods has to be applied with extreme precaution and always cross checked with several additional results in order to provide relevant, unambiguous information and, most importantly, avoid wrong data interpretation. As shown in this chapter, absolute scale calibration and the comparison of hydrodynamic properties of the obtained models with the ones obtained experimentally are two very useful tools for results checking and model validation.

5. Acknowledgment

The author acknowledges FAPESP for financial support (Proj. #2000/15087-4 and #2010/09277-7). University of São Paulo is acknowledged for the financial support of this book chapter. Prof. Carlos Bonafé is kindly acknowledged for his help and support to the preparation of the *G. Paulistus* samples. The author is grateful to Profa. Iris Torriani for several valuable discussions.

6. References

Andersen, E. S.; Dong, M.; Nielsen, M. M.; Jahn, K.; Subramani, R.; Mamdouh, W.; Golas, M. M.; Sander, B.; Stark, H.; Oliveira, C. L. P.; Pedersen, J. S.; Birkedal, V.; Besenbacher, F.; Gothelf, K. V.; Kjems, J. Self-assembly of a nanoscale DNA box with a controllable lid. *Nature (London)*, v. 459, p. 73-76, 2009. ISSN: 0028-0836.

Arndt, M. H. L. ; Oliveira, C. L. P. ; Regis, W. C. B. ; Torriani, I. L. and Santoro, M. M. . Small Angle X-Ray Scattering of the Hemoglobin from Biomphalaria glabrata. *Biopolymers (New York)*, v. 69, p. 470-479, 2003. ISSN (electronic): 1097-0282.

Bergmann, A; Fritz G; Glatter, O. (2000). Solving the generalized indirect Fourier transformation (GIFT) by Boltzmann simplex simulated annealing (BSSA). *J. Appl. Cryst.* (2000). 33, 1212-1216. ISSN (electronic): 1600-5767.

Bonafe, C. F. S.; Villas-Boas, M.; Suarez, M. C. and Silva, J. L. (1991). Reassembly of a Large Multisubunit Protein Promoted by Nonprotein Factors: Effects of Calcium and

Glycerol on the association of extracellular hemoglobin, *J. Biol. Chem.*, 266 (20), 13210-13216. ISSN 1083-351X

Chacon, P.; Morán. F.; Días, J. F.; Pantos, E. and Andreu J. M., (1998). "Low-resolution structures of proteins in solution retrieved from X-ray scattering whit a genetic algorithm". *Biophys. J.*, 74, 2760-2775. ISSN (electronic): 1542-0086

Chung, M. C. M. and Ellerton, H. D. (1979). "The physico-chemical and functional properties of extracellular respiratory haemoglobins and chlorocruorins". *Prog. Biophys. Molec. Biol.* 35, 33-102. ISSN: 0079-6107.

Ciccariello, S. (1985), Deviations from the Porod Law due to Parallel Equidistant Interfaces, *Acta Cryst., A*41 , 560-568; ISSN (electronic): 1600-5724.

Costa, M. C. P.; Bonafe, C. F. S.; Meirelles, N. C. and Galembeck, F., (1988). "Sedimentation Coefficient and Minimum molecular weight of extracellular hemoglobin of Glossoscolex paulistus (oligochaeta)". *Brazilian J. Med. Biol. Res.*, 21, 115-118. ISSN 1678-4510

Debye, P. (1915). Zerstreuung von Röntgenstrahlen. *Ann. Phys. (Leipzig)*, 46, 809–823. ISSN 00033804.

de La Torre, J. G.; Huertas, M. L. and Carrasco B. (2000). "Calculation of Hydrodynamic Properties of Globular Proteins from their Atomic-Level Structure". *Biophys. J.*, 78, 719-730. ISSN (electronic): 1542-0086

Feigin, A. and Svergun, D. I. (1987), Structure Analysis by Small-Angle X-Ray and Neutron Scattering, *Plenum Publishing Corporation – Plenum Press*, , ISBN: 0-306-42629-3 New York;

Fritz, G., Glatter O. (2006) Structure and interaction in dense colloidal systems: evaluation of scattering data by the generalized indirect Fourier transformation method. *Journal of Physics - Condensed Matter,* 18 (36) Issue 36, (13 September 2006), S2403-S2419. ISSN (electronic): 1361-648X.

Fushitani, K. and Riggs, A. F. (1991). "The extracellular hemoglobin of earthworm Lumbricus terrestris - oxygenation properties of isolated chains, trimer and a reassociated product". J. Biol. Chem., 266 (16), 10275-10281. ISSN (electronic): 1083-351X.

Glatter, O. (1977), A New Method for the Evaluation of Small-Angle Scattering Data, *J. Appl. Cryst.* 10, 415-421. ISSN (electronic): 1600-5767.

Glatter, O. (1979), The Interpretation of Real-Space Information from Small-Angle Scattering Experiments. *J. Appl. Cryst.* , 12, 166-175;

Glatter, O. (1980), Computation of Distance Distribution Functions and Scattering functions of Models for Small Angle Scattering Experiments . *Acta Physica Austriaca* , 52, 243-256. ISSN: 0001-6713.

Glatter, O. (1980) Determination of Particle-Size Distribution Functions from Small-Angle Scattering Data by Means of the Indirect Transformation Method, *J. Appl. Cryst.* 13, 7-11; ISSN (electronic): 1600-5767.

Glatter, O. (1981), Convolution Square Root of Band-Limited Symmetrical Functions and its Application to Small Angle Scattering Data, *J. Appl. Cryst.* 14, 101-108; ISSN (electronic): 1600-5767.

Glatter, O. (1984), Improvements in Real-Space Deconvolution of Small-Angle Scattering Data, *J. Appl. Cryst.* 17,435-441; ISSN (electronic): 1600-5767.

Glatter, O. (2002) Chapter 4. The Inverse Scattering Problem in Small-Angle Scattering, In: *Neutrons, X-Rays and Light: Scattering Methods Applied to Soft Condensed Matter*, P. Lindner and Th. Zemb (editors), pp. 73-124, Elsevier, ISBN: 0-444-51122-9, Amsterdam, The Netherlands.

Glatter, O. and O. Kratky, Eds. (1982). *Small Angle X Ray Scattering.* London, Academic Press. , ISBN: 0-12-286280-5, London, Great Britain.

Guinier, A. and Fournet, G. (1955). *Small-Angle Scattering of X-rays.* Wiley, New York, USA. ISBN: 9780598669933

Hansen, S. (2000). Bayesian estimation of hyperparameters for indirect Fourier transformation in small-angle scattering. *J. Appl. Cryst.* (2000). 33, 1415-1421; ; ISSN (electronic): 1600-5767.

Hirai, M.; Arai, S.; Iwase, H. and Takizawa, T. (1998) Small-Angle X-ray Scattering and Calorimetric Studies of Thermal Conformational Change of Lysozyme Depending on pH. *J. Phys. Chem. B* 102, 1308-1313. ISSN 1089-5647

Jackson, J. D., (1998). Classical Electrodynamics, Wiley, New York, USA. ISBN: 047130932x.

Kozin, B. Volkov, V. V. and Svergun, D. I. (1997), ASSA, a Program for Three-Dimensional Rendering in Solution Scattering from Biopolymers, *J. Appl. Cryst.*, 30, 811-815; ISSN (electronic): 1600-5767.

Lindner, P. and Zemb, Th. *Neutrons Eds, X-Rays and Light: Scattering Methods Applied to Soft Condensed Matter*, P. Lindner and Th. Zemb (editors), Elsevier, ISBN: 0-444-51122-9, Amsterdam, The Netherlands.

Mareuil, F.; Sizun, C.; Perez, J.; Schoenauer, M.; Lallemand, J.; and Bontems, F. (2007).A simple genetic algorithm for the optimization of multidomain protein homology models driven by NMR residual dipolar coupling and small angle X-ray scattering data. *Eur Biophys J*, 37:95–104. ISSN 0175-7571.

Nagle, J. F. and Wierner, M. C.(1989). Connection of electron density profiles to other structural quantities. *Biophysical J.*, 55, 309-313. ISSN (electronic): 1542-0086.

Oliveira, C. L. P. (2001),*Simulation of Small Angle X-Ray Scattering from 3D particles using low resolution models*, MSc thesis OL4s. State University of Campinas, Campinas, Brazil.

Oliveira, C. L. P.; Behrens, M. A. ; Pedersen, J. S. ; Erlacher, K. ; Otzen, D. ; Pedersen, J. S. A SAXS Study of Glucagon Fibrillation. *J. Mol. Biol.*, v. 387, p. 147-161, 2009. ISSN: 00222836

Oliveira, C. L. P.; Juul, S.; Joergensen, H. L.; Knudsen, B.; Tordrup, D.; Oteri, F.; Falconi, M.; Koch, J.; Desideri, A.; Pedersen, J. S.; Andersen, F. F.; Knudsen, B.R. Structure of Nanoscale Truncated Octahedral DNA Cages: Variation of Single-Stranded Linker Regions and Influence on Assembly Yields. *ACS Nano*, v. 4, p. 1367-1376, 2010.

Pape, E. H. and Kreutz, W. (1978). A Deconvolution Method for Evaluating Small-Angle X-ray Scattering from Lamellar Structures. *J. Appl. Cryst.* 11,421-429. ISSN (electronic): 1600-5767.

Patel, S. and Schimidt, P. W. (1971). Small-Angle X-ray Scattering Determination of the Electron Density of the Particles in a Colloidal Silica Suspension, *J. App. Cryst.*, 4 , 50. ISSN (electronic): 1600-5767.

Pedersen, J. S. (1997) Analysis of Small-angle Scattering Data from Polymeric and Colloidal Systems: Modelling and Least-squares Fitting. *Adv. Coll. Int. Sci.* 70, 171-201. ISSN 0001-868.

Pedersen, J. S.; Hansen, S.; Bauer, R., (1994) The Aggregation Behavior of Zinc-free Insulin Studied by Small-angle Neutron Scattering, *Eur. Biophys. J.* 22, 379-89. ISSN 0175-7571

Pedersen, J.S. (2001) Chapter 16. Modeling of Small-Angle Scattering Data, In: *Neutrons, X-Rays and Light: Scattering Methods Applied to Soft Condensed Matter*, P. Lindner and Th. Zemb (editors), pp. 171-201, Elsevier, ISBN: 0-444-51122-9, Amsterdam, The Netherlands.

Petoukhov, M.V. and Svergun, D.I. (2005) Global rigid body modelling of macromolecular complexes against small-angle scattering data. *Biophys. J.* 89 (2005) 1237-1250. . ISSN (electronic): 1542-0086

Rathgeber, S.; Monkenbusch, M.; Kreitschmann, M.; Urban, V., Brulet, A. (2002). Dynamics of star-burst dendrimers in solution in relation to their structural properties. *J. Chem. Phys.*, Vol. 117, No. 8, 22, 4047-4062.

Royer, W. E. Jr.; Strand, K.; van Heel, M. and Hendrickson, W. A. (2000). "Structural hierarchy in erythrocruorin, the giant respiratory assemblage of annelids", *PNAS*, 97 (13), 7107-7111. ISSN (electronic): 1091-6490.

Semenyuk, V. and Svergun, D. I. (1991) Gnom - A Program Package for Small-Angle Scattering Data-Processing. *J. Appl. Cryst.*, 24, 537-540; ISSN (electronic): 1600-5767.

Silva, J. L., Villas-Boas, M., Bonafe, C. F. S. and Meirelles, N. C. (1989). "Anomalous pressure dissociation of large protein agregates. Lack of concentration dependence and irreversibily at extreme degrees of dissociation of extracellular hemoglobin". *J. Biol. Chem.*, 25, 264(27): 15863-8. ISSN 1083-351X

Souza, C. F. (1990). *Espalhamento de Raios X a Baixo Ângulo: Estudo de Hemoglobinas Extracelulares em Solução*, PhD Thesis TDSo89e, State University of Campinas, Campinas, Brazil.

Sturhmann, H. B. (1973). Comparison of the Three Basic Scattering Functions of Myoglobin in Solution with those from the Known Structure in Crystalline State, *J. Mol. Biol.*, 77 , 363-369; ISSN: 00222836

Sturhmann, H. B. and Miller, A. (1978). Small-Angle Scattering of Biological Structures, *J. Appl. Cryst.*, 11 , 325-345; ISSN (electronic): 1600-5767.

Svergun, D. I. and Sturhmann, H. B. (1991) New Developments in Direct Shape Determination from Small-angle Scattering. 1. Theory and Model Calculations, *Acta Cryst.*, A47, 736-744; ISSN (electronic): 1600-5724.

Svergun, D. I., Barberato C. and Koch M.H.J. (1995) CRYSOL - a Program to Evaluate X-ray Solution Scattering of Biological Macromolecules from Atomic Coordinates. *J. Appl. Cryst.* , 28, 768-773. ISSN (electronic): 1600-5767.

Svergun, D. I. (1999). Restoring low resolution structure of biological macromolecules from solution scattering using simulated annealing. *Biophys. J.* 76, 2879-2886. ISSN (electronic): 1542-0086

Svergun DI, Petoukhov MV and Koch MHJ (2001). Determination of domain structure of proteins from X-ray solution scattering. *Biophys. J.* 80, 2946-2953. ISSN (electronic): 1542-0086

Svergun, D.I. (2007). Small-angle scattering studies of macromolecular solutions *J. Appl. Cryst.* 40, s10–s17. ISSN (electronic): 1600-5767.

Svergun, D. I. (2009) (coordinator). SAXIER: Small-Angle X-Ray Scattering Initiative for Europe. Available from: http://www.saxier.org/ .

Székely, P.; Ginsburg, A.; Ben-Nun, T.; Raviv, U. (2010). Solution X-ray Scattering Form Factors of Supramolecular Self-Assembled Structures. *Langmuir*, 26(16), 13110–13129. ISSN 0743-7463.

Voet, D.; Voet, J. G. and Pratt, C. W. (2006). *Fundamentals of Biochemistr: Life at the Molecular Level.* , Wiley, New York, USA. ISBN 0-471-21495-7

Volkov, V. V. and Svergun, D. I. (2003). Uniqueness of ab initio shape determination in smallangle scattering. *J. Appl. Cryst.* 36, 860-864. ISSN (electronic): 1600-5767.

Walther, D.; Cohen, F. E. Doniach, S. (2000). Reconstruction of low-resolution three-dimensional density maps from one-dimensional small-angle X-ray solution scattering data for biomolecules. *J. Appl. Cryst.* (2000). 33, 350-363. . ISSN (electronic): 1600-5767.

Weber, R. E. and Vinogradov, S. N. (2001). "Nonvertebrate Hemoglobins: Functions and Molecular Adaptations", *Phys. Rev.*, 81 (2), 569-628. ISSN (electronic): 1522-1210.

18

Complementary Use of NMR to X-Ray Crystallography for the Analysis of Protein Morphological Change in Solution

Shin-ichi Tate, Aiko Imada and Noriaki Hiroguchi
Department of Mathematical and Life Sciences, Hiroshima University
Japan

1. Introduction

A vast amount of protein structure data is going to pave new ways in protein structure research. They improved the quality of predicted protein structure from its primary sequence (Sanchez and Sali 2000). The possible protein interaction sites to small ligand and/or the other proteins could be predicted based on the protein complex structures in the Protein Data Bank (PDB) (Morris et al. 2009). The combined use of bioinformatics with the protein structure data has been frequently giving invaluable outcomes to facilitate the understanding on the experimental results in biochemistry and molecular biology. *In silico* protein structure analyses are now already essential approaches in protein research.

Protein structural data, most of which came from X-ray crystallography, are also useful to expand the protein structure analysis in solution, when combined with NMR. NMR chemical shift perturbation of a protein caused by the interaction with a compound, for example, allows sensitive identification of the interaction sites on protein (Shuker et al. 1996). This NMR derived binding site information with the protein structure facilitates drug design (Hajduk et al. 1997). This chemical shift-based approach is also applied to the protein-protein interaction, which enables to build a model protein complex structure (de Vries, van Dijk and Bonvin 2010, Dominguez, Boelens and Bonvin 2003). Although these approaches are now prevailingly used, there are limitations in their application. The approaches assume that the target protein retains the X-ray structure in solution and also negates the possible structural change caused by binding to a compound or a partner protein. Some of the proteins are known to have different domain arrangement from those in crystal (Skrynnikov et al. 2000a); it is often the case for the protein having domains linked by flexible linker. In addition, it is commonly found that proteins show structural change in response to compound binding or interaction with the other protein (Evenas et al. 2001). To expand the utility of protein structure data in the PDB in solution protein science, we need new NMR techniques to overcome the known limitations in the existing approaches, which could determine the structure changed from the one by X-ray in binding to a ligand or a partner protein to improve the modelled complex structure, for example.

Protein structure change caused by interaction with other molecules is primarily important in discussing protein functional regulation. The structural change is not limited in the region around the binding site. Sometimes, it also causes rather global change including domain

rearrangement or subunit rearrangement. Hereafter, we would mention to this global structural change as '*protein morphological change*' to emphasize its large amplitude in motion.
Protein morphological change is characterized by the considerable difference in the spatial arrangement of structural units from their original positions; structural units may include domain, subunit, or, sometimes, a small segment comprising of secondary structures. In the case, each unit mostly retains the original structure, except for the limited region engaged in the interaction (Fig. 1). The high-resolution unit structures from X-ray are therefore used as the templates in analyzing the morphological changes, for example, to see how the domains are rearranged upon ligand binding. As exemplified by Luciferase morphological change upon ligand binding, each domain shows little structural change, except for the intrinsically flexible loops (Nakatsu et al. 2006, Conti, Franks and Brick 1996) (Fig. 1). The bound form structure can be reconstituted from the apo-structure by rotating the small domain relative to the other (Fig. 1).

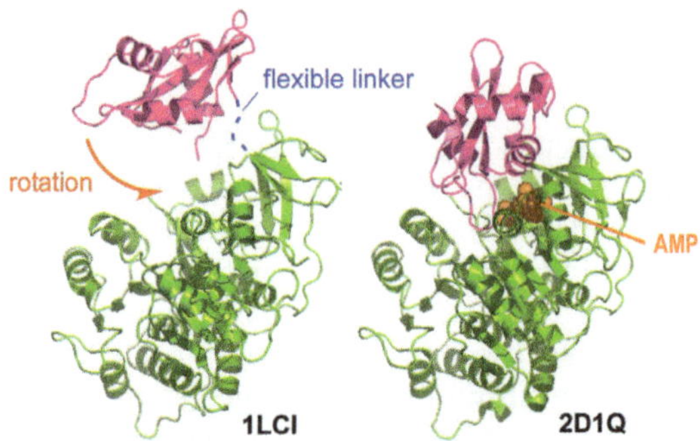

Fig. 1. Protein morphological change. Domain rearrangement of Luciferase upon binding to ATP. PDB codes 1LCI and 2D1Q for apo- and AMP-bound forms. Small domain is rotated over onto the large domain upon ligand binding. Both domains basically retain the structures in apo-form only with slight local changes.

The conventional NMR structure determination relying on NOEs and vicinal spin couplings, both of which give the only short-range structural information, does not properly determine the domain arrangement, particularly in the case that domains are connected by a linker and there are little inter-domain contacts. The accumulation of the less quantitative short-range structural information may even result in the erroneous determination of domain arrangement. It requires the global structural information that directly defines the domain arrangement.
In addition to the requirement for the global structural information, the molecular size limitation in solution NMR spectroscopy is another obstacle to work on the protein morphology by NMR. The conventional NMR approach has severe size limitation in determining protein structure, which is practically 30 kDa. Proteins having multiple domains tend to be greater than 50 kDa, thus the conventional NMR approach is not usable in protein morphology analysis.
Protein morphological change is of the great significance in biology, but there has not been any appropriate NMR techniques for this purpose, so far. With this concern, we have devised a new NMR approach, DIORITE (Determination of the Induced ORIentation by Trosy Experiments), that can overcome the obstacles in the conventional approaches, lacking global structural information and size limitation (Tate 2008, Tate, Shimahara and

Utsunomiya-Tate 2004). In this chapter, we are going to describe this NMR method from its theoretical backgrounds to applications to demonstrate how it works on the protein morphology analyses. This gives a complementary view on protein structure in solution, which is not seen by X-ray crystallography, although it requires high-resolution crystal structure as a template for the analysis. The combined use of high-resolution X-ray structure and its morphology analysis in solution by DIORITE may expand our understanding how protein works more vividly.

2. Residual anisotropic spin interactions

Conventional NMR structure analysis based on the short-range spin interactions does not give any global structural information required in protein morphology study. The residual anisotropic spin interactions, which become apparent for a *'weakly'* aligned protein, give molecular orientation information relative to the magnetic field. And the information gives the clues to analyze protein morphology.

Two different types of residual anisotropic spin interactions are observed in the amide ^{1}H-^{15}N spin pair on a peptide plane; one is the nuclear spin dipolar-dipolar interaction and the other is anisotropic shielding effect against the external magnetic field, which is called as chemical shift anisotropy (CSA) (Fig. 2).

These anisotropic spin interactions are not observed on a NMR spectrum for protein in an isotropic solution, where protein rapidly tumbles. Because of the rapid rotation that allows for protein to direct the entire angles against a magnetic field, anisotropic spin interactions are completely canceled for isotropic sample. In contrast, protein dissolved in a magnetically ordering liquid crystalline medium, for example, experiences some rotational restrictions by steric bump with the liquid crystalline molecules. Thus, it leads to incomplete cancellation of anisotropic spin interactions, which should give the residual anisotropic spin interactions observed on a spectrum. In a properly tuned liquid crystalline concentration, there is enough space to allow protein tumbling to some extent. The protein in an aligned liquid crystalline medium, accordingly, still can give high resolution NMR signals. In the higher liquid crystalline concentration, the protein tumbling becomes substantially limited and its NMR signals become severely broadened to prohibit their spectral observation. The condition that protein tumbling is slightly restricted to achieve incomplete cancellation of the anisotropic spin interactions, but it still gives narrow lines enough for the observation on a NMR spectrum is called as a *'weak'* alignment (Bax 2003).

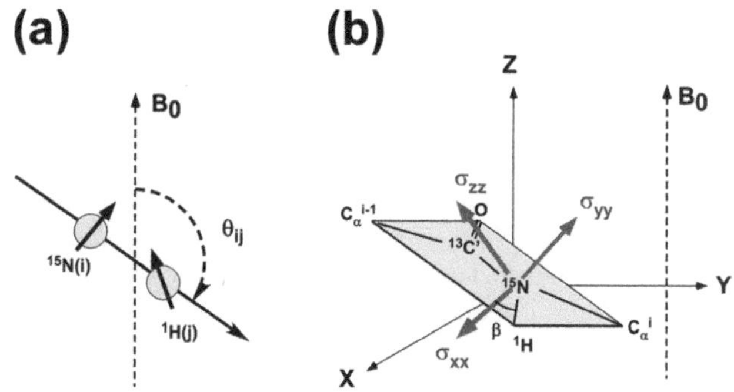

Fig. 2. Two anisotropic spin interactions observed for amide ^{1}H-^{15}N spin pair. (a) dipolar interaction. (b) ^{15}N chemical shift anisotropy (CSA) effect.

The spin interactions apparent only under a weakly aligning condition are the residuals by incomplete cancellation of the spatially anisotropic spin interactions. The effects are, therefore, called as *'residual'* anisotropic spin interactions. In the following text, we will focus on two types of the residual anisotropic spin interactions on a peptide plane prior to describe the DIORITE; they are the residual dipolar coupling (RDC) and the residual CSA (RCSA). DIORITE relies on these two anisotropic spin interactions, and their theoretical basics will be helpful in understanding DIORITE approach.

2.1 Residual dipolar coupling, RDC

Residual nuclear spin dipolar interaction (RDC) gives the direction of the internuclear vector against the magnetic field. In the peptide bond of a protein labeled with ^{15}N, the RDC for amide ^{15}N nucleus defines the angle between the NH bond vector and a magnetic field (Fig. 2a).

The RDCs amide ^{1}H-^{15}N spin pairs are the measurement of common choice, primarily due to the good spectral dispersion of the ^{1}H-^{15}N correlation signals on a 2D spectrum. The RDCs are obtained from a pair of ^{1}H-^{15}N HSQC spectra measured without ^{1}H decoupling during t_1 evolution period of ^{15}N spin; one spectrum is for the isotropic sample, and the other is for the aligned (Fig. 3). On a ^{1}H-coupled ^{1}H-^{15}N HSQC spectrum, each peak gives a pair of signals split along the ^{15}N axis (Fig. 3). The split width of signal for the isotropic sample corresponds to single-bond spin coupling ($^{1}J_{NH}$) between the covalently bonded ^{1}H and ^{15}N nuclei, around 93 Hz (Fig. 3a). Split width for the aligned sample gives the value deviated from $^{1}J_{NH}$ (Fig. 3b). The difference in the apparent $^{1}J_{NH}$ values between the isotropic and aligned sample is the RDC. In this section, we will describe how the RDCs can define the alignment tensor of protein, by describing its theoretical background.

2.2 Theoretical description on RDC

We focus on the amide ^{1}H-^{15}N spin pair here; i and j represent ^{1}H and ^{15}N nucleus, respectively. The dipolar Hamiltonian to describe the magnitude of dipolar split is written in the following equation in a laboratory frame under the high-field limit condition:

$$H_{ij}{}^{D}(t) = -\left(\frac{\mu_0}{4\pi}\right)\frac{\gamma_i \gamma_j h}{2\pi^2 r_{ij}{}^3} I_{iz} I_{jz} P_2(\cos\Theta(t)) \tag{1}$$

where r_{ij} is the distance between ^{1}H and ^{15}N atoms, γ_i and γ_j are the gyromagnetic ratios for ^{1}H and ^{15}N nuclear spins, respectively. As physical constants, h is Planck constant and μ_0 is vacuum permeability. I_{iz} and S_{jz} represent the spin angular-momentum operators for ^{1}H and ^{15}N spins, respectively. The angular part of the dipolar Hamiltonian is described by the second rank Legendre function, $P_2(\cos\Theta(t))$, here $\Theta(t)$ is a time-dependent angle between the magnetic field and the internuclear vector (NH bond vector in the present context) (Fig. 4a). In determining molecular orientation by RDC, bond libration effect is negligible due to its faster motion relative to molecular tumbling; bond libration occurs in a few hundred psec time regimes, whilst molecular tumbling happens in nsec range. In this sense, r_{ij} is defined as time-averaged effective internuclear distance, thus it is estimated to be slightly longer (1.04 Å) than the static bond length (1.02 Å).

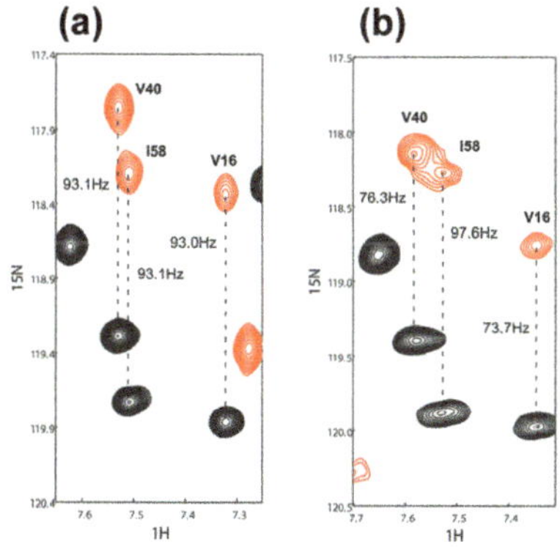

Fig. 3. RDC measurement with ^{1}H coupled HSQC. Isotropic sample (a) and aligned sample (b). Split with for each paired signals corresponds to $^1J_{NH}$. The apparent $^1J_{NH}$ is changed for aligned sample. The change in $^1J_{NH}$ from the isotropic value is RDC.

The experimentally measured residual dipolar couplings, D_{ij}^{res}, is represented as the time-averaged expectation value for the dipolar Hamiltonian, Eq. (1).

$$D_{ij}^{res.}(t) = -\left(\frac{\mu_0}{4\pi}\right)\frac{\gamma_i \gamma_j h}{2\pi^2 r_{ij}^3} < P_2(\cos\Theta(t)) > \tag{2}$$

The angle brackets denote the time-averaged resultant. The molecule in an isotropic tumbling gives no RDC, $< P_2(\cos\Theta(t)) >= 0$, due to the isotropic distribution of molecular orientation. On the other hand, the molecule in an anisotropic tumbling incompletely vanishes the term, $< P_2(\cos\Theta(t)) >\neq 0$. The magnitudes of the residual values are related to the molecular orientation angle and the alignment order against a magnetic field. The theoretical description to relate the RDCs to the orientation angles and orders will be given below.

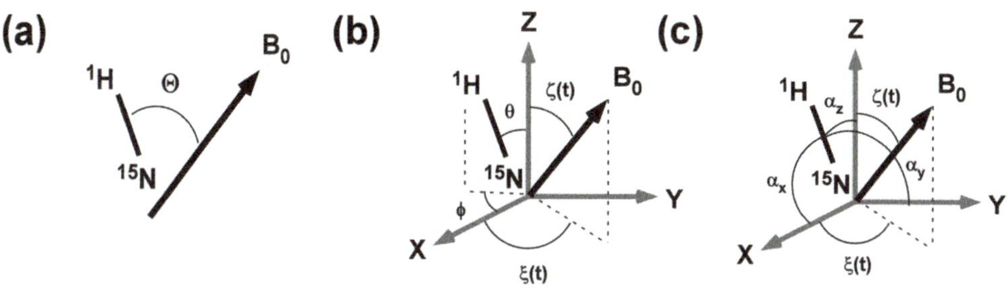

Fig. 4. Three different representations for NH bond vector against a magnetic field. (a) Angle dependency of the ^{1}H-^{15}N RDC is described by Θ, the angle between the NH bond vector and a magnetic field. (b) NH bond vector on a molecular coordinate system, whose direction is defined by polar angles (θ, ϕ). Molecular reorientation relative to a magnetic field is described by the time dependent polar angles ($\xi(t), \zeta(t)$). (c) NH bond vector is recast by direction cosines to three molecular axes.

Protein structure is represented by the atom positions in a Cartesian coordinate system. This coordinate system is referred as the molecular coordinate system here. The orientation of each NH bond vector in a protein is described by polar angles (θ, ϕ) in the molecular coordinate system (Fig. 4b). Now, we neglect the local NH bond libration, thus, the angles θ and ϕ are time-independent. Likewise, at any instant in time, the orientation of the magnetic field is described by the time dependent polar angles, ($\xi(t), \zeta(t)$) (Fig. 4b). The molecule reorients against a magnetic field in a time dependent manner; this molecular motion is represented by the directional change of the magnetic field on the molecular coordinate system (Fig. 4b). The polar angles defining a magnetic field direction, therefore, are described as time dependent parameters.

The Legendre function $P_2(\cos\Theta(t))$ is expanded by the spherical harmonics according to the spherical harmonic addition theorem:

$$P_2(\cos\theta(t)) = \frac{4\pi}{5}\sum_{q=-2}^{2} Y^*_{2q}(\theta,\varphi) Y_{2q}(\zeta(t),\xi(t)) \tag{3}$$

where the Y_{2q}'s are the normalized spherical harmonics. Using Eq. (3), the residual dipolar coupling (Eq. (2)) is expressed with spherical harmonics:

$$D_{ij}^{res} = -\left(\frac{\mu_0}{4\pi}\right)\frac{\gamma_i\gamma_j h}{2\pi^2 r_{ij}^3}\frac{4\pi}{5}\sum_{q=-2}^{2} Y^*_{2q}(\theta,\varphi)\left\langle Y_{2q}(\zeta(t),\xi(t))\right\rangle \tag{4}$$

In Eq. (4), the five time-averaged spherical harmonics $\left\langle Y_{2q}(\zeta(t),\xi(t))\right\rangle$ with q = -2, -1, 0, 1, 2 defines the molecular alignment angles and its magnitudes. Recasting these five terms makes more intuitively acceptable representation to describe the molecular alignment state, which is Saupe's order matrix.

The use of Saupe's order matrix reforms Eq. (4).

$$D_{ij}^{res} = -\left(\frac{\mu_0}{4\pi}\right)\frac{\gamma_i\gamma_j h}{2\pi^2 r_{ij}^3}\frac{4\pi}{5}\sum_{k,l=x,y,z} S_{kl}\cos(\alpha_k)\cos(\alpha_l) \tag{5}$$

where, S_{kl} is an element in Saupe's order matrix. The Saupe order matrix is a traceless, symmetric, 3 × 3 matrix, thus it comprises of five independent elements. In the representation using Saupe's order matrix, each bond vector orientation is defined by direction cosines against the molecular coordinate axes (Fig. 4c). The Saupe's order matrix elements that are made of the time-averaged spherical harmonics $\left\langle Y_{2q}(\zeta(t),\xi(t))\right\rangle$ are shown below:

$$\begin{aligned}
S_{xz} &= \sqrt{\tfrac{3}{8}}\sqrt{\tfrac{4\pi}{5}}\left(\left\langle Y_{21}(\xi(t),\zeta(t))\right\rangle - \left\langle Y_{2-1}(\xi(t),\zeta(t))\right\rangle\right)\\
S_{yz} &= i\sqrt{\tfrac{3}{8}}\sqrt{\tfrac{4\pi}{5}}\left(\left\langle Y_{21}(\xi(t),\zeta(t))\right\rangle + \left\langle Y_{2-1}(\xi(t),\zeta(t))\right\rangle\right)\\
S_{xy} &= i\sqrt{\tfrac{3}{8}}\sqrt{\tfrac{4\pi}{5}}\left(\left\langle Y_{22}(\xi(t),\zeta(t))\right\rangle - \left\langle Y_{2-2}(\xi(t),\zeta(t))\right\rangle\right)\\
S_{xx} - S_{yy} &= \sqrt{\tfrac{3}{2}}\sqrt{\tfrac{4\pi}{5}}\left(\left\langle Y_{22}(\xi(t),\zeta(t))\right\rangle + \left\langle Y_{2-2}(\xi(t),\zeta(t))\right\rangle\right)\\
S_{zz} &= \sqrt{\tfrac{4\pi}{5}}\left\langle Y_{20}(\xi(t),\zeta(t))\right\rangle
\end{aligned} \tag{6}$$

The Saupe order matrix describes the angle and the order for aligned protein relative to the magnetic field. Diagonalization of the Saupe order matrix gives the alignment tensor; its principal values represent the alignment order along each principal axis and the principal axes define the direction of the aligned molecule. In the alignment tensor, z-axis is defined as the most ordered axis. An asymmetry, $\eta = (S_{xx} - S_{yy}) / S_{zz}$, represents the deviation from axially symmetric ordering. The orientation of the alignment tensor axes (principal axes) is described on the molecular coordinate system by Euler angles.

In the alignment tensor axes system (the alignment tensor frame), the residual dipolar coupling D_{ij}^{res} is represented by the following two parameters, A_a and A_r, that include two non-zero average spherical harmonics:

$$\begin{aligned} \langle Y_{20}(\xi(t),\zeta(t))\rangle &= \sqrt{\tfrac{5}{4\pi}} S_{zz} = \sqrt{\tfrac{5}{4\pi}} A_a \\ \langle Y_{2\pm2}(\xi(t),\zeta(t))\rangle &= \sqrt{\tfrac{5}{24\pi}} \left(S_{xx} - S_{yy}\right) = \sqrt{\tfrac{15}{32\pi}} A_r \end{aligned} \quad (7)$$

Using the parameters, the D_{ij}^{res} is re-represented in a more intuitive form:

$$D_{ij}^{res} = -\left(\frac{\mu_0}{4\pi}\right)\frac{\gamma_i\gamma_j h}{2\pi^2 r_{ij}^3}\left[A_a(3\cos^2\theta' - 1) + \tfrac{2}{3}A_r \sin^2\theta' \cos 2\varphi'\right] \quad (8)$$

where the polar angles (θ', ϕ') define the orientation of NH bond vector in the alignment tensor frame. A_a and A_r represent the axial and rhombic components of the order magnitudes, respectively. The orientation of protein relative to a magnetic field is defined by the Euler angles (α, β, γ) that describes the position of the alignment tensor frame on the molecular coordinate system.

2.3 Residual chemical shift anisotropy, RCSA

Another anisotropic spin interaction associated with the amide ^{15}N nucleus comes from chemical shift anisotropy, CSA (Fig. 2b). This effect is observed as ^{15}N chemical shift change induced by a weak alignment (Fig. 5). The alignment induced chemical shift change is called as residual CSA, RCSA (Kurita et al. 2003).

The chemical shift for each nucleus is related to the shielding against the magnetic field by the electrons surrounding the nucleus, which is called as chemical shielding. In general, the spatial distribution of electrons around the nucleus is not uniform but rather spatially anisotropic. Therefore, the chemical shielding effect to the nucleus should change according to the orientation molecule against a magnetic field; the chemical shift for the nucleus can be orientation dependent. However, the molecule in an isotropic solution averages the orientation dependent shielding effects by its rapid isotropic tumbling, thus giving the orientation independent *'isotropic chemical shifts'* on a spectrum.

As in the case of the dipolar interaction, a weak alignment makes the orientation dependency of CSA observable, due to the incomplete averaging of anisotropic shielding effects. The amide ^{15}N has significantly large CSA, because of the substantial anisotropic distribution of the electron on a peptide plane. Protein in a weakly aligned state, thus, shows significant chemical shift changes along ^{15}N axis relative to its isotropic positions (Fig. 5). Although it is not described in detail in this manuscript, carbonyl ^{13}C has also large CSA and it shows significant RCSA under a weakly aligning state (Lundstrom, Hansen and Kay 2008).

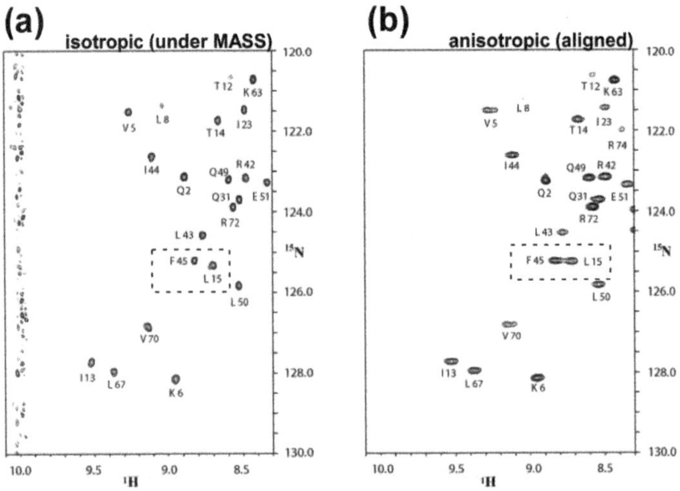

Fig. 5. ^{15}N chemical shift change induced by a weak alignment. ^{1}H-^{15}N HSQC spectra are compared for ^{15}N labeled ubiquitin dissolved in 7.5% (w/v) DMPC/DHPC/CTAB ternary bicelle solution between isotropic and aligned states: (a) under the magic angle sample spinning condition, giving an isotropic spectrum, (b) the aligned condition. Chemical shift changes along the ^{15}N axis caused by a weak alignment are clearly observed, for examples the signals in dotted square. These spectral changes come from the incomplete cancellation of ^{15}N chemical shift anisotropy (CSA) effect under a weak aligning condition. The change along the ^{15}N axis is the value for the residual CSA, RCSA.

The alignment induced chemical shift change, or RCSA, $\Delta\delta$, is expressed with Sauce's order matrix, S_{ij}, as for the RDC, Eq. (5). Based on this relation, we can also determine the alignment tensor of the molecule as done with the RDC.

$$\Delta\delta = \frac{2}{3}\sum_{i=x,y,z}\sum_{j=x,y,z}\sum_{k=x,y,z} S_{ij}\cos(\theta_{ik})\cos(\theta_{jk})\delta_{kk} \tag{8}$$

Here, $\cos\theta_{ij}$ denotes the direction cosine between *i*-axis in the molecular coordinate system and *j*-axis in the CSA tensor principal axis. The principal value for the CSA tensor axis *k* is represented by δ_{kk}. The spatially anisotropic distribution of the electrons around a nucleus is represented by a tensor (CSA tensor), where each principal axis represents the shielding magnitude. The ^{15}N CSA tensor axes are drawn on a peptide plane (Fig. 2b).

The alignment tensor is determined based on the Saupe order matrix obtained through the relation in Eq. (8) for a set of $\Delta\delta$ data. It is noted that in calculating the Saupe order matrix, we require the knowledge of the CSA tensor that is defined by the relative orientation of the CSA tensor frame on a peptide plane and the magnitudes of the principal values. In the case of the RDC, we assumed the effective interatomic distance, r_{ij}, to a fixed value. In contrast to the RDC based analysis that requires one pre-defined value for r_{ij}, the RCSA analysis requires five pre-defined parameters; two principal values δ_{zz}, δ_{xx}, with $\delta_{xx} + \delta_{yy} + \delta_{zz} = 0$, and Euler angles (α, β, γ) that define the CSA tensor orientation on a peptide plane. The RCSA based alignment tensor determination is therefore, much more knowledge demanding.

The ^{15}N CSA tensor is known to be significantly dependent on the local structure, particularly backbone torsion angle (Yao et al. 2010). Experimentally determined ^{15}N CSA tensors are reported with a various method (Fushman, Tjandra and Cowburn 1999, Cornilescu and Bax 2000, Boyd and Redfield 1999, Hiyama et al. 1988, Kurita et al. 2003). These data give consensus ^{15}N CSA tensor values for the residue in each type of secondary structure, α-helix, β-sheet and others. Practically, the use of the secondary structure specific ^{15}N CSA tensor values can determine the alignment tensor within experimental errors.
Experimental determination of ^{15}N CSA tensor for protein in solution is possible using the weak alignment technique. We previously proposed the method using magic-angle sample spinning to determine the accurate secondary structure specific ^{15}N CSA tensor, in which the bicellar media was used for a weak alignment (Kurita et al. 2003). In this experiment, we used only one aligned state, thus, only determined the ^{15}N CSA tensors in a secondary structure specific manner.
Recently, Bax and co-workers have applied this approach to determine the residue specific ^{15}N CSA tensors for a protein, where they used five more different aligning states to solve the Saupe order matrix for each residue (Yao et al. 2010). The residue specific ^{15}N CSA tensor determination that requires multiple aligned states of protein is rather demanding experiments, which require a various loop mutant to change aligning angle (Yao et al. 2010). However, the continuous effort to collect the residue specific ^{15}N CSA tensors in the similar way by Bax and co-workers will establish a clearer correlation between the ^{15}N CSA tensor and backbone torsion angles and also local interactions like hydrogen bonding, which may allow the prediction of the appropriate ^{15}N CSA tensor values from the structure. The refined ^{15}N CSA tensors will further improve the quality of alignment tensor analysis with the RCSA, although the present RCSA based approach gives an acceptable result.

3. Achieving weak alignment

In applying the residual anisotropic spin interactions described above, it is required to make a protein in a weakly aligned state. The aligning protein has to be carefully tuned to make the anisotropic interactions observable in a spectrum with keeping the spectral resolution and intensity. Alignment order is practically tuned to approximately 10^{-3}, giving about 20 Hz in maximum absolute magnitude for amide ^{1}H-^{15}N RDC. To achieve a weak alignment, some artificial medium has to be used, because the inherent magnetic susceptibility of a globular protein is too small to align to the desired extent, except for some heme-containing proteins having substantial magnetic susceptibility associated with a heme group. In this section, we will review some media for weak alignment.

3.1 Magnetically aligning liquid crystalline media

Magnetically ordering liquid crystalline media are commonly used. Discoidal phospholipid assembly, bicelle, is one of the prevailingly used materials for a weak alignment of protein (Ottiger and Bax 1999, Ottiger, Delaglio and Bax 1998, Tjandra and Bax 1997).. The bicelle is composed of a mixer of dimyristoylphosphatidylcholin (DMPC) and dihexynoyl-phosphatidylcholine (DHPC) in a ratio of 3:1. This phospholipid binary mixture forms lipid bilayers disks 30 nm - 40 nm in diameter. Bicelle has substantial magnetic susceptibility, and it spontaneously aligns under magnetic field with the normal of the bicelle surface staying perpendicular to the magnetic field (Fig. 6a).
In the experiments to measure the anisotropic spin interactions, an appropriate amount of bicelle is put into protein solution. In a high magnetic field, bicelles align and the aligned

discoidal liquid crystalline molecules limit the space for protein tumbling. Bicelle has flat surface, thus, the protein involved in the aligned bicellar media will be surrounded by flat walls. Because of the steric clash, protein does not rotate freely near the bicelle wall, which will hider some orientations of the protein. This hindrance on some orientations for the protein in the bicellar medium causes incomplete rotational averaging of the anisotropic interactions, which thus makes the residual anisotropic interactions observables (Berlin, O'Leary and Fushman 2009, Vijayan and Zweckstetter 2005, Zweckstetter 2008).

The aligning magnitude is readily tuned by the bicelle concentration; higher bicelle concentration induces the stronger order of alignment. The bicelle made of the binary phospholipid mixture involving DMPC and DHPC has neutral charge on its surface. Therefore, most proteins do not stick to bicelles. Protein aligns through collisional interaction described above.

Surface charge doping to the bicellar surface is possible. Incorporation of CTAB, cetyl trimethyl ammonium bromide, to DMPC/DHPC binary phospholipid bicelle generates the positively charged surface, while the addition of SDS, sodium dodecyl sulphate, makes it negatively charged. The surface charge doping to bicelle changes the aligning property from that by neutral bicelle. In charged bicellar solutoin, electrostatic interactions between the medium and protein become apparent. This makes it sometimes difficult to use the charged bicelle to acidic or basic proteins such as nucleic acid binding proteins.

There are some limitations in the bicelle application as aligning media. One is in the limited temperature range to keep bicelle in a liquid crystalline phase; it is typically 27°C - 45°C. Some proteins precipitate in this temperature range, and the bicelle is not used for such proteins. In addition, the bicelle is only stable around neutral pH; in acidic or basic solution, the bicellar medium tends to make phase-separation and loses aligning ability. To avoid the limitations associated with the physicochemical properties of the bicelle medium, other liquid crystalline media have been reported. By properly selecting the medium, we can now achieve weak alignment for rather various types of proteins, each of which requires its own optimal temperature, pH, and ionic strength (Prestegard, Bougault and Kishore 2004).

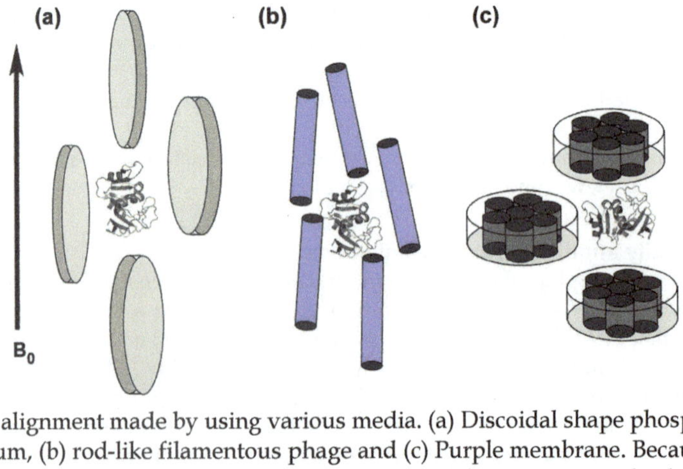

Fig. 6. Weak alignment made by using various media. (a) Discoidal shape phospholipid bicelle medium, (b) rod-like filamentous phage and (c) Purple membrane. Because of significant anisotropic magnetic susceptibility, the molecules spontaneously align in a magnetic field. Proteins within the aligning media are aligned through the collisional interaction or electrostatic interactions with the aligning molecules.

3.2 Naturally occurring materials that spontaneously align in a magnetic field

Some naturally available molecules that spontaneously align in a magnetic field are also used for weak alignment. Purple membrane is one example, which forms two-dimensional crystal lattice structure of bacteriorhodopsin (bR) that is rich in α-helices (Fig. 6b). Purple membrane constitutes of ordered α-helices and thus it has intrinsic high anisotropic magnetic susceptibility. Because of the structural characteristics, purple membrane can spontaneously align in a high magnetic field.

Suspension of purple membrane is added into protein solution to achieve a weak protein alignment. The alignment magnitude is tuned by the purple membrane concentration in a sample solution, as was the case for the bicelle medium (Sass et al. 1999). Purple membrane is the discoidal protein lipid complex, and it aligns with its normal parallel to the magnetic field. Purple membrane is rich in negative charge on its surface. Therefore, protein is aligned through the electrostatic interaction not collisional interaction.

Filamentous phage, which is made of a rod-like coat protein, is another example of the naturally occurring molecule used for weak alignment. Because of the highly anisotropic shape of the filamentous phage, it spontaneously aligns in a high magnetic field, with its rod axis parallel to a magnetic field (Fig. 6c). Filamentous phage has also negatively charged surface, thus it induces protein alignment through electrostatic interactions with protein as in the case of purple membrane. The alignment order can be tuned by adjusting the concentration of the phage suspension in protein solution.

The use of purple membrane and filamentous phage cannot be applied to basic proteins that are positively charged. The basic proteins tightly adsorb onto the negatively charged surfaces of the aligning molecules through the electrostatic interaction, which result in extreme ordering of proteins and thus prohibit observing high resolution NMR signals.

3.3 Compressed acryl amide gel

The other method uses anisotropically compressed acrylamide gel (Tycko, Blanco and Ishii 2000, Ishii, Markus and Tycko 2001, Sass et al. 1999). Acrylamide hydrogel has cavities to capture proteins within, whose cavity size is changed by altering the ratios of the composing chemicals, acrylamide and bis-acrylamide. Acrylamide forms a linear polymer chain and bis-acrylamide makes bridge to link acrylamide liner polymers, thus, the increased ratio of bis-acrylamide generates smaller cavity.

The acrylamide gel has a spatially isotropic cavity. Protein in the gel does not show any preferential orientation, and thus it shows no residual anisotropic spin interaction on its NMR spectrum. Protein alignment by the gel is achieved by inducing the spatial anisotropy to the cavities in the gel.

In a weakly alignment experiment using the gel, a cast gel including a protein solution is inserted into a NMR sample tube. There are two ways to make anisotropically compressed gel; one is to press the gel along the NMR tube (compressed gel) and the other is squeeze it in the lateral dimension, thus it becomes stretched vertically (stretched gel) (Fig. 7).

In vertical compression, gel is cast to have a shorter diameter than the inner diameter for NMR sample tube. After inserting the gel, vertically compressed with glass rod in the tube (Fig. 7a). On the other hand, stretched gel is made from the cast gel having a slightly larger diameter than that the inner diameter of the tube; the gel is inserted into the tube by using tapered device (Chou et al. 2001) (Fig. 7b). Because of the different pressing direction, these two preparations change the protein orientation to each other.

The charge doping in the anisotropically compressed gel is also possible. Negative charge is doped by replacing a part of acrylamide with acrylate, while a positive charge is introduced by adding DADMAC, diallyldimethylammonium chloride, in casting gel chip. These charge doping changes aligning property, because electrostatic interaction between protein and the gel becomes active.

Acrylamide gel is neutral and thus it induces protein alignment through collisional interaction between protein and walls having spatial anisotropy made by the compression. The non-charged acrylamide gel is readily used for any types of protein, independent on its pKa value. On the other hand, the charged gel is sometimes troublesome in achieving weak alignment of basic or acidic proteins, due to their adsorption onto the gel.

The aligning order magnitude is also tuned by changing the concentration of acrylamide and/or the ratio of the composing chemicals and also the diameter of the cased gel chip. In addition, the extent of the charge doping may have to be considered in some cases. There are more parameters to adjust than that for the liquid crystalline media. We, therefore, need some trial experiments to gain the optimal gel condition to have the anisotropic spin interactions observable in enough spectral resolution.

The great advantage in using the compressed gel is in the chemical stability of acrylamide gel. In using acrylamide gel for a weak alignment, we do not worry about the sample conditions including temperature, salt concentration, and solution pH. Under any sample conditions, we can align protein by the compressed gel. It is a keen difference from the limitation in applying liquid crystalline media and also in using the naturally occurring materials. The compressed gel is used as a universal media for weakly aligning protein.

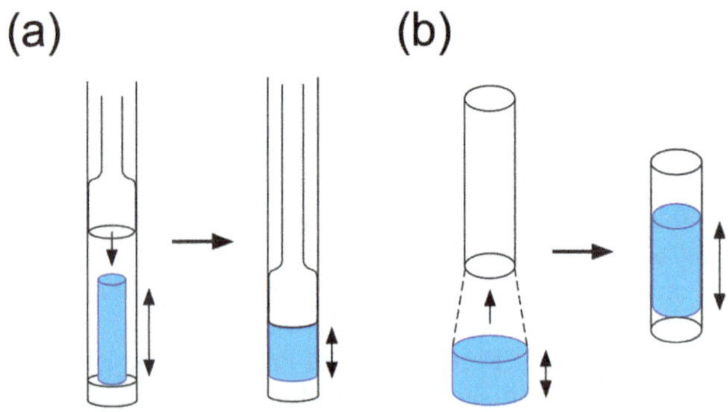

Fig. 7. Anisotropically compressed gels to achieve weak alignment. (a) Vertically compressed gel. (b) Laterally compressed, thus, stretched gel. Cast gel chip (blue) is inserted into a NMR glass tube. In the vertical compression, the cast gel is pressed by a glass rod in NMR tube. These compressions induce spatially anisotropic cavities within the acrylamide gel.

3.4 Protein structure in aligned media – Is it natural?

A weak alignment of protein is prerequisite for the analysis of protein morphology by NMR, because it relies on the anisotropic spin interactions that happen only under a weak aligned condition. Because protein is dissolved in an artificial medium like magnetically ordering liquid crystalline solution, one should worry about the structure under the condition; the domain orientation in such a medium represents the real state of the protein in an isotropic solution?

The mechanisms for aligning protein in magnetically aligned bicelle solution and rod-like filamentous phage suspension are well understood. The observed RDCs can be reproduced with using the protein crystal structure by the simulation based on the mechanism. The successful reproduction of the RDCs for the protein aligned by bicelle or filamentous phage supports that the proposed theory properly describes the state of protein in the aligning media.
According to the theory, the observed residual anisotropic spin interactions can be established by a small fraction of protein experiencing the rotational restriction through the collision between the protein and the aligning molecule. In using the charged filamentous phage, the electrostatic interactions are also active in inducing the rotational restriction on the protein near the medium. The fraction for that is around 0.1% of the total number of protein molecules in the sample. The interactions have to be weak and transient. The inter-molecule interactions, therefore, do not actively align the protein but just prohibit some fraction of the protein orientation near the medium. Under the condition, the aligning media do not induce artificial protein conformation or domain arrangement, because the prohibited orientation angles neat the medium is solely determined by the intrinsic molecular shape of protein in the bulk solution.
The structural perturbation by the medium is easily monitored by the NMR spectral changes. If the medium tightly interacts with protein to change the structure, some of the chemical shifts should change from those found in an isotropic state.
The bicellar medium is in the liquid crystalline state over its phase-transition temperature. Under the temperature, it becomes a micelle solution that does not align in a magnetic field, thus does not induce protein alignment. If protein in this micelle solution does not show apparent spectral changes from the solution, including no micelle, we exclude the possibility that the bicelle induces a structural change to the protein.
In the case of the charged filamentous phage, the increased concentration of cations, which shield the surface charge on the medium, impairs the ordering of the media in a magnetic field. The magnitudes of the anisotropic spin interactions for the protein in the phage suspension, therefore, are proportion to the cation concentration. If the spectral changes show linear dependency on the cation concentration, we could rule out the structural changes caused by the interaction with the medium.
Using compressed acrylamide gel allows the direct spectral comparison between the samples in the reference gel (non-compressed gel) and the isotropic solution without gel to see if any structure change happens through the interaction with the gel.
In general, under weakly aligning conditions, the spectral changes caused by the interaction with the media are rather small, ensuring that no apparent structural changes happen to protein in the media, even in the cases of multiple-domain proteins.

4. RDC-based domain orientation analysis – Basics and limitation

In this section, we will describe the experimental procedure to determine the domain orientation of a multiple-domain protein, from the RDC data collection to structure determination. In addition, the limitations in the RDC based analysis will be discussed to emphasize the necessity of our TROSY based DIORITE approach that follows.

4.1 Collecting the RDC data

RDC is measured on a pair of ^{1}H coupled HSQC spectra for the samples in isotropic and anisotropic states. The ^{1}H coupled HSQC spectrum gives a pair of split peaks along ^{15}N axis for each ^{1}H-^{15}N correlation. The doubled number of peaks on a ^{1}H coupled HSQC spectrum may increase signal overlaps that obstacle the accurate reading of peak positions. To avoid this drawback, a particular NMR technique is used to separate the up- and down-field components into different 2D spectra, IPAP-HSQC. IPAP-HSQC gives two separate spectra that have in-phase and anti-phase doubles, respectively. Addition or subtraction of the spectra will give two separate 2D spectra displaying only up-field or down-filed components of each doublet. This signal separation reduces signal overlap on a ^{1}H coupled HSQC spectrum, and keeps the spectral resolution to the same extent as in the original HSQC spectrum (Fig. 3).

The separation width between the up- and down-field components measured from IPAP-HSQC spectra gives ${}^{1}J_{NH}$ for isotropic sample and ${}^{1}J_{NH} + D_{NH}{}^{res}$ for aligned sample. Therefore, the RDC, $D_{NH}{}^{res}$, is obtained by their difference.

4.2 Domain orientation analysis based on the RDC data

Here, we describe the domain orientation analysis based on the RDCs. The domain orientation analysis should be done for the protein whose structure is already known by X-ray. The primal interest of the analysis is in exploring the domain reorientation upon interaction with the other protein or ligand. In the cases, each domain is assumed to retain the same structure as in crystal.

As described in the theory section, the alignment tensor for a weakly aligned protein is determined based by the RDCs and its structure coordinate, Eq. (5). In Eq. (5), direction cosines are calculated from the structure coordinate. Because the Saupe order matrix consists of five independent elements, we need more than five RDC data to determine the Saupe order matrix for the corresponding part. The singular value decomposition (SVD) to the matrix comprising of the relations for the observed RDCs will give the Saupe order matrix (Losonczi et al. 1999). Diagonalizing the Saupe order matrix gives the alignment tensor frame orientation relative to the molecular coordinate system and the magnate of the orders along each principal axis. As described in the theory part, the alignment tensor frame orientation is defined by the Euler angles (α, β, γ).

We consider a two-domain protein here. And we assume that the high resolution structure of each domain is available, and each domain structure is the same as in the crystal. Based on the collected RDCs, the alignment tensor for each domain is independently determined according to the above procedure. As schematically drawn in Fig. 8, the determined tensor frames for each domain are used as a guide to define solution domain orientation; one domain coordinate is rotated to make an overlay its tensor frame onto the other (Fig. 8). It is noted here, the RDCs do not provide any distance information between the domains. If the inter-domain segment has high flexibility, the additional distance restraints may be required to build the entire structure, which should come from the other experiments like paramagnetic relaxation enhancement (Clore and Schwieters 2002).

Alignment tensor determined by the RDCs has four possible orientations. Inversion around each principal axis gives the same RDCs values. Therefore, the inversion is not discriminated experimentally. To alleviate this ambiguity in orientation angle, additional alignment states using different aligning media, including charged bicelle, or charged acrylamide gel, will be used. In domain orientation analysis, however, the structural restrictions, which include the length of the inter-domain linker or possible inter-domain steric clash, may allow to define one inter-domain orientation even using a single aligning experiment.

The alignment tensor magnitudes along each principal axis represent the extent of the aligning order. If they differ between the domains, which may indicate that each rotates differently to each other. The aligning orders give the insight into the domain dynamics in a protein.

4.3 Significance of the domain orientation analysis by RDCs

The domain orientation analysis for protein in solution gives an invaluable outcome, even its high-resolution structure is available. There are some cases to show the different domain arrangements between solution and crystalline structures. The RDC analysis on maltose binding protein (MBP) in the complex with β-cyclodextrin has shown that the relative domain orientation in solution was different from that in the crystal (Skrynnikov et al. 2000b). This may indicate the crystal contact causes a subtle change in domain orientation. Bacteriophage T4 lysozyme in solution was shown to have a more open conformation relative to the crystal structure, which was also analyzed by the RDCs (Goto et al. 2001). This observation appears compatible with steric requirements for the ligand bindings.
These examples illustrate how the RDC based domain analysis complements X-ray crystallography in determining the relative domain orientation or protein morphology. The complementary role of the RDC based analysis is considerably emphasized in exploring the domain rearrangement upon binding to the other protein or ligand. Even if the complex structure cannot be solved by X-ray, the complex structure is determined by the RDCs in a solution state when the structure in a ligand-free form is available. This approach does not require tedious and time-consuming NOE analysis as required in conventional NMR structure determination, but just needs the backbone resonance assignments and a set of IPAP-HSQC spectra. The structure determination is, therefore, much more efficient over the conventional NMR structure determination.

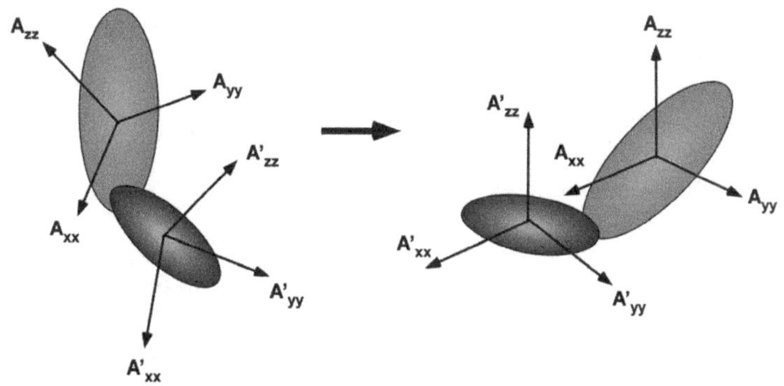

Fig. 8. Schematic representation for the procedure to determine the relative domain orientation based on the alignment tensors for each domain. Sets of RDC data determine the alignment tensor for each domain, independently. The domain orientation of a protein is established by rotating one domain coordinate to make an overlay of its tensor frame on the other.

4.4 Molecular size limitation in the RDC based approach

The domain orientation analysis with the RDCs is now recognized as a useful technique to elucidate overall protein morphology in solution, complementing the X-ray structure analysis. This approach has, however, severe size limitation. Here, this obstacle in the RDC application is discussed.

The size limitation comes from the rapid transverse relaxation rate of one of the split components observed on a 2D IPAP-HSQC spectrum. The high-filed component shows faster transverse relaxation rate than that of the other. This component has even faster relaxation rate than that of HSQC counterpart. This is due to cross-correlated relaxation interference to amide ^{15}N spin relaxation process; for the high-field component, the cross-correlated relaxation process additively affects, while for the low-field component, the interference reduces its relaxation rate. The transverse relaxation process of the HSQC signal is free from the interference.

In measuring the RDCs with IPAP-HSQC for high molecular weight protein, the high-field components of each amide spin pairs will broaden and severely reduce the signal intensities, thus, they will not be observed. In particular, the difficulty in observing the high-field component will be enhanced in an aligned state, due to the appearance of the residual dipolar interactions as relaxation causes. For proteins over 20 kDa, it is usually hard to observe the high-field component in an aligned state, thus making the RDC measurement impossible. The RDC based domain orientation analysis with IPAP-HSQC is practically limited up to around 20 kDa.

Simulation of the line broadening on each component of a double according to molecular size is shown in Fig. 9. In this figure, the low-field component that shows a longer transverse relaxation time and the other having a shorter transverse relaxation time are named as TROSY and anti-TROSY, respectively. The slower transverse relaxation associated with the low-field component is due to the mechanism used in TROSY (Transverse Relaxation Optimized SpetroscopY). As demonstrated on the simulation, the anti-TROSY component shows severe broadening even for the medium-size protein, 20 kDa.

The difference in line widths between the TROSY and anti-TROSY components will become considerable for higher molecular weight proteins. As seen in the simulation, protein over 150 kDa gives severely broadened anti-TROSY signal, which already hard to observe. Protein with 800 kDa never gives observable anti-TROSY signal. The size limitation in the RDC based approach is clearly demonstrated in this simulation.

It should be noted, the TROSY component can retain observable signal intensity even for 800 kDa protein (Fig. 9). This motivated us to devise an approach to determine an alignment tensor only using TROSY components.

4.5 Existing remedy for overcoming the size limitation in the RDC-based approach

Some remedies are proposed to overcome the size limitation the RDC application. They all rely on the TROSY.

The difference in the transverse relaxation rates between the TROSY (low-field) and anti-TROSY (high-field) components split along the ^{15}N axis are explained by the relaxation interference. The same effect is active in the split signals along ^{1}H dimension. In observing the ^{1}H-^{15}N single bond correlation spectrum without decoupling during t_1 and also t_2 durations, each spin pair gives a quartet on a spectrum; split signals in both ^{1}H and ^{15}N dimensions. The pure TROSY signal is the one having the longest transverse relaxation time among the quartets. In using the protein labeled with ^{15}N and ^{2}H, where an unwanted relaxation process is diminished by breaking the ^{1}H-^{1}H dipolar interaction network in a protein, TROSY effect is enhanced, and it allows ^{1}H-^{15}N correlation spectrum for proteins over 100 kDa (Pervushin et al. 1997).

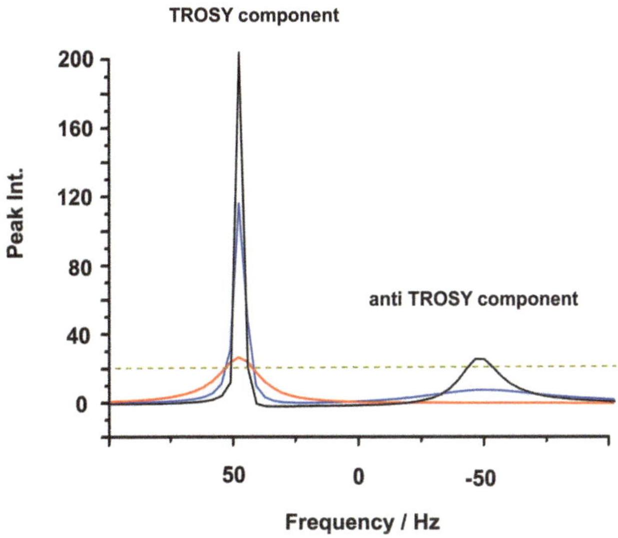

Fig. 9. Simulation of the molecular size dependency of the TROSY and anti-TROSY components observed on a IPAP-HSQC spectrum. The data represent a slice peak along the ^{15}N axis. This simulation assumed a 750 MHz experiment. Black, blue, and red lines are the simulated peaks for the sizes 20 kDa, 150 kDa, and 800 kDa, respectively. The dotted line is assumed the expected noise level.

One proposed remedy is the combinatorial use of TROSY and HSQC. The difference between TROSY and HSQC signals along the ^{15}N axis corresponds to a half of RDC. As discussed above, the transverse relaxation rate of the HSQC signal is slower than that for the anti-TROSY component in a IPAP-HSQC spectra. Therefore, the size limitation problem should be alleviated by replacing anti-TROSY signal with HSQC counterpart. For a 81.4 kDa protein, the transverse relaxation times for the TROSY, anti-TROSY, and HSQC signals are reported to be 65 msec, 10 msec, and 30 msec, respectively (Tugarinov and Kay 2003). In considering the difference between the transverse relaxation times between TROSY and HSQC signals, the combinatorial use does not fully solve the problem, but just alleviates it.
Another remedy is the use of J-scaled TROSY, which is also referred to as J-enhanced (JE) TROSY (Kontaxis, Clore and Bax 2000, Bhattacharya, Revington and Zuiderweg 2010). In this experiment, short J-evolution step is added in the standard TROSY, which induces J-dependent shift change from the standard TROSY shift. In the standard TROSY experiment, the shift difference between the signals along the ^{15}N axis on the same ^{1}H chemical shift corresponds to $^{1}J_{NH}$, whilst in the J-scaled TROSY, this shift difference is changed according to the additional duration for J-evolution. From the magnitude of the shift change induced by the applied J-evolution period, the apparent $^{1}J_{NH}$ coupling value is estimated. If the J-evolution period is applied to allow full recovery of $^{1}J_{NH}$ coupling, the observed signal position should be coincident with that of the HSQC signal. Usually, to gain the signal intensity for the observed signal on J-scaled TROSY, rather limited evolution time is set. In this J-evolution step, the coherences for TROSY and anti-TROSY are mixed; the equivalent

mixing of the two gives the coherence for observed as HSQC signal. The more increased the contribution of the anti-TROSY coherence to the observed signal leads to more broadened signals observed. Therefore, in the J-scaled TROSY experiment, partial recovery of the J-modulation is used to maintain the signal intensities on the J-scaled TROSY spectrum in the observable level.

Signals observed on a J-scaled TROSY spectrum have longer transverse relaxation times than those of the signals on a HSQC spectrum. Their transverse relaxation times, however, are still shorter than those for real TROSY counterparts. The combined use of TROSY and J-scaled TROSY is indeed advantageous over the TROSY/HSQC combination. In determining more accurate RDCs, J-scaled TROSY requires more extent of the mixing of the anti-TROSY coherences, which will result in the lesser sensitive J-scaled TROSY signals. The use of J-scaled TROSY is not the complete remedy for the problem we concern.

In spite of the limitations in the existing approaches, they expanded the RDC application up to 50 kDa protein (Jain, Noble and Prestegard 2003). However, it is also reported that the rapid transverse relaxation of the non-TROSY component is already an obstacle in measuring the RDCs for 81.4 kDa protein . The further expansion of the application limit is expected, and our DIORITE is one of the possible methods used for this purpose.

5. Alignment tensor determination using only TROSY

As discussed above, the molecular size limitation problem in the RDC based domain orientation analysis is not completely overcome, although the existing approaches have given some successful results. Most of the biologically interesting multi domain proteins tend to be over 100 kDa. The existing approaches are not thought to be applied to such higher molecular weight protein. This is because they do not take fully advantages of TROSY spectroscopy, which allows the longest transverse relaxation time for the observed signals. In contrast to the existing approaches, our approach, DIOIRTE, uses only TROSY spectra, where the signals having the longest transverse relaxation times are used. This may give considerable advantages over the existing methods in respect to the size limitation problem. In this section, we will describe the theoretical aspects of the TROSY based alignment tensor determination, which will allow the domain orientation analysis for higher molecular weight proteins ever.

5.1 Alignment induced TROSY shift changes

TROSY shift is changed when protein is transferred from isotropic to anisotropic states. This TROSY shift change along the ^{15}N axis contains the effects of two anisotropic spin interaction observed on a peptide plane; ^{1}H-^{15}N dipolar interaction and ^{15}N CSA. As depicted in Fig. 10, this alignment induced TROSY shift change, $\Delta\delta_{TROSY}$, contains a half of RDC and the full RCSA effects. In a Cartesian representation using the Saupe order matrix, the following relation should fold:

$$\begin{aligned}\Delta\delta_{TROSY} &= -\tfrac{1}{2}\left(\tfrac{\mu_0}{4\pi}\right)\frac{\gamma_i\gamma_j h}{2\pi^2 r_{ij}{}^3}\frac{4\pi}{5}\sum_{kl=x,y,z} S_{kl}\cos(\alpha_k)\cos(\alpha_l) \\ &\quad +\tfrac{2}{3}\sum_{kl=x,y,z}\sum_{j=x,y,z} S_{kl}\cos(\theta_{kj})\cos(\theta_{lj})\delta_{jj} \\ &= \sum_{kl} S_{lk}\{\tfrac{1}{2}D^0{}_{NH}\cos(\alpha_k)\cos(\alpha_l)+\tfrac{2}{3}\Delta_{kl}\}\end{aligned} \qquad (9)$$

$$\Delta_{kl} = \sum_{k,l=x,y,z} \sum_{j=x,y,z} S_{kl} \cos(\theta_{kj}) \cos(\theta_{lj}) \delta_{jj} \tag{10}$$

where, $\cos(\alpha_k)$ is the direction cosine for the NH bond vector relative the molecular axis k, k = x, y, z. D^0_{NH} is the static dipolar coupling constant, which equals 23.0 and 21.7 kHz for assumed NH bond length 1.02 Å and 1.04 Å, respectively; in solution, the effective NH bond length is estimated to be 1.04 Å, which values includes the bond libration effects. The tem $\cos(\theta_{kj})$ is the direction cosine of the CSA principal axis j relative to the molecular axis k. The principal value of the CSA tensor along j axis is denoted as δ_{jj}. As done for the RDC, the SVD calculation to the equations Eq. (9) for the residues in a protein gives the Saupe order matrix. At least, five $\Delta\delta_{TROSY}$ data are required to determine the Saupe order matrix constituted by five independent elements. The alignment tensor is obtained through the diagonalization of the Saupe matrix.

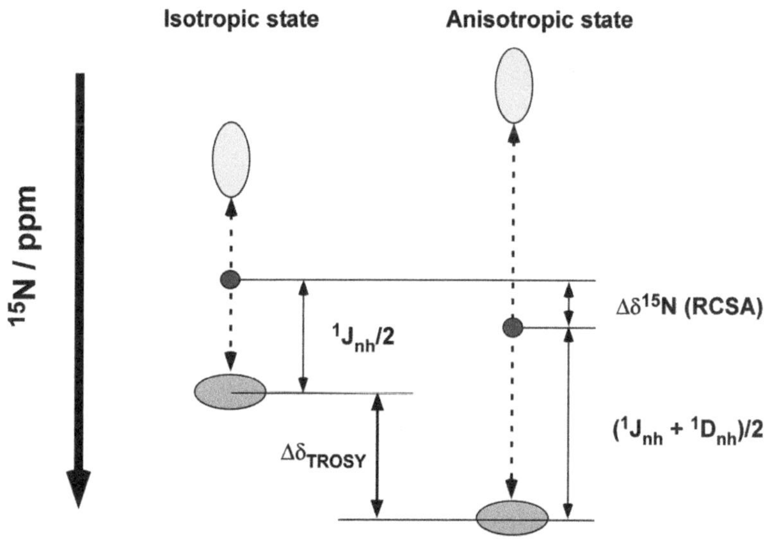

Fig. 10. Schematic drawing of the relationship among the signals for ^{1}H-^{15}N doublet in a ^{1}H coupled HSQC spectra observed for protein in isotropic and aligned states. $\Delta\delta_{TROSY}$ contains the contributions from a half of RDC and full RCSA.

DIORITE is the algorism to determine the alignment tensor solely from TRSOY spectra. It may be expected that DIORITE gives a more accurate alignment tensor over the ^{1}H-^{15}N RDCs, in particular, the case for huge protein over 50 kDa, due to the longest transverse relaxation time of the TROSY signal. DIORITE determines the alignment tensor based on two anisotropic spin interactions; RDC and RCSA. As described in theory part, RCSA contains the tensorial orientational information of the peptide plane against a magnetic field, while RDC gives only the bond vector orientation. RDC value does not change if the peptide plane is rotated along the NH bond axis, although RCSA should significantly change. Because of the inclusion of RCSA effect, DIORITE can discriminate the difference in the peptide plane orientation. Therefore, DIORITE should be more informative over the RDC based analysis in determining the domain orientation of a protein.

5.2 CSA tensor parameters used in DIOIRTE

DIORITE based alignment tensor determination requires ^{15}N CSA tensor for each residue. As discussed in the part for the RCSA, it is not trivial to get accurate ^{15}N CSA tensor for each residue in a protein, because it has significant local structure dependency; backbone torsion angle, hydrogen bonding, and so forth. In the DIORITE analysis, ^{15}N CSA tensors for every residue have to be known as input values.

In the domain orientation analysis with DIORITE, we use a high resolution domain coordinate from X-ray. Therefore, we know the detailed structure on each domain before the analysis. Some recent reports on the experimental determination of the residue specific ^{15}N CSA tensor for small proteins in solution had demonstrated that the ^{15}N CSA tensor value is primarily dependent on the backbone torsion angle. According to this correlation, we proposed the practical protocol for the DIORITE analysis that uses the secondary structure specific ^{15}N CSA tensors as inputs.

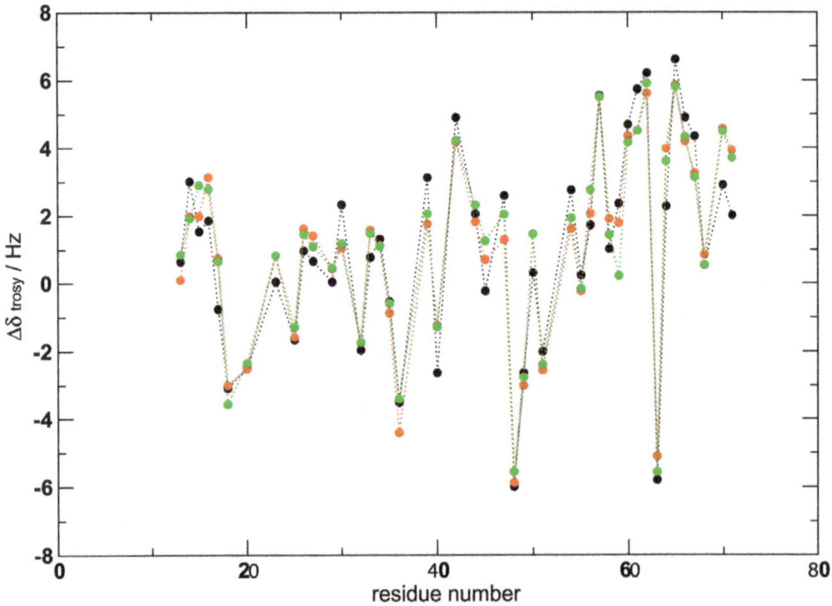

Fig. 11. Comparison of the back-calculated $\Delta\delta_{TROSY}$ obtained by DIORITE algorism using the secondary structure specific and the residue specific ^{15}N CSA tensors. Red and green circles are the values with secondary structure specific and residues specific tensors, respectively. Black circles are the observed $\Delta\delta_{TROSY}$.

The quality of the back calculated $\Delta\delta_{TROSY}$ values were assessed on ubiquitin. The values with the secondary structure specific ^{15}N CSA tensors and those with the residue specific tensors were compared (Fig. 11); the residue specific ^{15}N CSAs used here were determined by a set of elaborate spin relaxation analyses by Bodenhausen and co-workers (Loth, Pelupessy and Bodenhausen 2005). As demonstrated in the comparison, the use of the secondary structure specific ^{15}N CSA tensor gives consistent results with those with residue-specific values within an error range (Fig. 11). The DIORITE analysis using the secondary structure specific ^{15}N CSA tensors, therefore, can be a practical approach for exploring the domain orientation in a protein.

5.3 DIORITE analysis using different magnetic field strengths

TROSY effect reduces line width along ^{15}N dimension at a higher magnetic field; the original paper on TROSY has estimated the optimal frequency for obtaining the narrowest lines is around 1 GHz (Pervushin et al. 1997). In this section, the optimal magnetic-field strength for the DIORITE analysis will be discussed.

The alignment induced TROSY shift change, $\Delta\delta_{TROSY}$, show significant magnetic-field dependency due to the RCSA contribution. The magnitude of RCSA is proportional to the applied magnetic-field strength, while the RDC is independent on the applied magnetic-field strength. The field dependency of the RCSA gives a peculiar profile to the $\Delta\delta_{TROSY}$ value.

In a peptide plane, the least shielded ^{15}N CSA tensor axis, δ_{xx}, is close to the NH bond; the angle β ranges from 15 to 20 degrees (Fig. 2b). The small β angle result in the cancellation between RDC and RCSA values to each other, making the observed $\Delta\delta_{TROSY}$ value smaller than the corresponding RDC. In most of the residues in a protein, RDC and RCSA have an opposite sign to each other. Therefore, $\Delta\delta_{TROSY}$ should be roughly equivalent to that a half of RDC minus RCSA in their absolute values. The value $\Delta\delta_{TROSY}$ tends to be approximately one-third of the RDC in an absolute magnitude.

The inter-cancellation effect depends on the magnetic-field strength, but the dependency is not linear. Using ubiquitin, we simulated the root mean square (rms) $\Delta\delta_{TROSY}$ values on different magnetic-field strengths indicated by the protein resonance frequencies. In the simulation, the rms $\Delta\delta_{TROSY}$, which should correspond to the average magnitude of $\Delta\delta_{TROSY}$, decreases up to 800 MHz and then increases according to the field strength. In lower magnetic-field strength, where the RCSA contribution is rather small and the $\Delta\delta_{TROSY}$ is roughly approximated by a half of RDC. The increasing RCSA, which has an opposite sign to RDC, contribution overall decreases the RDC value but further enhancement of the RCSA contribution in a higher magnetic field becomes dominant in $\Delta\delta_{TROSY}$ value, which may explain the profile.

The condition for weak alignment is carefully tuned to avoid severe signal broadening. The alignment order is typically tuned to around 10^{-3}, giving the maximal absolute magnitude of RDC around 20 Hz. Under the condition, $\Delta\delta_{TROSY}$ is expected to give about 6 Hz. Even under the worst 800 MHz, the expected $\Delta\delta_{TROSY}$ should be 5.5 Hz. The accuracy in reading peak positions at a signal-to-noise ration of 40:1 was estimated to be 0.6 Hz. Therefore, the experimental error in the observed $\Delta\delta_{TROSY}$ is estimated to be 0.8 Hz, in considering the error propagation through subtracting the TROSY shift in an aligned state by that in an isotropic state. Accordingly, the expected $\Delta\delta_{TROSY}$ is well resolved within the error in peak picking. It is noted that the estimated reading error in RDC should be 1.2 Hz due to twice subtraction required.

The resolution on TROSY spectrum is also the factor to be considered in discussing the performance of DIORITE on different magnetic-field strength. Using the average ^{15}N CSA tensor value estimated from ubiquitin, the dependencies of ^{15}N TROSY line width and rms $\Delta\delta_{TROSY}$ value are plotted against proton resonance frequencies, where the values are in ppb (parts per billion) units instead of Hz to compare them in the context of spectral resolution

(Fig. 12a). In this simulation, the optimal frequency to minimize the ^{15}N line width of the observed TROSY signal is 972.5 MHz, close but slightly different from the value by the simpler estimation in the original paper. In comparing the line narrowing of TROSY signal, the rms $\Delta\delta_{TROSY}$ value shows less dependency on the magnetic-field strength. The effective accuracy in observing $\Delta\delta_{TROSY}$ value should be estimated by the value $\Delta\delta_{TROSY}$ divided by the ^{15}N line width of TROSY signal (Fig. 12b). In this simulated profile, the optimal magnetic-field strength for observing $\Delta\delta_{TROSY}$ is around 900 MHz for ^{1}H resonance frequency. Therefore, the magnetic-field strength for the maximal TROSY effect is almost optimal for the DIORITE analysis. DIORITE analysis, thus, can take full advantage of the TROSY effect on a 900 MHz NMR spectrometer, which is now commercially available.

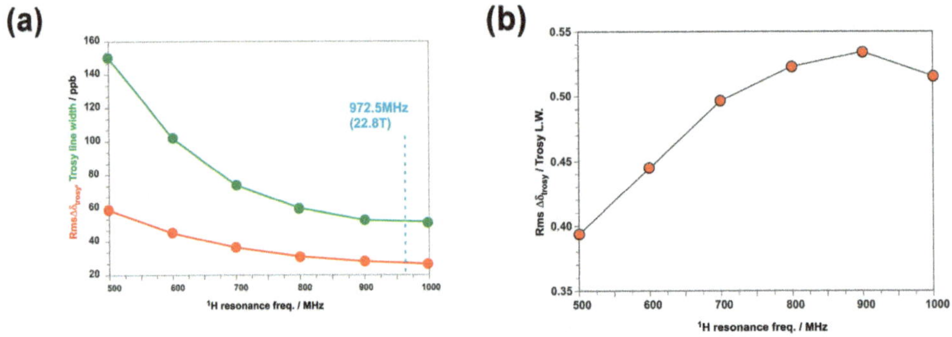

Fig. 12. Field dependency of the TROSY line width and the root-mean-square (rms) $\Delta\delta_{TROSY}$. (a) TROSY line width (green) and rms $\Delta\delta_{TROSY}$ (red) according to the magnetic field strength (b) Field strength dependency of the effective resolution in measuring $\Delta\delta_{TROSY}$.

5.4 Practical aspects of the DIORITE data collection

DIORITE analysis uses TROSY chemical shift changes $\Delta\delta_{TROSY}$ induced by a weak alignment of protein (Tate 2008, Tate et al. 2004). In general, chemical shift is very sensitive to the solution conditions, including temperature, sample concentration, pH, ionic strength and so no. In measuring $\Delta\delta_{TROSY}$, we have to exclude the other factors to change the chemical shift except for the alignment effects. To achieve this, anisotropically compressed polyacrylamide gel is the most appropriate medium for aligning protein. As described above, the anisotropically compressed gel (stretched gel) is made by inserting the cast gel chip having a little greater diameter than that of the inner diameter of NMR tube. The gel chip having the same diameter as that of NMR tube is not compressed after insertion, which can keep the isotropic cavity within. Protein in this non-compressed gel is not aligned and does not show any anisotropic spin interactions. The sample, therefore, can be used as the reference TROSY spectrum for measuring $\Delta\delta_{TROSY}$. Because the non-compressed gel consists of the same acrylamide composition, protein in this gel experiences the same chemical environments as in the compressed gel, which ensures that the observed TROSY shift changes solely come from the alignment effects. It is difficult in getting reference data with using the other

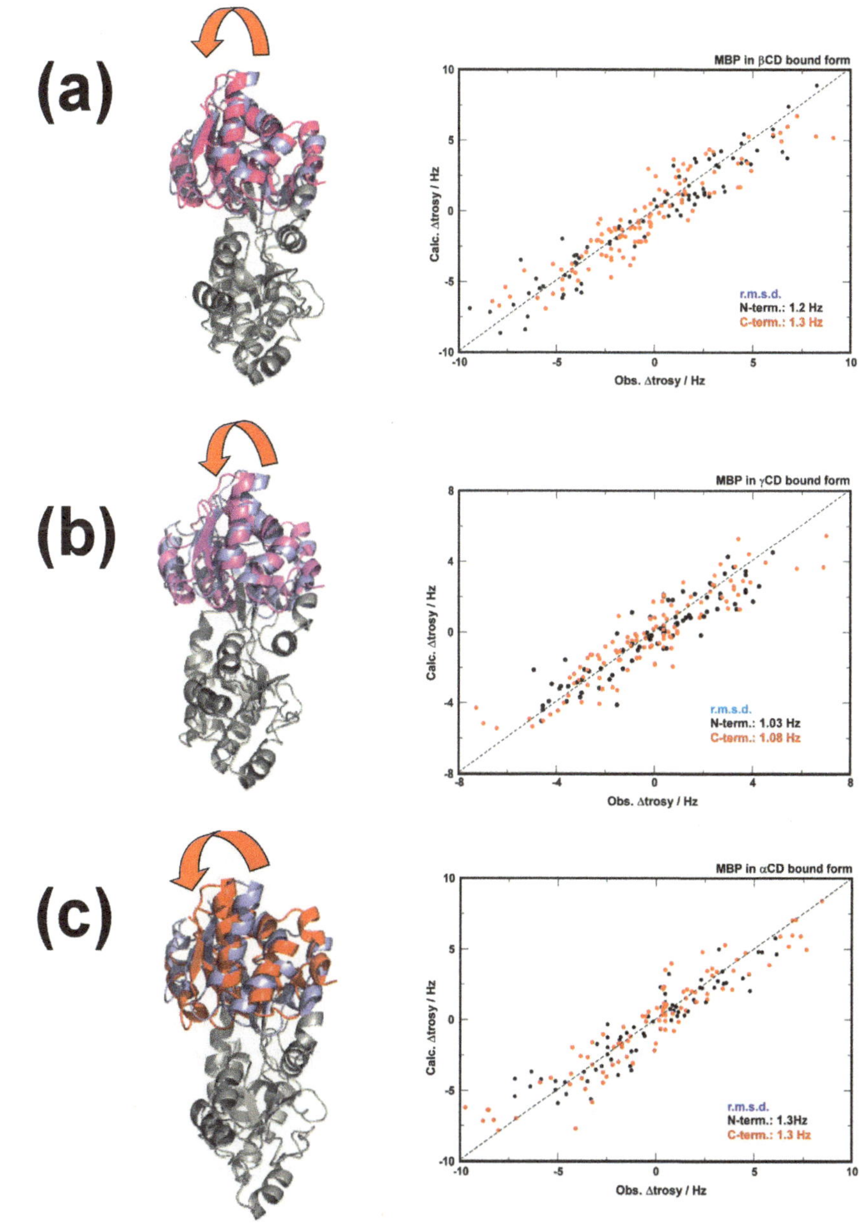

Fig. 13. Domain rotation angle change according to the ligand size in maltose binding protein (MBP). (a) β-CD, (b) γ-CD and (c) α-CD complex states. The hinge rotation angles induced by ligand binding were 13 deg., 14 deg. And 22 deg. for the cases of β-CD, γ-CD, and α-CD, respectively.

aligning media, including bicelle, filamentous phage. They cannot be the choice for DIORITE analysis.

In measuring $\Delta\delta_{TROSY}$ values, a set of TROSY spectra for the aligned and isotropic samples are used. The formation of the anisotropic cavity in the stretched gel can be monitored by the split of the deuteron signal from HOD; the deuteron in the water molecule within an anisotropic cavity shows a doublet, residual quadruplar coupling of the deuteron (Fig. 13a). The stronger anisotropy in a cavity gives a larger split. The water deuteron signal in the reference gel shows a singlet, confirming the cavity in the reference gel is isotropic.

In measuring NMR spectrum, water deuteron signal is used as a frequency lock to give the frequency standard. Because of the split of a deuteron signal in an aligned state, the resonance positions observed are biased; one of the doublet envelopes is used for the frequency lock, thus biasing the signals by a half of the split width. This offset should be subtracted from each signal position on the TROSY spectrum for protein in an aligned state.

5.5 DIORITE analyses on MBP in different ligand bound states

Here, we demonstrate the applications of the DIORITE analysis. Maltose binding protein (MBP) has two domains. From a series of X-ray analysis, MBP is known to show domain reorientation upon binding to ligand, and the domain rotation angle depend on size of the ligand molecule.

MBP binds to β-cyclodextrin (β-CD) comprising of seven glucose units, and it shows a slight change in the relative domain orientation from that in apo form as demonstrated by X-ray analysis. MBP also binds to different types of cyclodextrins comprising of the different number of glucose units; α–CD (six glucose units), γ-CD (eight glucose units). We expected to see the domain rotation angle changes according to the size of the three types of CDs.

Using the anisotropically compressed gel (stretched gel) and uncompressed reference gel, we collected a pair of TROSY spectra for each MBP in the complex with α-, β- and γ- CD. Using the apo-form MBP X-ray coordinate, we analyzed the relative domain orientation using DIORITE; the model structure was constructed based on the alignment tensors individually determined for N- and C-terminal domains. On each complex, the back-calculated and observed $\Delta\delta_{TROSY}$ values were well correlated, suggesting the alignment tensors for each domain were well determined (Fig. 13).

The DIORITE analyses on the MBP complexes demonstrated that for the smallest ligand, α-CD, the domain rotation angle was significantly larger than those for the β-CD complex structure. On the other hand, MBP in the complex with γ-CD retains almost the same domain orientation relative to the β-CD complex; the size over the β-CD does not change the domain rotation angle.

The example analyses demonstrated that the MBP showed significant domain reorientation according to the ligand size. It should be noted that the DIORITE analysis is very efficient to see this ligand-dependent domain reorientation; which requires just a pair of TROSY spectra collected for the sample in aligned and isotropic states.

6. Conclusion

X-ray structures are still exponentially accumulated every year. The huge collection of the protein coordinates should prompt the protein structure works using the existing protein coordinate data. In this review, we introduced protein structure analysis in solution with the assistance of protein structure data collected by X-ray crystallography.

There has been a lot of discussion on the significance of the solution structure determination by NMR. Most of the structures for single domain proteins or isolated domains have shown marginal structural deviations from the corresponding structures solved by X-ray. This diminishes the importance of NMR structure analysis, except for the case in which crystallization is hard. When X-ray structure is available on a type of protein behaving as an independent structural unit, the solution structure determination on the protein by NMR is not usually conducted, because the crystal structure should not largely differ from that in solution. Additionally, in most of the cases, the size limitation in NMR structure determination prohibits such solution structure analyses, even if they are required. NMR solution structure analysis, therefore, has been recognized as a complementary method in rather limited cases.

The situation seems different in the structure analysis of multiple domain proteins. There already appeared some examples to show the difference in the domain orientations of protein between solution and crystalline states. The kinds of example will be increased, because it is getting to know that many proteins have domains linked by seemingly flexible or unstructured linkers judged from the sequence. In the proteins, it is presumable that the domain arrangement tends to be defined artificially by crystal packing, thus not represent the protein morphology in a solution state.

The NMR techniques using a weak alignment have paved ways to directly determine the relative domain orientation in solution, which has not been ever done by the conventional NMR methods. Some variations of the methods were introduced in this review, with their limitations in practical applications. Our devised DIORITE approach has a significant advantage in the domain orientation analysis over the existing methods, when it is applied to higher molecular weight proteins. The domain orientation analysis by DIORITE will expand the X-ray structure assisted reach in exploring the protein morphological change in solution, which associates with the functional exertion.

The domain rearrangement in protein, or protein morphological change, upon binding to ligand or interaction with its partner protein will become much more important in discussing protein functional regulation, after getting the high-resolution crystal structure in a specific state, for example, apo-form. The combined use of DIORITE with X-ray structure data may give vivid views how protein works in solution by changing its morphology. NMR is now becoming a complementary partner to X-ray crystallography in protein structure analysis, in particular, on higher molecular weight proteins.

7. Acknowledgement

S.T. acknowledges financial support from PRESTO/JST and SENTAN/JST. We appreciate RIKEN for accessing high-field NMR instruments. This review article is dedicated to Prof. Mamoru Tamura who demised on August 7th, 2011. He has been encouraging for progressing the work described here as a supervisor in the PRESTO/JST project.

8. References

Bax, A. (2003) Weak alignment offers new NMR opportunities to study protein structure and dynamics. *Protein Science*, 12, 1-16.

Berlin, K., D. P. O'Leary & D. Fushman (2009) Improvement and analysis of computational methods for prediction of residual dipolar couplings. *Journal of Magnetic Resonance,* 201, 25-33.

Bhattacharya, A., M. Revington & E. R. P. Zuiderweg (2010) Measurement and interpretation of 15N–1H residual dipolar couplings in larger proteins. *Journal of Magnetic Resonance,* 203, 11-28.

Boyd, J. & C. Redfield (1999) Characterization of 15N Chemical Shift Anisotropy from Orientation-Dependent Changes to 15N Chemical Shifts in Dilute Bicelle Solutions. *Journal of the American Chemical Society,* 121, 7441-7442.

Chou, J., S. Gaemers, B. Howder, J. Louis & A. Bax (2001) A simple apparatus for generating stretched polyacrylamide gels, yielding uniform alignment of proteins and detergent micelles*. *Journal of Biomolecular NMR,* 21, 377-382.

Clore, G. M. & C. D. Schwieters (2002) Theoretical and computational advances in biomolecular NMR spectroscopy. *Current opinion in structural biology,* 12, 146-153.

Conti, E., N. P. Franks & P. Brick (1996) Crystal structure of firefly luciferase throws light on a superfamily of adenylate-forming enzymes. *Structure,* 4, 287-98.

Cornilescu, G. & A. Bax (2000) Measurement of Proton, Nitrogen, and Carbonyl Chemical Shielding Anisotropies in a Protein Dissolved in a Dilute Liquid Crystalline Phase. *Journal of the American Chemical Society,* 122, 10143-10154.

de Vries, S. J., M. van Dijk & A. M. Bonvin (2010) The HADDOCK web server for data-driven biomolecular docking. *Nat Protoc,* 5, 883-97.

Dominguez, C., R. Boelens & A. M. Bonvin (2003) HADDOCK: a protein-protein docking approach based on biochemical or biophysical information. *J Am Chem Soc,* 125, 1731-7.

Evenas, J., V. Tugarinov, N. R. Skrynnikov, N. K. Goto, R. Muhandiram & L. E. Kay (2001) Ligand-induced structural changes to maltodextrin-binding protein as studied by solution NMR spectroscopy. *Journal of Molecular Biology,* 309, 961-974.

Fushman, D., N. Tjandra & D. Cowburn (1999) An Approach to Direct Determination of Protein Dynamics from 15N NMR Relaxation at Multiple Fields, Independent of Variable 15N Chemical Shift Anisotropy and Chemical Exchange Contributions. *Journal of the American Chemical Society,* 121, 8577-8582.

Goto, N. K., N. R. Skrynnikov, F. W. Dahlquist & L. E. Kay (2001) What is the average conformation of bacteriophage T4 lysozyme in solution? A domain orientation study using dipolar couplings measured by solution NMR. *Journal of Molecular Biology,* 308, 745-764.

Hajduk, P. J., G. Sheppard, D. G. Nettesheim, E. T. Olejniczak, S. B. Shuker, R. P. Meadows, D. H. Steinman, G. M. Carrera, P. A. Marcotte, J. Severin, K. Walter, H. Smith, E. Gubbins, R. Simmer, T. F. Holzman, D. W. Morgan, S. K. Davidsen, J. B. Summers & S. W. Fesik (1997) Discovery of Potent Nonpeptide Inhibitors of Stromelysin Using SAR by NMR. *Journal of the American Chemical Society,* 119, 5818-5827.

Hansen, M. R., L. Mueller & A. Pardi (1998) Tunable alignment of macromolecules by filamentous phage yields dipolar coupling interactions. *Nature structural biology,* 5, 1065-1074.

Hiyama, Y., C. H. Niu, J. V. Silverton, A. Bavoso & D. A. Torchia (1988) Determination of 15N chemical shift tensor via 15N-2H dipolar coupling in Boc-glycylglycyl[15N glycine]benzyl ester. *Journal of the American Chemical Society,* 110, 2378-2383.

Ishii, Y., M. A. Markus & R. Tycko (2001) Controlling residual dipolar couplings in high-resolution NMR of proteins by strain induced alignment in a gel. *Journal of Biomolecular NMR,* 21, 141-151.

Jain, N. U., S. Noble & J. H. Prestegard (2003) Structural characterization of a mannose-binding protein-trimannoside complex using residual dipolar couplings. *Journal of Molecular Biology,* 328, 451-462.

Kontaxis, G., G. M. Clore & A. Bax (2000) Evaluation of cross-correlation effects and measurement of one-bond couplings in proteins with short transverse relaxation times. *J Magn Reson,* 143, 184-96.

Kurita, J., H. Shimahara, N. Utsunomiya-Tate & S. Tate (2003) Measurement of 15N chemical shift anisotropy in a protein dissolved in a dilute liquid crystalline medium with the application of magic angle sample spinning. *J Magn Reson,* 163, 163-73.

Losonczi, J. A., M. Andrec, M. W. Fischer & J. H. Prestegard (1999) Order matrix analysis of residual dipolar couplings using singular value decomposition. *Journal of magnetic resonance (San Diego, Calif.: 1997),* 138, 334-342.

Loth, K., P. Pelupessy & G. Bodenhausen (2005) Chemical Shift Anisotropy Tensors of Carbonyl, Nitrogen, and Amide Proton Nuclei in Proteins through Cross-Correlated Relaxation in NMR Spectroscopy. *Journal of the American Chemical Society,* 127, 6062-6068.

Lundstrom, P., D. F. Hansen & L. E. Kay (2008) Measurement of carbonyl chemical shifts of excited protein states by relaxation dispersion NMR spectroscopy: comparison between uniformly and selectively (13)C labeled samples. *J Biomol NMR,* 42, 35-47.

Morris, G. M., R. Huey, W. Lindstrom, M. F. Sanner, R. K. Belew, D. S. Goodsell & A. J. Olson (2009) AutoDock4 and AutoDockTools4: Automated docking with selective receptor flexibility. *Journal of Computational Chemistry,* 30, 2785-2791.

Nakatsu, T., S. Ichiyama, J. Hiratake, A. Saldanha, N. Kobashi, K. Sakata & H. Kato (2006) Structural basis for the spectral difference in luciferase bioluminescence. *Nature,* 440, 372-376.

Ottiger, M. & A. Bax (1999) Bicelle-based liquid crystals for NMR-measurement of dipolar couplings at acidic and basic pH values. *J Biomol NMR,* 13, 187-91.

Ottiger, M., F. Delaglio & A. Bax (1998) Measurement of J and dipolar couplings from simplified two-dimensional NMR spectra. *J Magn Reson,* 131, 373-8.

Pervushin, K., R. Riek, G. Wider & K. Wuthrich (1997) Attenuated T2 relaxation by mutual cancellation of dipole-dipole coupling and chemical shift anisotropy indicates an avenue to NMR structures of very large biological macromolecules in solution. *Proc Natl Acad Sci U S A,* 94, 12366-71.

Prestegard, J. H., C. M. Bougault & A. I. Kishore (2004) Residual dipolar couplings in structure determination of biomolecules. *Chemical reviews,* 104, 3519-3540.

Sanchez, R. & A. Sali (2000) Comparative protein structure modeling. Introduction and practical examples with modeller. *Methods Mol Biol,* 143, 97-129.

Sass, J., F. Cordier, A. Hoffmann, M. Rogowski, A. Cousin, J. G. Omichinski, H. Lowen & S. Grzesiek (1999) Purple Membrane Induced Alignment of Biological Macromolecules in the Magnetic Field. *Journal of the American Chemical Society,* 121, 2047-2055.

Shuker, S. B., P. J. Hajduk, R. P. Meadows & S. W. Fesik (1996) Discovering High-Affinity Ligands for Proteins: SAR by NMR. *Science,* 274, 1531-1534.

Skrynnikov, N. R., N. K. Goto, D. Yang, W. Y. Choy, J. R. Tolman, G. A. Mueller & L. E. Kay (2000a) Orienting domains in proteins using dipolar couplings measured by liquid-state NMR: differences in solution and crystal forms of maltodextrin binding protein loaded with beta-cyclodextrin. *J Mol Biol,* 295, 1265-73.

Skrynnikov, N. R., N. K. Goto, D. Yang, W. Y. Choy, J. R. Tolman, G. A. Mueller & L. E. Kay (2000b) Orienting domains in proteins using dipolar couplings measured by liquid-state NMR: differences in solution and crystal forms of maltodextrin binding protein loaded with beta-cyclodextrin. *Journal of Molecular Biology,* 295, 1265-1273.

Tate, S. (2008) Anisotropic nuclear spin interactions for the morphology analysis of proteins in solution by NMR spectroscopy. *Anal Sci,* 24, 39-50.

Tate, S., H. Shimahara & N. Utsunomiya-Tate (2004) Molecular-orientation analysis based on alignment-induced TROSY chemical shift changes. *Journal of magnetic resonance (San Diego, Calif.: 1997),* 171, 282-291.

Tjandra, N. & A. Bax (1997) Direct measurement of distances and angles in biomolecules by NMR in a dilute liquid crystalline medium. *Science,* 278, 1111-1114.

Tugarinov, V. & L. E. Kay (2003) Quantitative NMR studies of high molecular weight proteins: application to domain orientation and ligand binding in the 723 residue enzyme malate synthase G. *Journal of Molecular Biology,* 327, 1121-1133.

Tycko, R., F. J. Blanco & Y. Ishii (2000) Alignment of Biopolymers in Strained Gels: A New Way To Create Detectable Dipole−Dipole Couplings in High-Resolution Biomolecular NMR. *Journal of the American Chemical Society,* 122, 9340-9341.

Vijayan, V. & M. Zweckstetter (2005) Simultaneous measurement of protein one-bond residual dipolar couplings without increased resonance overlap. *Journal of Magnetic Resonance,* 174, 245-253.

Yao, L., A. Grishaev, G. Cornilescu & A. Bax (2010) Site-specific backbone amide (15)N chemical shift anisotropy tensors in a small protein from liquid crystal and cross-correlated relaxation measurements. *J Am Chem Soc,* 132, 4295-309.

Zweckstetter, M. (2008) NMR: prediction of molecular alignment from structure using the PALES software. *Nature Protocols,* 3, 679-690.

Permissions

The contributors of this book come from diverse backgrounds, making this book a truly international effort. This book will bring forth new frontiers with its revolutionizing research information and detailed analysis of the nascent developments around the world.

We would like to thank Dr. Annamalai Chandrasekaran ("Chandra"), for lending his expertise to make the book truly unique. He has played a crucial role in the development of this book. Without his invaluable contribution this book wouldn't have been possible. He has made vital efforts to compile up to date information on the varied aspects of this subject to make this book a valuable addition to the collection of many professionals and students.

This book was conceptualized with the vision of imparting up-to-date information and advanced data in this field. To ensure the same, a matchless editorial board was set up. Every individual on the board went through rigorous rounds of assessment to prove their worth. After which they invested a large part of their time researching and compiling the most relevant data for our readers. Conferences and sessions were held from time to time between the editorial board and the contributing authors to present the data in the most comprehensible form. The editorial team has worked tirelessly to provide valuable and valid information to help people across the globe.

Every chapter published in this book has been scrutinized by our experts. Their significance has been extensively debated. The topics covered herein carry significant findings which will fuel the growth of the discipline. They may even be implemented as practical applications or may be referred to as a beginning point for another development. Chapters in this book were first published by InTech; hereby published with permission under the Creative Commons Attribution License or equivalent.

The editorial board has been involved in producing this book since its inception. They have spent rigorous hours researching and exploring the diverse topics which have resulted in the successful publishing of this book. They have passed on their knowledge of decades through this book. To expedite this challenging task, the publisher supported the team at every step. A small team of assistant editors was also appointed to further simplify the editing procedure and attain best results for the readers.

Our editorial team has been hand-picked from every corner of the world. Their multi-ethnicity adds dynamic inputs to the discussions which result in innovative outcomes. These outcomes are then further discussed with the researchers and contributors who give their valuable feedback and opinion regarding the same. The feedback is then collaborated with the researches and they are edited in a comprehensive manner to aid the understanding of the subject.

Apart from the editorial board, the designing team has also invested a significant amount of their time in understanding the subject and creating the most relevant covers. They scrutinized every image to scout for the most suitable representation of the subject and create an appropriate cover for the book.

The publishing team has been involved in this book since its early stages. They were actively engaged in every process, be it collecting the data, connecting with the contributors or procuring relevant information. The team has been an ardent support to the editorial, designing and production team. Their endless efforts to recruit the best for this project, has resulted in the accomplishment of this book. They are a veteran in the field of academics and their pool of knowledge is as vast as their experience in printing. Their expertise and guidance has proved useful at every step. Their uncompromising quality standards have made this book an exceptional effort. Their encouragement from time to time has been an inspiration for everyone.

The publisher and the editorial board hope that this book will prove to be a valuable piece of knowledge for researchers, students, practitioners and scholars across the globe.

List of Contributors

David J. Hernández and Ivan Castillo
Universidad Nacional Autónoma de México, México

Sergey Pogodin, Shmuel Cohen, Tahani Mala'bi and Israel Agranat
Organic Chemistry, Institute of Chemistry, The Hebrew University of Jerusalem, Edmond J. Safra Campus, Jerusalem, Israel

Vânia André and M. Teresa Duarte
Centro de Química Estrutural, Dept. of Chemical and Biological Engineering, Instituto Superior Técnico, Universidade Técnica de Lisboa, Lisbon, Portugal

Angel Alvarez-Larena, Joan Farran and Joan F. Piniella
Universitat Autònoma de Barcelona, Spain

Anastasios D. Keramidas and Marios Stylianou
University of Cyprus, Cyprus

Chryssoula Drouza
Cyprus University of Technology, Cyprus

Aliakbar Dehno Khalaji
Department of Chemistry, Faculty of Science, Golestan University, Gorgan, Iran

Dan-Wei Zhang and Zhan-Ting Li
Department of Chemistry, Fudan University, Shanghai, China

Olga V. Kovalchukova
Peoples' Friendship University of Russia, Russian Federation

Soushi Miyazaki
Research Center for Materials Science, Nagoya University, Japan

Takahiko Kojima
Department of Chemistry, University of Tsukuba, Japan

Dai Oyama
Fukushima University, Japan

Viorel Cîrcu and Marin Micutz
University of Bucharest, Romania

Noriyuki Ishii
National Institute of Advanced Industrial Science and Technology (AIST), Japan

Nobuo N. Noda
Institute of Microbial Chemistry, Tokyo, Japan

Yoshinori Ohsumi
Tokyo Institute of Technology, Japan

Fuyuhiko Inagaki
Hokkaido University, Japan

Nathalie Colloc'h and Guillaume Marassio
CI-NAPS UMR 6232 Université de Caen Basse-Normandie - CNRS, Centre Cyceron, Caen, France

Thierry Prangé
LCRB, UMR 8015 - CNRS - Université Paris Descartes, Faculté de Pharmacie, Paris, France

Alejandro Giorgetti
Applied Bioinformatics Group, Dept. of Biotechnology, University of Verona, Italy
German Research School for Simulation Sciences, Jülich Research Center and RWTH-Aachen University, Jülich, Germany

Stefano Piccoli
Applied Bioinformatics Group, Dept. of Biotechnology, University of Verona, Italy

Antonello Merlino, Filomena Sica and Alessandro Vergara
Department of Chemistry, University of Naples "Federico II", Naples, Italy Istituto di Biostrutture e Bioimmagini, CNR, Naples, Italy

Cristiano Luis Pinto Oliveira
Department of Experimental Physics, Institute of Physics, University of São Paulo, Brazil

Shin-ichi Tate, Aiko Imada and Noriaki Hiroguchi
Department of Mathematical and Life Sciences, Hiroshima University
Japan